Modelling, Simulation and Data Analysis in Acoustical Problems

Modelling, Simulation and Data Analysis in Acoustical Problems

Special Issue Editors

Claudio Guarnaccia
Lamberto Tronchin
Massimo Viscardi

MDPI • Basel • Beijing • Wuhan • Barcelona • Belgrade • Manchester • Tokyo • Cluj • Tianjin

Special Issue Editors

Claudio Guarnaccia
University of Salerno
Italy

Lamberto Tronchin
University of Bologna
Italy

Massimo Viscardi
University of Naples Federico II
Italy

Editorial Office
MDPI
St. Alban-Anlage 66
4052 Basel, Switzerland

This is a reprint of articles from the Special Issue published online in the open access journal *Applied Sciences* (ISSN 2076-3417) (available at: https://www.mdpi.com/journal/applsci/special_issues/MSDAAP).

For citation purposes, cite each article independently as indicated on the article page online and as indicated below:

LastName, A.A.; LastName, B.B.; LastName, C.C. Article Title. *Journal Name* **Year**, *Article Number*, Page Range.

ISBN 978-3-03928-284-5 (Pbk)
ISBN 978-3-03928-285-2 (PDF)

Contents

About the Special Issue Editors

Claudio Guarnaccia is an associate professor of Applied Physics at the Department of Civil Engineering, University of Salerno. He has an M.Sc (2004) and Ph.D. (2008) in Physics from the Physics Department "E.R.Caianiello" of the University of Salerno. He has carried out post-doc research studies both in the Physics Department and in the Industrial Engineering Department of the University of Salerno. His research activity is focused on the application of physics to engineering and biomedical problems. The main focus of his research is acoustics. He works with several national and international universities and research institutes. He is a member of several national and international organizations, in the field of Acoustics and Applied Physics. He is an editor-in-chief, associate editor, member of the editorial board, and reviewer for several international journals, indexed in the most important international databases. He is the author of more than 100 papers published in international journals and the proceedings of international conferences with referees. He was invited to several conferences as a "plenary lecturer", to present the results of his research. Since 2005, he is working in teaching, and since 2007 he has been a teaching assistant in all the physics courses at the Engineering Faculty of Salerno University. He is a member of the Council of the Department and he is a delegate of the Director for student careers and mentorship/tutorship.

Lamberto Tronchin is Associate Professor in Environmental Physics from the University of Bologna and is recognised internationally as a leading authority on the subject of sound and acoustics. A pianist himself, with a diploma in piano from the Conservatory of Reggio Emilia, Dr Tronchin's principal area of research has been musical acoustics and room acoustics. He is the author of more than 200 papers and was Chair of the Musical Acoustics Group of the Italian Association of Acoustics from 2000 to 2008. Dr. Tronchin is the President of the Italian Section of AESDr. Tronchin is Associate Editor of the Journal of the Audio Engineering Society and a member of the Scientific Committee of the CIARM, the Inter- University Centre of Acoustics and Musical research. He has chaired sessions of architectural and musical acoustics during several international symposiums, been a referee for a number of International journals and is the Chair of the Organising and Scientific Committees of IACMA (International Advanced Course on Musical Acoustics). He was a visiting researcher at the University of Kobe in Japan, a visiting professor at the University of Graz in Austria and a special honoured International Guest at the International Workshop, 'Analysis, Synthesis and Perception of Music Signals', at Jadavpur University of Kolkata, India, in 2005. He has chaired the International Advanced Course on Musical Acoustics (IACMA), organised by the European Association of Acoustics, which was held in Bologna, in 2005. In 2008 and 2009, he gave plenary lectures at international congresses on acoustics in Cambridge (UK), Paris, Vancouver, Prague, Bucharest, Santander, Kos, Malta, Rodi, as well as at WIELS in Bruxelles. Dr Tronchin holds a Masters Degree in Building Engineering and a PhD in Applied Physics (Architectural Acoustics) from the University of Bologna. He has completed advanced courses on the Mechanics of Musical Instruments at CISM, Udine, Italy, and on Noise and Vibration at the University of Southampton in the UK where he has also worked as a visiting researcher. He was granted a post-doctoral scholarship in Room Acoustics. He designed theatres and other buildings, as an acoustic consultant, in collaboration with several architects, among them Richard Meier and Paolo Portoghesi. He is the inventor of an international patent with Alma Mater Studiorum: "Method for artificially reproducing an output signal of a non-linear time invariant system".

Massimo Viscardi graduated from the University of Naples "Federico II" with a degree in Aeronautical Engineering in 1994, and a Ph.D. in Aerospace Engineering in 1998. From 2006, he has been a confirmed researcher at the University of Naples "Federico II", and is Assistant Professor of Structural Testing and Assistant Professor of Experimental Vibroacoustic at the same university. His main research fields of interest are: acoustics and vibrations, structural testing, sensing and actuation systems, composite materials, natural fibers applications, NDT/Health monitoring activities and noise and vibration active control. He has been involved in many research activities within national (L297, PIA, Regional research program, PRIN) and international programs, like (EU Funded ASANCA, ASANCA II, MESA, FACE, DAMOCLES, PIROS, AERONEWS, ITEM B, SPAIN). From April 2002 to the end of 2006, he was a Member of the Scientific Committee on Ministry of Industry for the Research and Development program . He has developed specific competencies and has promoted industrial applications for innovative materials and relative testing facilities. He has been scientific coordinator of three international projects and more than 15 research and innovation projects within the national research framework. He is the author of more than 100 scientific works and more than 200 thesis degrees. He held more than 30 training courses in the reference research field at post-graduate and master's level.

Preface to "Modelling, Simulation and Data Analysis in Acoustical Problems"

Modelling and simulation in acoustics is currently gaining in importance. In fact, with the development and improvement of innovative computational techniques and with the growing need for predictive models, an impressive boost has been observed in this domain.

The design of a model requires a proper conversion of reality to functions and parameters. Once the model has been designed, an adequate simulation must be run in terms of modelling and computational parameters. Keeping in mind the limitations and approximations of any model, data analysis, both online and offline, is the last step of this process and can be extremely important to extract the required output from the process. These basic and general concepts can be applied in many acoustical problems. In acoustics, there is a large demand for modelling and simulation in several research and application areas, such as noise control, indoor acoustics, and industrial applications.

These factors led us to propose a special issue about "Modelling, Simulation and Data Analysis in Acoustical Problems", https://www.mdpi.com/journal/applsci/special_issues/MSDAAP, as we believe in the importance of these topics in modern acoustics' studies. In total, 81 papers were submitted and 33 of them were published, with an acceptance rate of 37.5%. Among the 33 papers published, two of them were classified as review papers, while the rest are classified as research papers.

According to the number of papers submitted, it can be affirmed that this is a trending topic in the scientific and academic community and this special issue will try to provide a future reference for the research that will be developed in the coming years.

Claudio Guarnaccia, Lamberto Tronchin, Massimo Viscardi
Special Issue Editors

Editorial

Special Issue on Modelling, Simulation and Data Analysis in Acoustical Problems

Claudio Guarnaccia [1,*], **Lamberto Tronchin** [2] and **Massimo Viscardi** [3]

1 Department of Civil Engineering, University of Salerno, 84084 Fisciano, Italy
2 Department of Architecture, University of Bologna, 40126 Bologna, Italy; lamberto.tronchin@unibo.it
3 Department of Industrial Engineering, University of Napoli, 80138 Napoli, Italy; massimo.viscardi@unina.it
* Correspondence: cguarnaccia@unisa.it

Received: 19 November 2019; Accepted: 23 November 2019; Published: 3 December 2019

1. Introduction

Modelling and simulation in acoustics is gathering more and more importance nowadays. In fact, with the development and improvement of innovative computational techniques and with the growing need of predictive models, an impressive boost has been observed in this domain. The design of a model needs a proper conversion of reality to functions and parameters. On the other hand, once the model has been designed, an adequate simulation must be run, in terms of modelling and computational parameters. Keeping in mind the limitation and the approximations of any model, the data analysis, both online and offline, is the last step of this process and can be extremely important to extract the required output from the process.

These basic and general concepts can be applied in many acoustical problems. In acoustics, in fact, there is a large demand for modelling and simulation, in several research and application areas, such as noise control, indoor acoustics, industrial applications, etc.

These motivations led us to the proposal of a Special Issue about "Modelling, Simulation and Data Analysis in Acoustical Problems", since we definitely believe in the importance of these topics in modern acoustics studies. In total, 81 papers were submitted and 33 were published, with an acceptance rate of 37.5%. Among the 33 papers published, two of them were classified as review papers, while the rest were classified as research papers. According to the number of papers submitted, it can be affirmed that this is a trendy topic in the scientific and academic community and this special issue will try to be a future reference for research to be developed in the next years.

2. Modelling, Simulation and Data Analysis in Acoustical Problems

As stated in the introduction, the need of models in acoustical problems is very large and can interest many subareas of acoustics. Withstanding this variety of possible applications and considering the interdisciplinary features of the Special Issue, several topics were studied in the papers submitted to the issue.

The noise control topic is studied by Wang et al. [1] and Zhang et al. [2], regarding respectively noise barrier insertion loss and car silencers.

Medical applications are presented in [3–6]. Hearing issues are studied by Cucis et al., comparing normal-hearing subjects and cochlear implant users, and Ito et al., regarding the effects of surgical instruments in ear surgery. High-intensity focused ultrasound (HIFU) non-invasive therapy is studied by Liu et al., Tan et al. and Gutierrez et al., concerning respectively the prediction of HIFU propagation in a dispersive medium, the influence of dynamic tissue properties on HIFU hyperthermia and acoustic field of focused ultrasound transducers.

In addition to HIFU, ultrasound waves are presented in several and various applications [7–10]. In particular, Choo et al. proposes a method to estimate the soil depth based on elastic wave velocity.

Some underwater applications are presented by Wang et al. and Wang et al., considering respectively underwater acoustic communication and channel modelling and estimation, and underwater acoustic sources estimation. Jin et al. [11] presents an application of hydro-elastic analysis of a submerged floating tunnel under extreme wave and seismic excitations.

The topic of the acoustic ultrasound emissions study for monitoring of fatigue crack growth in mooring chains [12] and for rail defect detection [13] is faced respectively by Angulo et al. and Shi et al., with very interesting results. Dobrzychi et al. applies the acoustic emissions technique to epoxy resin electrical treeing study [14]. Also, Teng et al. [15] proposes the evaluation of cracks in metallic material.

Vibration and vibroacoustic studies are presented in [16–19], by Chatterhee et al., Wu et al. and Qian et al., with applications to parabolic tapered annular circular plate (Chatterhee et al. 2018 and 2019) and sensorized prodder for landmine detection (Wu et al.). Also, Flückiger et al. studies the vibrations, but referred to piano keys and their influence on piano players' perception and performance [20].

Yin et al. [21] and Jiang et al. [22] present their results on transducers, respectively on a 3D model of electromagnetic acoustic transducers and balanced armature receiver optimal design.

Target speech with nonlinear soft masking is presented by Zou et al. [23] while Tronchin et al. [24] reports a study about spatial information on voice generation.

Propagation in a fluid-filled polyethylene pipeline is studied from an experimental point of view by Li et al. in [25].

Bo et al. [26] presents a study on the acoustics of the Syracuse open-air theatre, with experiments and simulations. A room acoustic experiment is proposed by Wang et al. in [27] to investigate fingerprinting acoustic localization indoors.

Noncontact audio recording and a multi-frame Principal Component Analysis (PCA) based stereo audio coding method are proposed respectively by Sato et al. [28] and Wang et al. [29].

Tarrazó-Serrano et al. [30] proposes a material for acoustic lenses compatible with magnetic resonance imaging.

An acoustic detection method for localization is proposed by Yin et al. in [31].

Kirkup [32] proposes a survey of the boundary element method in acoustics.

Sparse impulse response estimation is treated in [33] by Lim et al.

3. Conclusions

All the researches presented, published in this Special Issue, suggest that the topic of modelling and simulation in acoustic problems is extremely important and popular in the scientific community. The advances proposed by all the authors push further the knowledge in this area and open the way to new and interesting possible evolutions. We believe that this issue could become a reference in the near future of modelling and simulation in acoustics. The new horizons in this research area will be traced starting by state of the art innovations largely represented by the papers of this issue.

Acknowledgments: The success of this Special Issue is strongly related to the huge work and the great contributions of all the authors. In addition, we acknowledge the hard work and the professional support of the reviewers and of the editorial team of Applied Sciences. We are extremely grateful to all the reviewers involved in the issue, for their time and their knowledge. We congratulate the assistant editors from MDPI that collaborated with us and thank them for their tireless support. We hope that the editorial process, starting from the submission and focusing on the review, was appreciated by all the authors, despite the final decisions. The real value of the time and the work spent in this process must be traced in the help provided to the authors to improve their papers.

Conflicts of Interest: The authors declare no conflict of interest.

References

1. Wang, H.; Luo, P.; Cai, M. Calculation of Noise Barrier Insertion Loss Based on Varied Vehicle Frequencies. *Appl. Sci.* **2018**, *8*, 100. [CrossRef]
2. Zhang, H.; Fan, W.; Guo, L. A CFD Results-Based Approach to Investigating Acoustic Attenuation Performance and Pressure Loss of Car Perforated Tube Silencers. *Appl. Sci.* **2018**, *8*, 545. [CrossRef]

3. Cucis, P.; Berger-Vachon, C.; Hermann, R.; Millioz, F.; Truy, E.; Gallego, S. Hearing in Noise: The Importance of Coding Strategies—Normal-Hearing Subjects and Cochlear Implant Users. *Appl. Sci.* **2019**, *9*, 734. [CrossRef]

4. Ito, T.; Kubota, T.; Furukawa, T.; Matsui, H.; Futai, K.; Hull, M.; Kakehata, S. The Role of Powered Surgical Instruments in Ear Surgery: An Acoustical Blessing or a Curse? *Appl. Sci.* **2019**, *9*, 765. [CrossRef]

5. Liu, S.; Yang, Y.; Li, C.; Guo, X.; Tu, J.; Zhang, D. Prediction of HIFU Propagation in a Dispersive Medium via Khokhlov–Zabolotskaya–Kuznetsov Model Combined with a Fractional Order Derivative. *Appl. Sci.* **2018**, *8*, 609. [CrossRef]

6. Tan, Q.; Zou, X.; Ding, Y.; Zhao, X.; Qian, S. The Influence of Dynamic Tissue Properties on HIFU Hyperthermia: A Numerical Simulation Study. *Appl. Sci.* **2018**, *8*, 1933. [CrossRef]

7. Gutierrez, M.; Ramos, A.; Gutierrez, J.; Vera, A.; Leija, L. Nonuniform Bessel-Based Radiation Distributions on A Spherically Curved Boundary for Modeling the Acoustic Field of Focused Ultrasound Transducers. *Appl. Sci.* **2019**, *9*, 911. [CrossRef]

8. Choo, H.; Jun, H.; Yoon, H. Application of Elastic Wave Velocity for Estimation of Soil Depth. *Appl. Sci.* **2018**, *8*, 600. [CrossRef]

9. Wang, X.; Wang, X.; Jiang, R.; Wang, W.; Chen, Q.; Wang, X. Channel Modelling and Estimation for Shallow Underwater Acoustic OFDM Communication via Simulation Platform. *Appl. Sci.* **2019**, *9*, 447. [CrossRef]

10. Wang, F.; Chen, Y.; Wan, J. In-Depth Exploration of Signal Self-Cancellation Phenomenon to Achieve DOA Estimation of Underwater Acoustic Sources. *Appl. Sci.* **2019**, *9*, 570. [CrossRef]

11. Jin, C.; Kim, M. Time-Domain Hydro-Elastic Analysis of a SFT (Submerged Floating Tunnel) with Mooring Lines under Extreme Wave and Seismic Excitations. *Appl. Sci.* **2018**, *8*, 2386. [CrossRef]

12. Angulo, Á.; Tang, J.; Khadimallah, A.; Soua, S.; Mares, C.; Gan, T. Acoustic Emission Monitoring of Fatigue Crack Growth in Mooring Chains. *Appl. Sci.* **2019**, *9*, 2187. [CrossRef]

13. Shi, H.; Zhuang, L.; Xu, X.; Yu, Z.; Zhu, L. An Ultrasonic Guided Wave Mode Selection and Excitation Method in Rail Defect Detection. *Appl. Sci.* **2019**, *9*, 1170. [CrossRef]

14. Dobrzycki, A.; Mikulski, S.; Opydo, W. Using ANN and SVM for the Detection of Acoustic Emission Signals Accompanying Epoxy Resin Electrical Treeing. *Appl. Sci.* **2019**, *9*, 1523. [CrossRef]

15. Teng, X.; Zhang, X.; Fan, Y.; Zhang, D. Evaluation of Cracks in Metallic Material Using a Self-Organized Data-Driven Model of Acoustic Echo-Signal. *Appl. Sci.* **2019**, *9*, 95. [CrossRef]

16. Chatterjee, A.; Ranjan, V.; Azam, M.; Rao, M. Theoretical and Numerical Estimation of Vibroacoustic Behavior of Clamped Free Parabolic Tapered Annular Circular Plate with Different Arrangement of Stiffener Patches. *Appl. Sci.* **2018**, *8*, 2542. [CrossRef]

17. Chatterjee, A.; Ranjan, V.; Azam, M.; Rao, M. Comparison for the Effect of Different Attachment of Point Masses on Vibroacoustic Behavior of Parabolic Tapered Annular Circular Plate. *Appl. Sci.* **2019**, *9*, 745. [CrossRef]

18. Wu, Z.; Ma, H.; Wang, C.; Li, J.; Zhu, J. Numerical Analysis of a Sensorized Prodder for Landmine Detection by Using Its Vibrational Characteristics. *Appl. Sci.* **2019**, *9*, 744. [CrossRef]

19. Qian, C.; Ménard, S.; Bard, D.; Negreira, J. Development of a Vibroacoustic Stochastic Finite Element Prediction Tool for a CLT Floor. *Appl. Sci.* **2019**, *9*, 1106. [CrossRef]

20. Flückiger, M.; Grosshauser, T.; Tröster, G. Influence of Piano Key Vibration Level on Players' Perception and Performance in Piano Playing. *Appl. Sci.* **2018**, *8*, 2697. [CrossRef]

21. Yin, W.; Xie, Y.; Qu, Z.; Liu, Z. A Pseudo-3D Model for Electromagnetic Acoustic Transducers (EMATs). *Appl. Sci.* **2018**, *8*, 450. [CrossRef]

22. Jiang, Y.; Xu, D.; Jiang, Z.; Kim, J.; Hwang, S. Comparison of Multi-Physical Coupling Analysis of a Balanced Armature Receiver between the Lumped Parameter Method and the Finite Element/Boundary Element Method. *Appl. Sci.* **2019**, *9*, 839. [CrossRef]

23. Zou, Y.; Liu, Z.; Ritz, C. Enhancing Target Speech Based on Nonlinear Soft Masking Using a Single Acoustic Vector Sensor. *Appl. Sci.* **2018**, *8*, 1436. [CrossRef]

24. Tronchin, L.; Kob, M.; Guarnaccia, C. Spatial Information on Voice Generation from a Multi-Channel Electroglottograph. *Appl. Sci.* **2018**, *8*, 1560. [CrossRef]

25. Li, Q.; Song, J.; Shang, D. Experimental Investigation of Acoustic Propagation Characteristics in a Fluid-Filled Polyethylene Pipeline. *Appl. Sci.* **2019**, *9*, 213. [CrossRef]

26. Bo, E.; Shtrepi, L.; Pelegrín Garcia, D.; Barbato, G.; Aletta, F.; Astolfi, A. The Accuracy of Predicted Acoustical Parameters in Ancient Open-Air Theatres: A Case Study in Syracusae. *Appl. Sci.* **2018**, *8*, 1393. [CrossRef]
27. Wang, S.; Yang, P.; Sun, H. Fingerprinting Acoustic Localization Indoor Based on Cluster Analysis and Iterative Interpolation. *Appl. Sci.* **2018**, *8*, 1862. [CrossRef]
28. Sato, R.; Emoto, T.; Gojima, Y.; Akutagawa, M. Automatic Bowel Motility Evaluation Technique for Noncontact Sound Recordings. *Appl. Sci.* **2018**, *8*, 999. [CrossRef]
29. Wang, J.; Zhao, X.; Xie, X.; Kuang, J. A Multi-Frame PCA-Based Stereo Audio Coding Method. *Appl. Sci.* **2018**, *8*, 967. [CrossRef]
30. Tarrazó-Serrano, D.; Castiñeira-Ibáñez, S.; Sánchez-Aparisi, E.; Uris, A.; Rubio, C. MRI Compatible Planar Material Acoustic Lenses. *Appl. Sci.* **2018**, *8*, 2634. [CrossRef]
31. Yin, J.; Xiong, C.; Wang, W. Acoustic Localization for a Moving Source Based on Cross Array Azimuth. *Appl. Sci.* **2018**, *8*, 1281. [CrossRef]
32. Kirkup, S. The Boundary Element Method in Acoustics: A Survey. *Appl. Sci.* **2019**, *9*, 1642. [CrossRef]
33. Lim, J.; Lee, S. Regularization Factor Selection Method for l1-Regularized RLS and Its Modification against Uncertainty in the Regularization Factor. *Appl. Sci.* **2019**, *9*, 202. [CrossRef]

Article

Acoustic Emission Monitoring of Fatigue Crack Growth in Mooring Chains

Ángela Angulo [1,2,*], Jialin Tang [1], Ali Khadimallah [1], Slim Soua [1], Cristinel Mares [2] and Tat-Hean Gan [1,2]

[1] TWI Ltd., Condition and Structural Health Monitoring, Integrity Management Group. Granta Park, Great Abington, Cambridge CB21 6AL, UK; jialin.tang@twi.co.uk (J.T.); ali.khadimallah@twi.co.uk (A.K.); slim.soua@twi.co.uk (S.S.); tat-hean.gan@twi.co.uk (T.-H.G.)
[2] Brunel University London, Department of Mechanical Engineering, Kingston Lane, Uxbridge, Middlesex UB8 3PH, UK; cristinel.mares@brunel.ac.uk
* Correspondence: angela.angulo@twi.co.uk; Tel.: +44-0-1223-899-000

Received: 11 March 2019; Accepted: 20 May 2019; Published: 28 May 2019

Featured Application: Structural health monitoring experimental methodology for crack initiation and crack growth analysis for damage detection in mooring chains using acoustic emission.

Abstract: Offshore installations are subject to perpetual fatigue loading and are usually very hard to inspect. Close visual inspection from the turret is usually too hazardous for divers and is not possible with remotely operated vehicles (ROVs) because of the limited access. Conventional nondestructive techniques (NDTs) have been used in the past to carry out inspections of mooring chains, floating production storage and offloading systems (FPSOs), and other platforms. Although these have been successful at detecting and assessing fatigue cracks, the hazardous nature of the operations calls for remote techniques that could be applied continuously to identify damage initiation and progress. The aim of the present work is to study the capabilities of acoustic emission (AE) as a monitoring tool to detect fatigue crack initiation and propagation in mooring chains. A 72-day large-scale experiment was designed for this purpose. A detailed analysis of the different AE signal time domain features was not conclusive, possibly due to the high level of noise. However, the frequency content of the AE signals offers a promising indication of fatigue crack growth.

Keywords: structural health monitoring; acoustic emission; mooring chain; fatigue crack growth; structural integrity

---

1. Introduction

Offshore operators are constantly concerned by the safety and integrity of their assets, due to the high stakes involved. Ageing structures notably pose a significant threat to human lives and can incur exorbitant costs when unplanned shutdowns or catastrophic failures occur. Furthermore, in addition to the structural challenges that onshore structures experience, offshore assets withstand harsh marine environments as a result of severe storms, highly corroding sea water, seaquakes, and cyclic wave loading [1,2]. In this context, structural health monitoring (SHM) tools have been developed over the last few decades to mitigate such risks and offer continuous monitoring solutions that can be used in different industries.

One of the major problems in the design of offshore equipment is fatigue damage accumulation. Although this topic has been extensively studied in the literature, theoretically, numerically, and experimentally [3–9], the available inspection and monitoring technologies developed to date have not been able to fully overcome the severe environmental challenges associated with offshore service activities. Remotely operated vehicles (ROVs) have been widely used since the 1970s but face serious difficulties, despite technological advances, due to the highly unpredictable operating environment characterised by poor visibility and unstable conditions [10,11]. For fatigue damage detection in structural applications in general, several sensing techniques have been developed [12], including guided ultrasonic waves [13,14], fibre Bragg gratings [15–17], strain gauges [18], and piezoelectric sensors [19]. Acoustic emission (AE) has also been proposed as a potential solution to detect and monitor cracking in structures, such as vessels and pipelines [20] and bridges [21]. Roberts et al. [22] found a correlation between crack evolution rates and the AE count rates for a narrow range of loading in steel specimens subject to tension. Yu et al. [23] predicted the crack growth behaviour in compact tension steel specimens based on AE data. The AE data were filtered using specific techniques and obtained from AE sensors placed around the crack tip. The authors showed that an accurate life prediction model could be established when the absolute energy of the AE signals was analysed. Despite its limitations, it is now recognised that the AE technique is capable of monitoring fatigue crack initiation [24] and propagation [25] in steels and other metals.

Amongst the offshore assets that are vulnerable to corrosion-enhanced fatigue damage, mooring chains are one of the most crucial mooring components used in permanently anchored structures [26]. Despite their importance, limited experimental work on mooring chains exists in the literature. Studies have been carried out on the chains' material microstructural properties [27,28], whereas others have been numerical modelling oriented [29]. Few large-scale testing attempts have been made. Rivera et al. [30] conducted a 4-month feasibility study to establish the AE technique's capabilities in monitoring damage in mooring chain links subjected to stress corrosion cracking in artificial sea water. Although a metallurgical examination was not performed, the authors suggested that the AE technique has promising potential in monitoring fatigue cracking in mooring chains.

The present work is a continuation of a research programme [30–32] aimed at identifying the key AE signal features for the prediction of fatigue crack growth in mooring chain links. The primary goal of this study was to investigate the applicability of using ultrasonic guided waves (UGW) and AE approaches for detecting and monitoring crack initiation, location, and propagation on a mooring chain. The study shows preliminary modelling simulations using different finite element analysis (FEA) methods.

In this paper, in Section 2, a description of the full-scale test rig and the monitoring setup is presented. The results and discussion then follow before a brief conclusion. Due to the complexity of the experiment, a substantial amount of data was collected and analysed. For the purpose of simplicity and brevity, only the final outcomes relevant to the scope of this paper are presented.

2. Experimental Procedure

2.1. Test Rig Description

The experiment carried out was part of a large Joint Industry Project (JIP) research programme, which aimed to perform large-scale tests for the analysis of the fatigue performance of mooring chains in seawater. A full-scale fatigue test rig was arranged to perform the AE measurements. The setup was designed to test a short section of a chain made of 7 links in artificial seawater, with a link diameter of 127 mm and a maximum load capacity of 700 tons. The total length and width of the rig was 7.65 m and 2.2 m, respectively.

The chain was initially subjected to a tension of 1000 kN in order to calibrate the load cell. A cycling tensile loading ranging between 3113 kN and 3497 kN at 0.5 Hz frequency was applied during the 72-day experiment (Figure 1).

2.2. Hardware Selection and Sensor Deployment

The AE data acquisition was performed using a Vallen AMSY-6 (Vallen Systeme GmbH, Icking, Germany) digital multi-channel AE-measurement system (ASIP-2 dual channel acoustic signal processor). Different AE sensors with different resonance frequencies and frequency bandwidths were tested: Vallen-VS150-WIC-V01 (resonant frequency 150 kHz, bandwidth 100–450 kHz), VS375-WIC-V01 (resonant frequency 375 kHz, bandwidth 250–700 kHz), and VS900-WIC-V01 (resonant frequency 350 kHz, bandwidth 100–900 kHz). It has been found [33,34] that chain failure most likely occurs at the point of the intrados (KT point) and crown positions, due to higher localised stresses in these areas (Figure 2). Therefore, on each link, the sensor was placed 10 cm away from the weld on an accessible side (Figure 3b). Each sensor was equipped with an integrated 34 dB pre-amplification. Four sensors were used on Links 3, 4, 5, and 6 (Figure 3a, and Figure 3c): one VS150, one VS375, and two VS900. A water-based couplant was used to facilitate the transmission of the sound signal between the transducer and the link's surface. Links 1 and 7 were not monitored in the present setup.

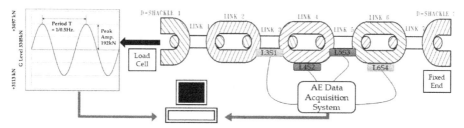

Figure 1. Test rig illustration: the chain is fixed at one end (right) and the loading (strain) is applied at the other end (left).

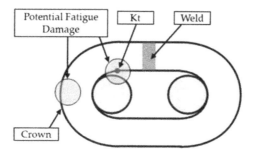

Figure 2. Chain link potential failure locations.

The sensors used in this study had already been calibrated according to the American Society for Testing and Materials standard, ASTM E1106 [35], prior to the experiment to ensure the reliability of the collected signals. Since the experiments were to be performed with the chain submerged in water, the reproducibility of the AE sensors response was verified in air and underwater, according to ASTM E976 [36], by carefully breaking a 0.5 mm pencil lead against the link's surface (the Pencil Lead Breakage (PLB) test). PLB, also known as the Hsu and Nielsen [37] pencil lead break, is a well-established technique and has long been used as a method to artificially generate reproducible AE signals [36]. Additionally, to verify the sensor coupling, the pulsing function was applied. Each sensor was used as a signal generator that sends signals to be intercepted by the rest of the sensors. These details are briefly summarised in Table 1.

Table 1. Signals reproducibility verification.

Pencil Lead Break	Sensor Pulsing
- A Pencil Lead Break (PLB) simulates Acoustic Emission (AE) events - PLBs at +−10 cm (4 at −10 cm, 4 at +10 cm) - PLBs at link inner face opposing weld (4 PLB per link) - Transmission across links was observed - Amplitude dropped from 85–90 dB to ~65 dB after transmission through two consecutive sensors (average) - PLB in air and underwater	- Emission of ultrasound pulses by the sensor - Pulsing from the sensor, four pulses per sensor - Transmission across links was observed - Amplitude dropped from 80 dB to ~70 dB after transmission through two consecutive sensors (average) - Pulsing in air and underwater

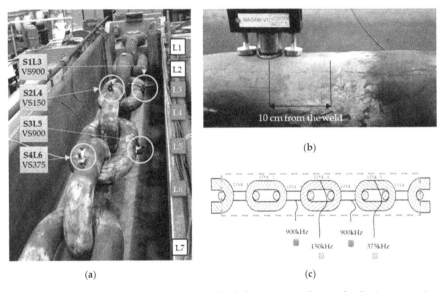

(a)   (c)

(b)

Figure 3. Sensor deployment on the chain in (**a**) the full-size test rig. Sensors fixed using magnetic holders 10 cm away from the link's weld (**b**). Illustrative drawing of the chain (**c**).

Figure 4 shows the results of the PLB calibration in air. Sensor 1 (S1), located on Link 3 (L3) and referred to as S1L3, and sensor S3L5 both correspond to the VS900 broadband AE sensor and seemed to receive higher dB levels of ~88 dB. For sensor S2L4 and sensor S4L6, corresponding respectively to VS150 and VS375, the dB levels received were approximately 10 dB lower at around 78 dB. From the figure, it is clear that the AE wave propagated through the links—the dB levels dropped to ~64 dB in the first adjacent link, ~54 dB in the second, and ~45 dB in the third (Figure 4a). Signal reproducibility was also demonstrated using the pulsing function. Each sensor generated four consecutive signals of peak amplitude equal to 81 dB (top rectangle in Figure 4b), which were intercepted by the rest of the sensors. As can be seen in Figure 4b, the wave was attenuated to a minimum of ~54 dB in the furthest sensor.

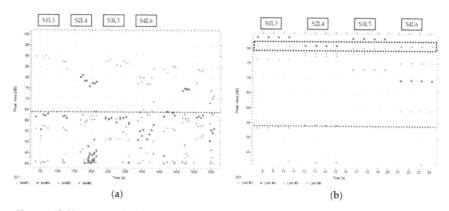

(a)   (b)

Figure 4. Calibration results for S1L3 (yellow), S2L4 (red), S3L5 (blue), and S4L6 (green) using (**a**) PLB and (**b**) the pulsing technique.

Figure 5 shows the behaviour of a wave generated in Link 3 from a PLB calibration event, where the peak amplitude of the signal ranges from ~90 dB in Link 3 to less than 45 dB in Link 6. The evolution of the other AE signal time domain features as the wave propagates through the chain links is shown in Figure 5. As the wave propagates through the links, the peak amplitude, number of counts, and average frequency decrease, whereas the rise time, duration, and time of arrival increase.

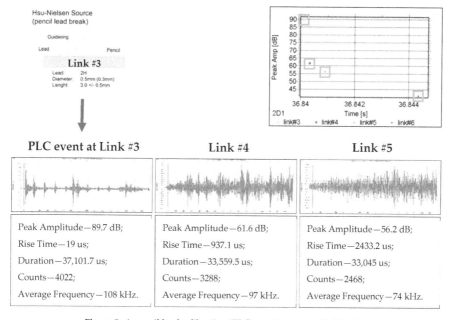

Figure 5. A pencil lead calibration (PLC) event generated in Link 3.

It should be noted that the signal reproducibility was also verified when the tank was being filled with water and the links were partially (almost completely) submerged (Figure 6).

In-air calibration was performed in order to understand in detail the propagation and resulting AE patterns across the links. The results obtained from underwater calibration during and after the tank filling was completed are similar to those obtained from the in-air calibration. However, due to the complexity of the setup, the values and conclusions acquired when full access was granted to the sensors' positions (in-air) were considered as the final calibration results.

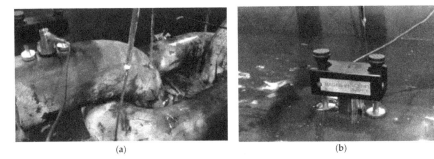

(a)   (b)

Figure 6. AE reproducibility was verified as the tank was being filled with water (**a**) before the onset of the experiment (**b**).

3. Results and Discussion

The total duration of the AE monitoring period was 72 days. Taking into account the findings of the reproducibility tests, the AE acquisition threshold was originally set at 45 dB. However, due to the large unsustainable level of events captured, mostly noise, the threshold was initially increased to 55 dB then finally to 60 dB within the first 24 h of the experiment. The data recorded in this experiment included the number of AE hits (number of emissions detected by the sensors) and AE time domain features, including the number of counts, energy, rise time, time of arrival, and duration, in addition to the load and displacement measured by the load cells. Sensors S1L3 and S4L6 failed during the first days of the experiment for unknown reasons and were not replaced. Only the data from S2L4 and S3L5 were considered in the present analysis.

The first time the pre-set limit displacement was exceeded occurred after 4,333,424 cycles. After the water tank was fully drained, a complete inspection of the chain using visual inspection and magnetic particle inspection (MPI) revealed a ~5 mm crack located at the weld area in Link 3 (Figure 7a). No other indication of cracking was found, and the experiment was resumed. At 4,923,552 cycles, close to the fatigue life (5×10^6 cycles when similar loading conditions are applied), the experiment was interrupted. A MPI inspection revealed that the same crack in Link 3 had grown to ~120 mm (Figure 7b). A complete inspection of the chain did not reveal any other cracks.

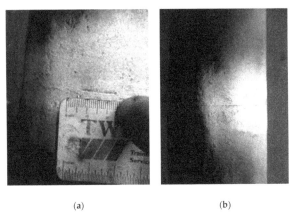

(a)   (b)

Figure 7. Crack indications after magnetic particle inspection (MPI) at (**a**) 4,333,424 cycles (5 mm) and (**b**) 4,923,552 cycles (120 mm).

The crack of interest was revealed 31 days before the end of the experiment. It should be noted that a few interruptions occurred at the beginning of the experiment. Fortunately, these were during the early stages of the test and the information collected after day 41 can be considered accurate. For these reasons, only the AE data that were acquired during the last 31 days were processed. The key AE signal features were examined in an effort to establish correlations that would link the AE activity to the fatigue crack evolution. Some of the relationships considered were:

- Peak amplitude (dB) vs. load (kN): It was desirable to establish a correlation between the AE amplitude of the events detected and the loading state of the chain links;
- Number of hits vs. average frequency (kHz): The frequency analysis was performed in order to measure the characteristic central frequency and characterise the damage evolution;
- Cumulative energy vs. number of hits: The cumulative AE energy is associated with the cumulative energy of the AE hits. An increase in the cumulative energy's slope may be related to the evolution of the damage [38].

The AE signal features detected during the experiment were studied in detail. An example of the peak amplitude vs. load plot is shown in Figure 8. As can be seen in the figure, the events detected by the sensor attached to Link 5 are exclusively located within certain load ranges: 3150–3250 kN, 3700–3780 kN, and 3800–3850 kN. The corresponding events detected in Link 4 cannot be distinguished due to their uniform distribution. The location pattern of the red events without a clear connection to any cracking suggests they are most likely due to a mechanical noise, possibly caused by the test setup or surface friction where the links meet. It is usually accepted that in fatigue cracking, the AE signals generated close to the high end of the loading range are related to crack growth. Roberts et al. [22] showed that any correlation attempt was unsuccessful when the complete loading range was considered. Attributing the top 10 and 5% of the load range to crack growth improved the correlation to AE significantly. Bhuiyan et al. [39] associated different AE signal groups to the crack behaviour based on a spectral analysis of the AE data. The authors discovered that the AE events generated in the top 75–85% of the load range were statistically different from those below 60% and were related to the fatigue crack growth. In the present study, due to the large amount of noise generated by the full-scale experiment, only data collected above 3820 kN are shown to offer good correlations.

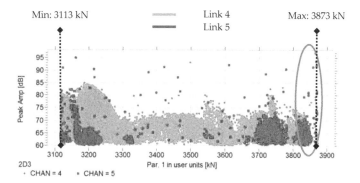

Figure 8. Peak amplitude (dB) vs. load (kN). Details of the minimum–maximum loading, and the 3820–3873 kN range is highlighted.

A closer look at the frequency content of the signals detected during the experiment, presented in Figure 9, shows a continuous shift of the average frequency towards higher values throughout the monitoring period.

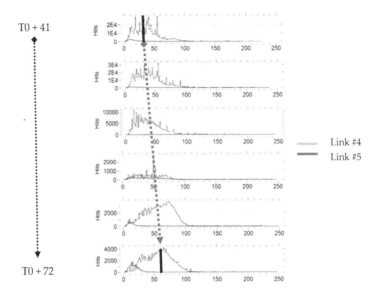

TO + 41

TO + 72

Link #4
Link #5

Figure 9. Hits vs. average frequency (kHz). Shift of the average frequency.

The shift in the AE signal frequency content may be indicative of a change in the damage mode. For example, in reinforcing steel bars used in reinforced concrete, a decrease of the average AE frequency is attributed to the onset of phase three of the corrosion loss phenomenological model [40]. Additionally, tensile cracking is associated with a higher AE signal frequency range when compared to shear cracking in the same class of materials [41]. In composites, the AE signal frequency can be correlated to the fracture mechanism [42].

Finally, the observation of the cumulative energy during the last 31 days of the experiment did not show any particular change except during the final cycles (yellow circle in Figure 10). The fast rise indicates a rapid release of energy corresponding to rapid crack propagation. No statistical change was noted in the slope of the cumulative energy after the crack was initially detected. A quick increase in the cumulative energy precedes the exposure of the 120 mm crack. The displacement measured by the gauge did not exceed the pre-set limit until 120 mm was reached. This indicates a sharp crack growth, as seen in Figure 10.

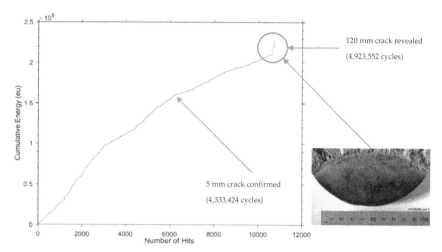

Figure 10. Cumulative energy (load > 3,820 kN) evolution as measured by the sensor fixed on Link 4. Photograph shows the fracture surface corresponding to the final crack.

4. Conclusions

The aim of this work was to assess the ability of AE to detect crack initiation and growth in mooring chains under realistic loading and environmental conditions. After careful evaluation of the different AE signal features and all possible correlations, it appears that the frequency content of the AE signals is the most promising parameter. An increase of the average frequency is observed with the growth of the crack in the chain link. However, due to the challenging environment and the high level of noise recorded, a more comprehensive frequency analysis will be needed.

Author Contributions: Conceptualization, Á.A.; Data curation, Á.A., J.T.; Formal analysis, Á.A., S.S.; Investigation, Á.A., A.K.; Methodology, Á.A., J.T.; Project administration, Á.A.; Resources, Á.A.; Software, Á.A., J.T.; Writing—original draft preparation, Á.A.; Supervision, C.M., S.S., T.-H.G.; Validation, C.M., S.S.; Writing—review & editing, Á.A., A.K., T.-H.G.

Funding: This research received no external funding.

Acknowledgments: The authors gratefully acknowledge the opportunity provided by TWI's JIP project (22116) sponsors to make use of the mooring chain full-scale test rig at TWI's facilities.

Conflicts of Interest: The JIP sponsors had no role in the design of the study; in the collection, analyses, or interpretation of data; in the writing of the manuscript; or in the decision to publish the results.

References

1. Moan, T. Life cycle structural integrity management of offshore structures. *Struct. Infrastruct. Eng.* **2018**, *14*, 911–927. [CrossRef]
2. Thorpe, T.; Scott, P.; Rance, A.; Silverster, D. Corrosion fatigue of BS 4360:50D structural steel in seawater. *Int. J. Fatigue* **1983**, *5*, 123–133. [CrossRef]
3. Agerskov, H.; Pedersen, N. Fatigue life of offshore steel structures under stochastic loading. *Struct. Infrastruct. Eng.* **1992**, *118*, 2101–2117. [CrossRef]
4. Alawi, H.; Ragab, A.; Shaban, M. Corrosion fatigue crack growth of steels in various environments. *J. Eng. Mater. Technol. (ASME)* **1989**, *111*, 40–45. [CrossRef]
5. Komai, K.; Noguchi, M.; Okamoto, H. Growth characteristics of surface fatigue cracks of high-tensile strength steel in synthetic seawater. *JSME Int. J.* **1988**, *31*, 613–618. [CrossRef]
6. Komai, K.; Minoshima, K.; Kinoshita, S.; Kim, G. Corrosion fatigue crack initiation of high-tensile-strength steels in synthetic seawater. *JSME Int. J.* **1988**, *31*, 606–612. [CrossRef]

7. Monsalve-Giraldo, J.S.; Dantas, C.M.S.; Sagrilo, L.V.S. Probabilistic fatigue analysis of marine structures using the univariate dimension-reduction method. *Mar. Struct.* **2016**, *50*, 189–204. [CrossRef]

8. Low, Y. A variance reduction technique for long-term fatigue analysis of offshore structures using Monte Carlo simulation. *Eng. Struct.* **2016**, *128*, 283–295. [CrossRef]

9. Zhang, Y.-H.; Dore, M. Fatigue crack growth assessment using BS 7910:2013—Background andrecommended developments. *Int. J. Press. Vessels Pip.* **2018**, *168*, 79–86. [CrossRef]

10. Shukla, A.; Karki, H. Application of robotics in offshore oil and gas industry—A review Part 2. *Robot. Auton. Syst.* **2016**, *75*, 508–524. [CrossRef]

11. Raine, G.A.; Lugg, M.C. ROV inspection of welds—A reality. *Insight Non-Destruct. Test. Cond. Monit.* **1996**, *38*, 346–350.

12. Papazian, J.M.; Nardiello, J.; Silberstein, R.P.; Welsh, G.; Grundy, C.C.D.; Evans, L.; Godfine, N.; Michaels, J.E.; Michaels, T.E.; Li, Y.; et al. Sensors for monitoring early stage fatigue cracking. *Int. J. Fatigue* **2007**, *29*, 1668–1680. [CrossRef]

13. Chan, H.; Masserey, B.; Fromme, P. High frequency guided ultrasonic waves for hidden fatigue crack growth monitoring in multi-layer model aerospace structures. *Smart Mater. Struct.* **2015**, *24*, 1–10. [CrossRef]

14. Cho, H.; Lissenden, C. Structural health monitoring of fatigue crack growth in plate structures with ultrasonic guided waves. *Struct. Health Monit.* **2012**, *11*, 393–404. [CrossRef]

15. Kuang, K.; Cantwell, W. Use of conventional optical fibers and fiber Bragg gratings for damage detection in advanced composite structures: A review. *Appl. Mech. Rev.* **2003**, *56*, 493–513. [CrossRef]

16. Bernasconi, A.; Carboni, M.; Comolli, L. Monitoring of fatigue crack growth in composite adhesively bonded joints using Fiber Bragg Gratings. *Procedia Eng.* **2011**, *10*, 207–212. [CrossRef]

17. Silva-Munoz, R.; Lopez-Anido, R. Structural health monitoring of marine composite structural joints using embedded fiber Bragg grating strain sensors. *Compos. Struct.* **2009**, *89*, 224–234. [CrossRef]

18. Deans, W.; Richards, C. A simple and sensitive method of monitoring crack and load in compact fracture mechanics specimens using strain gages. *J. Test. Eval.* **1979**, *7*, 147–154.

19. Ihn, J.-B.; Chang, F.-K. Detection and monitoring of hidden fatigue crack growth using a built-in piezoelectric sensor/actuator network: I. Diagnostics. *Smart Mater. Struct.* **2004**, *13*, 609. [CrossRef]

20. Fowler, T. Chemical industry application of acoustic emission. *Mater. Eval.* **1992**, *50*, 875–882.

21. Gong, Z.; Nyborg, E.; Oommen, G. Acoustic emission monitoring of steel railroad bridges. *Mater. Eval.* **1992**, *50*, 883–887. [CrossRef]

22. Roberts, T.; Talebzadeh, M. Acoustic emission monitoring of fatigue crack propagation. *J. Constr. Steel Res.* **2003**, *59*, 695–712. [CrossRef]

23. Yu, J.; Ziehl, P.; Zarate, B.; Caicedo, J. Prediction of fatigue crack growth in steel bridge components using acoustic emission. *J. Constr. Steel Res.* **2011**, *67*, 1254–1260. [CrossRef]

24. Berkovits, A.; Fang, D. Study of fatigue crack characteristics by acoustic emission. *Eng. Fract. Mech.* **1995**, *51*, 401–409. [CrossRef]

25. Lindley, T.; Palmer, I.; Richards, C. Acoustic emission monitoring of fatigue crack growth. *Mater. Sci. Eng.* **1978**, *32*, 1–15. [CrossRef]

26. Mathisen, J.; Larsen, K. Risk-based inspection planning for mooring chain. *J. Offshore Mech. Arct. Eng. (Trans. ASME)* **2004**, *126*, 250–257. [CrossRef]

27. Cheng, X.; Zhang, X.; Zhang, H. The influence of hydrogen on deformation under the elastic stress in mooring chain steel. *Mater. Sci. Eng. A* **2018**, *730*, 295–302. [CrossRef]

28. Cheng, X.; Zhang, H.; Li, H.; Shen, H. Effect of tempering temperature on the microstructure and mechanical properties in mooring chain steel. *Mater. Sci. Eng.* **2015**, *636*, 164–171. [CrossRef]

29. Cheng, Y.; Ji, C.; Zhai, G.; Oleg, G. Nonlinear analysis for ship-generated waves interaction with mooring line/riser systems. *Mar. Struct.* **2018**, *59*, 1–24. [CrossRef]

30. Rivera, F.G.; Edwards, G.; Eren, E.; Soua, S. Acoustic emission technique to monitor crack growth in a mooring chain. *Appl. Acoust.* **2018**, *139*, 156–164. [CrossRef]

31. Angulo, A.; Edwards, G.; Soua, S.; Gan, T.-H. Mooring integrity management: Novel approaches towards in situ monitoring. In *Structural Health Monitoring-Measurement Methods and Practical Applications*; Intechopen: London, UK, 2017.

32. Angulo, Á.; Allwright, J.; Mares, C.; Gan, T.H.; Soua, S. Finite element analysis of crack growth for structural health monitoring of mooring chains using ultrasonic guided waves and acoustic emission. *Procedia Struct. Integr.* **2017**, *5*, 217–224. [CrossRef]

33. Bastid, P.; Smith, S. Numerical analysis of contact stresses between mooring chain links and potential consequences for fatigue damage. In Proceedings of the ASME 2013 32nd International Conference on Ocean, Offshore and Arctic Engineering Volume 2B: Structures, Safety and Reliability, Nantes, France, 9–14 June 2013.

34. Perez, I.; Bastid, P.; Venugopal, V. Prediction of residual stresses in mooring chains and its impact on fatigue life. In Proceedings of the ASME 2017 36th International Conference on Ocean, Offshore and Arctic Engineering Volume 3A: Structures, Safety and Reliability, Trondheim, Norway, 25–30 June 2017.

35. ASTM E1106-12(2017). *Standard Test Method for Primary Calibration of Acoustic Emission Sensors*; ASTM Volume 03.03 Nondestructive Testing (I): C1331–E2373; ASTM International: West Conshohocken, PA, USA, 2018.

36. ASTM E976-15. *Standard Guide for Determining the Reproducibility of Acoustic Emission Sensor Response*; ASTM Volume 03.03 Nondestructive Testing (I): C1331–E2373; ASTM International: West Conshohocken, PA, USA, 2018.

37. Hsu, N.; Breckenridge, F. Characterization and calibration of acoustic emission sensors. *Mater. Eval.* **1981**, *39*, 60–68.

38. Bourchak, M.; Farrow, I.; Bond, I.; Rowland, C.; Menan, F. Acoustic emission energy as a fatigue damage parameter for CFRP composite. *Int. J. Fatigue* **2007**, *29*, 457–470. [CrossRef]

39. Bhuiyan, M.; Giurgiutiu, V. The signature of acoustic emission waveforms from fatigu crack advancing in thin metallic plates. *Smart Mater. Struct.* **2018**, *27*, 1–15. [CrossRef]

40. Kawasaki, Y.; Wakuda, T.; Kobarai, T.; Ohtsu, M. Corrosion mechanisms in reinforced concrete by acoustic emission. *Constr. Build. Mater.* **2013**, *48*, 1240–1247. [CrossRef]

41. Tomoda, Y.; Mori, K.; Kawasaki, Y.; Ohtsu, M. Monitoring corrosion-induced cracks in concrete by acoustic emission. In *Proceedings of the 7th International Conference on Fracture Mechanics of Concrete and Concrete Structures, Seoul, Korea, 23–28 May 2010*; Korea Concrete Institute: Seoul, Korea, 2010; ISBN 978-89-5708-181-5.

42. Groot, P.D.; Wijnen, P.; Janssen, R. Real-time frequency determination of acoustic emission for different fracture mechanism in carbon/epoxy composites. *Compos. Sci. Technol.* **1995**, *55*, 405–412. [CrossRef]

Article

Using ANN and SVM for the Detection of Acoustic Emission Signals Accompanying Epoxy Resin Electrical Treeing

Arkadiusz Dobrzycki [1,*], Stanisław Mikulski [1] and Władysław Opydo [2]

1   Faculty of Electrical Engineering, Poznań University of Technology, Piotrowo 3A str., 60-965 Poznań, Poland; stanislaw.mikulski@put.poznan.pl
2   Faculty of Telecommunications, Computer Science and Electrical Engineering, UTP University of Science and Technology, Al. prof. S. Kaliskiego 7, 85-796 Bydgoszcz, Poland; wladyslaw.opydo@put.poznan.pl
*   Correspondence: arkadiusz.dobrzycki@put.poznan.pl; Tel.: +48-061-665-2685

Received: 26 February 2019; Accepted: 8 April 2019; Published: 12 April 2019

Featured Application: From a practical point of view, the phenomenon of electrical treeing can be an exploitation problem, especially in the elements or places with locally increasing of electric field intensity, because it is an irreversible process. Such places are e.g., connecting points of the power network elements or parts of electrical devices in which there is constant insulation. The results of the carried out analyzes may complement the knowledge of identifying the type of PD occurring based on selected signal parameters.

Abstract: Electrical treeing is one of the effects of partial discharges in the solid insulation of high-voltage electrical insulating systems. The process involves the formation of conductive channels inside the dielectric. Acoustic emission (AE) is a method of partial discharge detection and measurement, which belongs to the group of non-destructive methods. If electrical treeing is detected, the measurement, recording, and analysis of signals, which accompany the phenomenon, become difficult due to the low signal-to-noise ratio and possible multiple signal reflections from the boundaries of the object. That is why only selected signal parameters are used for the detection and analysis of the phenomenon. A detailed analysis of various acoustic emission signals is a complex and time-consuming process. It has inspired the search for new methods of identifying the symptoms related to partial discharge in the recorded signal. Bearing in mind that a similar signal is searched, denoting a signal with similar characteristics, the use of artificial neural networks seems pertinent. The paper presents an effort to automate the process of insulation material condition identification based on neural classifiers. An attempt was made to develop a neural classifier that enables the detection of the symptoms in the recorded acoustic emission signals, which are evidence of treeing. The performed studies assessed the efficiency with which different artificial neural networks (ANN) are able to detect treeing-related signals and the appropriate selection of such input parameters as statistical indicators or analysis windows. The feedforward network revealed the highest classification efficiency among all analyzed networks. Moreover, the use of primary component analysis helps to reduce the teaching data to one variable at a classification efficiency of up to 1%.

Keywords: solid dielectrics; acoustic emission; artificial neural networks; electrical treeing; wavelets; non-destructive testing; high-voltage insulating systems

---

1. Introduction

The major factor that contributes to the deterioration of the insulation characteristics of dielectrics is partial discharges (PD), which occur under the influence of high-intensity electric fields. Partial discharges occur on the surface of or inside dielectrics, causing deterioration of their electrical insulating characteristics. PD in solid dielectrics often entails electrical treeing, which involves the formation of conductive or semi-conductive channels in the shape of trees inside the dielectric. Discharges in the channels result in further development of trees and finally lead to the low-resistance short-circuit of electrodes. That is why the forecasts for insulation in which treeing occurs are disastrous. Therefore, the information on whether treeing has been initiated in the insulation becomes of supreme importance for the purposes of high voltage (HV) insulation condition diagnostics.

Early studies on partial discharges (PD) occurring in dielectrics were carried out at the beginning of the Twentieth Century by Rayner, who investigated the impact of partial discharges on breakdown [1]. The development of the studies was stimulated by the development of power engineering and the high unreliability of paper and oil insulation used in HV cables, transformers, and generators. The studies were continued in the 1930s by Robinson [2] and Whitehead [3], who demonstrated that partial discharges in paper and oil insulation caused electrical treeing, which involves the formation of channels in the shape of trees or shrubs, or a sight in a dielectric, and are the main cause for the low durability of high-voltage cables with this type of insulation.

The main direction of studies in the second half of the Twentieth Century included the following: electrical treeing initiation [4], studies on the relationship between tree length, voltage impact time, and time to breakthrough [5,6], studies on the properties of cables with solid polyethylene insulation [7], and insulation breakthrough tests [8]. There have also been studies on partial discharge detection and measurement methods. Initially, mainly oscilloscopic methods were proposed by Tykociner [9]. Some studies on the development of insulation ageing models were also carried out [10].

In addition to detailed papers, several publications summarized and systematized the current knowledge status [11,12].

The turn of the Twenty-first Century saw studies related to forecasting treeing in dielectrics. Engineering progress contributed to the use of modern and sophisticated methods of partial discharge testing.

Shimizu [13,14] pointed out the possibility of examining treeing incubation by means of electroluminescence. Papers by Kudo [15] and Dissado [16] discussed the ways to predict treeing propagation using the theory of fractals and deterministic chaos. Since trees have a fractal-like nature, the authors used fractal size to identify the shape of the tree being formed. The theory of deterministic chaos was, in turn, used to identify the probable directions of tree development. One can observe a tendency for the development of non-invasive partial discharge methods, and the acoustic emission method in particular [17–22]. Advanced signal analysis methods, such as wavelet transformation or artificial neural networks, used for analysis parameters of solid insulation systems and identification of the discharge source and type have been employed [10,23–29].

Among the methods used for the detection of partial discharges (e.g., in the transformer tank), the acoustic emission method is common, focusing on the recording and analysis of an acoustic wave propagating in the material as a result of external impact exerted on it. The sources of the impact may include mechanical pressure (testing the stress inside the material) [30–34], electric field (partial discharge testing), or magnetic fields (used for Barkhausen noise analysis).

The purpose of the study was to try to automate the process of identifying the condition of insulation materials by developing a neural classifier allowing the detection of symptoms related to treeing in the recorded acoustic emission signals.

2. Test Setup and the Course of the Experiment

The first stage of the studies involved measurements of acoustic emission signals in electrically-stressed epoxy resin samples. The samples had cubicoid shapes of 25 mm × 10 mm × 4 mm. One sample surface of 25 mm × 4 mm was ground and coated with varnish conducive to electric current.

Since the process of tree channels forming in a dielectric may last very long, an electrode made of a T10 surgical needle with liquid resin poured inside the sample during sample formation was used, such that the distance between the sample bottom and the electrode end ranged from 1–3 millimeters. The procedure allowed obtaining the electric field intensity between the needle electrode and the electrode applied to the bottom of the sample, which was high enough to reduce the tree-forming duration to a few hours. The needle was connected to the neutral terminal of the transformer. The opposite base of the sample was adjacent to a plane copper electrode. The electrode was connected through a resistor of 0.5 MΩ resistance to a high-voltage terminal of a test transformer with the ratio of 220 V/30 kV and power of 10 kVA. The voltage was measured with an electrostatic voltmeter.

As part of the tests, acoustic signals were recorded for a dozen or so samples in which, under the influence of a high intensity of the electric field, the process of forming an electric tree began. The measurements were made for variable values of the supply voltage with a 50-Hz frequency and different distances between the electrodes.

The elastic waves of acoustic frequencies were emitted from the analyzed sample through a wave-guide made of a steel rod of 2 mm in diameter. One of its ends was put into a hole bored in the sample, while the other was connected to an electroacoustic converter.

As regards measurement time, the studied sample was placed in a methyl polymethacrylate vessel filled with electrical insulating oil. The voltage between the electrodes during the tests ranged from a few to several kV. The AE signals were measured by means of a Physical Acoustic Corporation (PAC) R3α electroacoustic converter and a filtering and amplifying system composed of 2/4/6 type pre-amplifier from PAC, a 20 ÷ 1000 kHz transmission band filter, and PAC AE5A amplifier. Upon amplification, the signal was recorded in the computer memory with an data acquisition (DAQ) NI–USB6251 card, which enabled signal recording with a sampling frequency up to 1 MS/s and 16-bit resolution. The test setup schema, the view of the measuring equipment, and the view a sample holder are presented respectively in Figure 1a–c.

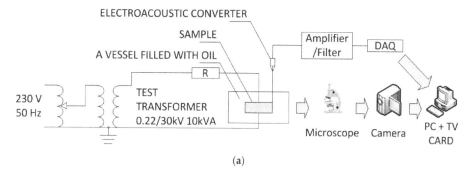

(a)

Figure 1. *Cont.*

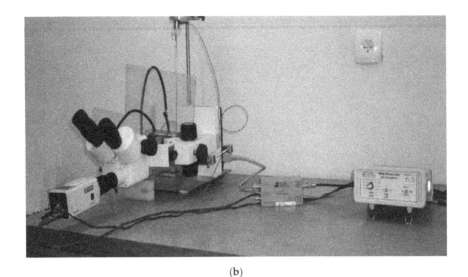

(b)

(c)

Figure 1. The test setup schema (**a**), view of the measuring equipment (**b**), and view of the sample holder (**c**) for measuring acoustic signals accompanying the treeing of solid dielectrics.

To ensure that the recorded signals concerned the electrical treeing while we were preparing the experiment, we also tested and observed under a microscope different states of the system, i.e., without any PDs, with corona, etc. We noticed that each of the PD types was connected with different waveforms of signals, and this can be found also in different researchers' results [35]. After that test, we rebuilt/reconfigured the setup and chose a voltage range to be sure that other types of PDs would not be present. Physically, electrical treeing is a combination of chemical and physical changes inside the specimen, so no one can be sure whether that particular signal is connected with the breaking polymer chain or PD inside the existing channel.

3. Method of Features' Extraction

3.1. Features' Definition

In order to extract the teaching data for neural classifiers, the analyzed signal fragment was divided into x blocks with N length of samples, where:

$$x = \{x_1, \ldots, x_n, \ldots, x_N\}. \tag{1}$$

For each x signal block, a set of statistical parameters typically used in the analysis of AE signals was identified [36,37], which was further used as input data for the neural network. The following parameters were used: signal energy (e), band power (p_{BP}), signal upper envelope (env), skewness ($skew$), and kurtosis ($kurt$).

The signal energy is identified based on the following relationship:

$$e = \sum\nolimits_{n=1}^{N} x_n^2 \tag{2}$$

Skewness, as the probability distribution asymmetry measure, is identified in the following way:

$$skew = \frac{\bar{x} - med(x)}{\sqrt{\sigma}} \tag{3}$$

where $\bar{x}$ is the mean value of samples, med (x) the block median, and σ the sample block variance.

Kurtosis, which is the measure of the results' concentration around the mean, amounts to:

$$kurt = \frac{\mu_4}{\sigma^4} - 3 \tag{4}$$

where μ_4 is the fourth central moment of samples in the block and σ the sample block variance. A teaching data vector d was identified for each x signal sample block, where:

$$d = \{e, p_{BP}, env, skew, kurt\} \tag{5}$$

In order to obtain training information y, which belongs to the set $\{0,1\}$, the recorded signal was blasted using the wavelet decomposition with the fifth-order Daubechies wave. Then, using the method of exceeding the trigger threshold, which was set at 20 mV for the presented problem, the information value y was allocated for each training set: 0 for no acoustic emission and 1 for the acoustic emission event (these are the red areas in Figure 2). Figure 2 presents one signal fragment with the sampling frequency of 1 MS/s selected for further studies. The red areas mark acoustic emission events determined on the basis of threshold crossing.

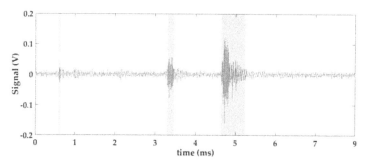

Figure 2. Fragment of the recorded signal where dielectric degradation occurred.

3.2. Passband Power

The selection of the band used for identifying band power p_{BP} was made based on the averaged spectrum for the recorded AE signals. Figure 3 presents an averaged spectrum for 1000 AE signals recorded and the spectrum of a selected signal fragment where no acoustic emission was observed. For both spectrums, the transfer function of the selected detector (R3α), given in the calibration datasheet, was taken into account.

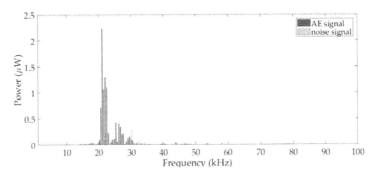

Figure 3. Averaged signal spectrum for 1000 recorded acoustic emission signals and measurement noise spectrum.

The presented spectrum was used to select a signal band with a frequency ranging from 15 kHz–30 kHz for further analysis because the power of the signal recorded within the band increased significantly when acoustic emission occurred.

3.3. Block Length

Selecting the right length of signal blocks to study is among the most important factors affecting classification quality. When performing an analysis using a taught network, we are not able to predict if the analyzed signal block starts before, during, or at the end of the acoustic emission signal, because it is necessary to know the influence of the signal window position against the acoustic signal on the values of selected network teaching features.

In order to identify the aforementioned relationships, window lengths of 10, 60, 200, and 300 samples were used. The windows were applied to a signal containing a single acoustic emission and moved by one sample (the window position against the AE signal was changed this way). A dataset was calculated for each position of the window. Figure 4 presents the values of each teaching feature as a function of the window position, at different window lengths, against the AE signal: kurtosis (a), skewness (b), and energy (c) respectively. Moreover, Figure 4c presents the identified upper envelope of the sample signal.

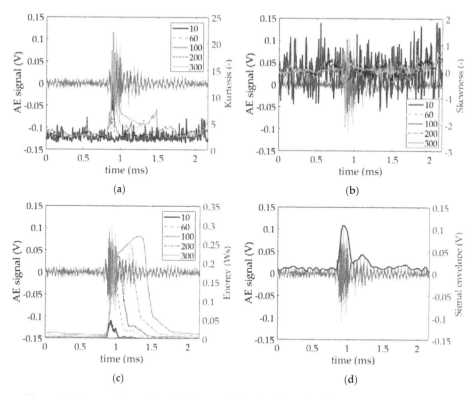

Figure 4. Influence of the position of different window lengths against the AE signal on the teaching properties' values: kurtosis (**a**); skewness (**b**); energy (**c**); upper envelope (**d**); the recorded signal wavelength is in the background.

3.4. Principal Component Analysis

Principal component analysis (PCA) is a statistical method for factor-based analysis. A collection *N* of *K* variable observations is analyzed as a collection of *N* points distributed in a *K*-dimensional space. Using the distribution of singular values for the deviation covariance matrix, it is possible to develop a new system that maximizes the variance of subsequent coordinates. This way, it is possible to reduce the space size (decrease the number of the analyzed input data) [30] and the size of the neural network, while maintaining the highest possible amount of information regarding the input process.

4. Artificial Neural Network Classifiers Used to Select EA Signals

4.1. Feedforward Neural Network

In a forward non-linear neural network (FNN), the flow of information (signals) is unidirectional. Its structure consists of the input layer, two or more hidden layers, and the output layer (Figure 5). All layer inputs can be linked only with the neurons in the preceding layer. Input signals *x* in a neuron are added to weights *w*. Neuron *y*'s output signal is calculated using a non-linear activation function φ, according to the relationship [38]:

$$y = \varphi\left(\sum_{m=1}^{M} w_m x_m\right) \tag{6}$$

The unipolar sigmoid function is among the most commonly-used activation functions:

$$\varphi(z) = \frac{1}{1 + e^{-z}} \tag{7}$$

alongside a bipolar function (hyperbolic tangent):

$$\varphi(z) = \frac{1 - e^{-z}}{1 + e^{-z}} \tag{8}$$

For learning feedforward neural network (FFN), the Levenberg–Marquardt algorithm was used. This method is one of the most effective learning algorithms. It has high convergence when network weights are near the optimal solution (as the Gauss–Newton method) and when the network is far from the optimal solution (as the descent gradient method). Detailed information about the Levenberg–Marquardt algorithm can be found in [39].

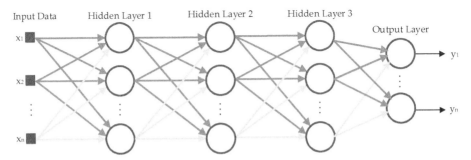

Figure 5. Sample structure of a feedforward network with three hidden layers.

4.2. Radial Basis Functions

A radial basis functions (RBF) network, similar to FNN, is a network with an oriented flow of signals. The network neurons have a radial activation function φ. The neuron output y is described by the relationship [40]:

$$y_j(x) = \sum_i^m w_{ij} \varphi(\|x - c_i\|) \tag{9}$$

where i is the hidden layer neuron number, j the radial network output number, w_{ij} the coefficient of the weight, x the input data vector, and c_i the center of the radial function for the ith neuron of the hidden layer.

In such a neuron, the value of the output signal is not proportional to the scalar product of x inputs and w neuron weights, but inversely proportional to the distance between x and the central point of radial function c located at the hyperspace of the network input parameters. The Gaussian function is among the most commonly-used radial functions. Figure 6 presents the structure of an RBF network.

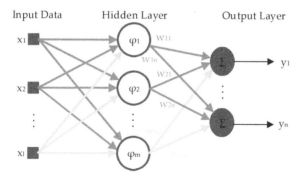

Figure 6. RBF network diagram.

4.3. Wavelet Neural Network

A wavelet neural network (WNN) structure consists of an input, output, and hidden layer of wavelet neurons (Figure 7).

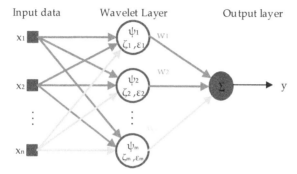

Figure 7. Wavelet neural network structure.

The function of wavelet neurons' activation φ is described by the relationship [41]:

$$\varphi(x) = \prod_i^m \psi\left(\frac{x_i - \zeta_i}{\varepsilon_i}\right). \tag{10}$$

where x is the input signal vector, ψ the wavelet function, ζ the translation parameter, and ε the scale parameter.

The process of such network teaching involves the selection of the scale parameters and shifting each wavelet ψ used inside a wavelet neuron nucleus. For the study, the authors used Mexican hat wavelet function.

Contrary to other network types, drawing of the initial values of the wavelet parameters ζ and ξ may result in the teaching algorithm being stuck in the local minimum. That is why network teaching is preceded by the initialization of initial parameters. A feature selection method developed by Oussar and Dreyfus was used for the study [42]. It consists of the following steps:

- Step 1: creating a library of wavelets with different parameters.
- Step 2: removing those wavelets not falling within the variability range of each input parameter.
- Step 3: developing a ranking of wavelets and iterative selection of the best wavelets using the Gram–Schmidt ortho-normalization algorithms.

Subsequently, fully-initialized WNN was learned with the backpropagation algorithm. A detailed description of the WNN's structure and learning algorithm used for the study can be found in [41].

4.4. Support Vector Machine

A support vector machine (SVM) was another network used in the studies. The purpose of the SVM network operation is to use a hyperplane spread in the hyperspace of the input parameters, which separates a collection of points (sets of input parameters) belonging to two different classes, with a certain error margin.

It is a type of binary classifier. Its operating principle can be presented on the example of a network with two input parameters $x = (x_1, x_2)$. Figure 8 presents the operating principle of such a classifier. The points lying in the plane (x_1, x_2) are divided into two classes marked in green and viloet. The SVM classifier searches for a straight line (black line in the drawing), which will maximize the error margin limited on both sides of the curve with support vectors, the brown lines in the drawing. The points exceeding the limits of the error margin will be misclassified. The purpose of SVM teaching is to select the coefficients of the straight line separating the points in such a way that the error margin is maximized (distance between the support vectors). Algorithms of non-linear optimization with constraints, and Lagrange's method in particular, were used for the network teaching [43].

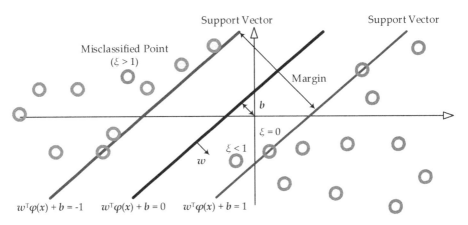

Figure 8. Principle of support vector machine operation.

5. Results

Each of the described networks was taught in order to classify the signals into two groups: that which contains and that which does not contain acoustic emission signals accompanying dielectric treeing. The previously-described properties were used to develop teaching data for the blocks with the following lengths: 10, 60, 100, 200, and 300 samples.

In the first part of the results, classification efficiency was specified for a test fragment of the signal other than the network teaching signal. In the second section, the signal classification efficiency was presented for the teaching data reduced by means of PCA.

Additionally, the optimum size of the network was achieved in the cross-validation, which consisted of dividing the set of reference data into equinumerous k subsets. The network with a given number of neurons was learned on the basis of k-1 subsets, and the unused subset was the validation data. This process was repeated k times. Thus, the obtained k learning outcomes networks were averaged.

5.1. Classifier Efficiency Analysis

Each network taught was subjected to a test developed based on a random signal fragment. Table 1 presents the classification efficiency for each network type at variable block lengths based on which, calculations were made.

Table 1. Classification efficiency test results for variable data block lengths. WNN, wavelet neural network.

Network Type	Block Length (Samples)	Neurons	Efficiency (%)	Network Type	Block Length (Samples)	Efficiency (%)
FFN	10	10	97.2	SVM	10	90.4
	60	14	97.6		60	87.8
	100	16	97.3		100	88.5
	200	15	97.5		200	95.6
	300	11	96.7		300	96.3
WNN	10	3	90.9	RBF	10	95.8
	60	3	88.2		60	95.3
	100	3	86.2		100	94.6
	200	2	94.7		200	94.8
	300	3	95.7		300	95.6

5.2. Classification Using PCA

Since the most stable efficiency results for each network were obtained for the data identified based on the block length of 300 samples, the data were subjected to PCA, and PCs were developed from them. Table 2 shows variances for different components and their percent share in the total variance.

Table 2. Classification efficiency test results for variable data block lengths.

PC	Variance (–)	Variance Share (%)
1st	3.6092	98.17
2nd	0.0606	1.65
3rd	0.0066	0.18
4th	6.3×10^{-5}	0
5th	8.7×10^{-11}	0

Based on the results presented in the table above, it can be concluded that the share of the first primary component in the whole signal amounted to over 98%, which denotes the possibility to reduce a system of five decisive variables to only one variable with no significant information loss. Table 3 presents the efficiency of the analyzed networks for teaching when only the first primary component was used.

Table 3. Classification efficiency test results for variable data block lengths.

PC Number	Efficiency (%)			
	FFN	WNN	RBF	SVM
1	96.6	95.5	95.7	96.2

6. Discussion

The paper proposed the use of ANN for the analysis of acoustic signals accompanying electrical treeing in epoxy resin.

The recorded signals were divided into blocks used to identify the values of the teaching data. Then, the features of the reference signal were analyzed using a moving window of a set length.

The features were divided into two groups whose properties depended on window length and position versus the beginning of the AE pulse. Hence, kurtosis and skewness, due to the high dynamics of changes, can be good indicators for the identification of the beginnings of AE pulses related to treeing. The values of the parameters, however, did not reflect the duration of AE signals, and their use required the application of an analysis window with a minimum length. In the analyzed case, the results were regarded as satisfactory and repeatable for window lengths over 200 samples.

The other group of analyzed parameters, which describes the signal by its power characteristics, i.e., band power and signal energy, demonstrated elevated values during the signal. The steepness of the signal energy increase did not depend on the analysis of the window length, but a relevant selection of the length helped to identify the place in the signal where the signal amplitude dropped significantly. This is seen in the diagram as an inflexion point.

Since AEs in electrical treeing are related to the breaking of polymer chains or the presence of partial discharge in the existing channels, the value and dynamics of signal energy changes during a single pulse can be linked to treeing intensity.

By analyzing the results for each neural network, one can see that FFN demonstrated the highest efficiency. Slightly poorer results were obtained for RBF, WNN, and SVM. Moreover, for FFN and RBF, the results for each window length applied to the signal were comparable. WNN and SVN rendered better results for the greater analysis of window lengths. Moreover, WNN in its realization used only two wavelet neurons, while the structure of FNN required at least 15 neurons.

In the case of networks taught by means of the first primary component, one can state that the efficiency of each network dropped by a value that did not exceed 0.5%.

Summing up, the high efficiency in detecting AE signals accompanying electrical treeing needs to be highlighted. In order, however, to obtain as much information as possible on the AE pulse course, it seems justified to use an algorithm based on the following three stages:

- signal envelope analysis,
- AE signals' detection,
- analysis of individual signals.

A signal envelope analysis could provide hints for identifying the optimum window length at the discrimination level assumed on an arbitrary basis.

The performed experiments revealed that ANN can be successfully used for signal detection and classification, with the kurtosis, skewness, and energy of a single signal as the classifying parameters.

The effectiveness of the applied method was confirmed by the results obtained for various electrical parameters, i.e., different inter-electrode distances and different intensity of the electric field in this space. Regardless of the values of the above parameters, the results of AE signal identification were characterized by equally high efficiency, which proves the correctness of the assumptions made. The next stage covered a detailed analysis of the signals classified in the previous step with regard to the criteria assumed by the experimenter.

The completed analyses justify the usefulness of further studies on using ANN for partial discharge analysis and electrical treeing for solid dielectrics.

This knowledge could be of particular value to persons making diagnoses on the condition that the proposed method could be used online. That is why future efforts should focus on developing algorithms that would satisfy this demand.

Author Contributions: A.D. and S.M. developed a concept for the use of neural networks for the analysis of AE signals, analyzed the results of measurements, edited text and prepare conclusions. A.D. author of the methodology, designed and built a measurement stand, planned a research program, carried out preliminary tests. S.M developed computational algorithms, program code and performed numerical calculations. W.O. analyzed the results of measurements and gave some valuable suggestions, improved the text of article.

Funding: This research was funded by Polish Government, Ministry of Science and Higher Education.

Conflicts of Interest: The authors declare no conflict of interest.

References

1. Rayner, E.H. High-voltage tests and energy losses in insulating materials. *Br. JIEE* **1912**, *49*, 3. [CrossRef]
2. Robinson, D.M. *Dielectric Phenomena in High Voltage Cables*; Chapman Hall Ltd.: London, UK, 1936.
3. Whitehead, S. *Breakdown of Solid Dielectrics*; Ernest Benn: London, UK, 1932.
4. Bahder, G.; Daking, T.W.; Lawson, J.H. Analysis of treeing type breakdown. In Proceedings of the International Conference on Large High Voltage Systems, Paris, France, 25 August–2 September 1976; pp. 709–714.
5. Grzybowski, S.; Dobroszewski, R. Influence of partial discharges of the development of electrical treeing in polyethylene insulated cables. In Proceedings of the International Symposium on Electrical Insulation, Philadelphia, PA, USA, 21–26 July 1978; pp. 122–125.
6. Noskov, M.D.; Malinovski, A.S.; Sack, M.; Schwab, A.J. Modelling of partial discharge development in electrical tree channels. *IEEE Trans. Dielectr. Electr. Insul.* **2003**, *10*, 425–434. [CrossRef]
7. Jocteur, T.; Osty, M.; Lemanique, H.; Terramorsi, G. Research and Development in France in the Field of Extruded Polyethylene Insulated High Voltage Cables. In Proceedings of the International Conference on Large High Tension Electric Systems, Reference 21-07, Paris, France, 28 August–6 Septembet 1972; pp. 1–22.
8. Kreuger, F.H.; Bentvelsen, P.A.C. Breakdown Phenomena in Polyethylene Insulated Cables. In Proceedings of the International Conference on Large High Tension Electric Systems, Reference 21-05, Paris, France, 28 August–6 Septembet 1972; pp. 1–7.
9. Tykociner, J.T.; Brown, H.A.; Paine, E.B. Oscillations due to ionization in dielectrics and methods of their detection and measurement. *Univ. Illinois Bull.* **1933**, *30*.
10. Tian, Y.; Lewin, P.L.; Davies, A.E.; Sutton, S.J.; Swingler, S.G. Application of acoustic emission techniques and artificial neural networks to partial discharge classification. In Proceedings of the Conference Record of the 2002 IEEE International Symposium on Electrical Insulation, Boston, MA, USA, 7–10 April 2002; pp. 119–123.
11. Dissado, L.A. Understanding electrical trees in solids: From experiment to theory. *IEEE Trans. Dielectr. Electr. Insul.* **2002**, *9*, 483–497. [CrossRef]
12. Eichhorn, R.M. Treeing in solid extruded electrical insulation. *IEEE Trans. Electr. Insul.* **1977**, *12*, 2–18. [CrossRef]
13. Shimizu, N.; Laurent, C. Electrical tree initiation. *IEEE Trans. Dielectr. Electr. Insul.* **1998**, *5*, 651–659. [CrossRef]
14. Shimizu, N.; Uchida, K.; Rasikawan, S. Electrical tree and deteriorated region in polyethylene. *IEEE Trans. Electr. Insul.* **1992**, *27*, 513–518. [CrossRef]
15. Kudo, K. Fractal analysis of electrical trees. *IEEE Trans. Dielectr. Electr. Insul.* **1998**, *5*, 713–727. [CrossRef]
16. Dissado, L.A.; Dodd, S.J.; Champion, J.V.; Williams, P.I.; Alison, J.M. Propagation of electrical tree structures in solid polymeric insulation. *IEEE Trans. Dielectr. Electr. Insul.* **1997**, *4*, 259–279. [CrossRef]
17. Opydo, W.; Dobrzycki, A. Detection of electric treeing of solid dielectrics with the method of acoustic emission. *Electr. Eng.* **2012**, *94*, 37–48. [CrossRef]
18. Dobrzycki, A.; Opydo, W. An attempt to appraise the progress of methyl polymethacrylate degradation induced by a strong electric field on the grounds of analysis of acoustic signal emission. *Pozn. Univ. Technol. Acad. J.* **2007**, *57*, 197–203.
19. Markalous, S.M.; Tenbohlen, S.; Feser, K. Detection and Location of Partial Discharges in Power Transformers using Acoustic and Electromagnetic Signals. *IEEE Trans. Dielectr. Electr. Insul.* **2008**, *15*, 1576–1583. [CrossRef]
20. Lundgaard, L.E. Partial discharge. XIV. Acoustic partial discharge detection-practical application. *IEEE Electr. Insul. Mag.* **1992**, *8*, 34–43. [CrossRef]
21. Casals-Torrens, P.; Gonzalez-Parada, A.; Bosch-Tous, R. Online PD detection on high voltage underground power cables by acoustic emission. *Procedia Eng.* **2012**, *35*, 22–30. [CrossRef]
22. Dobrzycki, A.; Opydo, W.; Zakrzewski, S. Acoustic emission signals associated with prebreakdown state in air high voltage insulating systems. *Comput. Appl. Electr. Eng.* **2015**, *13*, 278–286.
23. Boczar, T.; Borucki, S.; Cichon, A.; Zmarzly, D. Application Possibilities of Artificial Neural Networks for Recognizing Partial Discharges Measured by the Acoustic Emission Method. *IEEE Trans. Dielectr. Electr. Insul.* **2009**, *16*, 214–223. [CrossRef]
24. Candela, R.; Mirelli, G.; Schifani, R. PD recognition by means of statistical and fractal parameters and a neural network. *IEEE Trans. Dielectr. Electr. Insul.* **2000**, *7*, 87–94. [CrossRef]

25. Dobrzycki, A.; Mikulski, S.; Opydo, W. Analysis of acoustic emission signals accompanying the process of electrical treeing of epoxy resins. In Proceedings of the ICHVE 2014—2014 International Conference on High Voltage Engineering and Application, Poznan, Poland, 8–11 September 2014.

26. Dobrzycki, A.; Mikulski, S. Using of wavelet transform in the analysis of AE signals accompanying the process of epoxy resins electrical treeing. *Przegląd Elektrotechniczny* **2016**, *1*, 223–225. [CrossRef]

27. Mohanty, S.; Ghosh, S. Artificial neural networks modelling of breakdown voltage of solid insulating materials in the presence of void. *IET Sci. Meas. Technol.* **2010**, *4*, 278–288. [CrossRef]

28. Mathur, L.S.; Agrawal, A.; Singh, D.K. Modeling of Breakdown Voltage of Solid Insulating Materials by Artificial Neural Network. *Int. J. Eng. Sci. Res. Technol.* **2016**, *5*, 788–796.

29. Masood, A.; Zuberi, M.U. Correlation of Breakdown Strength Parameters of Solid Insulation using Artificial Neural Network (ANN). *Eur. J. Adv. Eng. Technol.* **2016**, *3*, 14–19.

30. Tsangouri, E.; Aggelis, D.G.; Matikas, T.E.; Mpalaskas, A.C. Acoustic Emission Activity for Characterizing Fracture of Marble under Bending. *Appl. Sci.* **2016**, *6*, 6. [CrossRef]

31. Świt, G. Acoustic Emission Method for Locating and Identifying Active Destructive Processes in Operating Facilities. *Appl. Sci.* **2018**, *8*, 1295. [CrossRef]

32. Ebrahimkhanlou, A.; Salamone, S. Single-sensor acoustic emission source localization in plate-like structures: A deep learning approach. In Proceedings of the Health Monitoring of Structural and Biological Systems XII, Denver, CO, USA, 5–8 March 2018; Kundu, T., Ed.; SPIE: Bellingham, WA, USA, 2018; p. 59.

33. Ebrahimkhanlou, A.; Salamone, S. Single-Sensor Acoustic Emission Source Localization in Plate-Like Structures Using Deep Learning. *Aerospace* **2018**, *5*, 50. [CrossRef]

34. Ebrahimkhanlou, A.; Choi, J.; Hrynyk, T.D.; Salamone, S.; Bayrak, O. Detection of the onset of delamination in a post-tensioned curved concrete structure using hidden Markov modeling of acoustic emissions. In Proceedings of the Sensors and Smart Structures Technologies for Civil, Mechanical, and Aerospace Systems 2018, Denver, CO, USA, 5–8 March 2018; Sohn, H., Ed.; SPIE: Bellingham, WA, USA, 2018; p. 74.

35. Sikorski, W. The detection and identification of partial discharges in power transformer with the use of the acoustic emission method. *Przegląd Elektrotechniczny* **2010**, *86*, 229–232.

36. Lin, L.; Xu, Q.; Zhou, Y. Extracting the Fault Features of an Acoustic Emission Signal Based on Kurtosis and Envelope Demodulation of Wavelet Packets. In *Advances in Acoustic Emission Technology*; Springer: Cham, Switzerland; pp. 101–111.

37. Mohammad, M.; Abdullah, S.; Jamaludin, N.; Nuawi, M.Z. Correlating Strain and Acoustic Emission Signals of Metallic Component Using Global Signal Statistical Approach. In Proceedings of the Materials and Manufacturing Technologies XIV, Istanbul, Turkey, 13–16 July 2011; Yigit, F., Hashmi, M.S.J., Eds.; Volume 445, p. 1064+.

38. Schmidhuber, J. Deep learning in neural networks: An overview. *Neural Netw.* **2015**, *61*, 85–117. [CrossRef]

39. Bishop, C.M. *Pattern Recognition and Machine Learning*; Springer: Berlin, Germay, 2006; ISBN 9780387310732.

40. Buhmann, M.D. *Radial Basis Functions*; Cambridge University Press: Cambridge, UK, 2003; ISBN 9780511543241.

41. Alexandridis, A.K.; Zapranis, A.D. Wavelet neural networks: A practical guide. *Neural Netw.* **2013**, *42*, 1–27. [CrossRef]

42. Oussar, Y.; Dreyfus, G. Initialization by selection for wavelet network training. *Neurocomputing* **2000**, *34*, 131–143. [CrossRef]

43. Cristianini, N.; Scholkopf, B. Support vector machines and kernel methods—The new generation of learning machines. *AI Mag.* **2002**, *23*, 31–41.

Article

An Ultrasonic Guided Wave Mode Selection and Excitation Method in Rail Defect Detection

Hongmei Shi [1,2], Lu Zhuang [1], Xining Xu [1,2,*], Zujun Yu [1,2] and Liqiang Zhu [1,2]

[1] School of Mechanical, Electronic, and Control Engineering, Beijing Jiaotong University, Beijing 100044, China; hmshi@bjtu.edu.cn (H.S.); 17121287@bjtu.edu.cn (L.Z.); zjyu@bjtu.edu.cn (Z.Y.); lqzhu@bjtu.edu.cn (L.Z.)
[2] Key Laboratory of Vehicle Advanced Manufacturing, Measuring and Control Technology, Beijing Jiaotong University, Ministry of Education, Beijing 100044, China
* Correspondence: xnxu@bjtu.edu.cn

Received: 27 February 2019; Accepted: 14 March 2019; Published: 19 March 2019

Abstract: Different guided wave mode has different sensitivity to the defects of rail head, rail web and rail base in the detection of rail defects using ultrasonic guided wave. A novel guided wave mode selection and excitation method is proposed, which is effective for detection and positioning of the three parts of rail defects. Firstly, the mode shape data in a CHN60 rail is obtained at the frequency of 35 kHz based on SAFE method. The guided wave modes are selected, combining the strain energy distribution diagrams with the phase velocity dispersion curves of modes, which are sensitive to the defects of the rail head, rail web and rail base. Then, the optimal excitation direction and excitation node of the modes are calculated with the mode shape matrix. Phase control and time delay technology are employed to achieve the expected modes enhancement and interferential modes suppression. Finally, ANSYS is used to excite the specific modes and detect defects in different rail parts to validate the proposed methods. The results show that the expected modes are well acquired. The selected specific modes are sensitive to the defects of different positions and the positioning error is small enough for the maintenance staff to accept.

Keywords: ultrasonic guided waves; SAFE; rail defect detection; mode excitation

---

1. Introduction

It is of great significance to discover the internal defects of rails in time for maintaining and ensuring the safety of trains. At present, the hand-push detection vehicle and large-scale detection vehicle are the main tools to detect the defect detection of Continuous Welded Rails [1–3]. These two kinds of detection vehicles are running in the maintenance time, generally from 12 a.m. to 4 a.m., and cannot be used for real-time monitoring of rail internal defects. Guided waves can propagate a long distance in the rail [4–6]. Based on the pulse-echo method [7–9], the on-line monitoring of rail internal defects can be implemented by detecting the echo signals generated at the rail defects during the propagation of the guided waves.

In the research of rail defect detection based on guided waves, some scholars selected the suitable modes for rail defect detection by a mode classification method. Hayashi [10] divided the modes of the rail base into three categories. By analyzing the dispersion characteristics of guided waves, the concept of dominant modes is proposed. In the frequency range of 60~200 kHz, the transverse and vertical vibration modes with smooth dispersion curves of rail base are more suitable for defect detection than the longitudinal vibration modes [11]. A rail base defect of 20 mm in length was successfully detected by exciting the vertical modes. Subsequently, Sheng Huaji [12] obtained the optimal excitation frequency and pulse period of the propagating modes at the rail base through simulation, and tested oblique cracks at different angles by numerical simulation. The results show that the detection effect of the vertical vibration modes is better than the transverse vibration modes.

Lu Chao et al. [13] excited the vertical vibration modes on the rail head with a kind of mode force hammer. The experimental results show that the vertical vibration modes can effectively detect the transverse crack of the rail head. In addition, some scholars chose the modes to detect the damage of different parts of the rail based on the mode energy distribution. Gharaibeh et al. [14] selected the mode with energy concentrated on the rail head to detect the rail head defect. And the signal was excited according to the experimental method. The transverse defect of rail head at a distance of 9 m from the excitation location was successfully identified. In order to distinguish large transverse cracks with a diameter of 30 mm from 6 mm welded seam, Long and Loveday [15] compared the symmetric mode 1 and the antisymmetric mode 2 focusing energy at the rail head, and the symmetric mode 1 successfully distinguished the crack and welded seam at the rail head. Mode 3 and mode 4 focused with energy at the rail web, mode 4 was suitable for detecting the welded seam as well as damage in the rail web. It can be seen that in the research of the excitation mode applied to the detection of rail defects, most of the scholars use the experimental methods to excite the expected modes without specific theoretical guidance and mathematical model. The excitation signal usually contains multiple modes, which also makes it difficult for echo signal analysis and defect location.

In view of the above problems, this paper will focus on the guided wave mode selection and excitation methods based on a mathematical model in rail defect detection, by which the modes can be selected and excited with different sensitivities to the defects of the rail head, rail web and rail base, and accordingly the defect position can be rapidly found in different parts of the rail. The paper is organized as follows: A mode selection method is described in Section 1, which is used to select the modes having the concentrated strain energy and small attenuation corresponded to the rail head, rail web and rail base. The mode excitation method is presented in Section 2, to acquire the optimum excitation direction and excitation node of guided wave modes. Section 3 introduces the results of the target modes excitation simulation and verification. The simulation and detection results of the rail defects are demonstrated and analyzed in Section 4.

2. Mode Selection Based on Semi-Analytical Finite Element (SAFE) Method

In order to study the mode selection method of guided waves in rail defect detection, it is necessary to obtain the mode shape data. Many studies have proved that SAFE method is suitable to solve the propagation characteristics of guided waves in waveguide, such as plates [16], cylinders [17], rails [18], etc. SAFE method is also used in this paper to solve the mode shape data of CHN60 rail, which is a 60 kg/m China rail. The CHN60 rail coordinate system is shown in Figure 1.

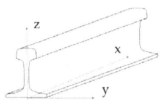

Figure 1. Coordinate system of a rail.

The displacement u, stress σ and strain ε of each particle in the rail can be expressed as Equation (1):

$$
\begin{aligned}
u &= \begin{bmatrix} u_x & u_y & u_z \end{bmatrix}^T \\
\sigma &= \begin{bmatrix} \sigma_x & \sigma_y & \sigma_z & \sigma_{yz} & \sigma_{xz} & \sigma_{xy} \end{bmatrix}^T \\
\varepsilon &= \begin{bmatrix} \varepsilon_x & \varepsilon_y & \varepsilon_z & \gamma_{yz} & \gamma_{xz} & \gamma_{xy} \end{bmatrix}^T
\end{aligned}
\tag{1}
$$

Based on SAFE theory, it is assumed that the displacement field of the guided wave propagation direction is harmonic [19]. The cross section of the rail is discretized into 255 triangular elements and 177 nodes, as shown in Figure 2.

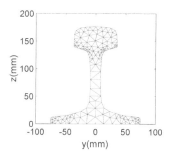

Figure 2. Discretization of triangle elements.

After discretization, the strain energy ϕ of each triangular element can be written as Equation (2):

$$\phi = \int_{V_e} \delta(\varepsilon^{(e)T})C\varepsilon^{(e)}dV_e = \delta q^{(e)T}(K_1 - i\zeta K_2 + \zeta^2 K_3)q^{(e)} \tag{2}$$

where $q^{(e)}$ is the nodal unknown displacement of each triangular element; ζ is the wavenumber; K_1, K_2, and K_3 are stiffness matrices; C is the elastic constant matrix of the rail.

By substituting strain energy and potential energy at any node in the rail cross section into Hamilton's formula, the general homogeneous wave equation of guided waves can be derived as Equation (3) [19]:

$$\left[K_1 + i\zeta K_2 + \zeta^2 K_3 - \omega^2 M\right]_M U = 0 \tag{3}$$

where M is the mass matrix; U is the nodal displacements; ω is frequency.

By solving the eigenproblem of Equation (3), the corresponding wavenumber ζ can be obtained for a given frequency ω. The triangular element nodal displacements U are calculated from the eigenvalue problem. And the strain fields are reconstructed from Equation (2).

At the frequency of 35 kHz, the nodal displacement $q^{(e)}$, stiffness matrix K_1, K_2, and K_3, and wavenumber ζ are substituted into the Equation (2), and the strain energy of each mode in each triangular element can be acquired. The strain energy distribution diagram of each mode is normalized as shown in Figure 3.

At the frequency of 35 kHz, there are 20 kinds of guided wave modes propagating in the rail. Our expected mode is that the strain energy is concentrated in a single part of the rail. From Figure 3, we can see that mode 1, mode 2, mode 3, mode 7, mode 10, and mode 20 have such characteristics while strain energy of the other modes is distributed at three rail parts.

In Figure 4a,b, the strain energy of mode 7 and mode 20 is concentrated on the rail head. In Figure 4c, it can be seen that the strain energy of mode 3 is focused on the rail web. In Figure 4d, Figure 4e,f, the strain energy of mode 1, mode 2, and mode 10 is concentrated on the rail base. Therefore, mode 7 and mode 20 can be selected to detect rail head defects, similarly, mode 3 for rail web and mode 1, mode 2 and mode 10 for rail base.

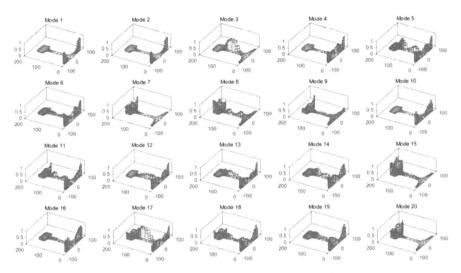

Figure 3. Strain energy distribution diagram of modes.

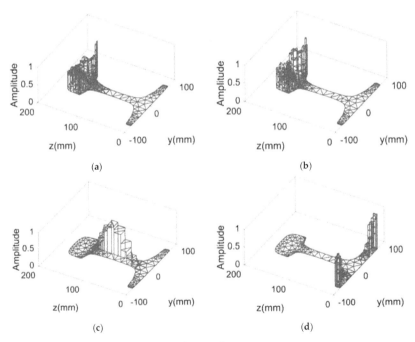

Figure 4. *Cont.*

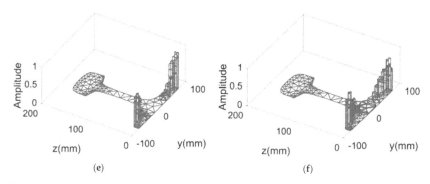

(e)   (f)

Figure 4. Normalized strain energy amplitude of modes. (**a**) mode 7; (**b**) mode 20; (**c**) mode 3; (**d**) mode 1; (**e**) mode 2; (**f**) mode 10.

The dispersion curves of the above modes are shown in Figure 5. When the frequency dispersion occurs, the wave packet composed of different frequency components will be dispersed, causing the waveform of the received signal to be distorted relative to the transmitted signal. Therefore, it is necessary to select modes with good non-dispersion characteristics. At the frequency of 35 kHz, the dispersion curves of mode 7 are gradual than that of mode 20. This means that the non-dispersion characteristic of mode 7 is better than that of mode 20. Hence mode 7 is selected to detect the rail head defect. Similarly, mode 3 is selected to detect the rail web defect. Mode 1, mode 2, and mode 10 are suitable to detect rail base defects.

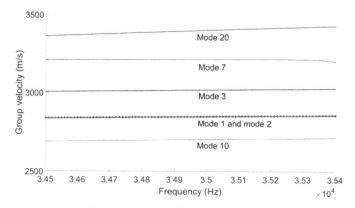

Figure 5. Group velocity dispersion curves.

3. Mode Excitation Based on Mode Matrix

In order to excite the expected modes, it is necessary to determine the optimal excitation direction and excitation node. The detailed excitation method is described in this section.

3.1. Determine Excitation Direction

Here Euclidean distance is used to describe the true distance between the vibration vectors of modes in the rail. The distances of displacement vector between mode m and other modes in x, y and z directions are calculated, and the direction with the largest Euclidean distance is chosen as the optimum excitation direction of mode m.

There are p nodes in the profile of rail cross section. Each node has 3 degrees of freedom. The displacements are represented as x_i, y_i, and z_i, respectively. The Euclidean distances X_{mn}, Y_{mn}, and Z_{mn} of the vibration vectors of mode m and mode n in x, y, and z directions can be defined as follows:

$$X_{mn} = \sqrt{\sum_{i=1}^{p}(x_{im} - x_{in})^2} \qquad (4)$$

$$Y_{mn} = \sqrt{\sum_{i=1}^{p}(y_{im} - y_{in})^2} \qquad (5)$$

$$Z_{mn} = \sqrt{\sum_{i=1}^{p}(z_{im} - z_{in})^2} \qquad (6)$$

At the frequency of 35 kHz, there are 20 kinds of guided wave modes in rails. Assuming the expected mode is mode m. Firstly, we can calculate the Euclidean distance between mode m and all other modes in x, y, and z directions respectively at the same frequency. Then 3 matrices of 20 × 20 can be obtained and the column m or the row m of the matrix represents the Euclidean distance between the mode m and other modes. Next, we calculate the mean of the distance between a given mode to all others in x, y, and z directions and call the value $\bar{x}_m$, $\bar{y}_m$, and $\bar{z}_m$. This is done separately for all three directions. By comparing the values of $\bar{x}_m$, $\bar{y}_m$, and $\bar{z}_m$, the direction with the largest value is selected as the optimum excitation direction.

According to the above method, the mode shape data of 20 modes at the frequency of 35 kHz are substituted into Equations (4)~(6). Then 3 matrices of 20 × 20 can be obtained and each column or each row of a matrix represents the Euclidean distance between the corresponding mode and other modes. This paper selects mode 1 to detect the rail base defect. Then we can calculate $\bar{x}_1$, $\bar{y}_1$, and $\bar{z}_1$ of mode 1. By comparing the 3 values of $\bar{x}_1$, $\bar{y}_1$, and $\bar{z}_1$, the z direction with the largest value is selected as the optimum excitation direction. The calculation process of mode 3 and mode 7 is the same and $\bar{x}$, $\bar{y}$, and $\bar{z}$ of mode 1, mode 3, and mode 7 in x, y, and z directions are shown in Table 1.

Table 1. The values of $\bar{x}$, $\bar{y}$, and $\bar{z}$.

Mode	$\bar{x}$ (mm)	$\bar{y}$ (mm)	$\bar{z}$ (mm)
1	0.063	0.054	0.074
3	0.067	0.077	0.054
7	0.068	0.053	0.080

According to the data in Table 1, the optimal excitation direction of mode 1, mode 3 and mode 7 is z, y, and z direction, respectively.

3.2. Determine Excitation Nodes

Covariance is employed to analyze the vibration displacement deviation of each node on the rail profile from the average displacement of all nodes. Then the optimal excitation point is selected according to the overall vibration displacement of the nodes and the node with the largest covariance is selected as the optimum positive excitation node in mode m. If there is a negative covariance value, the node with the minimum covariance value is the best inverse excitation node. Equation (7) describes the covariance of node t and node e of mode m in x, y, and z directions at the rail base.

$$RB\_COV_m(t, e) = \frac{\sum_{s=1}^{3}(t_s - \bar{t})(e_s - \bar{e})}{3-1} \qquad (7)$$

where t_s is the vibration displacement of node t in mode m. $\bar{t}$ is the average value of the vibration displacement of node t. e_s is the vibration displacement of node e of mode m. $\bar{e}$ is the average value of the vibration displacement of node e.

Similarly, Equations (8) and (9) respectively describes the covariance of node t and node e of mode m in x, y, and z directions at the rail web and rail head.

$$RW\_COV_m(t,e) = \frac{\sum_{s=1}^{3}(t_s - \bar{t})(e_s - \bar{e})}{3-1} \tag{8}$$

$$RH\_COV_m(t,e) = \frac{\sum_{s=1}^{3}(t_s - \bar{t})(e_s - \bar{e})}{3-1} \tag{9}$$

The mode shape data of mode 1, mode 3, and mode 7 is substituted into Equations (7)~(9), respectively. The optimal excitation nodes of each mode are calculated and the excitation method is selected according to mode shape characteristics. Figure 6 shows the optimal excitation nodes.

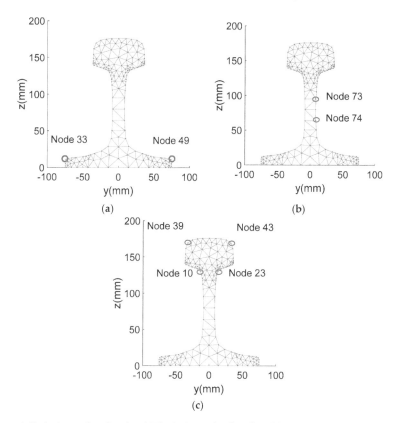

Figure 6. Excitation nodes of modes. (**a**) Excitation node of mode 1; (**b**) excitation nodes of mode 3; (**c**) excitation nodes of mode 7.

In summary, mode 1 selects the two-side symmetric excitation, which is excited on node 49 and node 33 in z positive direction. Node 33 receives the signal in z direction. Mode 3 selects the one-side symmetric excitation on node 73 and node 74 in y positive direction. Node 74 is selected to receive the signal in y direction. Mode 7 selects the symmetric excitation on nodes 43, node 39, node 10 and node 23 in z direction. Then node 43 receives the signal in z direction. The detailed selection process of excitation node can be referred to our previous work [20].

4. Mode Excitation Simulation and Verification

Mode excitation is simulated with ANSYS to verify the above proposed method. The excitation signal is a sine wave modulated by the Hanning window with a frequency of 35 kHz as shown in Figure 7.

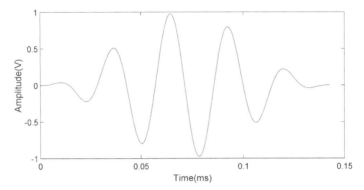

Figure 7. Excitation signal.

In the simulation process, the excitation signal is applied on the rail at 4 m from the right end of the rail as shown in Figure 8 in order to avoid the influence of the rail end echo on the simulation results. Between 0.8 and 2.7 m from the excitation node, a set of data acquisition array is set at intervals of 5 mm. There are 380 data acquisition nodes.

The two-dimensional Fast Fourier Transform(2D-FFT) is used on the acquired data to identify the phase velocity of the mode. The simulation results are compared with the frequency wavenumber dispersion curves calculated by SAFE method as shown in Figure 9.

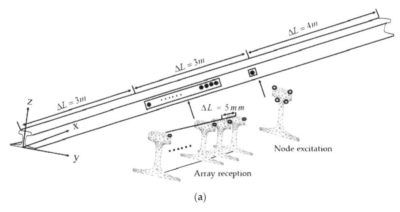

(a)

Figure 8. *Cont.*

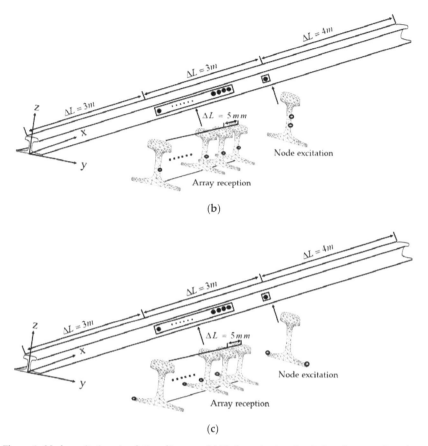

(b)

(c)

Figure 8. Node excitation simulation diagram. (**a**) Node excitation simulation diagram of mode 7; (**b**) node excitation simulation diagram of mode 3; (**c**) node excitation simulation diagram of mode 1.

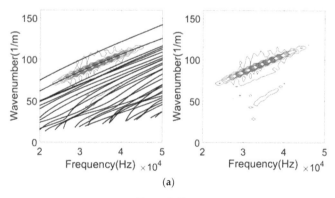

(a)

Figure 9. *Cont.*

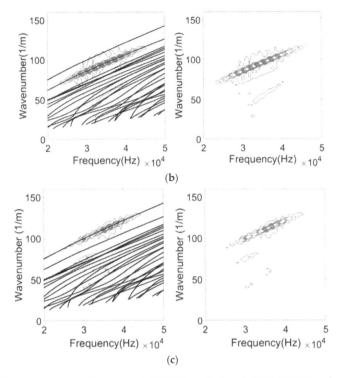

(b)

(c)

Figure 9. Frequency wavenumber curves. (a) 2D-FFT result of mode 7; (b) 2D-FFT result of mode 3; (c) 2D-FFT result of mode 1.

In Figure 9, the result of 2D-FFT coincides well with the frequency wavenumber curve of the corresponding mode. The ordinate value of the wavenumber ξ can be acquired from Figure 9 when the abscissa f is 35 kHz. The relationship between wavenumber ξ and frequency f is shown in Equation (10):

$$C_p = \frac{2\pi f}{\xi} \tag{10}$$

Then the frequency f and wavenumber ξ are substituted into the Equation (10) to calculate the simulation phase velocity of each mode. The calculation results are shown in Table 2.

Table 2. Phase velocity of modes.

Mode	Simulated Phase Velocity Value (m/s)	Theoretical Phase Velocity Value (m/s)	Phase Velocity Error
7	2669.4	2737.6	2.49%
3	2282.5	2286.1	0.16%
1	1982.5	1983.8	0.07%

It is found from Table 2 that the phase velocity errors of mode 7, mode 3 and mode 1 are 2.49%, 0.16%, 0.07% respectively. It is proved that the method of node excitation is effective, but there are still some interference modes. In order to excite a relatively single mode, it is necessary to study the method of mode enhancement and interference mode suppression.

5. Mode Enhancement and Defect Detection Simulation

5.1. Mode Enhancement Theory Based on Phase Control and Time Delay Technology

According to the theory of Rose [21], the installation interval and the excitation sequence of transducer array are set based on the expected guided wave mode, and the excited guided waves have the same phase angle. The excitation time interval is set as periodic T. According to the principle of wave superposition, the synthetic amplitude of two coherent waves with the same phase is the largest when the interval of transducers is offered based on the expected mode wavelength. As shown in Figure 10, 5 transducers are installed on the rail.

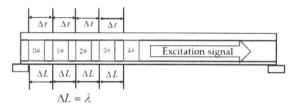

Figure 10. Phased array excitation diagram.

Transducer 0# is an excitation signal without time delay. The displacement function of each particle is $u_0(x, y, z, t)$.

$$u_0(x,y,z,t) = \begin{bmatrix} u_x(x,y,z,t) \\ u_y(x,y,z,t) \\ u_z(x,y,z,t) \end{bmatrix} = \begin{bmatrix} U_x(y,z) \\ U_y(y,z) \\ U_z(y,z) \end{bmatrix} e^{i(\xi x - \omega t)} \tag{11}$$

where ξ is the wavenumber.

The distance between transducer 1# and transducer 0# is λ, and the time delay is T. The displacement function of each particle is $u_1(x, y, z, t)$.

$$u_1(x,y,z,t) = \begin{bmatrix} u_x(x,y,z,t) \\ u_y(x,y,z,t) \\ u_z(x,y,z,t) \end{bmatrix} = \begin{bmatrix} U_x(y,z) \\ U_y(y,z) \\ U_z(y,z) \end{bmatrix} e^{i[\xi(x-\Delta L)-\omega(t-\Delta t)]} = \begin{bmatrix} U_x(y,z) \\ U_y(y,z) \\ U_z(y,z) \end{bmatrix} e^{i[\xi(x-\lambda)-\omega(t-T)]} \tag{12}$$

As we know:

$$\xi\lambda = 2\pi \tag{13}$$

$$\omega T = 2\pi \tag{14}$$

Substitute the Equations (13) and (14) into the Equation (12) to obtain the Equation (15).

$$u_1(x,y,z,t) = \begin{bmatrix} u_x(x,y,z,t) \\ u_y(x,y,z,t) \\ u_z(x,y,z,t) \end{bmatrix} = \begin{bmatrix} U_x(y,z) \\ U_y(y,z) \\ U_z(y,z) \end{bmatrix} e^{i[\xi(x-\lambda)-\omega(t-T)]} = \begin{bmatrix} U_x(y,z) \\ U_y(y,z) \\ U_z(y,z) \end{bmatrix} e^{i[\xi x - \omega t]} = u_0(x,y,z,t) \tag{15}$$

Similarly, the displacement and phase of transducer array 0#, 1#, 2#, 3#, and 4# can be derived as Equation (16):

$$u_0(x,y,z,t) = u_1(x,y,z,t) = u_2(x,y,z,t) = u_3(x,y,z,t) = u_4(x,y,z,t) \tag{16}$$

When the guided wave propagates along x direction, the expected mode is enhanced and the interference mode is suppressed. The transducer intervals of each mode are shown in Table 3.

Table 3. Transducer intervals.

Mode	Phase Velocity Value (m/s)	ΔL (mm)
7	2737.6	78
3	2286.1	65
1	1983.8	57

5.2. Mode Enhancement Simulation Results

The simulation process of mode enhancement is the same as Section 3, only the excitation method is changed to transducer array. The simulation results are compared with the frequency wavenumber dispersion curves as shown in Figure 11.

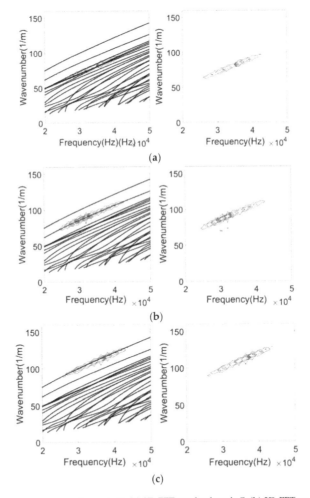

Figure 11. Frequency wavenumber curves. (**a**) 2D-FFT result of mode 7; (**b**) 2D-FFT result of mode 3; (**c**) 2D-FFT result of mode 1.

From Figure 11 we can see, the result of 2D-FFT coincides well with the frequency wavenumber curve of the corresponding mode and there is no interference mode. The wavenumber ξ and frequency *f* are substituted into Equation (10). The calculation results of simulated phase velocities are shown in Table 4, and compared with theoretical values. It can be seen that the maximum error is 0.68%.

Table 4. Phase velocities of modes.

Mode	Simulated Phase Velocity Value (m/s)	Theoretical Phase Velocity Value (m/s)	Phase Velocity Error
7	2719.0	2737.6	0.68%
3	2281.5	2286.1	0.20%
1	1982.0	1983.8	0.09%

The above discussion is about phase velocity, then the process of the group velocity calculation is described as following. The simulation data at 1 m and 2.2 m from the excitation nodes are extracted and the simulation results of mode 7, mode 3, and mode 1 are shown in Figure 12.

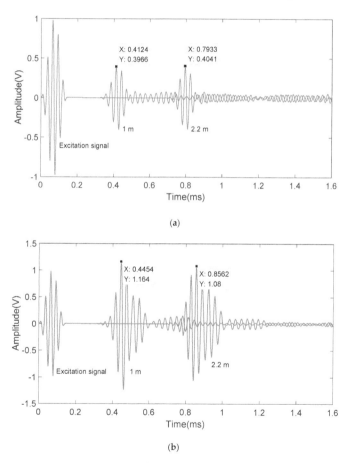

(a)

(b)

Figure 12. *Cont.*

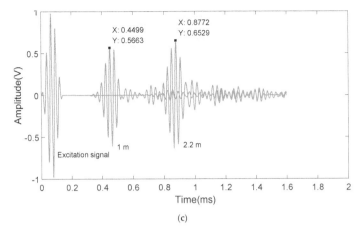

(c)

Figure 12. Mode simulation results. (**a**) Mode 7 simulation result; (**b**) Mode 3 simulation result; (**c**) Mode 1 simulation result.

The simulation group velocity can be calculated by calculating the peak time difference between 1 m and 2.2 m and the results are shown in Table 5.

Table 5. Group velocities of modes.

Mode	Simulated Group Velocity Value (m/s)	Theoretical Group Velocity Value (m/s)	Group Velocity Error
7	3150.4	3215.1	2.01%
3	2994.0	3019.9	0.86%
1	2818.2	2852.7	1.21%

As can be seen from Table 5, the group velocity error of each mode is relatively small. The results show that the method of array excitation can enhance the expected mode and suppress the interference mode, which indicates a relatively single mode is successfully excited.

5.3. Defect Detection Simulation

At present, rail head transverse defects are the most dangerous damage for rail operation. Figure 13a presents the simulation of detecting an internal transverse defect. The shape of the defect on the rail cross section is a circle with a diameter of 23 mm and a length of 5 mm in x direction. In Figure 13b, the rail web defect is assumed as a vertical crack. The depth of the crack in z direction is 30 mm and 5 mm in x direction. In Figure 13c, the rail base defect is a transverse defect, the length of which is 25 mm in the y direction and 10 mm in x direction. In defect detection simulation, the modes excitation and data acquisition method are the same as Section 5.2 and the rail defects are 3 m from the left end of the rail.

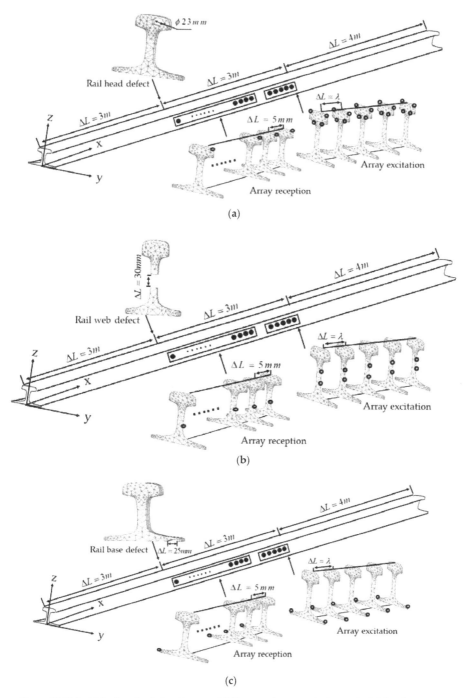

Figure 13. Defect detection simulation diagram. (**a**) diagram of array excitation mode 7 defect detection; (**b**) diagram of array excitation mode 3 defect detection; (**c**) diagram of array excitation mode 1 defect detection.

It is necessary to verify whether the received defect echo at the receiving nodes has occurred in mode conversion before calculating the defect location. The defect echoes extracted separately are transformed by 2D-FFT as shown in Figure 14.

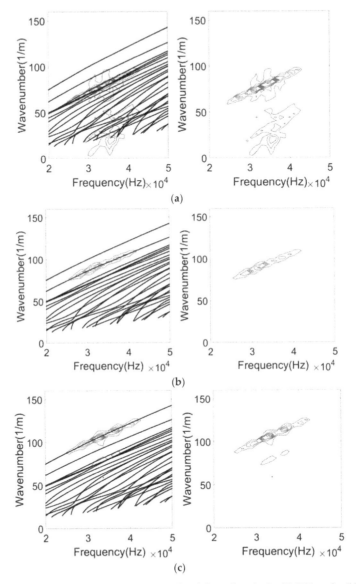

Figure 14. Frequency wavenumber curves. (**a**) mode 1 defect reflected echo 2D-FFT results; (**b**) mode 3 defect reflected echo 2D-FFT results; (**c**) mode 7 defect reflected echo 2D-FFT results.

Similarly, the phase velocity of each mode can be calculated from the frequency wavenumber curve. Table 6 shows the phase velocities of reflected echo modes.

Table 6. Phase velocities of reflected echo modes.

Mode	Simulated Phase Velocity Value (m/s)	Theoretical Phase Velocity Value (m/s)	Phase Velocity Error
7	2721.0	2737.6	0.61%
3	2262.5	2286.1	1.03%
1	1994.6	1983.8	0.54%

As can be seen from Table 6 that the error between the simulated phase velocity and the theoretical phase velocity of each defect echo can be accepted for application. The reflected echoes received at the corresponding nodes have not occurred mode conversion.

During the process of simulation, the receiving node is 1 m away from the defect. The position of the rail defect can be obtained by calculating the time difference between the received excitation signal and the reflected echo at the receiving node. As shown in Figure 15, the defect is 3 m from the left end of the rail. The receiving node is 1 m from the defect and 2 m from the excitation signal.

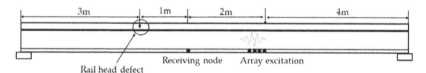

Figure 15. Diagram of the rail head defect detection.

When the defect is in the rail head, the detection results of mode 7, mode 3 and mode 1 are as shown in Figure 16. Only mode 7 has reflected echo. The time difference between the received excitation signal and the reflected echo is 0.6137 ms.

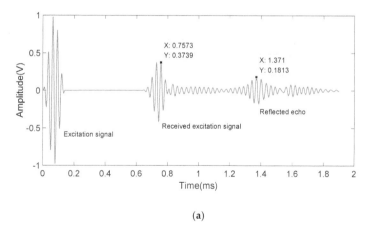

(a)

Figure 16. *Cont.*

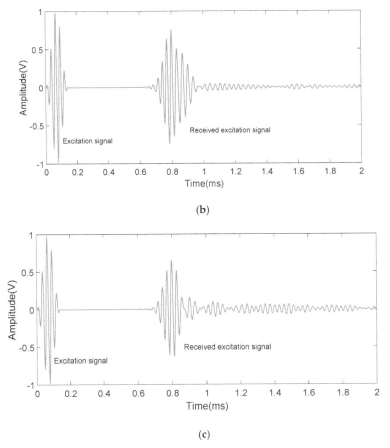

(b)

(c)

Figure 16. Rail head defect detection. (**a**) Mode 7 rail head defect detection; (**b**) mode 3 rail head defect detection; (**c**) mode 1 rail head defect detection.

When the defect is in the rail web, only mode 3 has reflected echo as shown in Figure 17 and the time difference is 0.6478 ms.

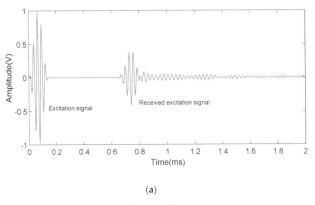

(a)

Figure 17. *Cont.*

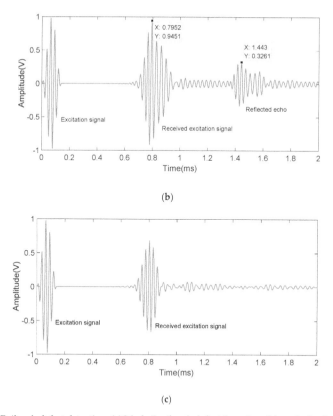

(b)

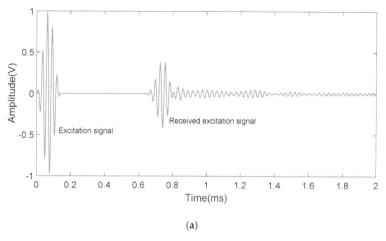

(c)

Figure 17. Rail web defect detection. (**a**) Mode 7 rail web defect detection; (**b**) mode 3 rail web defect detection; (**c**) mode 1 rail web defect detection.

Similarly, when the defect is in the rail base, only mode 1 has reflected echo as shown in Figure 18 and the time difference between is 0.6913 ms.

(a)

Figure 18. *Cont.*

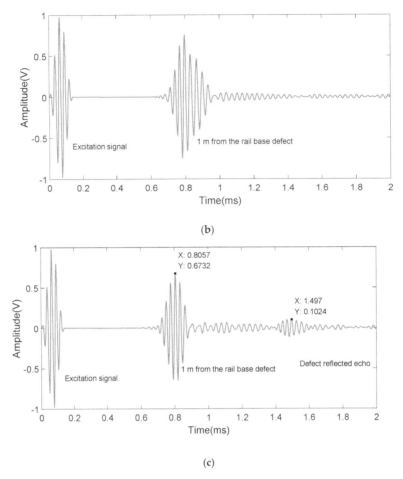

(b)

(c)

Figure 18. Rail base defect detection. (**a**) Mode 7 rail base defect detection; (**b**) mode 3 rail base defect detection; (**c**) mode 1 rail base defect detection.

The defect position can be calculated based on the time difference. Table 7 shows the detect results of the defect position. The results can be seen from the Table 7 that the defect positioning error of mode 7, mode 3 and mode 1 is 1.34%, 2.19%, and 1.40% respectively.

Table 7. Defect detection of each mode.

Mode	Time Difference (ms)	Theoretical Group Velocity Value (m/s)	Simulation Defect Position (m)	Positioning Error
7	0.6137	3215.1	0.9866	1.34%
3	0.6478	3019.9	0.9781	2.19%
1	0.6913	2852.7	0.9860	1.40%

6. Conclusions

In this paper, a novel method for mode selection and excitation in rail defect detection is proposed based on ultrasonic guided waves. The mode shape data in the CHN60 rail is obtained at the frequency of 35 kHz by using SAFE method. And the modes with good non-dispersion characteristics, which are respectively sensitive to the defects of the rail head, rail web and rail base, are selected combining the strain energy distribution diagrams with group velocity dispersion curves. The optimal excitation direction of the expected modes is solved by the Euclidean distance, and the optimal excitation node is calculated by covariance matrix. Then, phase control and time delay technology is applied to achieve the enhancement of expected modes and suppression of interferential modes. After that, ANSYS is used to simulate the expected modes excitation and defects detection in the different rail parts with phase velocity and group velocity to validate the proposed methods. The simulation results present an acceptable error and demonstrate the effectiveness of the guided wave mode selection and excitation method, which proposes a feasible scheme for the field test and application to rail defect detection.

Author Contributions: This paper is a result of the full collaboration of all the authors; H.S. and L.Z. (Lu Zhuang) conceived and proposed the main idea of the paper; X.X. combined with phase control and time delay technology to improve the mode excitation method; L.Z. (Liang Zhu) simulated the defect detection and analyzed the data; L.Z. (Liang Zhu) and Z.Y. put forward their views on data processing; H.S. and L.Z. (Lu Zhuang) wrote the paper.

Acknowledgments: The work is supported by the National Key Research and Development Program of China (2016YFB1200401), Supported by Foundation of Key Laboratory of Vehicle Advanced Manufacturing, Measuring, and Control Technology (Beijing Jiaotong University), Ministry of Education, China.

Conflicts of Interest: The authors declare no conflict of interest.

References

1. Zhao, Y.; Chen, J.W.; Sun, J.H.; Ma, J.; Sun, J.C.; Jia, Z.Q. The Application and Progress of Nondestructive Testing Technique and System for the Rail In-service. *Nondestruct. Test.* **2014**, *36*, 58–64.
2. Zhang, Y.H.; Xu, G.Y.; Li, P.; Shi, Y.S.; Huang, X.Y. Key Technology to Autonomous Ultrasonic Detection System of Rail Flaw Detection Car. *China Railw. Sci.* **2015**, *36*, 131–136.
3. Cerny, L. FRA Revises Defective Rail and Rail Inspection Standards, Adds New Standards for Detector Car Operators. *Railw. Track Struct.* **2014**, *110*, 2–19.
4. Ramatlo, D.A.; Wilke, D.N.; Loveday, P.W. Development of an optimal piezoelectric transducer to excite guided waves in a rail web. *NDT E Int.* **2018**, *95*, 72–81. [CrossRef]
5. Wilcox, P.; Pavlakovic, B.; Evans, M.; Vine, K.; Cawley, P.; Lowe, M.; Alleyne, D. Long Range Inspection of Rail Using Guided Waves. In *Review of Progress in Quantitative Nondestructive Evaluation, Proceedings of the AIP Conference*; American Institute of Physics: College Park, MD, USA, 2003; Volume 22, pp. 236–243.
6. Kijanka, P.; Staszewski, W.J.; Packo, P. Simulation of guided wave propagation near numerical Brillouin zones. In *Health Monitoring of Structural and Biological Systems, Proceedings of SPIE—The International Society for Optical Engineering*; SPIE: Bellingham, WA, USA, 2016; Volume 9805, p. 98050Q.
7. Lee, J.; Sheen, B.; Cho, Y. Multi-defect tomographic imaging with a variable shape factor for the RAPID algorithm. *J. Vis.* **2016**, *19*, 393–402. [CrossRef]
8. Escobar-Ruiz, E.; Wright, D.C.; Collison, I.J.; Cawley, P.; Nagy, P.B. Reflection Phase Measurements for Ultrasonic NDE of Titanium Diffusion Bonds. *J. Nondestruct. Eval.* **2014**, *33*, 535–546. [CrossRef]
9. Hill, S.; Dixon, S.; Sri, H.R.K.; Rajagopal, P.; Balasubramaniam, K. A new electromagnetic acoustic transducer design for generating torsional guided wave modes for pipe inspections. In *Review of Progress in Quantitative Nondestructive Evaluation, Proceedings of the AIP Conference*; American Institute of Physics: College Park, MD, USA, 2017; Volume 36, pp. 050003-1–050003-9.
10. Hayashi, T.; Miyazaki, Y.; Murase, M.; Abe, T. *Guided Wave Inspection for Bottom Edge of Rails*; American Institute of Physics: College Park, MD, USA, 2007; Volume 26, pp. 169–177.
11. Hayashi, T. Guided Wave Dispersion Curves Derived with a Semi Analytical Finite Element Method and Its Applications to Nondestructive Inspection. *Jpn. J. Appl. Phys.* **2014**, *47*, 3865–3870. [CrossRef]
12. Sheng, H.J. Research on Guided waves inspection for the Oblique Crack and Welding in the Rail bottom. Master's Thesis, Nanchang Hangkong University, Nanchang, China, June 2016.
13. Lu, C.; Liu, R.C.; Chang, J.J. Guided waves dispersion curves and wave structures of the rail's vertically vibrating modes and their application. *J. Vib. Eng.* **2014**, *79*, 25–37.
14. Gharaibeh, Y.; Sanderson, R.; Mudge, P.; Ennaceur, C.; Balachandran, W. Investigation of the behaviour of selected ultrasonic guided wave modes to inspect rails for long-range testing and monitoring. *Proc. Inst. Mech. Eng. Part F J. Rail Rapid Transit* **2011**, *1*, 1–14. [CrossRef]

15. Long, C.S.; Loveday, P.W. Prediction of guided wave scattering by defects in rails using numerical modelling. In *AIP Conference Proceedings*; AIP Publishing: College Park, MD, USA, 2014; Volume 1581, pp. 240–247.

16. Ahmad, Z.A.; Gabbert, U. Simulation of Lamb wave reflections at plate edges using the semi-analytical finite element method. *Ultrasonics* **2012**, *52*, 815–820. [CrossRef] [PubMed]

17. Marzani, A. Time–transient response for ultrasonic guided waves propagating in damped cylinders. *Int. J. Solids Struct.* **2008**, *45*, 6347–6368. [CrossRef]

18. Loveday, P.W. Guided Wave Inspection and Monitoring of Railway Track. *J. Nondestruct. Eval.* **2012**, *31*, 303–309. [CrossRef]

19. Bartoli, I.; Marzani, A.; Scalea, F.L.D.; Viola, E. Modeling wave propagation in damped waveguides of arbitrary cross-section. *J. Sound Vib.* **2006**, *295*, 685–707. [CrossRef]

20. Xu, X.N.; Zhuang, L.; Xing, B.; Yu, Z.J.; Zhu, L.Q. An Ultrasonic Guided Wave Mode Excitation Method in Rails. *IEEE Access* **2018**, *6*, 60414–60428.

21. Rose, J.L.; Pelts, S.P.; Quarry, M.J. A comb transducer model for guided wave NDE. *Ultrasonics* **1999**, *36*, 163–169. [CrossRef]

Article

Development of a Vibroacoustic Stochastic Finite Element Prediction Tool for a CLT Floor

Cheng Qian [1,*], Sylvain Ménard [1], Delphine Bard [2] and Juan Negreira [3]

[1] Department of Applied Sciences, University of Québec at Chicoutimi, Chicoutimi, QC G7H2B1, Canada; sylavin_menard@uqac.ca
[2] Division of Engineering Acoustics, Department of Construction Sciences, Lund University, 22362 Lund, Sweden; delphine.bard@construction.lth.se
[3] Saint-Gobain Ecophon Spain, 28020 Madrid, Spain; juan.negreira@saint-gobain.com
[*] Correspondence: cheng.qian1@uqac.ca

Received: 31 January 2019; Accepted: 8 March 2019; Published: 15 March 2019

Abstract: Low frequency impact sound insulation is a challenging task in wooden buildings. Low frequency prediction tools are needed to access the dynamic behavior of a wooden floor in an early design phase to ultimately reduce the low frequency impact noise. However, due to the complexity of wood and different structural details, accurate vibration predictions of wood structures are difficult to attain. Meanwhile, a deterministic model cannot properly represent the real case due to the uncertainties coming from the material properties and geometrical changes. The stochastic approach introduced in this paper aims at quantifying the uncertainties induced by material properties and proposing an alternative calibration method to obtain a relative accurate result instead of the conventional manual calibration. In addition, 100 simulations were calculated in different excitation positions to assess the uncertainties induced by material properties of cross-laminated-timber A comparison between the simulated and measured results was made in order to extract the best combination of Young's moduli and shear moduli in different directions of the CLT panel.

Keywords: wooden constructions; acoustics; low frequency noise; modelling

---

1. Introduction

Multi-story wooden constructions have increased in the market during the last two decades. However, this kind of constructions still face the challenge coming from low-frequency sound insulation. Although these types of buildings comply with the present regulations, subjective ratings of inhabitants have shown complaints [1–6] due to low-frequency impact noise. One of the important reasons resulting in this discrepancy between the subjective and objective evaluations is that the current standards are initially designed for heavy constructions, i.e., concrete constructions, but without an appropriate modification, they are directly applied to the wooden constructions. On the other hand, the low frequency impact noise is neglected in the standardized impact sound insulation evaluations. From recent research [3], it was shown that the subjective ratings are correlated better with the objective ratings with the help of the adaptation term $C_{I, 50-3000}$ following the standard ISO 717-2 [7]. But the correlation between the inhabitants' satisfaction (reported by means of questionnaires) and objective ratings can be further improved by taking the measurement from 20 Hz in comparison with the measurements performed from 50 Hz instead. Furthermore, the dynamic behavior of wood (as a natural material) is hard to predict due to its inhomogeneity. As a consequence, product development in the wooden industry is still based on empirical models and experimental tests, which could lead to an over-designed and expensive acoustical improvement solution. To end that, more knowledge about wooden construction is needed and accurate and handy prediction tools are called for in order to enable acoustic comfort in wooden constructions, especially in a low-frequency range.

The finite element (FE) method is a widely employed approach to develop numerical prediction models in wooden industry. By performing numerical simulations, experimental acoustical tests can be reduced, and parametric studies can also be carried out to investigate the influence of specific geometrical changes in construction as well as the influence of the variations/uncertainties in material properties, which always have a markable influence on the results. In Reference [8], experimental tests were conducted on a full-scale cross-laminated timber (CLT) floor. The material properties of the CLT were collected from the literature and then put into the established FE model to compare the simulation results with the measured ones. A better correlation between the testing results and the modelling results was attained after tuning the collected material properties of CLT. The latter points out the importance of knowing the material properties if a proper calculation of dynamic properties needs to be achieved. A similar conclusion was drawn in References [9–11].

However, one should know that the variety of the wood species, the variation in theoretical identical wood structure, as well as the lack of material properties data base, make the reliable material properties of wood difficult to obtain. Consequently, the calibration of a model is always tedious and time-consuming. Moreover, even with the calibrated model to predict the theoretical identical wooden structure, the prediction results may be different from the realistic case due to the uncertainties induced by wooden material properties, workmanship and geometrical details in structures. As a result, the deterministic model may not be representative enough in a realistic situation. Therefore, it is a necessary step to quantify the variations in the model's outputs in order to ultimately establish an accurate prediction tool. To achieve that, uncertainties of the wood material properties can be addressed in a model by introducing probabilistic parameters [6]. In Reference [12], a generalized probabilistic model was constructed to take into account the statistical fluctuation associated with the elastic properties in the model. The uncertainties of the mechanical constants of a wooden structure were also investigated in Reference [13]. A small data base of the Young's modulus in the longitudinal direction was established by means of vibration measurements. Then, by sampling the Young's modulus distribution established by the obtained data base, Monte Carlo simulations were performed in the FE model development to quantify the uncertainties induced by the elastic constants in modelling the vibro-acoustic behavior of wooden buildings in a low-frequency range. However, not only does the Young's modulus in the longitudinal direction have a great impact on the dynamic response of a wooden structure, but the Young's moduli in the other two directions as well as the shear moduli play an important role in its vibroacoustic performance. Regarding the uncertainties in structural dynamics, Shannon's maximum entropy principle [14] is an optimal choice to model the random data and the uncertainties [15]. In Reference [16,17], a probabilistic model was proposed to construct the probability distribution in high-dimension of a vector-valued random variable using the maximum entropy principle. This proposed probabilistic model was then extended to a tensor-valued coefficient (stiffness tensor) with different symmetric levels [15,18,19].

The research reported in this case aims at developing a stochastic FE prediction tool of a CLT slab in order to quantify the uncertainties induced by material properties and, subsequently, acquire the reliable material properties of the under investigated structure to accurately predict the vibroacoustic behavior of CLT in a low-frequency range. To achieve that, different mechanical constant distributions without available material property data base are constructed by means of the maximum entropy principle. The random elastic constants generated following the established distributions are considered as the inputs for the FE model to calibrate CLT in a low frequency range by comparing the simulated results with its dynamic responses obtained from experimental modal analysis (EMA) method. The best combination of the material properties of CLT panel is selected by minimizing different error metrics. The ultimate objective is to acquire the knowledge about the modeling method of CLT and to propose a generalized method to quantify the uncertainties coming from the orthotropic level material properties and to accurately calibrate the corresponding model by avoiding the repetitive manual calibration and, subsequently, to increase the prediction accuracy.

2. Measurements

2.1. Experimental Structure

The test structure was a 4×1.5 m^2 5-ply 175 mm thick CLT panel made of Canadian black spruce of machine stress rated grade 1950f-1.7E in parallel layers and visual grade No. 3 in perpendicular layers with density of 520 kg/m^3. At the initial stage, two CLT slabs were nailed together by a thin wooden panel (c.f. Figure 1a) and they were placed on a standardized sound insulation measurement console. However, this weak connection between two CLT panels resulted in a low stiffness coupling between two slabs and the imperfect simply supported boundary conditions of the measurement console makes the CLT floor easily "jump up" when there is an external excitation. In a real building construction, these problems can be resolved by adding a top layer on the CLT bare floor, which can add mass on CLT and, subsequently, enforce the connection between two CLT leaves. Consequently, due to the weak connection between two assembled CLT panels and the imperfect boundary conditions in the sound insulation laboratory, only the first eigen-mode can be extracted from the measurements, which cannot provide enough information to calibrate the established FE model. As a result, in order to easily extract more meaningful information, only one leaf of CLT floor was under investigation.

(a) (b)

Figure 1. (**a**) Two CLT panels connected by a thin wooden lath. (**b**) One CLT panel simply supported on two I-steel beams.

2.2. Measurement Procedure

2.2.1. Measurement Setup

EMA was performed on the CLT slab to characterize the dynamic properties of the CLT slab. The CLT panel was set free at two opposite long sides and simply supported (SFSF boundary condition) with two steel I-beams at two shorter edges (shown in Figure 1b) in order to simplify the boundary conditions. A predefined mesh was drawn on the CLT surface to determine the excitation positions and the measurement positions. In that way, the CLT was divided into five parts in the long direction and three parts in the short direction, which gives a total number of 16 excitation points (the nodes on the short edges were not excited). The size of mesh was decided based upon the highest frequency of interest, i.e., 200 Hz, to avoid the hammer exciting on the nodes of the eigen-modes.

A single input multiple output system (SIMO) can provide redundant data to better identify the eigen-frequencies and the local modes of complex structures based on different frequency response function (FRF) matrices [20,21]. Meanwhile, these different FRFs can be used as references to validate the FE model by comparing different simulated FRFs with different measured FRFs. Concerning the complexity and the inhomogeneity of CLT, the SIMO system was employed to characterize the dynamic properties of the CLT panel via different FRF matrices at different measurement positions. Five uni-axial accelerometers (Brüel & Kjær Accelerometer Type 4507 001, Brüel & Kjær Sound & Vibration Measurement A/S, Nærum, Denmark) were attached on the different nodes (red dots in Figure 2) with the help of Faber-Castell Tack-It Multipurpose Adhesive. An impact roving hammer (Brüel & Kjær Impact hammer Type 8208 serial No. 51994, Brüel & Kjær Sound & Vibration Measurement A/S, Nærum, Denmark) was used as excitation.

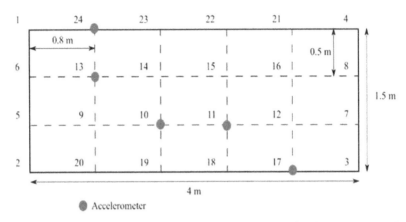

Figure 2. Mesh on the CLT panel. Five accelerometers (red dots) were placed at points 10, 11, 13, 17, and 24), whereas the hammer was moved around all nodes, except on the short edges.

Depending on the impulse shape and the force spectrum of the impact hammer shown in Figure 3, the medium hard hammer tip was selected for this measurement to ensure most of the input energy localizing within the frequency range of interest (up to 200 Hz). The roving hammer approach was employed. The accelerometers were kept fixing on the selected positions and the hammer were roved over the predefined mesh grid except for the points on boundaries. The accelerometers and the impact hammer were connected to Brüel & Kjær multi-channel front-end LAN-XI Type 3050. Then, the acceleration signals and the impact force signals were recorded by using the software Brüel & Kjær PULSE Labshop. There were three averages for each excitation position. The resolution of the FRF was 0.625 Hz.

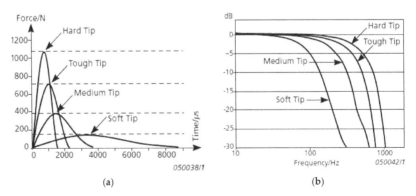

Figure 3. (a) Impulse shapes of the hammers showing the shape as a function of the used impact tip. (b) Force spectra of the hammers showing the frequency response as a function of used impact tip [22] Copyright © Brüel & Kjær.

2.2.2. Modal Parameters Extraction

The eigen-frequencies, the mode shapes, and the damping ratios were extracted by applying rational fraction polynomial—Z algorithm in Brüel & Kjær PULSE Connect. The frequency band of FRFs was divided into several parts, which means that identification of poles was restricted in a narrow frequency band each time in order to easily select the stable poles (eigen-frequencies) from the FRFs in each divided frequency range. An example of the stabilization diagram is shown in Figure 4.

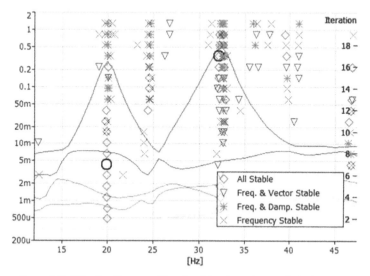

Figure 4. Stabilization diagram with rational fraction polynomial—Z algorithm.

Iteration times of rational fraction polynomial—Z algorithm was 40. The synthesized FRF is shown in Figure 5. The corresponding eigen-frequencies and the corresponding damping ratios are shown in Table 1.

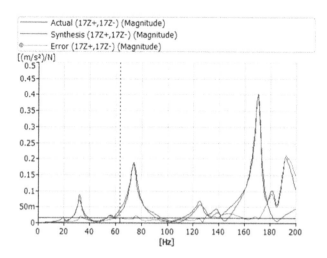

Figure 5. Synthesized FRF, measured FRF, and relative errors between the synthesized FRF and the measured FRF.

Table 1. Measured and simulated Eigen frequencies of the bending modes and the measured corresponding damping ratios.

Mode	Measured Eigen-Frequency/Simulated Eigen-Frequency	Damping Ratio
1	19.8 Hz/19 Hz	4.9%
2	32.2 Hz/33.2 Hz	3.6%
3	56.7 Hz/58.3 Hz	3.2%
4	73.8 Hz/72.5 Hz	3.2%
5	91.0 Hz/100.8 Hz	3.3%
6	125.5 Hz/117 Hz	2.8%
-	-/131 Hz	-
-	-/150 Hz	-
7	170.4 Hz/173.9 Hz	1.2%

Since the measurement system is the SIMO system. Different FRFs (measurement point 11/excitation point 11, measurement point 13/excitation point 13, measurement point 17/excitation point 17, measurement point 24/excitation point 24) can also be saved as references to calibrate the FE model.

3. Finite-Element Model

3.1. Model Description

A numerical model of the CLT slab was created in the commercial FE software Abaqus [23]. Five different layers were modelled by assigning different oriented material properties in different layers of the model. Therefore, there is no relative displacement between different layers in this model, which means that the CLT model is a complete model and each layer is fully tied together. The same material was assigned to each layer, except that the in-plan material orientation assignments were 90° oriented from the adjacent layer to mimic the cross-laminate layers of CLT panel. The material properties of CLT were gathered from the literature [24,25], reported in Table 2. The meshes of 20-node quadratic brick, reduced integration (C3D20R) quadratic type were assigned to the entire model. Since the shape of each layer is a simple rectangular so that there is no discontinuity between different layers. Details of mesh are shown in Figure 6. The mesh size was 0.1 to ensure the accuracy of the highest frequency interest. The eigen-frequencies and the eigen-modes were calculated by the linear perturbation frequency step. The FRF of the CLT was obtained by the Steady-state dynamics, Modal step. The damping extracted from the measurement was included in the model by means of the direct modal damping (c.f. Table 1). In this framework, the FRFs of CLT were first calculated to quantify the uncertainties of material properties. Afterward, the best FRF justified by different criterions was selected to extract the material properties of this under investigated CLT.

Table 2. Material Properties of CLT collected from the literature. Stiffness parameters have the unit of MPa, Poisson ratios are dimensionless, and the density is given in kg/m^3.

E_1	E_2	E_3	G_{12}	G_{13}	G_{23}	v_{12}	v_{13}	v_{23}	ρ
9200	4000	4000	900	90	63	0.3	0.3	0.4	520

In general, it is a challenge task to create appropriate constraints to describe the boundary conditions. Since dynamic responses of the structure are sensitive to the boundary conditions, slight changes in the FE model can lead to a big variation of eigen-frequencies and mode order. In order to mimic the simply supported boundary condition, all the displacements in three directions at one shorter edge were constrained and, at the other shorter edge, all the displacements were also constrained except the displacement in a vertical direction (U1 in Abaqus). This boundary set-up of the FE model was kept throughout the entire modelling process. More explanations about the choice of the boundary condition can be found in Section 3.3.

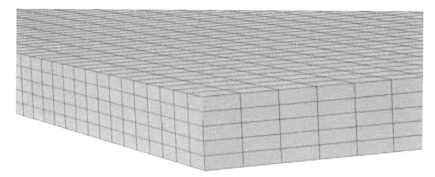

Figure 6. Meshes of FE CLT model.

3.2. Model Validation Criterion

The established model needs to be validated by two different metrics. One is the normalized relative frequency difference (NRFD), which characterizes the discrepancies between the simulated and the measured eigen-frequencies, defined by the equation below.

$$NRFD_i = \frac{\left|f_{ref_i} - f_i\right|}{f_{ref_i}}, \tag{1}$$

where f_{ref_i} is the measured resonance frequency and f_i is the simulated resonance frequency.

The other one is the modal assurance criterion (MAC), which quantifies the similarity of simulated and measured mode shapes, defined by the formula below.

$$MAC = \frac{|(\Phi_i^{sim})^T (\Phi_j^{exp})|^2}{\left(\Phi_i^{sim}\right)^T \left(\Phi_i^{sim}\right) \left(\Phi_j^{exp}\right)^T \left(\Phi_j^{exp}\right)} \tag{2}$$

where Φ_i^{sim} is the *i*-th simulated mode shape and Φ_j^{exp} is the *j*-th measured mode shape. The range of the MAC number is from 0 to 1. When the MAC number equals 1, the simulated eigen-mode perfectly correlates to the measured one. However, MAC equals 0, which implies irrelevant simulated and measured eigen-modes. Different from the conventional calibration procedure, stochastic process calibration begins with comparing eigen-frequencies (NRFDs) between the simulated FRFs and measured FRFs. Therefore, several FRFs of different positions are needed to make sure the calibrated model validating at different positions, which can increase the credibility of the model. However, the FRFs calibration can only ensure consistency of simulated and measured eigen-frequency but not the mode order. The simulated eigen-frequency may be the same as the measured eigen-frequency, but they could have different mode shapes. Therefore, both indictors are needed to ensure the simulated eigen-frequencies correlated with the measured ones to keep the simulated mode order corresponding to the measured one.

3.3. Preliminary Sensitivity Analysis

Wood as a kind of orthotropic material has nine different variables (three Young's moduli, three shear moduli, and three Poisson's ratios) to be calibrated. In order to decrease the complexity of calibration, a sensitivity analysis should be performed to investigate the effect of different elastic constants on simulated eigen-frequencies before the stochastic process is introduced to the FE model. In this section, Young's moduli in different directions were increased or decreased 25% when compared to the Young's moduli given by Table 2. Shear moduli in different directions were increased or decreased 15% when compared to the shear moduli given by Table 2. The Poisson ratios, v_{12}, v_{13}, v_{23} were set to be 0.25, 0.25, and 0.35, whereas the initial values were 0.3, 0.3, and 0.4. The reference eigen-frequencies employed in NRFDs were calculated, according to the elastic constants reported in Table 2. The measured eigen-frequencies were not selected as a reference since the objective of sensitivity analysis aims to investigate how different elastic constants affect the eigen-frequencies of an FE model.

From the NRFDs shown in Figures 7 and 8, it can be seen that the influence of Young's moduli and shear moduli on eigen-frequencies cannot be ignored. Among all the elastic constants, Young's modulus in the longitudinal direction has the most important influence on eigen-frequency changes. However, Young's modulus in a vertical direction barely changes the eigen-frequencies. Therefore, the variation of Young's modulus in the vertical direction was not reported in Figure 7. When looking at Figure 9, we could find that all the NRFDs are lower than 0.5%, which indicates that the influences of Poisson's ratios on eigen-frequencies are negligible. From this sensitivity analysis, it can be concluded that the Young's moduli and shear moduli have a more significant influence on eigen-frequency calculations than Poisson's ratios. As a result, the calibration of material properties can be reduced to six variables.

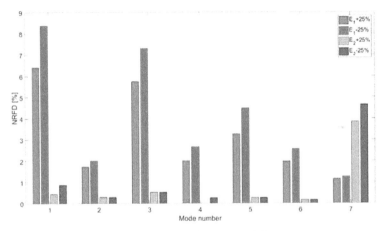

Figure 7. NRFDs of Young's moduli.

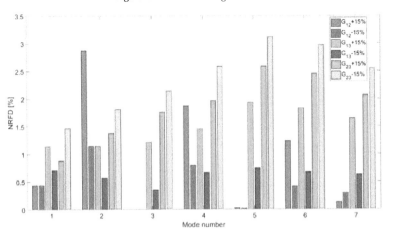

Figure 8. NRFDs of shear moduli.

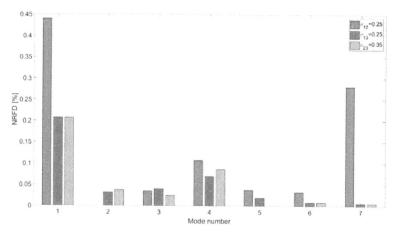

Figure 9. NRFDs of Poisson's ratios.

However, from this sensitivity analysis, it was also noticed that the first four simulated eigen-frequencies are always higher than the first four measured eigen-frequencies even decreasing different elastic constants. This may be caused by the over-stiffened boundary condition. Since the tested CLT is only placed on the top two steel I beams. It is difficult to have a perfect simply supported boundary conditions in reality. Therefore, restricting all the displacements at the boundaries of the FE model can create over stiffened boundary conditions, which results in over-estimated eigen-frequencies. Consequently, the displacement in a vertical direction at one boundary of the FE model is released to try to mimic the real boundary conditions.

4. Stochastic Process

Uncertainties of material properties are always assumed to follow the Gaussian distribution because of its simplicity and the lack of relevant experimental data, even though most material property distributions are non-Gaussian in nature [26,27]. The theory introduced in this case is about the probabilistic modeling of a random elasticity tensor in an orthotropic symmetric level within the framework of the maximum entropy principle under the constraint of the available information [18,19,28]. The established random elasticity tensor is considered as the inputs in the FE model to quantify the uncertainties induced by the CLT material properties and to seek the best combination of CLT material properties to calibrate the CLT model.

In this section, the elastic tensor is first decomposed in terms of random coefficients and tensor basis so that the fluctuation of different elastic constants can be characterized by the probability distribution functions (PDF). Next, construction of PDFs of different elastic constants in high-dimension [16] is shortly introduced. Lagrange multipliers associated with the explicit PDFs of random variables in high dimensions is estimated with the help of the Itô differential equation. The established PDFs is sampled by Metropolis-Hastings algorithm to obtain the random data to construct a random elasticity matrix [18,19] in order to derive the corresponding random combinations of elasticity constants to quantify the uncertainties of material properties and to calibrate the CLT model. A flow chart of the application of the stochastic procedure is shown in Figure 10.

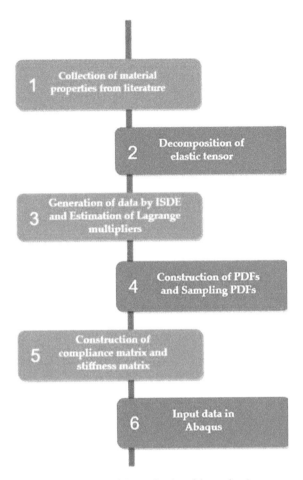

Figure 10. Flow chart of the application of the stochastic process.

4.1. Decomposition of the Random Elastic Tensor

Construction of PDF of a random value in a high-dimension approach can be applied to any arbitrary material symmetry class [15,18,19], such as isotropic symmetry, cubic symmetry, and transversely isotropic symmetry. In this work, this stochastic approach aims to seek a reasonable probability distribution of the random elastic tensor of the target material (CLT). Therefore, only the orthotropic symmetry case is considered in this section. The dimension of the random variable N is limited to 9.

Let C be a fourth-order random elastic tensor, which could be decomposed by using the equation below.

$$C = \sum_{i=1}^{N} c_i E_i, \tag{3}$$

where c_i is a set of random coefficients that can be described by its PDFs and E_i is the tensor basis of the random elastic tensor C, based on Walpole's derivation [29]. The CLT slab is modelled as orthotropic material in Abaqus so that the tensor basis E_i of the orthotropic symmetric elastic tensor are defined in the following form [18].

$$
\begin{cases}
E^{11} = a \otimes a \otimes a \otimes a, E^{12} = a \otimes a \otimes b \otimes b, E^{13} = a \otimes a \otimes c \otimes c, \\
E^{21} = b \otimes b \otimes a \otimes a, E^{22} = b \otimes b \otimes b \otimes b, E^{23} = b \otimes b \otimes c \otimes c, \\
E^{31} = c \otimes c \otimes a \otimes a, E^{32} = c \otimes c \otimes b \otimes b, E^{33} = c \otimes c \otimes c \otimes c, \\
E^4_{ijkl} = (a_i b_j + b_i a_j)(a_k b_l + b_k a_l)/2, \\
E^5_{ijkl} = (b_i c_j + c_i b_j)(b_k c_l + c_k b_l)/2, \\
E^6_{ijkl} = (c_i a_j + a_i c_j)(c_k a_l + a_k c_l)/2,
\end{cases} \tag{4}
$$

where a, b, and c are the unit orthogonal vectors, $\otimes$ is the Kronecker product.
The fourth-order elastic tensor C is decomposed as:

$$
C = c_1 E^{11} + c_2 E^{22} + c_3 E^{33} + c_4 \left(E^{12} + E^{21} \right) + c_5 \left(E^{23} + E^{32} \right) + c_6 \left(E^{31} + E^{13} \right)
$$
$$
+ c_7 E^4 + c_8 E^5 + c_9 E^6. \tag{5}
$$

4.2. Construction of Probability Distribution Function in High-Dimension Using the Maximum Entropy Principle

The objective of this section is to establish the PDFs of the random coefficients c_i, which control the statistical fluctuation of the fourth-order random elastic tensor. Let $c = (c_1, \ldots, c_N)$ be a vector in $\mathbb{R}^N$-valued second order random variable, which obeys certain unknown probability distribution $P_c(dc)$ with the Lebesgue measure $dc = dc_1 \ldots dc_N$. The element in vector $c = (c_1, \ldots, c_N)$ represent the random coefficient of the random elasticity matrix in the previous section (Equation (3)). The unknown probability distribution $P_c(dc)$ of the vector c can be presented by a probability density function $p_c(c)$, which satisfies the following normalization condition.

$$
\int p_c(c)dc = 1. \tag{6}
$$

The Maximum Entropy Principle applied here aims to construct the unknown probability distribution $P_c(dc)$ with the help of the available information. In this case, the probability density function (PDF) p_c could be written using the equation below.

$$
p_c = arg\, max S(p), \tag{7}
$$

where the entropy $S(p)$ is defined by the equation below.

$$
S(p) = - \int p(c) \log(p(c)) dc. \tag{8}
$$

In order to find an explicit probability density function $p_c(c)$, several constraints should be set as available information: (1) the mean values of the variables, (2) the log condition, and (3) the normalization condition.

$$
E\{C\} = \bar{c}, \text{ with } \bar{c} = (\bar{c}_1, \ldots, \bar{c}_9) \tag{9}
$$

$$
E\left\{ \log\left(\det\left(\sum_{i=1}^{N} c_i E \right) \right) \right\} = v_C, \text{ with } |v_C| < +\infty \tag{10}
$$

and

$$
\int p_c(c)dc = 1. \tag{11}
$$

Equation (9) indicates that the mean values of variables are supposed to be known and Equation (10) ensures that both the C and C^{-1} are second-order random variables. This equation also creates the statistical dependence between the different random variables.

To optimize the problem defined by Equation (7), Lagrange multipliers associated with Equations (9)–(11) are introduced. Let $\lambda^0 \in \mathbb{R}^+$, $\lambda^1 \in A_{\lambda_1}$, and $\lambda^2 \in A_{\lambda_2}$ be Lagrange multipliers associated with the constraints defined by Equations (9)–(11). It could be proven that the optimized Equation (7) could be written as [28]:

$$p_C(c) = k_0 exp\{- < \lambda_{sol}, g(c) >_{R^{N+1}}\}, \forall c \in \mathbb{R}^N, \tag{12}$$

where $k_0 = exp(-\lambda^0)$ is the normalizing constant, the operator $\langle,\rangle$ is the Euclidean inner product, the $c \to g(c)$ is the mapping defined on $S \times \mathbb{R}$, with the values in $\mathbb{R}^{N+1}$. $g(c)$ is defined by Equation (13) below.

$$g(c) = (c, \varphi(c)), \text{ with } \varphi(c) = \log\left(\det\left(\sum_{i=1}^N c_i E\right)\right), \tag{13}$$

where $\det(\sum_{i=1}^N c_i E) > 0$ is the support of Equation (13).

An $\mathbb{R}^N$-valued random variable B_λ parameterized by λ should be introduced to identify the Lagrange multipliers. Supposing the probability density function $b \to p_{B_\lambda}(b, \lambda)$ of the random variable B_λ is written as:

$$p_{B_\lambda}(b, \lambda) = k_\lambda \exp\{- < \lambda, g(b) >_{R^{N+1}}\}, \forall b \in \mathbb{R}^N, \tag{14}$$

where k_λ is the normalization constant parameterized by λ. Taking $k_0 = k_\lambda$, from Equations (12) and (14), it can be deduced that:

$$p_C(c) = p_{B_\lambda}(b, \lambda). \tag{15}$$

According to Equations (9), (10), (13), and (15), the calculation of Lagrange multipliers can be derived by evaluating the expectation of $g(b_\lambda)$:

$$E\{g(b_\lambda)\} = (\bar{c}, v_C). \tag{16}$$

As a result, the problem of the calculation of the Lagrange multipliers converts into generating the independent realizations of the random variable B_λ defined over $\mathbb{R}^N$ and then evaluating the left-hand side of Equation (16).

4.2.1. Calculation of Lagrange Multipliers

To derive Lagrange multipliers introduced in the previous section, there are several different methods to generate the independent the random variable B_λ with respect to the corresponding probability density function (Equation (15)), such as the Metropolis-Hastings method [30,31], Gibbs method [32]. However, it should be noticed that the Metropolis-Hastings method demands an appropriate proposal distribution, which is sometimes difficult to choose, and the Gibbs method requires us to know the conditional distributions. As a consequence, it could be intricate to give a robust calculation without an adequate initial guess, especially for a high-dimension case. Therefore, another alternative algorithm is introduced in Reference [16] to generate the independent random variable B_λ.

Random Number Generator

Let $u \to \Phi(u, \lambda)$ be the potential function defined as:

$$\Phi(u, \lambda) = < \lambda, g(u) >_{R^{N+1}}. \tag{17}$$

Let $\{(U(r), V(r))\}, r \in \mathbb{R}^+$ be the Markov stochastic process defined on the probability space (Θ, T, P) indexed by $\mathbb{R}^+$ with values in $\mathbb{R}^+ \times \mathbb{R}^+$, for $r > 0$, satisfying the following Itô stochastic differential equation (ISDE).

$$\begin{cases} dU(r) = V(r)dr, \\ dV(r) = -\nabla_u\Phi(U(r), \lambda)dr - \frac{1}{2}f_0V(r)dr + \sqrt{f_0}dW(r), \end{cases} \tag{18}$$

where $W(r)$ is the normalized Wiener process defined on $(\Theta, T,)$ indexed by $\mathbb{R}^+$ and with values in $\mathbb{R}^N$. The probability distribution of the initial condition $U(0)$ and $V(0)$ are supposed to be known. The parameter f_0 is a real positive number, which could dissipate the transition part of the response generated by the initial condition and ensures a reasonable fast convergence of the stationary solution corresponding to the invariant measure.

When r tends to infinity, the solution $U(r)$ of ISDE converges to the probability distribution of the random variable B_λ.

$$\lim_{r \to \infty}(U(r)) = B_\lambda. \tag{19}$$

n_s independent realization of B_λ is denoted as $B_\lambda(\theta_1)$, ..., $B_\lambda(\theta_{n_s})$. Let r_o be the iteration step that the solution of ISDE converge. When $r_k \geq r_0$, the ISDE could be written as the equation below.

$$\begin{cases} dU(r_k) = V(r_k)dr_k, \\ dV(r_k) = -\nabla_u \Phi(U(r_k), \lambda)dr_k - \frac{1}{2}f_0V(r_k)dr_k + \sqrt{f_0}dW(r_k). \end{cases} \tag{20}$$

Therefore, the independent realizations $B_\lambda(\theta_l)$ can be presented by the stationary solution of ISDE.

$$B_\lambda(\theta_l) = U(r_k, \theta_l). \tag{21}$$

It is worthwhile to mention that the Itô stochastic differential equation defined by Equation (18) can be discretized by the Explicit Euler scheme to obtain an approximate solution.

$$k = 1, \ldots, M - 1, \begin{cases} U^{k+1} = U^k + \Delta r V^k, \\ V^{k+1} = \left(1 - \frac{f_0}{2}\Delta r\right)V^k + \Delta r L^k + \sqrt{f_0}\Delta W^{k+1}, \end{cases} \tag{22}$$

with the initial conditions:

$$U^1 = u_0, \ V^1 = v_0, \tag{23}$$

where Δr is the iteration step and $\Delta W^{k+1} = W^{k+1} - W^k$ is a second-order Gaussian centered $\mathbb{R}^N$-valued random variable with a covariance matrix $E\left\{\Delta W^{k+1}\left(\Delta W^{k+1}\right)^T\right\} = \Delta r\{I_N\}$, where $W^1 = 0_N$. In Equation (22), L^k is an $\mathbb{R}^N$-valued random variable, which is the partial derivative of $\Phi(u, \lambda)$ defined by the equation below.

$$L_j^k \cong -\frac{\Phi\left(\Delta U^{k,j}, \lambda\right) - \Phi\left(U^k, \lambda\right)}{U_j^{k+1} - U_j^k}, \tag{24}$$

with

$$\Delta U^{k,j} = \left(U_1^k, \ldots, U_{j-1}^k, U_j^k + \Delta U_j^{k+1}, U_{j+1}^k, \ldots, U_N^k\right), \ \Delta U_j^{k+1} = U_j^{k+1} - U_j^k. \tag{25}$$

Mathematical Expectation Estimation

After the random number generator has been established and the ISDE has been discretized to obtain the random numbers, the expectation of these independent random numbers should be calculated to derive Lagrange multipliers. The mathematical expectation of the random variable B_λ can be estimated by using the Monte Carlo method. The evaluation of the mathematical expectation of the random variable B_λ is given by the equation below.

$$E\{g(B_\lambda)\} \cong \frac{1}{n_s}\sum_{=1}^{n_s} g(B_\lambda(\theta_l)). \tag{26}$$

After Lagrange multipliers are derived by evaluating the mathematical expectation of the random variable B_λ, the explicit PDFs of different elastic elements in the random elasticity matrix can be established. Depending on Equation (12), the PDF of the elastic tensor for the orthotropic symmetric class material could be defined by using the equation below.

$$p_C(c) = p_{C_1,\dots,C_6}(c_1,\dots,c_6)p_{C_7}(c_7)p_{C_8}(c_8)p_{C_9}(c_9), \tag{27}$$

with

$$p_{C_1,\dots,C_6}(c_1,\dots,c_6) = k\,det(Mat(c_1,\dots,c_6))exp\left(-\sum_{i=1}^{6}\lambda_i^{(1)}c_i\right), \tag{28}$$

with

$$Mat(c_1,\dots,c_6) = \begin{pmatrix} c_1 & c_4 & c_6 \\ c_4 & c_2 & c_5 \\ c_6 & c_5 & c_3 \end{pmatrix}, \tag{29}$$

and

$$p_{C_j}(c_j) = k_j \exp\left(-\lambda_i^{(1)}c_i\right)c_i^{-\lambda^{(2)}}, \; j = 7,8,9. \tag{30}$$

Remarks 1. *The random variables* $(C_1,\dots,C_6)$, C_7, C_8, *and* C_9 *are mutually independent of each other.* C_7, C_8, *and* C_9 *are Gamma-distributed and the k and* k_j *are the normalization constants.*

4.3. Numerical Application of the Orthotropic Symmetric Material (CLT)

The material properties of CLT gathered from the literature (c.f. Table 2) are regarded as an initial starting point (mean value) to procced with the stochastic approach presented in the previous sections. In Abaqus, the orthotropic materials compliance matrix can be defined by the engineering constants:

$$\begin{Bmatrix} \varepsilon_{11} \\ \varepsilon_{22} \\ \varepsilon_{33} \\ \gamma_{12} \\ \gamma_{13} \\ \gamma_{23} \end{Bmatrix} = \begin{bmatrix} \frac{1}{E_1} & -\frac{\nu_{21}}{E_2} & -\frac{\nu_{31}}{E_3} & 0 & 0 & 0 \\ -\frac{\nu_{12}}{E_1} & \frac{1}{E_2} & -\frac{\nu_{32}}{E_3} & 0 & 0 & 0 \\ -\frac{\nu_{13}}{E_1} & -\frac{\nu_{23}}{E_2} & \frac{1}{E_3} & 0 & 0 & 0 \\ 0 & 0 & 0 & \frac{1}{G_{12}} & 0 & 0 \\ 0 & 0 & 0 & 0 & \frac{1}{G_{13}} & 0 \\ 0 & 0 & 0 & 0 & 0 & \frac{1}{G_{23}} \end{bmatrix} \begin{Bmatrix} \sigma_{11} \\ \sigma_{22} \\ \sigma_{33} \\ \sigma_{12} \\ \sigma_{13} \\ \sigma_{23} \end{Bmatrix}. \tag{31}$$

The compliance matrix of the CLT panel can be derived depending on the elastic constants reported in Table 2 and Equation (31). The stiffness matrix is the inverse of the compliance matrix.

$$C = \begin{bmatrix} 10.58 & 2.3 & 2.3 & 0 & 0 & 0 \\ 2.3 & 5.2619 & 2.4028 & 0 & 0 & 0 \\ 2.3 & 2.4028 & 5.2619 & 0 & 0 & 0 \\ 0 & 0 & 0 & 0.9 & 0 & 0 \\ 0 & 0 & 0 & 0 & 0.09 & 0 \\ 0 & 0 & 0 & 0 & 0 & 0.063 \end{bmatrix} \times 10^9. \tag{32}$$

From Equation (32) and the corresponding orthotropic symmetric matrix basis (Equation (4)), the mean value of c_i defined in Equation (5) can be deduced by using the formulas below.

$$(\bar{c}_1,\dots,\bar{c}_9) = (10.58,\, 5.2619,\, 5.2619,\, 2.3,\, 2.4048,\, 2.3,\, 0.9,\, 0.09,\, 0.063) \times 10^9. \tag{33}$$

Let $f^{target} = (\bar{c}, \nu_C)$ be the target vector to compute the Lagrange multipliers.

$$f^{target} = (10.58,\, 5.2619,\, 5.2619,\, 2.3,\, 2.4048,\, 2.3,\, 0.9,\, 0.09,\, 0.063,\, 5.3059). \tag{34}$$

And let $f^{est}(\lambda) = (\bar{c}(\lambda), \nu_C(\lambda))$ be the estimated vector to compare with the target vector so that the optimal solution of λ can be obtained by solving the following optimization function.

$$J(\lambda^{opt}) = argmin\left((1-\alpha)(\bar{c} - \bar{c}(\lambda))^2 + \alpha(\nu_C - \nu_C(\lambda))^2\right), \tag{35}$$

where $\alpha \in [0,1]$ is a free parameter. In this scenario, α is set to be 0.5 to give a robust estimation.

Lagrange multipliers associated with $\lambda^{(1)}$ can be further expressed by means of $\lambda^{(2)}$ [15].

$$\begin{cases} \lambda_1^{(1)} = -\lambda^2 \frac{c_2 c_3 - c_5^2}{\triangle}, \lambda_2^{(1)} = -\lambda^{(2)} \frac{c_1 c_3 - c_6^2}{\triangle}, \lambda_3^{(1)} = -\lambda^{(2)} \frac{c_1 c_2 - c_4^2}{\triangle}, \\ \lambda_4^{(1)} = -\lambda^{(2)} \frac{(c_5 c_6 - c_3 c_4)}{\triangle}, \lambda_5^{(1)} = -\lambda^{(2)} \frac{(c_4 c_6 - c_1 c_5)}{\triangle}, \lambda_6^{(1)} = -\lambda^{(2)} \frac{(c_4 c_5 - c_2 c_6)}{\triangle}, \\ \lambda_7^{(1)} = -\lambda^{(2)} \frac{1}{c_7}, \lambda_8^{(1)} = -\lambda^{(2)} \frac{1}{c_8}, \lambda_9^{(1)} = -\lambda^{(2)} \frac{1}{c_9}, \end{cases} \tag{36}$$

with

$$\triangle = c_1 c_2 c_3 + 2c_4 c_5 c_6 - c_1 c_5^2 - c_2 c_6^2 - c_3 c_4^2. \tag{37}$$

From the sensitivity analysis, the initial guess of $\lambda^{(2)}$ is set to be -2. The target optimization function (Equation (35)) is evaluated by using the interior-point method (*fmincon* function) in Matlab [33].

From Figure 11, it could be observed that the optimization algorithm converges fast to a small value and the corresponding optimal values of the Lagrange multipliers are found to be:

$$\lambda^{(1)} = (0.1263, \ 0.2904, \ 0.2905, \ -0.0758, -0.2324, -0.0758, \ 12.9089, \ 18.4413, \ 1.2909),$$
$$\lambda^{(2)} = -1.1618. \tag{38}$$

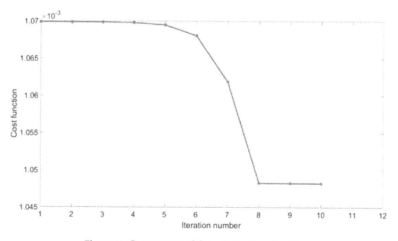

Figure 11. Convergence of the optimization algorithm.

The estimated vector $f^{est}(\lambda)$ is evaluated by the Monte Carlo method.

$$conMC(n_s) = \left| n_s^{-1} \sum_{i=1}^{n_s} g(C(\theta_i)) \right|. \tag{39}$$

Therefore, the estimated vector $f^{est}(\lambda)$ yields the following.

$$f^{est} = (10.5882, \ 5.2677, \ 5.2675, \ 2.3035, \ 2.4121, \ 2.2860, \ 0.8997, \ 0.0904, \ 0.0636, \ 5.3088). \tag{40}$$

The cost function of the target vector $J(\lambda^{opt})$ is 1.048211×10^{-3}, which implies good agreement of the estimated values with the reference values.

4.4. Sampling the Defined Probability Distribution Function by Metropolis-Hastings Alforithm

Following the process introduced in the previous section, the PDFs of the random elasticity tensor of the CLT were constructed. The objective of this section is to generate the random numbers that obey the defined PDFs. The corresponding random stiffness matrix of CLT can be constructed following the generated data. Subsequently, the compliance matrix of CLT can be derived by inversing the stiffness matrix. The random elastic constants (engineer constants in Abaqus) of CLT can be determined according to Equation (31). Lastly, the generated elastic constants should be imported into Abaqus to analyze the dynamic response of the CLT panel. The Markov Chain Monte Carlo method is wildly used to sample the high-dimension PDFs. A specific algorithm, called the Metropolis-Hastings algorithm (MHA), is used in this case to sample the target the function. The proposed distribution is a conventional multivariate Gaussian distribution. The mathematical support of $p_{C_1,...,C_6}(c_1,...,c_6)$ is $det(\sum_{i=1}^{N} c_i E) > 0$ and the mathematical support of $p_{C_j}(c_j)$ is $c_j > 0$, $j = 7, 8, 9$. When sampling the target function, the generated data should stay in the supports of the sampled function. A total of 50,000 combinations of $(C_1,...,C_9)$ are realized by performing the MHA and by obeying the mathematical constraints (supports) of the target functions. The marginal distributions of the different mechanical constants reconstructed by *ksdensity* function in Matlab are shown in Figure 12.

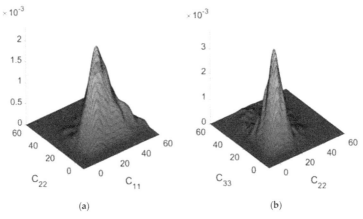

(a)   (b)

Figure 12. (a) Joint probability density function of random variables C_{11} and C_{22}. (b) Joint probability density function of random variables C_{22} and C_{33}.

5. Implementation of Stochastic Data in Abaqus

A total of 50,000 generated random numbers $(C_1,...,C_9)$ have been generated. The corresponding random stiffness tensor could be determined. The random compliance tensor can also be derived by inversing the random stiffness tensor. Since Poisson's ratios have a very slight influence on the eigen-frequencies and the mode shape order of the CLT numerical model, the variation of the Poisson ratios will not be taken into account in the FE model. 50,000 generated random elastic constants are only satisfied with the mathematical constraints. The constraints associated with the physical meaning are not considered when generating the random numbers. In order to ensure the date implemented in the FE model has physical meanings, several physical constraints are set: (1) Variation of Young's moduli and shear moduli should be in a reasonable range of wood. The ranges of Young's modulus and the Shear modulus are assumed in $\pm 50\%$ of the respective mean values.

$$E_1 \in [4600MPa, \ 13800MPa], \ E_2 \in [2000MPa, \ 6000MPa], E_3 \in [2000MPa, \ 6000MPa], \qquad (41)$$

$$G_{12} \in [450MPa, \ 1350MPa], \ G_{13} \in [45MPa, \ 135MPa], \ G_{23} \in [45MPa, \ 135MPa]. \qquad (42)$$

(2) Young's modulus (Shear modulus) in a principle direction should be larger than the Young's moduli (Shear moduli) in the other two directions.

$$E_1 > E_2; \ E_1 > E_3; \qquad (43)$$

$$G_{12} > G_{13}; \ G_{12} > G_{23}. \qquad (44)$$

(3) Young's modulus (Shear modulus) in direction 2 should be larger or equal to Young's modulus (Shear modulus) in direction 3.

$$E_2 \geq E_3; \ G_{13} \geq \ G_{23}. \qquad (45)$$

Only the generated random elastic constants fulfilled with the above requirements (from Equation (41) to Equation (45)) could be imported in Abaqus. In the work reported in this case, due to limited time and limited computer calculation capacity, only 100 different combinations of elastic constants were selected to calibrate the model and to investigate the influence of material properties on the dynamic response of CLT. This large sum of Abaqus input files with different input mechanical constants were realized with the help of Python scripts.

6. Results and Discussion

6.1. Quantification of Uncertainties

One of the objectives of this article is to investigate the effects of uncertainties induced by material properties on the model output and to ultimately obtain a reliable model to predict the vibro-acoustic behavior of different designs. To achieve that, 100 steady-state simulations with different combinations of material properties were carried out in Abaqus.

In Figure 13, each subfigure has 100 FRF simulations and the measurement results are shown in blue. Figure 13 shows that there is an obvious envelope overlapped around the first four peaks lower than 100 Hz. The large variation envelope range is due to varying five mechanical constants in one time since five elastic constants variations have a larger impact on the dynamic response of CLT than only changing one elastic constant in one time.

On the contrary of the frequency range lower than 100 Hz, the simulated FRFs begin to scatter above 100 Hz. No clear envelope peaks can be found around the measured resonances higher than 100 Hz. One possible reason of these scattering curves in a relative higher frequency range is the complexity of the mode shapes. It is known that the lower the eigen-frequency is, the simpler the mode shape is and the longer the wave length is. Long wavelength are not sensitive to the small details in the CLT panel, such as the non-uniform air gaps throughout the laminas and the edge bonding (c.f. Figure 14). It implies that the dynamic behavior of CLT in a low frequency range can be mimicked by a simplified homogeneous orthotropic laminated FE model. However, when the mode shapes become more complex in a higher frequency range, the wavelength becomes smaller. As a consequence, small details of the CLT panel begin to affect the vibration of the CLT panel. In this case, the homogeneous orthotropic laminated FE model could not properly describe the dynamic response of the CLT panel in the higher frequency range. In order to increase the accuracy of the FE in a high frequency range, non-homogeneous laminate layer, such as an account for the irregular air gap in laminas, should be modeled. Nevertheless, it should be aware that the calculation time will become longer, when more details are taken into account in the model. The stochastic method needs a large number of calculations to quantify the uncertainties induced by the material properties so that a compromise should be carefully made between the accuracy of the model and the calculation time.

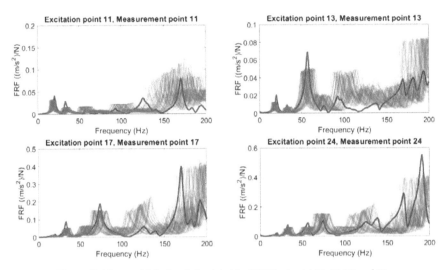

Figure 13. Measured (blue) and simulated (red) FRFs at point 11, 13, 17, and 24.

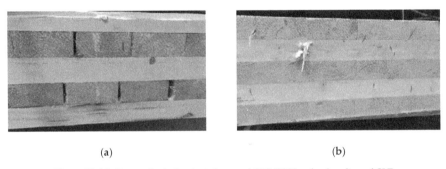

(a)   (b)

Figure 14. (a) Air gaps in the laminate layers of CLT. (b) No edge-bonding of CLT.

6.2. Calibration of the CLT Panel

In this section, the best combination of elastic constants of CLT should be identified by selecting the best NRFDs and MACs among 100 simulations. To select the best combination of mechanical constants, the NRFDs of the first four resonances at point 11, point 13, point 17, and point 24 were calculated. The simulations with the smallest NRFDs of the first four resonances at each point were selected from 100 simulations at each point. The NRFDs of the simulated and measured eigen-frequencies at four excitation positions are shown in Figure 15. It can be seen from the NRFDs of each excitation position that the NRFDs of the first four resonances are lower than 5%. However, the NRFDs of the 5th and 6th resonances are relatively high when compared to the first four resonances. This result emphasizes that more structural details should be involved in the FE model to calibrate the dynamic behavior of CLT in the frequency range higher than 100 Hz. However, only NRFD values are not enough to justify the best combination of elastic constants. Since the NRFDs can only represent the simulated eigen-frequency shifts when compared with the experimental results, the mode order can be different even with a low NRFD. Therefore, MAC numbers are needed to validate the model by ensuring the modes in the same order with reference even if there are low NRFDs. The simulated eigen-frequencies and mode shape are reported in Table 1 and Figure 17. The corresponding material properties of CLT are reported in Table 3. Figure 18 shows that the first six simulated modes are in the same order with the measured ones. However, there exist two extra modes in the simulation results.

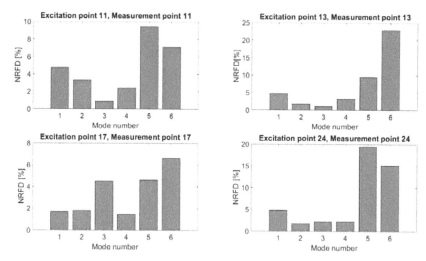

Figure 15. NRFDs of the first six resonances at points 11, 13, 17, and 24.

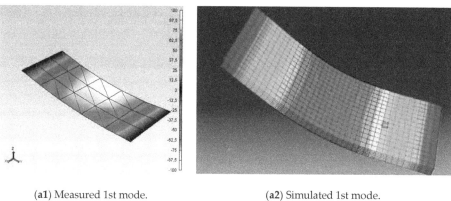

(**a1**) Measured 1st mode.　　　　　　　(**a2**) Simulated 1st mode.

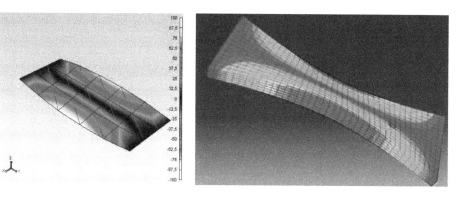

(**b1**) Measured 2nd mode.　　　　　　　(**b2**) Simulated 2nd mode.

Figure 16. *Cont.*

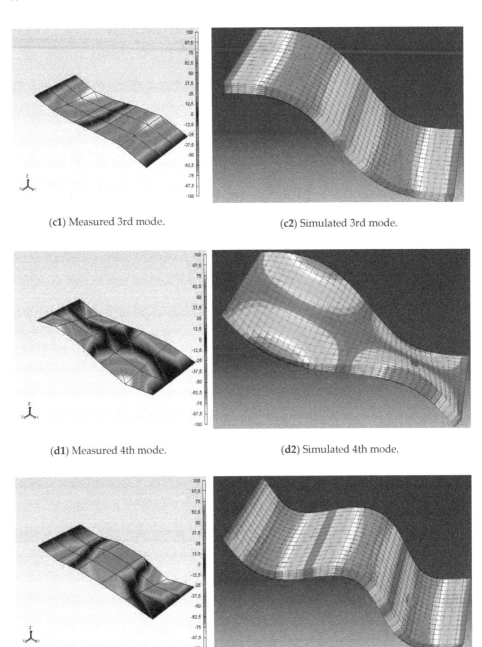

(**c1**) Measured 3rd mode.

(**c2**) Simulated 3rd mode.

(**d1**) Measured 4th mode.

(**d2**) Simulated 4th mode.

(**e1**) Measured 5th mode.

(**e2**) Simulated 5th mode.

Figure 17. *Cont.*

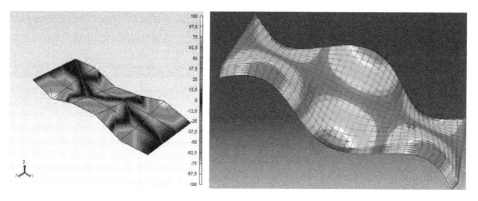

(**f1**) Measured 6th mode.   (**f2**) Simulated 6th mode.

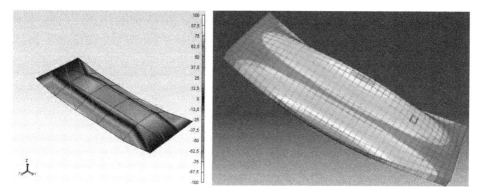

(**g1**) Measured 7th mode.   (**g2**) Simulated 7th mode.

Figure 17. Measured and simulated modes.

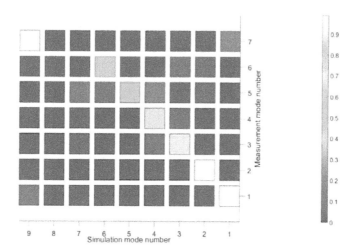

Figure 18. Cross-MAC.

Table 3. Material Properties of CLT used in the calibrated FE model. Stiffness parameters have the unit of MPa and the density is given in kg/m³.

E_1	E_2	E_3	G_{12}	G_{13}	G_{23}	v_{12}	v_{13}	v_{23}	ρ
13,396	4712.5	4681.6	974.6	63.64	60.46	0.4	0.4	0.3	520

The same results can also be seen in the mobility of different excitation points of the lowest NRFD values (c.f. Figure 19). The simulated FRFs at these four different excitation points correlate better with the measured ones, while there are extra peaks and eigen-frequency shifts in the simulated FRFs in the frequency range from 110 Hz to 170 Hz. We suspect that these discrepancies are higher than the 110 Hz result from the over-simplified homogenous laminated FE model, which ignores the geometrical details contained in the real CLT panel. Yet, the boundary condition set-up in the model could not describe the real measurement boundary conditions.

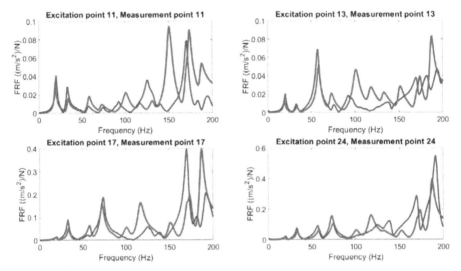

Figure 19. Magnitude of the complex mobility in the vertical direction of point 11, 13, 17, and 24. Simulated FRFs in red, measured FRFs in blue.

The dynamic properties of wooden structures highly depend on the material properties of the structure, the geometry details, and the workmanship. Consequently, the deterministic model may not be able to represent the dynamic response of the wooden structures in a realistic way. A calibrated model may not be able to accurately predict the dynamic behavior of the theoretical identical wooden structure due to the uncertainties. The stochastic method is applied in this case to quantify the uncertainties induced by material properties so that this model can estimate the dynamic response of a class of wooden structures, instead of only one structure. Moreover, the influence of material properties on the vibration of CLT is the coupling effect of Young's moduli and shear moduli in all directions, so that calibration is always time consuming and tedious work to find the appropriate combination of elastic constants in different directions. The calibration employing the stochastic approach could start from the material properties collected from the literature and set the mathematical and physical constraints to generate the input data to find the best combination of the mechanical constants of the under investigated structure. This method could automate the calibration step to avoid the repetitive manual calibrations. However, we should pay attention to the mathematical and physical constraints before generating the input elastic constant data. Because the generated elastic constants should be in a reasonable range of the under investigated material. Otherwise, the input elastic constants may not have a physical meaning, even though the calibration results fit well with the reference. The stochastic method uses a large amount of data to describe an unknown problem (the data base of CLT in our case). More calculations are made and more accurate calibration can be achieved. However, a trade-off between the calculation time and the accuracy of the result should be made in order to keep the calculation time in a reasonable range. This method can not only be applied to CLT but also can be employed to calibrate the other wooden structures whose stiffness constants are difficult to obtain.

Furthermore, one of the objectives of this stochastic approach is to calibrate the material properties of the target structure. It would be better to decrease the influence of other influence factors, such as boundary conditions. Therefore, it is suggested to hang up the under investigated structure (free-free boundary condition) or fix the structure boundary to the ground (perfect simply supported condition) to eliminate the influence of boundary conditions as far as possible. In the work reported in this case, due to a lack of support materials, the CLT panel just laid on top of the I-steel beam and it was not screwed into the ground. Consequently, when the CLT is excited, the deformation of the I-steel beam can affect the vibration of the CLT slab. Furthermore, the laboratory boundary conditions are always different from the in-situ boundary conditions [34]. Thus, it would be necessary to investigate the dynamic response of CLT in a real building. To achieve that, the FRFs could be first measured from a CLT bare floor in real mounting conditions. Then, the same CLT bare floor could be set in the simplified laboratory conditions to compare the relative differences between different FRFs under different boundary conditions.

From the FE CLT modelling perspective, the model validation criterions (NRFD and MAC) and the simulated FRFs suggest that dynamic behavior of the CLT panel can be modelled by the homogenous orthotropic laminated FE model in the frequency range lower than 100 Hz. In a higher frequency range, as the inhomogeneity of the laminated layers of CLT slab begins to pronounce in the vibration of CLT panel, more geometrical details in the CLT panel should be taken into account in the FE CLT model to obtain more accurate results.

7. Conclusions

Low frequency sound insulation is always a challenge for the wooden constructions, especially for the multi-family dwellings. Even though the wooden constructions are satisfied with the standards in force, acoustic comfort is not always met. Since the evaluation frequency range even with the adaptation term of the current standards is from 50 Hz to 3150 Hz, however, the first few resonance frequencies of the wooden floor, which are believed to cause most annoyances, are left out of the evaluation scope. Low frequency prediction tools are needed to access the vibratory performance of wooden buildings at the early design stage due to complaints coming from the inhabitants in wooden buildings. Accessing an accurate low frequency prediction tool requires involving the structure details in the model. Moreover, material properties are another important factor, which can induce a remarkable change in the modelling output.

In this paper, we introduced the stochastic process into the FE model to quantify the uncertainties generated by the material properties. By performing Monte Carlo simulations, variation of Young's moduli and shear moduli in different directions were taken into account in FE model to investigate the coupling effect of different elastic moduli on the dynamic response of structure. Furthermore, 100 simulations were calculated at four different driving points. Clear envelopes can be observed from the simulations lower than 100 Hz. However, the simulations begin to scatter in the frequency range higher than 100 Hz. The best combination of material properties is selected from 100 different combinations of elastic constants to calibrate the FE CLT model. It was noticed that the simulated dynamic response that was lower than 100 Hz was correlated better with the measured dynamic response of CLT. From the promising results, it was concluded that the stochastic method can be applied to a deterministic model (FE model) to quantify the uncertainties of the structures. Furthermore, this method can be employed to calibrate the FE model to acquire the material properties of the under-investigated structure.

Author Contributions: Data curation, C.Q. Investigation, C.Q. Methodology, C.Q. and J.N. Supervision, S.M. and D.B. Validation, C.Q. Visualization, C.Q. Writing—original draft, C.Q. Writing—review & editing, C.Q. and J.N.

Funding: The authors are grateful to FPInnovation and Nordic Structure and Industrial Chair on Eco-responsible Wood Construction (CIRCERB).

Conflicts of Interest: The authors declare no conflict of interest.

References

1. Späh, M.; Hagberg, K.; Barlome, O.; Weber, L.; Leistner, P.; Liebl, A. Subjective and Objective Evaluation of Impact Noise Sources in Wooden Buildings. *Build. Acoust.* **2013**, *20*, 193–213. [CrossRef]
2. Sipari, P. Sound Insulation of Multi-Storey Houses—A Summary of Finnish Impact Sound Insulation Results. *Build. Acoust.* **2000**, *7*, 15–30. [CrossRef]
3. Ljunggren, F.; Simmons, C.; Hagberg, K. Correlation between sound insulation and occupants' perception –Proposal of alternative single number rating of impact sound. *Appl. Acoust.* **2014**, *85*, 57–68. [CrossRef]
4. Vardaxis, N.G.; Bard, D.; Persson, W. Review of acoustic comfort evaluation in dwellings—part I: Associations of acoustic field data to subjective responses from building surveys. *Build. Acoust.* **2018**, *25*, 151–170. [CrossRef]
5. Forssén, J.; Kropp, W.; Brunskog, S.; Ljunggren, S.; Bard, D.; Sandberg, G.; Ljunggren, F.; Agren, A.; Hallstrom, O.; Dybro, H.; et al. *Acoustics in Wooden Buildings: State of the Art 2008*; Vinnova Project 2007-01653; SP Technical Research Institute of Sweden: Stockholm, Sweden, 2008.
6. Flodén, O.; Persson, K.; Sandberg, G. Numerical methods for predicting vibrations in multi-story wood buildings. In Proceedings of the World Conference On Timber Engineering, Vienna, Austria, 22–25 August 2016.
7. ISO717-2. Acoustics—Rating of sound insulation in buildings and of building elements. In *Part 2: Impact Sound Insulation*; International Organization for Standardization: Geneva, Switzerland, 2013.
8. Bolmsvik, A.; Linderholt, A.; Jarnerö, K. FE modeling of a lightweight structure with different junctions. In Proceedings of the European Conference on Noise Control, Prague, Czech Republic, 10–13 June 2012; pp. 162–167.
9. Wang, P.; Van Hoorickx, C.; Dijckmans, A.; Lombaert, G.; Reynders, E. Numerical prediction of impact sound in dwelling from low to high frequencies. In Proceedings of the INTER-NOISE 2018—47th International Congress and Exposition on Noise Control Engineering: Impact of Noise Control Engineering, Chicago, IL, USA, 26–29 August 2018.
10. Negreira, J.; Sjöström, A.; Bard, D. Low frequency vibroacoustic investigation of wooden T-junctions. *Appl. Acoust.* **2016**, *105*, 1–12. [CrossRef]
11. Flodén, O.; Persson, K.; Sandberg, G. A multi-level model correlation approach for low-frequency vibration transmission in wood structures. *Eng. Struct.* **2018**, *157*, 27–41. [CrossRef]
12. Coguenanff, C. Robust Design of Lightweight Wood-Based Systems in Linear Vibroacoustics. Doctoral Thesis, Université Paris-Est, Paris, France, October 2015.
13. Persson, P.; Flodén, O. Towards Uncertainty Quantification of Vibrations in Wood Floors. In Proceedings of the 25th International Congress on Sound and Vibration, Hiroshima, Japan, 8–12 Junly 2018.
14. Shannon, C.E. A Mathematical Theory of Communication. *Bell Syst. Tech. J.* **1948**, *27*, 379–423. [CrossRef]
15. Staber, B.; Guilleminot, J. Approximate Solutions of Lagrange Multipliers for Information-Theoretic Random Field Models. *SIAM/ASA J. Uncertain. Quantif.* **2015**, *3*, 599–621. [CrossRef]
16. Soize, C. Construction of probability distributions in high dimension using the maximum entropy principle: Applications to stochastic processes, random fields and random matrices. *Int. J. Numer. Method. Eng.* **2008**, *76*, 1583–1611. [CrossRef]
17. Soize, C. *Stochastic Models of Uncertainties in Computational Mechanics*; American Society of Civil Engineer: Reston, VA, USA, 2012; ISBN 978-0-7844-1223-7.
18. Guilleminot, J.; Soize, C. On the Statistical Dependence for the Components of Random Elasticity Tensors Exhibiting Material Symmetry Properties. *J. Elast.* **2013**, *111*, 109–130. [CrossRef]
19. Guilleminot, J.; Soize, C. Generalized stochastic approach for constitutive equation in linear elasticity: A random matrix model. *Int. J. Numer. Method. Eng.* **2012**, *90*, 613–635. [CrossRef]
20. Martini, A.; Troncossi, M.; Vincenzi, N. Structural and Elastodynamic Analysis of Rotary Transfer Machines by Finite Element Model. *J. Serb. Soc. Comput. Mech.* **2017**, *11*. [CrossRef]

21. Manzato, S.; Peeters, B.; Osgood, R.; Luczak, M. Wind turbine model validation by full-scale vibration test. In Proceedings of the European Wind Energy Conference (EWEC) 2010, Warsaw, Poland, 20–23 April 2010.

22. Brüel & Kjær. *Product data—Heavy Duty Impact Hammers—Type 8207, 8208 and 8210*; Brüel & Kjær Sound and Vibration Measurement A/S: Nærum, Denmark, 2012.

23. ABAQUS/CAE, 2017. Dassault Systèmes SIMULIA: Vélizy-Villacoublay, France, 2017.

24. Zhou, J.H.; Chui, Y.H.; Gong, M.; Hu, L. Elastic properties of full-size mass timber panels: Characterization using modal testing and comparison with model predictions. *Compos. Part Eng.* **2017**, *112*, 203–212. [CrossRef]

25. Ussher, E.; Arjomandi, K.; Weckendorf, J.; Smith, I. Prediction of motion responses of cross-laminated-timber slabs. *Structures* **2017**, *11*, 49–61. [CrossRef]

26. Shang, S.; Yun, G.J. Stochastic finite element with material uncertainties: Implementation in a general purpose simulation program. *Finite Elem. Anal. Des.* **2013**, *64*, 65–78. [CrossRef]

27. Stefanou, G. The stochastic finite element method: Past, present and future. *Comput. Methods Appl. Mech. Eng.* **2009**, *198*, 1031–1051. [CrossRef]

28. Soize, C. Random matrix theory for modeling uncertainties in computational mechanics. *Comput. Methods Appl. Mech. Eng.* **2005**, *194*, 1333–1366. [CrossRef]

29. Walpole, L.J. Fourth-rank tensors of the thirty-two crystal classes: Multiplication tables. *Proc. R. Soc. Lond. Math. Phys. Sci.* **1984**, *391*, 149–179. [CrossRef]

30. Chib, S.; Greenberg, E. Understanding the Metropolis-Hastings Algorithm. *Am. Stat.* **1995**, *49*, 327–335.

31. Robert, P.C.; Casella, G. *Monte Carlo Statistical Methods*; Springer: New Yourk, NY, USA, 2013.

32. Kroese, P.D.; Taimre, T.; Botev, Z. *Handbook of Monte Carlo Methods*; John Wiley & Sons: Hoboken, NJ, USA, 2010.

33. Matlab, R2017b. The MathWorks Inc.: Natick, MA, USA, 2017.

34. Martini, A.; Troncossi, M. Upgrade of an automated line for plastic cap manufacture based on experimental vibration analysis. *Case Stud. Mech. Syst. Signal Process.* **2016**, *3*, 28–33. [CrossRef]

applied
sciences

Article

Nonuniform Bessel-Based Radiation Distributions on A Spherically Curved Boundary for Modeling the Acoustic Field of Focused Ultrasound Transducers

Mario Ibrahin Gutierrez [1], Antonio Ramos [2], Josefina Gutierrez [3,*], Arturo Vera [4] and Lorenzo Leija [4]

1   CONACYT—Instituto Nacional de Rehabilitación, Subdirección de Investigación Biotecnológica, División de Investigación en Ingeniería Médica, Calz. Mexico-Xochimilco 289, Tlalpan, Mexico City 14389, Mexico; m.ibrahin.gutierrez@gmail.com
2   Consejo Superior de Investigaciones Científicas, CSIC, Instituto de Tecnologías Físicas y de la Información, R&D Group. Ultrasonic Signals, Systems and Technologies, C/Serrano 144, 28006 Madrid, Spain; aramos@ia.cetef.csic.es
3   Instituto Nacional de Rehabilitación, Subdirección de Investigación Biotecnológica, División de Investigación en Ingeniería Médica, Calz. Mexico-Xochimilco 289, Tlalpan, Mexico City 14389, Mexico
4   Centro de Investigación y de Estudios Avanzados del IPN, Cinvestav-IPN, Department of Electrical Engineering, Bioelectronics Section, Av. IPN 2508, Gustavo A. Madero, Mexico City 07360, Mexico; arvera@cinvestav.mx (A.V.); lleija@cinvestav.mx (L.L.)
*   Correspondence: jgutierrez@inr.gob.mx; Tel.: +52-555-999-1000

Received: 14 December 2018; Accepted: 19 February 2019; Published: 4 March 2019

Abstract: Therapeutic focused ultrasound is a technique that can be used with different intensities depending on the application. For instance, low intensities are required in nonthermal therapies, such as drug delivering, gene therapy, etc.; high intensity ultrasound is used for either thermal therapy or instantaneous tissue destruction, for example, in oncologic therapy with hyperthermia and tumor ablation. When an adequate therapy planning is desired, the acoustic field models of curve radiators should be improved in terms of simplicity and congruence at the prefocal zone. Traditional ideal models using uniform vibration distributions usually do not produce adequate results for clamped unbacked curved radiators. In this paper, it is proposed the use of a Bessel-based nonuniform radiation distribution at the surface of a curved radiator to model the field produced by real focused transducers. This proposal is based on the observed complex vibration of curved transducers modified by Lamb waves, which have a non-negligible effect in the acoustic field. The use of Bessel-based functions to approximate the measured vibration instead of using plain measurements simplifies the rationale and expands the applicability of this modeling approach, for example, when the determination of the effects of ultrasound in tissues is required.

Keywords: focused transducer; acoustic field; nonuniform radiation distribution; Bessel radiation distribution; spherically curved uniform radiator; rim radiation; Lamb waves; finite element modeling

1. Introduction

In recent years, the use of focused ultrasounds (FUS) has been increased in biological applications for both high intensity and low intensity modalities [1–5]. A high-intensity focused ultrasound is used for the rapid destruction of tissues by thermal ablation [2,3,6], for example in oncology, while low-intensity applications are based on producing midterm hyperthermia and nonthermal ultrasonic therapy [3,7], with multiple possible applications [4,8,9]. Among the non-ablating FUS applications reported in literature can be mentioned the low-intensity pulsed ultrasound (LIPUS) using focused transducers [8], drug delivery in deep tissues (as blood–brain barrier disruption) [10], gene transfer therapy [11], and sonothrombolysis [12,13], among others. In all these applications, the control of the dose in the target volume should be precise during long periods of time to avoid cell death in non-treated zones [12,14]. Noninvasive (and non-expensive) technologies to monitor the temperature in the treated zone [12], more precise and simpler calibration techniques for FUS transducers to determine effective radiating parameters [15,16], in conjunction with accurate and simple computational models capable of effectively representing the acoustic fields of real FUS transducers are required. This could provide more information to adequately study the produced effects along the ultrasound pathway and to quantify undesired consequences in surrounding tissues, previous the application of the therapy. However, in medical applications, the use of simple, but inefficient, ideal models for the acoustic field of FUS transducers is a common and not very questioned practice [17–19].

One of the first approaches to calculate the acoustic field of curved radiators is the O'Neil solution proposed in 1949 [20]. This solution is based on the Rayleigh integral, and it assumes a spherically curved uniform radiator (SCUR) oscillating with a uniform velocity distribution. Although this approximation could be appropriate for many modern transducers at the focus, in the regions where the pressure amplitudes are highly affected by diffraction, e.g., before the focus, the discrepancies are very evident [17]. These variations between "ideal" theory and experiments occur because the assumption of a normal uniform velocity distribution is very conservative and, usually, unreal [17,21,22]. The rather complex vibration of piezoelectric plates not only is composed by a thickness-extensional (TE) vibration mode but also includes contributions of radial modes [23,24], edge waves [17], and Lamb waves [21,25], whose effects are more noticeable under a continuous regime [17]; this vibration can be more complicated if we consider that the piezoelectric plate is not vibrating freely, but it is somehow clamped by its edge producing a higher vibration amplitude at the center of the plate with an attenuated vibration at the edge [26,27]. This complex vibration occurs in both planar and concave plates, and it should be accounted for when producing models of acoustic fields for more accurate results.

In this paper, we are proposing an approach to model the acoustic field produced by FUS transducers in a low-intensity regime (considering linear propagation) using polynomial-Bessel based functions as nonuniform radiating distributions on a curved surface. The reason of proposing these functions is based on the reported vibration patterns produced in piezoelectric curved disks composed of a main vibrating thickness-extensional (TE) mode generating a wave in the thickness direction and a second component of Lamb waves propagating radially [17,28,29]. These two components are modified when the disk is fixed to the transducer case, which produced a combined vibration pattern that can be approximated by a polynomial-Bessel based function, in accordance with the measurements for any specific transducer. With this approach, we got better results than the widely used SCUR, which can be comparable with the results using the intuitive approach for modeling the acoustic field using the velocity distribution measured on the radiating surface [29,30]. Using analytical functions instead of measurements will permit to propose mathematical algorithms to produce realistic acoustic fields of actual FUS transducers; however, these demonstrations are beyond the scope of this paper. The future applications of these new models are open, since the assumptions taken for this work are not particular.

2. Materials and Methods

In this paper, the acoustic field of a FUS transducer has been modeled. A 2 MHz spherically focused transducer with a 20 mm nominal focal length and a 20 mm of nominal aperture (Onda Corporation, Sunnyvale, CA, USA) was used for the experiments; these values usually differ from the measured ones as it will be discussed later [15]. This kind of transducer, designed for high-intensity applications but used here for low-intensity measurements, has its negative terminal exposed, i.e., it does not have a nonconductive front-layer. This aspect will be useful for the explained below curvature measurements.

2.1. Measurement Setup

Three different sets of measurements were carried out for this work using the setup shown in Figure 1. For the first set, the curvature of the FUS transducer was determined by taking advantage of its exposed negative terminal on the radiating surface; for this, a multimeter (Fluke 289, Fluke Corporation, Everett, WA, USA) and a 3D positioning system (Onda Corporation, Sunnyvale, CA, USA) driven in manual mode were used. The transducer was placed at the final position for acoustic field measurements (to be made after) in which the transducer will emit in positive z-direction; this first measurement was made in the air. Then, the multimeter was set to measure electric continuity, and one of its probes was attached to the positioning system while the other was connected to the negative terminal of the transducer. The measurement probe was moved step-by-step (step resolution of 6.36 µm) towards the transducer surface, down in the z-direction (relative to the direction of the ultrasound emission), for a fixed x-coordinate until the probe touched the transducer conductive surface, indicated by a "beep" from the multimeter. This procedure was repeated diametrically for every x-coordinate from −14.573 mm to 14.573 mm with a step resolution of 0.3312 mm, covering the transducer case and the curved piezoelectric plate. The used step-resolution was smaller than a half of a wavelength of 2 MHz ultrasound in water at 20 °C (approx. 0.38 mm). The determined z-coordinates for all the "x" were saved for further use. These final coordinates were used for the second below explained measurement.

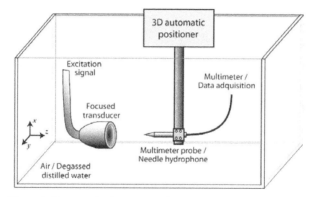

Figure 1. The setup for the curvature/acoustic pressure measurements.

For the second set, the transducer was immersed in a tank filled with degassed distilled water carefully keeping intact the transducer setup used for the first set (see Figure 1). This second set of measurements was carried out using the previous obtained coordinates for the curvature to determine the radiation pattern "very close to the transducer surface". The radiation measured at this region would closely represent the vibration distribution of the curved piezoelectric disk, assuming the effect of other regions of the radiating surface is negligible. For this, a wideband needle hydrophone PZTZ44-0400 (Onda Corporation, Sunnyvale, CA, USA) with a 40 µm aperture and sensitivity of −260 dB referred to 1 V/µPa was mounted on the positioning system, replacing the multimeter probe. The measurements were made at the positions determined previously but 1 mm in front of the transducer surface (in z-direction). The transducer was driven with a wave generator (Array 3400, Array Electronic Co., Taiwan, China) using s 20 Vpp sine tone-burst; the received data were recorded in a PC. The ultrasound signals were recorded for 1, 5, 10, and 20 sine cycles to determine the radiation pattern under different excitation conditions but, more specifically, to know in which number of cycles the vibration presents a quasi-stationary pattern. The data were post-processed in MATLAB (R2017a, MathWorks, Natick, MA, USA) to determine the peak-to-peak values at each spatial point and the full radiation distribution for each excitation condition.

The third set of measurements was performed to determine the full acoustic fields of the focused transducer. This was measured using the previously described setup. The acoustic fields were obtained by saving only the peak-to-peak acoustic pressure at each point in the radiated volume. It was recorded on an XZ plane covering the transducer dimensions starting at $z = 2$ mm and finishing at $z = 50$ mm from the transducer case, with a step resolution of 0.3312 mm in the x-direction and 0.5000 mm in the z-direction. The XY planes were measured at two depths, 0.2 cm and 1.7 cm (at the focus), using a step resolution of 0.3312 mm in both directions; these planes were obtained to verify the symmetry of the acoustic radiation. As formerly, the acoustic fields were captured for 1, 5, 10, and 20 sine cycles to determine the conditions for a nearly continuous emission pattern. The data were postprocessed in MATLAB to reconstruct the fields.

2.2. Acoustic Field Modeling Using FEM

The acoustic field was modeled using the finite element method (FEM) based on the geometry shown in Figure 2. The software used for the FEM processing was COMSOL Multiphysics (COMSOL AB, Stockholm, Sweden) working in a PC of 8-core 3 GHz microprocessor and 64 RAM (Dell, Round Rock, TX, USA). Based on the cylindrical symmetry of the transducer, the problem was assumed axisymmetric. The validity of this assumption was verified with acoustic field measurements in which the profile of the radiation along the azimuthal coordinate is similar for any angle (data not shown). The mesh in the rectangular part of Figure 2 consisted of 10 square elements per wavelength, i.e., more than 530,000 elements; triangular elements were used only for the zone created with the boundaries 1 and 5 to simplify meshing. The requirements to mesh using squares are very specific, and this kind of element cannot be used in certain curved geometries; triangular elements are sometimes the only option for these complicated zones. The mesh convergence was verified by increasing the mesh resolution, which indicated an error of 0.01% at the focus amplitude between meshing using 9 and 10 elements per wavelength. Using square elements instead of triangular permitted to increase the spatial resolution with the same interpolation functions because this kind of mesh has a larger number of nodes (and therefore more degrees of freedom) than the triangular one with the same number of elements (more nodes per element area); this can be noticed with the reduction of the solution time compared to the time required for solving the problem using the same number of triangular elements. The largest solution time registered for our main model, which was barely the most demanding of computational resources, was 45 s.

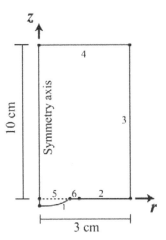

Figure 2. The finite element method (FEM) axisymmetric geometry for modeling the focused acoustic field.

It was assumed the linear ultrasound propagation (i.e., rather low acoustic intensity) was in an attenuation-free homogeneous media (degassed distilled water). The ultrasound propagation is determined with FEM based on the homogeneous Helmholtz wave equation for harmonic radiation, assuming a purely harmonic source and a no-frequency dispersion. This equation can be written as

$$\nabla^2 p + k^2 p = 0 \tag{1}$$

where the wavenumber is $k = \omega/c$, ω is the angular frequency (rad), and c is the speed of sound in water (m/s) assumed constant. Boundary 1 was set with the harmonic normal acceleration. The amplitude of this normal acceleration was adjusted for the different radiation conditions presented in this paper. For instance, in the uniform radiation model of the next section, the amplitude of the harmonic acceleration was set constant along the radius. Usually, the radiation of a boundary is expressed in terms of particle velocity [20,26,31]. In harmonic conditions [32], the particle acceleration a_0 on the radiator surface can be obtained by time-deriving the constant-amplitude particle velocity $v_0 e^{j\omega t}$ as

$$a_0 = \frac{d}{dt}\left(v_0 e^{j\omega t}\right) = v_0 \omega e^{j\left(\omega t + \frac{\pi}{2}\right)} = \frac{\omega}{\rho c} p_0 e^{j\left(\omega t + \frac{\pi}{2}\right)}, \tag{2}$$

where ρ is the medium density (kg/cm^3). The last term was determined by considering $p_0 = \rho c v_0$, which is true in the field very close to the radiator surface [27]. The term $\frac{\pi}{2}$ in Equation (2) is a time phase shift between the velocity and the acceleration, and this is not related to the spatial profile along the transducer surface [32]. Then, under harmonic simulations, using either acceleration or velocity for the radiation distributions of the plate is not relevant if adequate amplitude considerations are taken.

Boundaries 2–4 were configured to match the acoustic impedance of water and to reduce the ultrasound reflections at the walls [27]. Then, $Z = \rho c = 1.5$ MRayls at 25 °C, where Z is the acoustic impedance (MRayls) given by the product of the media density and the speed of sound; the walls were considered to be perfectly flat. However, the dimensions of the FEM geometry (10 cm × 3 cm) were big enough to not have residual reflected waves affecting the region of interest for this application, i.e., a region of 5.0 cm depth and 1.5 cm width after the transducer. Boundary 5 was set with a continuity condition, and it was used only to simplify the meshing. Boundary 6 represents the transducer rim, which was set, accordingly to the measurements, to uniformly radiate a relative pressure of 8% of the average pressure radiated by the curved surface in the effective radiating area; this area was determined as the area producing 95% of the transducer's radiation, as defined in other applications for planar radiators [27]. The use of this effective area instead of the nominal area improved the model results as it will be explained later; this was already proposed by other means in Reference [15]. The rim radiation was included because it represents an important contribution in the field for this transducer, more evident in the post-focus field but with some little effects in the pressure amplitude on the propagation axis (along z-axis) in the prefocus zone.

2.2.1. Radiator with Uniform Vibration Distribution

Conventionally, the acoustic field produced by planar ultrasound transducers has been represented as the product of the radiation coming from a uniform vibrating surface (the piston in a Baffle and the Rayleigh equation) [33]. This assumption produces adequate results for wideband transducers, in which the vibration is usually damped by the backing material and, thus, produces quasi-uniform displacements along the transducer radiating surface [34–36]. However, narrowband transducers often do not have a backing material (air-backed) which makes the vibration of their piezoelectric components less uniform [27]. These nonuniformities have an important effect in the field.

The acoustic field of focused transducers has been historically modeled following the same supposition of planar plates, in which it is assumed the plate is a slightly curved uniform radiator (SCUR) [20]. For this, the curvature of the piezoelectric plate of the transducer is usually assumed spherical; conversely, this curvature is not often reported in datasheets. The most well-known theoretical proposal to determine the acoustic field produced by curved surfaces was made by O'Neil in 1949 [20]. This model works adequately for low-power wideband transducers with a backing material, and it has produced poor results when used to model the acoustic field of air-backed narrowband radiators [17,36]. In order to compare our proposal with the most used ideal approach of curved transducers, the acoustic field produced by a SCUR was determined. For this, a constant amplitude harmonic normal acceleration was used in the curved boundary 1, in Figure 2.

2.2.2. Radiator with "Classical" Nonuniform Vibration

Nonuniform vibration distributions have been proposed to generate consistent acoustic field models of some planar radiators but with a limited range of applicability. These are based on the assumption that, under certain conditions, a radiator can behave not only as a piston but also as a membrane and a clamped plate [26,31,37]. The general equation of "classical" nonuniform radiation distributions include the two more common theoretical conditions for fixing the edge of a rigid piezoelectric plate [38]: 1) a plate with simply supported edges that restrict edge movement in any direction but allow rotation by the edge (simply supported radiator) and 2) a plate with clamped edges that restricts the movement and rotation in any direction (clamped radiator). The general equation for the nonuniform radial acceleration $a_0(r)$ of the surface of a plate radiator can be expressed by [26,31,37]

$$a_0(r) = \sum_{mn} A_{mn} \left[1 - \left(\frac{r}{R} \right)^{2m} \right]^{n+1}, \tag{3}$$

where R is the fixed radius of the plate (m), A_{mn} is the adjustable amplitude of the acceleration, and constants m and n will determine the shape of the distribution. However, only two cases have been reported to represent an actual meaning in Equation (3), specifically, when $n = 0$, the radiation distribution corresponds to the simply supported radiator (SSR) for $m = 1, 2, 3, \ldots$; when $m = 1$, the radiation distribution is that of a clamped radiator (CR) for $n > 0$. This radial acceleration is set in boundary 1 shown in Figure 2. The summation in Equation (3) can be reduced after assuming that the first vibration mode dominates [31].

2.2.3. Radiator with Bessel Acceleration

The vibration distribution of a piezoelectric plate is closer to a combination of Bessels than a uniform distribution [38–40]. This behavior depends on the material's piezoelectric properties, the constraint conditions, the coupling layers, and the shape. When the plate is concave, the vibration of the concave-shape piezoelectric plates can also be composed of Bessel-like vibration distributions due to the Lamb waves [17]. The components affecting the vibration of an ultrasound transducer are vast and difficult to quantify. Then, for the proposal of this paper, the vibration was determined by measuring the acoustic pressure very close to the radiating surface [27]. If we suppose the ultrasound behaves as a plane wave in the region very close to the radiating surface (when $r \rightarrow 0$), the acoustic pressure is related with the acceleration as shown in Equation (2).

Thus, the proposal of this paper is to use, as the nonuniform radiation distribution, a Bessel-based function composed by two sub-functions [27]. For this, the acceleration on the curved radiating surface will be expressed by

$$a_0(r) = A_0 f(r) \cdot g(r), \tag{4}$$

where r is the variable radius related to each point in the curved boundary and A_0 is the amplitude of the emitted acoustic pressure, provided that the amplitude of the product of $f(r)$ and $g(r)$ is 1 at $r = 0$. The function $f(r)$ is given by the acceleration of Equation (3), and it represents the effect of attaching the piezoelectric plate to the transducer case (edge clamping), then $f(r) = \sum_{mn} A_{mn} \left[1 - (r/R)^{2m} \right]^{n+1}$. This function can be determined by adjusting the parameters to get a correct representation of the measured profile, which can be evaluated by correlation.

The function $g(r)$ represents the radial measured peak distribution (MPD) shown in Figure 3. This was approximated using the Bessel-based function

$$g(r) = C_1 J_0 \left(\beta_{2N} \frac{r}{R} \right) + C_2.$$ (5)

Equation (5) was proposed after considering the reports and simulations about Bessel patterns in the vibration distribution on some transducer surfaces [17,28,29]. For the profile measured for the transducer used for this paper, we used a pure negative Bessel J_0 mounted over either an SSR or CR function to approximate the experimental MPD. Then, the Bessel-SSR and Bessel-CR distributions were obtained by combining Equations (5) and (3) into Equation (4), with adequate constants. Here, C_1 and C_2 are the coefficients that control the amplitude and the offset of Equation (5), respectively, and β_{2N} is the $2N$ zero of J_0, where N is the number of peaks in the MPD. When C_1 is close to zero, the acoustic field is close to the ideal uniform radiation (SCUR) for $C_2 \neq 0$; for practical concerns, the average amplitude into the effective radiating radius of the function $a_0(r)$, after selecting adequate functions $f(r)$ and $g(r)$, should be 1, which can be adjusting with an adequate value of A_{mn}. The effective radius was determined by calculating the effective area containing 95% of the total radiation. Then, the amplitude of the emitted pressure can be controlled with A_0. In Figure 3, $C_1 = -0.3$, $C_2 = 0.9$, $R = 1$ cm, $m = 1$, $n = 0.3$, $N = 5$, $A_{mn} = 11.7$, and $\beta_{2N} = 30.67$, which is the tenth zero of Bessel J_0. The values of m and n of Equation (3) were determined by the values producing the maximum correlation with the measured data. The Bessel-SSR of Figure 3 requires $n = 0$; Bessel-MOD (MOD, modified Bessel-SSR) have the same constants as the Bessel-SSR, but $n = 0.3$. These values are highly dependent on the transducer construction, not only on the piezoelectric constants and disk geometry; they are dependent on the way the disk was attached to the case, the thickness and properties of extra layers (electrodes, glue, etc.), and the mechanical properties of the case. Because of these reasons, it would not be possible to specify any rule to determine the parameters of Equations (3) and (5), neither by simple transducer inspection nor even after knowing the transducers' electrical characteristics. Actual measurements of the emitted field very close to the transducer should be made.

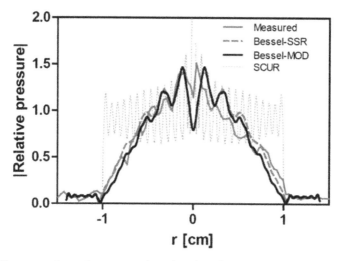

Figure 3. The measured acoustic pressure at 1 mm from the radiating concave surface along its radius compared with the uniform and proposed Bessel distributions. The graphs are normalized with the average radiated pressure at the radiator surface.

3. Results

In this paper, we propose a Bessel-based nonuniform vibration distribution for the surface of a spherically curved radiator to get a modeled acoustic field closer to a real measured field. Three classical approaches based on uniform (SCUR) and nonuniform distributions are included to contrast our main results. Figure 3 shows the measured pressure profile at 1 mm from the piezoelectric curved plate that is composed of local peaks distributed along the radius. This distribution can be represented with Equation (4), combining an adequate Bessel function (Equation (5)) and a function representing the radiator clamping using the traditional nonuniform approaches (SSR or CR with Equation (3)). Using a radiating function instead of the measured data for determining the acoustic field could permit the generalization of the radiation coming from this kind of transducer and eventually simplify the calculations by proposing analytic solutions for certain conditions. Although SSR and CR nonuniform distributions have been suggested since the 1960s decade for planar radiators, these have not been formally proposed for curved radiators [26,37]; their inclusion in this paper is not just for comparison purposes but as two valid alternatives for producing certain types of acoustic fields. Then, the use of SSR/CR nonuniform distribution for curved radiators is not discarded by this work, so this can be an effective approach if the operation conditions of the transducer produce these kinds of vibration profiles.

In Figure 4, it is shown the relative pressure profile along the propagation axis (z-direction) of classical SCUR and two "traditional" nonuniform vibration distributions compared with the measured field. The improvement in the fields using nonuniform distributions on the radiator can be noticed. When the vibration of a curved radiator is uniform, the field has very sharp local peaks along the propagation axis that are smoothed when the emitted radiation is gradually reduced at the plate rim, as in the proposed nonuniform approaches for $r \to R$. The concordances of these fields and the measurements are much better, except in the prefocus zone in which the diffraction patterns differ. In both nonuniform approaches, the post-focus zone is smoother, following the same tendency as the measured profile. There is an after-lobe at about 3 cm, which is well-represented by both nonuniform approaches but only when those included the rim radiation. Without this component, this lobe disappears [41], keeping the most other components of the field. This rim radiation was also included in the main results of this paper.

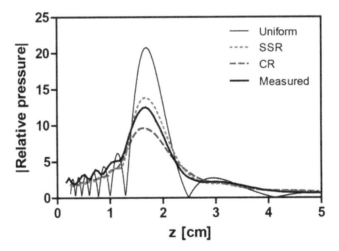

Figure 4. The amplitude of the relative acoustic pressure along the propagation axis of the spherically curved radiator with uniform distribution (SCUR), simply supported radiator (SSR), and clamped radiator (CR) conditions vs. measured field. The pressure is relative to the average pressure on the radiator surface of each condition.

The results of using the Bessel-based radiation distributions are shown in Figure 5. For these proposals, Equations (3) to (5) were used to set the acceleration of the curved radiating boundary; also, the radiation from a 3.5 mm rim was included (boundary 6). For the Bessel-SSR approach, it can be noticed that the acoustic field at the prefocal zone has a peak distribution very close to the measured field, and the post-focal zone has the after-lobe produced by the rim radiation with an amplitude equivalent to that of the measurements. The Bessel-CR approach did not provide a good result in the prefocal zone, preserving only one peak of the three located at the measured field. The after-lobe in the far field was not present for this approximation, even when it included the rim radiation, which could indicate this approach is useless for this particular transducer. However, controlling the value of n in Equation (3) can reduce the amplitude of the focus of the pure SSR condition to make the model fit with the measurements if correlation does not provide an acceptable result. For this paper, correlation was useful, and the value of m and n were easily found by this method, which permitted to find the constants for the modified SSR profile (Bessel-MOD) in Figure 5, with a more accurate result comparable with the measurements.

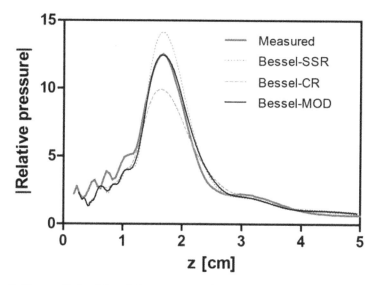

Figure 5. The amplitude of the relative pressure on the propagation axis using the Bessel-based nonuniform approximations with SSR and CR conditions on the curved radiating surface. Bessel-MOD is the Bessel-SSR with $n = 0.3$, which emulates correctly the measured field distribution. The pressure is relative to the average emitted pressure. The three models include an 8% of rim radiation, equivalent to the radiation measured at that zone.

The graphs of Figure 6 show the radial profiles at different depths. Important regions were chosen to plot these graphs: Figure 6a is the graph obtained at the first depth measured for the XZ plane; Figure 6b,c is at two equally space depths before the focus; Figure 6d is at the focus; Figure 6e is at the first minimum after the focus; and Figure 6f is at the after-lobe (after the focus). In these figures, the radial profiles at different z-distances for the Bessel-MOD are closer to the measurements than the uniform approach, not only in amplitude but also in shape. Other approaches presented in this paper were omitted from this figure for clarity, even if they also would have shown an improvement compared with the uniform approach. For instance, the pure SSR produced in the prefocal zone a profile like the Bessel-MOD without the peaks (a smoother profile), while after the focus, the SSR profiles were the same as those presented for the Bessel-MOD in Figure 6e,f.

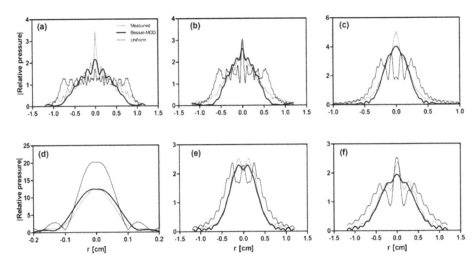

Figure 6. The radial relative pressure at different depths of the measured acoustic field compared with the SCUR distribution and the Bessel-MOD distribution, i.e., Bessel-SSR with $n = 0.3$ in Equation (3). Other analyzed distributions did not fully match the measured profile, and they were not included in these graphs for clarity, namely, SSR, CR, and Bessel-CR. The depths are (**a**) 0.2 cm, (**b**) 0.5 cm, (**c**) 1.0 cm, (**d**) 1.7 cm (focus); (**e**) 2.7 cm (down peak); and (**f**) 3.1 cm (lobe after focus). All the graphs have the same type of lines as in (**a**) and the same axis labels. The pressure is relative to the average pressure of Figure 3 of each distribution.

In Figure 7, the full acoustic fields of the more representative models of this work are shown. The Bessel-based approaches with modified SSR (SSR with $n = 0.3$) and CR components are presented in Figure 7c,d, respectively, in order to be compared with the measured field of Figure 7a and the most used classical SCUR distribution of Figure 7b. This ideal SCUR distribution is composed of a very characteristic diffraction pattern before the focus, which is not present in the measured profile. The Bessel-MOD proposal provides a better result in the prefocal zone, with a similar diffraction pattern to the measured field. The Bessel-CR still has a similar diffraction pattern but with a narrower acoustic field and larger focus size, probably because of the reduced effective radiating area [41]. The focus locations in the Bessel-based models were more in agreement with the measured field than the SCUR; this was probably because of the reduction of the effective radiating areas, which were determined from each data set (models and measurements) to normalize the average emitted radiation that was the reference of the relative pressure used in the field comparisons.

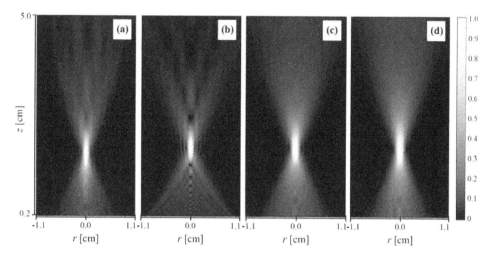

Figure 7. The normalized acoustic field of the focused transducer at a central longitudinal plane: the transducer would be at the bottom of each figure. The linear color scale: white = 1 and black = 0. (**a**) The measured acoustic field; (**b**) the modeled field using SCUR; (**c**) the modeled field using a modified Bessel-SSR ($m = 1$, $n = 0.3$); and (**d**) the modeled field using Bessel-CR ($m = 1$, $n = 1$).

4. Discussion

A proposal to effectively model the acoustic field of focused transducers has been presented. This was based on another previously proposed nonuniform distribution for planar radiators named Bessel-based nonuniform radiation distributions (shown in Figure 3). In this work, two other more general proposals are also used as comparison: the simply supported radiator (SSR) and clamped radiator (CR). The three proposals have shown that can be used to generate acoustic fields in good agreement with the field produced by focused radiators. The distributions were set in a curved boundary, and the results were obtained with a FEM commercial software. Only low-power linear behavior was considered in this work, but the main rationale presented here could be applied for high-power nonlinear measurements. The emission was considered to be purely harmonic operating at the transducer's nominal frequency of 2 MHz; the measured transducer emission bandwidth was 100 kHz, which represents about 5% of the central operation frequency. Because of that narrow bandwidth, the effect of this parameter in the acoustic field was considered negligible. The use of an analytical expression in the radiating boundary instead of the raw measurements, for acoustic field processing, would open the possibility not only to reduce computation but also to find a simpler way to calculate analytically the acoustic field in a closed form under certain conditions, for instance using the Hankel transform [42] or the direct integration of the Rayleigh integral [22,24]. However, this process is beyond the scope of this paper.

Figure 4 shows the axial acoustic fields for the SCUR approach and the two classical nonuniform distributions SSR and CR already proposed for planar radiators. It can be noticed that the latter produce good results in the overall field amplitude when they are set in this curved radiating boundary but with poor agreement in the prefocal zone. At this region, the diffraction effects are quite different between the measured profiles in both the SSR and CR models. Actually, the diffraction profile of pure SSR and CR fields are similar to the SCUR field but with a reduced amplitude and an "offset". This indicates that reducing the emitted field amplitude at the edges of the radiator provokes also a smooth overall acoustic field. The SSR and CR fields are closer in amplitude to the measured field, and this could occur because the piezoelectric plate of the transducer is fixed at the edges, which produces an emitted radiation closer to a SSR/CR distribution than a SCUR, as seen in the measured field in Figure 3. This is also noticeable in the focus locations, which were closer to the measurements in the

SSR and CR distributions than the SCUR; this was improved probably because of the reduction of the effective radiating area in the nonuniform distributions [15]

Although these classical nonuniform distributions are closer in amplitude to the measured field, their results are not fully satisfactory. Then, it is proposed the use of Bessel-based distributions is closer to the real vibration distribution produced in concave transducers. The results of combining a Bessel with SSR and CR edge conditions are shown in Figure 5. From those results, it can be noticed that the prefocal zone of the acoustic field is better modeled after including the Bessel behavior in the radiation distribution. The peak-distribution in the modeled prefocal zone using Bessel-SSR has the same number of peaks as the measured field and almost at the same position. The focus amplitude was improved when adapting the equation with $n = 0.3$, for the Bessel-MOD, which produced a field more congruent with the measurements. The focus locations of these representations were also more in agreement with the measurements than the classical models presented before, probably due to the reduction of the effective radiating areas. In the post-focus zone, the model can also produce the after-lobe observed in the measurements but with a slightly lower amplitude. This after-lobe was only present in the Bessel-SSR and Bessel-MOD simulations with 8% of rim radiation, which was based on the measurements (see rim radiation in Figure 3). The Bessel-CR condition produced a similar profile at the prefocal zone but with less pronounced peaks. In this case, the after-lobe in the post-focal zone was practically unnoticeable, even with the inclusion of the rim radiation, probably because of the total dominance of the acoustic intensities coming from the main radiating surface.

Figure 6 shows the radial distributions at different depths of the three more significant cases for our comparison purposes, namely, the SCUR, the modified Bessel-SSR (Bessel-MOD), and the measurements. Before the focus, in Figure 6a–c, the radial profile using the Bessel-MOD follows adequately the measured field with some variations in amplitude in the central peak. In spite of the complexity of adequately simulating this region, the Bessel-MOD approach correctly produces acceptable results. A similar agreement was obtained in other less complicated regions as the focus and after the focus in both amplitude and field width. In the former, shown in Figure 6d, the Bessel-MOD produced almost the same amplitude as the measurements, with a 0.4% relative error; the focus with the SCUR is 40% larger than the measured one. After the focus (Figure 6e,f), our model still has a similar tendency as the measured field, with a good match of the graphs in Figure 6e and little variation in the central peak in Figure 6f. From these graphs, the SCUR does not represent an adequate model for this kind of radiators, i.e., those with nonuniform Bessel-based vibration patterns.

The acoustic fields presented in Figure 7 show the differences among each approach in a detailed manner using a linear color scale. The prefocal zones of both Bessel approaches are similar but with a larger focus for the Bessel-CR approach. Focus locations were also improved using our proposed models. The SCUR approach differs at practically any region from the measured field. For this transducer, the Bessel-MOD presents better results than the other nonuniform distributions studied here, with a clear good agreement in most of the graphs shown in this paper. However, this does not mean that the applicability of this model is universal for curved radiators. Some transducers could still behave as ideal pistons, simply supported curved disks, or clamped curved disks that could require the use of any of the other approaches (SCUR, SSR, CR, and Bessel-CR). The use of any of these models should depend on the real radiation profile measured very close to the radiating surface. This paper has presented a new modeling approach to effectively simulate the acoustic field of focused transducers that can be adapted to most of the devices used in different medical applications. Eventually, using the proposed functions as radiation distributions of this kind of transducers could potentially permit to find more analytical solutions of this kind of radiators that could better match the measurements.

5. Conclusions

In this paper, we have presented four proposals of nonuniform vibration distributions on the radiating surface of a curved transducer to obtain more realistic simulated acoustic fields very congruent with field measurements. From the results here shown, it was possible to conclude that these models provide better approximations of the vibration distribution capable of producing accurate representations of acoustic field for focused applications. When using Bessel-based functions for the vibration of the curved radiator, the prefocal zone of the transducer was correctly simulated. For the post-focal zone, our Bessel proposal had to include the radiation coming from the transducer rim, which allowed the incorporation of the central "after-lobe" observed in the measured field. In the other proposed approaches, the amplitude of the focus significantly varied with respect to the measurements. This happens possibly because of the differences in the proportion of the emitted average radiation used for determining the relative pressure and because of the effective radiating area in each condition, which is a parameter rarely considered for this kind of transducers [15].

Having a model to correctly simulate the acoustic field in the prefocal zone for focused radiators is a very important improvement in the field. This will permit to increase the accuracy in therapy planning, to improve the prediction of thermal increments in tissues outside the focus, to produce better thermal models in hyperthermia, and in general, to have a better dose control. The use of more realistic but still simple models of acoustic field for focused radiators will help to control the undesired effects out of the treatment zone, i.e., before and after the focus, and to easily incorporate this model proposal to therapy planning. In this work, it was proven that our models represent better alternatives for focused radiators than the widely used ideal uniform approaches.

Author Contributions: Conceptualization, M.I.G. and A.R.; data curation, M.I.G.; formal analysis, M.I.G.; funding acquisition, M.I.G., A.R., A.V., and L.L.; investigation, M.I.G.; methodology, M.I.G. and A.R.; project administration, M.I.G., A.V., and L.L.; resources, M.I.G., J.G., A.V., and L.L.; software, M.I.G.; supervision, M.I.G., A.R., J.G., A.V., and L.L.; validation, M.I.G.; visualization, M.I.G.; writing—original draft, M.I.G.; writing—review and editing, M.I.G. and A.R.

Funding: This research was funded by CONACyT, grant number 257966; CSIC, grant number COOPB20166; ERAnet-EMHE CSIC, grant number 200022; Spanish P.N RETOS, grant number DPI2017-90147-R; and CYTED-Ditecrod Network Ref. 218RT0545.

Acknowledgments: The authors would like to thank Rubén Pérez Valladares for his technical support during the acoustic field measurements.

Conflicts of Interest: The authors declare no conflict of interest.

References

1. Bystritsky, A.; Korb, A.S.; Douglas, P.K.; Cohen, M.S.; Melega, W.P.; Mulgaonkar, A.P.; DeSalles, A.; Min, B.-K.; Yoo, S.-S. A review of low-intensity focused ultrasound pulsation. *Brain Stimul.* **2011**, *4*, 125–136. [CrossRef] [PubMed]

2. Miller, D.L.; Smith, N.B.; Bailey, M.R.; Czarnota, G.J.; Hynynen, K.; Makin, I.R.S.; Bioeffects Committee of American Institute of Ultrasound in Medicine. Overview of Therapeutic Ultrasound Applications and Safety Considerations. *J. Ultrasound Med.* **2012**, *31*, 623–634. [CrossRef] [PubMed]

3. Diederich, C.J.; Hynynen, K. Ultrasound technology for hyperthermia. *Ultrasound Med. Biol.* **1999**, *25*, 871–887. [PubMed]

4. Huang, T.-H.; Tang, C.-H.; Chen, H.-I.; Fu, W.-M.; Yang, R.-S. Low-Intensity Pulsed Ultrasound-Promoted Bone Healing Is Not Entirely Cyclooxygenase 2 Dependent. *J. Ultrasound Med.* **2008**, *27*, 1415–1423. [CrossRef] [PubMed]

5. Fishman, P.S.; Frenkel, V. Focused Ultrasound: An Emerging Therapeutic Modality for Neurologic Disease. *Neurotherapeutics* **2017**, *14*, 393–404. [CrossRef] [PubMed]

6. Zhou, Y. Generation of uniform lesions in high intensity focused ultrasound ablation. *Ultrasonics* **2013**, *53*, 495–505. [CrossRef] [PubMed]

7. Hynynen, K.; Roemer, R.; Anhalt, D.; Johnson, C.; Xu, Z.X.; Swindell, W.; Cetas, T. A scanned, focused, multiple transducer ultrasonic system for localized hyperthermia treatments. 1987. *Int. J. Hyperth.* **2010**, *26*, 1–11. [CrossRef] [PubMed]

8. Jung, Y.J.; Kim, R.; Ham, H.-J.; Park, S.I.; Lee, M.Y.; Kim, J.; Hwang, J.; Park, M.-S.; Yoo, S.-S.; Maeng, L.-S.; et al. Focused low-intensity pulsed ultrasound enhances bone regeneration in rat calvarial bone defect through enhancement of cell proliferation. *Ultrasound Med. Biol.* **2015**, *41*, 999–1007. [CrossRef] [PubMed]

9. Feng, Y.; Tian, Z.; Wan, M. Bioeffects of Low-Intensity Ultrasound In Vitro. *J. Ultrasound Med.* **2010**, *29*, 963–974. [CrossRef] [PubMed]

10. Frenkel, V. Ultrasound mediated delivery of drugs and genes to solid tumors. *Adv. Drug Deliv. Rev.* **2008**, *60*, 1193–1208. [CrossRef] [PubMed]

11. Hynynen, K. Ultrasound for drug and gene delivery to the brain. *Adv. Drug Deliv. Rev.* **2008**, *60*, 1209–1217. [CrossRef] [PubMed]

12. Jenne, J.W.; Preusser, T.; Günther, M. High-intensity focused ultrasound: Principles, therapy guidance, simulations and applications. *Z. Med. Phys.* **2012**, *22*, 311–322. [CrossRef] [PubMed]

13. Yamashita, T.; Ohtsuka, H.; Arimura, N.; Sonoda, S.; Kato, C.; Ushimaru, K.; Hara, N.; Tachibana, K.; Sakamoto, T. Sonothrombolysis for intraocular fibrin formation in an animal model. *Ultrasound Med. Biol.* **2009**, *35*, 1845–1853. [CrossRef] [PubMed]

14. Arora, D.; Cooley, D.; Perry, T.; Skliar, M.; Roemer, R.B. Direct thermal dose control of constrained focused ultrasound treatments: Phantom and in vivo evaluation. *Phys. Med. Biol.* **2005**, *50*, 1919–1935. [CrossRef] [PubMed]

15. Zhang, S.; Hu, P.; Li, X.; Jeong, H. Calibration of focused circular transducers using a multi-Gaussian beam model. *Appl. Acoust.* **2018**, *133*, 182–185. [CrossRef]

16. Li, X.; Zhang, S.; Jeong, H.; Cho, S. Calibration of focused ultrasonic transducers and absolute measurements of fluid nonlinearity with diffraction and attenuation corrections. *J. Acoust. Soc. Am.* **2017**, *142*, 984–990. [CrossRef] [PubMed]

17. Cathignol, D.; Sapozhnikov, O.A.; Zhang, J. Lamb waves in piezoelectric focused radiator as a reason for discrepancy between O'Neil's formula and experiment. *J. Acoust. Soc. Am.* **1997**, *101*, 1286–1297. [CrossRef]

18. Rosnitskiy, P.B.; Yuldashev, P.V.; Sapozhnikov, O.A.; Maxwell, A.D.; Kreider, W.; Bailey, M.R.; Khokhlova, V.A. Design of HIFU Transducers for Generating Specified Nonlinear Ultrasound Fields. *IEEE Trans. Ultrason. Ferroelectr. Freq. Control* **2017**, *64*, 374–390. [CrossRef] [PubMed]

19. Nell, D.M.; Myers, M.R. Thermal effects generated by high-intensity focused ultrasound beams at normal incidence to a bone surface. *J. Acoust. Soc. Am.* **2010**, *127*, 549–559. [CrossRef] [PubMed]

20. O'Neil, H.T. Theory of Focusing Radiators. *J. Acoust. Soc. Am.* **1949**, *21*, 516–526. [CrossRef]

21. Sapozhnikov, O.A.; Pishchal'nikov, Y.A.; Morozov, A.V. Reconstruction of the normal velocity distribution on the surface of an ultrasonic transducer from the acoustic pressure measured on a reference surface. *Acoust. Phys.* **2003**, *49*, 354–360. [CrossRef]

22. Lucas, B.G.; Muir, T.G. The field of a focusing source. *J. Acoust. Soc. Am.* **1982**, *72*, 1289–1296. [CrossRef]

23. Baboux, J.C.; Lakestani, F.; Perdrix, M. Theoretical and experimental study of the contribution of radial modes to the pulsed ultrasonic field radiated by a thick piezoelectric disk. *J. Acoust. Soc. Am.* **1984**, *75*, 1722–1731. [CrossRef]

24. Maréchal, P.; Levassort, F.; Tran-Huu-Hue, L.-P.; Lethiecq, M. Effect of Radial Displacement of Lens on Response of Focused Ultrasonic Transducer. *Jpn. J. Appl. Phys.* **2007**, *46*, 3077–3085. [CrossRef]

25. Riera-Franco de Sarabia, E.; Ramos-Fernandez, A.; Rodriguez-Lopez, F. Temporal evolution of transient transverse beam profiles in near-field zones. *Ultrasonics* **1994**, *32*, 47–56. [CrossRef]

26. Laura, P.A. Directional Characteristics of Vibrating Circular Plates and Membranes. *J. Acoust. Soc. Am.* **1966**, *40*, 1031–1033. [CrossRef]

27. Gutierrez, M.I.; Calas, H.; Ramos, A.; Vera, A.; Leija, L. Acoustic Field Modeling for Physiotherapy Ultrasound Applicators by Using Approximated Functions of Measured Non-Uniform Radiation Distributions. *Ultrasonics* **2012**, *52*, 767–777. [PubMed]

28. Delannoy, B.; Bruneel, C.; Lasota, H.; Ghazaleh, M. Theoretical and Experimental-Study of the Lamb Wave Eigenmodes of Vibration in Terms of the Transducer Thickness to Width Ratio. *J. Appl. Phys.* **1981**, *52*, 7433–7438. [CrossRef]

29. Canney, M.S.; Bailey, M.R.; Crum, L.A.; Khokhlova, V.A.; Sapozhnikov, O.A. Acoustic characterization of high intensity focused ultrasound fields: A combined measurement and modeling approach. *J. Acoust. Soc. Am.* **2008**, *124*, 2406–2420. [CrossRef] [PubMed]

30. Fan, X.B.; Moros, E.G.; Straube, W.L. Ultrasound field estimation method using a secondary source-array numerically constructed from a limited number of pressure measurements. *J. Acoust. Soc. Am.* **2000**, *107*, 3259–3265. [CrossRef] [PubMed]

31. Dekker, D.L.; Piziali, R.L.; Dong, E. Effect of Boundary-Conditions on Ultrasonic-Beam Characteristics of Circular Disks. *J. Acoust. Soc. Am.* **1974**, *56*, 87–93. [CrossRef] [PubMed]

32. Gutierrez, M.I.; Lopez-Haro, S.A.; Vera, A.; Leija, L. Experimental Verification of Modeled Thermal Distribution Produced by a Piston Source in Physiotherapy Ultrasound. *Biomed. Res. Int.* **2016**, *2016*, 1–16. [CrossRef] [PubMed]

33. Rayleigh, L. *The Theory of Sound*; Dover Publications: New York, NY, USA, 1945.

34. Khokhlova, V.A.; Souchon, R.; Tavakkoli, J.; Sapozhnikov, O.A.; Cathignol, D. Numerical modeling of finite-amplitude sound beams: Shock formation in the near field of a cw plane piston source. *J. Acoust. Soc. Am.* **2001**, *110*, 95–108. [CrossRef]

35. Sapozhnikov, O.A.; Khokhlova, V.A.; Cathignol, D. Nonlinear waveform distortion and shock formation in the near field of a continuous wave piston source. *J. Acoust. Soc. Am.* **2004**, *115*, 1982–1987. [CrossRef]

36. Chapelon, J.-Y.; Cathignol, D.; Cain, C.; Ebbini, E.; Kluiwstra, J.-U.; Sapozhnikov, O.A.; Fleury, G.; Berriet, R.; Chupin, L.; Guey, J.-L. New piezoelectric transducers for therapeutic ultrasound. *Ultrasound Med. Biol.* **2000**, *26*, 153–159. [CrossRef]

37. Miner, G.W.; Laura, P.A. Calculation of Nearfield Pressure Induced by Vibrating Circular Plates. *J. Acoust. Soc. Am.* **1967**, *42*, 1025–1030. [CrossRef]

38. Thomas, O.; Bilbao, S. Geometrically nonlinear flexural vibrations of plates: In-plane boundary conditions and some symmetry properties. *J. Sound Vib.* **2008**, *315*, 569–590. [CrossRef]

39. Guo, N.; Cawley, P. Transient-Response of Piezoelectric Disks to Applied Voltage Pulses. *Ultrasonics* **1991**, *29*, 208–217. [CrossRef]

40. Shaw, E.A.G. On the Resonant Vibrations of Thick Barium Titanate Disks. *J. Acoust. Soc. Am.* **1956**, *28*, 38–50. [CrossRef]

41. Gutierrez, M.I.; Vera, A.; Leija, L.; Ramos, A.; Gutierrez, J. Acoustic field modeling of focused ultrasound transducers using non-uniform radiation distributions. In Proceedings of the 2017 14th International Conference on Electrical Engineering, Computing Science and Automatic Control (CCE), Mexico City, Mexico, 20–22 September 2017; pp. 1–4.

42. Marechal, P.; Levassort, F.; Tran-Huu-Hue, L.P.; Lethiecq, M. Electro-acoustic response at the focal point of a focused transducer as a function of the acoustical properties of the lens. In Proceedings of the World Congress on Ultrasonics 2003, Paris, France, 7–10 September 2003; pp. 535–538.

Article

Comparison of Multi-Physical Coupling Analysis of a Balanced Armature Receiver between the Lumped Parameter Method and the Finite Element/Boundary Element Method

Yuan-Wu Jiang [1], Dan-Ping Xu [2], Zhi-Xiong Jiang [1], Jun-Hyung Kim [1] and Sang-Moon Hwang [1,*]

[1] School of Mechanical Engineering, Pusan National University, Busan 609-735, Korea;
    evan.jiang.pnu@gmail.com (Y.-W.J.); jzx20180902@gmail.com (Z.-X.J.); joonyng7@gmail.com (J.-H.K.)
[2] School of Mechatronics Engineering and Automation, Shanghai University, Shanghai 200-072, China;
    sallyxu45@gmail.com
*   Correspondence: shwang@pusan.ac.kr; Tel.: +82-051-510-3204

Received: 24 January 2019; Accepted: 25 February 2019; Published: 27 February 2019

Abstract: The balanced armature receiver (BAR) is a product based on multiphysics that enables coupling between the electromagnetic, mechanical, and acoustic domains. The three domains were modeled using the lumped parameter method (LPM) that takes advantage of an equivalent circuit. In addition, the combined finite element method (FEM) and boundary element method (BEM) was also applied to analyze the BAR. Both simulation results were verified against experimental results. The proposed LPM can predict the sound pressure level (SPL) by making use of the BAR parts dimension and material property. In addition, the previous analysis method, FEM/BEM, took 36 h, while the proposed LPM takes 1 h. So the proposed LPM can be used to check the BAR parts' dimension and material property influence on the SPL and develop the BAR product efficiently.

Keywords: balanced armature receiver; lumped parameter method; finite element method and Boundary element method

---

1. Introduction

In terms of acoustic transducers, there are the MEMS receiver [1,2], the dynamic receiver [3], the speaker box with passive radiator [4], and the balanced armature receiver [5–20]. The Balance armature structure was devized by Olsen [5]. Three types of structure are described: the unpolarized armature, the polarized reed, and the polarized balanced armature. A new magnetic circuit balanced armature structure transducer has been developed for use in hearing aids [6,7]. In the balanced armature structure transducer, the coil is moved outside the magnetic structure. Stationary gaps are located at the two side legs. Based on the proposed polarized balanced armature microactuator, an implantable hearing device was developed [8]. The balanced electromagnetic separation transducer used in a bone-anchored hearing aid is presented in ref. [9]. To minimize harmonic distortions, quadratic distortion forces and static forces are counterbalanced. A closed loop armature is proposed in the structure design of BAR [10,11].

Nowadays, the balanced armature receiver (BAR) is widely used in hearing aids and earphones because of its small size and high sensitivity. The BAR is a product that follows the principles of multiphysics. When current is passed through a coil, the flux density in the upper and lower air gap is different, which contributes to the generated force on the armature's end. With the input force, the armature and pin vibrate. The vibration of the pin contributes to the diaphragm's vibration, which leads to sound radiation through the spout. The BAR consists of electromagnetic, mechanical, and acoustic physical domains, which are coupled with each other.

To construct the electromagnetic mathematic modeling of the BAR, the lumped parameter method (LPM) was proposed in previous research [12,13] which does not investigate the acoustic domain and cannot predict the SPL result, the key performance criteria of BAR. The multimode of BAR is investigated by using the lumped parameter in refs. [14,15]. The electromagnetic domain and acoustic domain is not involved. To obtain the SPL result, the acoustic domain is considered [16–18] while the input electromagnetic parameter (resistance, inductance, force factor) and mechanical parameters (mass, stiffness, force factor) come from experiment, which means the input parameter cannot be obtained in the analysis if there is no sample or experiment.

This paper is organized in the following structure. In the first part, to predict the SPL and develop a new structure BAR product, LPM is proposed to analyze the performance of the BAR by modeling the electromagnetic, mechanical, and acoustic domains along with their coupling effect according to the modeling dimension and material property, which is defined as the proposed method. Second, BAR is analyzed by FEM (electromagnetic, mechanical domain) and BEM (acoustic domain) which is defined as the previous method [19,20]. Consequently, the samples were manufactured according to the analyzed modeling. Based on the samples, the SPL experiment result was obtained. Finally, the result shows that LPM was verified by experiment, as the previous method (FEM and BEM). The difference is that the proposed method (LPM) takes 1 hour, while the previous method (FEM and BEM) took 36 hours. In conclusion, the contribution of the paper is that the proposed LPM method makes use of the BAR modeling dimension and material property to predict the SPL result without parameter identification by experiment and is more efficient than the previous method. The proposed LPM method can be used to design a new structure BAR and shorten the development cycle of the BAR. The method detail is listed in Table 1.

Table 1. Comparison of simulation methods in the paper.

Domain	Proposed Method	Previous Method
Electromagnetic	LPM	FEM
Mechanical	LPM	FEM
Acoustic	LPM	BEM

2. Analysis by LPM

2.1. Electromagnetic Analysis

The electromagnetic modeling can be described by an equivalent circuit, which is demonstrated in Figure 1. By solving Kirchhoff's voltage and current laws, the flux density in the armature can be expressed as a dimension of the electromagnetic circuit and material property, which includes the B-H curve and coercive force. The B-H curve indicates that the permeability changes with flux density in the armature and magnet house. When the position of the armature end changes, the reluctance of the air gap between armature and permanent magnet is changed, which means the different position contributes to a different flux density and force on the armature. Consequently, the electromagnetic characteristic such as inductance and force factor are nonlinear. Hence, to handle the nonlinear characteristic property of the B-H curve, the under-relaxed Newton–Raphson method is adopted to solve the Kirchhoff's current and voltage laws [21–23].

By post processing flux density, the cogging force, force factor, and inductance can be obtained which are depicted in Figure 2.

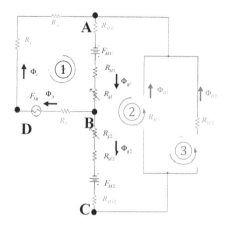

Figure 1. The equivalent circuit of the electromagnetic part.

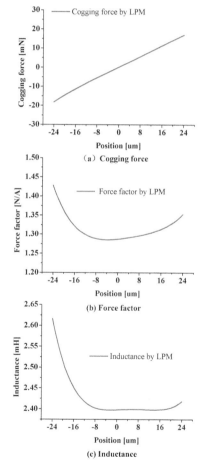

Figure 2. Electromagnetic parameters vs displacement.

The cogging force, force factor, and inductance expressions are listed in the following equations:

$$F_{cogging}(x) = 2\mu_0 A_g F_M^2 \frac{(D_{eff2}+D_{eff1}+4D_{eff})(D_{eff2}-D_{eff1}+2x)}{[(D_{eff1}-x)(D_{eff2}+x)+D_{eff}(D_{eff1}+D_{eff2})]^2}$$

$$L_E(x) = \mu_0 A_g N^2 \frac{(D_{eff2}+D_{eff1})}{(D_{eff1}-x)(D_{eff2}+x)+D_{eff}(D_{eff2}+D_{eff1})} \tag{1}$$

$$Bl(x) = \frac{\mu_0 A_g}{2} \frac{F_M N \left[(D_{eff2}+D_{eff1}+4D_{eff})(D_{eff2}+D_{eff1})+(D_{eff2}-D_{eff1}+2x)^2\right]}{[(D_{eff1}-x)(D_{eff2}+x)+D_{eff}(D_{eff2}+D_{eff1})]^2}$$

where

$$D_{eff1} = \mu_0 A_g (R_M + R_{Hi1}) + D$$

$$D_{eff2} = \mu_0 A_g \left(R_M + R_{Hi2} + \tfrac{1}{2}R_H\right) + D$$

$$D_{eff} = \mu_0 A_g (R_A + R_i + R_{ii})$$

where $F_{cogging}$, Bl and L_E are cogging force, force factor, and voice coil inductance, respectively, which are expressed by the armature position x. R_A, R_H are the reluctance of the armature and magnet house. F_M, A_g, N and D are the magnetomotive force of the magnet, the area of the magnet, the coil turns, and the air gap width respectively.

The mathematic equation for the electrical part is given in the following equation:

$$Z(s) = R_E + sL_E \tag{2}$$

where R_E is the electrical voice coil resistance at DC and L_E is the voice coil inductance.

2.2. Mechanical Analysis

The modeling of the mechanical simulation is demonstrated in Figure 3. The structure contains the armature, pin, and diaphragm. Magnetic force is generated on the armature end. The diaphragm vibrates through the pin connection. The edge of the diaphragm is fixed.

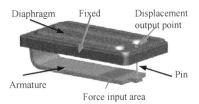

Figure 3. Mechanical simulation modeling.

To describe the mechanical system, one degree-of-freedom of the vibration system's governing equation is adopted as follows:

$$F = M_{ms}\ddot{x} + R_{ms}\dot{x} + \frac{1}{C_{ms}}x \tag{3}$$

where M_{ms} is the mechanical mass of the driver diaphragm, R_{ms} denotes the mechanical resistance of the total driver losses, and C_{ms} denotes the mechanical compliance of the driver suspension.

In the modeling, if the input force is 0.01 N, there will be 6.38×10^{-3} mm displacement on the output point. If the input force is 0.02 N, the displacement becomes 12.76×10^{-3} mm. Hence, the stiffness of the mechanical system is calculated by dividing the difference of the force with the displacement, which is 1570 N/m and defined as original stiffness ($K_{original}$). The cogging force stiffness is 720 N/m, which can be treated as negative stiffness. Therefore, the modified stiffness is

850 N/m, which means C_{ms} is 1.176 m/N. By conducting a modal analysis of the mechanical system, the resonance frequency is obtained, which is 3169.9 Hz. According to the following equation:

$$f_0 = \frac{1}{2\pi} \sqrt{\frac{K_{original}}{M_{ms}}}$$ (4)

The mass is calculated as 3.95×10^{-6} kg.

2.3. Acoustic Analysis

The acoustic domain is modeled as described in the following sections:

There are four tubes in the acoustic modeling which are depicted in Figures 4 and 5. These tubes can be treated as transmission line models which are depicted in Figure 6.

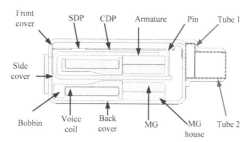

Figure 4. Acoustic modeling in the BAR.

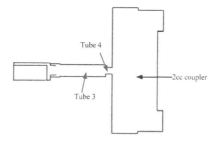

Figure 5. Acoustic modeling in the test jig.

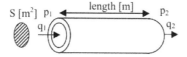

Figure 6. Tube model.

The governing equation is listed below:

$$\begin{pmatrix} p_1 \\ q_1 \end{pmatrix} = \begin{bmatrix} \cos(kl) & jZ_w \sin(kl) \\ (j/Z_w)\sin(kl) & \cos(kl) \end{bmatrix} \begin{pmatrix} p_2 \\ q_2 \end{pmatrix}$$ (5)

and

$$p_1 = \cos(kl) \times p_2 + jZ_w \sin(kl) \times q_2$$
$$q_1 = (j/Z_w)\sin(kl) \times p_2 + \cos(kl) \times q_2$$ (6)

The parameters are $k = \omega/c$, where $\omega = 2\pi f$; l = tube length; and $Z_w = \rho c/S$, where ρ is the air density and c is the speed of sound propagation.

The 2-cc coupler is modeled as the acoustic capacity, which is shown in Figure 7.

Figure 7. Equivalent circuit of cavity.

The related equation defining the acoustic capacity is given as follows:

$$C_a = \frac{V}{\rho c^2} \tag{7}$$

The SPL (Sound Pressure Level) can be calculated as below:

$$SPL = 20log\left(p/20 \times 10^{-6}\right) \tag{8}$$

2.4. Electromagnetic-Mechanical–Acoustic Coupling Factors

The force factor is the electromagnetic–mechanical coupling factor and treated as a gyrator which generates a back EMF in the electromagnetic domain and a driving force in the mechanical domain. The effective area is the mechanical–acoustic coupling factor and is treated as the transformer which changes the vibration velocity of the diaphragm into the volume velocity of the air motion. The effective area of the diaphragm is calculated as half of the area of the diaphragm, because it vibrates just like a cantilever beam. The total simulation tool is depicted in Figure 8 which contains the electromagnetic, mechanical, and acoustic domains along with the coupling factor.

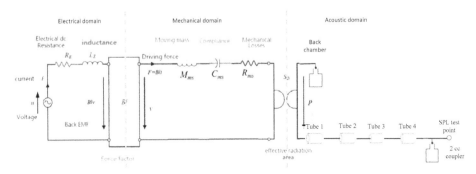

Figure 8. Equivalent circuit of BAR.

3. Analysis by FEM and BEM

To check the efficacy of LPM, the BAR is also analyzed by FEM and BEM. In the electromagnetic simulation part, the BAR is modeled with a different deformation, i.e., different displacement. Varying current is then input to the coil. The magnetic vector potential is defined as zero on the surface of the surrounded air box which is shown in Figure 9. Finally, the flux density is solved through FEM simulation. The electromagnetic FEM simulation governing equation is demonstrated by the following equation:

$$\nabla \times \left(\frac{1}{\mu_r}\nabla \times \mathbf{A}\right) = \mathbf{J}\mu_0 + \mu_0\nabla \times \mathbf{H}_c \tag{9}$$

where **A** is the magnetic vector potential, **J** is the current density, $\mathbf{H}_c$ is the permanent magnetic intensity, μ_0 is the permeability in the vacuum. After solving the governing equation, the flux density of every node in the modeling can be obtained. By post process, the force factor, inductance, and cogging force are obtained, which become the input of the vibro-acoustic simulation.

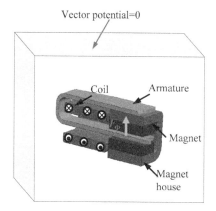

Figure 9. Modeling of electromagnetic FEM simulation.

The governing equation of the mechanical FEM simulation is listed in equation

$$[M]\ddot{u} + [C]\dot{u} + [K]u = [F_i] \tag{10}$$

where $[M]$, $[C]$, $[K]$, $\{F_i\}$, and u denote the mass matrix, damping matrix, stiffness matrix, vector of current force, and displacement. The surround part of the diaphragm is fixed which means that the displacement is zero. With a given input force, the displacement of every node can be solved.

In the acoustic domain, the velocity of the tube surface is defined as zero, which means a rigid wall and is depicted in Figure 10. The sound pressure on the test point is solved by solving the Helmholtz governing equation.

$$\nabla^2 p - k^2 p = -j\rho_0 wq \tag{11}$$

where p, k, ρ_0, w, and q are sound pressure, wave number air density, angular frequency, and volume velocity. The volume velocity is from the displacement result obtained in the mechanical FEM simulation. The details are outlined in Tables 2 and 3.

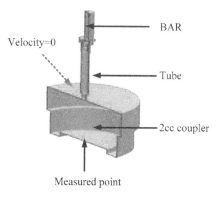

Figure 10. Modeling of acoustic BEM simulation.

Table 2. Simulation detail by FEM and BEM.

Domain	Package	Elements Order	Number of Nodes
Electromagnetic	ANSYS	3	146817
Mechanical	ANSYS	3	525006
Acoustic	Virtual lab	2	1329

Table 3. Model detail.

Domain	Value	Unit
Total dimension	$5 \times 3 \times 2.6$	mm
Coil resistance	23	Ohm
Input voltage	0.115	V

Table 2 shows the simulation detail. ANSYS is taken advantage of in the electromagnetic and mechanical system and Virtual Lab is adopted to simulate the acoustic system. In order to obtain an accurate result, the numbers of nodes in the three systems are 146817, 525006, and 1329. Based on the same computer and the listed node number in Table 2, the computation time by FEM and BEM is 36 h while, the time by the LPM method is 1 hour. So the computation time is improved.

4. Experiment

Furthermore, according to 3D modeling, the balanced armature sample is manufactured and assembled. The sample is shown in Figure 11. The experimental condition is demonstrated in Figure 12. The sweep input source ranges from 100 Hz to 20,000 Hz. The SPL of the BAR is tested using a microphone through the 2-cc coupler jig. In the experimental result, the SPL can maintain 100 dB because the BAR is tested in an enclosed tube and chamber. The first SPL peak is due to resonance of the mechanical structure. The second SPL peak is due to the tube and front chamber. Figure 13 shows the comparison of SPL between the experiment and simulation. The LPM simulation results were verified through experimental results, using FEM and BEM.

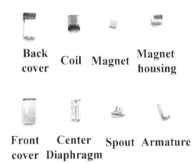

Back cover   Coil   Magnet   Magnet housing

Front cover   Center Diaphragm   Spout   Armature

Figure 11. Parts of BAR.

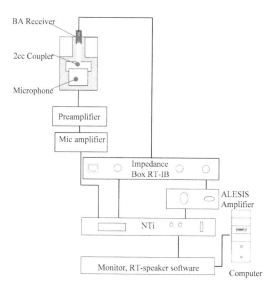

Figure 12. Experimental condition.

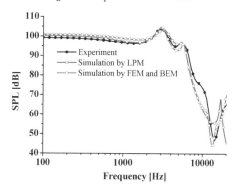

Figure 13. Comparison of experiment and simulation.

5. Conclusions

The BAR is a product based on multiphysics that considers electromagnetic, mechanical, and acoustic domains. These physical domains are coupled with each other. This study derived LPM simulation, which modeled the multiphysics characteristic and considered the coupling effect. The SPL performance of the BAR was predicted through LPM, which provided the same validation as FEM and BEM. Additionally, it was found that LPM was more efficient and could be used to develop the BAR. In the future, the proposed LPM method will be discussed in relation to the BAR dimension and material property influence on SPL. For example, in the electromagnetic domain, the sensitivity analysis of the magnetomotive force of the magnet will be done to check its influence on SPL. Finally, according to the design target, every dimension can be determined.

Author Contributions: Conceptualization, Y.-W.J. and D.-P.X.; methodology, Y.-W.J.; software, Y.-W.J.; validation, Y.-W.J., Z.-X.J. and J.-H.K.; formal analysis, Y.-W.J.; D.-P.X. investigation, Y.-W.J.; D.-P.X. resources, Y.-W.J.; D.-P.X. data curation, Y.-W.J.; D.-P.X. writing—original draft preparation, Y.-W.J.; writing—review and editing, Y.-W.J., Z.-X.J. and J.-H.K.; visualization, Y.-W.J.; supervision, S.-M.H.; project administration, S.-M.H.

Funding: This research received no external funding.

Conflicts of Interest: The authors declare no conflict of interest.

Abbreviations

The following abbreviations are used in this manuscript:

SPL	Sound pressure level
MEMS	Micro-Electro-Mechanical System
LPM	Lumped parameter method
FEM	Finite element method
BEM	Boundary element method
BAR	Balanced armature receiver
EMF	Electromotive force
2-cc	2 cubic centimeter

References

1. Zhao, C.; Knisely, K.E.; Grosh, K. Design and fabrication of a piezoelectric MEMS xylophone transducer with a flexible electrical connection. *Sens. Actuators A Phys.* **2018**, *275*, 29–36. [CrossRef]
2. Zhao, C.; Knisely, K.E.; Grosh, K. Modeling, fabrication, and testing of a MEMS multichannel aln transducer for a completely implantable cochelar implant. In Proceedings of the 2017 19th International Conference on Solid-State Sensors, Actuators and Microsystems (TRANSDUCERS), Kaohsiung, Taiwan, 18–22 June 2017; pp. 16–19.
3. Hwang, S.M.; Lee, H.J.; Kim, J.H.; Hwang, G.Y.; Lee, W.Y.; Kang, B.S. New development of integrated microspeaker and dynamic receiver used for cellular phones. *IEEE Trans. Magn.* **2003**, *39*, 3259–3261. [CrossRef]
4. Jiang, Y.W.; Kwon, J.H.; Kim, H.K.; Hwang, S.M. Analysis and Optimization of Micro Speaker-Box Using a Passive Radiator in Portable Device. *Arch. Acoust.* **2017**, *42*, 753–760. [CrossRef]
5. Olson, H.F. *Dynamical Analogies*; Van Nostrand: New York, NY, USA, 1943; pp. 126–148.
6. Bauer, B.B. Magnetic Translating Device. U.S. Patent 2,454,425, 23 November 1948.
7. Bauer, B. A miniature microphone for transistorized amplifiers. *Trans. IRE Prof. Group Audio* **1953**, *1*, 5–7. [CrossRef]
8. Bernhard, H.; Stieger, C.; Perriard, Y. New implantable hearing device based on a micro-actuator that is directly coupled to the inner ear fluid. In Proceedings of the 2006 International Conference of the IEEE Engineering in Medicine and Biology Society, New York, NY, USA, 30 August–3 September 2006; pp. 3162–3165.
9. Håkansson, B.E.V. The balanced electromagnetic separation transducer: A new bone conduction transducer. *J. Acoust. Soc. Am.* **2003**, *113*, 818–825. [CrossRef] [PubMed]
10. Jayanth, V.; Nepomuceno, H.G. Balanced Armature Bone Conduction Shaker. U.S. Patent 7,869,610, 11 January 2011.
11. Blanchard, M.A.; Geswein, B.C.; Hruza, E.A.; Geschiere, O. Balanced Armature with Acoustic Low Pass Filter. U.S. Patent 8,135,163, 13 March 2012.
12. Jensen, J.; Agerkvist, F.T.; Harte, J.M. Nonlinear time-domain modeling of balanced-armature receivers. *J. Audio Eng. Soc.* **2011**, *59*, 91–101.
13. Ziolkowski, M.; Kwiatkowski, W.; Gratkowski, S.; Ziolkowski, M. Static analysis of a balanced armature receiver. *COMPEL Int. J. Comput. Math. Electr. Electron. Eng.* **2018**, *37*, 1392–1404. [CrossRef]
14. Sun, W.; Hu, W. Lumped element multimode modeling of balanced-armature receiver using modal analysis. *J. Vib. Acoust.* **2016**, *138*, 061017. [CrossRef]
15. Sun, W.; Hu, W. Lumped Element Multimode Modeling for a Simplified Balanced-Armature Receiver. In Proceedings of the 23rd International Congress on Sound and Vibration, Athens, Greece, 10–14 July 2016.
16. Kim, N.; Allen, J.B. Two-port network analysis and modeling of a balanced armature receiver. *Hear. Res.* **2013**, *301*, 156–167. [CrossRef] [PubMed]
17. Tsai, Y.-T.; Huang, J.H. A study of nonlinear harmonic distortion in a balanced armature actuator with asymmetrical magnetic flux. *Sens. Actuators A Phys.* **2013**, *203*, 324–334. [CrossRef]
18. Bai, M.R.; You, B.-C.; Lo, Y.-Y. 'Electroacoustic analysis, design, and implementation of a small balanced armature speaker. *J. Acoust. Soc. Am.* **2014**, *136*, 2554–2560. [CrossRef] [PubMed]

Appl. Sci. **2019**, *9*, 839

19. Xu, D.P.; Lu, H.W.; Jiang, Y.W.; Kim, H.K.; Kwon, J.H.; Hwang, S.M. Analysis of Sound Pressure Level of a Balanced Armature Receiver Considering Coupling Effects. *IEEE Access* **2017**, *5*, 8930–8939. [CrossRef]

20. Jiang, Y.W.; Xu, D.P.; Hwang, S.M. Electromagnetic-Mechanical Analysis of a Balanced Armature Receiver by Considering the Nonlinear Parameters as a Function of Displacement and Current. *IEEE Trans. Magn.* **2018**, *99*, 1–4.

21. Furlani, E.P. *Permanent Magnet and Electromechanical Devices: Materials, Analysis, and Applications*; Academic Press: Cambridge, MA, USA, 2001; pp. 335–345.

22. Campbell, P. *Permanent Magnet Materials and Their Application*; Cambridge University Press: Cambridge, UK, 1996.

23. Asghari, B.; Dinavahi, V. Novel transmission line modeling method for nonlinear permeance network based simulation of induction machines. *IEEE Trans. Magn.* **2011**, *47*, 2100–2108. [CrossRef]

applied
sciences

Article

Comparison for the Effect of Different Attachment of Point Masses on Vibroacoustic Behavior of Parabolic Tapered Annular Circular Plate

Abhijeet Chatterjee [1,*], Vinayak Ranjan [2], Mohammad Sikandar Azam [1] and Mohan Rao [3]

[1] Department of Mechanical Engineering, Indian Institute of Technology (ISM), Dhanbad 826004, India; mdsazam@gmail.com

[2] Department of Mechanical Engineering, Bennett University, Greater Noida 201310, India; vinayakranjan@gmail.com

[3] Department of Mechanical Engineering, Tennessee Tech University, Cookeville, TN38505, USA; mrao@tntech.edu

* Correspondence: abhijeet.ism@gmail.com

Received: 19 November 2018; Accepted: 14 January 2019; Published: 20 February 2019

Abstract: In this paper, a comparison for the effect of different arrangement of point masses on vibroacoustic behavior of parabolic tapered annular circular plate with different taper ratios are analyzed by keeping the total mass of the plate plus point masses constant. Three different arrangement of thickness variation are considered. The mathematical tool FEM using ANSYS is used to determine the vibration characteristic and both FEM and Rayleigh integral is used to determine the acoustic behavior of the plate. Further, Case II plate (parabolic decreasing increasing thickness variation) for all combination of point masses is found to have reduction in natural frequency parameter in comparison to other cases of parabolic tapered plate. In terms of acoustic behavior, sound power levels of different cases of plate with different point mass combination are observed. It is observed that the Case II plate with two point masses combination shows the highest sound power and the Case III plate for all cases of point mass combination is least prone to acoustic behavior. Furthermore, It is observed that at low forcing frequency average radiation efficiency of parabolic tapered plate for different arrangement of point masses is almost same, but at high forcing frequency average radiation increases for higher taper ratio. Finally, a brief discussion of peak sound power reduction and actuation for different arrangement of point masses with different taper ratios are provided.

Keywords: thick annular circular plate; finite element modeling; Rayleigh integral; point mass; taper ratio; parabolic thickness variation

---

1. Introduction

Plate with tapered annular circular plates with different combinations of point masses has many engineering applications. They are used in many structural components, i.e., building, design, diaphragms, and deck plates in launch vehicles, diaphragms of turbines, aircraft and missiles, naval structures, nuclear reactors, optical systems, construction of ships, automobiles and other vehicles, the space shuttle, etc. These tapered plates with different combinations of point masses are found to have greater resistance to bending, buckling, and vibration in comparison to plates of uniform thickness. It is interesting to know that tapered plates with different thickness variation have drawn the attention of most of the researchers in this field. However, tapered plates with different combination of point masses can alter the dynamic characteristic of structures with a change in stiffness. Hence, for practical design purposes, the vibration and acoustic characteristics of such tapered plates are equally important. In comparison to the present study, several existing works are presented where the researchers have investigated the vibration response [1–9] of circular or annular plates of tapered or uniform thickness. However, in terms of acoustic behavior, many researchers have contributed most. Lee and Singh [10] used the thin and thick plate theories to determine the sound radiation from out-of-plane modes of a uniform thickness annular circular plate. Thompson [11] used the Bouwkamp integral to determine the mutual and self-radiation impedances both for annular and elliptical pistons. Levine and Leppington [12] analyzed the sound power generation of a circular plate of uniform thickness using exact integral representation. Rdzanek and Engel [13] determined the acoustic power output of a clamped annular plate using an asymptotic formula. Wodtke and Lamancusa [14] minimized the acoustic power of circular plates of uniform thickness using the damping layer placement. Wanyama [15] studied the acoustic radiation from linearly-varying circular plates. Lee and Singh [16] used the flexural and radial modes of a thick annular plate to determine the self and mutual radiation. Cote et al. [17] studied the vibro acoustic behavior of an unbaffled rotating disk. Jeyraj [18] used an isotropic plate with arbitrarily varying thickness to determine its vibro-acoustic behavior using the finite element method (FEM). Ranjan and Ghosh [19] studied the forced response of a thin plate of uniform thickness with attached discrete dynamic absorbers. Bipin et al. [20] analyzed an isotropic plate with attached discrete patches and point masses with different thickness variation with different taper ratios to determine its vibro acoustic response. Lee and Singh [21] investigated the annular disk acoustic radiation using structural modes through analytical formulations. Rdzanek et al. [22] investigated the sound radiation and sound power of a planar annular membrane for axially-symmetric free vibrations. Doganli [23] determined the sound power radiation from clamped annular plates of uniform thickness. Nakayama et al. [24] investigated the acoustic radiation of a circular plate for a single sound pulse. Hasegawa and Yosioka [25] determined the acoustic radiation force used on the solid elastic sphere. Lee and Singh [26] used a simplified disk brake rotor to investigate the acoustic radiation through a semi-analytical method. Thompson et al. [27,28] analyzed the modal approach for different boundary conditions to calculate the average radiation efficiency of a rectangular plate. Rayleigh [29] determined the sound radiation from flat finite structures. Maidanik [30] analyzed the total radiation resistance for ribbed and simple plates using a simplified asymptotic formulation. Heckl [31] used the wave number domain and Fourier transform to analyze the acoustic power. Williams [32] determined the wave number as a series in ascending power to estimate the sound radiation from a planar source. Keltie and Peng [33] analyzed the sound radiation using the cross-modal coupling from a plane. Snyder and Tanaka [34] demonstrated the importance of cross-modal contributions for a pair of modes through total sound power output using modal radiation efficiency. Martini et al. [35] investigated the structural and elastodynamic analysis of rotary transfer machines by a finite element model. Croccolo et al. [36] determined the lightweight design of modern transfer machine tools using the finite element model. Martini and Troncossi [37] determined the upgrade of an automated line for plastic cap manufacture based on experimental vibration analysis. Pavlovic et al. [38] investigated the modal analysis and stiffness optimization: the case of a tool machine for ceramic tile surface finishing using FEM.

In this paper, a numerical simulation has been proposed for the vibroacoustic behavior of parabolic tapered plate with different attachment of point masses. The review of the literature suggested that the vibroacoustic behavior of plate with different point masses is not reported. Hence taking consideration of all these facts, this paper is focused on the vibroacoustic behavior of parabolic tapered annular circular plate keeping the mass of the plate plus point masses constant for all cases of parabolic thickness variation. Therefore, the numerical simulation is provided in this paper by taking consideration of different plates with different taper ratios under time varying harmonic excitations.

2. Materials and Methods

2.1. Free Vibration of Plate

In this simulation process, the vibration of the plate is performed and its modal characteristic is determined. The natural frequency along with the modes shape of the plate during modal analysis is obtained as

$$\left([K] - \omega^2 [M] \right) \psi_{mn} = 0 \tag{1}$$

In the above formulation, $[M]$ is said to be mass matrix and $[K]$ is said to be the stiffness matrix. The mode shape is represented by ψ_{mn} and the corresponding natural frequency of the plate is represented by ω denoted as rad/sec. Further, λ^2 known as the non-dimensional frequency parameter which is obtained as

$$\lambda^2 = \omega a^2 \sqrt{\frac{\rho h}{D}} \tag{2}$$

where, D is said to be the flexure rigidity $= \frac{Eh^3}{12(1-v^2)}$, a is said to be the outer radius, E is said to be the Young's modulus of elasticity, v is said to be Poisson's ratio, h is said to be the thickness of the plate, and ρ is said to be the density of plate.

2.2. Acoustic Radiation Formulation of Plate with Point Masses

In this numerical simulation, we consider an annular circular plate in flexural vibration is set on flat rigid baffle having infinite extent as reported in Figure 1. We are neglecting the acoustic scattering of the edges of a vibrating structure in this investigation. Further, if P be considered as sound pressure amplitude, S_s be considered as the surface of the sound source, q be considered as the Green methods function in free field. Furthermore, if l_s and l_p be considered as the position vectors of source and receiver and If the surface normal vector at l_s is taken as f, then using by Rayleigh integral [10], structure sound radiation can be obtained by Equation (3)

$$P(l_p) = \int_{S_s} \left(\frac{\partial q}{\partial f} (l_p, l_s) P(l_p) - \frac{\partial P}{\partial f} (l_s) q (l_p, l_s) \right) ds(l_s) \tag{3}$$

Consideration of the plane wave approximation to determine the sound pressure radiated from non-planar source in far and free field environment can be obtained by Equation (4)

$$P(l_p) = \frac{\rho_0 c_0 B}{4\pi} \int_{S_s} \frac{e^{iB|l_p - l_s|} U(l_s)}{|l_p - l_s|} (1 + \cos \eta) dS \tag{4}$$

Further for our consideration, if ρ_0 is considered as the mass density of air, c_0 is considered as the speed of sound in air, B is considered as the corresponding acoustic wave number, and U and $\dot{u}$ is considered as both the corresponding vibratory velocity amplitude and spatial dependent vibratory velocity amplitude in the z direction at l_s, then the modal sound pressure P_{mn} from a normal plane [10], for an annular plate with (m, n)th mode is obtained from simplifying above Equation (4) with Hankel transform and is obtained Equations (5) and (6)

$$P_{mn}(R, \alpha, \beta) = \frac{\rho_0 c_0 B e^{iB_{mn} R_d}}{2R_d} \cos n\beta (-i)^{n+1} A_n \left[\dot{u}(l) \right] (1 + \cos \eta) \tag{5}$$

$$A_f \left[\dot{u}(l) \right] = \int_0^\infty \dot{u}(l) J_n(B_l l) l dl, Bl = B \sin \theta; R_d = |l_p - l_s| \tag{6}$$

where J_n is considered as Bessel function of order n, (α, β) are considered as the cone and azimuthal angles of the observation positions, η is considered as angle between the surface normal vector and the vector from source position to receiver position, and A is considered as Hankel transform. Further from the far field condition, R_d in the denominator is approximated by R where $R = |l_p|$ is considered to be radius of the sphere. The observation positions are represented by some points having equal angular increments ($\Delta\varphi$, $\Delta\alpha$) on a sphere S_v. Then, at all of the observation positions, the sound pressure is obtained by the above Equations (4)–(6). Where '$\Delta\varphi$' represents the small increment in the circumferential direction of the plate. In far field for the (m, n)th mode, the modal sound power S_{mn} [10,16] is obtained by Equation (7)

$$S_{mn} = (D_{mn}S_v)_s = \frac{1}{2}\int_0^{2\pi}\int_0^{\pi}\frac{P_{mn}^2}{\rho_0 c_0}R^2\sin\alpha\, d\alpha\, d\beta \tag{7}$$

where, we considered D_{mn}, as the acoustic intensity and we considered S_v as area of the control surface. Furthermore, if σ_{mn} is considered as radiation efficiency of the plate, then radiation efficiency [10] is obtained by Equation (8)

$$\sigma_{mn} = \frac{S_{mn}}{\left|\dot{u}_{mn}^2\right|ts}, \left|\dot{u}_{mn}^2\right|ts = \frac{1}{2\pi(a^2 - b^2)}\int_b^a\int_0^{2\pi}u^2 d\varphi\, dl \tag{8}$$

where, $\left|\dot{u}_{mn}^2\right|ts$ is considered to be the spatially average r.m.s velocity for the two normal surfaces of the plate. If the plate thickness (h) effect is considered, then from the two normal surfaces of the plate at ($Z = 0.5h$ and $-0.5h$), the modal sound power [16] due to the sum of two sound radiations is obtained by Equations (9)–(11)

$$P_{mn}(R, \alpha, \beta) = (1 + \cos\alpha)P_{mn}^s(R, \alpha, \beta) + (1 - \cos\alpha)P_{mn}^o(R, \alpha, \beta) \tag{9}$$

$$P_{mn}^s(R, \alpha, \beta) = \frac{\rho_0 c_0 B_{mn}e^{iB_{mn}R}}{2R}e^{-iB_{mn}(\frac{h}{2})\cos\alpha}\cos n\beta(-i)^{n+1}A_n\left[U(l)\right] \tag{10}$$

$$P_{mn}^o(R, \alpha, \beta) = \frac{\rho_0 c_0 B_{mn}e^{iB_{mn}R}}{2R}e^{-iB_{mn}(\frac{h}{2})\cos\alpha}\cos n(\beta + \phi)(-i)^{n+1}A_n\left[U(l)\right] \tag{11}$$

where, for (m, n)th mode, B_{mn} is considered to be the corresponding acoustic wave number whereas s and o represent the source side and opposite to source side.

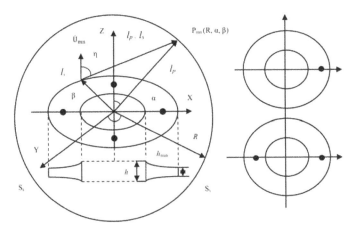

Figure 1. Acoustic radiation due to the vibration modes of unbaffled tapered plate in Z direction with different combination of attached point masses enclosed in a sphere.

2.3. Thickness Variation of the Plate

In this numerical simulation processes, we considered three different parabolic thickness variation of plate for vibration analysis and are reported in Figure 2. The radial direction is taken for thickness variation and the total mass of the plate plus point masses are kept constant. The thickness is varied in radial direction and is given by $h_x = h [1 - T_x \{f(x)\}^n]$, where '$h$' is the maximum thickness of the plate.

$$\text{where, } f(x) = \{^{0, x=b}_{1, x=a} \text{ and } f(x) = \frac{x - b}{a - b} \text{ where } b < x < a \tag{12}$$

In this study, the taper ratio (T_x) is obtained by Equation (13)

$$T_x = \left(1 - \frac{h_{min}}{h}\right) \tag{13}$$

The Case I plate (parabolically decreasing) thickness variation, Case II plate (parabolically decreasing–increasing) thickness variation, and Case III plate (parabolically increasing–decreasing) thickness variation of (Figure 2) are obtained by Equations (14)–(16)

$$h_x = h\left\{1 - T_x\left(\frac{x - b}{a - b}\right)^n\right\} \tag{14}$$

$$h_x = h\left\{1 - T_x\left(1 - abs\left(1 - 2\frac{(x - b)}{(a - b)}\right)\right)^n\right\} \tag{15}$$

$$h_x = h\left\{1 - T_x abs\left(1 - 2\frac{(x - b)}{(a - b)}\right)^n\right\} \tag{16}$$

where, $n = 2$ for parabolic thickness variation. The total volume of the plate plus point masses as well as unloaded plate is kept constant and is given by the Equation (17)

$$\text{Volume} = \pi(a^2 - b^2)h = \int_b^a (a^2 - b^2)h_x dx \tag{17}$$

The process is based on the numerical simulation technique using FEM. The plate is modeled in ANSYS with Plane 185 with 8 brick nodes and having three degrees of freedom at each node. A Structural Mass 21 in ANSYS is added to locate the point mass at the nodes. The number of element and nodes for uniform unloaded plate comes out to be 5883 and 1664, respectively. The numerical results obtained using FEM are compared with the other existing literature. The structure is modeled as such that the volume of the uniform unloaded plate is equal to the volume of the plate with point mass and as a result the whole mass of the plate with point mass remains constant. For plates with different cases of thickness variation with different point masses, we tried to keep the mesh as close to the mesh of the uniform unloaded plate. For vibration analysis and for Case I plate with one point mass combination, the modal structure consists of 5726 elements with 1575 nodes. Similarly, for Case I plate with two point masses combinations, the modal structure will be consists of 5720 elements with 1537 nodes where as for Case I plate with four point masses combination, the modal structure will be consists of 5712 elements with 1517 nodes respectively. In this numerical simulation technique, the meshes of different cases of plate with different plate thickness and with different combination of point masses are not exactly equal to the uniform unloaded plate. However for plate with other combination of point masses with different parabolic thickness, variations of 1–2% of mesh as that of uniform unloaded plate are considered. Furthermore, around the plate for creating the acoustic medium environment FLUID 30 and FLUID 130 are used. FLUID 30 is used for fluid structure interaction. FLUID 130 elements are created by imposing a condition of infinite space around the source and to prevent the back reflection of sound waves to the source. For acoustic calculation, the number of elements and nodes for uniform unloaded plate comes out to be 14,680 and 3639, respectively. For Case I plate with one point mass the number of elements and nodes after proper convergence comes out to be 14,662 and 3627, respectively. Further, after proper convergence, for Case I plate with two point masses the number of element and nodes after proper convergence comes out to be 14,644 and 3615 respectively. For Case I plate with four point masses, the number of element and nodes after proper convergence found to be 14,620, and 3605 respectively. For other cases of plate with different point masses combination, again a variation of 1–2% of mesh as that of uniform unloaded plate is taken in this numerical simulation technique. Consider the plate is vibrating in air medium with air density $\rho_0 = 1.21$ kg/m^3. At 20 °C, the speed of sound c_0 of air is assumed as 343 m/s. The plate with structural damping coefficient is taken as 0.01. Rayleigh integral is applied to determine acoustic power calculation and ANSYS is used as a tool for numerical simulation. In this paper, we considered a plate with outer radius 'a' and inner radius 'b' as shown in Figure 2.

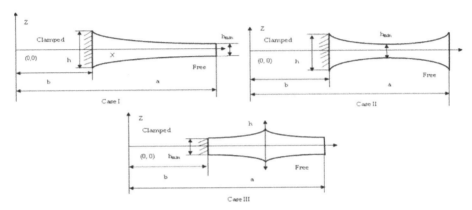

Figure 2. Plate with different parabolic varying thickness variations.

In this numerical simulation processes, a comparison for the effect of natural frequency parameter, effect of sound power level, effect of average radiation efficiency, and effect of peak sound power level for different cases of parabolic tapered plate with different combinations of point masses are obtained. For all cases of tapered plate with point masses, the mass of the plate is kept constant. The out of plane (m, n)th modes are considered and the different taper ratios are varied in the range of 0.00–0.75 for this process of simulation. The plate is clamped at inner and free at the outer boundary. We are considered the three arrangements of plate with different combination of point masses as shown in Figure 3. The selection of different combinations of point masses are such that the mass of uniform unloaded plate is equal to mass of plate + point mass and in all cases total mass of the plate with point masses remains constant. The dimension and the material properties of an annular circular plate with point masses are reported in Table 1.

Table 1. Different specific dimension and material properties of plate with point mass consider in this literature.

Dimension of the Plate with Point Mass	Isotropic Annular Circular Plate
Outer radius (a) m	0.1515
Inner radius (b) m	0.0825
Radii ratio, (b/a)	0.54
Thickness (h) m	0.0315
Thickness ratio, (h/a)	0.21
Density, ρ (kg/m^3)	7905.9
Young's modulus, E (GPa)	218

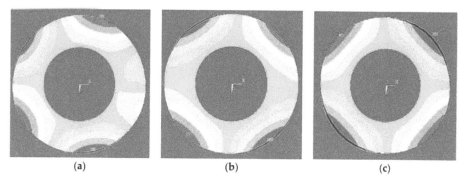

(a)	(b)	(c)

Figure 3. Plate with different combinations of point masses with (0, 2) modes. (a) Plate with one point mass. Mass of point mass = 0.1 times mass of the plate. (b) Plate with two point masses. Mass of each point mass = 0.05 times mass of the plate. (c) Plate with four point masses. Mass of each point mass = 0.025 times mass of the plate.

3. Validation of the Present Study

In this process of numerical simulation technique, the validation of modal frequency of thick annular isotropic plate is done with the published result of Lee et al. [10] as shown in Table 2. In [10], Lee et al. used the thick and thin plate theories to provide the solution for the natural frequency parameter of uniform annular circular plate. In our study, we have calculated our result using FEM by taking the same dimension of plate as that of Lee et al. Therefore, our result of simulation in this study has a good agreement with the published results [10] as reported in Table 2. For the acoustic power calculation, the computed analytical, numerical, and published experimental results [10] are considered as reported in Figure 4. From Figure 4, good agreements of computed acoustic results are seen to be obtained analytically and numerically in line with published experimentally results [10].

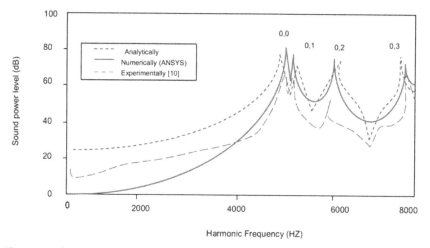

Figure 4. Analytical, experimental and numerical comparison of sound power level of unloaded plate having uniform thickness for taper ratio T_x = 0.00.

Table 2. Comparison and validation of natural frequency parameter λ^2 of clamped-free uniform annular circular plate with that of Lee et al. [10] at $T_x = 0.00$.

Plate	Mode	Non Dimensional Frequency Parameter, λ^2	
		H. Lee et al. [10]	Present Work
Uniform plate $b/a = 0.54$ $h/a = 0.21$	(0, 0)	13.61	13.49
	(0, 1)	13.43	13.50
	(0, 2)	15.28	14.12
	(0, 3)	16.81	16.67

4. Result and Discussion

4.1. Effect of Natural Frequency Parameter (λ^2) of Plate with Different Combinations of Point Masses with Different Taper Ratios

In this paper, the simulation results for the effect of natural frequency parameter (λ^2) due to different combination of point mass are considered. The different parabolic thickness variations are taken where the analysis of the plate is done by keeping mass of the plate + point mass constant. We have considered the first four frequency parameter and hence the numerical comparison is done between the uniform unloaded plate and plate with different combination of point masses for uniform thickness at $T_x = 0$ as reported in Table 3. We find from Table 3 that the effect of natural frequency parameter for the unloaded plate and the plate with different combination of point masses is almost same. For further simulation results, a comparison of percentage variation of frequency parameter with the modes are investigated for different cases of plate with point masses combination as reported in Figure 5. It is clear from Figure 5 that the (0,1) mode increases for two point mass and four point mass combinations but decreases for one point mass combinations. In our numerical simulation results, we see that there is abrupt decrease of (0, 3) mode for all point masses combinations due to more stiffness associated with these modes. In our numerical simulation we compare λ^2 with all modes both for unloaded plate and plate with four point masses combinations as reported in Figure 6. We find that the effect of natural frequency parameter due to four point masses shows the little decrease in the frequency parameter. This may happen due to more stiffness associated with this plate. Further, in this numerical simulation process, we compare Tables 4–6 for natural frequency parameter (λ^2) of plate with different combinations of point masses combinations for different cases of tapered plate. It is observed from the Tables 4–6 that Case II plate (parabolically decreasing—increasing thickness variation) reports the reduction in natural frequency parameter for all cases of thickness variations with different combinations of point masses in respect to Case I plate (parabolic decreasing thickness variation). This reduction of natural frequency parameter for Case II plate may be due to the less stiffness associated than that of Case I plate. It is found that due to more stiffness associated with Case III plate (parabolic increasing—decreasing thickness variation), it shows the almost equal effect of natural frequency parameter as that of uniform unloaded plate for all combination of point masses. However, for plate with different parabolically thickness variations with all cases of four point mass combinations, alteration of modes are observed at higher taper ratios.

Table 3. Numerical comparison of different frequency parameter λ^2 with different modes of uniform unloaded plate for taper ratio $T_x = 0.00$ with that of different combinations of point masses.

Mode	Un-Loaded Plate	Plate with One Point Mass	% λ^2	Plate with Two Point Masses	% λ^2	Plate with Four Point Masses	% λ^2
(0,0)	13.49	13.46	0.223	13.45	0.296	13.35	1.033
(0,1)	13.50	13.48	0.148	13.44	0.444	13.32	1.333
(0,2)	14.12	14.08	0.283	14.06	0.424	14.02	0.708
(0,3)	16.67	16.64	0.017	16.64	0.017	16.62	0.299

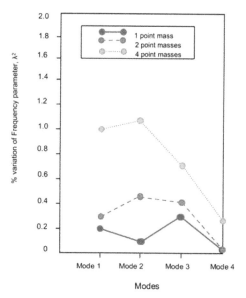

Figure 5. Comparison of % variation of natural frequency parameter with modes for uniform plate with different combinations of point masses.

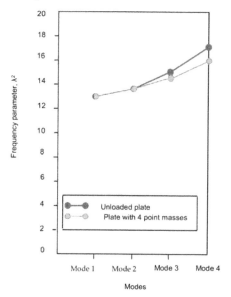

Figure 6. Comparison of variation of natural frequency parameter with modes for unloaded plate and for plate with four point masses.

Table 4. Numerical comparison of different frequency parameter λ^2 with different modes of plate with one point mass combinations for different parabolic thickness variations and for different taper parameters (T_x).

Case	Mode	Natural Frequency Parameter, λ^2			
		$T_x = 0.00$	$T_x = 0.25$	$T_x = 0.50$	$T_x = 0.75$
I	(0,0)	13.4802	12.9904	12.4703	11.9305
	(0,1)	13.4942	12.9745	12.4442	11.8918
	(0,2)	14.1132	13.6135	13.0936	12.5502
	(0,3)	16.6624	16.0733	15.4611	14.8254
II	(0,0)	13.4811	12.8952	12.2718	11.6202
	(0,1)	13.4932	12.8774	12.2414	11.5752
	(0,2)	14.1134	13.5187	12.8943	12.2384
	(0,3)	16.6624	15.9610	15.2277	14.4605
III	(0,0)	13.4812	13.4891	13.4902	13.4904
	(0,1)	13.4943	13.4784	13.4808	13.4804
	(0,2)	14.1136	14.1109	14.1134	14.1125
	(0,3)	16.6625	16.6600	16.6624	16.6618

Table 5. Numerical comparison of different frequency parameter λ^2 with different modes of plate with two point masses combinations for different parabolic thickness variations and for different taper parameters (T_x).

Case	Mode	Natural Frequency Parameter, λ^2			
		$T_x = 0.00$	$T_x = 0.25$	$T_x = 0.50$	$T_x = 0.75$
I	(0,0)	13.4753	12.9768	12.4592	11.9285
	(0,1)	13.4825	12.9682	12.4392	11.8825
	(0,2)	14.0925	13.6092	13.0878	12.5325
	(0,3)	16.6532	16.0691	15.4592	14.8125
II	(0,0)	13.4768	12.8825	12.2685	11.6125
	(0,1)	13.4832	12.8785	12.2386	11.5624
	(0,2)	14.0965	13.5085	12.9186	12.2252
	(0,3)	16.6582	15.9528	15.2582	14.4518
III	(0,0)	13.4793	13.4758	13.5076	13.4721
	(0,1)	13.4825	13.4768	13.5195	13.4821
	(0,2)	14.0968	14.0952	14.1392	14.0926
	(0,3)	16.6582	16.6592	16.7002	16.6523

Table 6. Numerical comparison of different frequency parameter λ^2 with different modes of plate with four point masses combinations for different parabolic thickness variation and for different taper parameter (T_x).

Case	Mode	Natural Frequency Parameter, λ^2			
		$T_x = 0.00$	$T_x = 0.25$	$T_x = 0.50$	$T_x = 0.75$
I	(0,0)	13.4852	12.9877	12.4701	11.9301
	(0,1)	13.4902	12.9765	12.4443	11.8937
	(0,2)	14.1025	13.6040	13.0838	12.5412
	(0,3)	16.6635	16.0712	15.4619	14.8258
II	(0,0)	13.4842	12.8924	12.2715	11.6202
	(0,1)	13.4902	12.8805	12.2418	11.5762
	(0,2)	14.1035	13.5102	12.8858	12.2304
	(0,3)	16.6638	15.9618	15.2302	14.4612
III	(0,0)	13.4852	13.4820	13.4850	13.4852
	(0,1)	13.4906	13.4853	13.4902	13.4902
	(0,2)	14.1037	14.1008	14.1022	14.1032
	(0,3)	16.6638	16.6612	16.6639	16.6635

4.2. Acoustic Radiation of Tapered Annular Circular Plate with Different Attachment of Point Masses with Different Taper Ratios

In this numerical simulation processes, the sound power level (dB, reference = 10^{-12} watts) of annular circular plate with different combinations of point masses is considered. The plate with different parabolic thickness variation is analyzed due to transverse vibration. The different taper ratios are taken as range from (0.00–0.25). A concentrated load of 1N is considered under time-varying harmonic excitations which are acted at different excitation location at different nodes. A harmonic frequency range of 0–8000 Hz is taken to determine the sound radiation characteristic. We considered the Case I plate with parabolic decreasing thickness variation as a convergence study. Figure 7 compares analytically and numerically the sound power level for Case I plate with four point masses combination for taper ratio, T_x = 0.75 and for different modes. On comparison of sound power, we observed a good agreement of computed results as depicted from Figure 7. In this numerical simulation, the numerical comparison of sound power level for Case I plate with different combinations of point masses for different taper ratios are reported in Figures 8–10. From Figures 8–10, it is investigated that for sound power level up to 20 dB, we do not get any design options for different taper ratios for plate with both one point mass and for two point masses combinations. However, for four point masses combination, we do not find any sound power level upto 30 dB. However, for sound power level up to 30 dB, we get all taper ratios, T_x = 0.00, 0.25, 0.50, and 0.75 as design options in frequency bands A and B for plate with one point mass combinationas reported in Figure 8. It is noteworthy that, for sound power level up to 50 dB, we get more design options for sound power levels in different frequency bands, i.e., C, D, and E as reported in Figure 8. From Figure 9, it is apparent that for sound power level up to 30 dB, then in frequency band A only taper ratio T_x = 0.00, 0.25, 0.50, and 0.75 are available design alternative for plate with two point mass combination. However, for sound power level up to 50 dB, we get wider frequency bands, B, C, and D for different taper ratios as reported in Figure 9. From Figure 10, it is investigated that for sound power level up to 40 dB is possible only in frequency bands A, B and C only with all taper ratios T_x = 0.00, 0.25, 0.50, and 0.75 and therefore is the available design alternative for plate with four point mass combination. However, for sound power level up to 50 dB, we get broader range of frequency denoted as D, E and F for all taper ratios as reported in Figure 10. It may be inferred from Figures 8–10 that plate with different combinations of point masses plays a significant role in sound power reduction in different frequency bands. For plates with four point masses combinations, the lowest sound power is observed in comparison to one point mass and two point mass combinations. However stiffness contribution due to various taper ratios have very limited impact on sound power level reduction in comparison to that of modes and excitation locations of plate with different combination of point masses. From Figures 8–10, it is observed that for excitation frequencies up to 2000 Hz, the effect of different combinations of point masses and stiffness variation due to different taper ratios do not have a significant effect on sound power radiation for clamped-free forcing boundary condition. However, when the excitation frequency increases beyond 2000 HZ and up to the first peak, Case I plate with one point mass combinationreports the higher sound power level only for a higher taper ratio. However, for Case I plate with two point masses and four point masses combinations, there are variations of the sound power level. This is due to variation of peaks due to different taper ratios at this forcing region. Beyond 2000 HZ, Case II plate with two point massescombinations is seen to have the highest sound power level. However, the sound power for Case III plate is found to be decreased for all combination of point masses. Different modes do influence the sound power peaks as evident from Figures 8–10. Sound power level peak obtained for different modes (0, 0) and (0, 1) is investigated and it is observed that the dissimilar peak for (0, 0) and (0, 1) is observed for plate with different point masses. However, with increasing taper ratio, sound power levels do shift towards lower frequency for all combinations of point masses. It is observed that at higher forcing frequency beyond 4000 Hz, different taper ratios alter its stiffness for different cases of thickness variations. It is needless to mention that for higher frequency beyond 4000 Hz up to 8000 HZ, plate with different combination of point mass alter its stiffness at higher forcing frequency. The acoustic power curve is seen to intersect each other at this high forcing region. Table 7 compares the peak sound power level of different parabolic tapered plate with different combinations of point masses for taper ratio $T_x = 0.75$. It is interesting to note that the lowest sound power of 76 dB is observed for plate with four point mass combinations among all different thicknesses and the highest power of 82 dB is observed for plate with one point mass combination. Figures 11–13 compares Case I, Case II, and Case III for sound power level numerically for different combinations of point mass for taper ratio, $T_x = 0.75$. From Figures 11–13 it is investigated that for excitation frequency up to 2000 Hz, plate with different parabolic thickness variations does not have contribute much on sound power radiation. However, beyond excitation frequency of 2000 HZ and up to the first peak, it is investigated that Case II plate with two point masses combination is very good sound radiator of sound power 83 dB in comparison to 82 dB of plate with one point mass combination. Case III plates with all combinations of point masses are seen to have poor sound radiation. Figure 14 compares radiation efficiency (σ_{mn}) analytically and numerically for Case I plate with four point masses combination having parabolically decreasing thickness variation and for taper ratio $T_x = 0.75$. It is observed that on comparison the results obtained for radiation efficiency matches well with each other as reported in Figure 14. Figure 15 compares the variation of radiation efficiency for Case I plate (parabolic decreasing thickness variation) with different arrangement of point masses for different taper parameter T_x. In this numerical simulation, it is found that for exciting frequencies up to 1000 HZ, the effect of radiation efficiency with different arrangement of point masses and for different taper ratios is independent of excitation frequency. However, at a given forcing frequency beyond 1000 HZ, higher taper ratios cause higher radiation efficiency as evident from Figure 15. It can also be seen that with increasing taper ratio sound power level peaks do shift towards lower frequency as reported in Figure 15. Moreover, beyond 2000 HZ, different taper ratios alter its stiffness at higher frequency and radiation efficiency curve tends to intersect each other at this high forcing region. It can also be seen that all radiation curves due to all combination of point masses tends converge in a frequency range of 6800–7200 HZ and clear peaks are seen at this frequency band. From Figure 15, it is noted that with increasing taper ratio the radiation efficiency increases for all combination of point masses. Among these radiation combinations, the highest radiation efficiency is shown by Case II plate with two point mass combinztions. The moderate radiation efficiency is seen to be observed for Case I plate with one point mass and two point masses combinations as reported in Table 7. However, at higher forcing frequencies, different parabolic tapered plate (Cases I, II, and III) with four point masses combination shows the least radiation efficiency as evident from Table 7. It is interesting to note that the lowest radiation efficiency (σ_{mn}) is shown by Case III plate. Thus, Case III plate may be considered a poor radiator among all the thickness variation with different combinations of and point masses. Figure 16 compares the radiation efficiency numerically for plates with different parabolic thickness variation two point masses combination for taper ratio $T_x = 0.75$. It is investigated that all cases of parabolic tapered plate contribute almost the same radiation efficiency as depicted from Figure 16. Figure 17 shows the numerical comparison of sound power level for plate with two point masses combination for different parabolic thickness variation and for taper ratio $T_x = 0.75$. It is observed that almost equal and increasing peak sound power level is seen for all cases of parabolic tapered plate. Hence, the stiffness variation due to different taper ratios has negligible effect on acoustic radiation as evident from Figures 16 and 17.

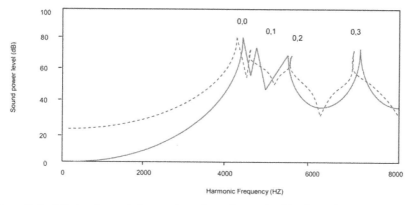

Figure 7. Analytical and numerical comparison of sound power level for Case I plate with four point masses having parabolic decreasing thickness variation for taper ratio $T_x = 0.75$.

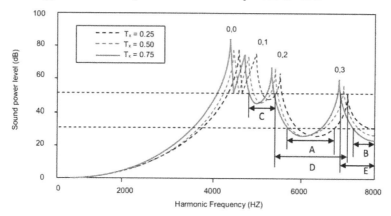

Figure 8. Numerical comparison of sound power level for Case I plate with one point mass having parabolic decreasing thickness variation with different taper ratio T_x.

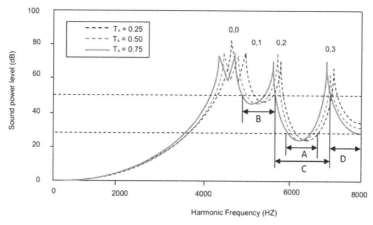

Figure 9. Numerical comparison of sound power level for Case I plate with two point masses having parabolic decreasing thickness variation with different taper ratio T_x.

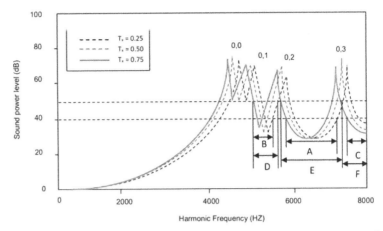

Figure 10. Numerical comparison of sound power level for Case I plate with four point masses having parabolic decreasing thickness variation with different taper ratio T_x.

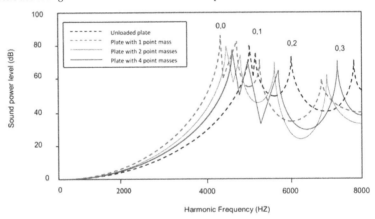

Figure 11. Numerical comparison of sound power level for Case I plate having parabolic decreasing thickness variation for different combinations of point masses for taper ratio $T_x = 0.75$.

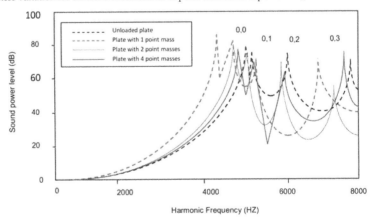

Figure 12. Numerical comparison of sound power level for Case II plate having parabolic decreasing increasing thickness variation for different combinations of point masses for taper ratio $T_x = 0.75$.

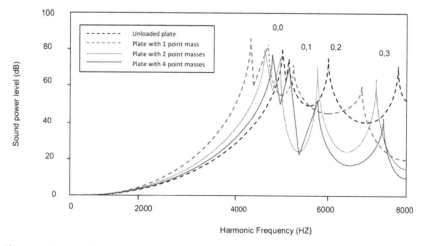

Figure 13. Numerical comparison of sound power level for Case III plate having parabolic increasing decreasing thickness variation for different combinations of point masses for taper ratio $T_x = 0.75$.

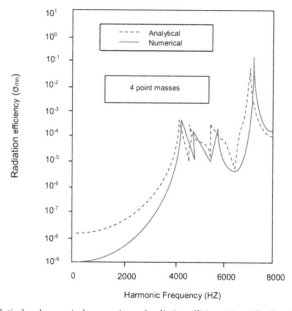

Figure 14. Analytical and numerical comparison of radiation efficiency (σ_{mn}) for Case I plate with four point masses having parabolic decreasing thickness variation for taper ratio $T_x = 0.75$.

Table 7. Comparison of peak sound power level and radiation efficiency of plate having different parabolically varying thickness with different combinations of point masses for $T_x = 0.75$.

Type	Plate Thickness Variation	Plate with One Point Mass		Plate with Two Point Masses		Plate with Four Point Masses	
		SPL (dB)	RE (σ_{mn})	SPL (dB)	RE (σ_{mn})	SPL (dB)	RE (σ_{mn})
Point masses	Case I	82	1.058	78	1.007	77	0.994
	Case II	81	1.045	83	1.079	77	0.994
	Case III	79	1.020	77	0.994	76	0.935

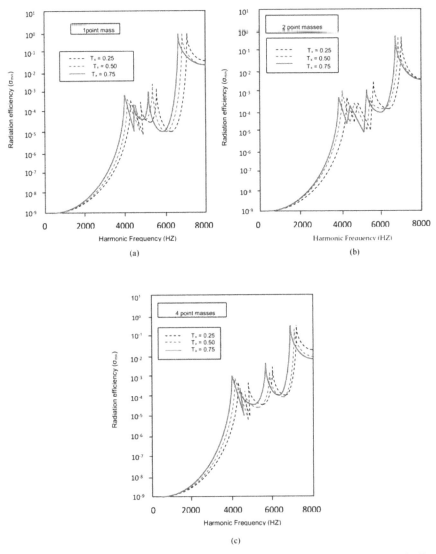

Figure 15. Numerical comparison of radiation efficiency (σ_{mn}) of Case I plate for parabolically decreasing thickness variation with different combination of point masses combinations for taper ratio $T_x = 0.75$.

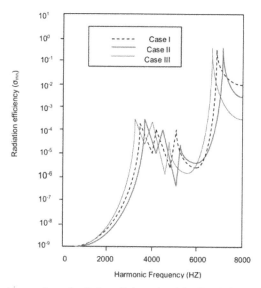

Figure 16. Numerical comparison of radiation efficiency (σ_{mn}) for Case I plate with two point masses combinations having different parabolic thickness variation for taper ratio $T_x = 0.75$.

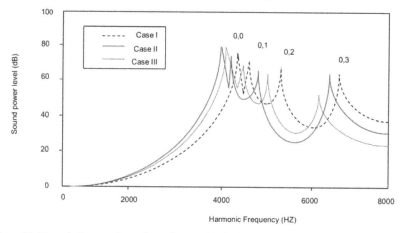

Figure 17. Numerical comparison of sound power level (dB) for Case I plate with two point masses combinations having different parabolic thickness variation for taper ratio $T_x = 0.75$.

4.3. Peak Sound Power Level Variation with Different Taper Ratios for All Combinations of Point Masses Attached to a Plate

In this numerical simulation, the peak sound power level is calculated. The different cases of plate with different parabolically varying thickness are considered. The different combinations of point mass are taken and different taper ratios are considered as shown in Figure 18. We are aiming at the highest peak sound power level for different combinations of point mass attached to a plate which is reported at first peak which corresponds to (0, 0) mode of the plate. In this numerical simulation, different cases of plate are investigated with point mass combination. For the Case I plate with one point mass combination, it is seen that peak sound power level increases for increasing value of taper ratio. For two point masses and four point masses combinations, there seems to be the variation of peak sound power levels for increasing value of taper ratio as evident from Figure 18. For Case I plate it is seen that peak is maximum for taper ratio, $T_X = 0.75$ for plate with one point mass combination and peak is minimum for taper ratio, $T_X = 0.75$ for four point masses combination. It is further noticed that for Case II plate the highest peak is seen for taper ratio, $T_X = 0.75$ for two point masses combination. Similarly, it is seen that for Case III plate lowest peak is observed for taper ratio, $T_X = 0.75$ for plate with four point masses combination. Thus from the simulation result, it is quite obvious that peak sound power level corresponds to (0, 0) mode is deeply affected by different combinations of point masses. It is observed that plate with different combinations of point masses with different taper ratios provide us design options for peak sound power level. As for example, for peak sound power reduction, taper ratios, $T_X = 0.75$ with four point mass combination and taper ratio, $T_X = 0.75$ with two point masses combination, for Case III plate may be the options. Similarly, for sound power actuation, taper ratio $T_X = 0.75$ with one point mass combination for Case I plate and two point masses combination for Case II plate may be the another alternative solution.

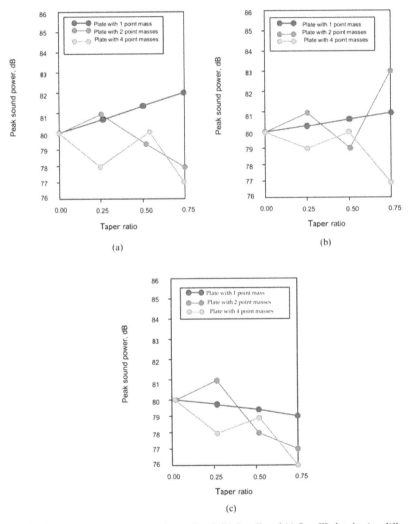

Figure 18. Peak sound power level (dB) for (**a**) Case I, (**b**) Case II, and (**c**) Case III plate having different parabolic thickness variation with different combination of point masses.

5. Conclusions

This paper represents a comparison of vibroacoustic behavior of different cases of parabolic tapered annular circular plate. The different combinations of point masses of tapered plate are considered. The clamped free boundary condition of the plate is taken where the mass of the unloaded plate and the mass of the plate plus point masses is kept constant. In this numerical simulation, it is investigated that Case II plate for all combination of point masses shows reduction in natural frequency parameter in comparison to Case I plate. This may happen perhaps due to less stiffness associated with the Case II plate. However, the natural frequency parameter for Case III plate is found to be same as that of uniform unloaded plate. For acoustic radiation behavior, it is noted that mode variation and all cases of parabolic tapered plate with different combinations of pointmasses have significant impact on sound power level. Whereas the sound power level is less contributed by the stiffness variation due to different taper ratios. Up to 50 dB, we get abroad range of frequencies as design options for all taper ratios, $T_x = 0.00, 0.25, 0.50$, and 0.75. This includes all cases of parabolic tapered plate with different combinations of point massesin different frequency bands. The numerical simulation results in minimum sound power level for all cases of thickness variation of plate with four point masses combination. On other hand, Case II plate reports the highest sound power level with two point masses combinations. It is interesting to note that Case III plate with all combination of point masses is seen to have the lowest sound power level among all variations and may be considered as the lowest sound radiator. Finally, design options for peak sound power level different combinations of point masses with different taper ratios are considered. For example, for peak sound power reduction, taper ratios $T_x = 0.75$ with four point masses combination and taper ratio $T_x = 0.75$ with two point masses combination for Case III plate may be the options. Similarly, for sound power actuation, taper ratio $T_x = 0.75$ with one point mass for Case I plate and two point masses combination for Case II plate may be the another solution.

Author Contributions: V.R. and M.R. supervised the research. A.C. and M.S.A. developed the research concept, developed the theory and performed the analysis. M.S.A. collects the data. A.C. wrote the paper. V.R. and M.R. revised the manuscript, and made important technical and grammatical suggestions. A.C. provided the APC funding.

Funding: The work is carried out in Indian Institute of Technology (ISM) Dhanbad, India. The APC will be funded by corresponding author only.

Conflicts of Interest: The authors declare no conflict of interest.

References

1. Wang, C.M.; Hong, G.M.; Tan, T.J. Elasting buckling of tapered circular plates. *Comput. Struct.* **1995**, *55*, 1055–1061. [CrossRef]
2. Gupta, A.P.; Goyal, N. Forced asymmetric response of linearly tapered circular plates. *J. Sound Vib.* **1999**, *220*, 641–657. [CrossRef]
3. Vivio, F.; Vullo, V. Closed form solutions of axisymmetric bending of circular plates having non-linear variable thickness. *Int. J. Mech. Sci.* **2010**, *52*, 1234–1252. [CrossRef]
4. Sharma, S.; Lal, R.; Neelam, N. Free transverse vibrations of non-homogeneous circular plates of linearly varying thickness. *J. Int. Acad. Phys. Sci.* **2011**, *15*, 187–200.
5. Wang, C.Y. The vibration modes of concentrically supported free circular plates. *J. Sound Vib.* **2014**, *333*, 835–847. [CrossRef]
6. Liu, T.; Kitipornchai, S.; Wang, C.M. Bending of linearly tapered annular Mindlin plates. *Int. J. Mech. Sci.* **2001**, *43*, 265–278. [CrossRef]
7. Duana, W.H.; Wang, C.M.; Wang, C.Y. Modification of fundamental vibration modes of circular plates with free edges. *J. Sound Vib.* **2008**, *317*, 709–715. [CrossRef]
8. Gupta, U.S.; Lal, R.; Sharma, S. Vibration of non-homogeneous circular Mindlin plates with variable thickness. *J. Sound Vib.* **2007**, *302*, 1–17. [CrossRef]

9. Kang, J.H. Three-dimensional vibration analysis of thick circular and annular plates with nonlinear thickness variation. *Comput. Struct.* **2003**, *81*, 1663–1675. [CrossRef]
10. Lee, H.; Singh, R. Acoustic radiation from out-of-plane modes of an annular disk using thin and thick plate theories. *J. Sound Vib.* **2005**, *282*, 313–339. [CrossRef]
11. Thompson, W., Jr. The computation of self- and mutual-radiation impedances for annular and elliptical pistons using Bouwkamp integral. *J. Sound Vib.* **1971**, *17*, 221–233. [CrossRef]
12. Levine, H.; leppington, F.G. A note on the acoustic power output of a circular plate. *J. Sound Vib.* **1988**, *21*, 269–275.
13. Rdzanek, W.P., Jr.; Engel, W. Asymptotic formula for the acoustic power output of a clamped annular plate. *Appl. Acoust.* **2000**, *60*, 29–43. [CrossRef]
14. Wodtke, H.W.; Lamancusa, J.S. Sound power minimization of circular plates through damping layer placement. *J. Sound Vib.* **1998**, *215*, 1145–1163. [CrossRef]
15. Wanyama, W. Analytical Investigation of the Acoustic Radiation from Linearly-Varying Circular Plates. Doctoral Dissertation, Texas Tech University, Lubbock, TX, USA, 2000.
16. Lee, H.; Singh, R. Self and mutual radiation from flexural and radial modes of a thick annular disk. *J. Sound Vib.* **2005**, *286*, 1032–1040. [CrossRef]
17. Cote, A.F.; Attala, N.; Guyader, J.L. Vibro acoustic analysis of an unbaffled rotating disk. *J. Acoust. Soc. Am.* **1998**, *103*, 1483–1492. [CrossRef]
18. Jeyraj, P. Vibro-acoustic behavior of an isotropic plate with arbitrarily varying thickness. *Eur. J. Mech. A/Solds* **2010**, *29*, 1088–1094. [CrossRef]
19. Ranjan, V.; Ghosh, M.K. Forced vibration response of thin plate with attached discrete dynamic absorbers. *Thin Walled Struct.* **2005**, *43*, 1513–1533. [CrossRef]
20. Kumar, B.; Ranjan, V.; Azam, M.S.; Singh, P.P.; Mishra, P.; PriyaAjit, K.; Kumar, P. A comparison of vibro acoustic response of isotropic plate with attached discrete patches and point masses having different thickness variation with different taper ratios. *Shock Vib.* **2016**, *2016*, 8431431.
21. Lee, M.R.; Singh, R. Analytical formulations for annular disk sound radiation using structural modes. *J. Acoust. Soc. Am.* **1994**, *95*, 3311–3323. [CrossRef]
22. Rdzanek, W.J.; Rdzanek, W.P. The real acoustic power of a planar annular membrane radiation for axially-symmetric free vibrations. *Arch. Acoust.* **1997**, *4*, 455–462.
23. Doganli, M. Sound Power Radiation from Clamped-Clamped Annular Plates. Master's Thesis, Texas Tech University, Lubbock, TX, USA, 2000.
24. Nakayama, I.; Nakamura, A.; Takeuchi, R. Sound Radiation of a circular plate for a single sound pulse. *Acta Acust. United Acust.* **1980**, *46*, 330–340.
25. Hasegawa, T.; Yosioka, K. Acoustic radiation force on a solid elastic sphere. *J. Acoust. Soc. Am.* **1969**, *46*, 1139–1143. [CrossRef]
26. Lee, H.; Singh, R. Determination of sound radiation from a simplified disk brake rotor using a semi-analytical method. *Noise Control Eng. J.* **2000**, *52*. [CrossRef]
27. Squicciarini, G.; Thompson, D.J.; Corradi, R. The effect of different combinations of boundary conditions on the average radiation efficiency of rectangular plates. *J. Sound Vib.* **2014**, *333*, 3931–3948. [CrossRef]
28. Xie, G.; Thompson, D.J.; Jones, C.J.C. The radiation efficiency of baffled plates and strips. *J. Sound Vib.* **2005**, *280*, 181–209. [CrossRef]
29. Rayleigh, J.W. *The Theory of Sound*, 2nd ed.; Dover: New York, NY, USA, 1945.
30. Maidanik, G. Response of ribbed panels to reverberant acoustic fields. *J. Acoust. Soc. Am.* **1962**, *34*, 809–826. [CrossRef]
31. Heckl, M. Radiation from plane sound sources. *Acustica* **1977**, *37*, 155–166.
32. Williams, E.G. A series expansion of the acoustic power radiated from planar sources. *J. Acoust. Soc. Am.* **1983**, *73*, 1520–1524. [CrossRef]
33. Keltie, R.F.; Peng, H. The effects of modal coupling on the acoustic power radiation from panels. *J. Vib. Acoust. Stress Reliab. Des.* **1987**, *109*, 48–55. [CrossRef]
34. Snyder, S.D.; Tanaka, N. Calculating total acoustic power output using modal radiation efficiencies. *J. Acoust. Soc. Am.* **1995**, *97*, 1702–1709. [CrossRef]
35. Martini, A.; Troncossi, M.; Vincenzi, N. Structural and elastodynamic analysis of rotary transfer machines by Finite Element model. *J. Serb. Soc. Comput. Mech.* **2017**, *11*, 1–16. [CrossRef]

36. Croccolo, D.; Cavalli, O.; De Agostinis, M.; Fini, S.; Olmi, G.; Robusto, F.; Vincenzi, N. A Methodology for the Lightweight Design of Modern Transfer Machine Tools. *Machines* **2018**, *6*, 2. [CrossRef]

37. Martini, A.; Troncossi, M. Upgrade of an automated line for plastic cap manufacture based on experimental vibration analysis. *Case Stud. Mech. Syst. Signal Process.* **2016**, *3*, 28–33. [CrossRef]

38. Pavlovic, A.; Fragassa, C.; Ubertini, F.; Martini, A. Modal analysis and stiffness optimization: The case of a tool machine for ceramic tile surface finishing. *J. Serb. Soc. Comput. Mech.* **2016**, *10*, 30–44. [CrossRef]

Article

Numerical Analysis of a Sensorized Prodder for Landmine Detection by Using Its Vibrational Characteristics

Zhiqiang Wu [1,2], Hui Ma [1], Chi Wang [1,2,*], Jinhui Li [2] and Jun Zhu [2,*]

[1] Deptment of Precision Mechanical Engineering, Shanghai University, Shanghai 200072, China; iwuwilliam@shu.edu.cn (Z.W.); mahui123456@shu.edu.cn (H.M.)
[2] Science and Technology on Near-Surface Detection Laboratory, Wuxi 214035, China; xin5star@shu.edu.cn
*   Correspondence: wangchi@shu.edu.cn (C.W.); ericbi@shu.edu.cn (J.Z.);
    Tel.: +86-137-6473-9726 (C.W.); +86-183-2173-5257 (J.Z.)

Received: 31 December 2018; Accepted: 5 February 2019; Published: 20 February 2019

check for
updates

Featured Application: This research is related to fields such as landmine detection (humanitarian landmine mine detection especially) and nondestructive testing.

Abstract: Prodders are widely used devices in landmine detection. A sensorized prodder has been developed to detect shallow buried landmines by their vibrational characteristics. However, the influencing mechanisms of prodder's components on the measured vibrational characteristics are not clear, and the vibration intensity of the buried landmine decreases with burial depth. A numerical analysis method is proposed to investigate the effects of parameters of prodder-object coupling system on the measured vibrational characteristics. The calculated main resonance frequency is 109.2 Hz, which corresponds well with the published analogy result of 110 Hz, and the mathematical method is also validated by the previous experimental results. Based on the proposed analysis method, an optimized prodder is designed, whereby the signal strength can theoretically increase 122.78%, which means that a greater depth of detection can be acquired. This optimal design is verified by the simulation experiment that was conducted with the optimization function of Adams software.

Keywords: landmine detection; lumped parameter model; prodder; resonance frequency

---

1. Introduction

As cheap but effective defensive weapons, landmines have been widely deployed since World War I. About 100 million landmines remain to be removed in more than 80 countries. These landmines kill or injure approximately 20,000 people every year. More people get poorer because the contaminated land is not suitable for agricultural or industrial production [1–4]. However, landmine detection has been a historically difficult problem.

Landmines are 5–30 cm in diameter (including anti-personnel and anti-tank mines), their casing materials vary from plastic to metal, and the burial environments are very complicated. Under these circumstances, even though many techniques have been developed for landmine detection, current researches indicate that no single device can detect landmine to a real performance level. Metal detection based on electromagnetic induction (EMI) can only detect landmines containing a certain amount of metal components, and it may lead to a high rate of false alarm due to metal shreds that are very common on battlefields [5,6]. Ground penetrating radar (GPR) images geophysical subsurface by acquiring reflected radar signals that are related to underground dielectric variations and it has a larger detection depth when compared to EMI method, but it also may cause a high false-positive detection rate due to the existences of big rocks, animal burrows, and roots [7,8].

The nuclear quadrupole resonance method observes the presence of abundant nitrogen in the explosives (TNT and DNT) of the mine, is why this device is very expensive and difficult to operate [9]. Dogs and rats have also been trained to sense the explosive vapor that is diffused from the mine for their remarkable sense of smell, but the training time is long and the detection result is dependent on individual animal's work status [10]. Other methods, like X-ray backscatter and electric impedance tomography (EIT), are limited by the lack of safety involved in X-ray radiation [11] and detection depth [12].

Mine prodders can be traced to World War I and they are still most common supplementary tools in landmine detection because of their adaptability on detecting metal or nonmetal landmines that are buried in complex circumstances [13–18]. Generally, prodding a landmine with a force of 25 N is unlikely to detonate a mine, though 50 N is regarded to be "safe" by some authorities [19]. A conventional prodder is usually a steel stick that is 30 cm in length. After the rain has softened the soil, deminers use the prodder to inspect subsurface of suspect area by poking the ground gently to contact the buried object to sense whether it is a landmine by their senses of touch and sound. Therefore, in recent years, many novel prodders with force feedback have been developed to explore new solutions for landmine detection by combining sensors with prodders [20–25].

Giovanni Borgioli et.al. introduced a sensorized prodder to detect landmines that are buried in shallow soil based on their mechanical feature of big compliance when compared to stone, root, and other clutters [26]. This prodder consists of three main parts: an electromechanical actuator, a thin steel rod, and an accelerometer that is fixed on it. After the steel rod is inserted in to the soil to obtain contact with the buried object to form a prodder-object coupling system, an external electronic unit provides power supply and excitation signals to drive the actuator to stimulate the buried object, and the vibration signal that is measured by the accelerometer mounted on the rod is transmitted to the signal processing unit for characteristic analysis. The tests that were carried out in the laboratory and an outdoor test bed validated that the buried objects' vibration responses are well distinguishable both in the frequency range and vibration intensity due to the landmine having a larger compliance when compared to roots, stones, and other rigid clutters for the landmine structure contains an air cavity between the explosive cavity and upper casing.

Since the vibration response is the key feature in distinguishing landmine from other rigid buried objects. When compared to the Ref. [26], we further studied the theoretical model of the prodder-object coupling system to investigate the effects of system parameters on the measured response signal features in this paper. A numerical analysis method is proposed to quantitatively investigate the resonance behavior of the prodder-object system and their influencing factors. The calculation result met well with the simulation and experimental data published in [26], which verifies the feasibility of this proposed numerical method. Based on the analysis results regarding system sensitivity, an optimal design for a better performing prodder is proposed, and the measured signal strength can increase 122.78% theoretically, which means greater depth of detection can be acquired, and a simulation experiment is conducted to verify the proposed optimal design with the optimization function of Adams software. This analysis procedure provides a reference for developing a good performing prodder.

2. Modeling of the Prodder-Object System and the Numerical Analysis Method

As shown in Figure 1a, the prodder that is described in Ref. [26] is composed of seven components, namely, (1) electromechanical actuator; (2) adjusting load spring; (3) aluminum cylinder for mechanical connection between springs; (4) aluminum cylinder for mechanical connection between spring and steel rod; (5) steel rod; (6) triaxial accelerometer with the y-axis aligned with the steel rod; and, (7) shell of the prodder.

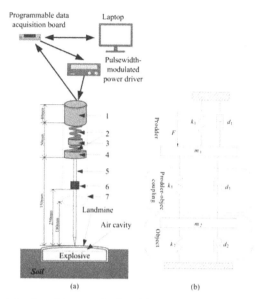

Figure 1. (a) Mechanical sketches of the sensorized prodder; (b) Kinetic model of "prodder-object" system.

The air cavity between a thin upper casing and the explosive makes the landmine structure more compliant than other buried clutters, such as rocks, tree roots, metal shreds, etc. The prodder is designed to distinguish landmines from those clutters by sensing buried objects' resonance behavior, including vibration intensity and frequency response.

In the detection procedure, the rigid steel rod is inserted into soil to get direct contact with the buried object to form a prodder-object coupling system. Subsequently, a programmable data acquisition board generates a chrip signal to control a pulsewidth-modulated (PWM) power driver that excites the electromechanical actuator of the prodder to vibrate. The steel rod stimulates the buried object with the mechanical force output from the actuator resultantly. For safety consideration, the output force of the prodder should be controlled below 10 N by adjusting the PWM power driver current. Meanwhile, the acceleration of the contact point is measured by a light silicon MEMS accelerometer that is mounted on the steel rod and then transmitted to the acquisition board for data analysis. After Analogue-to-Digital (A/D) conversion, filtering and Fast Fourier Transformation (FFT) processing of the measured data, the frequency response of the buried object can be acquired. Landmine structure is typically much more compliant than those rigid clutters, thus the vibration intensity of landmine would be much larger and its natural frequency would focus on the low frequency band when compared to other rigid clutters. Since these two resonance behaviors are the key features in distinguishing landmines from other rigid clutters, the resonance mechanism of the prodder-object coupling system should be further studied to make the influencing factors and their influencing mechanism clear.

The kinetic model of this prodder-object system can be obtained by lumped parameter method; this method has been used to investigate acoustic landmine detection [27,28]. As indicated in Figure 1b, k_1 and c_1 represent the stiffness and damping constants of adjusting spring respectively; and, k_2 and c_2 represent the viscoelastic parameters of the moving part of buried object's upper casing; Stiffness constant k_3 and damping constant c_3 of the prodder-object coupling part are equivalent to the viscoelastic constants of steel rod. While m_1 and m_2 represent the mass of the moving parts of prodder and buried object respectively.

According to the lumped model that is shown in Figure 1b and linear vibration theory, we can obtain the following kinetic equations:

$$m_1\ddot{x}_1 = F - k_1 x_1 - c_1 \dot{x}_1 - k_3(x_1 - x_2) - c_3(\dot{x}_1 - \dot{x}_2)$$
$$m_2\ddot{x}_2 = k_3(x_1 - x_2) + c_3(\dot{x}_1 - \dot{x}_2) - (k_2 x_2 + c_2 \dot{x}_2) \tag{1}$$

The above equations in matrix form can be written, as follows:

$$KX + C\dot{X} + M\ddot{X} = F \tag{2}$$

where $X = \text{diag}(x_1, x_2)$ is the displacement matrix dependent on the moving part of prodder and buried object, thus $\dot{X}$ and $\ddot{X}$ are the corresponding velocity and acceleration matrixes, respectively; while

$$K = \begin{bmatrix} k_1 + k_3 & -k_3 \\ -k_3 & k_2 + k_3 \end{bmatrix}, C = \begin{bmatrix} c_1 + c_3 & -c_3 \\ -c_3 & c_2 + c_3 \end{bmatrix}, M = \begin{bmatrix} m_1 \\ m_2 \end{bmatrix}, F = \begin{bmatrix} F \\ 0 \end{bmatrix}.$$

After being Fourier transformed, the matrix Equation (2) can be written, as follows:

$$\left(K - w^2 M + jwC\right)X(w) = F(w) \tag{3}$$

Accordingly, $Z(w)$, the impedance matrix of Equation (3), can be written as the following equation:

$$Z(w) = K - w^2 M + jwC \tag{4}$$

Subsequently, $\det Z(w)$ and $adjZ(w)$, the determinant and adjoint matrixes of $Z(w)$, respectively, can be written, as follows:

$$adjZ(w) = \begin{bmatrix} k_1 + k_2 - w^2 m_2 + jw[c_1 + c_2] & k_1 + jwc_1 \\ k_1 + jwc_1 & k_1 + k_3 - w^2 m_1 + jw[c_1 + c_3] \end{bmatrix} \tag{5}$$

$$\det Z(w) = \left[k_1 + k_3 - w^2 m_1 + jw[c_1 + c_3]\right]\left[k_1 + k_2 - w^2 m_2 + jw[c_1 + c_2]\right] - (k_1 + jwc_1)^2 \tag{6}$$

Afterwards, the system frequency response function $H(w)$ can be written, as follows:

$$H(w) = \frac{adjZ(w)}{\det Z(w)} = \begin{bmatrix} H_{11}[w] & H_{12}[w] \\ H_{21}[w] & H_{22}[w] \end{bmatrix} \tag{7}$$

In the 2·2 matrix $H(w)$, $H_{lp}(w)$ is the transfer function between the l th and p th component of the system. In this case, $H_{12}(w)$, the transfer function between prodder and buried object, can be written, as follows:

$$|H_{12}|w|| = \sqrt{\frac{C^2 + D^2}{E^2 + F^2}} \tag{8}$$

where

$A = k_1 + k_3 - w^2 m_1;$
$B = w(c_1 + c_3);$
$C = k_2 + k_3 - w^2 m_2;$
$D = w(c_2 + c_3);$
$E = AC - BD - k_3^2 + w^2 c_3^2;$ and,
$F = AD + BC - 2wc_3 k_3.$

3. Numerical Analysis of the Influencing Mechanisms of Prodder-Object System Parameters

3.1. Verification of the Mathematical Analysis Method

For testing the numerical analysis method, the results of mathematical analysis and experiments are compared. The data reported in Table 1 are the measured mechanical parameters of the prodder-object system, a plastic box with a diameter of 95 mm is used as the non-metallic landmine simulant, since the effects of a hollow plastic cap that simulates the upper casing of a landmine [26].

Table 1. Mechanical parameters of the lumped model.

Component	Mass (kg)	Elastic Coefficient (N/m)	Damping Coefficient (Ns/m)
Moving part of the prodder	$m_1 = 0.054$	$k_1 = 1818$	$c_1 = 0.1$
Plastic box	$m_2 = 0.013$	$k_2 = 30\cdot10^3$	$c_2 = 4$
Steel rod	N.A.	$k_3 = 3213\cdot10^3$	$c_3 = 0$

After substituting all of the above data into Equation (8), the calculated frequency response of the plastic box is shown in Figure 2. The resultant main resonance frequency is 109.2Hz, meeting well with the result 110 Hz that is published in Ref. [26], which validates our mathematical analysis method.

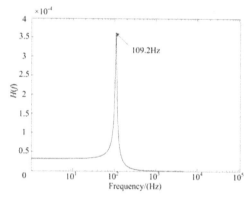

Figure 2. Viewgraph of the frequency response of the plastic box.

3.2. Effects of Prodder Components on the Vibrational Characteristics of Buried Object

The resonance behavior of prodder-object system is the most important feature in distinguishing a buried landmine. However, the influencing mechanisms of prodder's each component on the buried object's vibrational characteristics are not clear. On the other hand, the amplitude at resonance frequency is equivalent to the measured signal strength, but the object's vibration intensity usually diminishes with the increase of burial depth, which indicates that a better performing prodder should acquire greater signal strength under the same condition. Therefore, in the following sections, the effects of prodder's each component on the measured signal are quantitatively analyzed with the proposed mathematical method to explore the optimal design parameters for a better performing prodder.

3.2.1. Effects of m_1, the Mass of the Moving Part of Prodder

Parameter m_1 represents the mass of the moving part of prodder, which includes the total mass of steel rod, accelerometer, actuator's mobile part, adjusting springs, and aluminum cylinders. As shown in Figure 3, the resulting vibration intensity at the main resonance frequency of the plastic box increases by adjusting the value of m_1 from 10 g (mass of accelerometer) up to 100 g, while the other parameter values did not change, which indicates that the value of m_1 should be reasonably large to obtain greater signal strength.

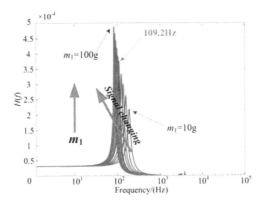

Figure 3. Effects of m_1 on the frequency response of the plastic box.

3.2.2. Effects of k_1, the Stiffness of Adjusting Load Spring

Figure 4 shows that both the intensity and frequency features of the plastic box vibration change little when the value of k_1 is variable from 600 N/m to 30 kN/m. It demonstrates that, with adjusting spring, which as an energy storage element, the effect of its stiffness on the measured signal could be ignored.

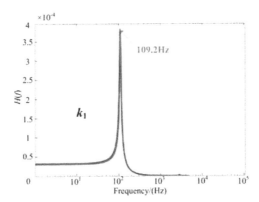

Figure 4. Effect of k_1 on the frequency response of the plastic box.

3.2.3. Effects of c_1, the Damping Coefficient of the Adjusting Spring

With the value of damping coefficient c_1 increasing from 0.1 Ns/m to 9 Ns/m, the resonance frequency of the plastic box remains unchanged, while the vibration intensity decreases a lot, as shown in Figure 5.

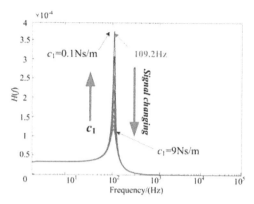

Figure 5. Effects of c_1 on the frequency response of the plastic box.

Ideally, the adjusting spring is used as an energy storage element without changing the magnitude of the output force that is exported from electromechanical actuator. However, in practice, there is still a part of energy consumed by the damping effect of the spring. Thus, damping coefficient of the adjusting spring should be small enough to acquire better detection performance.

3.3. Effects of the Prodder-Object Coupling Part on Plastic Box Vibrational Characteristics

The lumped model shown in Figure 1b is a two-degree freedom mechanical model that has two resonance frequencies, namely the main resonance frequency and a higher second resonance frequency, as introduced by the coupling part (steel rod). The values of k_3 and c_3 are uncertain when the rod and the measured object are not in good contact. Moreover, the value of c_3 is difficult to measure. Therefore, it is necessary to analyze the effects of the coupling part on the measured signal characteristics.

3.3.1. Effects of k_3, the Stiffness of Steel Rod

As shown in Figure 6, when the value of k_3 is increasing from 100 kN/m to 3213 kN/m, the main resonance frequency of the plastic box increases slightly while the related amplitude decreases, which means that a steel rod with smaller stiffness may lead to greater measured signal strength. Thus, steel rods for a better performing prodder should have smaller stiffness.

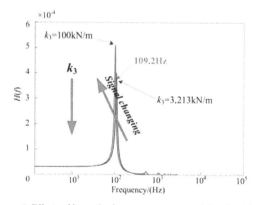

Figure 6. Effects of k_3 on the frequency response of the plastic box.

3.3.2. Effects of c_3, the Damping Coefficient of the Steel Rod

Figure 7 shows that the main resonance frequency and vibration intensity of the plastic box remain the same, while the amplitude at the second resonance frequency increases with the value of c_3 increasing from 0 up to 90 Ns/m. It can be seen that the damping effect of the coupling part only introduces a second resonance frequency without changing the amplitude of the measured signal, and when the prodder contacts well with buried object, the smallest damping effect of the coupling part can be achieved.

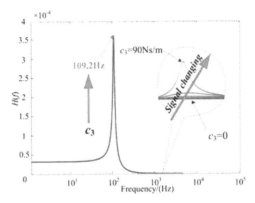

Figure 7. Effects of c_3 on the frequency response of the plastic box.

Figure 8 represents the one-degree freedom model that ignores the coupling effects of the steel rod. In this case, the output force from the actuator is directly applied to the measured object.

Figure 8. Mechanical model of one-degree freedom prodder-object system.

The dynamic equation of the one-degree freedom model can be illustrated as:

$$(k_1+k_2)x+(c_1+c_2)\dot{x}+(m_1+m_2)\ddot{x} = F \qquad (9)$$

Subsequently, the frequency response function of this one-degree freedom system can be written, as follows:

$$|H_0|\omega|| = \frac{1}{\sqrt{[k_1+k_2-\omega^2[m_1+m_2]]^2+\omega^2(c_1+c_2)^2}} \qquad (10)$$

As shown in Figure 9, the theoretical frequency response of the plastic box can be easily obtained by calculating Equation (10). The calculated main resonance frequency is 110 Hz, which is consistent with the main resonance frequency of the two-degree freedom mechanical model that is shown in Figure 1b. It indicates that the mechanical model shown in Figure 1b has two degrees of freedom is due to the coupling effects of the steel rod, and it would attenuate to a one-degree freedom model as the decrease of the rod's damping coefficient.

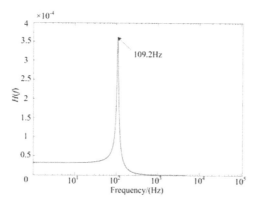

Figure 9. The frequency response of the one-degree freedom model of the plastic box.

3.4. Effects of Buried Object'S Parameters on the Measured Signal Features

Since the landmine is buried in soil, the actual value of stiffness k_2 and damping coefficient c_2 might be significantly influenced by the varying environmental conditions, such as burial depth, moisture, particle size, and porosity of soil. Meanwhile, different landmines may have different viscoelastic coefficients. Obviously, the measured signal characteristics would be influenced by parameters of landmine and soil conditions. In the following part, the viscoelastic coefficients of buried object are analyzed quantitatively with the same mathematical method.

3.4.1. Effects of k_2, the Stiffness of the Moving Part of Plastic Box

Figure 10 shows the effects of landmine upper casing stiffness k_2 on the measured signal features, the amplitude of the measured signal decrease, while the related resonance frequency increases as the value of k_2 increases from 30 kN/m up to 66 kN/m. Since the casing stiffness of the plastic landmines are smaller than the stiffness of metal landmines [26], according to the analysis results above, the prodder may have better performance in the detection of non-metallic landmines that cannot be detected by the common device, for example, metal detector.

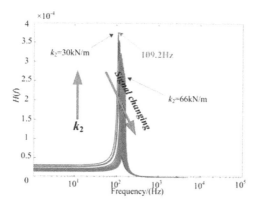

Figure 10. Effects of k_2 on the frequency response of the plastic box.

3.4.2. Effects of c_2, the Damping Coefficient of the Moving Part of Plastic Box

Figure 11 shows the upper casing damping effects on the measured signal characteristics; the amplitude at the main resonance frequency decreases significantly as the value of c_2 increases from 4 Ns/m up to 24 Ns/m. This is due to the significant amount of vibrational energy that can be consumed by damping effects. However, the actual value of damping coefficient c_2 increases with the burial depth. As a result, the signal strength, namely the amplitude at main resonance frequency of buried landmine, would decrease with burial depth. These results are consistent with the experimental results in Ref. [26]. It highlights the importance of designing an optimized prodder that can obtain larger signal strength to achieve a greater depth of detection.

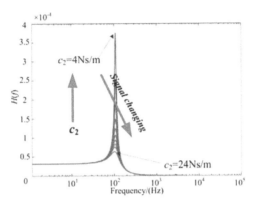

Figure 11. Effects of c_2 on the frequency response of the plastic box.

According to the analysis above, the effects of each prodder's component on buried landmine's vibrational characteristics can be summarized in Table 2 as the relevant parameter values increase. The notation "↑" represents an increase, "↓" represents a decrease, and "—" means no influence.

Table 2. Effects of Parameters of the System on Its Vibrational characteristics.

Component	Prodder			Steel Rod		Buried Object	
	m_1	k_1	c_1	k_3	c_3	k_2	c_2
Parameter value	↑	↑	↑	↑	↑	↑	↑
The main resonance frequency	↓	—	—	↑	—	↑	—
Vibrational amplitude	↑	—	↓	↓	—	↓	↓

4. Evaluation of the Optimized Design

On the basis of the above analysis results, an optimized prodder with parameter values of $m_1 = 0.1$ kg, $k_1 = 1818$ N/m, $c_1 = 0.1$ Ns/m, and $k_3 = 100$ kN/m can obtain greater signal strength and larger detection depth. As shown in Figure 12, the plastic box's theoretical resonance vibration intensity, as measured with the optimized prodder, is 8.07×10^{-4}, while the same value that was measured with the original prodder is $3.60 \cdot 10^{-4}$, thus the signal strength increases by approximately 122.78%.

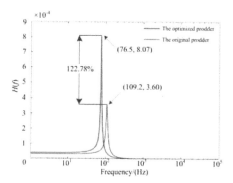

Figure 12. Performances comparison between the optimized prodder and the original prodder.

Finally, Adams, which is a widely used multi-body dynamics simulation software, is used to conduct a simulation experiment to verify the numerical analysis results. The optimization function of this software allows for users to obtain the optimal parameter values for the design objectives within certain constraints. As shown in Figure 13a, a parametric dynamics model explaining the working principle of prodder is built in the Adams software. According to the above analysis results, k_1 has few influence on the measured signal, smaller values for the damping coefficients c_1 and c_3 tend to enhance performance. Therefore, k_3 and m_1, the viscoelastic coefficients of steel rod and the mass of the moving part of prodder, are set as key design variables. The objective function is the mean value of the variance between the time domain vibration data measured with the original and new designed prodders. It is obvious that the maximum value of the objective function corresponds to the strongest difference of vibration intensity, as measured by these two prodders. The corresponding parameter values are optimal in the design of better performing prodder. The optimization result is shown in Figure 13b and the theoretical main resonance frequencies of vibration signal measured by the original and optimized prodders are 109.27Hz and 76.5Hz, respectively, which perfectly meets with the numerical analysis results that are shown in Figure 12. Due to Adams software taking earth gravity into account, the vibration intensity of simulation results increases 161.56%, which is larger than the numerical analysis result of 122.78%, but they are of the same order. The corresponding values for k_3 and m_1 are 100 kN/m and 0.1 kg, respectively, which are equal to the results deduced by the mathematical method, which verifies this numerical analysis method and the optimal design. It should be noted that the value ranges of the above parameters are selected experientially, mainly for studying the parameter's effect on the prodder's performance. This paper principally provides an optimization analysis method for developing a better performance prodder used for landmine detection.

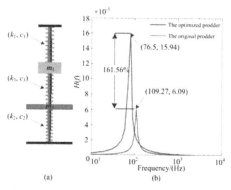

Figure 13. (a) Parametric dynamics model of prodder-object system; and, (b) Performance comparison between the original prodder and the prodder optimized with the function of Adams software.

5. Conclusions

Based on the working principle of a sensorized prodder that is used for landmine detection, a mathematical analysis method is proposed to analyze the influences of each of the prodder's components on the measured signal. The calculated results meet well with the published electromechanical analogy simulation results and experimental results respectively, proving the feasibility of the proposed numerical analysis method. According to the analysis results, we can draw the following conclusions:

If the mass of the moving part of prodder increases, the main resonance frequency of the measured signal would decrease while the corresponding vibration intensity would increase. Therefore, the mass of the moving part of prodder should be reasonable large to acquire greater signal strength to obtain larger detection depth.

The stiffness of prodder's adjusting spring has no effect on the vibrational characteristics of the measured object, but the damping effect of spring can attenuate the vibration intensity. Thus, the spring of an optimized prodder should have a smaller damping coefficient.

The steel rod's coupling effects generate a second resonance of the measured signal, while its influence on the main resonance vibration could be ignored. However, the measured signal strength would decrease with the increase of the stiffness of the steel rod. As a result, a better performing prodder should adopt a steel rod with smaller stiffness, for example, a hollow steel rod with conic-shaped tip.

The analysis results of the effects of the measured object's stiffness and damping coefficient show that the theoretical signal strength for the plastic mine is larger than the related value for metal mine, which indicates that the prodder might be better suited for non-metallic landmine detection.

According to the above analysis results, when the mass of the prodder's moving part is 0.1 kg, the adjusting spring is still the same, and the stiffness of the rod is 100 kN/m, the measured signal strength would increase approximately 122.78% resultantly, which means that a greater depth of detection could be achieved, and this optimal design is verified by the simulation experiment that was conducted with the optimization function of Adams software. It should be mentioned that, although many prodders have been proposed, prodder with force feedback is more of an exploration solution for landmine detection and is under consideration by some mine clearance organizations at this stage for its safety as a contact method. However, this optimized prodder could be of considerable interest in landmine detection training, other shallow buried landmine detection, and nondestructive testing for structural health. In our future work, we will develop an optimized prodder and dedicate it to these fields.

Author Contributions: Conceptualization, J.Z. and Z.W.; methodology, Z.W.; writing-original draft preparation, Z.W.; writing-review and editing, J.L.; project administration, C.W.; simulation experimental validation, H.M.

Funding: This research was funded, in part, by the NATIONAL NATURAL SCIENCE FOUNDATION OF CHINA, grant number 61773249; the NATURAL SCIENCE FOUNDATION OF SHANGHAI, grant number 16ZR1411700; the SCIENCE AND TECHNOLOGY ON NEAR-SURFACE DETECTION LABORATORY, grant numbers 6142414090117, TCGZ2017A006; and the NATIONAL YOUTH FUNDATION OF CHINA, grant number 41704123.

Acknowledgments: Thanks are due to Ziyang Song and Zhiwen Yuan for their valuable discussion and help.

Conflicts of Interest: The authors declare no conflict of interest.

References

1. Kowalenko, K. Saving lives, one land mine at a time. *IEEE Inst.* **2004**, *28*, 10–11.
2. Geoff, H. The economics of landmine clearance: Case study of Cambodia. *J. Int. Dev.* **2000**, *12*, 219–225.
3. Meier, R., III; Smith, W.K. Landmine injuries and rehabilitation for landmine survivors. *Phys. Med. Rehabil. Clin. N. Am.* **2002**, *13*, 175–187. [CrossRef]
4. Gibson, J.; Barns, S.; Cameron, M.; Lim, S.; Scrimgeour, F.; Tressler, J. The value of statistical life and economics of landmine clearance in developing countries. *World Dev.* **2007**, *35*, 512–531. [CrossRef]

5. Kim, B.; Yoon, J.W.; Lee, S.E.; Han, S.H.; Kim, K. Pulse-induction metal detector with time-domain bucking circuit for landmine detection. *Electron. Lett.* **2015**, *51*, 159–161. [CrossRef]

6. Collins, L.; Gao, P.; Tantum, S. Model-based statistical signal processing using electromagnetic induction data for landmine detection and classification. In Proceedings of the 11th IEEE Signal Processing Workshop on Statistical Signal Processing, Singapore, 8 August 2001.

7. Carin, L. Ground-Penetrating Radar (Paper I). Alternatives for Landmine Detection. Available online: https://www.rand.org/pubs/monograph_reports/MR1608.html (accessed on 18 February 2003).

8. Takahashi, K.; Igel, J.; Preetz, H.; Sato, M. Influence of heterogeneous solis and clutter on the performance of ground-penetrating radar for landmine detection. *IEEE Trans. Geosci. Electron.* **2014**, *52*, 3464–3472.

9. Tan, Y.Y.; Tantum, S.L.; Collins, L.M. Landmine detection with nuclear quadrupole resonance. In Proceedings of the IEEE International Geoscience and Remote Sensing Symposium, Toronto, ON, Canada, 24–28 June 2002.

10. Bach, H.; Ian, M.G.; Conny, A.; Rebecca, S. Improving mine detection dogs: An overiew of the GICHD dog program. In Proceedings of the International Conference on Requirements and Technologies for the Detection, Removal and Neutralization of Landmines and UXO, Brussels, Belgium, 15–18 September 2003.

11. Harding, G. X-ray scatter tomography for explosives detection. *Radiat. Phys. Chem.* **2004**, *71*, 869–881. [CrossRef]

12. Church, P.; McFee, J.E.; Gagnon, S.; Wort, P. Electrical impedance tomographic imaging of buried landmines. *IEEE Trans. Geosci. Remote Sens.* **2006**, *44*, 2407–2420. [CrossRef]

13. Gasser, R.; Thomas, T.H. Prodding to detect mines: A technique with a future. In Proceedings of the Second International Conference on Detection of Abandoned Land Mines, Edinburgh, UK, 12–14 October 1998.

14. Hussein, E.M.; Waller, E.J. Landmine detection: The problem and the challenge. *Appl. Radiat. Isot.* **2000**, *53*, 557–563. [CrossRef]

15. Machler, P. Detection technologies for anti-personnel mines. In Proceedings of the Symposium on Autonomous Vehicles in Mine Countermeasures, Monterey, CA, USA, 4–6 April 1995.

16. Maki, H.K. Mine detection and sensing technologies-new development potentials in context of humanitarian demining. In Proceedings of the 27th Annual Conference of the IEEE Industrial Electronics Society, Denver, CO, USA, 29 November–2 December 2001.

17. Frigui, H.; Gader, P.; Keller, J. Fuzzy clustering for land mine detection. In Proceedings of the 1998 Conference of the North American Fuzzy Information Processing Society, Pensacola Beach, FL, USA, 20–21 August 1998.

18. Russell, K. Contact Methods. Alternatives for Landmine Detection. Available online: https://www.rand.org/content/dam/rand/pubs/monograph_reports/MR1608/MR1608.appw.pdf (accessed on 18 February 2003).

19. Russell, G. Technology for Humanitarian Landmine Clearance. Ph.D. Thesis, University of Warwick, Coventry, UK, 2000; p. 70.

20. Ishikawa, J.; Iino, A. A study on prodding detection of antipersonnel landmine using active sensing prodder. In Proceedings of the International Symposium "Humanitarian Demining 2010", Sibenik, Croatia, 27–29 April 2010.

21. Gallagher, P.J. Characteristic Discriminating Landmine Hand Prodder. U.S. Patent 5,754,494, 19 May 1998.

22. Uno, M.A.; Kimura, T.; Kato, M. Method and Apparatus for Detecting and Discriminating Objects under the Ground. U.S. Patent 5,672,825, 30 September 1997.

23. Borza, M.A. Prodder with Force Feedback. U.S. Patent 6,386,036, 14 May 2003.

24. Steinway, W.; Scott, W. Mine Detection using Radar Vibrometer. U.S. Patent 7,183,964, 27 February 2007.

25. Fernandez, R.; Montes, H.; Armada, M. Intelligent multisensory prodder for training operators in humanitarian demining. *Sensors* **2016**, *16*, 965. [CrossRef] [PubMed]

26. Borgioli, G.; Bulletti, A.; Calzolai, M.; Capineri, L. A new sensorized prodder device for the detection of vibrational characteristics of buried objects. *IEEE Trans. Geosci. Remote Sens.* **2014**, *52*, 3440–3452. [CrossRef]

27. Donskoy, D.; Ekimov, A.; Sedunov, N.; Tsionskiy, M. Nonlinear seismo-acoustic land mine detection and discrimination. *J. Acoust. Soc. Am.* **2002**, *111*, 2705–2713. [CrossRef]

28. Donskoy, D.; Reznik, A.; Zagrai, A.; Ekimov, A. Nonlinear vibrations of buried landmines. *J. Acoust. Soc. Am.* **2004**, *117*, 690–700. [CrossRef]

applied sciences

Article

Hearing in Noise: The Importance of Coding Strategies—Normal-Hearing Subjects and Cochlear Implant Users

Pierre-Antoine Cucis [1,2,*], **Christian Berger-Vachon** [3,4], **Ruben Hermann** [1,2], **Fabien Millioz** [5], **Eric Truy** [1,2] and **Stéphane Gallego** [6,7]

[1] Integrative, Multisensory, Perception, Action and Cognition Team (IMPACT), INSERM U1028, CNRS UMR 5292, Claude-Bernard Lyon1 University, 69676 Bron, France; ruben.hermann@chu-lyon.fr (R.H.); eric.truy@chu-lyon.fr (E.T.)

[2] ENT and Cervico-Facial Surgery Department, Edouard Herriot Hospital, 69003 Lyon, France

[3] Brain Dynamics and Cognition (DYCOG), INSERM U1028, CNRS UMR 5292, Claude-Bernard Lyon1 University, 69676 Bron CEDEX, France; christian.berger-vachon@univ-lyon1.fr

[4] Biomechanics and Impact Mechanics Laboratory (LBMC), French Institute of Science and Technology for Transport, Development and Networks (IFSTTAR), CNRS UMR T9406, Claude-Bernard Lyon1 University, 69675 Bron CEDEX, France

[5] Centre for Research into the Acquisition and Processing of Images for Healthcare (CREATIS), INSERM U1206, CNRS UMR5220, INSA, Claude-Bernard Lyon1 University, F69621 Lyon, France; fabien.millioz@univ-lyon1.fr

[6] Sensory and Cognitive Neuroscience Laboratory (LNSC), CNRS UMR 7260, Aix-Marseille University, 13331 Marseille, France; sgallego@hotmail.fr

[7] Institute of Rehabilitation Sciences and Techniques (ISTR), Claude-Bernard Lyon1 University, 69373 Lyon CEDEX 08, France

[*] Correspondence: cucis.pa@gmail.com; Tel.: +33-472-110-0518

Received: 7 December 2018; Accepted: 11 February 2019; Published: 20 February 2019

Abstract: Two schemes are mainly used for coding sounds in cochlear implants: Fixed-Channel and Channel-Picking. This study aims to determine the speech audiometry scores in noise of people using either type of sound coding scheme. Twenty normal-hearing and 45 cochlear implant subjects participated in this experiment. Both populations were tested by using dissyllabic words mixed with cocktail-party noise. A cochlear implant simulator was used to test the normal-hearing subjects. This simulator separated the sound into 20 spectral channels and the eight most energetic were selected to simulate the Channel-Picking strategy. For normal-hearing subjects, we noticed higher scores with the Fixed-Channel strategy than with the Channel-Picking strategy in the mid-range signal-to-noise ratios (0 to +6 dB). For cochlear implant users, no differences were found between the two coding schemes but we could see a slight advantage for the Fixed-Channel strategies over the Channel-Picking strategies. For both populations, a difference was observed for the signal-to-noise ratios at 50% of the maximum recognition plateau in favour of the Fixed-Channel strategy. To conclude, in the most common signal-to-noise ratio conditions, a Fixed-Channel coding strategy may lead to better recognition percentages than a Channel-Picking strategy. Further studies are indicated to confirm this.

Keywords: cochlear implant; coding strategy; Fixed-Channel; Channel-Picking; vocoder simulation; normal-hearing

---

1. Introduction

In 2012, cochlear implants (CIs) had successfully restored partial hearing to over 324,200 deaf people worldwide [1]. In most cases, users of modern CIs perform well in quiet listening conditions.

Four CI manufacturers are presently on the French market: Cochlear® and Neurelec®/Oticon Medical® for Channel-Picking (CP) strategies and Advanced Bionics® and Med-El® for Fixed-Channel (FC) strategies. For most CI users, performances for speech perception significantly decrease in noisy environments [2].

All modern sound coding strategies are based on the analysis of acoustic information by a bank of band-pass filters and each strategy has its own philosophy [3].

Two coding schemes are mainly in use. FC strategies transmit all available channels to the electrodes, which usually stimulate at a high rate. CP strategies (sometimes called n-of-m strategies) use various stimulation rates (a high, medium or low rate), estimate the outputs of all the available channels (m) and select a subset of channels (n) with the largest amplitudes.

The present study focuses on the relative contribution of FC strategies and CP strategies on syllable recognition in noise. We wish to compare the efficiency of the FC and CP coding strategies, first in simulation and secondly with CI users.

1.1. Sound Coding Strategies

In practice, a wide variation of outcomes is observed amongst implanted patients, which is probably linked to the duration of deafness, the age at implantation, the age at onset of deafness, the duration of implant use and the patient's social environment [4].

Some studies showed a superiority of an FC strategy over a CP strategy particularly in noise [5]. Others like Skinner et al. and Kiefer et al. [6,7] showed a positive difference, for speech recognition, in favour of the Advanced Combination Encoder (ACE) (CP strategy) over Continuous Interleaved Sampling (CIS) (FC strategy) and Spectral Peak Picking (SPEAK) (CP strategy) [6]. Brockmeier et al. [8] compared the musical activities and perception of cochlear implant users and concluded that CIS, SPEAK and ACE did not differ significantly. No clear advantage for a particular coding scheme has been identified yet.

The number of spectral channels required for speech recognition depends on the difficulty of the listening situation for both FC and CP strategies [9,10]. For FC strategies, all channels are transmitted to the corresponding electrodes, usually between 12 and 16, leading to a relatively large amount of spectral information that may be blurred by the current spread in the cochlea. When the stimulation rate is high, some results suggest that this rate may be beneficial to speech perception [11]. Another feature of the strategies lies in the non-overlapping (interleaved) pulse delivery; pulses are brief with a minimum delay between them and rapid variations in speech can be tracked [12].

1.2. Influence of Noise

The assessment of the performance of CI users in noise has become of great interest as it is considered to be representative of daily listening conditions.

In noise, the natural gaps in speech are filled and speech envelopes are distorted making speech recognition more difficult. The CP coding strategies may select noise-dominated channels, instead of the dominant speech channels, at low signal-to-noise ratios (SNRs) [13]. Unlike the CP strategies, FC strategies transmit the information of all available channels leaving the task of selecting the informational cues to the auditory system.

The presence of noise reduces the effective dynamic range for CI users by compressing the region of audibility into the upper position of the dynamic range [14]. Good speech perception in noise is a target in the management of deafness [15–18] and this aspect is also of great importance when CI coding strategies are concerned. Thus, tests in noise are more sensitive to changes in the fitting parameters and more ecological than tests in quiet conditions.

1.3. Simulation with Normal-Hearing Subjects

Considering the heterogeneity in a group of CI users, it is usually difficult to draw strong conclusions. Additionally, as the FC and CP strategies are fitted to each group of CI users, the heterogeneity of the populations is increased.

On the contrary, a simulation work, which can be done with NH subjects, allows greater homogeneity of the participants. In this case, the same subject can face different situations [19], such as coding schemes and SNRs, allowing one to focus on different controllable features such as time and amplitude cues and to ensure efficient paired comparisons. However, the results observed with NH listeners cannot be directly extrapolated to CI users and many studies have been conducted on this subject. Dorman extensively studied this matter in 2000 [20,21] and stated that "performance of the NH listeners established a benchmark for how well implant recipients could perform if electrode arrays were able to reproduce, by means of the cochlea, the stimulation produced by auditory stimulation of the cochlea and if patients possessed neural structures capable of responding to the electrical stimulation" [22]. They also indicated that the best CI users achieved scores that were within the ranges of scores observed with NH subjects. On the contrary, other authors point out the limitations of using vocoders to simulate electric hearing and the importance of making experiments with CI users [23].

Consequently, both approaches (with CI and NH subjects) seem necessary; with NH subjects, we can evaluate the consequences of the coding strategies and with CI users we can evaluate the real aspect on a clinical point of view. Practically, for a given strategy, several fitting procedures are recommended by the manufacturers and each CI is fitted to the patient.

2. Material & Methods

2.1. Participants

The work presented in this paper follows a previous pilot study [24] and was approved by the French Ethics Committee "Sud-Est 2" (August, 27, 2014, ID-RCB: 2014-A00888-39), under the supervision of the HCL (Civil Hospitals of Lyon). The participants were recruited between November 2014 and April 2016. They were all informed at least a week before entering the study, verbally and in writing and they filled out a consent form.

2.1.1. Normal-Hearing Subjects

Twenty NH subjects participated in this experiment. Their age ranged from 18 to 33 years old, with an average of 25 years. They were recruited among the students of the Claude Bernard Lyon 1 University, through a recruitment notice sent via email. An otologic examination was performed before entering the study in order to exclude subjects with previous pathologies or deafness. All these subjects were considered to have normal hearing according to recommendations of the International Bureau for Audio-Phonology, as their auditory thresholds were below 20 dB HL for all frequencies between 250 and 8000 Hz.

2.1.2. Cochlear Implant Subjects

Forty-five CI users were included in this study. Their ages ranged from 18 to 60 years old, with an average of 37 years. They were recruited in the general population of CI users who have their classical follow-up examination in our tertiary referral centre. Nineteen subjects were fitted with an FC strategy (Advanced Bionics® and Med-El®) and twenty-six had a CP strategy (Cochlear® and Neurelec®/Oticon Medical®); the CI population was constituted of two groups (one for each coding scheme). CI users included in the experiment were people implanted unilaterally and bilaterally. In the case of people with bilateral implantation, only one implant was tested: the one giving the best outcomes according to the patient. Demographic details are indicated in Appendix A.

2.2. Stimuli

The acoustic material incorporates Fournier's lists mixed in with a cocktail-party noise.

2.2.1. Fournier's Disyllabic Lists

These lists were adapted to test the participants. They were created by Jean-Etienne Fournier in 1951 and are approved by the French National College of Audiology (C.N.A.). Forty lists with a male voice are available and each list is constituted of 10 two-syllable common French words (e.g., le bouchon = the cork), leading to 20 syllables per list. They are a French equivalent to the American Spondee lists (e.g., baseball). The recognition step was one syllable (5%).

2.2.2. Noise

In this study, we used cocktail-party noise. It was a voice mix of eight French-speaking people, four males and four females. This kind of noise was sufficiently heterogeneous for the task and the masking was rather invariable throughout a session.

2.3. Hardware

Stimuli were recorded on a CD (44.1 kHz sampling frequency, 16-bit quantization) and presented using a PHILIPS CD723 CD player connected to a Madsen orbiter 922® Clinical audiometer to control the general volume and the SNR. The sound was delivered in free field with two JBSYSTEMS ISX5 loudspeakers for CI users and with a TDH 39 headset for NH subjects. Devices used in our experiment are regularly calibrated and checked according to the NF EN ISO 8253 standard.

2.4. Experimental Conditions and Procedures

For the two groups of subjects, the experiment consisted in speech audiometry in noise with one syllable as the error unit. For a fixed speech level of 60 dB SPL, the maximum level delivered was below 65 dB SPL. According to the conditions requested by the ethics committee, it did not exceed the 80 dB SPL limitation recommended for professional noise exposure.

2.4.1. Normal-Hearing Subjects

Processed stimuli were delivered to only one ear, as in the experiment conducted with CI users. Furthermore, we chose to test the subjects in the right ear considering a lateralization of the treatment of sounds and especially that speech understanding seems associated with the left hemisphere activity [25,26].

For a fixed speech level of 60 dB SPL, five SNR were tested for each sound-coding scheme [FC and CP (8 out of 20)]. The lower SNR was −3 dB and the higher was +9 dB with 3 dB steps between each tested SNR. For the SNR of +9 dB, the recognition percentage was 100%. Each combination (coding scheme + SNR) was assigned to a Fournier's list so that the lists were not repeated. Each session started with a short training period to help the listener understand the instructions. Then the 10 noise and coding scheme conditions were randomly presented to each subject (1 list per condition: 5 SNRs and 2 coding schemes). The sessions lasted about 15 min (plus half an hour for the auditory check).

2.4.2. Cochlear Implant Users

The procedure was slightly different with the CI users as the task was more difficult for them than for the NH subjects. The speech level was fixed to 60 dB SPL. Most of the CI users did not reach the 100% recognition score; the percentage regularly increased with the SNR. The SNRs were presented from +18 dB to −3 dB with 3 dB steps. Only one strategy (corresponding to their CI) could be tested with a patient. Lists were presented in increasing order of difficulty (from +18 dB to −3 dB of SNR) to avoid discouragement effects; this procedure was the same for both coding schemes.

CI users were tested at the beginning of their periodical clinical check-up and device setting, which occurs at the "CRIC" (Cochlear Implant Setting Centre) located in the ORL department of the Edouard-Herriot hospital. The patient follow-up consists of an appointment with a speech therapist, a setting of the implant parameters by an audiologist and a clinical examination by a physician.

The following tasks were realized in our work:

- verification of the patient's medical file;
- a short training session to help the patient understand the instructions.

2.5. Implant Simulation

For the simulation of "CI like" speech processing, we used a vocoder implemented in Matlab® (MathWorks, Natick, MA) to simulate an FC and a CP coding strategy. We did not simulate channel interaction in this study.

A diagram representing the signal processing performed by the vocoder is shown in Figure 1. The different steps of the signal processing are as follows:

- The input signal goes through a pre-emphasis filter, which is a high-pass filter (cut-off frequency 1.2 kHz and slope 3 dB/octave).
- The signal is then sampled (16 kHz sampling frequency, 16 bit quantization). A short-term fast Fourier transform (STFFT) is applied to the samples and the frame length is 128 points (8 ms). There is a frame overlap of 6 ms (75% overlap) and a set of pulses is calculated for each frame. Sixty-four spectral bins are extracted in each frame (amplitude and phase). The step between two bins is 125 Hz.
- The spectral bins are then grouped into frequency bands that are logarithmically distributed, according to ear physiology [27]. Considering the usual values taken in CI, we used 20 bands (leading to 20 channels). The corresponding mapping is shown in Table 1.
- In each band, the energy is calculated using the Parseval's formula (the squares of the amplitude of each beam are added). In the FC coding, all the channels were taken. For the CP coding strategy, only the eight most energetic channels were kept. The value n = 8 is a standard in CIs [28].
- Each channel is represented by a narrowband spectrum coming from a white noise spectrum. The amplitude of the narrowband follows the energy detected in the corresponding channel. The synthesis filters covered the corresponding analysis bands but were 70 Hz narrower (35 Hz less on each side). Moreover, filters used here were 20th order Butterworth bandpass filters to avoid channel interaction. The first two-channels were represented by sine waves.
- The output signal is obtained by summing the selected channels (8 for the CP strategy; 20 for the FC strategy).

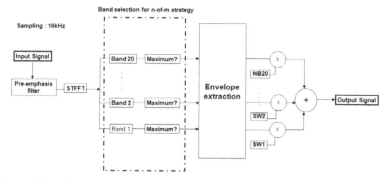

Figure 1. Block diagram representing the signal processing performed by the n-of-m simulator.

Table 1. Centre and cut-off frequencies of the vocoder coding.

Channel	Centre Frequency (Hz)	Analysis		Synthesis		Carrier
		Lower Cut-off	Higher Cut-off	Lower Cut-off	Higher Cut-off	
1	250	190	310	225	275	sine
2	375	315	435	350	400	sine
3	500	440	560	475	525	noise band
4	625	565	685	600	650	noise band
5	750	690	810	725	775	noise band
6	875	815	935	850	900	noise band
7	1000	940	1060	975	1025	noise band
8	1125	1065	1185	1100	1150	noise band
9	1313	1250	1375	1285	1340	noise band
10	1563	1500	1625	1535	1590	noise band
11	1813	1750	1875	1785	1840	noise band
12	2125	2000	2250	2035	2215	noise band
13	2500	2375	2625	2410	2590	noise band
14	2938	2750	3125	2785	3090	noise band
15	3438	3250	3625	3285	3590	noise band
16	4000	3750	4250	3785	4215	noise band
17	4688	4375	5000	4410	4965	noise band
18	5500	5125	5875	5160	5840	noise band
19	6438	6000	6875	6035	6840	noise band
20	7438	7000	7875	7035	7840	noise band

2.6. Mathematical Analysis of the Data

2.6.1. Comparison of the Percentages

The score for each test was the number of correctly repeated syllables (20 syllables per condition) expressed as a percentage.

In the case of NH subjects, we used a two-way repeated-measure ANOVA (coding scheme × SNR). For CI users, we used a two-way mixed model ANOVA [coding scheme × SNR] on intelligibility scores. Because the groups were relatively small and the data were not normally distributed, all the post-hoc analyses were performed with non-parametrical tests: Mann–Whitney's test for unpaired data and Wilcoxon's test for paired data.

We also calculated the Cohen's d term as an effect size index for each average score tested [29]. Cohen's d is a quantitative measure of the magnitude of a phenomenon: a large absolute value indicates a strong effect. Cohen's *d* is defined as the difference between two means divided by a standard deviation for the data.

2.6.2. Curve Fitting with a Sigmoid Function

The recognition percentages versus the SNR can be classically represented by a sigmoid curve regression (Figure 2).

Three parameters were considered on this curve:

- the SNR corresponding to 50% of the maximum recognition denoted here by $x_{50\%}$;
- the "slope" (SNR interval, given in dB, between 25 and 75% of the maximum recognition) which is denoted here by $\Delta_{25-75\%}$;
- the top asymptote y_{max} showing the maximum recognition score.

These analytical values are represented on the sigmoid curve. The minimum recognition is 0% (measured for SNR = −3 dB). Thus, the sigmoid equation is

$$y = \frac{a}{1 + e^{-b(x-c)}}$$

where

- y is the recognition percentage,
- x is the SNR,
- a is y_{max},
- c is $x_{50\%}$, and
- b is linked to the slope: $b = 2.2/\Delta_{25-75\%} \Rightarrow \Delta_{25-75\%} = 2.2/b$.

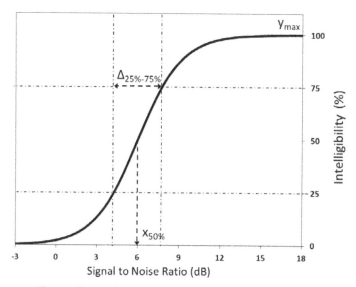

Figure 2. Fitting of the recognition percentages by a sigmoid curve.

2.6.3. Bonferroni Correction

We considered the Bonferroni correction as an indicator but we did not adjust our probability (p) thresholds because of the small number of comparisons and the indicative orientation of this work [30]. The main objective was to look for clues that will need to be further investigated in the future. Streiner et al. [31] "advise against correcting in these circumstances but with the warning that any positive results should be seen as hypothesis generating, not as definitive findings." Consequently, to avoid overcorrection, we used the Holm–Bonferroni method, which adjusts the rejection criteria of each of the individual comparisons. The lowest p-value is evaluated first with a Bonferroni correction involving all tests. The second is evaluated with a Bonferroni correction involving one less test and so on for the other tests. Holm's approach is more powerful than the Bonferroni approach but it still keeps control on the Type 1 error.

3. Results

3.1. Normal-Hearing Subjects

3.1.1. Recognition Percentages

The results of syllable recognition versus the SNR are shown in Figure 3. Significant differences are indicated by an asterisk.

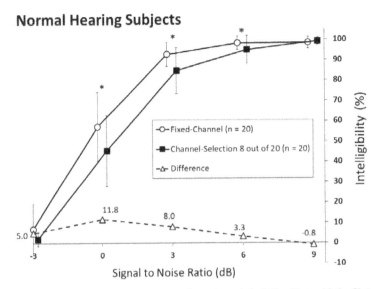

Figure 3. Syllable recognition function of the signal-to-noise ratio by NH subjects with the CI simulator using both strategies. Bars indicate the standard deviation. The asterisks indicate the significant differences (5% threshold).

3.1.2. Statistical Analysis

The ANOVA showed a significant effect of the SNR [$F_{(4,95)} = 519$; $p < 10^{-4}$] and of the coding scheme [$F_{(1,95)} = 16$; $p < 10^{-4}$]; there was no significant interaction between them [$F_{(4,95)} = 1.95$; $p = 0.11$]. Consequently, a post-hoc analysis was performed for the coding scheme.

For each SNR, comparisons were made with paired Wilcoxon's tests (on the 20 subjects who participated in the experiment). In the whole experiment, we had five paired series (one for each SNR); for each paired series we had 20 pairs of values (one per subject). For the extreme SNR values (−3 dB and +9 dB), the recognition percentages were not significantly different between FC and CP (Table 2). *P*-values were below 5% for the SNRs 0 dB, +3 dB and +6 dB.

Using the Holm–Bonferroni correction, the first corrected decision threshold was 1% and differences become not significant since the lowest *p*-value was 0.019. For SNRs +3 and 0 dB, differences were close to significance and worth further investigation; additionally, the Cohen's effect sizes were respectively strong (0.89) and medium (0.68). This coheres with the general ANOVA results.

Table 2. Percentage comparisons for normal-hearing subjects between the simulation strategies.

Normal-Hearing Subjects		Fixed-Channel (n = 20)	Channel-Picking (n = 20)	*p* (Wilcoxon)	Effect Size (Cohen's d)
SNR +9 dB	m	98.50	99.25	0.374	0.31 (small)
	σ	2.86	1.83		
SNR +6 dB	m	98.25	95.00	0.046	0.60 (medium)
	σ	3.35	6.88		
SNR +3 dB	m	92.75	84.75	0.019	0.89 (strong)
	σ	5.95	11.29		
SNR 0 dB	m	57.25	45.50	0.020	0.68 (medium)
	σ	16.97	17.54		
SNR −3 dB	m	7.00	2.00	0.065	0.53 (medium)
	σ	12.50	3.40		

3.1.3. Sigmoid Parameters

The comparison of the sigmoid parameters (Table 3) showed that the $x_{50\%}$ values were different between FC and CP ($p = 0.038$). No differences were found for the slope and the plateau. Considering the Holm–Bonferroni correction, the first adjusted decision threshold was 1.7%. The effect size for $x_{50\%}$ was strong (0.85).

Table 3. Comparison of the sigmoid parameters, for normal-hearing listeners.

Normal-Hearing Subjects		Fixed-Channel (n = 20)	Channel-Picking (n = 20)	p (Wilcoxon)	Effect Size (Cohen's d)
$x_{50\%}$	m	−0.28	0.39	0.038	0.85 (strong)
	σ	0.85	0.73		
$\Delta_{25-75\%}$	m	2.19	2.54	0.287	0.27 (small)
	σ	1.38	1.23		
y_{max}	m	98.5	99.25	0.374	0.31 (small)
	σ	2.86	1.83		

3.2. Cochlear Implant Users

3.2.1. Recognition Percentages

CI users with FC stimulations and with CP have been gathered into two groups (FC and CP); percentages are shown in Figure 4.

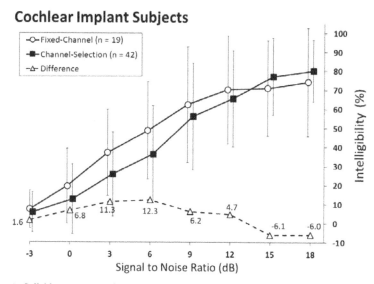

Figure 4. Syllable recognition function of the signal-to-noise ratio by CI users. Bars indicate the standard deviation. The asterisks indicate the significant differences (5% threshold).

3.2.2. Statistical Analysis

The ANOVA indicated a significant effect of SNR [F (1,301) = 146; $p < 10^{-4}$] but not for the coding scheme [F (1,43) = 0.66; $p = 0.42$]. A significant interaction was seen between them [F (1,301) = 2.23; $p = 0.032$], which may need further investigation.

The recognition percentages are indicated in Figure 4 and Table 4. We can see that the plateau was not reached for high SNRs and an inversion of the performances may be noticed between CP and FC at +15 dB.

Table 4. Percentage comparisons for cochlear implant users between the coding strategies.

Cochlear-Implant Users		Fixed-Channel (n = 19)	Channel-Picking (n = 26)	Effect Size (Cohen's d)
SNR +18 dB	m	74.21	80.19	0.28 (small)
	σ	28.44	16.09	
SNR +15 dB	m	71.05	77.12	0.27 (small)
	σ	25.14	20.16	
SNR +12 dB	m	70.26	65.58	0.17 (very small)
	σ	28.31	25.39	
SNR +9 dB	m	62.37	56.15	0.21 (small)
	σ	30.38	28.01	
SNR +6 dB	m	48.68	36.35	0.48 (small)
	σ	25.43	25.56	
SNR +3 dB	m	37.11	25.77	0.50 (medium)
	σ	22.69	22.08	
SNR 0 dB	m	19.47	12.69	0.35 (small)
	σ	19.71	18.4	
SNR −3 dB	m	7.37	5.77	0.15 (very small)
	σ	9.91	10.46	

3.2.3. Sigmoid Parameters

Gathering the four implants according to their coding schemes (FC and CP), Mann–Whitney's tests indicated a significant difference only for $x_{50\%}$ ($p = 0.042$) (Table 5). After considering the Holm–Bonferroni correction, this difference needs to be discussed. The effect size was medium (0.73).

Table 5. Comparison of the analytical values, for cochlear implant users.

Cochlear-Implant Users		Fixed-Channel (n = 19)	Channel-Picking (n = 26)	p (Mann–Whitney)	Effect Size (Cohen's d)
$x_{50\%}$	m	3.95	6.17	0.042	0.73 (medium)
	σ	2.43	3.20		
$\Delta_{25-75\%}$	m	7.52	6.09	0.189	0.40 (small)
	σ	3.61	3.48		
y_{max}	m	89.78	86.24	0.460	0.28 (small)
	σ	10.66	14.80		

We also looked for a possible link between $x_{50\%}$ and y_{max} (Figure 5). The scatter plot indicates that all the situations can be observed with every implant. No correlation was seen for any manufacturer ($p_{Med-El} = 0.62$, $p_{Advanced\ Bionics} = 0.47$, $p_{Cochlear} = 0.055$, $p_{Neurelec} = 0.55$).

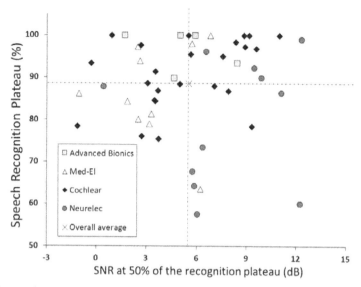

Figure 5. Speech recognition plateau versus the $x_{50\%}$ parameter for cochlear implant users.

4. Discussion

Several items have been considered in this study: the coding scheme, the influence of noise and the simulation of CI coding in NH subjects.

4.1. On the Coding Strategy

The choice of a coding strategy is a delicate matter and some studies have shown that CI users have a subjective preference for a particular strategy that is not always the one that yields the best performances [32].

Additionally, many technical parameters concerning the coding scheme, such as stimulation rate, gain control, update rate and filter settings, influence the final results and have an effect on the performances related to the coding strategy [28,33].

It is interesting to note that the four manufacturers have taken different stimulation strategies and the results are dispersed; it is therefore difficult to draw definitive conclusions. For any manufacturer, all coding strategies can be implemented within the processor.

From our results with NH listeners, the Bonferroni correction lowered the significance limit but the Cohen analysis indicated that the differences were reliable. The FC strategy presented better recognition percentages than CP, particularly in the SNR range 0 to +6 dB with effect sizes medium (0 and +6 dB) and strong (+3 dB). Moreover, the comparison between the $x_{50\%}$ highlighted a strong effect size in favour of the FC strategy.

Of course, our results only stand for a CP strategy with an extraction of 8 channels out of 20 and an FC strategy with 20 channels. However, this seems to be an interesting hint, as the conditions and the subjects were the same for both strategies (CP or FC), used the same random approach, had the same SNRs and had the same signal processing (window length, sampling rate, channel band-pass, quantization, etc.), the subjects were of the same type (range of age, education, etc.) and we were able to use paired comparisons. However, this simulates an "ideal case" where there is no channel interaction and no pitch-shift due to the insertion depth of the electrode array and where all the channels are functional. This is why these results should not be taken on their own without taking into account the experiment conducted with the CI subjects.

With the CI subjects, considering the non-saturation of the percentages, we raised the SNR up to +18 dB and FC led to, albeit not significant, higher scores than CP in the 0–12 dB range. A small inversion of the results was also observed above +12 dB SNR, which can be linked to the significant interaction between SNR and coding scheme. Nevertheless, results are to be taken with caution, considering the wide dispersion of the results and the comparison of relatively small and unbalanced groups. Because of the difference in the number of included patients in each group, we used nonparametric tests to compare them; they are well adapted for this kind of comparisons. While they are generally less powerful than parametric tests, they are more robust. However, like in the experiment with NH subjects, the comparison between the $x_{50\%}$ showed an interesting difference in favour of the FC strategies. With a medium effect size, this result would be worth investigating further in another study.

Our results are consistent with studies that showed a superiority of the FC strategy over the CP strategy, particularly in noise [5]. This was less true when CP strategies, such as ACE, with a high stimulation rate were introduced [6,7]. No clear advantage for a particular coding scheme can be identified taking all the literature on the subject.

In many studies, when the FC strategy was used, the stimulation rate was an important factor as the possibility to follow the quick changes in the signal helps the recognition performances mostly for consonants [34,35].

4.2. Cochlear Implant Users and Normal-Hearing Subjects

Despite the fact that the groups of CI users were heterogeneous, the general recognition behaviour was the same for CI users and NH subjects, whatever implant was used. With NH listeners, for a SNR of +9 dB, the 100% recognition level was reached.

With CI users, the plateau was not always reached with a SNR of +18 dB; additionally, it was below the 100% measured with NH subjects. For a +9 dB SNR (maximum tested in simulation), the CI users' performances were below the scores observed with NH subjects; the mean scores with CI users ranged from 50 to 75%; this is consistent with previous studies [36].

The same results were also seen with the $x_{50\%}$ (sigmoid fitting), which was better in NH subjects than with CI users.

With the CI users, an inversion occurred between +12 and +15 dB SNR and it was also seen for +18 dB SNR; performances observed with a CP strategy were higher than the performances with an FC strategy.

The reliability of the data obtained from CI users is a real issue. Is there a link between the plateau and the $x_{50\%}$? The scatter point diagram of the four CI user populations is shown in Figure 5. It shows that, for each manufacturer, all possibilities exist, either with a good plateau and a poor $x_{50\%}$ or vice versa. All intermediate situations were found and the correlation coefficients were not significantly different from zero.

As the very goal is to provide an opportunity for every CI user to hear in everyday life [28], the work ahead is important. The efficiency of a CI is affected by many factors such as the recognition and linguistic skills, the degrees of nerve survival and the technical choices that are made when fitting the device and the variations are wide with every subject.

4.3. Listening in Noise

Listening in noise is a clear challenge, which is not handled in the same way by CI users and NH people. Noise flattens the spectrum and the subsequent structures in the auditory system do not react identically [37]. The study of speech recognition in noise has become of great interest as it is present in daily listening conditions. Additionally, we can see the coding behaviour for different SNRs (floor and ceiling effect and intermediate situation).

In this study, the CI user group was older on average than the NH group. In general, older people have lower speech perception scores in noise, even with normal or age-related hearing, compared to young people. However, the purpose of the study was to test both groups and see if a similar trend (between FC and CP coding schemes) could emerge and not to compare CI users with NH subjects.

Another finding was the effect of noise on performances for each strategy, which makes this study interesting in that each manufacturer can set any coding scheme on their devices. Consequently, it is worthwhile to evaluate the results through different approaches. Our work may suggest that the strategy is noise-dependent.

The number of channels needed to understand speech in noise or in quiet is an important issue. Studies have indicated that, when speech is processed in the same manner as in CIs and presented in quiet to NH listeners, sentence recognition scores are higher than 90% with as few as 4 channels [38,39]. In the literature, results show that more channels are needed to understand speech in noise than in quiet [10] but selecting more than 12 channels may not yield significant improvements on the recognition performances [21]. These considerations orientated the choice of our parameters.

In noise, performances of CI users reach a plateau as the number of channel increases; for NH subjects performances continue to increase (up to 100%), suggesting that, CI subjects could not fully utilize the spectral information provided by the number of electrodes, possibly because of the channel interaction [38]. As indicated above, trends are similar for NH and CI listeners but results are not interchangeable. It is sensible to say that more channels imply more information but this also implies more overlap between the electrodes. This conflict needs to be studied in the future; we can simulate channel interaction with NH subjects.

The acoustical material (in our case the Fournier's lists and the cocktail party noise) seemed to be well adapted to the situation.

5. Conclusions

A simulation study of NH listeners measured syllable recognition in a noisy environment, using both Fixed-Channel and Channel-Picking coding schemes. The results were also compared with CI users' performances. CI users were divided into two groups corresponding to the coding schemes available. Twenty NH subjects and 45 CI users participated in this experiment. The acoustic material was the Fournier French dissyllabic lists mixed with a cocktail-party noise.

The results obtained in the simulation with the NH subjects indicated an advantage of the fixed-channel strategy over the channel-picking coding in a middle SNR range (from 0 to +6 dB); parameters (patients, technology and protocol) were well controlled in this approach. This trend was confirmed using the sigmoid curve regression. The results seemed similar with the CI users.

Nevertheless, results were less reliable with CI users, probably due to the wide dispersion in the patients' results. Additionally, an inversion of the coding strategy was seen with high SNRs, with CI users. This aspect should be examined in the future, considering its practical application and we need to consider the physiological and electrical phenomena involved in a multichannel stimulation such as channel interaction. Simulation and tests with CI users are useful as they give two complementary insights into the difficult task of determining an "optimal" sound coding strategy to enhance the auditory performance of CI users.

Author Contributions: Conceptualization, C.B.-V., E.T. and S.G.; formal analysis, P.A.C. and S.G.; investigation, P.A.C.; methodology, P.A.C., C.B.-V. and S.G.; resources, E.T.; software, P.A.C. and F.M.; supervision, C.B.-V, E.T. and S.G.; visualization, P.A.C. and C.B.-V; writing—original draft preparation, P.A.C. and C.B.-V; writing—review and editing, P.A.C., C.B.-V, F.M., R.H., E.T. and S.G.

Funding: This research received no external funding.

Acknowledgments: The authors would like to thank the people and institutions who participated in this study: Kevin Perreault who initiated the work, Charles-Alexandre Joly and Fabien Seldran for their scientific contribution, Evelyne Veuillet for contacts with the ethic-committee, the members of the CRIC team of the Edouard Herriot University hospital of Lyon for their collaboration, the normal-hearing subjects and the cochlear implant users who entered the study and the Hospitals of Lyon and the Polytechnic School of Lyon for their administrative support.

Conflicts of Interest: The authors declare no conflict of interest.

Appl. Sci. **2019**, *9*, 734

Appendix A

Table A1. Demographic details for cochlear implant users.

Characteristic	N
Gender	
Male	23
Female	22
Ear	
Right	32
Left	13
Origin of deafness	
Congenital	17
Acquired	18
Unknown	10
Age in years at implantation	
1–5 years	9
6–10 years	3
11–20 years	6
>20 years	27
Duration in years of implant use	
1–5 years	14
6–10 years	14
11–15 years	7
16–20 years	9
>20 years	1
Duration of deafness in years	
1–10 years	4
11–20 years	18
21–30 years	4
31–40 years	8
>40 years	5
Unknown	7
Cochlear implant	
Cochlear	13
Med-El	12
Advanced Bionics	7
Neurelec/Oticon Medical	13
Coding strategy	
Channel-picking (SPEAK, ACE …)	26
Fixed-channel (FS4, HiRes …)	19

References

1. NIDCD. Available online: https://www.nidcd.nih.gov/ (accessed on 14 June 2017).
2. Fetterman, B.L.; Domico, E.H. Speech recognition in background noise of cochlear implant patients. *Otolaryngol. Head Neck Surg.* **2002**, *126*, 257–263. [CrossRef] [PubMed]
3. Clark, G. *Cochlear Implants: Fundamentals and Applications*; Springer Science & Business Media: New York, NY, USA, 2006; ISBN 978-0-387-21550-1.
4. Blamey, P.; Artieres, F.; Başkent, D.; Bergeron, F.; Beynon, A.; Burke, E.; Dillier, N.; Dowell, R.; Fraysse, B.; Gallégo, S.; et al. Factors affecting auditory performance of postlinguistically deaf adults using cochlear implants: An update with 2251 patients. *Audiol. Neurotol.* **2013**, *18*, 36–47. [CrossRef] [PubMed]
5. Kiefer, J.; Müller, J.; Pfennigdorff, T.; Schön, F.; Helms, J.; von Ilberg, C.; Baumgartner, W.; Gstöttner, W.; Ehrenberger, K.; Arnold, W.; et al. Speech understanding in quiet and in noise with the CIS speech coding strategy (MED-EL Combi-40) compared to the multipeak and spectral peak strategies (nucleus). *ORL J. Otorhinolaryngol. Relat. Spec.* **1996**, *58*, 127–135. [CrossRef] [PubMed]

6.   Kiefer, J.; Hohl, S.; Stürzebecher, E.; Pfennigdorff, T.; Gstöettner, W. Comparison of speech recognition with different speech coding strategies (SPEAK, CIS and ACE) and their relationship to telemetric measures of compound action potentials in the nucleus CI 24M cochlear implant system. *Audiology* **2001**, *40*, 32–42. [CrossRef] [PubMed]

7.   Skinner, M.W.; Holden, L.K.; Whitford, L.A.; Plant, K.L.; Psarros, C.; Holden, T.A. Speech recognition with the nucleus 24 SPEAK, ACE and CIS speech coding strategies in newly implanted adults. *Ear Hear* **2002**, *23*, 207–223. [CrossRef] [PubMed]

8.   Brockmeier, S.J.; Grasmeder, M.; Passow, S.; Mawmann, D.; Vischer, M.; Jappel, A.; Baumgartner, W.; Stark, T.; Müller, J.; Brill, S.; et al. Comparison of musical activities of cochlear implant users with different speech-coding strategies. *Ear Hear* **2007**, *28*, 49S–51S. [CrossRef] [PubMed]

9.   Dorman, M.F.; Loizou, P.C.; Spahr, A.J.; Maloff, E. A comparison of the speech understanding provided by acoustic models of fixed-channel and channel-picking signal processors for cochlear implants. *J. Speech Lang. Hear. Res.* **2002**, *45*, 783–788. [CrossRef]

10.   Shannon, R.V.; Fu, Q.-J.; Galvin, J. The number of spectral channels required for speech recognition depends on the difficulty of the listening situation. *Acta Otolaryngol. Suppl.* **2004**, *124*, 50–54. [CrossRef]

11.   Verschuur, C.A. Effect of stimulation rate on speech perception in adult users of the Med-El CIS speech processing strategy. *Int. J. Audiol.* **2005**, *44*, 58–63. [CrossRef]

12.   Wilson, B.S.; Finley, C.C.; Lawson, D.T.; Wolford, R.D.; Eddington, D.K.; Rabinowitz, W.M. Better speech recognition with cochlear implants. *Nature* **1991**, *352*, 236–238. [CrossRef]

13.   Qazi, O.U.R.; van Dijk, B.; Moonen, M.; Wouters, J. Understanding the effect of noise on electrical stimulation sequences in cochlear implants and its impact on speech intelligibility. *Hear. Res.* **2013**, *299*, 79–87. [CrossRef] [PubMed]

14.   Garnham, C.; O'Driscoll, M.; Ramsden, R.; Saeed, S. Speech understanding in noise with a Med-El COMBI 40+ cochlear implant using reduced channel sets. *Ear Hear* **2002**, *23*, 540–552. [CrossRef] [PubMed]

15.   Hu, Y.; Loizou, P.C. A new sound coding strategy for suppressing noise in cochlear implants. *J. Acoust. Soc. Am.* **2008**, *124*, 498–509. [CrossRef] [PubMed]

16.   Jeanvoine, A.; Gnansia, D.; Truy, E.; Berger-Vachon, C. Contribution of Noise Reduction Algorithms: Perception Versus Localization Simulation in the Case of Binaural Cochlear Implant (BCI) Coding. In *Emerging Trends in Computational Biology, Bioinformatics and System Biology*; Elsevier Inc.: Amsterdam, The Netherlands, 2015; pp. 307–324.

17.   Wang, Q.; Liang, R.; Rahardja, S.; Zhao, L.; Zou, C.; Zhao, L. Piecewise-Linear Frequency Shifting Algorithm for Frequency Resolution Enhancement in Digital Hearing Aids. *Appl. Sci.* **2017**, *7*, 335. [CrossRef]

18.   Kallel, F.; Frikha, M.; Ghorbel, M.; Hamida, A.B.; Berger-Vachon, C. Dual-channel spectral subtraction algorithms based speech enhancement dedicated to a bilateral cochlear implant. *Appl. Acoust.* **2012**, *73*, 12–20. [CrossRef]

19.   Seldran, F.; Gallego, S.; Thai-Van, H.; Berger-Vachon, C. Influence of coding strategies in electric-acoustic hearing: A simulation dedicated to EAS cochlear implant, in the presence of noise. *Appl. Acoust.* **2014**, *76*, 300–309. [CrossRef]

20.   Dorman, M.F.; Loizou, P.C.; Fitzke, J.; Tu, Z. Recognition of monosyllabic words by cochlear implant patients and by normal-hearing subjects listening to words processed through cochlear implant signal processing strategies. *Ann. Otol. Rhinol. Laryngol. Suppl.* **2000**, *185*, 64–66. [CrossRef]

21.   Loizou, P.C.; Dorman, M.F.; Tu, Z.; Fitzke, J. Recognition of sentences in noise by normal-hearing listeners using simulations of speak-type cochlear implant signal processors. *Ann. Otol. Rhinol. Laryngol. Suppl.* **2000**, *185*, 67–68. [CrossRef]

22.   Dorman, M.F.; Loizou, P.C.; Fitzke, J. The identification of speech in noise by cochlear implant patients and normal-hearing listeners using 6-channel signal processors. *Ear Hear* **1998**, *19*, 481–484. [CrossRef]

23.   Winn, M.B.; Rhone, A.E.; Chatterjee, M.; Idsardi, W.J. The use of auditory and visual context in speech perception by listeners with normal hearing and listeners with cochlear implants. *Front. Psychol.* **2013**, *4*, 824. [CrossRef]

24.   Perreaut, K.; Gallego, S.; Berger-Vachon, C.; Millioz, F. Influence of Microphone Encrusting on the Efficiency of Cochlear Implants Preliminary Study with a Simulation of CIS and "n-of-m" Strategies. *AMSE J. Ser. Model. C* **2014**, *75*, 199–208.

25. Hornickel, J.; Skoe, E.; Kraus, N. Subcortical Laterality of Speech Encoding. *Audiol. Neurotol.* **2009**, *14*, 198–207. [CrossRef]

26. Zatorre, R.J.; Belin, P.; Penhune, V.B. Structure and function of auditory cortex: Music and speech. *Trends Cogn. Sci. (Regul. Ed.)* **2002**, *6*, 37–46. [CrossRef]

27. Traunmüller, H. Analytical expressions for the tonotopic sensory scale. *J. Acoust. Soc. Am.* **1990**, *88*, 97–100.

28. Wouters, J.; McDermott, H.J.; Francart, T. Sound Coding in Cochlear Implants: From electric pulses to hearing. *IEEE Signal Process. Mag.* **2015**, *32*, 67–80. [CrossRef]

29. Sullivan, G.M.; Feinn, R. Using Effect Size—Or Why the P Value Is Not Enough. *J. Grad. Med. Educ.* **2012**, *4*, 279–282. [CrossRef]

30. Armstrong, R.A. When to use the Bonferroni correction. *Ophthalmic Physiol. Opt.* **2014**, *34*, 502–508. [CrossRef]

31. Streiner, D.L.; Norman, G.R. Correction for multiple testing: Is there a resolution? *Chest* **2011**, *140*, 16–18. [CrossRef]

32. Skinner, M.W.; Arndt, P.L.; Staller, S.J. Nucleus 24 advanced encoder conversion study: Performance versus preference. *Ear Hear* **2002**, *23*, 2S–17S. [CrossRef]

33. Kallel, F.; Laboissiere, R.; Ben Hamida, A.; Berger-Vachon, C. Influence of a shift in frequency distribution and analysis rate on phoneme intelligibility in noisy environments for simulated bilateral cochlear implants. *Appl. Acoust.* **2013**, *74*, 10–17. [CrossRef]

34. Riss, D.; Hamzavi, J.-S.; Blineder, M.; Flak, S.; Baumgartner, W.-D.; Kaider, A.; Arnoldner, C. Effects of stimulation rate with the fs4 and hdcis coding strategies in cochlear implant recipients. *Otol. Neurotol.* **2016**, *37*, 882–888. [CrossRef] [PubMed]

35. Wilson, B.S.; Sun, X.; Schatzer, R.; Wolford, R.D. Representation of fine structure or fine frequency information with cochlear implants. *Int. Congr. Ser.* **2004**, *1273*, 3–6. [CrossRef]

36. Dorman, M.F.; Loizou, P.C. The identification of consonants and vowels by cochlear implant patients using a 6-channel continuous interleaved sampling processor and by normal-hearing subjects using simulations of processors with two to nine channels. *Ear Hear* **1998**, *19*, 162–166. [CrossRef] [PubMed]

37. Aguiar, D.E.; Taylor, N.E.; Li, J.; Gazanfari, D.K.; Talavage, T.M.; Laflen, J.B.; Neuberger, H.; Svirsky, M.A. Information theoretic evaluation of a noiseband-based cochlear implant simulator. *Hear. Res.* **2015**, *333*, 185–193. [CrossRef] [PubMed]

38. Friesen, L.M.; Shannon, R.V.; Baskent, D.; Wang, X. Speech recognition in noise as a function of the number of spectral channels: Comparison of acoustic hearing and cochlear implants. *J. Acoust. Soc. Am.* **2001**, *110*, 1150–1163. [CrossRef] [PubMed]

39. Loizou, P.C.; Dorman, M.; Tu, Z. On the number of channels needed to understand speech. *J. Acoust. Soc. Am.* **1999**, *106*, 2097–2103. [CrossRef] [PubMed]

Article

In-Depth Exploration of Signal Self-Cancellation Phenomenon to Achieve DOA Estimation of Underwater Acoustic Sources

Fang Wang [1,*], Yong Chen [1] and Jianwei Wan [2]

[1] College of Physics and Communication Electronics, Jiangxi Normal University, Nanchang 330022, China; yongchen@jxnu.edu.cn

[2] College of Electronic Science, National University of Defense Technology, Changsha 410073, China; kermitwjw@139.com

* Correspondence: fangwang@jxnu.edu.cn

Received: 9 December 2018; Accepted: 6 February 2019; Published: 8 February 2019

Abstract: In the ocean environment, the minimum variance distortionless response beamformer usually has the problem of signal self-cancellation, that is, the acoustic signal of interest is erroneously suppressed as interference. By exploring the useful information behind the signal self-cancellation phenomenon, a high-precision direction estimation method for underwater acoustic sources is proposed. First, a pseudo spatial power spectrum is obtained by performing unit circle mapping on the beam response in the direction interval. Second, the online calculation process is given for reducing the computational complexity. The computer simulation results show that the proposed algorithm can obtain satisfactory direction estimation accuracy under the conditions of low intensity of acoustic source, strong interference and noise, and less array snapshot data.

Keywords: minimum variance distortionless response; signal self-cancellation; direction estimation; underwater acoustic source; spatial power spectrum

---

1. Introduction

Underwater acoustic source localization determines the altitude or depth, range, and bearing angle of the underwater target, that is, the three coordinates of the underwater target in the elliptical coordinate system [1]. The estimation of bearing angle (or direction of arrival (DOA)) of underwater acoustic source is an important and indispensable step in underwater acoustic source localization. In fact, in some underwater acoustic source localization methods, it is the target direction information that is used to estimate the target distance [2]. Using a vector hydrophone (or vector hydrophone array) is perhaps one of the simplest and most straightforward methods of underwater acoustic source DOA estimation. The vector hydrophone is capable of simultaneously measuring sound pressure and particle velocity along one to three orthogonal directions. Therefore, only a single vector hydrophone can generate a directional beam pattern. A Directional Autonomous Seafloor Acoustic Recorder (DASAR) system consisting of several vector hydrophones has been reported [3]. By using a vector hydrophone, directional industrial noise is effectively suppressed, and weak marine mammal sounds can be successfully detected.

Conventional beamforming (or delay-and-sum beamforming) is also one of the common DOA estimation methods [4–6]. Based on the ray-path approximation for the sound channel's impulse response, the frequency-difference beamforming method for the sparse hydrophone array is proposed in [7,8], which can estimate the signal phase difference by using the conventional delay-and-sum beamforming of the field product at the difference frequency. In contrast, by determining the array weight coefficients in a nonlinear manner, the minimum variance distortionless response

(MVDR) method can achieve higher angular resolution than conventional beamforming [4]. In order to cope with the model mismatch problems in the uncertain ocean environment, such as the signal look direction mismatch, the signal spatial signature mismatch due to local scattering or wavefront distortion [9,10], many robust MVDR-based adaptive beamforming algorithms have been proposed [10–12].

Similar to the MVDR-based methods, the subspace-based high-resolution DOA estimation techniques also use information carried by the covariance matrix. The most representative subspace-based DOA estimation methods may be the Multiple Signal Classification (MUSIC) algorithm [13], the Estimation of Signal Parameters via Rotational Invariance Technique (ESPRIT) [14], and the Propagator Method (PM) [15]. The key to the subspace-based DOA estimation methods is the estimation of the signal subspace (or noise subspace). To achieve this purpose, one can first perform eigendecomposition on the sample covariance matrix, then construct the signal subspace with the eigenvectors corresponding to the larger eigenvalues, and form the noise subspace with the eigenvectors corresponding to the smaller eigenvalues. Another more sophisticated approach is to reconstruct the sample covariance matrix according to the Toeplitz structure of the covariance matrix [16], and then obtain the signal subspace in a similar way as above. In contrast to the above eigenvalue-based methods, the eigenvector pruning algorithm implements the estimation of the signal subspace by using the statistical properties of eigenvectors of signal-free sample covariance matrix [17].

In many underwater acoustic source DOA estimation scenarios, the number of acoustic sources distributed in the underwater far field is much smaller than the number of hydrophones in the observation array. This is the intrinsic basis of the popular sparse-based DOA estimation methods. In the least absolute shrinkage and selection operator (LASSO) method [18], the signal amplitude vector is obtained by solving an l_1-norm regularized least-squares problem. The LASSO method contains the l_1-norm constraint on the solution vector, thus making the result of the solution vector sparse [19]. By linear transformation of the solution vector, the weighted LASSO method [20] imposes certain structural constraint on the solution vector to achieve efficient processing of spatially extended sources (e.g., underwater embedded objects in acoustic imaging [21]). The total variation norm regularization method [22] for DOA estimation of spatially extended sources can be seen as a special case of the weighted LASSO method, which uses the band matrix to realize the linear transformation of the solution vector, so that the solution vector has block sparsity [19,23]. Besides, using the information contained in the covariance matrix, the sparse spectrum fitting (SpSF) algorithm [24] first performs a vectorization operation on the covariance matrix, and then fits the estimated covariance matrix and the ideal covariance matrix under the l_2-norm. At the same time, considering the sparsity of the source, l_1-norm penalization is imposed on the source strength vector. However, SpSF algorithm is based on the assumption that ambient noise is white Gaussian noise. Therefore, Yang L., Yang Y. X., and Wang Y. proposed the directional noise field sparse spectrum fitting (DN-SpSF) algorithm [25], which uses the slowly varying characteristics of the noise spectral density function to derive the general expression of the covariance matrix of underwater directional ambient noise, and takes an optimization process similar to the SpSF algorithm.

The naturally occurring ambient noise in the ocean is generally considered to be a nuisance [1,26]. Therefore, one of the purposes of the sonar signal processing algorithms is to distinguish the desired signal from the ambient noise and suppress the ambient noise as much as possible. However, recent studies have shown that ocean ambient noise actually contains a lot of useful information [27,28], and could be used for the underwater imaging [26], the geoacoustic inversion [29,30], and the determination of seabed sub-bottom layer profile [31–33]. A similar situation is that when the presumed steering vector is mismatched with the actual steering vector, MVDR-based beamformers exhibit a so-called signal self-cancellation phenomenon, that is, the signal of interest is erroneously treated as interference, thereby being greatly suppressed. Therefore, signal self-cancellation is commonly regarded as a nuisance, and the existing related algorithms are intended to reduce the effect of signal self-cancellation. To the best of our knowledge, there is currently no research on how to use the information contained in the signal self-cancellation phenomenon. Therefore, this paper does not take signal self-cancellation as a troublesome thing, but explores the potential information behind the signal self-cancellation phenomenon and uses it to achieve high-precision DOA estimation of underwater acoustic sources.

The main contributions of this paper are: (1) Treating the signal self-cancellation problem of the MVDR-based beamformers from a new perspective, that is, although for the MVDR-based beamformers, signal self-cancellation is a nuisance, it also contains favorable information and can be used for DOA estimation of underwater acoustic sources. (2) A novel unit circle mapping method is proposed, which effectively correlates the signal self-cancellation and the beam response curves by uniformly mapping all beam response sample values in the direction interval to a unit circle. (3) DOA estimation performance of the proposed method is analyzed in the underwater acoustic propagation simulation environment, and the performance comparisons with existing DOA methods is also completed.

2. Signal Self-Cancellation of MVDR Beamformer

Assume that the sensor array is a horizontal linear array composed of M omnidirectional hydrophones. In addition, suppose that there are N far-field narrowband underwater acoustic signals impinging on the hydrophone array, and their directions of arrival are equal to $\theta_1, \theta_2, ..., \theta_N$. Let $\mathbf{x}(k)$ be the array snapshot vector at time k, then it can be expressed as

$$\mathbf{x}(k) = [\mathbf{a}(\theta_1), \mathbf{a}(\theta_2), \cdots, \mathbf{a}(\theta_N)] \begin{bmatrix} s_1(k) \\ s_2(k) \\ \cdots \\ s_N(k) \end{bmatrix} + \mathbf{n}(k), \tag{1}$$

where $s_i(k), i = 1, 2, \cdots, N$ represents the amplitude of the ith received underwater acoustic signal at time k, $\mathbf{n}(k)$ denotes the array noise vector at time k, where the data element on the jth row corresponds to the recorded noise of the jth hydrophone, $j = 1, 2, \cdots, M$. Besides, $\mathbf{a}(\theta)$ is the array manifold (or steering vector) towards direction θ, which can be formulated as

$$\mathbf{a}(\theta) = [1, e^{j(2\pi/\lambda)d\sin(\theta)}, \cdots, e^{j(M-1)(2\pi/\lambda)d\sin(\theta)}]^\mathrm{T}, \tag{2}$$

where the superscript T represents the transposition operation, λ denotes the signal wavelength, and d is the distance between adjacent hydrophones.

Beamformer can preserve the received signal impinging from a specific direction while suppressing signals in other directions, that is, with spatial filtering capability. The MVDR beamformer achieves the above objects by solving the convex optimization problem as follows,

$$\min_{\mathbf{w}} \mathbf{w}^H \mathbf{R}_{i+n} \mathbf{w} \quad s.t. \quad \mathbf{w}^H \mathbf{a}(\theta_d) = 1, \tag{3}$$

where θ_d is the specified direction (i.e., the direction of the desired signal), $\mathbf{w}$ denotes the weight vector of MVDR beamformer, and $\mathbf{R}_{i+n}$ represents the interference-plus-noise covariance matrix. In Equation (3), the objective function is equal to the power of the interference and noise passing through the beamformer, and the constraint guarantees that the gain of the signal in the specified direction is 1.

It should be pointed out that in some practical applications, such as passive sonar detection, the array received snapshot data contains both interference, noise, and the desired signal. Therefore, in this case, it is difficult to directly obtain a covariance matrix only for interference and noise. A simple solution is to replace the interference-plus-noise covariance matrix directly with the sample covariance matrix. Therefore, the optimization problem about the MVDR beamformer needs to be re-expressed as

$$\min_{\mathbf{w}} \mathbf{w}^H \hat{\mathbf{R}} \mathbf{w} \quad s.t. \quad \mathbf{w}^H \mathbf{a}(\theta_d) = 1, \tag{4}$$

where $\hat{\mathbf{R}}$ represents the sample covariance matrix, which can be directly calculated from several array snapshot data,

$$\hat{\mathbf{R}} = \frac{1}{K} \sum_{k=1}^{K} \mathbf{x}(k)\mathbf{x}^{H}(k), \tag{5}$$

where K is the number of array snapshots that can be used to calculate the sample covariance matrix. In Equation (4), the objective function is equal to the total power of the desired signal, interference, and noise passing through the beamformer. Please note that when the array snapshot number K is large enough, the sample covariance matrix is approximately equal to the theoretical signal covariance matrix (i.e., $\mathbf{R} = E\{\mathbf{x}(k)\mathbf{x}^{H}(k)\}$). Furthermore, when the specified direction is completely equal to the DOA of the desired signal, it is easy to prove that the optimization problems in Equations (3) and (4) are equivalent.

However, in the practical applications, such as underwater acoustic source localization, even if the number of available array snapshots is sufficient, there are still many unfavorable factors that make the performance of the MVDR beamformer significantly degraded. For example, the specified direction is usually difficult to accurately equal the DOA of the desired signal, which leads to a direction error. In addition, when the acoustic signal propagates a long distance in the inhomogeneous ocean medium, the wavefront of the acoustic wave will no longer be a theoretical plane wave, and a so-called random wavefront fluctuation occurs. Other negative factors include, errors in mounting positions of the hydrophones, errors in the amplitude and phase gain of the hydrophones. Thus, the given steering vector is equal to the sum of the true steering vector and the steering vector error, that is,

$$\bar{\mathbf{a}} = \mathbf{a} + \mathbf{a_e}, \tag{6}$$

where $\bar{\mathbf{a}}$ is the given steering vector with respect to the desired signal, $\mathbf{a}$ represents the actual steering vector for the desired signal, and $\mathbf{a_e}$ denotes the steering vector error caused by the aforementioned unfavorable factors. From Equations (4) and (6), it can be found that when the steering vector error in Equation (6) is not equal to zero, the constraint in Equation (4) will become $\mathbf{w}^{H}\bar{\mathbf{a}} = 1$, which means that the MVDR beamformer will retain a certain signal corresponding to the given steering vector $\bar{\mathbf{a}}$, and Not the desired signal. Even worse, the minimization of the objective function in Equation (4) will result in the power of the desired signal being greatly reduced as it passes through the MVDR beamformer. This phenomenon in which the desired signal is cancelled is often referred to as the signal self-cancellation of the MVDR beamformer.

3. SSC-MVDR Algorithm for DOA Estimation

According to the analysis in the previous section, we already know that in the ideal case, that is, when there is no steering vector error, the steering vector model of the linear array can be expressed by Equation (2). Through further analysis, we can also find that the direction error, which is one of the unfavorable factors in practical applications, only changes the DOA of the desired signal without changing the steering vector model. However, other unfavorable factors, including random wavefront fluctuations, hydrophone position errors, and hydrophone amplitude phase errors, will affect the representation of the steering vector model.

In this paper, it is assumed that the MVDR beamformer has a certain degree of direction error, and the expression of the steering vector model is known. Specifically, the actual steering vector has the following form,

$$\bar{\mathbf{a}}(\theta) = [1, \alpha_1 e^{j((2\pi/\lambda)d\sin(\theta)+\phi_1)}, \cdots, \alpha_{M-1} e^{j((M-1)(2\pi/\lambda)d\sin(\theta)+\phi_{M-1})}]^{\mathrm{T}}, \tag{7}$$

where $\alpha_i, i = 1, 2, \cdots, M-1$ and $\phi_i, i = 1, 2, \cdots, M-1$ are known constants, the former representing the amplitude deviation of the steering vector and the latter representing the phase deviation of the steering vector.

To propose the SSC-MVDR algorithm for DOA estimation, the following two cases are specifically analyzed. First, when the direction error is not equal to zero, the MVDR beamformer appears to cancel the desired signal, that is, the beam response produces sharp nulls for the desired signal; second, when the direction error is exactly equal to zero, the beam response of the MVDR beamformer produces a main lobe of a certain width for the desired signal. Furthermore, we also assume that the approximate interval of the DOA of the desired signal is known (in fact, a coarse estimate of the DOA of the desired signal can be obtained by conventional beamformer. Even in complex uncertain ocean environment, conventional beamformer still exhibits sufficient robustness). If the direction interval of the desired signal is sufficiently narrow, we will find that in the above two cases, the beam response in the direction interval appears as two different shapes. Specifically, in the first case, the beam response in the direction interval is a curve containing a steep null; and in the second case, a relatively flat curve is obtained by the beam response in the direction interval.

Inspired by the above analysis, we first define the direction interval of the desired signal as Θ, and discretely sample the direction interval to get L direction samples, that is, $\vartheta_i \in \Theta, i = 1, 2, \cdots, L$. Then, the beam response of the MVDR beamformer on the above L direction samples is calculated.

$$B(\vartheta_i) = |\frac{\bar{\mathbf{a}}^H(\theta_d)\hat{\mathbf{R}}^{-1}\bar{\mathbf{a}}(\vartheta_i)}{\bar{\mathbf{a}}^H(\theta_d)\hat{\mathbf{R}}^{-1}\bar{\mathbf{a}}(\theta_d)}|, \quad i = 1, 2, \cdots, L, \tag{8}$$

where $B(\vartheta_i)$ is the beam response of the MVDR beamformer on the direction ϑ_i.

Although the shape of the beam response curve is intuitively distinguishable, how to quickly distinguish the shape of the beam response curve in the direction interval by calculation is a major problem faced by this algorithm. In this paper, we present a unit circle mapping method, whose main idea is to uniformly map all beam response sample values in the direction interval to a unit circle,

$$B_m(\vartheta_i) = |\frac{\bar{\mathbf{a}}^H(\theta_d)\hat{\mathbf{R}}^{-1}\bar{\mathbf{a}}(\vartheta_i)}{\bar{\mathbf{a}}^H(\theta_d)\hat{\mathbf{R}}^{-1}\bar{\mathbf{a}}(\theta_d)}|e^{j(2\pi/L)i}, \quad i = 1, 2, \cdots, L, \tag{9}$$

where $B_m(\vartheta_i)$ represents the unit circle mapped value of the beam response on the direction ϑ_i. The essence of the unit circle mapping method is to convert a series of scalars into directional vectors. Specifically, the amplitude of the i-th vector is equal to the i-th beam response sample value, and the phase of the i-th vector is equal to $(2\pi/L)i$. If all beam response sample values are equal to 1, then the converted vectors are exactly on the unit circle, so the above method is called the unit circle mapping method. Next, all the unit circle mapped values are summed. In the summation process, the beam response curves of the two different shapes will correspond to significantly different results. Specifically, for a beam response curve segment containing a steep null, the magnitude of the summation result is related to the depth of the null, and the deeper the null, the greater the magnitude of the summation result. The phase of the summation result differs from the phase of the null on the unit circle by approximately 180 degrees. For the relatively flat beam response curve segment, the unit circle mapped values have almost cancelled each other during the summation process, so that the summation result is approximately equal to zero. Therefore, we define the pseudo spatial power spectrum as follows,

$$P_{\text{SSC-MVDR}}(\theta) = 1/|\sum_{i=1}^{L}|\frac{\bar{\mathbf{a}}^H(\theta)\hat{\mathbf{R}}^{-1}\bar{\mathbf{a}}(\vartheta_i)}{\bar{\mathbf{a}}^H(\theta)\hat{\mathbf{R}}^{-1}\bar{\mathbf{a}}(\theta)}|e^{j(2\pi/L)i}|, \quad \theta \in \Theta. \tag{10}$$

It should be noted that the pseudo spatial power spectrum $P_{\text{SSC-MVDR}}(\theta)$ is only defined in the direction interval Θ. When the direction θ is exactly equal to the DOA of the desired signal, the pseudo spatial power spectrum is expected to achieve a maximum. In practical applications, to achieve the desired estimation performance of the SSC-MVDR algorithm, a reasonable direction interval should be selected. Generally, the center of the direction interval can be set as the coarse estimate of the DOA of the acoustic source, and the width of the direction interval should be set to a suitable value to ensure that the true DOA of the acoustic source always falls within the direction interval, and meanwhile, the beam response segment in the direction interval contains only the null generated by the signal self-cancellation, and does not contain other unrelated nulls. Therefore, the direction interval width is not only related to the error of the coarse estimation of the DOA, but also to the specific position of the nulls in the beam response.

The implementation principle and detailed steps of the SSC-MVDR algorithm are shown in Figure 1. First, the sample covariance matrix is calculated according to the array snapshots, then the beam response of the MVDR beamformer is calculated, and the beam response on the direction interval is mapped to the unit circle, and finally the pseudo spatial power spectrum is calculated. Meanwhile, in Figure 1, examples of the beam response of the MVDR beamformer and examples of the unit circle mapping of the beam response curve segment on the direction interval are also given. Specifically, Figure 1a,b are the results when there is no signal self-cancellation. In the Figure 1a, the direction interval is indicated by two dashed lines, and the beam response curve segment in this direction interval is relatively flat. Therefore, in the Figure 1b, all the mapping vector amplitudes are approximately equal. Moreover, since the phases of the mapping vectors are uniformly distributed in the range of 0 to 360 degrees, the magnitude of the sum of the mapping vectors will be very small. Therefore, the pseudo spatial power spectrum for this case will be very large, as shown by the higher red dashed line in Figure 1e. Figure 1c,d correspond to the case where there is signal self-cancellation. In the Figure 1c, the direction interval is also indicated by two dashed lines, and the beam response curve segment in this direction interval contains a steep null. Therefore, in the Figure 1d, the magnitudes of some mapping vectors are much smaller than the amplitudes of other mapping vectors. Similarly, the phases of the mapping vectors are uniformly distributed in the range of 0 to 360 degrees, thus the magnitude of the sum of the mapping vectors will be large, and the pseudo spatial power spectrum for this case will be very small, as shown by the lower red dashed line in Figure 1e.

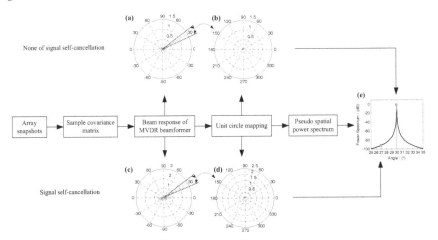

Figure 1. Implementation principle and detailed steps of the SSC-MVDR algorithm. (**a**) example of the beam response of the MVDR beamformer when there is no signal self-cancellation. (**b**) example of the unit circle mapping of the beam response curve segment on the direction interval when there is no signal self-cancellation. (**c**) example of the beam response of the MVDR beamformer when there is signal self-cancellation. (**d**) example of the unit circle mapping of the beam response curve segment on the direction interval when there is signal self-cancellation. (**e**) example of the pseudo spatial power spectrum.

4. Online Computation of SSC-MVDR Algorithm

The computational complexity of an algorithm is one of the important factors we need to consider when applying the algorithm to the actual project. In the SSC-MVDR algorithm (i.e., Equation (10)), the inverse of the sample covariance matrix and the corresponding matrix operations are the main calculation steps. In the following, the online computation process of the SSC-MVDR algorithm is given and a way to reduce the amount of calculation is provided.

First, the iterative calculation process of the sample covariance matrix is as follows

$$\hat{\mathbf{R}}(k+1) = \gamma \hat{\mathbf{R}}(k) + \frac{1}{k+1} \mathbf{x}(k+1) \mathbf{x}^H(k+1), \tag{11}$$

where $\hat{\mathbf{R}}(k+1)$ and $\hat{\mathbf{R}}(k)$ represent the sample covariance matrices at times k and $k+1$, respectively. γ is a constant less than 1 but very close to 1. When the array snapshots are non-stationary, the coefficient γ is used to ensure that the SSC-MVDR algorithm still works reliably. Using the Woodbury matrix identity, the inverse of the sample covariance matrix at time $k+1$ can be expressed as,

$$\hat{\mathbf{R}}^{-1}(k+1) = \gamma^{-1}(\hat{\mathbf{R}}^{-1}(k) - \frac{\mathbf{y}(k+1)\mathbf{y}^H(k+1)}{(k+1)\gamma + \mathbf{x}^H(k+1)\mathbf{y}(k+1)}), \tag{12}$$

where $\mathbf{y}(k+1)$ represents the product of the inverse of the sample covariance matrix at time k and the array snapshot at time $k+1$, that is, $\mathbf{y}(k+1) = \hat{\mathbf{R}}^{-1}(k)\mathbf{x}(k+1)$.

Second, the symbol g is introduced to represent the generalized inner product of the steering vectors with respect to the inverse of the sample covariance matrix,

$$g_{k+1}(\theta, \vartheta_i) = \bar{\mathbf{a}}^H(\theta)\hat{\mathbf{R}}^{-1}(k+1)\bar{\mathbf{a}}(\vartheta_i), \tag{13}$$

Substituting Equation (12) into Equation (13), the generalized inner product g can be equivalently expressed as,

$$g_{k+1}(\theta, \vartheta_i) = \gamma^{-1}(g_k(\theta, \vartheta_i) - q_{k+1}(\theta, \vartheta_i)), \tag{14}$$

where $q_{k+1}(\theta, \vartheta_i)$ is defined as

$$q_{k+1}(\theta, \vartheta_i) = \frac{\bar{\mathbf{a}}^H(\theta)\mathbf{y}(k+1)\mathbf{y}^H(k+1)\bar{\mathbf{a}}(\vartheta_i)}{(k+1)\gamma + \mathbf{x}^H(k+1)\mathbf{y}(k+1)}. \tag{15}$$

Finally, the pseudo spatial power spectrum at time $k+1$ can be calculated by substituting Equation (14) into Equation (10), that is,

$$P_{SSC-MVDR}^{k+1}(\theta) = 1/|\sum_{i=1}^{L}|\frac{g_k(\theta, \vartheta_i) - q_{k+1}(\theta, \vartheta_i)}{g_k(\theta, \theta) - q_{k+1}(\theta, \theta)}|e^{j(2\pi/L)i}|, \quad \theta \in \Theta. \tag{16}$$

It can be seen from Equation (16) that the calculation of the pseudo spatial power spectrum at time $k+1$ depends on the results of two functions, which are the generalized inner product at time k and the function q at time $k+1$, respectively. Please note that the former is known during the calculation process at time $k+1$ (because it has been obtained in the previous calculation), and the latter is calculated as Equation (15). Since it only involves vector operations, its computational complexity is relatively small.

5. Simulation Results and Analysis

In computer simulations, it is assumed that the linear array consists of 10 omnidirectional hydrophones, and the spacing of adjacent hydrophones is set to half the wavelength of the narrowband acoustic signal. In addition, assuming that there are two underwater acoustic sources in the far field, their directions of arrival are set to 30 degrees and 60 degrees, respectively. In the simulations below, the intensity of the acoustic source at 30 degrees is set to be variable for testing the DOA estimation performance of the SSC-MVDR algorithm, while the intensity of the acoustic source at 60 degrees is set to always be 30 dB (relative to noise) for testing the algorithm performance in a strong interference environment. Meanwhile, the received noise of the linear array is assumed to be spatially white Gaussian noise.

It should be noted that various unfavorable factors that may be encountered in the complex uncertain ocean environment are also considered in the simulations, including direction error, random wavefront fluctuations, hydrophone position errors, and hydrophone amplitude phase errors. Therefore, the steering vector model in Equation (7) is used while assuming that the amplitude deviation coefficients of the steering vector and the phase deviation coefficients of that are known.

In Figures 2 and 3, the pseudo spatial power spectrums of the SSC-MVDR algorithm are given under different acoustic source intensities and different snapshot numbers (i.e., the number of snapshots used to calculate the sample covariance matrix in Equation (5)). Specifically, Figure 2 corresponds to a friendly simulation environment where the signal-to-noise ratio (SNR) of the acoustic source in the direction of 30 degrees is set to 20 dB, the number of snapshots is set to 100, while Figure 3 corresponds to a poor simulation environment where the SNR of the same acoustic source is only −10 dB, and the number of snapshots is only 30. For performance comparison, the spatial power spectrums of the MVDR method [4], the MUSIC method [13] and the PM method [15] under the same simulation conditions are also given in Figures 2 and 3. Compared to other algorithms, the SSC-MVDR algorithm exhibits sharper peaks near the true DOA of the acoustic signal in Figures 2 and 3.

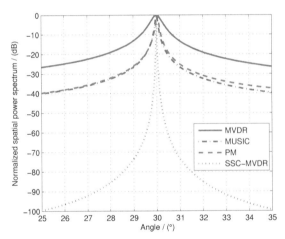

Figure 2. Normalized spatial power spectrums obtained by the MVDR method, the MUSIC method, the PM method and the SSC-MVDR method when the acoustic source intensity is 20 dB and the snapshot number is 100.

Figures 4 and 5 show the DOA estimation accuracy results of the SSC-MVDR algorithm. In Figure 4, the number of snapshots is fixed at 100, and the intensity of the acoustic source is gradually increased from −20 dB to 20 dB. In Figure 5, the acoustic source intensity is fixed at −10 dB, and the number of snapshots is gradually increased from 10 to 100. The DOA estimation accuracy is evaluated by the root mean square error (RMSE) of DOA estimation, which is defined as follows,

$$\text{RMSE} = 20\log\sqrt{\frac{1}{S}\sum_{i=1}^{S}(\hat{\theta}_i - \theta_a)^2} \qquad (17)$$

In Equation (17), RMSE is actually the logarithm of the root mean square error, so the unit of RMSE is dB. The number of independent computer simulations is represented by S. In the following simulations (i.e., from Figures 4–7), $S = 100$ is set, which means that each simulation result requires 100 independent runs. $\hat{\theta}_i$ represents the DOA estimate obtained by the ith computer simulation, and θ_a is the corresponding true DOA value. Meanwhile, the results of the DOA estimation accuracy of the MVDR method [4], the MUSIC method [13], the Root-MUSIC method [34], the TLS-ESPRIT method [35] and the PM method [15] are also given in Figures 4 and 5. It can be seen from Figure 4 that the RMSE results of the SSC-MVDR algorithm is significantly lower than that of other methods when the acoustic source intensity is in the range of -20 dB to -10 dB. When the number of snapshots is between 20 and 90, the RMSE results of the SSC-MVDR algorithm in Figure 5 is also significantly lower than that of other algorithms.

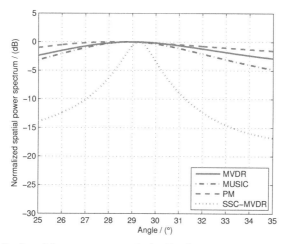

Figure 3. Normalized spatial power spectrums obtained by the MVDR method, the MUSIC method, the PM method and the SSC-MVDR method when the acoustic source intensity is -10 dB and the snapshot number is 30.

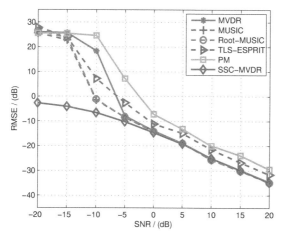

Figure 4. DOA estimation accuracy results obtained by the MVDR method, the MUSIC method, the Root-MUSIC method, the TLS-ESPRIT method, the PM method and the SSC-MVDR method when the number of snapshots is fixed at 100, and the intensity of the acoustic source is gradually increased from -20 dB to 20 dB.

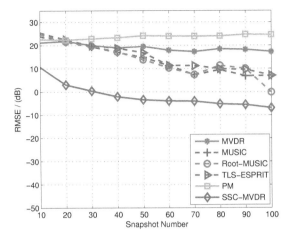

Figure 5. DOA estimation accuracy results obtained by the MVDR method, the MUSIC method, the Root-MUSIC method, the TLS-ESPRIT method, the PM method and the SSC-MVDR method when the acoustic source intensity is fixed at −10 dB, and the number of snapshots is gradually increased from 10 to 100.

Although the interference acoustic source in the far field has been considered in the above simulation (that is, an interference acoustic source with an intensity of 30 dB and a direction of 60 degrees is always included in the simulation settings), it is still necessary to perform simulation and analysis for the case of multiple interference acoustic sources. The results of the direction estimation accuracy under multiple interference acoustic sources are shown in Figure 6. In Figure 6, the intensities of the multiple interference acoustic sources are set to be the same, and the interference-to-noise ratio (INR) of each interference is gradually increased from 0 dB to 50 dB. Besides, N_I represents the number of interference acoustic sources. For example, $N_I = 3$ means that there are 3 interference acoustic sources in the far field at the same time, their directions are 60 degrees, 10 degrees, and −20 degrees, respectively. As can be seen from Figure 6, the three curves corresponding to different numbers of interferences are almost coincident.

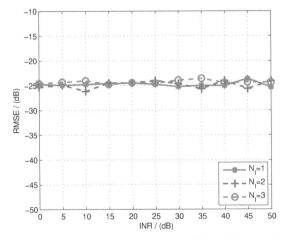

Figure 6. DOA estimation accuracy results obtained by the SSC-MVDR algorithm when there are multiple interfering sound sources with the source intensity gradually increased from 0 dB to 50 dB.

Finally, the influence of the direction interval, which is one of the important parameters of the SSC-MVDR algorithm, on the performance of the proposed algorithm is analyzed by simulation. In Figure 7, the horizontal axis is marked as the window width, that is, the width of the direction interval Θ. The three curves in Figure 7 are the accuracy results of DOA estimation obtained by the SSC-MVDR algorithm when the parameter θ_e takes different values, where θ_e represents the deviation of the center of the direction interval from the true DOA. It can be seen from Figure 7 that when the width of the direction interval is in the range of 10 degrees to 20 degrees, the RMSEs obtained by the SSC-MVDR algorithm is significantly lower than that when different window widths out of the above range is used. The reason for the above results is that when the center of the direction interval deviates from the true DOA and the selected direction interval is too narrow, the actual DOA of the acoustic source will fall outside the selected direction interval, thus the obtained DOA estimation must be wrong. However, when the selected direction interval is too wide, it will cause the beam response segment in the direction interval to contain some unwanted or even unfavorable information, such as other nulls in the beam response that are independent of signal self-cancellation.

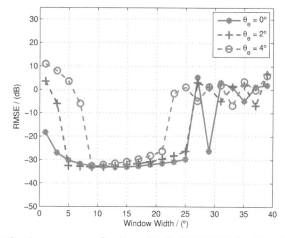

Figure 7. DOA estimation accuracy results obtained by the SSC-MVDR algorithm when the width of the direction interval varies between 1 and 40 degrees, and the deviation of the center of the direction interval from the true DOA is equal to 0, 2, and 4 degrees, respectively.

6. Conclusions

The MVDR beamforming-based underwater acoustic source localization techniques often encounter many unfavorable factors in the ocean environment, such as direction error, random wavefront fluctuation, hydrophone position error, hydrophone gain error, etc. These unfavorable factors lead to signal self-cancellation problems and severe performance degradation of the MVDR beamformer. Therefore, the signal self-cancellation problem is generally considered to be a disadvantage of the MVDR beamformer and is suppressed. On the contrary, by exploiting the signal self-cancellation phenomenon, this paper proposes a high-precision DOA estimation method for the underwater acoustic sources. First, the beam response of the MVDR beamformer in the direction interval is calculated according to the steering vector model. Then, the pseudo spatial power spectrum is calculated using the unit circle mapping technique. Finally, to reduce the computational complexity, the online calculation process of the pseudo spatial power spectrum is given. The computer simulation results show that the SSC-MVDR algorithm can obtain satisfactory direction estimation accuracy under the conditions of low intensity of acoustic source, strong interference and noise, and less array snapshot data. At the same time, computer simulation also gives reasonable suggestions for the width of the direction interval.

Appl. Sci. **2019**, *9*, 570

Author Contributions: Conceptualization, J.W.; Methodology, F.W., Y.C.; Software, F.W.; Writing—original draft, F.W.; Writing—review and editing, F.W.

Funding: This research was funded by National Natural Science Foundation of China (61601209), Natural Science Foundation of Jiangxi Province, P.R. China (20171BAB202003), and Science and Technology Research Project of Education Department, Jiangxi Province, P. R. China (8140).

Conflicts of Interest: The authors declare no conflict of interest.

References

1. Dowling, D.R.; Sabra, K.G. Acoustic Remote Sensing. *Annu. Rev. Fluid Mech.* **2015**, *47*, 221–243. [CrossRef]
2. Wang, L.; Yang, Y.; Liu, X. A distributed subband valley fusion (DSVF) method for low frequency broadband target localization. *J. Acoust. Soc. Am.* **2018**, *143*, 2269–2278. [CrossRef] [PubMed]
3. Thode, A.M.; Kim, K.H.; Norman, R.G.; Blackwell, S.B.; Greene, C.R. Acoustic vector sensor beamforming reduces masking from underwater industrial noise during passive monitoring. *J. Acoust. Soc. Am.* **2016**, *139*, EL105–EL111. [CrossRef] [PubMed]
4. Van Trees, H.L. *Optimum Array Processing: Part IV of Detection, Estimation And Modulation Theory*; Wiley: Hoboken, NJ, USA, 2002; pp. 293–303.
5. Jensen, F.B.; Kuperman, W.A.; Porter, M.B.; Schmidt, H. *Computational Ocean Acoustics*; Springer Science & Business Media: Berlin/Heidelberg, Germany, 2011.
6. Trucco, A.; Traverso, F.; Crocco, M. Broadband performance of superdirective delay-and-sum beamformers steered to end-fire. *J. Acoust. Soc. Am.* **2014**, *135*, EL331–EL337. [CrossRef]
7. Abadi, S.H.; Song, H.C.; Dowling, D.R. Broadband sparse-array blind deconvolution using frequency-difference beamforming. *J. Acoust. Soc. Am.* **2012**, *132*, 3018–3029. [CrossRef]
8. Douglass, A.S.; Song, H.C.; Dowling, D.R. Performance comparisons of frequency-difference and conventional beamforming. *J. Acoust. Soc. Am.* **2017**, *142*, 1663–1673. [CrossRef]
9. Bai, M.R.; Chen, C.C. Regularization using Monte Carlo simulation to make optimal beamformers robust to system perturbations. *J. Acoust. Soc. Am.* **2014**, *135*, 2808–2820. [CrossRef]
10. Vorobyov, S.A.; Gershman, A.B.; Luo, Z.Q. Robust adaptive beamforming using worst-case performance optimization: a solution to the signal mismatch problem. *IEEE Trans. Signal Process.* **2003**, *51*, 313–324. [CrossRef]
11. Cox, H.; Zeskind, R.; Owen, M. Robust adaptive beamforming. *IEEE Trans. Acoust. Speech Signal Process.* **1987**, *35*, 1365–1376. [CrossRef]
12. Guo, X.; Miron, S.; Yang, Y.; Yang, S. Second-order cone programming with probabilistic regularization for robust adaptive beamforming. *J. Acoust. Soc. Am.* **2017**, *141*, EL199–EL204. [CrossRef]
13. Schmidt, R. Multiple emitter location and signal parameter estimation. *IEEE Trans. Antennas Propag.* **1986**, *34*, 276–280. [CrossRef]
14. Roy, R.; Kailath, T. ESPRIT-estimation of signal parameters via rotational invariance techniques. *IEEE Trans. Acoust. Speech Signal Process.* **1989**, *37*, 984–995. [CrossRef]
15. Marcos, S.; Marsal, A.; Benidir, M. The propagator method for source bearing estimation. *Signal Process.* **1995**, *42*, 121–138. [CrossRef]
16. Quijano, J.E.; Zurk, L.M. Beamforming using subspace estimation from a diagonally averaged sample covariance. *J. Acoust. Soc. Am.* **2017**, *142*, 473–481. [CrossRef]
17. Quijano, J.E.; Zurk, L.M. Eigenvector pruning method for high resolution beamforming. *J. Acoust. Soc. Am.* **2015**, *138*, 2152–2160. [CrossRef] [PubMed]
18. Tibshirani, R. Regression Shrinkage and Selection via the Lasso. *J. R. Stat. Soc. Ser. B (Methodol.)* **1996**, *58*, 267–288. [CrossRef]
19. Xenaki, A.; Fernandez-Grande, E.; Gerstoft, P. Block-sparse beamforming for spatially extended sources in a Bayesian formulation. *J. Acoust. Soc. Am.* **2016**, *140*, 1828–1838. [CrossRef]
20. Tibshirani, R.J.; Taylor, J. The solution path of the generalized lasso. *Ann. Stat.* **2011**, *39*, 1335–1371. [CrossRef]
21. Palmese, M.; Trucco, A. Acoustic imaging of underwater embedded objects: Signal simulation for three-dimensional sonar instrumentation. *IEEE Trans. Instrum. Meas.* **2006**, *55*, 1339–1347. [CrossRef]

22. Sidky, E.Y.; Pan, X. Image reconstruction in circular cone-beam computed tomography by constrained, total-variation minimization. *Phys. Med. Biol.* **2008**, *53*, 4777–4807. [CrossRef]

23. Tibshirani, R.; Saunders, M.; Rosset, S.; Zhu, J.; Knight, K. Sparsity and smoothness via the fused lasso. *J. R. Stat. Soc. Ser. B (Stat. Methodol.)* **2005**, *67*, 91–108. [CrossRef]

24. Zheng, J.; Kaveh, M. Sparse Spatial Spectral Estimation: A Covariance Fitting Algorithm, Performance and Regularization. *IEEE Trans. Signal Process.* **2013**, *61*, 2767–2777. [CrossRef]

25. Yang, L.; Yang, Y.; Wang, Y. Sparse spatial spectral estimation in directional noise environment. *J. Acoust. Soc. Am.* **2016**, *140*, EL263–EL268. [CrossRef] [PubMed]

26. Buckingham, M.J.; Berknout, B.V.; Glegg, S.A.L. Imaging the ocean with ambient noise. *Nature* **1992**, *356*, 327–329. [CrossRef]

27. Harrison, C.H. The ocean noise coherence matrix and its rank. *J. Acoust. Soc. Am.* **2018**, *143*, 1689–1703. [CrossRef] [PubMed]

28. Harrison, C.H. Separation of measured noise coherence matrix into Toeplitz and Hankel parts. *J. Acoust. Soc. Am.* **2017**, *141*, 2812–2820. [CrossRef] [PubMed]

29. Quijano, J.E.; Dosso, S.E.; Dettmer, J.; Zurk, L.M.; Siderius, M. Trans-dimensional geoacoustic inversion of wind-driven ambient noise. *J. Acoust. Soc. Am.* **2013**, *133*, EL47–EL53. [CrossRef]

30. Yardim, C.; Gerstoft, P.; Hodgkiss, W.S.; Traer, J. Compressive geoacoustic inversion using ambient noise. *J. Acoust. Soc. Am.* **2014**, *135*, 1245–1255. [CrossRef]

31. Harrison, C.H. Performance and limitations of spectral factorization for ambient noise sub-bottom profiling. *J. Acoust. Soc. Am.* **2005**, *118*, 2913–2923. [CrossRef]

32. Siderius, M.; Harrison, C.H.; Porter, M.B. A passive fathometer technique for imaging seabed layering using ambient noise. *J. Acoust. Soc. Am.* **2006**, *120*, 1315–1323. [CrossRef]

33. Harrison, C.H.; Siderius, M. Bottom profiling by correlating beam-steered noise sequences. *J. Acoust. Soc. Am.* **2008**, *123*, 1282–1296. [CrossRef] [PubMed]

34. Yan, F.G.; Shuai, L.; Wang, J.; Shi, J.; Jin, M. Real-valued root-MUSIC for DOA estimation with reduced-dimension EVD/SVD computation. *Signal Process.* **2018**, *152*, 1–12. [CrossRef]

35. Kalgan, A.; Bahl, R.; Kumar, A. Studies on underwater acoustic vector sensor for passive estimation of direction of arrival of radiating acoustic signal. *Indian J. Geo-Mar. Sci.* **2015**, *44*, 213–219.

Article

Channel Modelling and Estimation for Shallow Underwater Acoustic OFDM Communication via Simulation Platform

Xiaoyu Wang, Xiaohua Wang, Rongkun Jiang, Weijiang Wang, Qu Chen and Xinghua Wang *

School of Information and Electronics, Beijing Institute of Technology, Beijing 100081, China;
wangxiaoyu@bit.edu.cn (X.W.); wangxiaohuabit@163.com (X.W.); jiangrongkun@bit.edu.cn (R.J.);
wangweijiang@bit.edu.cn (W.W.); chenqu@bit.edu.cn (Q.C.)
* Correspondence: wangxinghuabit@163.com

Received: 30 December 2018; Accepted: 23 January 2019; Published: 28 January 2019

Abstract: The performance of underwater acoustic (UWA) communication is heavily dependent on channel estimation, which is predominantly researched by simulating UWA channels modelled in complex and dynamic underwater environments. In UWA channels modelling, the measurement-based approach provides an accurate method. However, acquirement of environment data and simulation processes are scenario-specific and thus not cost-effective. To overcome such restraints, this article proposes a comprehensive simulation platform that combines UWA channel modelling with orthogonal frequency division multiplexing (OFDM) channel estimation, allowing users to model UWA channels for different ocean environments and simulate channel estimation with configurable input parameters. Based on the simulation platform, three independent simulations are conducted to determine the impacts of receiving depth, sea bottom boundary, and sea surface boundary on channel estimation. The simulations show that UWA channel estimation is greatly affected by underwater environments. The effect can be mainly attributed to changing acoustic rays tracing which result in fluctuating time delay and amplitude. With 10 m receiving depth and flat sea bottom, the channel estimation achieves optimal performance. Further study indicates that the sea surface has stochastic effects on channel estimation. As the significant wave height (SWH) increases, the average performance of channel estimation shows improvements.

Keywords: UWA communication; channel modelling; OFDM; channel estimation; simulation platform

1. Introduction

In recent years, underwater acoustic (UWA) communication has been widely employed in military affairs [1], ocean exploration [2], pollution monitoring [3], offshore oil drilling [4], etc. In view of these applications, UWA communication technology shows great potential as an area of research. Nevertheless, the nature of UWA channel (such as low available bandwidth, time-varying multipath, and the speed of sound) hinders the efficiency of communication devices [5]. Orthogonal frequency division multiplexing (OFDM), a robust method of encoding digital data on multiple carrier frequencies, is generally used in UWA time-dispersive channels and has the ability to render inter-symbol interference (ISI) negligible by embedding a cyclic prefix [6]. Furthermore, in order to suppress inter-channel interference (ICI) [7] introduced by Doppler spread in the OFDM system, the knowledge of channel impulse response (CIR) is essential, which could be acquired with the help of pilot signals [8]. Therefore, the estimation of channel is the key factor in the performance of UWA communication.

Although the estimation of UWA channel still remains a sophisticated problem due to complex underwater environments, channel modelling is an efficient way to perform preliminary evaluation and decide parameters such as estimation algorithm, pilot interval, and number of carriers for the OFDM communication device [9]. UWA channel modelling can be classified to two generic approaches: geometry-based and measurement-based [10].

Geometry-based UWA channel modelling relies on mathematical analysis in physical characteristics. Zajic [11] proposed a geometry-based model for multiple-input multiple-output (MIMO) mobile-to-mobile (M-to-M) shallow UWA channel by taking both macro- and microscattering effects into account. Qarabaqi and Stojanovic [12] developed a statistical model of UWA channels that incorporates physical aspects of acoustic propagation with the effects of inevitable random channel variations. Naderi et al. [13] constructed a stochastic channel model for wideband single-input single-output (SISO) shallow UWA channels under the assumption of rough ocean surface and bottom. These geometry-based models generally display stochastic channel responses in consideration of the time-varying characteristic of UWA channels collected through measurement campaigns. Although geometry-based channels could achieve a high level of accuracy, the physical features of the channels are limited in specific cases. For example, the probability distribution functions (PDF) of UWA channel envelope was revealed to be Rice distribution in [14], Rayleigh distribution in [15], log-normal distribution in [16], and K-distribution in [17].

The differential distributions above imply that the precise UWA channels should be modelled with real environment data. BELLHOP [18,19] is a beam tracing model that considers ocean environment properties data, such as sea surface boundary, sea bottom boundary, and sound speed profile (SSP). Utilizing the precision of BELLHOP, numerous measurement-based UWA channels were built through the model. In [20], Tomasi et al. compared real UWA channel data measured from an experiment with the channel obtained through the BELLHOP model. The result showed a great agreement in the conditions of calm ocean, which confirmed the feasibility of BELLHOP model. With strong winds, the measured channels were slightly worse than the result of the BELLHOP model. The deviation was caused by the lack of consideration on the sea surface. It also revealed the limitation of the BELLHOP model when dealing with stochastic parameters. In [21–23], shallow UWA channels were modelled in the eastern shore of Johor and the Taiwan Strait through the BELLHOP model and environment data were sourced from various databases. These approaches of measurement-based channel modelling provided an accurate way to analyse the channel characteristics such as transmission loss, CIR and ray tracing. However, environment data was scenario-specific and hard to acquire. In [24], Gul et al. designed a graphical user interface (GUI) with configurable parameters to model UWA channels and showed channel effect over an acoustic signal. The GUI brought much convenience to constructing environment files and visualizing simulation results. Nevertheless, the accuracy of modelling might be deteriorated as empirical data, including Munk SSP, was used in the GUI.

In this paper, three contributions are presented:

1. A comprehensive simulation platform is designed to model and estimate realistic UWA channels. To maintain the accuracy of UWA channels, measured data such as SSP, sea bottom boundary, and sea surface boundary are interfaced from open databases. Therefore, the realistic UWA channels in most areas of the ocean can be modelled through BELLHOP model. Furthermore, the simulation platform is realized through MATLAB GUI, making building BELLHOP files and processing the result of simulation with configurable inputs user-friendly. In addition, the simulation platform provides a method of fixing the defect of the BELLHOP model when dealing with stochastic parameters, to simulate stochastic channels. In conclusion, the simulation platform is a powerful and efficient tool to simulate and analyse the UWA channels.

2. The modelling of UWA channels is combined with OFDM channel estimation. To study the modelling of UWA channels, a great number of researchers focus on analysing the transmission loss, CIR, and ray tracing, which show the transmission characteristic of UWA channels. Furthermore, bit error rate (BER) and mean square error (MSE) are key parameters in judging the performance of channel estimation. Hence the normalized CIRs of channels are computed in UWA channel modelling part and the comparison of performance of BER and MSE is made in

the OFDM channel estimation part. Based on this framework, users can analyse the estimation performance of different modelling channels, modulation schemes, and estimation algorithms to assess the implementation scheme of the UWA communication device.

3. Based on the simulation platform, three simulations are conducted. Realistic UWA channels in the East China Sea are modelled to analyse the factors that could influence the performance of channel estimation. On the one hand, deterministic channels are modelled with different receiving depth and types of sea bottom to compare the performance of channel estimation. The result shows that channels with 10 m receiving depth and flat sea bottom yield optimal performance of channel estimation. On the other hand, a batch of channels is modelled with stochastic sea surface in different significant wave heights (SWH) and the result of channel estimation performance is synthesized. Overall, the channels with high SWH have a good average performance.

The rest of this article is structured as follows. In Section 2, the article provides a brief overview of channel estimation in OFDM communication system. The implementation of simulation platform is demonstrated in Section 3. In Section 4, the effects of receiving depth, sea bottom boundary, and sea surface boundary on UWA channel estimation are analysed by modelling and comparison. Finally, a concise conclusion is made in Section 5.

2. Channel Estimation in OFDM

The main principle of OFDM is to transmit data stream in low rate through numerous orthogonal subcarriers simultaneously [25]. Figure 1 shows a typical end-to-end configuration of OFDM communication system. The transmitted data is modulated on N subcarriers and passed through the UWA channels.

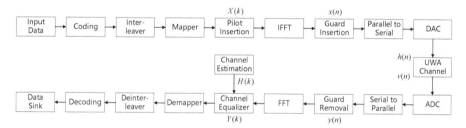

Figure 1. Block diagram of orthogonal frequency division multiplexing (OFDM) system transceiver.

The CIR of UWA channel can be expressed as [26],

$$h(n) = \sum_{l=0}^{L-1} h_l \delta(n - l) \tag{1}$$

where L is the length of channel and h_l is the coefficient of lth tap. The L can be calculated by $L = \tau_{max} / T_s$ where τ_{max} is the maximum delay of the channel and T_s is the length of an OFDM symbol.

Now we suppose the guard interval (CP) length is no shorter than the maximum delay of UWA channel, which means the current received OFDM symbol is not affected by the previous symbol. Then the received signal processed by FFT can be expressed as,

$$\mathbf{Y} = \mathbf{XH} + \mathbf{V} \tag{2}$$

where $\mathbf{Y}$, $\mathbf{X}$, $\mathbf{H}$, and $\mathbf{V}$ denote the matrix of received symbol, transmitted symbol, channel frequency response, and noise in frequency domain, respectively. It can also be presented as,

$$
\begin{bmatrix} Y[0] \\ Y[1] \\ \vdots \\ Y[N-1] \end{bmatrix} = \begin{bmatrix} X[0] & 0 & \cdots & 0 \\ 0 & X[1] & & \vdots \\ \vdots & & \ddots & 0 \\ 0 & \cdots & 0 & X[N-1] \end{bmatrix} \begin{bmatrix} H[0] \\ H[1] \\ \vdots \\ H[N-1] \end{bmatrix} + \begin{bmatrix} V[0] \\ V[1] \\ \vdots \\ V[N-1] \end{bmatrix} \tag{3}
$$

In the channel estimation problem of OFDM system, the matrix of transmitted pilot symbol X_p and the matrix of received pilot symbol Y_p are usually given and the estimate $\hat{H}_p$ of channel frequency response H_p needs to be figured out.

Least square (LS), minimum mean square error (MMSE), and linear minimum mean square error (LMMSE) are the most traditional algorithms used in OFDM channel estimation. LS algorithm focuses on finding the estimate $\hat{H}_p$ to minimize the value of following cost function,

$$
J\left(\hat{H}_p \right) = \left\| Y_p - \hat{Y}_p \right\| \tag{4}
$$

The solution to the LS channel estimation is as follows [26,27],

$$
\hat{H}_{LS} = (X_p^H X_p)^{-1} X_p^H Y_p = X_p^{-1} Y_p \tag{5}
$$

MMSE algorithm aims at finding the estimate $\hat{H}_p$ to minimize the value of mean square, and can be expressed as,

$$
J\left(\hat{H}_p \right) = E\left\{ \left\| H_p - \hat{H}_p \right\|^2 \right\} \tag{6}
$$

The solution to the MMSE channel estimation is [26,27],

$$
\hat{H}_{MMSE} = R_{H_p H_p} \left[R_{H_p H_p} + \left(X_p X_p^H \right)^{-1} N\sigma_n^2 \right]^{-1} \hat{H}_{LS} \tag{7}
$$

where $R_{H_p H_p}$ is the autocorrelation matrix of H_p. N is number of subcarriers. σ_n is the variance of noise.

MMSE algorithm is much more complicated than LS algorithm because of matrix inversion. Suppose that the output of code modulation is equiprobable, then $\left(X_p X_p^H \right)^{-1}$ can be replaced with $E\left\{ \left(X_p X_p^H \right)^{-1} \right\}$ and the solution of LMMSE is presented as [28,29],

$$
\hat{H}_{LMMSE} = R_{H_p H_p} \left[R_{H_p H_p} + \frac{\beta}{SNR} I \right]^{-1} \hat{H}_{LS} \tag{8}
$$

where $\beta = E\left\{ |X_p|^2 \right\} E\left\{ 1/|X_p|^2 \right\}$ is a coefficient associated with code modulation scheme and $SNR = E\left\{ |X_p|^2 \right\} / N\sigma_n^2$ is signal-to-noise ratio. Since LMMSE algorithm has similar performance and lower complexity than MMSE algorithm, LS and LMMSE algorithms are used to estimate UWA channels in this paper.

3. Implementation of Simulation Platform

3.1. Basic Framework

Based on the GUI of MATLAB, the simulation platform is mainly composed of two parts, namely, UWA channel modelling and OFDM channel estimation. User interface of the simulation platform is illustrated in Figure 2. Figure 2a shows the interface of UWA channel modelling, which is divided into four major components. In the zone of modelling parameters, users can set the position and depth of transmitter and receiver, as well as signal frequency, number of beams, and SWH. Figures of normalized CIR and eigenrays trace are presented in the middle. The status indicator displays the consequences of script function in progress. Furthermore, the channel list is used to save or delete modelled channels. The interface of OFDM channel estimation is illustrated in Figure 2b. The parameters, such as modulation type, number of carriers, simulated channel, and algorithm, are freely set in the zone of estimation parameters. Figures of the simulation results are plotted in the lower part and can be saved with specified names if necessary.

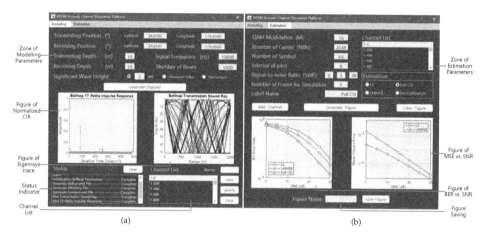

Figure 2. The user interface of simulation platform. (**a**) underwater acoustic (UWA) channel modelling. (**b**) OFDM channel estimation.

Figure 3 shows the flow chart of using the simulation platform in this paper. When conducting a simulation, first, users should enter proper channel parameters and click 'Generate Channel' button repeatedly until enough channels are saved in the list. Then, parameters of OFDM channel estimation and UWA channels need to be configured. For deterministic channels, a UWA channel needs to be selected from the channel list. While simulating with stochastic channels, users should choose a group of channels by clicking 'Add Channel' button. After that, curves of BER and MSE will be plotted by clicking the 'Generate Figure' button. Repeating the progress of modifying parameters and plotting curve, result of performance comparison will be produced. Eventually, the simulation figures can be saved in high-resolution by clicking 'Save Figure' button.

The design of simulation platform puts forward a high-efficiency approach to the research on UWA channels. One of the distinct advantages lies in the reduction of workload for finding real data, building input files, and processing output data. The users using simulation platform without any previous experience in BELLHOP model or MATLAB programing can still perform simulation for research. Besides, the GUI of simulation platform makes using BELLHOP model more intuitive and provides users with immediate visual feedback about the results of the program. Moreover, based on the real environment data acquired from open databases, the simulation platform provides users a flexible way to choose parameters. In sight of the accuracy and convenience of simulation platform, users can modify position, depth, frequency, SWH, and even the number of beams to study the UWA channel. At the same time, the factors of OFDM modulation, pilot pattern, and estimation algorithms can also be analysed by changing these values. In addition, by combining UWA channel modelling with OFDM channel estimation, the simulation platform helps users analyse characteristics of UWA channel and pick the appropriate OFDM implementation scheme for UWA communication device.

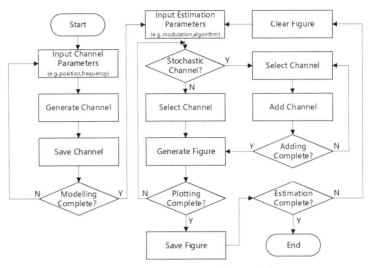

Figure 3. Flow chart of using simulation platform.

3.2. Channel Modelling

The most common models for solving the problem of UWA communication are propagation models. Wave equation, derived from the equations of state, continuity, and motion, is the theoretical base for all mathematical models of acoustic propagation. The simplified wave equation is as follows,

$$\nabla^2\Phi=\frac{1}{c^2}\frac{\partial^2\Phi}{\partial t^2} \qquad (9)$$

where ∇^2 is the Laplacian operator. ϕ is the potential function. c is the speed of sound. t is the time.

Furthermore, there are five canonical models for solving wave equation [9]: ray theory, normal mode, multipath expansion, fast field, and parabolic equation model. Among the five models above, ray theory model is both applicable and practical in high frequency, and is therefore widely used in UWA channel simulation. Based on the theory of Gaussian beams [30], BELLHOP is one of the most effective implementations of ray model for solving the ray equations with cylindrical symmetry [31],

$$\frac{dr}{ds} = c\xi(s), \quad \frac{d\xi}{ds} = -\frac{1}{c^2}\frac{\partial c}{\partial r},$$
$$\frac{dz}{ds} = c\zeta(s), \quad \frac{d\zeta}{ds} = -\frac{1}{c^2}\frac{\partial c}{\partial z}, \qquad (10)$$

where $r(s)$ and $z(s)$ are the ray coordinates in cylindrical coordinates and $c(s)[\xi(s),\zeta(s)]$ is the tangent versor along the ray. With the initial conditions $(r(0) = r_s, z(0) = z_s, \xi(0) = \frac{\cos\theta_s}{c_s}$, and $\zeta(0) = \frac{\sin\theta_s}{c_s}$. θ_s is the launching angle. $[r_s, z_s]$ is the source position. c_s is the sound speed at the source position), the coordinates of ray can be obtained by

$$\tau = \int_\Gamma \frac{ds}{c(s)} \qquad (11)$$

The BELLHOP model is designed to simulate two-dimensional acoustic ray tracing for a specific SSP in waveguides with flat or variable absorbing boundaries [18,19,32]. The program of BELLHOP offers various output options, including ray coordinates, eigenray coordinates, acoustic pressure, travel time, and amplitudes. In order to generate a valid output of UWA channels, the parameters for modelling should be set correctly in BELLHOP files (environment file, altimetry file, bathymetry file, etc.) according to the user's guide [32].

3.3. Data Interface

This article aims at modelling realistic UWA channels through BELLHOP model, which means the simulation platform should implement the constructing of input files with real data. Although modelling in measurement-based approaches is a complex process with numerous parameters that need to be determined, there are lots of open databases available on the Internet that can be interfaced through script functions. The data flow diagram of simulation platform is illustrated in Figure 4.

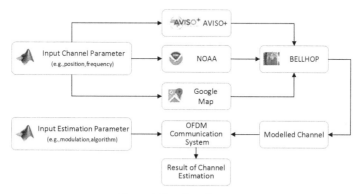

Figure 4. Data flow diagram.

In the channel modelling, users enter basic channel parameters such as position, frequency, and number of beams to the GUI. After getting the data of position, the simulation platform will be able to read real data about sea surface boundary, SSP, and sea bottom boundary from the databases. Sea surface boundary is a stochastic parameter that can be generated from the Gauss-Lagrange model. The function of Gauss-Lagrange is as follows,

$$x(t, u) = \sum_{j=0}^{N} \sqrt{S_j \Delta \omega \rho_j} R_j \cos(k_j u - \omega_j t + \Theta_j + \theta_j) \qquad (12)$$

where ρ_j is the amplitude response and θ_j is the phase response. WafoL [33], a toolbox of MATLAB, is used to solve the function and generate Gauss-Lagrange waves. In order to make the waves closer to reality, amplitudes of Gauss-Lagrange waves should be adjusted according to real SWH, which can be obtained from altimetry satellite missions in Aviso+ [34]. The speed of sound in sea water is the basic variable of acoustic channel, which is affected by many factors such as water temperature, salinity, and depth. World Ocean Atlas 2013 (WOA2013) [35] is a data product of National Oceanic and Atmospheric Administration (NOAA) where the ocean properties can be accessed with the index of time and position. With the real data, SSP in any geographical location can be calculated by the UNESCO equation [36]. Sea bottom boundary can reflect and scatter the sound ray. According to Google Map API [37], the simulation platform gets a set of discrete depth samples in order to interpolate the sea bottom terrain.

After getting the values of sea surface boundary, SSP, and sea bottom boundary, the simulation platform will be able to construct BELLHOP files (environment file, altimetry file, bathymetry file, etc.) and calculate the eigenray coordinates, travel time and amplitudes of UWA channel. Therefore, the eigenrays tracing is plotted on the simulation platform with the coordinates. Furthermore, the CIR is calculated by accumulating travel time and amplitudes for each multipath. For the sake of convenient simulation in OFDM communication system, the CIR is normalized and saved in channel list. In OFDM channel estimation, the modelled channels are simulated with Monte Carlo method based on specific input parameters of channel estimation and the BER and MSE curves are plotted for analyses.

4. Simulations and Analysis

Control signals being transmitted from sea surface to underwater in shallow sea is one of the most commonly used applications of UWA communication. This article focuses on this application and analyses the factors in the estimation of UWA channel. There are two scenarios of shallow sea illustrated in Figure 5a with a range of 2 km from the southeast of Meizhou island in the East China Sea, which are selected to simulate on the proposed platform. Scenario A is located between 119.6590° E, 24.8385° N and 119.6689° E, 24.8541° N with an approximate flat sea bottom at the depth of about 70 m. Scenario B is located between 119.5708° E, 24.3453° N and 119.5609° E, 24.3297° N where the depth of sea bottom increases from 50 m to 80 m. The time of databases is set at noon on June 22. According to WOA2013, with an increase of sea depth, the temperature gradually decreases from 27.9 °C to 16.7 °C and the salinity slowly increases from 33.5 psu to 34.2 psu. The SSP calculated by temperature and salinity is shown in Figure 5b, varying from 1513 m/s to 1540 m/s. In addition, the average SWH can also be found in Aviso+ with a value of 1.29 m.

Three simulations are set to model channels in different receiving depth, sea bottom boundaries, and sea surface boundaries, respectively, and the influence on channel estimation performance is analysed afterwards. The detailed simulation parameters are tabulated in Table 1.

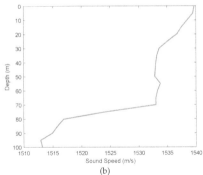

(a)   (b)

Figure 5. (**a**) Geographical location of two scenarios in simulations. (**b**) sound speed profile (SSP) curve generated from WOA2013.

Table 1. Parameters of simulations.

Parameter Class	Parameters	Values
UWA Channel Modelling	Transmitting Depth	10 m
	Receiving Depth	10 m, 30 m, 50 m, 70 m
	Significant Wave Heitht	0 m (flat), 0.5 m, 1.29 m (measured), 2 m
	Sea Bottom Boundaries	flat, rising, falling, protruding, sagged
	Signal Frequency	10,000 Hz
	Number of Beams	1000
OFDM Channel Estimation	Modulation Scheme	16 QAM
	Symbol Order	Grey
	Algorithm	LS, LMMSE
	Number of Carriers	2048
	Size of FFT	2048
	Length of Cyclic Prefix	512
	Pilot Interval	8
	Number of Symbols per frame	64

4.1. Receiving Depth

When sending control signals, transmitting device is usually located below the surface of the sea while the depth of receiving device is not fixed. The first simulation is designed to model UWA channels with different receiving depths and examine the influence to OFDM channel estimation. In this simulation, the energy converter is 10 m in depth and the acoustic signal is evenly emitted from a 30-degree emission angle. The depth of receiving device is set to 10 m, 30 m, 50 m, and 70 m, respectively. In order to control the effect of sea surface and sea bottom in this simulation, UWA channels are modelled in the sea area of scenario A (flat bottom boundary) with flat sea surface boundary.

Figure 6 illustrates the eigenrays path of modelled channels during propagation. The color of lines indicates whether eigenrays hit surface boundary or bottom boundary of the sea, and basically there are four cases: The black line represents eigenray hitting both boundaries; the blue line represents eigenray hitting bottom only; the green line represents eigenray hitting surface only; and the red line represents eigenray hitting neither bottom nor surface. As can be seen in the eigenrays figure, the main eigenray type is black line. In shallower depth, the number of blue lines is greater than that in deeper depth.

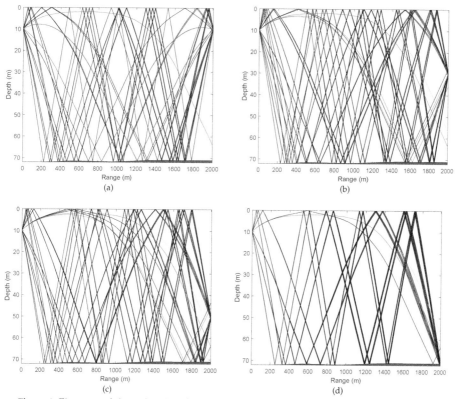

Figure 6. Eigenrays of channel with different receiving depth. The color of lines indicates whether eigenrays hit surface boundary or bottom boundary of the sea. Receiving depth of each figure: (**a**) 10 m; (**b**) 30 m; (**c**) 50 m; (**d**) 70 m.

Different types and lengths of eigenrays lead to different transmission losses and time delay. The normalized CIR is calculated by accumulating all eigenrays. Figure 7 shows the normalized CIRs of channel with different receiving depth. Comparing the four normalized CIRs, the normalized CIR with 10 m receiving depth is more concentrated in low time delay and has a maximum amplitude of 0.9415. Whereas the other normalized CIRs are scattered and have maximum amplitudes under 0.5.

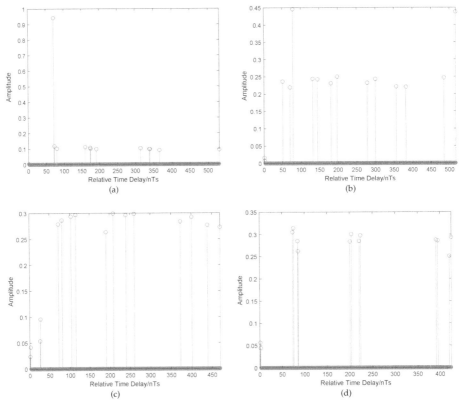

Figure 7. Normalized channel impulse responses (CIRs) of channel with different receiving depth. Receiving depth of each figure: (**a**) 10 m; (**b**) 30 m; (**c**) 50 m; (**d**) 70 m.

Figure 8a shows BER performance of channels in different receiving depths. It is obvious that the performance of channel estimation in 10 m depth channel is much better than others. The gap of performance between 10 m depth channel and other channels is gradually increasing as SNR increases. With a 30 dB SNR, the gap in LS algorithm reaches around 20 dB and the gap in LMMSE algorithm reaches around 30 dB. Comparing the performance of LS and LMMSE algorithms in different channels, the difference between LS and LMMSE algorithms is larger in 10 m depth channel. MSE performance of channels in different receiving depth is illustrated in Figure 8b. The curves of LS algorithm have the same performance due to the principle of algorithm. Furthermore, the performance of MSE has similar features to that of BER in LMMSE algorithm.

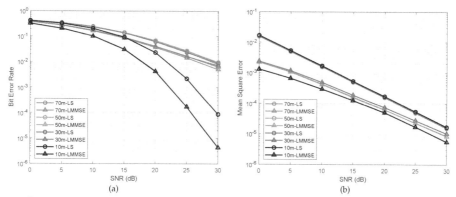

Figure 8. (**a**) bit error rate (BER) performance of channels in different receiving depth, estimated by LS and LMMSE algorithms, respectively. (**b**) MSE performance of channels in different receiving depth, estimated by LS and LMMSE algorithms, respectively.

4.2. Sea Bottom Boundary

Seabed terrain has some structures that result from common physical phenomena. In the shallow sea, the depth of seabed usually changes slowly and the sea bottom boundary can be considered as flat in short distances, which is modelled in previous simulation. Furthermore, the seabed may rise or fall rapidly in the littoral area and have protruding or sagged structures under specific conditions. The four special types of seabed above will be modelled in this simulation to analyse the influence of sea bottom boundary on channel estimation. The rising and falling sea boundaries are sampled from scenario B with opposite transmitting and receiving positions. The protruding and sagged sea boundaries are rare to find in real bathymetry data. Therefore, the sea bottom boundary of scenario A is adjusted to produce the boundaries artificially. In this simulation, the receiving depth is fixed at 10 m, which has been confirmed to yield the best channel estimation performance among other depths in previous simulation. The sea surface boundary is set as flat to control the stochastic effect.

Figure 9 shows the eigenrays of channels with different sea bottom boundaries. The channel with rising bottom has more eigenrays than others because rising bottom increases the length of eigenrays transmission path and disperses the eigenrays gradually in the process of transmission. The falling bottom can also disperse the eigenrays. Since falling bottom decreases the length of eigenrays transmission path, the number of eigenrays in channel with falling bottom reduces a lot. The channel with protruding bottom has shorter length of eigenray transmission path than the channel with sagged bottom as the impact of seabed terrain.

Figure 10 illustrates the normalized CIRs of channels above. Overall, the length of normalized CIRs of channels with rising and sagged bottom are larger than that of channels with falling and protruding bottom. It is obvious that the CIR of channel with rising bottom has more multipaths than that with falling bottom. Owing to the seabed terrain, the relative time delay of path with maximum amplitude in protruding channel bottom is less than that in sagged channel.

Figure 11 is plotted to compare the BER and MSE performance between the four channels. The BER performance of different channels in LS algorithm is all around 0.0075 when SNR is 30 dB. The value is similar to the worse BER value in Figure 8a, implicating that there may exist a worst-case performance bound of UWA channel estimation through LS algorithm. In addition, the BER performance of different channels in LMMSE algorithm has a few differences. The channel with falling bottom is estimated in the best BER performance while the channel with sagged bottom is estimated in the worst BER performance. The performance of MSE also has similar characteristics to that of BER in LMMSE algorithm, shown in Figure 11b.

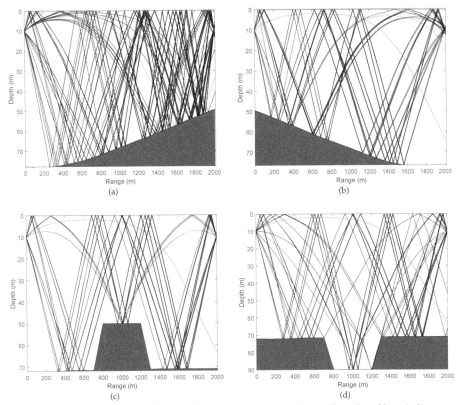

Figure 9. Eigenrays of channel with different sea bottom boundaries. The color of lines indicates whether eigenrays hit surface boundary or bottom boundary of the sea. Sea bottom boundaries of each figure: (**a**) rising bottom; (**b**) falling bottom; (**c**) protruding bottom; (**d**) sagged bottom.

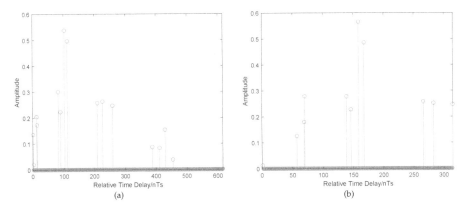

Figure 10. *Cont.*

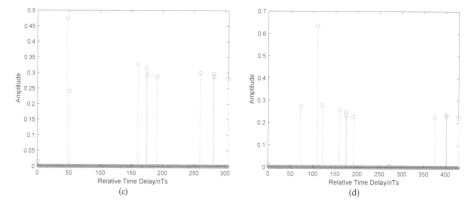

Figure 10. Normalized CIRs of channel with different sea bottom boundaries. Sea bottom boundaries of each figure: (**a**) rising bottom; (**b**) falling bottom; (**c**) protruding bottom; (**d**) sagged bottom.

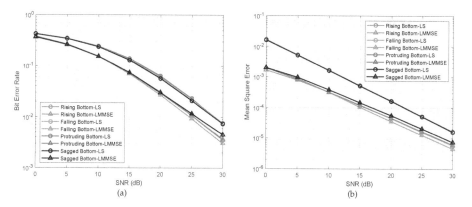

Figure 11. (**a**) BER performance of channels in different sea bottom boundaries, estimated by LS and LMMSE algorithms, respectively. (**b**) MSE performance of channels in different sea bottom boundaries, estimated by LS and LMMSE algorithms, respectively.

4.3. Sea Surface Boundary

In a real ocean environment, the sea surface boundaries are usually not flat. The last simulation is designed to model UWA channels with different sea surface boundaries and analyse the impact on OFDM channel estimation. In this simulation, the receiving depth is fixed at 10 m and UWA channels are modelled in the sea area of scenario A with flat sea bottom boundary, according to the experience of previous simulations. For the sake of high complexity in stochastic channel simulation, LS algorithm is used to estimate UWA channel in this simulation.

First, four UWA channels are modelled with different sea surface boundaries, which are generated by Gauss-Lagrange wave model with measured SWH (1.29 m). Figure 12 shows the eigenrays of different sea surface boundaries. As can be seen from the figures, the stochastic wave boundaries have variable influence on reflection angle, which leads to the number of reflection changes.

Furthermore, Figure 13 shows the normalized CIRs of channels with different sea surface boundaries. In this figure, the effect of sea surface appears as the change of time delay and response amplitude. The CIRs of channels with sea surface 2, sea surface 3, and sea surface 4 are more concentrated than that with sea surface 1. Furthermore, the CIR of channel with sea surface 4 has the maximum amplitude (0.9702).

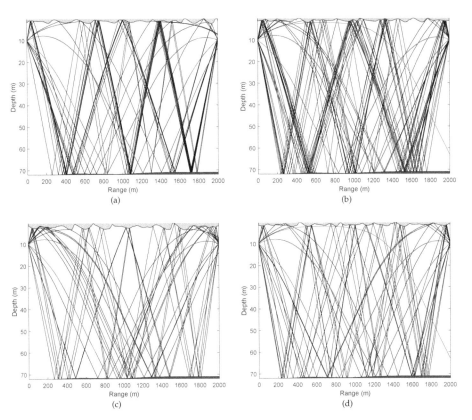

Figure 12. Eigenrays of channel with different sea surface boundaries. The color of lines indicates whether eigenrays hit surface boundary or bottom boundary of the sea. Sea surface boundaries of each figure: (**a**) Sea Surface 1; (**b**) Sea Surface 2; (**c**) Sea Surface 3; (**d**) Sea Surface 4.

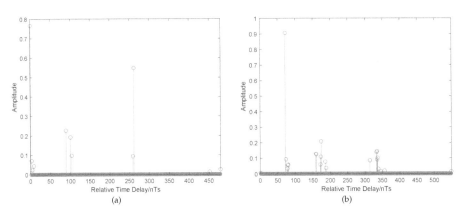

Figure 13. *Cont.*

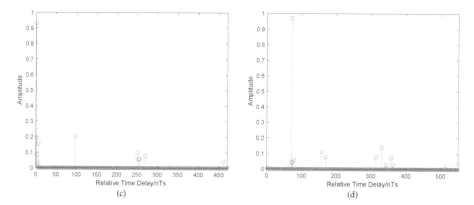

Figure 13. Normalized CIRs of channel with different sea surface boundaries in 1.29 m SWH. Sea surface boundaries of each figure: (**a**) Sea Surface 1; (**b**) Sea Surface 2; (**c**) Sea Surface 3; (**d**) Sea Surface 4.

Figure 14a shows BER performance in different sea surface boundaries. Comparing with the BER performance in UWA channel with flat sea surface boundary channel in Figure 8a, the Gauss-Lagrange waves can improve or deteriorate the BER performance of OFDM channel estimation to some extent. The BER performance of channels with sea surface 1, sea surface 2 and sea surface 3 has a range of deterioration from 1.50 dB to 19.38 dB, while the BER performance of channel with sea surface 4 has an improvement of 10.86 dB. The MSE performance of the four channels has the same value as that in previous simulations, which are illustrated in Figure 14b.

From the figures above, the result of different BER value is mainly caused by the stochastic effect of sea surface on the reflection angle of interfering eigenrays. The main energy of CIRs comes from the blue eigenrays (eigenrays hitting bottom only) that are the same in the four channels. Owing to the stochastic effect of sea surface on the reflection angle of interfering eigenrays which are mainly composed of black eigenrays that stand for the eigenrays hitting both boundaries, the amplitudes and delay of interfering eigenrays are different. In the channel with sea surface 1, the interfering eigenrays have more rebounds than that in the channel with sea surface 4. As a result, the interfering eigenrays in the channel with sea surface 1 have shorter time delay and higher amplitudes, which leads to a worse channel estimation performance.

The result above reveals the stochastic effect of sea surface and explains the reason for the deviation between modelled channels and real channels in [20]. In order to further investigate the statistical effect on OFDM channel estimation, batch of channels with 0.5 m, 1.29 m, and 2 m SWH are modelled and compared in terms of BER performance.

Figure 15a shows the BER curves of channels with different SWH. The blue, green, and red curves represent the BER performance of the modelled channel with 0.5 m, 1.29 m, and 2 m SWH, respectively. For each color, there are 20 curves. When SNR is 30 dB, the worst curves of the three colors have similar BER value with the worst curves in Figure 8a and Figure 11a, which confirms the existence of worst-case performance bound of UWA channel estimation through LS algorithm. Considering the distribution characteristic of curves in different colors, the blue, green, and red curves are mainly concentrated in the high, middle, and low BER performance part of the distribution range, respectively. Figure 15b shows the quantitative result by plotting the mean channel estimation BER curves for each value of SWH. It is obvious that the mean BER performance of stochastic channels is worse than that of channel with flat sea surface. As the SWH rises, the average performance gradually increases. When SNR is 30 dB, UWA channel with 2 m SWH is 7.95 dB better than that with 0.5 m SWH and 8.06 dB worse than that with flat sea surface.

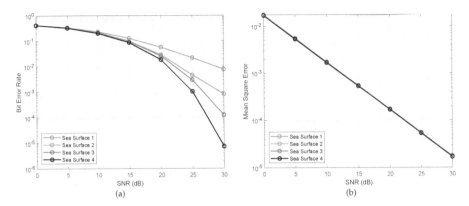

Figure 14. (a) BER performance of channels in different sea surfaces boundaries. (b) MSE performance of channels in different sea surfaces boundaries.

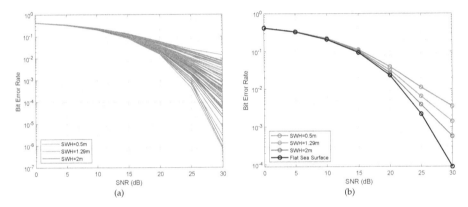

Figure 15. (a) Distribution of BER curves with difference SWH. There are 20 curves for each SWH value. (b) Mean BER performance of channels in different SWH.

The result is mainly caused by the effect of sea surface and the worst-case performance bound. According to the analysis of Figures 12–14, sea surface can affect the time delay and amplitudes of interfering eigenrays, which leads to the variation of estimation performance. As SWH rises, the variation becomes bigger and makes channel estimation more likely to get better or worse performance while the worst-case performance bound limits the performance from getting worse. Consequently, the channels with high SWH can get better average performance of channel estimation. In addition, the possibility of negative effect of sea surface is more than the positive effect, which makes the rough surface channel get worse average performance of channel estimation than the flat surface channel.

5. Conclusions

In this article, we design a comprehensive simulation platform combining UWA channel modelling with OFDM channel estimation. The simulation platform is presented in a GUI and interfaced from various databases, allowing the user to model realistic UWA channels in most areas of the ocean and estimate channels with configurable inputs. Three simulations are conducted based on the simulation platform. Realistic UWA channels in the East China Sea are modelled to study the influence of receiving depth, sea bottom boundary, and sea surface boundary on OFDM channel estimation. The simulations present that different environmental factors have specific effects on rays tracing, which result in the change of time delay and amplitudes, causing the specific effect on the performance of channel estimation. The results show that: (1) The UWA channel with 10 m receiving depth has more concentrated normalized CIR and better channel estimation performance than the other channels with deeper receiving depth. When SNR is 30 dB, the gap of performance reaches around 20 dB in LS algorithm and 30 dB in LMMSE algorithm. (2) The UWA channels with complicated sea bottom boundaries yield poor channel estimation. The BER performance of the channels in LS algorithm is around 0.0075, which is similar to the worse BER value in the first simulation. A worst-case performance bound exists in LS algorithm in which the UWA channels can hardly get worse performance of channel estimation. (3) The sea surface modelled in Gauss-Lagrange waves only affects the interfering eigenrays and has a stochastic effect on the performance of channel estimation. With the increase of SWH, the average performance gradually gets better because of the worst-case performance bound and increasing range of stochastic effect. Though the effect is stochastic, the rough surface channels get worse average performance than the flat one. When SNR is 30 dB, the 2 m SWH channels get 7.95 dB better mean BER performance than the 0.5 m SWH channels and 8.06 dB worse mean BER performance than the flat surface channel in the LS algorithm.

Author Contributions: Conceptualization, X.W. (Xiaoyu Wang) and R.J.; Formal analysis, X.W. (Xiaohua Wang) and Q.C.; Methodology, X.W. (Xiaohua Wang) and X.W. (Xinghua Wang); Resources, R.J. and W.W.; Software, X.W. (Xiaoyu Wang); Writing—original draft, X.W. (Xiaoyu Wang); Writing—review & editing, Q.C. and X.W. (Xinghua Wang).

Funding: This research received no external funding.

Conflicts of Interest: The authors declare no conflict of interest.

References

1. Headrick, R.; Freitag, L. Growth of underwater communication technology in the U.S. Navy. *IEEE Commun. Mag.* **2009**, *47*, 80–82. [CrossRef]
2. Baptista, A.; Howe, B.; Freire, J.; Maier, D.; Silva, C. Scientific Exploration in the Era of Ocean Observatories. *Comput. Sci. Eng.* **2008**, *10*, 53–58. [CrossRef]
3. Shin, D.; Na, S.Y.; Kim, J.Y.; Baek, S.J. Fish Robots for Water Pollution Monitoring Using Ubiquitous Sensor Networks with Sonar Localization. In Proceedings of the IEEE 2007 International Conference on Convergence Information Technology (ICCIT 2007), Hydai Hotel Gyeongui, Korea, 21–23 November 2007. [CrossRef]
4. Dalbro, M.; Eikeland, E.; In't Veld, A.J.; Gjessing, S.; Lande, T.S.; Riis, H.K.; Søråsen, O. Wireless Sensor Networks for Off-shore Oil and Gas Installations. In Proceedings of the IEEE 2008 Second International Conference on Sensor Technologies and Applications (sensorcomm 2008), Cap Esterel, France, 25–32 August 2008. [CrossRef]
5. Stojanovic, M.; Preisig, J. Underwater acoustic communication channels: Propagation models and statistical characterization. *IEEE Commun. Mag.* **2009**, *47*, 84–89. [CrossRef]
6. Wang, X.; Ho, P.; Wu, Y. Robust channel estimation and ISI cancellation for OFDM systems with suppressed features. *IEEE J. Sel. Areas Commun.* **2005**, *23*, 963–972. [CrossRef]
7. Mostofi, Y.; Cox, D.C. ICI mitigation for pilot-aided OFDM mobile systems. *IEEE Trans. Wirel. Commun.* **2005**, *4*, 765–774. [CrossRef]
8. John Proakis, M.S. *Digital Communications*; IRWIN: Boston, MA, USA, 2007.
9. Etter, P.C. *Underwater Acoustic Modeling and Simulation*; Spon Press (Tay & Francis Group): Boca Raton, FL, USA, 2013; pp. 351–383.
10. Zajic, A. *Mobile-to-Mobile Wireless Channels*; Artech House Books: Norwood, MA, USA, 2013.
11. Zajic, A.G. Statistical Modeling of MIMO Mobile-to-Mobile Underwater Channels. *IEEE Trans. Veh. Technol.* **2011**, *60*, 1337–1351. [CrossRef]
12. Qarabaqi, P.; Stojanovic, M. Statistical Characterization and Computationally Efficient Modeling of a Class of Underwater Acoustic Communication Channels. *IEEE J. Ocean. Eng.* **2013**, *38*, 701–717. [CrossRef]

13. Naderi, M.; Patzold, M.; Zajic, A.G. A geometry-based channel model for shallow underwater acoustic channels under rough surface and bottom scattering conditions. In Proceedings of the 2014 IEEE Fifth International Conference on Communications and Electronics (ICCE), Danang, Vietnam, 30 July–1 August 2014. [CrossRef]
14. Radosevic, A.; Proakis, J.G.; Stojanovic, M. Statistical characterization and capacity of shallow water acoustic channels. In Proceedings of the OCEANS 2009-EUROPE, Bremen, Germany, 11–14 May 2009. [CrossRef]
15. Galvin, R.; Coats, R. A stochastic underwater acoustic channel model. In Proceedings of the OCEANS 96 MTS/IEEE Conference Proceedings, The Coastal Ocean—Prospects for the 21st Century, Fort Lauderdale, FL, USA, 23–26 September 1996. [CrossRef]
16. Tomasi, B.; Casari, P.; Badia, L.; Zorzi, M. A study of incremental redundancy hybrid ARQ over Markov channel models derived from experimental data. In Proceedings of the Fifth ACM International Workshop on UnderWater Networks—WUWNet'10, Woods Hole, MA, USA, 30 September–1 October 2010. [CrossRef]
17. Yang, W.B.; Yang, T.C. High-frequency channel characterization for M-ary frequency-shift-keying underwater acoustic communications. *J. Acoust. Soc. Am.* **2006**, *120*, 2615–2626. [CrossRef]
18. Ocean Acoustics Library, Rays. Available online: http://oalib.hlsresearch.com/ (accessed on 12 December 2018).
19. Bellhop Gaussian Beam/Finite Element Beam Code. Available online: http://oalib.hlsresearch.com/Rays (accessed on 12 December 2018).
20. Tomasi, B.; Zappa, G.; McCoy, K.; Casari, P.; Zorzi, M. Experimental study of the space-time properties of acoustic channels for underwater communications. In Proceedings of the OCEANS'10 IEEE SYDNEY, Sydney, Australia, 24–27 May 2010. [CrossRef]
21. Jiang, R.; Cao, S.; Xue, C.; Tang, L. Modeling and analyzing of underwater acoustic channels with curvilinear boundaries in shallow ocean. In Proceedings of the 2017 IEEE International Conference on Signal Processing, Communications and Computing (ICSPCC), Xiamen, China, 22–25 October 2017. [CrossRef]
22. Bahrami, N.; Khamis, N.H.H.; Baharom, A.B. Study of Underwater Channel Estimation Based on Different Node Placement in Shallow Water. *IEEE Sens. J.* **2016**, *16*, 1095–1102. [CrossRef]
23. Bahrami, N.; Khamis, N.H.H.; Baharom, A.; Yahya, A. Underwater Channel Characterization to Design Wireless Sensor Network by Bellhop. *Telecommun. Comput. Electron. Control* **2016**, *14*, 110–118. [CrossRef]
24. Gul, S.; Zaidi, S.S.H.; Khan, R.; Wala, A.B. Underwater acoustic channel modeling using BELLHOP ray tracing method. In Proceedings of the 2017 14th International Bhurban Conference on Applied Sciences and Technology (IBCAST), Islamabad, Pakistan, 10–14 January 2017. [CrossRef]
25. Nee, R.V. *Ofdm for Wireless Multimedia Communications*; Artech House Inc.: Norwood, MA, USA, 1999.
26. Van de Beek, J.J.; Edfors, O.; Sandell, M.; Wilson, S.; Borjesson, P. On channel estimation in OFDM systems. In Proceedings of the 1995 IEEE 45th Vehicular Technology Conference, Countdown to the Wireless Twenty-First Century, Chicago, IL, USA, 25–28 July 1995. [CrossRef]
27. Sutar, M.B.; Patil, V.S. LS and MMSE estimation with different fading channels for OFDM system. In Proceedings of the 2017 International conference of Electronics, Communication and Aerospace Technology (ICECA), Coimbatore, India, 20–22 April 2017. [CrossRef]
28. Aida, Z.; Ridha, B. LMMSE channel estimation for block—Pilot insertion in OFDM systems under time varying conditions. In Proceedings of the 2011 11th Mediterranean Microwave Symposium (MMS), Hammamet, Tunisia, 8–10 September 2011. [CrossRef]
29. Noh, M.; Lee, Y.; Park, H. Low complexity LMMSE channel estimation for OFDM. *IEE Proc. Commun.* **2006**, *153*, 645. [CrossRef]
30. Porter, M.B.; Bucker, H.P. Gaussian beam tracing for computing ocean acoustic fields. *J. Acoust. Soc. Am.* **1987**, *82*, 1349–1359. [CrossRef]
31. Jensen, F.B.; Kuperman, W.A.; Porter, M.B.; Schmidt, H. *Computational Ocean Acoustics, AIP Series in Modern Acoustics And Signal Processing*; American Institute of Physics: New York, NY, USA, 1994; Volume 80.
32. Porter, M.B. *The Bellhop Manual and User's Guide: Preliminary Draft*; Heat, Light, and Sound Research, Inc.: La Jolla, CA, USA, 2011.
33. Lindgren, G. *Prev—A Wafo module for Analysis of Random Lagrange Waves, Mathematical Statistics*; Lund University: Lund, Sweden, 2015.
34. AVISO+, GRIDDED WIND/WAVE PRODUCTS. Available online: https://www.aviso.altimetry.fr/en/data/products/windwave-products/mswhmwind.html (accessed on 12 December 2018).

35. World Ocean Atlas 2013. Available online: https://www.nodc.noaa.gov/OC5/woa13/woa-info.html (accessed on 12 December 2018).
36. Wong, G.S.K.; Zhu, S.-M. Speed of sound in seawater as a function of salinity, temperature, and pressure. *J. Acoust. Soc. Am.* **1995**, *97*, 1732–1736. [CrossRef]
37. Google Maps Platform. Available online: https://cloud.google.com/maps-platform/ (accessed on 12 December 2018).

Article

Experimental Investigation of Acoustic Propagation Characteristics in a Fluid-Filled Polyethylene Pipeline

Qi Li [1,2,3], Jiapeng Song [1,2,3] and Dajing Shang [1,2,3],*

1   Acoustic Science and Technology Laboratory, Harbin Engineering University, Harbin 150001, China;
    leechi319@163.com (Q.L.); sjp91872@sina.com (J.S.)
2   Key Laboratory of Marine Information Acquisition and Security (Harbin Engineering University),
    Ministry of Industry and Information Technology, Harbin 150001, China
3   College of Underwater Acoustic Engineering, Harbin Engineering University, Harbin 150001, China
*   Correspondence: shangdajing@hrbeu.edu.cn; Tel.: +86-133-0450-3268

Received: 16 November 2018; Accepted: 4 January 2019; Published: 9 January 2019

Abstract: Fluid-filled polyethylene (PE) pipelines have a wide range of applications in, for example, water supply and gas distribution systems, and it is therefore important to understand the characteristics of acoustic propagation in such pipelines in order to detect and prevent pipe ruptures caused by vibration and noise. In this paper, using the appropriate wall parameters, the frequencies of normal waves in a fluid-filled PE pipeline are calculated, and the axial and radial dependences of sound fields are analyzed. An experimental system for investigating acoustic propagation in a fluid-filled PE pipeline is constructed and is used to verify the theoretical results. Both acoustic and mechanical excitation methods are used. According to the numerical calculation, the first-, second-, and third-order cutoff frequencies are 4.6, 10.4, and 16.3 kHz, which are close to the experimentally determined values of 4.7, 10.6, and 16 kHz. Sound above a cutoff frequency is able to propagate in the axial direction, whereas sound below this frequency is attenuated exponentially in the axial direction but can propagate along the wall in the form of vibrations. The results presented here can provide some basis for noise control in fluid-filled PE pipelines.

Keywords: fluid-filled polyethylene (PE) pipeline; noise control; acoustic propagation; cutoff phenomenon

1. Introduction

In pipeline systems, a number of different materials can be used for the pipe walls, with the most common being steel and thermoplastics such as polyethylene (PE). PE pipelines have a wide range of applications, including transportation of liquids on ships and aircraft, long-distance transportation of natural gas, storage and transportation of liquids in the chemical industry, and urban water supply. They are of particular importance in the last of these applications owing to their advantages of high strength, high resistance to corrosion and wear, good stability over a wide range of temperatures, and lack of toxicity [1]. However, fluid-filled PE pipelines often suffer from problems caused by excessive vibration and noise [2,3]. Vibration can cause long-term fatigue damage to the pipeline system, while noise not only reduces the stability and safety of the entire pipeline system, but can also have a deleterious effect on the environment for people in the vicinity of the pipeline. Burst water supply pipelines not only result in losses of large amounts of water and consequent serious disruption to daily life, but can also lead to secondary consequences such as traffic jams and even disasters such as landslides if they are not repaired rapidly. Therefore, it is of great importance to investigate the acoustic transmission characteristics of fluid-filled PE pipelines and to develop methods to control the noise and vibration that are generated in these systems.

Before investigating acoustic propagation in fluid-filled pipelines, it is useful to consider the problem of elastic vibrations of a cylindrical shell. In the nineteenth century, Rayleigh [4] investigated the vibration of cylindrical shells and obtained the free-vibration frequency of an infinite cylindrical shell in vacuum. More recently, the dispersion characteristics of pipelines have been extensively investigated [5–9]. Junger [10–12] and Muggeridge [13] investigated vibration problems for cylindrical shells in liquids. The first to investigate acoustic propagation in a fluid-filled pipeline was Lamb [14], who came to the conclusion that this propagation is influenced by the strength of longitudinal waves in the wall compared with that of bending waves. Later, Lin and Morgan [15] investigated the dispersion properties of a sound field in a fluid-filled pipeline, and analyzed the first four normal modes of waves in an axisymmetric rigid pipeline.

Using a short-pulse signal, Kwun et al. [16] experimentally investigated the dispersion of longitudinal waves in a liquid-filled cylindrical shell and found that the liquid in the pipeline slightly reduced the group velocity and cutoff frequency of the longitudinal mode in the tube wall. Horne et al. [17] conducted an experimental investigation of acoustic propagation in a liquid-filled pipeline, examining the effects of different pipe-wall materials on the sound field in the pipe. However, their experiment suffered from the limitations that the end of the pipeline was not muffled and the sound pressure was measured at only one point at a given time. Pan et al. [18] investigated acoustic propagation in a fluid-filled pipeline both experimentally and numerically. In their experiment, sound field measurements could not be obtained throughout the fluid, because they used only two PZT (Lead zirconate titanate piezoelectric ceramic) circular transducers mounted on the pipe wall, one at the transmitting end and the other at the receiving end. Lafleur [19], Aristegui et al. [20] and Baik et al. [21] conducted systematic theoretical and experimental investigations of acoustic propagation in liquid-filled elastic pipelines, but their experimental methods and equipment were not very different from those used in previous studies. To date, there have been few theoretical and experimental studies focusing on sound below the cutoff frequency. Many engineers are not aware of the low-frequency cutoff effect and the propagation path of low-frequency noise. Most pipeline mufflers are limited to liquid noise reduction alone, and do not deal with wall vibration [22,23], which is a serious omission.

There have been a number of systematic investigations of acoustic propagation in gas-filled pipelines [24,25], and the results obtained on noise in such systems can provide some guidance for studies of noise in liquid-filled pipelines. However, these results cannot be carried over completely to the liquid case because of the differences in the characteristic impedance and speed of motion of the respective media [26,27]. There have also been a few theoretical studies of noise generated in fluid-filled elastic pipelines by supersonic flow [28], wall vibration [29], and bubble oscillations [30], but there is a lack of corresponding experimental investigations.

The present study aims to improve on the results of previous work by taking full account of the existence of a low-frequency cut-off phenomenon in a fluid-filled pipeline such that sound below a cut-off frequency is mainly propagated through the pipeline wall. It thereby also aims to remedy some shortcomings of previous attempts at pipeline noise reduction. Both theoretical and experimental investigations of acoustic propagation in a fluid-filled PE pipeline are conducted. The remainder of the paper is organized as follows. In Section 2, the eigenequation for the sound field in an elastic PE pipeline is obtained from a theoretical analysis, and the cutoff frequencies of a normal wave in the PE tube are calculated. Section 3 describes the experimental system and the scheme for determining the acoustic propagation characteristics of the fluid-filled PE pipeline. Section 4 discusses the experimental results. The general distribution law of the sound field and the propagation path of noise in the fluid-filled PE pipeline are analyzed. Finally, Section 5 presents the conclusions of this study.

2. Theoretical Analysis

2.1. Eigenequation in the Pipeline

It should first be noted that although calculations in pipe acoustics have generally been performed under the assumption of absolute soft or absolute hard boundary conditions, the characteristic impedance of a liquid is large compared with that of air, and cannot be ignored, and so ideal boundary conditions are no longer applicable.

The infinitely long straight pipeline considered here has outer diameter a and inner diameter b, as shown in Figure 1 [31].

Figure 1. Infinitely long straight fluid filled pipeline model.

In the following, the displacement scalar potential function is denoted by ϕ, the vector potential function by $\vec{\Psi}$, the longitudinal-wave velocity in the wall by c_l, the shear-wave velocity by c_s, the shear modulus of the pipe wall material by μ, the wave velocity in the liquid in the pipe by c_0, and the liquid density in the pipe by ρ_1. Some previous studies have assumed an axisymmetric source in a plate [32] or in a pipe, using a cylindrical coordinate system [33], while some have assumed a point source [34,35]. In the present study, the assumption of axisymmetric excitation is an important one.

Under axisymmetric excitation, with $\vec{\Psi} = (0, \psi_\theta, 0)$, the wave equation in the wall can be represented by the following equations for two scalar potential functions:

$$\nabla^2 \phi = \frac{1}{c_l^2} \frac{\partial^2 \phi}{\partial t^2},$$
$$\left(\nabla^2 - \frac{1}{r^2} \right) \psi_\theta = \frac{1}{c_s^2} \frac{\partial^2 \psi_\theta}{\partial t^2}, \tag{1}$$

where

$$\nabla^2 = \frac{\partial^2}{\partial r^2} + \frac{1}{r} \frac{\partial}{\partial r} + \frac{\partial^2}{\partial z^2}.$$

The radial and axial components of the displacement are

$$u_r = \frac{\partial \phi}{\partial r} - \frac{\partial \psi_\theta}{\partial z},$$
$$u_z = \frac{\partial \phi}{\partial z} + \frac{1}{r} \frac{\partial (r \psi_\theta)}{\partial r}, \tag{2}$$

where u_r is the radial components of the displacement in the wall, u_z is the axial components of the displacement in the wall, and the normal and tangential components of the stress are

$$\delta_{rr} = \lambda \Delta + 2\mu \frac{\partial u_r}{\partial r},$$
$$\delta_{rz} = \mu \left(\frac{\partial u_r}{\partial z} + \frac{\partial u_z}{\partial r} \right), \tag{3}$$

where δ_{rr} is the normal component of the stress in the wall, δ_{rz} is the tangential components of the stressin the wall, λ and μ are the Lame coefficients.

Under the assumption of a simple harmonic vibration in the z direction, the displacement potential function can be expressed as

$$\phi = \Phi e^{i(k_z z - \omega t)}, \psi_\theta = \Psi e^{i(k_z z - \omega t)}.$$

With the time dependence ignored, by substitution of these expressions into the wave equation, the relationship between the displacement potential function and the displacement and stress in the pipe wall can be obtained.

When $b \leq r \leq a$, the formal solution for the potential function in the wall is

$$\begin{aligned}
\phi(r,z) &= [A J_0(k_l r) + B Y_0(k_l r)] e^{i k_z z}, \ k_l^2 + k_z^2 = (\omega/c_l)^2, \\
\psi_\theta(r,z) &= [C J_0(k_t r) + D Y_0(k_t r)] e^{i k_z z}, \ k_t^2 + k_z^2 = (\omega/c_s)^2,
\end{aligned} \tag{4}$$

where A, B, C, and D are constants.

The wave equation satisfied by the water potential function is

$$\nabla^2 \phi_1 = \frac{1}{c_0^2} \frac{\partial^2 \phi_1}{\partial t^2}. \tag{5}$$

The radial and axial components of the displacement in the water are

$$\begin{aligned}
u_{rf} &= \frac{\partial \phi_1}{\partial r}, \\
u_{zf} &= \frac{\partial \phi_1}{\partial z},
\end{aligned} \tag{6}$$

where u_{rf} is the radial component of the displacement in the water, u_{rr} is the axial component of the displacement in the water.

The normal stress in the water is

$$\delta_{rrf} = \rho_1 \omega^2 \phi_1, \tag{7}$$

where δ_{rrf} is the normal stress in the water.

Similarly, under the assumption of a simple harmonic vibration in the z direction, the displacement potential function can be expressed as

$$\phi_1 = \Phi_1 e^{i(k_z z - \omega t)}.$$

With the time dependence ignored, on substitution of these expressions into the wave equation, the relationship between the displacement potential function and the displacement and stress in the water can be obtained.

When $0 \leq r \leq b$, the formal solution for the potential function in the water is

$$\phi_1(r,z) = E J_0(k_r r) e^{i k_z z}, \ k_r^2 + k_z^2 = (\omega/c_0)^2, \tag{8}$$

where E is a constant. The boundary conditions are

$$\left\{ \begin{array}{l} \delta_{rr}|_b = \delta_{rrf}|_{b'} \\ \delta_{rz}|_b = 0, \\ u_r|_b = u_{rf}|_{b'} \end{array} \right. \qquad \left\{ \begin{array}{l} \delta_{rr}|_a = 0, \\ \delta_{rz}|_a = 0. \end{array} \right. \tag{9}$$

Substitution of the formal solutions from Equations (4) and (10) into the expressions for the stress and displacement, and substitution into the boundary conditions (9), then gives the eigenequation

$$
\begin{bmatrix}
P(a) & Q(a) & R(a) & S(a) & 0 \\
P(b) & Q(b) & R(b) & S(b) & -\frac{\rho_1 \omega^2}{2\mu} J_0(k_r b) \\
M J_1(k_l a) & M Y_1(k_l a) & G J_1(k_t a) & G Y_1(k_t a) & 0 \\
M J_1(k_l b) & M Y_1(k_l b) & G J_1(k_t b) & G Y_1(k_t b) & 0 \\
k_l J_1(k_l b) & k_l Y_1(k_l b) & i k_z k_t J_1(k_t b) & i k_z k_t Y_1(k_t b) & k_r J_1(k_r b)
\end{bmatrix}
\begin{bmatrix} A \\ B \\ C \\ D \\ E \end{bmatrix} = 0, \tag{10}
$$

where

$$
\begin{cases}
P(r) = -T J_0(k_l r) + \frac{k_l}{r} J_1(k_l r), & R(r) = N\left[J_0(k_t r) - \frac{1}{k_t r} J_1(k_t r)\right], \\
Q(r) = -T Y_0(k_l r) + \frac{k_l}{r} Y_1(k_l r), & S(r) = N\left[Y_0(k_t r) - \frac{1}{k_t r} Y_1(k_t r)\right], \\
T = \frac{1}{2}(k_t^2 - k_z^2), & G = k_t(k_t^2 - k_z^2), \quad N = -i k_z k_t^2, \quad M = 2 i k_z k_l.
\end{cases}
$$

2.2. Calculation of the Normal Frequency

For Equation (10) to have a nonzero solution, the determinant of the coefficient matrix must vanish, i.e.,

$$
\begin{vmatrix}
P(a) & Q(a) & R(a) & S(a) & 0 \\
P(b) & Q(b) & R(b) & S(b) & -\frac{\rho_1 \omega^2}{2\mu} J_0(k_r b) \\
M J_1(k_l a) & M Y_1(k_l a) & G J_1(k_t a) & G Y_1(k_t a) & 0 \\
M J_1(k_l b) & M Y_1(k_l b) & G J_1(k_t b) & G Y_1(k_t b) & 0 \\
k_l J_1(k_l b) & k_l Y_1(k_l b) & i k_z k_t J_1(k_t b) & i k_z k_t Y_1(k_t b) & k_r J_1(k_r b)
\end{vmatrix} = 0. \tag{11}
$$

This equation is the dispersion relation.

If $k_z = 0$, then $k_r = (\omega/c_0)^2$, $k_l = (\omega/c_l)^2$, and $k_t = (\omega/c_s)^2$, and there is no sound propagation in the axial direction of the pipeline. Then, ω can be obtained by substituting these values of k_r, k_l, and k_t into Equation (11), and the corresponding frequency is the normal frequency of the corresponding order of vibration of the fluid-filled elastic pipeline.

Table 1 shows an example of the normal frequencies of the first four orders of vibration calculated using Newton's iterative method with the wall parameters of the experimental liquid-filled PE pipeline (also shown in the table).

Table 1. Wall material parameters of the experimental polyethylene (PE) pipeline and an example of the numerically calculated normal frequencies.

Wall Material Parameters			
a (m)	b (m)	ρ_0 (kg/m^3)	ρ_1 (kg/m^3)
0.125	0.116	1000	940
c_0 (m/s)	c_1 (m/s)	μ (GPa)	
1470	1640	0.377	
Numerically calculated values of the normal frequency			
First order (kHz)	Second order (kHz)	Third order (kHz)	Fourth order (kHz)
4.6	10.4	16.3	22.2

2.3. Axial and Radial Dependence of the Sound Field

The sound field in the tube can be analyzed in terms of the formal solution in Equation (8) for the displacement potential function in the water. If the radial wavenumber k_r is negligible, then only the axial dependence of the wave needs be considered. When k_z is real, $e^{ik_z z}$ is a periodic function, and the sound wave is able to propagate for long distances along the axial direction. When k_z is imaginary, $e^{ik_z z}$ is an exponential function, and the normal wave is transformed into a nonuniform wave attenuated according to an exponential law along the axial direction, and thus has very little influence on the sound field far from the pipeline axis: The sound wave cannot propagate for long distances in the pipeline.

If the axial wavenumber k_z is negligible, then only the radial dependence of the wave needs be considered. It can be seen that this is given by the zeroth-order Bessel function $J_0(k_r r)$, and so the sound pressure is greatest close to the axis.

3. Experimental Apparatus and Procedure

To verify the theoretical results, experiments were carried out using the system shown in Figure 2. These experiments focused on the distribution of the sound field and the acoustic propagation behavior in the pipeline for different excitation sources when the liquid in the pipeline was stationary. The experimental conditions are listed in Table 2, and photographs of the experimental conditions and experiment apparatus are shown in Figure 3.

Table 2. Experimental conditions.

Working Condition	Source	Frequency (kHz)	Remarks
1	White noise signal	0–20	
2	Single-frequency acoustic signal 1	4.2	Single-frequency signal below the cutoff frequency
3	Single-frequency acoustic signal 2	5.2	Single-frequency signal above the cutoff frequency
4	Single-frequency mechanical signal 1	4.2	The excitation point is outside the tube wall, directly below the sound source
5	Single-frequency mechanical signal 2	5.2	The excitation point is outside the tube wall, directly below the sound source

Normal waves in the pipeline can be analyzed under white noise conditions. The white noise frequency range was selected as 0–20 kHz according to the sampling frequency of the collector and the theoretically calculated normal frequency. The variation of the sound field along the axial direction in the pipeline can be analyzed under a single-frequency-signal condition. The two frequencies below and above the cutoff frequency were selected in experimental conditions 2 and 3, respectively, to verify the cutoff effect of the sound in the pipeline. Single-frequency mechanical excitation corresponds to the acoustic signal experimental conditions, and the propagation path of the sound was analyzed along the axial direction, and therefore experimental conditions 4 and 5 involved transmission of two single-frequency mechanical force signals of the same frequency as the sound source.

The maximum sampling rate of the B&K pulse collector was 131,072 Hz. The sensors used in this experiment were a B&K8103 hydrophone and B&K4371 vibration sensor. Their specifications are given in Table 3.

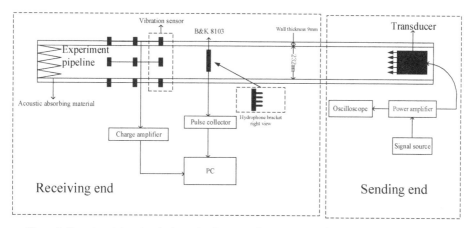

Figure 2. Experimental system for investigating acoustic propagation characteristics of a fluid-filled PE pipeline.

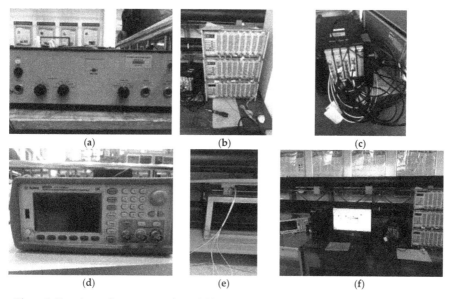

Figure 3. Experimental apparatus and acquisition system: (**a**) B&K2713 power amplifier; (**b**) YE5859 charge amplifier; (**c**) B&K pulse collector; (**d**) Agilent 33522A signal source; (**e**) hydrophone bracket; (**f**) experimental acquisition system.

Table 3. Specifications of sensors.

Hydrophone (B&K 8103)		
Voltage sensitivity	Frequency range	Maximum operating static pressure
30 μV/Pa	0.1–180 kHz	4×10^6 Pa
Vibration sensor (B&K 4371)		
Charge sensitivity	Frequency range	Maximum operational level
1 pC/ms^2	0.1–25 200 Hz	6000g

As mentioned in references [10–13], there have been many experimental investigations of acoustic propagation in fluid-filled elastic pipelines; however, these pipelines were rather short, and there was no special treatment of the end of the pipeline other than, in some cases, the simple addition of a flange, which caused inverse superposition of the sound field in the axial direction. These previous experiments used a single hydrophone, measuring the sound pressure spectrum at a single point only, and therefore it was not possible to obtain the distribution of the sound field along the entire pipeline.

To avoid the above problems, in the present experiment, an 18-m-long PE pipeline with two layers of anechoic tips of different lengths installed at the end was used, which completely eliminated echo and prevented inverse superposition of the sound field in the axial direction. As the source, a piston transducer was mounted on the front of the pipe through a flange, and vibration isolation material was interposed between the flange and the pipeline, thereby preventing direct excitation of the pipe wall. The wall parameters were the same as those used in the theoretical calculations (see Table 1). Five slots, each of length 1 m, in the axial direction were cut in the pipe wall at a distance of 1 m from the transducer, with a 1 m gap between them (see Figure 4).

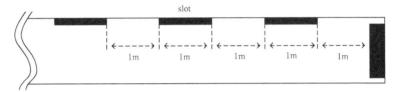

Figure 4. Slots of the pipeline.

Four 8103 hydrophones were mounted on a bracket, and so the sound pressure at four different radial positions could be measured at the same time, as shown in Figure 2. The hydrophone bracket was made from polyvinyl chloride (PVC), which has characteristic impedance similar to that of water, greatly reducing the scattering of sound waves as they passed through the bracket. Three other hydrophones measured the near-field signal at 0, 0.05, 0.1, and 0.15 m from the axial center of the transducer. After the pipeline was filled with water, it was allowed to stand for more than 30 h to eliminate the effect of bubbles on the experiment.

After the fluid column (water) was excited by the transducer, the hydrophones were moved along the axial slots away from the source, and recordings were taken at 10 cm intervals; thus, each slot had 9 or 10 recording points. The use of the hydrophone bracket ensured that the radial positions of the hydrophones remained unchanged when the axial position was changed. In this way, the sound pressure distributions of the fluid column along both the axial and radial directions were measured.

The analysis bandwidth of the collector was set to 0–25 600 Hz, in accordance with the theoretical normal frequencies, and the corresponding sampling frequency was set to 65,536 Hz by the Nyquist law, with a sampling time of 10 s. The power spectrum of the corresponding working condition was obtained through fast Fourier transform (FFT) processing of the time-domain signal. To determine the propagation path of the sound in the experimental system and compare it with the acoustic signal of the fluid in the pipeline at the same time, the vibration of the wall was also measured in this experiment. There were three rows of vibration sensors encircling the outside of the wall, each with four sensors, with a separation of 0.33 m between each row.

The deployment of the hydrophones and vibration sensors and the corresponding labels are shown in Figure 5.

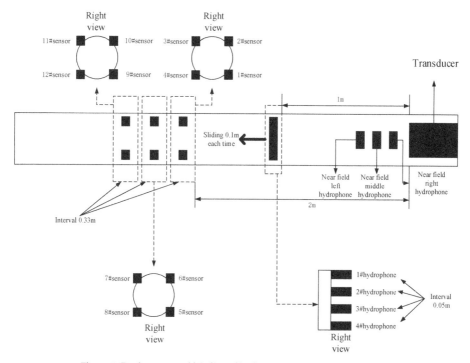

Figure 5. Deployment and labeling of hydrophones and vibration sensors.

4. Results and Discussion

4.1. Behavior of Normal Waves in the Pipeline

For working condition 1, the transmitting voltage level response of the transducer is shown in Figure 6. The measurement results from the hydrophones at different positions along the axial direction in the pipeline are shown in Figure 7. It can be seen that the acoustic signal in the pipeline exhibits a significant cutoff phenomenon, with a cutoff frequency of 4.7 kHz, which is close to the theoretically determined cutoff frequency of 4.6 kHz.

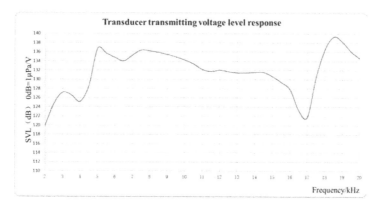

Figure 6. Transmitting voltage level response of transducer.

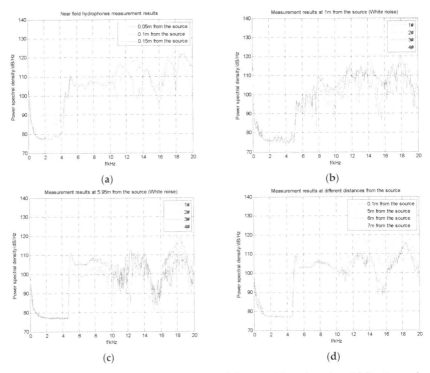

Figure 7. Measurement results from hydrophones at different positions along the axial direction under white noise excitation: (**a**) 0.05, 0.1, and 0.15 m from the source; (**b**) 1.1 m from the source; (**c**) 5.95 m from the source; (**d**) 0.1, 5, 6, and 7 m from the source.

The sound energy of the far field can be divided into four intervals from the spectrum: (1) Below 4.7 kHz; (2) 4.7–10.6 kHz; (3) 10.6–16 kHz; (4) 16 kHz and above. The boundary points of these intervals are the frequencies of the normal wave in the pipeline, which are basically consistent with the calculated frequencies of the corresponding orders, as shown in Table 4.

Table 4. Comparison of calculated and experimental normal frequencies.

Order	1	2	3
Calculated frequency (kHz)	4.6	10.4	16.3
Experimental frequency (kHz)	4.7	10.6	16
Relative error (%)	2.17	1.92	1.84

As mentioned before, the experimental pipeline is slotted. PE is not a very rigid material and, as a result of the slotting process, the tube is deformed radially. This is the main reason for the error between the actual measured frequency and the theoretically calculated frequency. In addition, the parameters in Table 1 are the material elastic parameters of standard high-density PE, which are not necessarily exactly the same as those of the wall of the experimental pipeline, which also leads to an error between the theoretical and experimental results.

In interval (1), the curve of the power spectrum is very close to the background noise. Close to 4.7 kHz, however, the curve suddenly rises, exhibiting a cutoff phenomenon. Then, in interval (2), the curve changes relatively gently, which is basically consistent with the corresponding frequency band of the transmitting transducer frequency response curve. As the distance between the measurement points and the source becomes greater, it can be seen that the curve decreases monotonically; in interval (3), the curve changes more sharply, and many resonance peaks appear. As the frequency increases, the distance between adjacent peaks also increases, and the appearance of the curve in interval (4) is similar to that in interval (3).

In terms of the behavior of the normal wave, interval (1) is below the cutoff frequency, and the normal wave is attenuated exponentially in the axial direction. The curve of the power spectrum in this frequency band is basically the same as the background, and it can be seen that the power spectrum decays exponentially with frequency in the near field.

Interval (2) lies between the first-order and second-order normal wave frequencies; only the first-order normal wave can propagate, and the curve of the power spectrum hardly changes with frequency. The trend in this section of the curve is related to the transmitting response of the source, with the curve monotonically decreasing along the axial direction as a result of absorption of sound waves by the tube wall.

The first and second orders of the normal wave propagate simultaneously in interval (3), and the first, second, and third orders propagate in interval (4), where there is strong interference leading to large fluctuations in the curve and to the appearance of many resonant peaks.

The results of the vibration signal measurements are shown in Figure 8. The frequency response of the vibration signal is similar to that of the acoustic signal in its overall trend, with a cutoff phenomenon, and the division of the modal frequencies of each order is obvious. The vibration signal in the frequency band above the cutoff frequency is transmitted from the acoustic signal in the water.

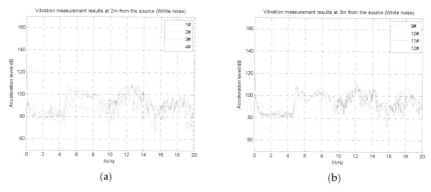

Figure 8. Measurement results from vibration sensors on the outside of the pipeline wall: (**a**) 2 m from the source; (**b**) 3 m from the source.

4.2. Variation of the Sound Field along the Axial Direction

The variation of the sound field along the axial direction can be seen more clearly from analysis of the response to a single-frequency source compared with the response to the white noise in working condition 1. In working conditions 2 and 3, two representative single-frequency signals were used, 4.2 and 5.2 kHz, which are respectively below and above the cutoff frequency. The main frequency power spectrum was obtained as an average of the measurements by the four hydrophones on the bracket, and its variation along the axial distance is shown in Figure 9. It can be seen from Figure 9a that for a source frequency below the cutoff frequency, the power spectrum of the main frequency is attenuated very rapidly, indeed exponentially, as the distance increases. Acoustic signals below the cutoff frequency cannot propagate axially over long distances in the pipeline. For a source frequency above the cutoff frequency, as shown in Figure 9b, the power spectrum of the main frequency hardly changes with distance. There is only 4 dB attenuation from 3 to 10 m, and this attenuation is a result of acoustic absorption by the pipeline wall.

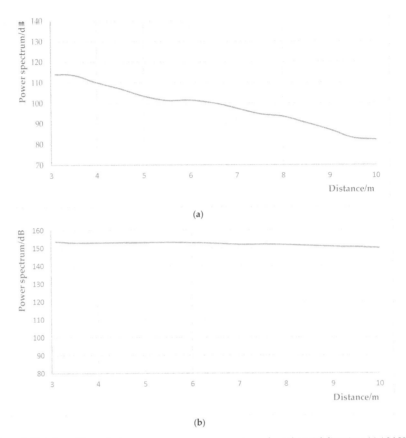

(a)

(b)

Figure 9. Variation of the main frequency average power spectrum along the axial direction: (**a**) 4.2 kHz source; (**b**) 5.2 kHz source.

4.3. Variation of the Sound Field along the Radial Direction

To explore the variation of the sound field in the radial direction, measurements were performed before and after the hydrophone bracket was raised by 1.5 cm, as shown schematically in Figure 10. The results of these measurements are shown in Figure 11.

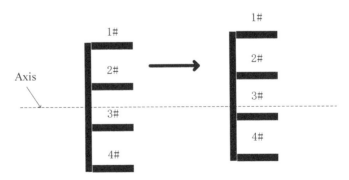

Figure 10. Raising of hydrophone bracket.

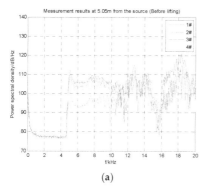
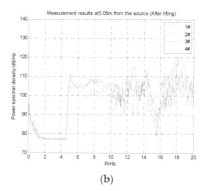

(a)   (b)

Figure 11. Results of hydrophone measurements at different depths 5.05 m from the source: (**a**) Before and (**b**) after lifting of hydrophone bracket.

The frequency band between the first- and second-order normal frequencies was analyzed. Before lifting, the 2# and 3# hydrophones were at the same distance from the axis, and similarly for the 1# and 4# hydrophones. Therefore, in Figure 11a, the curves of the power spectra from the 2# and 3# hydrophones are the same, as are those from the 1# and 4# hydrophones. After lifting, the 3# hydrophone is nearest to the axis, followed in order by the 2#, 4#, and 1# hydrophones, which is consistent with the increasing strengths of the respective power spectra. This is in accordance with the theoretical radial dependence on the Bessel function $J_0(k_r r)$ in Equation (8).

The following is a quantitative analysis of the radial distribution. The distance between each pair of hydrophones is 50 mm, and the hydrophone bracket is initially at the radial center position. After lifting, the distances of the 3#, 2#, 4#, and 1# hydrophones from the axis are 10, 40, 60, and 90 mm, respectively. At the cutoff frequency, $k_r = k_0$, depending on the distance r from the axis, the theoretical differences between the sound pressure self-spectrum measured by the 3# hydrophone and those measured by the other three hydrophones can be calculated, and the results are compared with the experimental measurements in Table 5. For convenience of exposition, the distance between the 3# hydrophone and the axis is set as r_0, and the differences between the sound pressure power spectrum of the 1#, 2#, and 4# hydrophones and the 3# hydrophone as X_1, X_2, and X_3, respectively.

Table 5. Comparison of hydrophone measurements and theoretical values at different radial locations.

Power Spectral Difference	Difference between r_0 and r	Theoretical Power Spectral Difference (dB)	Experimental Power Spectral Difference (dB)
X_1	80	12.1	11.1
X_2	30	1.6	2
X_3	40	3.8	3.3

There is good agreement between theory and experiment, and the radial distribution of the normal wave in the tube is quantitatively confirmed to follow the Bessel function behavior.

The reasons for the error are as follows. In the experiment, the magnitude of the lifting was controlled manually, and not very accurately, which is the main source of error: If the lifting range were slightly larger, and the 3# and 4# hydrophones closer to the axis, the amplitude would be higher. If the 1# and 2# hydrophones were further away from the axis, the amplitude would be smaller, which would lead to an increase in X_2; according to the properties of the Bessel function, the closer the value of the function is to the axis, the slower is its rate of change. Therefore, the power spectrum at the 4# hydrophone would increase more if it were lifted, so the value of X_3 would be reduced. If the 1# hydrophone were closer to the upper slot and the outside medium were air (which can be regarded as an absolutely soft boundary), the sound would be totally reflected, so the amplitude would become higher, leading to a decrease in X_1. In addition, the radial deformation of the pipeline due to slotting, scattering by the hydrophone bracket, and the fact that the orientation of the hydrophone was not strictly in the axial direction are all possible sources of error.

4.4. Measurements under Mechanical Excitation

Working conditions 4 and 5 use mechanical force excitation from outside the pipeline wall in a position directly under that of the sound source in the previous working conditions, as shown in Figure 12. Corresponding to working conditions 2 and 3, the single frequencies of excitation applied in working conditions 4 and 5 are 4.2 and 5.2 kHz, respectively. The measurement results from the hydrophones and vibration sensors are shown in Figures 13–16. In contrast to the response of the acoustic signal, the excitation of the exciter to the pipe wall was a single-point excitation. Therefore, when the sound propagated mainly along the wall, the sound source excitation and the exciter excitation had completely different radial distribution laws. When the sound propagated mainly along the liquid in the tube, both had the same radial distribution. This is an important basis for judging the propagation path of sound in a pipeline.

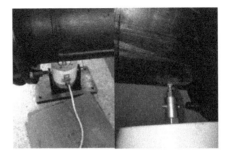

Figure 12. Exciter and excitation point in working conditions 4 and 5.

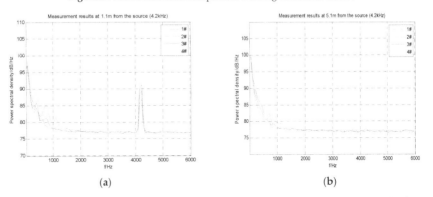

Figure 13. Mechanical excitation at a single frequency of 4.2 kHz. Measurement results from hydrophones at different distances from the excitation point: (**a**) 1.1 m; (**b**) 5.1 m.

Even if the pipeline wall is excited by a mechanical force, an acoustic signal below the cutoff frequency cannot propagate a long distance in the case of weak excitation. It can be seen from Figure 13b that the hydrophones 5.1 m from the excitation point have difficulty in picking up a signal. In Figure 13a, the acoustic signal at the main frequency exhibits a radial dependence that is different from the Bessel function: The hydrophone near the lower outer wall close to the excitation point receives a stronger signal.

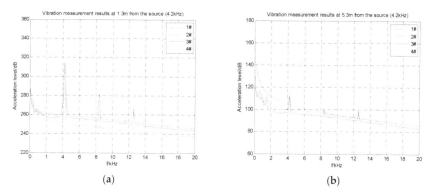

Figure 14. Mechanical excitation at a single frequency of 4.2 kHz. Measurement results from vibration sensors at different distances from the excitation point: (**a**) 1.3 m; (**b**) 5.3 m.

From the vibration measurement results in Figure 14, it can be seen that a vibration signal at the main frequency can still be measured on the wall at 5.3 m, but it is attenuated compared with the measurement from the sensor at 1.3 m, and the hydrophones in the pipeline are unable to detect any signal at 5.3 m. It can be deduced that the signal at 4.2 kHz is propagated mainly through the pipe wall in the form of vibrations. The second and third peaks in Figure 14 result from frequency doubling. The exciter generates frequency doubling when transmitting a single-frequency signal, which is a consequence of its own physical structure and has no effect on the results of this experiment.

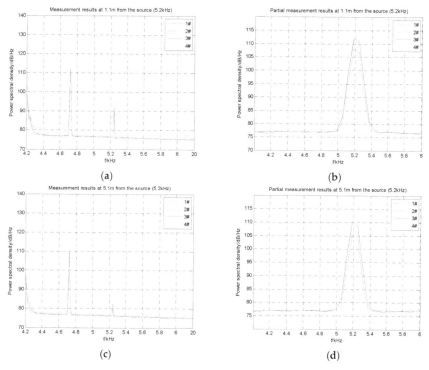

Figure 15. Mechanical excitation at a single frequency of 5.2 kHz. Measurement results from hydrophones at different distances from the excitation point: (**a**) 1.1 m; (**b**) 1.1 m (local); (**c**) 5.1 m; (**d**) 5.1 m (local).

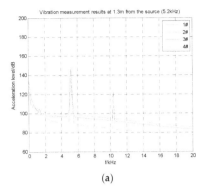
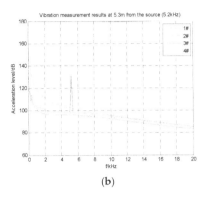

(a)   (b)

Figure 16. Mechanical excitation at a single frequency of 5.2 kHz. Measurement results from vibration sensors at different distances from the excitation point: (**a**) 1.3 m; (**b**) 5.1 m.

For excitation at 5.2 kHz, which is above the cutoff frequency, Figure 15a,b shows the measurement results from hydrophones 1.1 m from the excitation point, and Figure 15c,d those from hydrophones 5.1 m from the excitation point, with Figure 15b,d being partial displays of the frequency band near the main frequency shown in Figure 15a,c, respectively. Figure 16 shows the measurement results from the wall vibration sensors 1.3 and 5.3 m from the excitation point.

The sound power at the main frequency suffers almost no attenuation in the axial direction from 1.1 to 5.1 m, and conforms to the Bessel function dependence in the radial direction. It can be deduced that the signal at 5.2 kHz is propagated mainly in the form of sound through the fluid in the pipeline.

5. Conclusions

The first four orders of normal frequencies in a fluid-filled PE pipeline were calculated, and the distributions of sound in the axial and radial directions were analyzed. The acoustic propagation characteristics of such a pipeline were also studied in an experimental system.

Both the theoretical and experimental investigations have revealed the following:

1. Sound in a fluid-filled PE pipeline propagates through the pipeline with the normal frequency at each order.
2. Sound above a certain cutoff frequency can propagate in the axial direction of the pipeline for long distances, whereas sound below the cutoff frequency is attenuated exponentially in the axial direction and cannot propagate over long distances.
3. In the fluid in the pipeline, the sound power is highest at the axial center and decreases with radial distance from the axial center according to a Bessel function dependence $J_0(k_r r)$.
4. Sound above the cutoff frequency is propagated mainly through the fluid, while sound below the cutoff frequency propagates in the form of vibrations along the pipe wall.
5. Controlling and reducing the vibration of the pipe wall is the most effective way to reduce low-frequency noise in a fluid-filled pipeline system.

Author Contributions: Conceptualization, Q.L. and J.S.; methodology, Q.L.; software, J.S.; data validation, J.S.; formal analysis, D.S.; writing—original draft preparation, J.S.; writing—review and editing J.S. and D.S.; supervision and project administration, Q.L.

Funding: This research was funded by Acoustic Science and Technology Laboratory, Harbin Engineering University (SSJSWDZC2018010) and by the National Science Foundation of China (11874131).

Conflicts of Interest: The authors declare no conflict of interest.

References

1. Serebrennikov, A.; Serebrennikov, D. Mathematical model of polyethylene pipe bending stress state. *Mater. Sci. Eng.* **2017**, *327*, 1–6. [CrossRef]
2. Gao, Y.; Liu, Y.Y. Theoretical and experimental investigation into structural and fluid motions at low frequencies in water distribution pipes. *Mech. Syst. Signal Process.* **2017**, *90*, 126–140. [CrossRef]
3. Wen, J.H.; Shen, H.J.; Yu, D.L.; Wen, S.X. Theoretical and experimental investigation of flexural wave propagating in a periodic pipe with fluid-filled loading. *Chin. Phys. Lett.* **2010**, *27*, 11.
4. Rayleigh, J.W.S. *The Theory of Sound*; Dover Publications: New York, NY, USA, 1945.
5. Scott, J.F.M. The free modes of propagation of an infinite fluid-loaded thin cylindrical shell. *J. Sound Vib.* **1988**, *125*, 241–280. [CrossRef]
6. Sinha, B.K.; Piona, T.J.; Kostek, S. Axisymmetric wave propagation in fluid-loaded cylindrical shells. I. Theory. *J. Acoust. Soc. Am.* **1992**, *92*, 1132–1143. [CrossRef]
7. Sinha, B.K.; Piona, T.J.; Kostek, S. Axisymmetric wave propagation in fluid-loaded cylindrical shells. II. Theory versus experiment. *J. Acoust. Soc. Am.* **1992**, *92*, 1144–1155. [CrossRef]
8. Guo, Y.P. Approximate solutions of the dispersion equation for fluid-loaded cylindrical shells. *J. Acoust. Soc. Am.* **1994**, *95*, 1435–1440. [CrossRef]
9. Xi, Z.C.; Liu, G.R.; Lam, K.Y. Dispersion and characteristic surfaces of waves in laminated composite circular cylindrical shells. *J. Acoust. Soc. Am.* **2000**, *108*, 2179–2186. [CrossRef]
10. Junger, M.C. Radiation loading of cylindrical and spherical surfaces. *J. Acoust. Soc. Am.* **1952**, *24*, 288–298. [CrossRef]
11. Junger, M.C. Vibration of elastic shells in a fluid medium and the associated radiation of sound. *J. Appl. Mech.* **1952**, *19*, 439–445.
12. Junger, M.C. The physical interpretation of the expression for an outgoing wave in cylindrical coordinates. *J. Acoust. Soc. Am.* **1953**, *25*, 40–53. [CrossRef]
13. Muggeridge, D.B.; Buckley, T.J. Flexural vibration of orthotropic cylindrical shells in a fluid medium. *AIAA J.* **1979**, *17*, 1019–1022. [CrossRef]
14. Lamb, H. On the velocity of sound in a tube, as affected by the elasticity of the walls. *Manch. Mem.* **1898**, *42*, 1–16.
15. Lin, T.C.; Morgan, G.W. Wave propagation through fluid contained in a cylindrical elastic shell. *J. Acoust. Soc. Am.* **1956**, *28*, 1165–1176. [CrossRef]
16. Kwun, H.; Bartels, K.A.; Dynes, C. Dispersion of longitudinal waves propagating in liquid-filled cylindrical shells. *J. Acoust. Soc. Am.* **1999**, *105*, 2601–2611. [CrossRef]
17. Horne, M.P.; Hansen, R.J. Sound propagation in a pipe containing a liquid of comparable acoustic impedance. *J. Acoust. Soc. Am.* **1982**, *71*, 1400–1405. [CrossRef]
18. Pan, H.; Koyano, K.; Usui, Y. Experimental and numerical investigations of axisymmetric wave propagation in cylindrical pipe filled with fluid. *J. Acoust. Soc. Am.* **2003**, *113*, 3209–3214. [CrossRef]
19. Lafleur, L.D.; Shields, F.D. Low-frequency propagation modes in a liquid-filled elastic tube waveguide. *J. Acoust. Soc. Am.* **1994**, *97*, 1435–1445. [CrossRef]
20. Aristegui, C.; Lowe, M.J.S.; Cawley, P. Guided waves in fluid-filled pipes surrounded by different fluids. *Ultrasonics* **2001**, *39*, 367–375. [CrossRef]
21. Baik, K.; Jiang, J.; Leighton, T.G. Acoustic attenuation, phase and group velocities in liquid-filled pipes: Theory, experiment, and examples of water and mercury. *J. Acoust. Soc. Am.* **2010**, *128*, 2610–2624. [CrossRef]
22. Liu, B.Y.; Liu, J.C. Suppression of low frequency sound transmission in fluid-filled pipe systems through installation of an anechoic node array. *AIP Adv.* **2018**, *8*, 11. [CrossRef]
23. Wang, M.; Qiao, G.; He, Y.A. Noise control of fluid-filled pipes. *Appl. Acoust.* **2003**, *22*, 35–38.
24. Zhu, Z.X.; Lee, P.C.; Wang, Z.G. Numerical solution compared with experimental result for sound propagation in ducts. *Acta Acust.* **1988**, *13*, 1–8. (In Chinese)
25. Zhu, Z.X.; Guo, R. A study of noise reduction by anti-sound using bypass in ducts with flow. *Acta Acust.* **1997**, *22*, 1–10. (In Chinese)
26. Zhang, Y.M.; Tang, R.; Li, Q.; Shang, D.J. The low-frequency sound power measuring technique for an underwater source in a non-anechoic tank. *Meas. Sci. Technol.* **2018**, *29*, 035101. [CrossRef]

27. Shang, D.J.; Shang, Q.L.D.J.; Lin, H. Experimental investigation on flow-induced noise of the underwater hydrofoil structure. *Acta Acust.* **2012**, *37*, 416–423. (In Chinese)

28. Lan, K.; He, X.T.; Lai, D.X.; Li, S.G. Radiative energy flux of diffusive supersonic wave in a cylinder. *Acta Phys. Sin.* **2016**, *55*, 3789–3795. (In Chinese)

29. Gao, G.J.; Deng, M.X.; Li, M.L. Modal expansion analysis of nonlinear circumferential guided wave propagation in a circular tube. *Acta Phys. Sin.* **2015**, *64*, 184303. (In Chinese)

30. Wang, C.H.; Cheng, J.C. Nonlinear forced oscillations of gaseous bubbles in elastic microtubules. *Acta Phys. Sin.* **2013**, *62*, 114301. (In Chinese)

31. Tang, W.L.; Wu, Y. Interior noise field of a viscoelastic cylindrical shell excited by the TBL pressure fluctuations: I. Production mechanism of the noise. *Acta Acust.* **1997**, *22*, 60–69. (In Chinese)

32. Haider, M.F.; Giurgiutiu, V. Analysis of axis symmetric circular crested elastic wave generated during crack propagation in a plate: A Helmholtz potential technique. *Int. J. Solids Struct.* **2018**, *134*, 130–150. [CrossRef]

33. Rose, J.L. *Ultrasonic Guided Waves in Solid Media*; Cambridge University Press: Cambridge, UK, 2014.

34. Mostafapour, A.; Davoodi, S. A theoretical and experimental study on acoustic signals caused by leakage in buried gas-filled pipe. *Appl. Acoust.* **2015**, *87*, 1–8. [CrossRef]

35. Achenbach, J.A.; Achenbach, J.D. *Reciprocity in Elastodynamics*; Cambridge University Press: Cambridge, UK, 2003.

Article

Regularization Factor Selection Method for l_1-Regularized RLS and Its Modification against Uncertainty in the Regularization Factor

Junseok Lim [1],* and Seokjin Lee [2]

[1] Department of Electrical Engineering, College of Electronics and Information Engineering, Sejong University, Gwangjin-gu, Seoul 05006, Korea

[2] School of Electronics Engineering, College of IT Engineering, Kyungpook National University, Daegu 41566, Korea; sjlee6@knu.ac.kr

* Correspondence: jslim@sejong.ac.kr; Tel.: +82-2-3408-3299

Received: 5 December 2018; Accepted: 4 January 2019; Published: 8 January 2019

Featured Application: This algorithm can be applied to various kinds of sparse channel estimations, e.g., room impulse response, early reflection, and underwater channel response.

Abstract: This paper presents a new l_1-RLS method to estimate a sparse impulse response estimation. A new regularization factor calculation method is proposed for l_1-RLS that requires no information of the true channel response in advance. In addition, we also derive a new model to compensate for uncertainty in the regularization factor. The results of the estimation for many different kinds of sparse impulse responses show that the proposed method without a priori channel information is comparable to the conventional method with a priori channel information.

Keywords: l_1-regularized RLS; sparsity; room impulse response; total least squares; regularization factor

---

1. Introduction

Room impulse response (RIR) estimation is a problem in many applications that use acoustic signal processing. The RIR identification [1] is fundamental for various applications such as room geometry related spatial audio applications [2–5], acoustic echo cancellation (AEC) [6], speech enhancement [7], and dereverberation [8]. In [9], the RIR has relatively large magnitude values during the early part of the reverberation and fades to smaller values during the later part. This indicates that most RIR entries have values close to zero. Therefore, the RIR has a sparse structure. The sparse RIR model is useful for estimating RIRs in real acoustic environments when the source is given a priori [10]. There has been recent interest in adaptive algorithms for sparsity in various signals and systems [11–22]. Many adaptive algorithms based on least mean square (LMS) [11,12] and recursive least squares (RLS) [14–17] have been reported with different penalty functions. Sparse estimation research, such as that done by Eksioglu and Tanc [17], has proposed a sparse RLS algorithm, l_1-RLS, which is fully recursive like the plain RLS algorithm. The algorithm of l_1-RLS in [17] proposed a proper calculation method for the regularization factor. These recursive algorithms have the potential for sparse RIR estimation; however, the regularization factor should be established prior to applying these algorithms. The regularization factor calculation method requires information about a true sparse channel response for a good performance. The authors in [18,19] have also proposed recursive regularization factor selection methods; however, these methods still need the true impulse response in advance.

In this paper, we propose a new regularization factor calculation method for l_1-RLS algorithm in [17]. The new regularization factor calculation needs no information for the true channel response in advance. This makes it possible to apply l_1-RLS algorithm in various room environments. In addition, we derive a new model equation for l_1-RLS in [17] with uncertainty in the regularization factor and show that the new model is similar to the total least squares (TLS) model that compensates for uncertainty in the calculated regularization factor without the true channel response. For the performance evaluation, we simulate four different sparse channels and compare channel estimation performances. We show that, without any information of the true channel impulse response, the performance of the proposed algorithm is comparable to that of l_1-RLS with the information of the true channel impulse response.

This paper is organized as follows. In Section 2, we summarize l_1-RLS in [17]. In Section 3, we summarize the measure of sparsity. In Section 4, we propose a new method for the regularization calculation. In Section 5, we show that l_1-RLS with uncertainty in the regularization factor can be modeled as the TLS model. In Section 6, we summarize l_1-RTLS (recursive total least squares) algorithm as a solution for l_1-RLS with uncertainty in the regularization factor. In Section 7, we present simulation results to show the performance of the proposed algorithm. Finally, we give the conclusion in Section 8.

2. Summarize l_1-RLS

In the sparse channel estimation problem of interest, the system observes a signal represented by an $M \times 1$ vector $\mathbf{x}(k) = [x_k, \cdots, x_{k-M+1}]^T$ at time instant n, performs filtering, and obtains the output $y(i) = \mathbf{x}^T(k)\mathbf{w}_o(k)$, where $\mathbf{w}_o(k) = [w_k, \cdots, w_{k-M+1}]^T$ is the M dimensional actual system with finite impulse response (FIR) type. For system estimation, an adaptive filter system applies with M dimensional vector $\mathbf{w}(k)$ to the same signal vector $\mathbf{x}(k)$ and produces an estimated output $\hat{y}(k) = \mathbf{x}^T(k)\mathbf{w}(k)$, and calculates the error signal $e(k) = y(k) + n(k) - \hat{y}(k) = \tilde{y}(k) - \hat{y}(k)$, where $n(k)$ is the measurement noise, $y(k)$ is the output of the actual system, and $\hat{y}(k)$ is the estimated output. In order to estimate the channel impulse response, an adaptive algorithm minimizes the cost function defined by

$$\mathbf{w} = \operatorname*{argmin}_{\mathbf{w}} \frac{1}{2} \sum_{m=0}^{k} \lambda^{k-m}(e(m))^2. \tag{1}$$

From the gradient based minimization, Equation (1) becomes

$$\mathbf{R}(k)\mathbf{w}(k) = \mathbf{r}(k), \tag{2}$$

where $\mathbf{R}(k) = \sum_{m=0}^{k} \lambda^{k-m}\mathbf{x}(m)\mathbf{x}^T(m)$ and $\mathbf{r}(k) = \sum_{m=0}^{k} \lambda^{k-m}\tilde{y}(m)\mathbf{x}(m)$. This equation is the normal equation for the least squares solution. Especially, $\mathbf{w}_o(k)$ is considered as a sparse system when the number of nonzero coefficients K is less than the system order of M. In order to estimate the sparse system, most estimation algorithms exploit non-zero coefficients in the system [11–17]. In [17], Eksioglu proposed a full recursive l_1-regularized algorithm by the minimization of the object function as shown in Equation (3).

$$J_k = \frac{1}{2}\varepsilon_k + \gamma_k\|\mathbf{w}\|_1, \tag{3}$$

where $\varepsilon_k = \sum_{m=0}^{k} \lambda^{k-m}(e(m))^2$. From the minimization of Equation (3), a modified normal equation was derived as shown in Equation (4).

$$\mathbf{R}(k)\mathbf{w}(k) = \mathbf{r}(k) - \gamma_k\nabla^s\|\mathbf{w}(k)\|_1 = \hat{\mathbf{p}}(k). \tag{4}$$

When we solve Equation (4), we should select the regularization factor as shown in Equation (5).

$$\gamma_k = \frac{2^{\frac{tr(\mathbf{R}^{-1}(k))}{M}}}{\|\mathbf{R}^{-1}(k)\nabla^s f(\mathbf{w}(k))\|_2^2} \times [(f(\mathbf{w}(k)) - \rho) + \nabla^s f(\mathbf{w}(k))\mathbf{R}^{-1}(k)\varepsilon(k)], \tag{5}$$

where $f(\mathbf{w}(k)) = \|\mathbf{w}(k)\|_1$ and the subgradient of $f(\mathbf{w})$ is $\nabla^s\|\mathbf{w}\|_1 = \text{sgn}(\mathbf{w})$. In Equation (5), the regularization factor has the parameter, ρ, which should be set beforehand. In [17], the parameter was set as $\rho = f(\mathbf{w}_{true}) = \|\mathbf{w}_{true}\|_1$ with $\mathbf{w}_{true}$ indicating the impulse response of the true channel. There was no further discussion about how to set ρ. However, it is not practical to know the true channel in advance.

3. Measure of Sparseness

In [20], the sparseness of a channel impulse response is measured by Equation (6).

$$\chi = \frac{L}{L - \sqrt{L}}\left(1 - \frac{\|\hat{\mathbf{w}}\|_1}{\sqrt{L}\|\hat{\mathbf{w}}\|_2}\right),\tag{6}$$

where $\|\hat{\mathbf{w}}\|_p$ is the p-norm of $\hat{\mathbf{w}}$ and L is the dimension of $\hat{\mathbf{w}}$. The range of χ is $0 \le \chi \le 1$. That is dependent on the sparseness of $\hat{\mathbf{w}}$. As $\hat{\mathbf{w}}$ becomes sparser, the sparsity, χ, comes close to 1, and as $\hat{\mathbf{w}}$ becomes denser, χ comes close to 0. We often have small and none-zero value of χ, even in a dense channel. For example, Figure 1 shows the relation of the value of χ and the percentage of none-zero components in $\hat{\mathbf{w}}$ with $L = 215$. In Figure 1, we consider all possible cases of none-zero components in $\hat{\mathbf{w}}$.

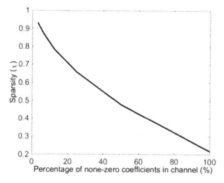

Figure 1. Sparsity (χ) vs. the percentage of none zero coefficients in the channel impulse response.

4. New ρ Selection Method in the Sparsity Regularization Constant γ_k

Section 2 shows that the regularization constant γ_k in Equation (5) needs ρ to be set as $\rho = \|\text{true system impulse response}\|_1 = \|\mathbf{w}_{true}\|_1$. However, we need a new method in the constant selection because Equation (5) is not practical. Therefore, Section 4 proposes a new method to set this constant.

For a practical method for the constant selection, we can consider using the estimated vector $\hat{\mathbf{w}}$ instead of using the true vector $\mathbf{w}_{true}$ because $\hat{\mathbf{w}}$, the solution with l_1-norm, will be closer to the sparse true vector than the solution of the conventional RLS. The more iteration is repeated, the more $\hat{\mathbf{w}}$ converges to the true value. Conventional RLS also converges to the true value; however, the solution with l_1-norm, is closer to the sparse true value. Therefore, we can use sparse estimate $\hat{\mathbf{w}}$ instead of $\mathbf{w}_{true}$ when we set ρ, and the uncertainty arising from this is compensated through a TLS solution in the next section. When we determine ρ using the estimated $\hat{\mathbf{w}}$, we choose between the average ρ and the current estimate $\|\hat{\mathbf{w}}\|_1$. Table 1 summarizes the ρ selection steps.

The determination method for ρ value shown in Table 1 is as follows. In Step 1, the sparsity of the estimated $\hat{\mathbf{w}}$ is calculated. The sparsity represents the sparseness of $\hat{\mathbf{w}}$ as a number [23]. In Step 2, l_1-norm of the estimated $\hat{\mathbf{w}}$ is scaled and the value is averaged with the previous ρ value. The scaling value approaches 1 as the sparsity, χ, gets close to 1. However, the scaling value gets close to $e^{-1} \simeq 0.37$ as the sparsity, χ, gets close to 0. Therefore, the scaling does not change l_1-norm of $\hat{\mathbf{w}}$ for the sparse $\hat{\mathbf{w}}$.

Instead the scaling changes the l_1-norm smaller for the dense $\hat{\mathbf{w}}$. In Step 3, the smaller one between the averaged ρ and the l_1-norm of the estimated $\hat{\mathbf{w}}$ is selected as the new ρ value. In this case, the ρ value becomes completely new if the l_1-norm of the estimated $\hat{\mathbf{w}}$ is selected, otherwise the previous trend is maintained. In Figure 1, the reference value 0.75 used in Step 3 means that less than 16% of all the impulse response taps are not zero.

Table 1. ρ selection method in the sparsity regularization constant γ_k.

Step 1	Sparsity: $\chi = \frac{L}{L-\sqrt{L}}\left(1 - \frac{\|\hat{\mathbf{w}}\|_1}{\sqrt{L}\|\hat{\mathbf{w}}\|_2}\right)$ [20] where L is the length of the impulse response.
Step 2	$\rho(k) = 0.99\rho(k-1) + 0.01\left(e^{\chi-1} \times \|\hat{\mathbf{w}}\|_1\right)$
Step 3	$\rho(k) = \begin{cases} min(\rho(k), 0.98\|\hat{\mathbf{w}}\|_1), \text{if } \chi > 0.75 \\ min(\rho(k), 0.999\|\hat{\mathbf{w}}\|_1), \text{otherwise} \end{cases}$

5. New Modeling for l_1-RLS with Uncertainty in the Regularization Factor

If we set $\rho = $ constant, the regularization factor becomes

$$
\begin{aligned}
\tilde{\gamma}_k &= \frac{2\frac{tr\left(\mathbf{R}^{-1}(k)\right)}{M}}{\left\|\mathbf{R}^{-1}(k)\nabla^s f(\mathbf{w}(k))\right\|_2^2} \times \left[(f(\mathbf{w}(k)) - \text{constant}) + \nabla^s f(\mathbf{w}(k)) \times \mathbf{R}^{-1}(k)\varepsilon(k)\right] \\
&= \frac{2\frac{tr\left(\mathbf{R}^{-1}(k)\right)}{M}}{\left\|\mathbf{R}^{-1}(k)\nabla^s f(\mathbf{w}(k))\right\|_2^2} \times (f(\mathbf{w}(k)) - \|\mathbf{h}\|_1 + \|\mathbf{h}\|_1 - \text{constant}) + \frac{2\frac{tr\left(\mathbf{R}^{-1}(k)\right)}{M}\nabla^s f(\mathbf{w}(k))\mathbf{R}^{-1}(k)\varepsilon(k)}{\left\|\mathbf{R}^{-1}(k)\nabla^s f(\mathbf{w}(k))\right\|_2^2}.
\end{aligned}
\tag{7}
$$

Then,

$$
\tilde{\gamma}_k = \gamma_k + \frac{2\frac{tr\left(\mathbf{R}^{-1}(k)\right)}{M}(\|\mathbf{h}\|_1 - \text{constant})}{\left\|\mathbf{R}^{-1}(k)\nabla^s f(\mathbf{w}(k))\right\|_2^2} = \gamma_k + \Delta\gamma.
\tag{8}
$$

Using Equation (8), Equation (4) becomes

$$
\mathbf{R}(k)\mathbf{w}(k) = \mathbf{r}(k) - (\gamma_k + \Delta\gamma)\nabla^s\|\mathbf{w}(k)\|_1 \quad .
\tag{9}
$$

$\nabla^s\|\mathbf{w}\|_1 = \text{sgn}(\mathbf{w})$ is represented as

$$
\nabla^s\|\mathbf{w}(k)\|_1 = \begin{bmatrix} \ddots & & \\ & \frac{1}{|\mathbf{w}_i|} & \\ & & \ddots \end{bmatrix}\mathbf{w}(k).
\tag{10}
$$

By applying Equation (10) to Equation (9),

$$
\left(\mathbf{R}(k) + \Delta\gamma\begin{bmatrix} \ddots & & \\ & \frac{1}{|\mathbf{w}_i|} & \\ & & \ddots \end{bmatrix}\right)\mathbf{w}(k) = \mathbf{r}(k) - \gamma_k\|\mathbf{w}(k)\|_1,
\tag{11}
$$

where $\mathbf{w}_i$ is i-th element of $\mathbf{w}(k)$. Then it is simplified as

$$
\left(\mathbf{R}(k) + \Delta\gamma\begin{bmatrix} \ddots & & \\ & \frac{1}{|\mathbf{w}_i|} & \\ & & \ddots \end{bmatrix}\right)\mathbf{w}(k) = \hat{\mathbf{p}}(k).
\tag{12}
$$

Equation (12) is very similar to the system model in Figure 2 that is contaminated by noise both in input and in output. Suppose that an example of the system in Figure 2 is represented as

$$
\begin{bmatrix}
x_k + n_{i,k} & \cdots & x_{k-N+1} + n_{i,k-N+1} \\
x_{k-1} + n_{i,k-1} & \cdots & x_{k-N} + n_{i,k-N} \\
\vdots & \ddots & \vdots \\
x_{k-N+1} + n_{i,k-N+1} & \cdots & x_{k-2N+2} + n_{i,k-2N+2}
\end{bmatrix}
\times \mathbf{w}(k) =
\begin{bmatrix}
y_k + n_{o,k} \\
y_{k-1} + n_{o,k-1} \\
\vdots \\
y_{k-N+1} + n_{o,k-N+1}
\end{bmatrix},
\tag{13}
$$

where x_k is $x(k)$, $n_{i,k}$ is $n_i(k)$, and $n_{o,k}$ is $n_o(k)$. Equation (13) is simplified as

$$
\mathbf{A}\mathbf{w}(k) = \mathbf{b}.
\tag{14}
$$

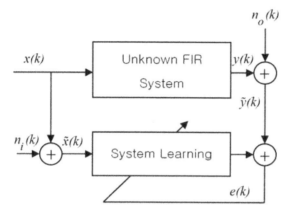

Figure 2. The model of a noisy input and noisy output system.

If we multiply Equation (14) by $\mathbf{A}^H$ and average it, we get

$$
E\left(\mathbf{A}^H \mathbf{A}\right)\mathbf{w}(k) = E\left(\mathbf{A}^H \mathbf{b}\right).
\tag{15}
$$

We can rewrite Equation (15) as follows

$$
\begin{bmatrix}
r_{xx}(0) + \sigma_n^2 & r_{xx}(1) & \cdots & r_{xx}(N-1) \\
r_{xx}(1) & r_{xx}(0) + \sigma_n^2 & \cdots & r_{xx}(N-2) \\
\vdots & \vdots & \ddots & \vdots \\
r_{xx}(N-1) & r_{xx}(N-2) & \cdots & r_{xx}(0) + \sigma_n^2
\end{bmatrix}
\mathbf{w}(k) =
\begin{bmatrix}
r_{xy}(0) \\
r_{xy}(1) \\
\vdots \\
r_{xy}(N-1)
\end{bmatrix}
\tag{16}
$$

Then, it can be represented as

$$
\left(\mathbf{R} + \sigma_n^2 \mathbf{I}\right)\mathbf{w}(k) = \tilde{\mathbf{p}}(k).
\tag{17}
$$

When we compare Equation (12) with Equation (17), the two system models have almost the same form. Therefore, it is feasible that the TLS method can be applied to Equation (12) [24–30]. Therefore, we expect to obtain almost the same performance as l_1-RLS with the true channel response if we apply the TLS method by the regularization factor with the new ρ in Table 1. In the next section, we summarize l_1-RTLS (recursive total least squares) algorithm in [29].

6. Summarize l_1-RTLS for the Solution of l_1-RLS with Uncertainty in the Regularization Factor

Lim, one of the authors of this paper, has proposed the TLS solution for l_1-RLS known as l_1-RTLS [30]. In this section, we summarize l_1-RTLS in [30] for the solution of Equation (11).

The TLS system model assumes that both input and output are contaminated by additive noise as Figure 2. The output is given by

$$\tilde{y}(k) = \tilde{\mathbf{x}}^T(k)\mathbf{w}_o + n_o(k), \tag{18}$$

where the output noise $n_o(k)$ is the Gaussian white noise with variance σ_o^2. The noisy input vector in the system is modeled by

$$\tilde{\mathbf{x}}(k) = \mathbf{x}(k) + \mathbf{n}_i(k) \in C^{M \times 1}, \tag{19}$$

where $\mathbf{n}_i(k) = [n_i(k), n_i(k-1), \cdots n_i(k-M+1)]^T$ and the input noise $n_i(k)$ is the Gaussian white noise with variance σ_i^2. For the TLS solution, we set the augmented data vector as

$$\overline{\mathbf{x}}(k) = \left[\tilde{\mathbf{x}}^T(k), \tilde{y}(k) \right]^T \in R^{(M+1) \times 1}. \tag{20}$$

The correlation matrix is represented as

$$\overline{\mathbf{R}} = \begin{bmatrix} \tilde{\mathbf{R}} & \mathbf{p} \\ \mathbf{p}^T & c \end{bmatrix}, \tag{21}$$

where $\mathbf{p} = E\{\tilde{\mathbf{x}}(k)y(k)\}$, $c = E\{y(k)y(k)\}$, $\mathbf{R} = E\{\mathbf{x}(k)\mathbf{x}^T(k)\}$ and $\tilde{\mathbf{R}} = E\left\{\tilde{\mathbf{x}}(k)\tilde{\mathbf{x}}^T(k)\right\} = \mathbf{R} + \sigma_i^2 \mathbf{I}$. In [27,28], the TLS problem becomes to find the eigenvector associated with the smallest eigenvalue of $\overline{\mathbf{R}}$. Equation (22) is the typical cost function to find the eigenvector associated with the smallest eigenvalue of $\overline{\mathbf{R}}$.

$$J(k) = \frac{1}{2}\tilde{\mathbf{w}}^T(k)\overline{\mathbf{R}}(k)\tilde{\mathbf{w}}(k), \tag{22}$$

where $\overline{\mathbf{R}}(k)$ is a sample correlation matrix at k-th instant, and $\tilde{\mathbf{w}}(k) = \left[\hat{\mathbf{w}}^T(k), -1\right]^T$ in which $\hat{\mathbf{w}}(k)$ is the estimation result for the unknown system at k-th instant. We modify the cost function by adding a penalty function in order to reflect prior knowledge about the true sparsity system.

$$J(k) = \frac{1}{2}\tilde{\mathbf{w}}^T(k)\overline{\mathbf{R}}(k)\tilde{\mathbf{w}}(k) + \lambda\left(\tilde{\mathbf{w}}^T(k)\tilde{\mathbf{w}}(k-1) - 1\right) + \gamma_k f(\tilde{\mathbf{w}}(k)), \tag{23}$$

where λ is the Lagrange multiplier and γ_k is the regularized parameter in [13]. We solve the equations by $\nabla_{\hat{\mathbf{w}}}J(k) = 0$ and $\nabla_\lambda J(k) = 0$ simultaneously. $\nabla_{\hat{\mathbf{w}}}J(k) = 0$:

$$2\overline{\mathbf{R}}(k)\tilde{\mathbf{w}}(k) + \lambda\tilde{\mathbf{w}}(k-1) + \gamma_k\nabla^s f(\tilde{\mathbf{w}}(k)) = 0, \tag{24}$$

$$\nabla_\lambda J(k) = 0: \quad \tilde{\mathbf{w}}^T(k)\tilde{\mathbf{w}}(k-1) = 1, \tag{25}$$

where the subgradient of $f(\tilde{\mathbf{w}}) = \|\tilde{\mathbf{w}}\|_1$ is $\nabla_{\tilde{\mathbf{w}}}^s\|\tilde{\mathbf{w}}\|_1 = \text{sgn}(\tilde{\mathbf{w}})$. From (24), we obtain

$$\tilde{\mathbf{w}}(k) = -\frac{\lambda}{2}\overline{\mathbf{R}}^{-1}(k)\tilde{\mathbf{w}}(k-1) - \gamma_k\overline{\mathbf{R}}^{-1}(k)\nabla^s f(\tilde{\mathbf{w}}(k)). \tag{26}$$

Substituting Equation (26) in Equation (25), we get

$$\left(-\frac{\lambda}{2}\overline{\mathbf{R}}^{-1}(k)\tilde{\mathbf{w}}(k-1) - \gamma_k\overline{\mathbf{R}}^{-1}(k)\nabla^s f(\tilde{\mathbf{w}}(k))\right)^T \times \tilde{\mathbf{w}}(k-1) = 1, \tag{27}$$

or

$$\lambda = -2\frac{1 + \gamma_k \nabla^s f(\widetilde{\mathbf{w}}(k))^T \overline{\mathbf{R}}^{-1}(k) \widetilde{\mathbf{w}}(k-1)}{\widetilde{\mathbf{w}}^T(k-1) \overline{\mathbf{R}}^{-1}(k) \widetilde{\mathbf{w}}(k-1)}. \tag{28}$$

Substituting λ in Equation (26) by Equation (28) leads to

$$\widetilde{\mathbf{w}}(k) = \frac{1 + \gamma_k \nabla^s f(\widetilde{\mathbf{w}}(k))^T \overline{\mathbf{R}}^{-1}(k) \widetilde{\mathbf{w}}(k-1)}{\widetilde{\mathbf{w}}^T(k-1) \overline{\mathbf{R}}^{-1}(k) \widetilde{\mathbf{w}}(k-1)} \times \overline{\mathbf{R}}^{-1}(k) \widetilde{\mathbf{w}}(k-1) - \gamma_k \overline{\mathbf{R}}^{-1}(k) \nabla_{\widetilde{\mathbf{w}}} f(\widetilde{\mathbf{w}}(k)). \tag{29}$$

Equation (29) can be expressed in a simple form as

$$\widetilde{\mathbf{w}}(k) = \alpha \overline{\mathbf{R}}^{-1}(k) \widetilde{\mathbf{w}}(k-1) - \gamma_k \overline{\mathbf{R}}^{-1}(k) \nabla^s f(\widetilde{\mathbf{w}}(k)), \tag{30}$$

where $\alpha = \frac{1 + \gamma_k \nabla^s f(\widetilde{\mathbf{w}}(k))^T \overline{\mathbf{R}}^{-1}(k) \widetilde{\mathbf{w}}(k-1)}{\widetilde{\mathbf{w}}^T(k-1) \overline{\mathbf{R}}^{-1}(k) \widetilde{\mathbf{w}}(k-1)}$. Because asymptotically $\|\widetilde{\mathbf{w}}(k)\| = 1$ as $k \rightarrow \infty$, Equation (29) can be approximated as the following two equations.

$$\widetilde{\mathbf{w}}(k) \simeq \overline{\mathbf{R}}^{-1}(k) \widetilde{\mathbf{w}}(k-1) - \gamma_k \left(\widetilde{\mathbf{w}}^T(k-1) \overline{\mathbf{R}}^{-1}(k-1) \widetilde{\mathbf{w}}(k-1) \right) \overline{\mathbf{R}}^{-1}(k) \nabla^s f(\widetilde{\mathbf{w}}(k-1)). \tag{31}$$

$$\widetilde{\mathbf{w}}(k) = \widetilde{\mathbf{w}}(k) / \|\widetilde{\mathbf{w}}(k)\|. \tag{32}$$

Finally, we obtain the estimated parameter of the unknown system as

$$\hat{\mathbf{w}}(k) = -\widetilde{\mathbf{w}}_{1:M}(k) / \widetilde{\mathbf{w}}_{M+1}(k). \tag{33}$$

For Equation (23), we can use the modified regularization factor γ_k in [30]

$$\gamma_k = \frac{2^{\frac{tr\left(\overline{\mathbf{R}}^{-1}(k)\right)}{M}}}{\left\| \overline{\mathbf{R}}^{-1}(k) \nabla^s f(\hat{\mathbf{w}}_{aug}(k)) \right\|_2^2} \times [(f(\hat{\mathbf{w}}_{aug}(k)) - \rho) + \nabla^s f(\hat{\mathbf{w}}_{aug}(k)) \overline{\mathbf{R}}^{-1}(k) \varepsilon(k)], \tag{34}$$

where $\hat{\mathbf{w}}_{aug}(k) = \left[\hat{\mathbf{w}}^T(k), -1 \right]^T$, $\hat{\mathbf{w}}_{aug,RLS}(k) = \left[\hat{\mathbf{w}}_{RLS}^T(k), -1 \right]^T$, $\varepsilon(k) = \hat{\mathbf{w}}_{aug}(k) - \hat{\mathbf{w}}_{aug,RLS}(k)$, and $\hat{\mathbf{w}}_{RLS}(k)$ is the estimated parameter by recursive least squares (RLS). As $f(\hat{\mathbf{w}}) = \|\hat{\mathbf{w}}\|_1$, the subgradient of $f(\hat{\mathbf{w}}_{aug}(k))$ is

$$\nabla^s \|\hat{\mathbf{w}}_{aug}(k)\|_1 = \text{sgn}(\hat{\mathbf{w}}_{aug}(k)). \tag{35}$$

As mentioned in Section 4, we apply new constant ρ in Table 1, to the regularization factor γ_k in Equation (34) instead of $\|\mathbf{w}_{true}\|_1$, where $\mathbf{w}_{true}$ is the true system impulse response.

7. Simulation Results

This section confirms the performance of the proposed algorithm in sparse channel estimation. In the first experiment, the channel estimation performance is compared with other algorithms using randomly generated sparse channels. In this simulation, we follow the same scenario in the experiments as [17]. The true system vector $\mathbf{w}_{true}$ is 64 dimensions. In order to generate the sparse channel, we set the number of the nonzero coefficients, S, in the 64 coefficients and randomly position the nonzero coefficients. The values of the coefficients are taken from an $N(0, 1/S)$ distribution, where $N(\ )$ is the normal distribution. In the simulation, we estimate the channel impulse response by the proposed algorithms that are l_1-RLS using the ρ in Table 1 and l_1-RTLS using the ρ in Table 1. For the comparison, we estimate the channel impulse response by l_1-RLS using the true channel response; in addition, we also execute the regular RLS algorithm in an oracle setting (oracle-RLS) where the positions of the true nonzero system parameters are assumed to be known. For the estimated channel results, we calculate the mean standard deviation (MSD), where MSD $= E\left(|\hat{\mathbf{w}} - \mathbf{w}_{true}|^2\right)$, $\hat{\mathbf{w}}$ is the estimated channel response and $\mathbf{w}_{true}$ is the true channel response. For the performance evaluation, we simulate the algorithms in the sparse channels for S = 4, 8, 16, and 32.

Figure 3 illustrates the MSD curves. For S = 4, Figure 3a shows that the estimation performance of l_1-RTLS using the regularization factor with the ρ in Table 1 is almost the same as the l_1-RLS using regularization with a true channel impulse response. However, the performance of l_1-RLS using the regularization factor with the ρ in Table 1 is gradually degraded and shows a kind of uncertainty accumulation effect. In the other cases of S, we can observe the same trend in the MSD curves. Therefore, we can confirm that the new regularization factor selection method and the new modeling for l_1-RLS can estimate the sparse channel as good as l_1-RLS using the regularization with the true channel impulse response. In all the simulation scenarios, oracle RLS algorithm produces the lowest MSD as expected.

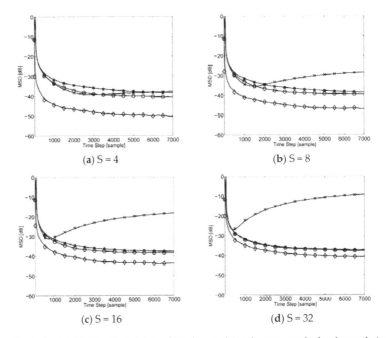

Figure 3. Steady-state MSD for S = 4, 8, 16, and 32 when applying the new ρ method to the regularization factor (-o-: l_1-RLS with the true channel response, -×-: l_1-RLS with the new ρ method, -*-: proposed l_1-RTLS with the new ρ method, -◇-: oracle-RLS).

Table 2 summarizes the steady-state MSD values as varying S from 4 to 32. The results show that the proposed l_1-RTLS with the new ρ is comparable to l_1-RLS with the true channel.

In the second experiment, we compare channel estimation performance using room impulse response. The size of the room is (7.49, 6.24, 3.88 m). The position of the sound source is (1.53, 0.96, 1.12 m) and the position of the receiver is (1.81, 5.17, 0.71 m), respectively. T60 is set to 100 ms and 400 ms. The impulse response of the room is generated using the program in [31]. We focus on the direct reflection part and the early reflection part in the RIR because the direct reflection and early reflection part of the RIR has a sparse property. This is the part that is estimated in the AEC applications [32]. This part is also related to localization and clarity in room acoustics [33–35]. Comparing the impulse response (IR) generated by setting T60 = 100 ms to the channel with 65 coefficients used in the first experiment, it is equivalent to S = 4 in the channel with 65 coefficients. In the same manner, the IR generated by setting T60 = 400 ms is equivalent to S = 10.

Table 2. MSD (mean square deviation) comparison.

Sparsity (S)	Algorithm	MSD
4	l_1-RLS with the true channel	-40.6 dB
	l_1-RLS with the new ρ method	-37.8 dB
	proposed l_1-RTLS with the new ρ method	-38.5 dB
	Oracle-RLS	-50.4 dB
8	l_1-RLS with the true channel	-39.5 dB
	l_1-RLS with the new ρ method	-28.4 dB
	proposed l_1-RTLS with the new ρ method	-38.5 dB
	Oracle-RLS	-46.9 dB
16	l_1-RLS with the true channel	-38.4 dB
	l_1-RLS with the new ρ method	-18.2 dB
	proposed l_1-RTLS with the new ρ method	-37.6 dB
	Oracle-RLS	-43.6 dB
32	l_1-RLS with the true channel	-37.6 dB
	l_1-RLS with the new ρ method	-9.1 dB
	proposed l_1-RTLS with the new ρ method	-37.3 dB
	Oracle-RLS	-40.6 dB

Table 3 summarizes the steady-state MSD values. The results also show the same trend as Table 2. In RIR estimation, the proposed l_1-RTLS with the new ρ is also comparable to l_1-RLS with the true channel.

Table 3. MSD (mean square deviation) comparison in sparse RIR estimations.

Reverberation Time (T60)	Algorithm	MSD
100 ms	l_1-RLS with the true channel	-38.5 dB
	l_1-RLS with the new ρ method	-34.7 dB
	proposed l_1-RTLS with the new ρ method	-35.4 dB
	Oracle-RLS	-45.3 dB
400 ms	l_1-RLS with the true channel	-32.1 dB
	l_1-RLS with the new ρ method	-20.9 dB
	proposed l_1-RTLS with the new ρ method	-30.1 dB
	Oracle-RLS	-36.0 dB

8. Conclusions

In this paper, we have proposed the regularization factor for recursive adaptive estimation. The regularization factor needs no prior knowledge of the true channel impulse response. We have also reformulated the recursive estimation algorithm as l_1-RTLS type. This formulation is robust to the uncertainty in the regularization factor without a priori knowledge of the true channel impulse response. Simulations show that the proposed regularization factor and l_1-RTLS algorithm provide good performance comparable to l_1- RLS with the knowledge of the true channel impulse response.

Author Contributions: Conceptualization, J.L.; Methodology, J.L.; Validation, J.L. and S.L.; Formal analysis, J.L.; Investigation, J.L. and S.L.; Writing—original draft preparation, J.L.; Writing—review and editing, S.L.; Visualization, J.L.; Project administration, J.L.; Funding acquisition, J.L. and S.L.

Funding: This research received no external funding.

Acknowledgments: This research was supported by Agency for Defense Development (ADD) in Korea (UD160015DD).

Conflicts of Interest: The authors declare no conflict of interest.

References

1. Benichoux, A.; Simon, L.; Vincent, E.; Gribonval, R. Convex regularizations for the simultaneous recording of room impulse responses. *IEEE Trans. Signal Process.* **2014**, *62*, 1976–1986. [CrossRef]
2. Merimaa, J.; Pulkki, V. Spatial impulse response I: Analysis and synthesis. *J. Audio Eng. Soc.* **2005**, *53*, 1115–1127.
3. Dokmanic, I.; Parhizkar, R.; Walther, A.; Lu, Y.M.; Vetterli, M. Acoustic echoes reveal room shape. *Proc. Natl. Acad. Sci. USA* **2013**, *110*, 12186–12191. [CrossRef] [PubMed]
4. Remaggi, L.; Jackson, P.; Coleman, P.; Wang, W. Acoustic reflector localization: Novel image source reversion and direct localization methods. *IEEE Trans. Audio Speech Lang. Process.* **2017**, *25*, 296–309. [CrossRef]
5. Baba, Y.; Walther, A.; Habets, E. 3D room geometry interference based on room impulse response stacks. *IEEE Trans. Audio, Speech Lang. Process.* **2018**, *26*, 857–872. [CrossRef]
6. Goetze, S.; Xiong, F.; Jungmann, J.O.; Kallinger, M.; Kammeyer, K.; Mertins, A. System Identification of Equalized Room Impulse Responses by an Acoustic Echo Canceller using Proportionate LMS Algorithms. In Proceedings of the 130th AES Convention, London, UK, 13 May 2011; pp. 1–13.
7. Yu, M.; Ma, W.; Xin, J.; Osher, S. Multi-channel l_1 regularized convex speech enhancement model and fast computation by the split Bregman method. *IEEE Trans. Audio Speech Lang. Process.* **2012**, *20*, 661–675. [CrossRef]
8. Lin, Y.; Chen, J.; Kim, Y.; Lee, D.D. Blind channel identification for speech dereverberation using l1-norm sparse learning. In Proceedings of the Twenty-First Annual Conference on Neural Information Processing Systems, Vancouver, BC, Canada, 3–6 December 2007; pp. 921–928.
9. Naylor, P.A.; Gaubitch, N.D. *Speech Dereverberation*; Springer: London, UK, 2010; pp. 219–270.
10. Duttweiler, D.L. Proportionate normalized least-mean-squares adaptation in echo cancelers. *IEEE Trans. Speech Audio Process.* **2000**, *8*, 508–518. [CrossRef]
11. Gu, Y.; Jin, J.; Mei, S. Norm Constraint LMS Algorithm for Sparse System Identification. *IEEE Signal Process. Lett.* **2009**, *16*, 774–777.
12. Chen, Y.; Gu, Y.; Hero, A.O. Sparse LMS for system identification. In Proceedings of the IEEE International Conference on Acoustics, Speech and Signal Processing, Taipei, Taiwan, 19–24 April 2009; pp. 3125–3128.
13. He, X.; Song, R.; Zhu, W.P. Optimal pilot pattern design for compressed sensing-based sparse channel estimation in OFDM systems. *Circuits Syst. Signal Process.* **2012**, *31*, 1379–1395. [CrossRef]
14. Babadi, B.; Kalouptsidis, N.; Tarokh, V. SPARLS: The sparse RLS algorithm. *IEEE Trans. Signal Process.* **2010**, *58*, 4013–4025. [CrossRef]
15. Angelosante, D.; Bazerque, J.A.; Giannakis, G.B. Online adaptive estimation of sparse signals: Where RLS meets the l_1-norm. *IEEE Trans. Signal Process.* **2010**, *58*, 3436–3447. [CrossRef]
16. Eksioglu, E.M. Sparsity regularised recursive least squares adaptive filtering. *IET Signal Process.* **2011**, *5*, 480–487. [CrossRef]
17. Eksioglu, E.M.; Tanc, A.L. RLS algorithm with convex regularization. *IEEE Signal Process. Lett.* **2011**, *18*, 470–473. [CrossRef]
18. Sun, D.; Liu, L.; Zhang, Y. Recursive regularisation parameter selection for sparse RLS algorithm. *Electron. Lett.* **2018**, *54*, 286–287. [CrossRef]
19. Chen, Y.; Gui, G. Recursive least square-based fast sparse multipath channel estimation. *Int. J. Commun. Syst.* **2017**, *30*, e3278. [CrossRef]
20. Kalouptsidis, N.; Mileounis, G.; Babadi, B.; Tarokh, V. Adaptive algorithms for sparse system identification. *Signal Process.* **2011**, *91*, 1910–1919. [CrossRef]

21. Candes, E.J.; Wakin, M.; Boyd, S. Enhancing sparsity by reweighted minimization. *J. Fourier Anal. Appl.* **2008**, *14*, 877–905. [CrossRef]
22. Lamare, R.C.; Sampaio-Neto, R. Adaptive reduced-rank processing based on joint and iterative interpolation, decimation, and filtering. *IEEE Trans. Signal Process.* **2009**, *57*, 2503–2514. [CrossRef]
23. Petraglia, M.R.; Haddad, D.B. New adaptive algorithms for identification of sparse impulse responses—Analysis and comparisons. In Proceedings of the Wireless Communication Systems, York, UK, 19–22 September 2010; pp. 384–388.
24. Golub, G.H.; Van Loan, C.F. An analysis of the total least squares problem. *SIAM J. Numer. Anal.* **1980**, *17*, 883–893. [CrossRef]
25. Dunne, B.E.; Williamson, G.A. Stable simplified gradient algorithms for total least squares filtering. In Proceedings of the 34th Annual Asilomar Conference on Signals, Systems, and Computers, Pacific Grove, CA, USA, 29 October–1 November 2000; pp. 1762–1766.
26. Feng, D.Z.; Bao, Z.; Jiao, L.C. Total least mean squares algorithm. *IEEE Trans. Signal Process.* **1998**, *46*, 2122–2130. [CrossRef]
27. Davila, C.E. An efficient recursive total least squares algorithm for FIR adaptive filtering. *IEEE Trans. Signal Process.* **1994**, *42*, 268–280. [CrossRef]
28. Soijer, M.W. Sequential computation of total least-squares parameter estimates. *J. Guid. Control Dyn.* **2004**, *27*, 501–503. [CrossRef]
29. Choi, N.; Lim, J.S.; Sung, K.M. An efficient recursive total least squares algorithm for raining multilayer feedforward neural networks. *Lect. Notes Comput. Sci.* **2005**, *3496*, 558–565.
30. Lim, J.S.; Pang, H.S. l_1-regularized recursive total least squares based sparse system identification for the error-in-variables. *SpringerPlus* **2016**, *5*, 1460–1469. [CrossRef] [PubMed]
31. Lehmann, E. Image-Source Method: MATLAB Code Implementation. Available online: http://www.eric-lehmann.com/ (accessed on 17 December 2018).
32. Gay, S.L.; Benesty, J. *Acoustic Signal Processing for Telecommunication*; Kluwer Academic Publisher: Norwell, MA, USA, 2000; pp. 6–7.
33. Swanson, D.C. *Signal Processing for Intelligent Sensor Systems with MATLAB*, 2nd ed.; CRC Press: Boca Raton, FL, USA, 2012; p. 70.
34. Kuttruff, H. *Room Acoustics*, 6th ed.; CRC Press: Boca Raton, FL, USA, 2017; p. 168.
35. Bai, H.; Richard, G.; Daudet, L. Modeling early reflections of room impulse responses using a radiance transfer method. In Proceedings of the IEEE Workshop on Applications of Signal Processing to Audio and Acoustics, New Paltz, NY, USA, 20–23 October 2013; pp. 1–4.

Article

Evaluation of Cracks in Metallic Material Using a Self-Organized Data-Driven Model of Acoustic Echo-Signal

Xudong Teng [1,2,†], Xin Zhang [3,†], Yuantao Fan [4] and Dong Zhang [1,*]

1   Key Laboratory of Modern Acoustics (Nanjing University), Ministry of Education, Institute of Acoustics, Nanjing 210093, China; txd19@163.com
2   School of Electronic and Electric Engineering, Shanghai University of Engineering Science, Shanghai 201620, China
3   Nanjing Manse Acoustics Technology Co. Ltd., Nanjing 210017, China; yusheng.yan@hotmail.com
4   Center for Applied Intelligent Systems Research (CAISR), Halmstad University, SE-30118 Halmstad, Sweden; fanyuantao@gmail.com
*   Correspondence: dzhang@nju.edu.cn; Tel.: +86-25-83597324
†   These authors contributed equally to this work.

Received: 24 November 2018; Accepted: 22 December 2018; Published: 28 December 2018

Abstract: Non-linear acoustic technique is an attractive approach in evaluating early fatigue as well as cracks in material. However, its accuracy is greatly restricted by external non-linearities of ultra-sonic measurement systems. In this work, an acoustical data-driven deviation detection method, called the consensus self-organizing models (COSMO) based on statistical probability models, was introduced to study the evolution of localized crack growth. By using pitch-catch technique, frequency spectra of acoustic echoes collected from different locations of a specimen were compared, resulting in a Hellinger distance matrix to construct statistical parameters such as z-score, p-value and T-value. It is shown that statistical significance p-value of COSMO method has a strong relationship with the crack growth. Particularly, T-values, logarithm transformed p-value, increases proportionally with the growth of cracks, which thus can be applied to locate the position of cracks and monitor the deterioration of materials.

Keywords: crack growth; acoustic echo; COSMO; p-value

1. Introduction

Nonlinear ultrasonic behaviors, such as harmonics, mix frequencies, and the resonance frequency shift, have been proven to be sensitive to structure imperfections and early degradation of materials [1–6]. In the early stage of damage, material fatigue can induce a number of micro-cracks with a typical length of 1–100 µm by continuous loading cycles, then the micro-cracks further grow, coalesce with other micro-cracks and eventually form macro-cracks [7]. Since the fatigue cracks are localized [8], not uniformly distributed in the structure, the generated nonlinear response is basically dependent on the configuration of the crack area related to the localized hysteretic deformation. Clapping between the contacting surface and dissipative mechanism due to frictional sliding and so on [7,8], leads to a much stronger nonlinearity than the surrounding material behaving linearly [8]. However, non-linear effects induced by the localized cracks in the materials are not obvious enough to conveniently be measured and analyzed [9,10]. In addition, the use of power amplifier, transducers, and coupling media in ultrasonic testing system also bring about external non-linear change. Since it is difficult to separate structure-induced non-linearity in materials from external non-linearity, non-linear ultrasonic technology is not applied widely to accurately evaluate and locate the structure imperfections in practical applications [11].

In terms of the uncertainties in real-life testing conditions, "big data" sets, collected from acoustic echo-signals of vast amounts of damaged material, show different but distinguishable statistical characteristics compared with intact material [12–17]. Some statistical models, also called data-driven models, relate the degree of damage to the probability of detection (PoD). They were first introduced by the National Aeronautics and Space Administration (NASA) in 1973 and were soon accepted as a standard method [18,19]. Later Lu and Meeker further developed statistical methods to estimate a time–to–failure distribution for a broad class of degradation structures [16]. Gebraeel employed Bayesian to update a method to develop a closed-form residual-life distribution for the monitored device [20]. Gang Qi et al. proposed a framework to meet the challenge by systematically evaluating material damage based on large data sets collected by using acoustic emission (AE) [12]. Zhou et al. investigated AE relative energy, amplitude distribution as well as amplitude spectrum to discern the delamination damage mechanism of the composites [21]. Kûs et al. employed the model-based Clustering (Hellinger divergence) method to classify certain attributes of the original pure data obtained directly from the acoustic emission signals and form normed frequency spectra to perform physical separation tasks of AE random signals [22].

However, most data-driven methods are probabilistic and obtained historical degradation data or empirical knowledge [23,24], which require a large amount of experimental data to construct the reference curves or the preset feature threshold. Hence, in our previous work [25], we first introduced consensus self-organizing model (COSMO), neither any domain knowledge nor supervision to extract useful features, detecting a single flaw located at a fixed positon of steel specimens based on acoustical echo-signals. Nevertheless, as the contacting surfaces of the flaw produced by an electrical discharge machine were totally separated in [25], evaluating the actual fatigue cracks by applying COMSO method would potentially be problematic. In this paper, four cracks distributed along the length direction of a steel specimen were investigated, and the crack's growth (produced using fatigue testing [26]) was further discussed in details, which is probably accord with the mechanisms involved in contact acoustic nonlinearity and hysteretic nonlinearity. Both Numerical simulation and experimental measurements showed that COSMO models are effective in NDT inspection, as well as health monitoring for regular metallic structures.

2. The Consensus Self-Organizing Models (COSMO)

The COSMO method identifies the typical variability within a group of systems and to evaluate the likelihood of any individual being significantly different from the majority. In an ultrasonic testing system, a group of acoustic echo-signals were collected using pitch-catch method from N different locations (i.e., the total amount of samples) on a measured object. The spectral density of the acoustic signals were then obtained so that the difference between spectral density of two acoustic signals were compared [27,28], and Hellinger distance $d_{i,j}$ was employed [22–24]:

$$d_{i,j} = \frac{1}{\sqrt{2}} \sqrt{\sum_{i,j=1}^{m} \left(\sqrt{p_i} - \sqrt{q_j} \right)^2}, \tag{1}$$

where p_i and q_j are normalized histograms, representing spectral density of two acoustic signals. m is the number of sampled points of each acoustic signal. Two acoustic signals with different spectral density will yield a greater value of Hellinger distance than two similar ones. From the perspective of clustering analysis, samples with larger sum of distances to all peers are prone to be outliers.

For all pairs of these histograms, Hellinger distance was computed, resulting in a symmetric distance matrix D:

$$D = \begin{pmatrix} 0 & d_{1,2} & \cdots & d_{1,N} \\ d_{2,1} & 0 & \cdots & d_{2,N} \\ \vdots & \vdots & \vdots & \vdots \\ d_{N,1} & \cdots & d_{N,N-1} & 0 \end{pmatrix}.$$

Within D, the row with minimum sum of distances is chosen as the most central pattern *c*, representing the most typical testing sample within this group. Based on the most central pattern *c*, the *z*-score is computed for each sample *k*, representing the percentage of all samples that are further away from *c* than sample *k*, that is,

$$z(k) = \left| \frac{\{i=1,2,\dots,N:d_{i,c}<d_{k,c}\}}{N} \right| \quad (k = 1,2,\dots,N;) ,$$ (2)

where $d_{k,c}$ is the distance between sample *k* and *c*; $d_{i,c}$ denotes the distance between sample *i* and *c*; N is the total amount of samples. A sample with z-score close to 0 indicates a large distance to its peers. If there is no micro-crack at a particular position, the z-scores of a set of testing samples from this position should be uniformly distributed between (0, 1). We approximate the distribution of the average based on *n* samples using a normal distribution with mean 0.5 and variance 1/12n, i.e., [23,24]

$$z \sim N[\frac{1}{2}, (1/12n)^{1/2}].$$ (3)

Therefore, the one-sided *p*-value, which is the probability that a single observation *z* picked from a normal distribution with parameters $(1/2, (1/12n)^{1/2})$ will fall in the interval $(-\infty, z]$, can be computed as:

$$p = \frac{2}{\sqrt{2\pi}} \int_{-\infty}^{x} e^{-\frac{(z-1/2)^2}{2(1/\sqrt{12n})^2}} dz.$$ (4)

Based on a uniformity test of z-scores over an area, the resulting *p*-value of this test is obtained to estimate whether the inspected contains any micro-flaws.

3. Crack Identification Based on COSMO

In this section, we employed the COSMO model to identify and locate the cracks to steel specimens by numerical simulation and experimental tests.

3.1. Numerical Simulation

A two-dimensional model of a steel board embedded with a single crack was simulated by using a commercial software (Comsol Multiphysics V4.3a. COMSOL, Inc., Palo Alto, CA, USA). Figure 1 shows the schematic illustration of an ultrasonic measurement system in the simulation. Two longitudinal transducers with 60° wedge as the transmitter were typically used to carry out the ultrasonic inspection on the top surface of the steel board with the length of 220 mm and the height of 30 mm. A fixed distance between the two transducers was kept at a constant interval to make sure the first back wall echo was fully collected by the receiver. The single crack with a depth of *d* was located at the middle of the steel board along the x direction, i.e., x = 110 mm.

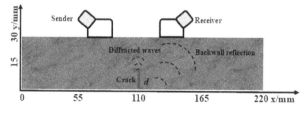

Figure 1. Schematic illustration of scanning on a simulated specimen with a single crack.

A 0.5 MHz continuous sinusoidal signal with signal-noise ratio (SNR) of 15 dB was applied as the exciting signal. Both the transmitter and the receiver were moved simultaneously to scan the steel board along the x direction. The receiving waveforms, at eight different positions, were spaced by

20 mm on the top of the simulated steel board, recorded, and the corresponding spectrums were then analyzed. Figure 2a depicts spectrums at eight positions of the simulated steel board with crack length $d = 2$ mm.

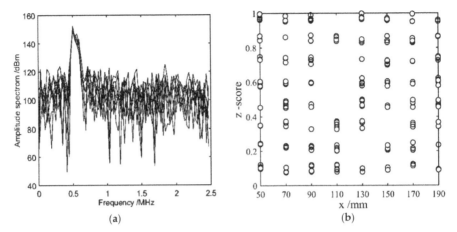

(a)   (b)

Figure 2. (a) Spectrums and (b) z-score distribution of echoes at eight monitoring points (depth of crack d = 2 mm) along x direction.

According to the COSMO algorithm model, the corresponding spectrums, at eight different observation positions, were saved then a group of z-scores were calculated by Equation (2) after every scanning process, finally 30 groups of z-score were obtained by scanning repeatedly 30 times. It is clearly shown in Figure 2b that z-score of observation points are distributed almost evenly between 0.3 and 1 except $x = 110$ mm, while most of z-score at $x = 110$ mm are mainly distributed below 0.4, just right at the crack's position. It is shown that the distribution of z-scores could be used to locate and identify cracks or defects in materials, i.e., z-score of damaged regions might be below 0.4. However, the conclusions need to be subjected to hypothesis tests to reach statistical significance, which determines whether a null hypothesis can be rejected or retained.

Figure 3a shows the calculated level of significance testing for crack depth of 2, 4, and 5 mm by Equation (4). It could be seen from Figure 3a that the p-value has a much smaller value than 0.1 around the crack region from 90 to 150 mm, which suggests that the imperfect structure of this region is significant. To make the comparisons and analysis clearly, an indicator called deviation level is defined as,

$$T = -\lg(p),\qquad(5)$$

i.e., logarithm transformed p-value, obviously the small T-value indicates little significant probability of crack. The T-value curve of significance testing is shown in Figure 3b. Obviously, the maximum T-value occurs around the position of crack ($x = 110$ mm) for crack depth of $d = 2$, 4 and 5 mm, respectively. Furthermore, the maximum T-value increases simultaneously as crack growth, e.g., the maximum T-value is close to 4 when crack depth is 2 mm, and the peak of T-value up to 10 when crack depth equal to 5. This result indicates that the higher level T-value is strongly correlated with the crack depth, which could become an index to exhibit the evolution of crack growth inside materials.

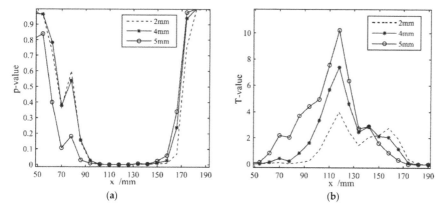

Figure 3. The calculated level of significance testing (**a**) *p*-value curve and (**b**) deviation level T-value curve can be plotted for crack depth of 2, 4 and 5 mm by Equations (4) and (5) along *x* direction on the simulated board.

When the crack depth changed from 0 mm to 5 mm, the peaks of T-value around the cracks were obtained and thus the relationship curve between the maximum T-value and crack depth d was plotted, as shown Figure 4. It can be seen that as the crack depth increased from 0 to 1, the slope of the curve sharply increased. When crack depth was less than 1 mm, the T-value was not larger than 5, which is basically considered the formation stage of crack, due to the relative small change of crack depth, thus this phase is called stage I. As the crack depth gradually expanded form 1.5 mm to 3.7 mm, the maximum T-value increased slowly from 5 to 7, at stage II. When crack depth was larger than 4 mm, at stage III, the value of the curve increased rapidly up to 10, and as high as 2 times than that of stage I, which means that the small cracks had already expanded to macro-cracks. Therefore, the peaks of T-value might track the progression of damage and evaluate the evolution of crack growth.

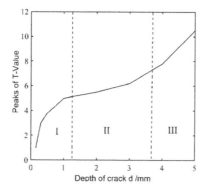

Figure 4. Deviation level T-value grows monotonically as depth of crack increases.

3.2. Experimental Measurement

A specimen made of Q235 (See Table 1) with dimensions 800 mm × 250 mm × 20 mm was used in the experimental measurement, as shown in Figure 5. Four sections embedded with cracks with average depth of 6 mm, 2 mm, 1 mm, and 0.5 mm were manufactured in the specimen, mainly located at 150 mm, 300 mm, 450 mm, and 600 mm, respectively, denoted by B, C, D and E. Additionally, Section A and F represented as undamaged regions located in the two ends of the specimens. A portable TOFD ultrasonic detector (PXUT-920, Nantong Union Digital Tech., China) was used to excite a narrow-pulse acoustic signal with 200 ns in width and stored the echo signals from inspected cracked region. Scanning was manually carried out by a scanner unit with one pair of 5 MHz normal transducer, i.e., the transmitter and the receiver, with 60° wedges for longitudinal waves. Two transducers, spaced 62 mm apart, were located at equidistant over the crack region center, and scanning was done by moving the scanner in the length direction of steel plate parallel to the crack region. The echo signal sampled by the detector, containing 1496 points, was acquired every 0.5 mm along the length direction. After a scanning, a total of 1600 echo signals (A-scan) were obtained and stored in the ultrasonic detector.

Table 1. Mechanical characteristics of Q235 carbon steel. (Provided by HBO Windpower Equipment Co., Ltd, Nantong, China).

Grade	Elements (%)					Yield Strength (MPa)	Tensile Strength (MPa)	Elongation (%)
	C	Mn	Si	P	S			
Q235A	0.14~0.22	0.30~0.65	0.30	0.045	0.030	235	375~460	21~26

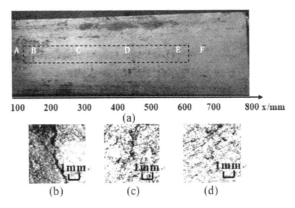

Figure 5. (a) A specimen with four cracks with average depth of (b) 6 mm, (c) 2 mm, (d) 1 mm, and 0.5 mm located at Section B, C, D, E, respectively. Section A and F without cracks.

COSMO method was applied to analyze this dataset of echo-signal recorded by TOFD ultrasonic detector. Firstly, the Hellinger distance matrix D was constructed using Equation (1), and the row with minimum sum was chosen in the metric D so that the z-score could be determined using Equation (2). Figure 6a shows the z-score distribution for 30 scanning. It is shown that the z-score in undamaged sections is much larger than those in the region with cracks. For example, the z-score for positions A and F are about 0.4~1, while those for positions B, C, D and E are below 0.4. The results suggest that the z-score is closely related to cracks of specimen, just as the simulated results.

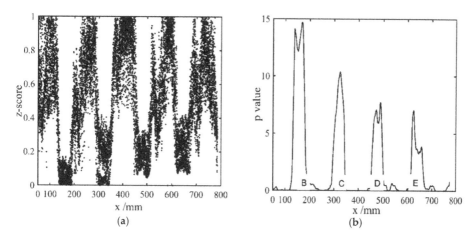

Figure 6. (**a**) z-score distribution (**b**) the curve of T-value of specimen with cracks and along x direction.

By using Equations (1)~(4) and (5), the deviation level T-Value is calculated to make a significant analysis. It can be observed from Figure 6b that the T-value at B, C, D and E are of high level compared to those in uncracks region. For instance, the T-value at $x = 150$ up to 13, and T- value at $x = 600$ sharply increasing from 0 to 7, but T-value of sections without cracks almost equal to 0, far less than T-value at crack region. In addition, T-value increase almost linearly with the depth of cracks. The relationship curve between the peaks of T-value and crack depth is plotted in Figure 7. It is shown that the peak of T value increases with crack growth from 0 to1.5 mm quickly up to 5, which is exactly in stage I. When the crack depth is larger than 2, the slope of curve go slow but still faster than the simulated results. It is worth noting that the change of slope is not distinct enough to easily recognize stage II or III when depth of crack exceeding 1.5 mm, different from the simulated curve, which might attributed to the result of multi-physical mechanisms.

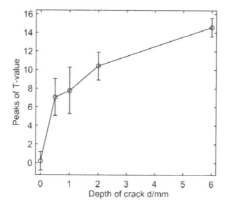

Figure 7. The relationship curve of specimen between T-value peaks and crack depth.

4. Conclusions

Defects or cracks can significantly increase acoustic non-linearity, and the nonlinear acoustical parameters, and thereby can be exploited to evaluate the state of material damage. However, the harmonics are usually too weak to be detected in early fatigue. Therefore, the non-linear ultrasonic technique is rarely used to qualify crack growth. In this work, the COSMO method was applied to compare the spectrum of different positions by ultra-sonic scanning in order to obtain the distribution

Appl. Sci. **2019**, *9*, 95

of z-scores as well as the corresponding significance level in every scanning position. The results show that: (1) the z-scores in the location with cracks are distributed below 0.4, while the z-scores in the location without cracks are above 0.4; (2) the deviation level T - value at locations with crack are much larger than those at locations without cracks, and the T-value would get larger with the increase of crack depth; (3) based on the quantitative relations between the T values and the crack depth, we can evaluate and monitor the online state of the structural health by COSMO model. However, it is noted that the COSMO model is still a simple model that does not consider some other factors, such as the shape and mechanical properties of structure, as well as the requirements on data size. Therefore, the reliability of COSMO needs to be further optimized to reach a solution for non-destructive evaluation in future.

Author Contributions: D.Z., conceived and designed the experiments. X.T., and X.Z., performed the experiments. X.T., X.Z., and Y.F., analyzed the data. X.T., X.Z., and D.Z., wrote the manuscript.

Funding: This research was funded by the National Natural Science Foundation of China (Grant no's., 81627802, 11674173 and 11874216), QingLan Project, and the Fundamental Research Funds for the Central Universities.

Conflicts of Interest: The authors declare no conflict of interest.

References

1. Broda, D.; Staszewski, W.J.; Martowicz, A.; Uhl, T.; Silberschmidt, V.V. Modelling of nonlinear crack–wave interactions for damage detection based on ultrasound—A review. *J. Sound Vib.* **2014**, *333*, 1097–1118. [CrossRef]

2. Novak, A.; Bentahar, M.; Tournat, V.; Guerjouma, R.; Simon, L. Nonlinear acoustic characterization of micro-damaged materials through higher harmonic resonance analysis. *NDT E Int.* **2012**, *45*, 1–8.

3. Donskoy, D.; Sutin, A.; Ekimov, A. Nonlinear acoustic interaction on contact interfaces and its use for nondestructive testing. *NDT E Int.* **2001**, *34*, 231–238.

4. Nagy, P.B. Fatigue damage assessment by nonlinear ultrasonic materials characterization. *Ultrasonics* **1998**, *36*, 375–381. [CrossRef]

5. Dos Santos, S.; Vejvodova, S.; Prevorovsky, Z. Nonlinear signal processing for ultrasonic imaging of material complexity. *Proc. Est. Acad. Sci.* **2010**, *59*, 108–117. [CrossRef]

6. Jhang, K.Y. Nonlinear ultrasonic techniques for nondestructive assessment of micro damage in material: A review. *Int. J. Precis. Eng. Manuf.* **2009**, *10*, 123–135. [CrossRef]

7. Guo, X.; Zhang, D.; Zhang, J. Detection of fatigue-induced micro-cracks in a pipe by using time-reversed nonlinear guidedwaves: A three-dimensional model study. *Ultrasonics* **2012**, *52*, 912–919. [CrossRef]

8. Blanloeuil, P.; Rose, L.F.; Veidt, M.; Wang, C.H. Time reversal invariance for a nonlinear scatterer exhibiting contactacoustic nonlinearity. *J. Sound Vib.* **2018**, *417*, 413–431. [CrossRef]

9. Ostrovsky, L.A.; Johnson, P.A. Dynamic nonlinear elasticity in geomaterials. *La Rivista Del Nuovo Cimento* **2008**, *24*, 1–46.

10. Hall, D.A. Review Nonlinearity in piezoelectric ceramics. *J. Mater. Sci.* **2001**, *36*, 4575–4601. [CrossRef]

11. Zhenggan, Z.; Siming, L. Nonlinear Ultrasonic Techniques Used in Nondestructive Testing: A Review. *J. Mech. Eng.* **2011**, *47*, 2–9.

12. Gang, Q.; Steven, F.W. A Framework of Data-Enabled Science for Evaluation of Material Damage Based on Acoustic Emission. *J. Nondestruct. Eval.* **2014**, *33*, 597–615.

13. Sollier, T.; Blain, C. IRSN preliminary analysis on statistical methods for NDE performances assessment. In Proceedings of the 12th International Conference on Non-Destructive Evaluation in Relation to Structural Integrity for Nuclear and Pressurized Components, Dubrovnik, Croatia, 4–6 October 2016; pp. 1–10.

14. Keprate, A.; Ratnayake, R.M.C. Probability of Detection as a Metric for Quantifying NDE Reliability: The State of The Art. *J. Pipeline Eng.* **2015**, *14*, 199–209.

15. Schneider, C.R.A.; Rudlin, J.R. Review of statistical methods used in quantifying NDT reliability. Insight-Non-Destruct. *Test. Cond. Monit. (INSIGHT)* **2004**, *46*, 77–79.

16. Lu, C.J.; Meeker, W.Q. Using Degradation Measures to Estimate a Time-to-Failure Distribution. *Technometrics* **1993**, *35*, 161–174. [CrossRef]

17. Wu, S.; Tsai, T. Estimation of time-to-failure distribution derived from a degradation model using fuzzy clustering. *Qual. Reliab. Eng. Int.* **2015**, *16*, 261–267. [CrossRef]

18. Batzel, T.D.; Swanson, D.C. Prognostic Health Management of Aircraft Power Generators. *IEEE Trans. Aerosp. Electron. Syst.* **2009**, *45*, 473–482. [CrossRef]
19. Lakhtakia, A. Transition from Nondestructive Testing (NDT) to Structural Health Monitoring (SHM): Potential and challenges. *SPIE Smart Struct. Nondestruct. Eval.* **2014**, *9055*, 90550Z.
20. Gebraeel, N.Z.; Lawley, M.A.; Li, R. Residual-Life Distributions from Component Degradation Signals: A Bayesian Approach. *IIE Trans.* **2005**, *37*, 543–557. [CrossRef]
21. Zhou, W.; Zh, L.V.; Wang, Y.R. Acoustic Response and Micro-Damage Mechanism of Fiber Composite Materials under Mode-II Delamination. *Chin. Phys. Lett.* **2015**, *32*, 046201. [CrossRef]
22. Kůs, V.; Tláskal, J.; Farová, Z.; Santos, D.S. Signal detection, separation & classification under random noise background. In Proceedings of the 13th Biennial Baltic Electronics Conference, Tallinn, Estonia, 3–5 October 2012; pp. 287–290.
23. RÖgnvaldsson, T.; Norrman, H.; Byttner, S.; Järpe, E. Estimating *p*-values for deviation detection. In Proceedings of the IEEE Eighth International Conference on Self-Adaptive and Self-Organizing Systems, London, UK, 8–12 September 2014; pp. 100–109.
24. Fan, Y.T.; Nowaczyk, S.; RÖgnvaldsson, T. Evaluation of Self-Organized Approach for Predicting Compressor Faults in a City Bus Fleet. *Procedia Comput. Sci.* **2015**, *53*, 447–456. [CrossRef]
25. Teng, X.; Fan, Y.; Nowaczyk, S. Evaluation of micro-flaws in metallic material based on a self-organized data-driven approach. In Proceedings of the 2016 IEEE International Conference on Prognostics and Health Management (ICPHM), Ottawa, ON, Canada, 20–22 June 2016; pp. 1–5.
26. Kim, J.Y.; Yakovlev, V.A.; Rokhlin, S.I. Parametric modulation mechanism of surface acoustic wave on a partially closed crack. *Appl. Phys. Lett.* **2003**, *82*, 3203–3205. [CrossRef]
27. Krüger, S.E.; Rebello, J.M.A.; Camargo, P.C. Hydrogen damage detection by ultrasonic spectral analysis. *NDT E Int.* **1999**, *32*, 275–281.
28. Hillis, A.J.; Neild, S.A.; Drinkwa, B.W.; Wilcoxter, P.D. Global crack detection using bispectral analysis. *Proc. R. Soc. A* **2006**, *462*, 1515–1530. [CrossRef]

Article

Influence of Piano Key Vibration Level on Players' Perception and Performance in Piano Playing

Matthias Flückiger *, Tobias Grosshauser and Gerhard Tröster

Electronics Laboratory, ETH Zurich, Gloriastrasse 35, 8092 Zurich, Switzerland; tobias@grosshauser.de (T.G.); troester@ife.ee.ethz.ch (G.T.)
* Correspondence: matthias.flueckiger@gmail.com

Received: 29 October 2018; Accepted: 16 December 2018; Published: 19 December 2018

Abstract: In this study, the influence of piano key vibration levels on players' personal judgment of the instrument quality and on the dynamics and timing of the players' performance of a music piece excerpt is examined. In an experiment four vibration levels were presented to eleven pianists playing on a digital grand piano with grand piano-like key action. By evaluating the players' judgment of the instrument quality, strong integration effects of auditory and tactile information were observed. Differences in the sound of the instrument were perceived by the players, when the vibration level in the keys was changed and the results indicate a sound-dependent optimum of the vibration levels. By analyzing the influence of the vibration levels on the timing and dynamics accuracy of the pianists' musical performances, we could not observe systematic differences that depend on the vibration level.

Keywords: piano playing; vibrotactile feedback; interaction; musical performance; auditory perception; sensors; actuators

---

1. Introduction

Playing the piano is a complex multi-modal task, where the pianist controls the instrument through his or her intention and perceived instrument feedback. There are four main musician-musical instrument interaction modalities: visual feedback, auditory feedback, force feedback, and vibrotactile feedback. The interaction with a musical instrument can be modeled as a feedback controller [1], where the musician's brain controls his or her body, arms, and fingers to modify the instrument's behavior based on changes in sensory inputs. This closed-loop model implies that if the instrument's feedback is altered, the pianist will adapt his or her playing to compensate for and retain the desired instrument behavior. Vibrotactile feedback can support the precise control of finger force, as shown by Ahmaniemi [2] with a basic force repetition experiment on a rigid sensor box. Furthermore, Goebl and Palmer [3] demonstrated that tactile sensations from the finger-key surface interaction support some pianists to improve timing accuracy and precision of finger movements.

In piano playing, vibrotactile feedback is perceived through the fingers in contact with the keys and the feet in contact with the pedals. The keybed and soundboard vibrations excite the piano keys and pedals [4]. Askenfelt and Jansson [5] measured the vibrations of a depressed piano key and a depressed piano pedal. Piano key vibrations comprise broadband and tonal parts [4,6]. The tonal parts come from the string vibrations, and the broadband parts come from mechanical impacts (e.g., hammer-string impact and key–keybed impact) of the piano action when the piano key is played [4,6].

The levels of the vibrations' tonal part can rise to the micrometer range [7]; the vibrations are close to the limits of human vibration perception and are often sensed subconsciously [8]. However, piano key vibrations can be detected up to the middle octave of the keyboard [9]. Further, the vibration levels vary considerably among different pianos, but it remains unclear if pianists can perceive these differences [7].

Keane and Dodd [8] found that the ratio of broadband and tonal parts of piano key vibrations influenced the instrument's perceived sound. An upright piano was mechanically modified to reduce the broadband parts' amplitude, with the expectation to improve the instrument's quality. The pianists preferred the modification with regard to tone and loudness in an evaluation study. Interestingly, the participants did not report differences with regard to touch or vibrations.

Fontana et al. [10] showed by ratings of evaluation criteria that realistic piano key vibrations rendered on a digital keyboard are preferred to a no-vibration condition. In the same study, key vibrations did not show a significant effect on pianists' timing and dynamics accuracy during a scale playing task.

In addition to the state-of-the-art, the influence of four piano key vibration levels on pianists' personal judgment of an instrument's sound, control, and feel is investigated in this study. We designed the experiment, such that the control of vibration levels was independent of the sound of the instrument and aimed to explore connections between vibrotactile feedback and the perceived quality of the instrument. To test if pianists adapt their playing to vibrotactile feedback and to analyze if the vibrations support the control of finger forces, the effect of the vibration levels on timing and dynamics accuracy in pianists' performances is also studied in this paper.

2. Methods

2.1. Equipment

The pianists played on an AvantGrand N3X, a digital hybrid grand piano from Yamaha. This instrument was chosen because it simulates piano key vibrations, features state-of-the-art grand piano sound-rendering algorithms, and has a piano action resembling that of acoustic grand pianos. Musical instrument digital interface (MIDI) messages and the headphone audio output of the instrument were recorded. The pianists played with closed-back headphones to block the small amount of sound that vibrating keys radiate.

2.2. Experiment Design

The target was to control the key vibration levels independent of the sound and to cover the level range of piano key vibrations of acoustic concert grand pianos. Independent control could not be achieved with the built-in vibrotactile feedback rendering system of the AvantGrand N3X; therefore, it was extended as illustrated in Figure 1.

The mono audio output signal of the AvantGrand N3X was processed with a digital signal processor (DSP). Through a combined approach of vibrometer measurements and subjective evaluation by playing on the instrument, the DSP's filter stage was tuned to create vibration level V_3 (see Figure 1), which approached the maximum vibration levels previously measured on acoustic grand pianos [7]. After implementation of V_3, vibration levels V_2 and V_1 were created by attenuating the signal in steps of 6 dB. The no-vibration condition V_0 completed the levels of the experiment. As shown in Figure 1, the vibration levels cover the range of acoustic grand pianos for notes A2, A3, and A4. For notes A0 and A1, the levels are more than 10 dB lower. The chosen music piece avoided the lowest notes. The deviations of the vibration level curves in Figure 1 are due to non-idealities of the excitation system.

The experiment was created to study the influence of four vibration levels on players' personal judgment of the instrument quality and on the dynamics and timing of the players' performance of a music piece excerpt. The experiment was designed so that the participants were unaware of the independent variable, and the session was split into three parts to steer the pianists' attention to different instrument properties. Free verbalizations were used to assess the players' judgment, allowing for unrestrained and possibly unexpected answers. Since a small influence of the piano key vibrations on musical performance and the players' judgments was assumed—as natural levels are close to the threshold of vibration sensation [7–9]—numerous repetitions were included in the protocol, and participants with high levels of playing experience were selected.

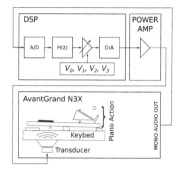

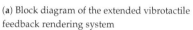

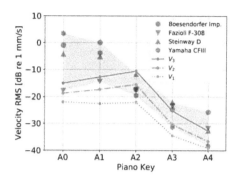

(**a**) Block diagram of the extended vibrotactile feedback rendering system

(**b**) Comparison of the measured tonal part of the vibration levels to levels measured on acoustic grand pianos

Figure 1. (**a**) Block diagram of the vibrotactile feedback rendering system to generate the key vibrations; the mono audio output of the N3X was filtered and attenuated with a DSP. Thereafter the signal was power amplified to drive the transducer of the built-in vibrotactile rendering system of the N3X. (**b**) Comparison of the tonal part of the vibration levels V_1, V_2, V_3 to vibration levels of four acoustic concert grand pianos. The comparison is based on vibrometer measurements of *forte* keystrokes [7]. (V_0 is not shown because it corresponds to no vibrations.)

2.3. Participants

Eleven pianists participated in the study: seven piano students, 22–26 years of age, with an average playing experience of 17 years; and four professional pianists, 31–40 years of age, with an average playing experience of 26 years. None of them reported having auditory or tactile impairments.

2.4. Procedure

The session for each participant lasted around 1.5 h. The participants were asked to prepare an interpretation of a music piece excerpt. The excerpt was 15 bars long and the participants were instructed to adhere to the tempo, dynamics, accents, and pedaling information. The excerpt was taken from Klage by Gretchaninov [11] and was edited to cover the dynamic range from *pianissimo* to *fortissimo* (see Figure A1 in the Appendix A). The participants were not informed about the purpose of the experiment beforehand. The participants were only told that their judgments of various settings of the instrument will be evaluated.

The experiment comprised a warm-up (with a duration of around 5–10 min), questionnaires, and three parts (A, B, and C) with a duration of roughly 20 min each.

During the warm-up, the pianists were free to play whatever they wanted and were instructed to evaluate the instrument in a way comparable to choosing an instrument for a concert or for purchase. After this familiarization, the pianists were asked to express their first impression by answering a set of questions about the sound and touch of the AvantGrand N3X and by comparing the instrument to their main instrument.

During the main parts of the experiment (A, B, and C) the pianists were asked to repeat the excerpted music piece accurately in 12 direct comparisons of different instrument settings. After each trial the pianists were asked to indicate a personal preference ("better", "worse", or "similar") of the current setting relative to the previous setting and to describe their impression in a few words. The participants were told that the differences between the comparisons can be small and that some might be perceptually irrelevant. In part A the pianists were told that a slight adjustment of the instrument (not further specified in order to not suggest an answer or category) was made between each repetition, in part B it was claimed that a small adjustment to the sound was made between each trial, and in part C the pianists' attention was directed to the keyboard by asking about the instrument's control and feel.

In fact, the only independent variable throughout the experiment was the key vibration level (V_0, V_1, V_2, V_3). The sequence of parts (A, B, and C) was the same for all participants. Each part had a different randomized sequence of vibration levels. All pair of levels were compared twice in each part—once for each order. In total, nine trials per vibration level and per participant were recorded.

At the very end of each experiment session, personal information was collected and the participant was asked about his or her experience and preference of piano key vibrations in piano playing, before we disclosed and explained the purpose of the experiment.

3. Results

3.1. Influence of Vibration Levels on Perceived Instrument Sound, Control, and Feel

In the analysis of the preference ratings ("better", "worse", or "similar"), a high variance and for some participants also controversial ratings were observed. We decided to present the ratings across all parts and participants here, because it was not possible to draw conclusions from the ratings per participant or per part. The result is presented in Figure 2.

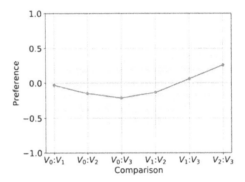

Figure 2. Analysis of the preference ratings ("better" (1), "worse" (−1), or "similar" (0)) of the vibration levels across all parts and all participants. The ratings are based on direct comparisons between all pairs of levels. The ratings are relative. For example, a positive value for V_0:V_1 indicates a preference of V_0 over V_1 and a negative value a preference of V_1 over V_0. The shaded area marks the standard deviation of the ratings.

The high variance of the ratings in Figure 2 reflects the closeness of the key vibration levels to the limits of human perception. Additional factors that can have disturbed the ratings are the mood and fatigue of the player. Also a self-evaluation of the playing, the difficulty of the task, or the imposed expectation of a difference between the settings might have disturbed the ratings. However, a visual comparison of the mean values in Figure 2 indicates a tendency in the preference of the players toward vibration level V_2. The preference of vibration level V_2 is confirmed by the evaluation of the player's verbal self-reports presented hereafter.

The free verbalizations were analyzed with an approach presented by Pate et al. [12], where concepts by Dubois [13] were applied to musical instrument evaluation.

Based on the context and for each participant, the meaning, category, and preference of each statement was identified via linguistic tools such as reformulations, oppositions, and comparatives. Thereafter, the statements were classified into positive and negative statements for three categories: sound, control, and feel. Statements covering multiple categories were split before classification. Statements not indicating a preference or a perceived difference were counted as "no difference".

The results are summarized in Table 1. Significance was evaluated with Pearson's χ^2-tests at a confidence level of 95%. The test was performed on all statements (positive and negative) per category.

Table 1. Evaluation per vibration level derived from the free verbalizations of the pianists. The number of positive and negative statements per category was counted for each vibration level (V_0, V_1, V_2, V_3). The frequency counts were evaluated with χ^2-tests. The χ^2-statistics and p-values are given for each category; $p < 0.05$ is highlighted in bold.

Category	Examples	V_0	V_1	V_2	V_3	χ^2	p
sound pos. s_p	"round", "balanced"	21	26	39	24		
sound neg. s_n	"harsh", "artificial"	28	19	12	19		
relative number of positive statements $s_p/(s_p+s_n)$		0.4	0.6	0.8	0.6	11.86	**0.0079**
control pos. c_p	"reactive", "controllable"	25	24	25	24		
control neg. c_n	"limited", "hard to create dynamics"	15	11	13	16		
relative number of positive statements $c_p/(c_p+c_n)$		0.6	0.7	0.7	0.6	0.69	0.88
feel pos. f_p	"comfortable", "grand piano feeling"	9	8	7	11		
feel neg. f_n	"exhausting", "tedious"	5	4	1	4		
relative number of positive statements $f_p/(f_p+f_n)$		0.6	0.7	0.9	0.7	1.52	0.68
no difference	"similar", "somewhat different"	12	24	11	13		

Although we did not alter the sound throughout the experiment, Table 1 shows that the vibration levels (V_0, V_1, V_2, V_3) have an influence on the pianists' sound perception. Pairwise testing with Bonferroni correction showed that the significance of the χ^2-test in the sound category arises from the difference between V_0 and V_2.

The phenomenon of vibrotactile feedback causing a difference in sound perception is known as integration of auditory and tactile information [14–16] or weak synesthesia [17]. The preference of V_2 over V_3, confirming the evaluation of the preference ratings in Figure 2, was surprising because vibration level V_3 is closer to the levels of acoustic instruments (see Figure 1). An explanation lies in five statements about vibration level V_3, which were classified as negative. These statements criticized the balance of the perceived sound as having "too much bass" or being "unbalanced".

The results for control and feel are not significant according to the χ^2-tests. For vibration level V_1 there were twice as many "no difference" statements than for all other levels, which indicates that V_1 is most difficult to differentiate.

Only two participants consciously noticed a change in vibration levels during the experiment, when vibration level V_3 was compared to V_0 and vice versa. Both recognized the vibrations during the last part, when the keyboard was the focal point.

To find possible explanations for the above presented differences, we analyzed the verbal self-reports of the participants in more specific categories. We observed that the key vibrations influence the timbre and the perceived loudness of the bass keys. Also, the timbre of treble notes was judged more pleasant when playing with V_2 or V_3. Some participants also noted a sensation of space when playing with higher vibration levels (V_2, V_3) and described it as room or reverb effect of the sound. In contrast when pianists played with vibration levels V_0 or V_1 the sound was sometimes described as dry. Comparisons to acoustic instruments and e-pianos also align with this observation.

A critical aspect for discussion is that the sequence of parts (A, B, and C) was the same for all participants, which might have influenced the results. We designed the experiment protocol to steer the players' attention to different multi-modal aspects to discover unexpected connections between vibrotactile feedback and the players' judgment of the instrument quality.

In part A the participants could freely describe their impressions and we did not suggest any quality criteria for the comparisons. Unbiased comparisons are only possible within the first part of the experiment. Ten out of eleven participants naturally made statements about sound and control in the first part, which justifies the suggestion of these criteria in the following parts.

We decided to put part C at the end of the experiment session, because we did not want to risk that the participants are already consciously aware of the vibrations, when comparing the levels with regard to sound. In part C the participants focused on the keyboard. Therefore we expected that it is most likely that the participants recognize the vibrations in this part (which happened in two cases).

Finally, also the difficulty of the evaluation task might have altered the judgments of the pianists. In each trial, the participant played the music piece excerpt for a duration of around 30 s, communicated his or her impression with regard the previous setting and sometimes also answered clarifying questions from the experimenter. Thereafter he or she performed the music piece for the next comparison.

3.2. Influence of Piano Key Vibration Levels on Musical Performance

To analyze the MIDI-based performance data, a custom data structure was used. The structure groups notes played at the same time ($\pm$40 ms) into clusters, removes accidentally played wrong notes, and assigns if the note was played by the left or right hand.

Key velocity v, a measure of a keystroke's excitation strength, was directly extracted from the MIDI messages. For the calculation of the inter-onset interval τ—the time interval between two subsequent note onsets—only notes played by the left hand were considered. The tempo of each trial was normalized.

To compare trials by the distribution of key velocity v (analogously for inter-onset interval τ) histogram intersection was used. Histogram intersection was introduced by Swain and Ballard [18] to identify objects by color similarity in computer vision. Histogram intersection is defined as [18]

$$H_1(v) \cap H_2(v) = \sum_{i=1}^{n} \min\left(h_{1i}(v), h_{2i}(v)\right), \tag{1}$$

where H_1 and H_2 represent two trials by normalized discrete distributions of key velocity v with n bins h_{1i}, h_{2i}. Equation (1) measures the overlap of two histograms in the range $[0,1]$. The number '1' corresponds with perfect overlap; '0' means no overlap. In contrast to an evaluation based on mean values only, histogram intersection also identifies differences in the distributions' shape or offset.

To judge significance, two tests were demanded to reject the null hypothesis: non-parametric Friedman analysis of variance with a 95% confidence level in combination with pairwise Wilcoxon signed-rank tests with Bonferroni correction. We used non-parametric tests, because the evaluated quantities do not necessarily follow a normal distribution and because of the sample size.

Two approaches were used to compare the distributions of key velocity v and inter-onset interval τ by histogram intersection. The distribution of both parameters was calculated for each trial and was analyzed for each participant separately.

The first approach considered if the pianists adapted their playing to the vibration levels (e.g., if a pianist perceived an overemphasis of bass notes and therefore played the bass notes with less finger force than before). This force adaption manifests in the shape of the distribution of key velocity v. An increase in the "amount of adaption" was expected with increasing vibration levels. The "amount of adaption" for key velocity v was measured as follows.

Let $H_{P_k,V_\ell,i}(v)$ denote the normalized histograms describing the distribution of key velocity v for trial $i \in \{1,\dots,9\}$, vibration level V_ℓ with $\ell \in \{0,1,2,3\}$, and pianist P_k with $k \in \{1,\dots,11\}$. Then the "amount of adaption" A_{P_k,V_n} of pianist P_k to vibration levels V_n with $n \in \{1,2,3\}$ was estimated as the histogram difference relative to V_0 condition $H_{P_k,V_0,i}(v) \cap H_{P_k,V_n,j}(v)$ for all combinations of trials $i,j \in \{1,\dots,9\}$ and key velocity v. For the inter-onset interval τ the same procedure was conducted.

Figure 3 shows the "amount of adaption" of the pianist's playing to the feedback levels for both performance parameters. The differences in Figure 3 are not significant. There is no general tendency that the participants adapt their playing to the key vibration level. Nonetheless, by analyzing the influence of the vibration levels per participant individually, three participants showed significant differences for key velocity v and three for inter-onset interval τ.

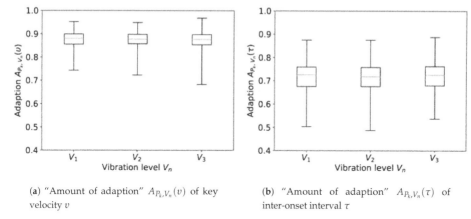

(a) "Amount of adaption" $A_{P_k,V_n}(v)$ of key velocity v

(b) "Amount of adaption" $A_{P_k,V_n}(\tau)$ of inter-onset interval τ

Figure 3. "Amount of adaption" to the feedback levels V_n ($n = \{1, 2, 3\}$) relative to the no-vibration condition V_0 across all pianists P_k and for both performance parameters. The differences in the amount of adaption for all vibration levels are not significant for both parameters. The line in the center of the box-plot marks the median, the box extends from the first to the third quartile, and the whiskers mark the value range.

Possible explanations for the majority of pianists not adapting their playing to the feedback levels include that the combined task of playing, judging the impression, and adapting their playing was too difficult, that the levels were too small to cause a reaction, or that the method was not accurate enough to unveil such differences.

The second approach investigated how accurately the pianists could repeat the music piece excerpt when playing with different vibration levels. If key vibrations support the precise control of finger forces, a lower variance in the distribution of key velocity v could be expected. In consequence the shape of the distribution of key velocity v would be altered and hence a difference in repeatability could be detected. Likewise, if the pianist's tempo was more stable, a different shape of the distribution of the inter-onset interval τ would occur. Indirect causes are also possible (e.g., the pianist feels more comfortable to play and therefore plays with higher repeatability). The time-point of the trials during the experiment was not taken into account. We decided to analyze and present the data for each pianist individually hereafter, because we observed a strong dependency of the repeatability on the player.

The repeatability $R_{P_k,V_0}(v)$ for participant P_k playing with vibration level V_0 was computed by comparing the distributions of key velocity v by $H_{P_k,V_0,i}(v) \cap H_{P_k,V_0,j}(v)$ for all combinations of trials $i, j \in \{1, \ldots, 9\}$, where $i \neq j$. For vibration levels V_1, V_2, and V_3, and for inter-onset interval τ similar procedures were conducted.

The resulting repeatability per vibration level and per participant is presented in Figure 4 for key velocity v and inter-onset interval τ. The differences in repeatability were significant for a majority of the participants. However, no consistent tendency or pattern in repeatability occurred among the pianists in Figure 4.

Therefore, the vibration levels of our experiments do not have a conclusive influence on the pianists' repeatability, and the measured MIDI data do not support the hypothesis that key vibrations assist the precise control of finger force. Consequently the observations do not confirm the results of Ahmaniemi [2] or Galica et al. [19], for the piano playing case. Galica et al. [19] showed that even unconscious vibratory stimulation applied to the soles of the feet can cause lower variance in kinematic interactions.

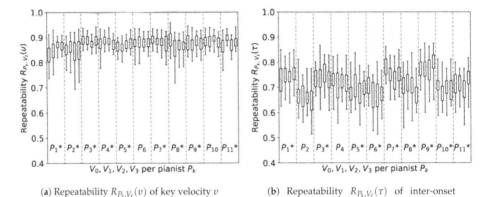

(a) Repeatability $R_{P_k,V_\ell}(v)$ of key velocity v (b) Repeatability $R_{P_k,V_\ell}(\tau)$ of inter-onset interval τ

Figure 4. Estimated repeatabilities per participant P_k and vibration level V_ℓ for both performance parameters. No consistent tendency occurred among the pianists, but the vibration levels had a significant influence (marked with *) on repeatability for a majority of the participants.

In summation, the repeatability estimates $R_{P_k,V_\ell}(v)$, $R_{P_k,V_\ell}(\tau)$ for vibration levels (V_0, V_1, V_2, V_3) depended on the player. The pianists in this study were more accurate in repeating key velocity v (median of $R_{P_k,V_\ell}(v) \in [0.85, 0.88], \forall k, \ell$) than in repeating inter-onset interval τ (median of $R_{P_k,V_\ell}(\tau) \in [0.66, 0.77], \forall k, \ell$). For the inter-onset interval τ the intra-individual variance was also considerably larger (see Figure 4), although we normalized the tempo of each trial before the analysis.

As a concluding aspect of interest, no categorical differences (in repeatability or playing adaption to feedback levels) were found between the group of students (P_1 to P_7 in Figure 4) and the group of professional pianists (P_8 to P_{11} in Figure 4).

4. Conclusions

By systematically investigating the players' personal judgment of the instrument quality of the vibration level in the keys, we observed strong integration effects of auditory and tactile information. The results give an illustration of the strong multi-modal effects in piano playing. The subjects perceived differences in the sound of the instrument when the vibration level in the keys was changed. The preference of vibration level V_2 over V_3 indicates an optimum or a "sweet spot" of piano key vibration levels, which depends on the instrument's sound and sound balance.

In line with the results of Keane and Dodd [8], the vibration levels in this experiment significantly affected the instrument's judged sound quality but not its control and feel. However, in contrast to the design of the present experiment, Keane and Dodd [8] reduced the level of the broadband part of piano key vibrations of an acoustic instrument.

An interesting direction for future research is to determine the vibration level differences that pianists can differentiate. This would help to understand if an instrument can be identified based on its vibrotactile feedback only, while the instrument's auditory and force feedback are kept constant. Some participants in this experiment perceived a spatial sensation and described it as room or reverb effect on the sound when playing with higher vibration levels (V_2, V_3). For several applications it could be interesting to understand the conditions that can cause such an illusion.

We did not find systematic differences by analyzing the influence of the vibration levels on the timing and dynamics accuracy of the pianists' musical performances. We can not exclude that such an influence exists but with the proposed measures, "amount of adaption" and repeatability, we could not measure such a relation. Furthermore, the basic results of Ahmaniemi [2], that vibrotactile feedback assists the precise control of finger forces could not be confirmed in our case. For future studies, we suggest to include a larger number of participants. This could help to identify groups reacting similarly to key vibrations. Future studies might also include multiple experiments over a certain range of time to exclude influences of physical and mental state on the day of testing. Finally, an analysis on a note-by-note basis could clarify if, for example, the vibrotactile feedback of a long-lasting bass note helps the precise control of the dynamics in subsequent keystrokes. We could not generalize such a relation with the data of the presented experiment.

If our results can be confirmed on acoustic instruments, our findings of the perception part of the experiment suggest that piano manufacturers should design the vibrations in the piano keys in balance with the sound of the lower notes of the instrument. Furthermore, it would be interesting to investigate the just-noticeable difference of piano key vibration levels, which might possibly be around 6 dB. Further research in this area could help to answer the question, if the tonal parts of the Steinway and Sons and the Yamaha concert grand pianos (the tonal parts for notes A2, A3, and A4 differ by more than 6 dB [7]) can be differentiated based on their vibrotactile feedback only by the player. Of course in such an experiment the vibrotactile feedback should be rendered on the same instrument, otherwise cues from the auditory or kinematic sensations might dominate the perceived impression.

Author Contributions: M.F. designed, accomplished, and evaluated this study. T.G. and G.T. contributed as consultants. All discussed the results.

Funding: This research has been pursued as part of the "Musician's behavior based on multi-modal real-time feedback" project, Grant No. 166588, funded by the Swiss National Science Foundation (SNSF).

Acknowledgments: The authors are indebted to Anders Askenfelt for contributing to the experiment's design, for inspiring discussions, and for offering advice about the topics of the presented study. They would also like to thank Yamaha for generously providing the AvantGrand N3X.

Conflicts of Interest: The authors declare no conflict of interest.

Appendix A

Figure A1. Music sheet of the study. The excerpt was taken from Klage by composer Gretchaninov [11]. The excerpt was edited to cover a broad dynamic range and also to include accents.

References

1. O'Modhrain, S.; Gillespie, R.B. Once more, with feeling: Revisiting the role of touch in performer-instrument interaction. In *Musical Haptics*; Springer: Berlin, Germany, 2018; pp. 11–27.

2. Ahmaniemi, T. Effect of dynamic vibrotactile feedback on the control of isometric finger force. *IEEE Trans. Haptics* **2013**, *6*, 376–380. [CrossRef] [PubMed]

3. Goebl, W.; Palmer, C. Tactile feedback and timing accuracy in piano performance. *Exp. Brain Res.* **2008**, *186*, 471–479. [CrossRef] [PubMed]

4. Askenfelt, A. Observations on the transient components of the piano tone. In Proceedings of the Stockholm Musical Acoustics Conference, Stockholm, Sweden, 28 July–1 August 1993; pp. 297–301.

5. Askenfelt, A.; Jansson, E.V. On vibration sensation and finger touch in stringed instrument playing. *Music Percept.* **1992**, *9*, 311–349. [CrossRef]

6. Keane, M. Separation of piano keyboard vibrations into tonal and broadband components. *Appl. Acoust.* **2007**, *68*, 1104–1117. [CrossRef]
7. Flückiger, M.; Grosshauser, T.; Tröster, G. Evaluation of Piano Key Vibrations among Different Acoustic Pianos and Relevance to Vibration Sensation. *IEEE Trans. Haptics* **2018**, *11*, 212–219. [CrossRef] [PubMed]
8. Keane, M.; Dodd, G. Subjective assessment of upright piano key vibrations. *Acta Acust. United Acust.* **2011**, *97*, 708–713. [CrossRef]
9. Fontana, F.; Papetti, S.; Järveläinen, H.; Avanzini, F. Detection of keyboard vibrations and effects on perceived piano quality. *J. Acoust. Soc. Am.* **2017**, *142*, 2953–2967. [CrossRef] [PubMed]
10. Fontana, F.; Järveläinen, H.; Papetti, S.; Avanzini, F.; Klauer, G.; Malavolta, L.; di Musica, C.; Pollini, C. Rendering and subjective evaluation of real vs. synthetic vibrotactile cues on a digital piano keyboard. In Proceedings of the International Conference on Sound and Music Computing, Maynooth, Ireland, 26 July–1 August 2015; pp. 161–167.
11. Gretchaninov, A. *Klage*; Julia Suslin, International Musikverlage Hans Sikorski: Hamburg, Germany, 2009.
12. Paté, A.; Carrou, J.L.L.; Navarret, B.; Dubois, D.; Fabre, B. Influence of the electric guitar's fingerboard wood on guitarists' perception. *Acta Acust. United Acust.* **2015**, *101*, 347–359. [CrossRef]
13. Dubois, D. Categories as acts of meaning: the case of categories in olfaction and audition. *Cogn. Sci. Q.* **2000**, *1*, 35–68.
14. Wilson, E.C.; Braida, L.D.; Reed, C.M. Perceptual interactions in the loudness of combined auditory and vibrotactile stimuli. *J. Acoust. Soc. Am.* **2010**, *127*, 3038–3043. [CrossRef] [PubMed]
15. Deas, R.; Adamson, R.B.; Garland, P.; Bance, M.L.; Brown, J. Combining auditory and tactile inputs to create a sense of auditory space. *Proc. Meet. Acoust.* **2010**, *11*, 015003. [CrossRef]
16. Merchel, S.; Altinsoy, M.E. Auditory-tactile music perception. *Proc. Meet. Acoust.* **2013**, *19*, 015030.
17. Parncutt, R. Piano touch, timbre, ecological psychology, and cross-modal interference. In Proceedings of the International Symposium on Performance Science in Vienna, Austria, published by Association Européenne des Conservatoires, Brussels, Belgium, 28–31 of August 2013; pp. 763–768.
18. Swain, M.J.; Ballard, D.H. Color indexing. *Int. J. Comput. Vis.* **1991**, *7*, 11–32. [CrossRef]
19. Galica, A.M.; Kang, H.G.; Priplata, A.A.; D'Andrea, S.E.; Starobinets, O.V.; Sorond, F.A.; Cupples, L.A.; Lipsitz, L.A. Subsensory vibrations to the feet reduce gait variability in elderly fallers. *Gait Posture* **2009**, *30*, 383–387. [CrossRef] [PubMed]

Article

MRI Compatible Planar Material Acoustic Lenses

Daniel Tarrazó-Serrano [1], Sergio Castiñeira-Ibáñez [1], Eugenio Sánchez-Aparisi [2],
Antonio Uris [1] and Constanza Rubio [1,*]

[1]   Centro de Tecnologías Físicas, Universitat Politècnica de València, Camí de Vera s/n, 46022 València, Spain;
dtarrazo@fis.upv.es (D.T.-S.); sercasib@upvnet.upv.es (S.C.-I.); auris@fis.upv.es (A.U.)
[2]   Hospital Francesc de Borja, 46702 Gandia, València, Spain; sanchez_eug@gva.es
*   Correspondence: crubiom@fis.upv.es

Received: 10 October 2018; Accepted: 13 December 2018; Published: 15 December 2018

Abstract: Zone plate lenses are used in many areas of physics where planar geometry is advantageous in comparison with conventional curved lenses. There are several types of zone plate lenses, such as the well-known Fresnel zone plates (FZPs) or the more recent fractal and Fibonacci zone plates. The selection of the lens material plays a very important role in beam modulation control. This work presents a comparison between FZPs made from different materials in the ultrasonic range in order to use them as magnetic resonance imaging (MRI) compatible materials. Three different MRI compatible polymers are considered: Acrylonitrile butadiene styrene (ABS), polymethyl methacrylate (PMMA) and polylactic acid (PLA). Numerical simulations based on finite elements method (FEM) and experimental results are shown. The focusing capabilities of brass lenses and polymer zone plate lenses are compared.

Keywords: MRI; Zone Plates; ultrasonic lenses

1. Introduction

The development of modulating and focusing energy systems has been a field of study of great interest for scientist and engineers. The lens is a devices that is able to perform this energy modulation. Lenses allow beam forming, control propagation and focusing the energy that impinges on them. These effects are produced by refractive and diffractive phenomena. Transmission efficiency is one of the most important aspects, particularly when low impedance contrast is presented between the lens and the host medium. Due to the wide versatility of the lenses, they have been used in different areas. For example, they have been applied in sonochemistry [1], construction [2] and the pharmaceutical industry [3].

The acoustic lenses, depending on the physics involved in the beam formation, can be divided into different groups, including refractive lenses and diffractive lenses. One example of lenses based on the refraction phenomenon are sonic crystal lenses made of periodic distributions of rigid cylinders [4]. Due to the subsonic sound speed inside the crystal, these lenses act similar to those in optical systems. Another example of this typology of acoustic lenses are those which modify the refractive index using labyrinths. These type of lenses are the so-called Gradient-Index lenses [5–7].

The other subtype of lenses, based on the diffractive phenomenon, conducts its behavior on the constructive interferences of the pressure field. An example of these types of lenses is the fractal lenses, which are able to generate different foci depending on their fractal geometrical properties [8]. Fresnel Zone Plates (FZP) have an improved focusing capacity. Among the different ways to implement FZPs, one of the most common and easiest is to alternate transparent and blocking zones, which results in a Soret type FZP [9]. To obtain these blocking areas, materials that are opaque to sound are required. This fact is accomplished by selecting materials that have a high impedance contrast with the host medium. There are studies that have implemented Soret FZP (SZP) by ultrasounds based on these type of lenses [10].

A material that has a high impedance with respect to water and that allows for the creation of opaque zones to achieve a Soret type lens is brass. However, this type of material has limitations, especially when used in fields such as bioengineering. The use of acoustic lenses in medicine for high intensity focused ultrasounds (HIFU) treatment is one of the current lines of research. magnetic resonance imaging (MRI) is the technique that is most used for guiding HIFU treatment [11].

MRI is a technique used for soft tissue structure imaging in a non-invasive way. The image is obtained by aligning and relaxing the magnetic moments of the atoms of the introduced elements in the MRI. Tissues are exposed to a strong external time-independent magnetic field. Thus, metallic elements cannot be introduced in the resonance zone due to their interference in the image and because they could damage MRI-systems. To avoid interaction with the electromagnetic field, non-metallic materials should be used in the construction of lenses. One of these materials is polylactic acid (PLA) [12]. The MRI environment requires materials such as PLA for medical instruments and patient supports. Recently, PLA has been used for this purpose and its reliability has been shown [13]. The HIFU transducer is embedded within a specially designed table that fits into the MRI device. This integrated system, has a degassed water bath where the transducer is located. The patient lies over this system on [14]. Although the transducer and the lens must be immersed in this water bath, degradation of the PLA will occur over long-term immersion. PLA degrades in water after a period ranging from months to a year [15]. Therefore it must be taken into account that, in MRI systems, the lens and the transducer are not permanently submerged. After 20 to 25 min, the system is extracted from the water bath, and for this reason, the time of degradation due to being immersed in water can be prolonged considerably.

In this work, three lenses with three types of compatible materials with MRI environments are compared. In this sense, acrylonitrile butadiene styrene (ABS), polymethyl methacrylate (PMMA) and polylactic acid (PLA) materials are used. Furthermore, a SZP built in brass is compared. Although, this material is not MRI compatible, it is the nearest to the ideal SZP that can be implemented in real projects. In the comparison, a not compatible with MRI lens built in brass and an ideal Soret lens will be added. Results are obtained and compared both numerically and experimentally. Numerical results have been obtained using the commercial software COMSOL Multiphysics 4.3a by COMSOL Inc. (Sweden) [16]. In this work, it has been verified that the ratio of the transmission capacity that is related to the ratio of impedances of the medium and the lens, directly influences the focusing capacity.

2. Methodology and Theoretical Analysis

Fresnel zone plates are circular concentric structures, which are known as Fresnel regions. Every consecutive region has a π phase shift between them. This fact makes a coherent contribution to obtain high intensity levels at focal length (F_L), which is the location in the axial coordinate where the focus is placed. The number of Fresnel regions is defined as N, this includes both opaque and transparent acoustic sections. The working frequency is defined as f_0 and radial distances (r_n) of each Fresnel zone can be obtained using Equation (1) valid for plane wave incidence.

$$r_n = \sqrt{n\lambda F_L + \left(\frac{n\lambda}{2}\right)^2} \qquad n = 1, 2, ..., N \tag{1}$$

In this work, underwater transmission is considered and lenses are designed for ultrasound applications. Therefore, piston sources have to be considered when FZPs are implemented. Due to spherical wave incidence consideration, Equation (2) has been used where d is the separation between the point source and the lens.

$$d + F_L + \frac{n\lambda}{2} = \sqrt{d^2 + r_n^2} + \sqrt{F_L^2 + r_n^2} \tag{2}$$

The acoustic wave has to propagate through the host medium, then cross Fresnel regions and afterwards continue through the host medium. A three-layer configuration has to be considered (Figure 1). Acoustic impedance (Z) is defined as the product of the medium density (ρ) and the sound propagation velocity (c) in it. Therefore, it is necessary to consider the input (Z_{in}) and output (Z_{out}) acoustic impedance and the transmission pressure coefficients must be calculated (t). This coefficient is a clear indicator of the blocking capacity of the elements of the FZP. Hence, t is defined as the relation between the transmitted field and the incident field. Density (ρ), sound propagation velocity (c) and acoustic impedance (Z_{mat}) values have been shown in Table 1. Using these values in Equation (3), Z_{in} could be obtained [17].

$$Z_{in} = Z_{mat} \frac{Z_{out} + jZ_{mat}\tan(k_{mat}d)}{Z_{mat} + jZ_{out}\tan(k_{mat}d)} \tag{3}$$

where k_m is the wave number, defined as $k_m = w/c$. Considering $w = 2\pi f_0$. Once Z_{in} is obtained, reflection coefficient is defined in Equation (4).

$$r_{in} = \frac{Z_{in} - Z_{water}}{Z_{in} + Z_{water}} \tag{4}$$

The equation that relates the field balance as a function of the impedance and reflection coefficient of the system is defined in Equation (5) and gives t values depending on the material.

$$|t| = \frac{|p_t^+|}{|p_{in}^+|} = \sqrt{(1 - |r_{in}|^2)} \tag{5}$$

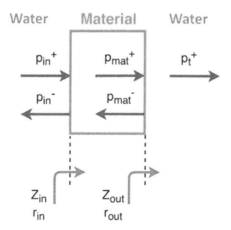

Figure 1. Transmission diagram of the implemented lenses.

Table 1. Density and sound speed values. Acrylonitrile butadiene styrene (ABS), polylactic acid (PLA), and polymethyl methacrylate (PMMA).)

Material	ρ (kg/m^3)	c (m/s)	Z_{mat} (Rayls)	Z_{mat}/Z_{water}
ABS	1050	2250	$2.4 \cdot 10^6$	1.58
PLA	1240	2220	$2.8 \cdot 10^6$	1.84
PMMA	2690	1191	$3.2 \cdot 10^6$	2.14
Brass	8400	4700	$39.5 \cdot 10^6$	26.32
PLA-Air-PLA	398	944	$3.6 \cdot 10^5$	0.25

Considering the transmission coefficient values obtained (0.23 for brass, 0.51 for PLA-Air-PLA and more than 0.95 for ABS, PLA and PMMA), it can be affirmed that full implemented MRI compatible material lenses will focus less energy at the F_L if it is compared to brass FZP or ideal SZP. Therefore, one solution is proposed to obtain the desired impedance contrast. A FZP that includes an air chamber inside the structure has been implemented by using a 3D-printer. Thus, both lenses, full-PLA and air-chamber, have been compared.

2.1. Numerical Model

The finite elements method (FEM) has been used to obtain a numerical solution of the physical problem. The finite elements method allows us to study the physical phenomena involved in the interaction of waves with FZPs. Therefore, a mathematical model that replicates the conditions of the problem has been implemented. This method also allows us to determine the pressure distribution of the diffracted fields generated by the FZP when there is a piston emitter, causing interference phenomena. From the mesh generated by FEM, a partial differential equation solution is obtained for each node [18]. In this case, acoustic Helmholtz equation is considered (Equation (6)). To solve the Helmholtz equation, standard values of water such as density of the medium (ρ =1000 kg/m^3) and sound propagation velocity (c = 1500 m/s) have been considered. The working frequency of the FZPs is 250 kHz and it can be found by its relation with the angular velocity (ω). Finally, p corresponds to the acoustic pressure.

$$\nabla \cdot \left(-\frac{1}{\rho_0}(\nabla p) \right) = \frac{\omega^2 p}{\rho_0 c^2} \tag{6}$$

If a 3D model is considered, this will require high computational resources. To simplify the model and reduce this computational cost, as shown in previous works [19,20], the geometrical properties of the model are used taking advantage of its axisymmetry. Therefore, the model is simplified by implementing a semi-lens only. A complete solution is obtained by rotating it from its symmetry axis. This procedure achieves a reduction of the degrees of freedom necessary to obtain the results of the numerical simulation and thus significantly diminishing the calculation time.

The boundary conditions defined in the numerical models are explained below as seen in Figure 2. The contours of the model are defined as wave radiation condition boundary to emulate an infinitely large medium and therefore the Sommerfeld condition is satisfied. Acoustic impedance domain definition has been used for all opaque Fresnel regions for each lens. In the case of the SZP lens, the contours are considered infinitely rigid, applying the Neumann condition (the sound velocity in the contour is zero).

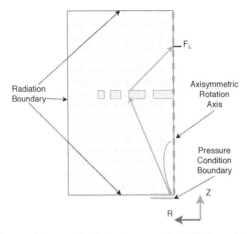

Figure 2. Scheme of the finite element method (FEM) conditions.

3. Experimental Set-Up

It is required to validate the results obtained from the theoretical models with other solutions such as numerical models and experimental measurements. In this sense, obtaining experimental results is fundamental to validate the numerical models. A complex measurement and acquisition system is needed to perform the experiments given the technical difficulties to control the underwater devices. The Center for Physics Technologies: Acoustics, Materials and Astrophysics of the Universitat Politècnica de València has a robotized and automated system for high precision ultrasound measurements. The robot is built based on the size of the immersion tank where the tests and experiments are carried out, which contains distilled and degassed water, with dimensions of 0.5 m wide by 0.5 m high by 1 m long. These dimensions suppose that the immersion tank must contain around 200 L of distilled water to be functional, and allow both transducers and devices to be completely submerged, and avoid reflections due to the impedance changes produced by the change medium.

The measurement system is composed by a fixed emitter and a hydrophone coupled to the robotic system. This system obtains reliable and precise results that allow for the evaluation of the acoustic phenomena involved in these types of lenses. A plane immersion piston transducer built by Imasonic with 250 kHz of central working frequency and an active diameter of 32 mm has been used as the emitter. Also, a Precision Acoustics hydrophone, model 1.0 mm Needle Hydrophone is used as the receiver. This hydrophone is capable of measuring high frequencies, even if they have a very weak signal level. The sensitivity of the hydrophone is 850 nV/Pa (-241.4 dB 1V/μPa) with a tolerance of ± 3 dB. The frequency response is flat ± 2 dB between 3 and 12 MHz and ± 4 dB between 200 kHz and 15 MHz. The bandwidth ranges from 5 kHz to 15 MHz. Figure 3 shows an experimental set-up in a measurement. Two types of different configurations are used to generate and amplify the signals. The first one is to use an external function generator connected to a high power amplifier. The second configuration is to use a pulse generator (5077PR of Panametrics) with integrated amplifier. This generator and amplifier allows generating pulses with frequencies between 100 kHz and 20 MHz, a pulse repetition frequency (PRF) from 100 Hz to 5 kHz and a pulse amplitude between 100 and 400 V.

Figure 3. Experimental set-up.

All the results shown below are obtained for a working frequency of 250 kHz. For the experimental comparison, three lenses have been implemented, two made of PLA and one of brass. Every lens considered in this work was designed with 11 Fresnel zones and an outer radius of 88.8 mm. The thickness of the brass lens was 1 mm. For manufacturing reasons, the rest of the lenses had a total thickness of 5 mm. Figure 4 shows both PLA and brass lenses. Both PLA lenses are identical, the only difference being an inner air chamber to achieve a higher impedance contrast. As described in the previous section, it is not possible to differentiate them by the naked eye.

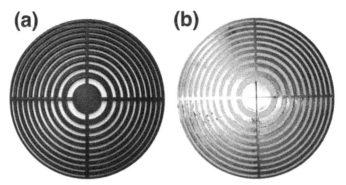

Figure 4. Implemented lenses, (**a**) PLA and (**b**) brass.

4. Results

Intensity gain for longitudinal axis cuts and maps have been calculated to compare all the lenses coherently. One parameter, which can be used to evaluate the focusing capacity of a lens is the intensity gain (G). The intensity gain is related with the intensity with both the intensity with lens (I) and intensity without lens (I_0), as shown in Equation (7).

$$G(dB) = 10 \cdot \log_{10}(I/I_0) \tag{7}$$

Intensity gain values have been calculated from Equation (7). Figure 5, shows the intensity gain for longitudinal cuts on the Z axis for both, numerical and experimental results. It can be seen from Figure 5a, that higher impedance contrast, as in the case of the ideal SZP or brass FZP lens, gives rise to a higher gain levels. As expected, the lower gains are obtained with those materials with impedance contrast values between 1 and 2 and for impedance contrast values lower than 1, the intensity gain increases. ABS, PLA, PLA-Air-PLA, and PMMA polymers, according with the values showed in Table 1, are not able to achieve enough intensity gain as brass FZP. By comparing the experimental results obtained for brass and PLA (see Figure 5b) with numerical ones (see Figure 5a) it can be seen that there is a good agreement. From Figure 5b, it is observed that the air chamber PLA lens has higher intensity gain than full PLA lens. This fact can be explained by the introduction of an air layer. This layer, due to its low acoustic impedance, can block the transmission of the ultrasonic waves approaching its behavior to an ideal SZP. Nevertheless, a focal length displacement of 1.66λ is observed in the FZP lens built with air chamber and PLA. In this case, the displacement is due to the new three-layer configuration (PLA-Air-PLA). The resolution of the 3D-printer and the wall width needed to avoid porosities means that there is an interface between the host medium and the air chamber.

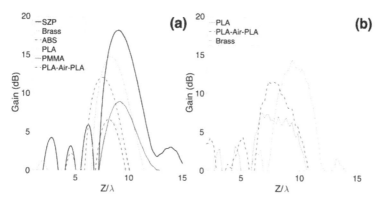

Figure 5. Intensity gain longitudinal cuts for (**a**) FEM results and (**b**) experimental results.

Figure 6 shows four intensity gain maps, the first three obtained experimentally and the fourth numerically. The experimental ones correspond to PLA, PLA-Air-PLA and brass, while the numerical one has been obtained using an ideal SZP. It has been verified how the results obtained with brass resembles the ideal SZP. This is due to the rigidity of the material. On the other hand, in PLA lens results, a diminishing intensity gain is observed. This intensity gain level can be improved using a PLA-Air-PLA lens. All the lenses are designed with a focal length located at 8.33λ for a working frequency of 250 kHz. When the lens is able to block destructive interference, it is possible to locate the focus in F_L. This occurs in brass and the ideal SZP case. Resulting from the lack of blocking capability, the full PLA FZP could not impede the incident pressure wave transferal generating aberrations in the F_L.

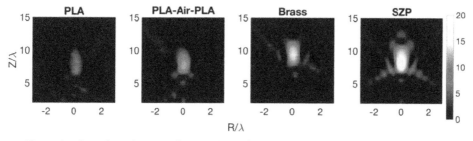

Figure 6. Intensity gain maps for experimental measurements and ideal SZP numerically obtained (FEM).

5. Conclusions

Non-metallic materials can be used for the construction of acoustic lenses. Three alternative materials, compatible with magnetic resonance, have been proposed instead of brass lenses. It has been possible to verify that the higher the impedance contrast of the materials, the higher the intensity gain levels. The PMMA lens has higher intensity level than ABS and PLA ones, because it has a slightly higer impedance contrast value than ABS or PLA. However, the use of an air chamber inside the PLA lens increases the intensity gain levels, due to the fact that values of impedance contrast less than one means blocking of the waves. PLA is a biocompatible material and is cheaper than PMMA. 3D printers give open field of new lens design MRI compatible. Moreover, since PLA is a biodegradable material, it is a environmental friendly material. This point is important in procedures that generate waste. Nevertheless, the manufacture of PLA lenses require great care because of microporosities that could appear. The appearance of pores can cause water to enter into the lens, drastically reducing the blocking capacity. In addition, polymers such as PLA, ABS or PMMA are more affordable than metal plates. This will lower the costs in the production of HIFU treatment devices based on acoustic lenses. For this reason, PLA is proposed as an MRI compatible material with great potential for therapeutic applications of ultrasound focusing.

Author Contributions: A.U. and C.R. coordinated the theoretical development, participating in the establishment of the theory principles used in this work, as well as in the drafting of the manuscript. D.T.-S. coordinated experimental development. S.C.-I. developed part of the theory used and designed some characterization. E.S.-A. participated in the analysis of the state of art.

Funding: This research was funded by spanish Ministerio de Economía y Competitividad (MINECO) TEC2015-70939-R.

Acknowledgments: This work has been supported by Spanish MINECO (TEC2015-70939-R).

Conflicts of Interest: The authors declare no conflict of interest.

References

1. Li, J.T.; Han, J.F.; Yang, J.H.; Li, T.S. An efficient synthesis of 3, 4-dihydropyrimidin-2-ones catalyzed by NH_2SO_3H under ultrasound irradiation. *Ultrason. Sonochem.* **2003**, *10*, 119–122. [CrossRef]

2. McCann, D.M.; Forde, M.C. Review of NDT methods in the assessment of concrete and masonry structures. *NDT E Int.* **2001**, *34*, 71–84. [CrossRef]

3. Albu, S.; Joyce, E.; Paniwnyk, L.; Lorimer, J.P.; Mason, T.J. Potential for the use of ultrasound in the extraction of antioxidants from Rosmarinus officinalis for the food and pharmaceutical industry. *Ultrason. Sonochem.* **2004**, *11*, 261–265. [CrossRef] [PubMed]

4. Cervera, F.; Sanchis, L.; Sánchez-Pérez, J.V.; Martínez-Sala, R.; Rubio, C.; Meseguer, F.; López, C.; Caballero, D.; Sánchez-Dehesa, J. Refractive acoustic devices for airborne sound. *Phys. Rev. Lett.* **2002**, *88*, 023902. [CrossRef] [PubMed]

5. Li, Y.; Liang, B.; Tao, X.; Zhu, X.F.; Zou, X.Y.; Cheng, J.C. Acoustic focusing by coiling up space. *Appl. Phys. Lett.* **2012**, *101*, 233508. [CrossRef]

6. Welter, J.T.; Sathish, S.; Christensen, D.E.; Brodrick, P.G.; Heebl, J.D.; Cherry, M.R. Focusing of longitudinal ultrasonic waves in air with an aperiodic flat lens. *J. Acoust. Soc. Am.* **2011**, *130*, 2789–2796. [CrossRef] [PubMed]

7. Peng, P.; Xiao, B.; Wu, Y. Flat acoustic lens by acoustic grating with curled slits. *Phys. Lett. A* **2014**, *378*, 3389–3392. [CrossRef]

8. Castiñeira-Ibáñez, S.; Tarrazó-Serrano, D.; Fuster, J.; Candelas, P.; Rubio, C. Polyadic cantor fractal ultrasonic lenses: Design and characterization. *Appl. Sci.* **2018**, *8*, 1389. [CrossRef]

9. Soret, J. Ueber die durch Kreisgitter erzeugten Diffractionsphänomene. *Ann. Phys.* **1875**, *232*, 99–113. [CrossRef]

10. Calvo, D.C.; Thangawng, A.L.; Nicholas, M.; Layman, C.N. Thin Fresnel zone plate lenses for focusing underwater sound. *Appl. Phys. Lett.* **2015**, *107*, 014103. [CrossRef]

11. McDannold, N.; Hynynen, K.; Wolf, D.; Wolf, G.; Jolesz, F. MRI evaluation of thermal ablation of tumors with focused ultrasound. *J. Magn. Resonance Imaging* **1998**, *8*, 91–100. [CrossRef]

12. Drumright, R.E.; Gruber, P.R.; Henton, D.E. Polylactic acid technology. *Adv. Mater.* **2000**, *12*, 1841–1846. [CrossRef]

13. Herrmann, K.H.; Gärtner, C.; Güllmar, D.; Krämer, M.; Reichenbach, J.R. 3D printing of MRI compatible components: Why every MRI research group should have a low-budget 3D printer. *Med. Eng. Phys.* **2014**, *36*, 1373–1380. [CrossRef] [PubMed]

14. Köhler, M.O.; Mougenot, C.; Quesson, B.; Enholm, J.; Le Bail, B.; Laurent, C.; Moonen, C.T.; Ehnholm, G.J. Volumetric HIFU ablation under 3D guidance of rapid MRI thermometry. *Med. Phys.* **2009**, *36*, 3521–3535. [CrossRef] [PubMed]

15. Rocca-Smith, J.R.; Whyte, O.; Brachais, C.H.; Champion, D.; Piasente, F.; Marcuzzo, E.; Sensidoni, A.; Debeaufort, F.; Karbowiak, T. Beyond biodegradability of poly (lactic acid): physical and chemical stability in humid environments. *ACS Sustain. Chem. Eng.* **2017**, *5*, 2751–2762. [CrossRef]

16. COMSOL-Multiphysics. *COMSOL-Multiphysics User Guide (Version 4.3a)*; COMSOL User Guide (Version 4.3a); COMSOL Inc.: Stockholm, Sweden, 2012; pp. 39–40.

17. Kinsler, L.E.; Frey, A.R.; Coppens, A.B.; Sanders, J.V. *Fundamentals of Acoustics*, 4th ed.; John Wiley and Sons: New York, NY, USA, 1999; p. 560.

18. Zienkiewicz, O.C.; Taylor, R.L.; Zienkiewicz, O.C.; Taylor, R.L. *The Finite Element Method*; McGraw-Hill: London, UK, 1977; Volume 36.

19. Castiñeira-Ibáñez, S.; Tarrazó-Serrano, D.; Rubio, C.; Candelas, P.; Uris, A. An ultrasonic lens design based on prefractal structures. *Symmetry* **2016**, *8*, 28. [CrossRef]

20. Tarrazó-Serrano, D.; Rubio, C.; Minin, O.V.; Candelas, P.; Minin, I.V. Manipulation of focal patterns in acoustic Soret type zone plate lens by using reference radius/phase effect. *Ultrasonics* **2019**, *91*, 237–241. [CrossRef] [PubMed]

Article

Theoretical and Numerical Estimation of Vibroacoustic Behavior of Clamped Free Parabolic Tapered Annular Circular Plate with Different Arrangement of Stiffener Patches

Abhijeet Chatterjee [1,*], Vinayak Ranjan [2], Mohammad Sikandar Azam [1] and Mohan Rao [3]

[1] Department of Mechanical Engineering, Indian Institute of Technology (ISM), Dhanbad 826004, India; mdsazam@gmail.com

[2] Department of Mechanical Engineering, Bennett University, Greater Noida 201310, India; vinayakranjan@gmail.com

[3] Department of Mechanical Engineering, Tennessee Tech University, Cookeville, TN 38505, USA; mrao@tntech.edu

* Correspondence: abhijeet.ism@gmail.com

Received: 13 September 2018; Accepted: 2 November 2018; Published: 8 December 2018

Abstract: This paper compares the vibroacoustic behavior of a tapered annular circular plate having different parabolic varying thickness with different combinations of rectangular and concentric stiffener patches keeping the mass of the plate and the patch constant for a clamped-free boundary condition. Both numerical and analytical methods are used to solve the plate. The finite element method (FEM) is used to determine the vibration characteristic and both Rayleigh integral and FEM is used to determine the acoustic behavior of the plate. It is observed that a Case II plate with parabolic decreasing–increasing thickness variation for a plate with different stiffener patches shows reduction in frequency parameter in comparison to other cases. For acoustic response, the variation of peak sound power level for different combinations of stiffener patches is investigated with different taper ratios. It is investigated that Case II plate with parabolic decreasing–increasing thickness variation for an unloaded tapered plate as well as case II plate with 2 rectangular and 4 concentric stiffeners patches shows the maximum sound power level among all variations. However, it is shown that the Case III plate with parabolically increasing–decreasing thickness variation with different combinations of rectangular and concentric stiffeners patches is least prone to acoustic radiation. Furthermore, it is shown that at low forcing frequency, average radiation efficiency with different combinations of stiffeners patches remains the same, but at higher forcing frequency a higher taper ratio causes higher radiation efficiency, and the radiation peak shifts towards the lower frequency and alters its stiffness as the taper ratio increases. Finally, the design options for peak sound power actuation and reduction for different combinations of stiffener patches with different taper ratios are suggested.

Keywords: thick annular circular plate; Rayleigh integral; finite element modeling; rectangular and concentric stiffener patches; taper ratio; thickness variation

---

1. Introduction

Tapered annular circular plates with different combinations of rectangular and concentric patches has many engineering applications. They are used in many structural components i.e., building, design, diaphragms and deck plates in launch vehicles, diaphragms of turbines, aircraft and missiles, naval structures, nuclear reactors, optical systems, construction of ships, automobiles and other vehicles, the space shuttle etc. These tapering plates with different combinations of rectangular and concentric patches are found to have greater resistance to bending, buckling and vibration in comparison to plates of uniform thickness. It is interesting to know that tapered plates with different thickness variation have drawn the attention of most of the researchers in this field. However, tapered plates with different combination of rectangular and concentric patches can alter the dynamic characteristic of structures with a change in stiffness. Hence, for practical design purposes, the vibration and acoustic characteristics of such tapered plates are equally important. In comparison to the present study, several existing works are presented where the researchers have investigated the vibration response [1–9] of circular or annular plates of tapered or uniform thickness. But in terms of acoustic behavior, many researchers have contributed most. Lee and Singh [10] used the thin and thick plate theories to determine the sound radiation from out-of-plane modes of a uniform thickness annular circular plate. Thompson [11] used the Bouwkamp integral to determine the mutual and self-radiation impedances both for annular and elliptical pistons. Levine and Leppington [12] analyzed the sound power generation of a circular plate of uniform thickness using exact integral representation. Rdzanek and Engel [13] determined the acoustic power output of a clamped annular plate using an asymptotic formula. Wodtke and Lamancusa [14] minimized the acoustic power of circular plates of uniform thickness using the damping layer placement. Wanyama [15] studied the acoustic radiation from linearly-varying circular plates. Lee and Singh [16] used the flexural and radial modes of a thick annular plate to determine the self and mutual radiation. Cote et al. [17] studied the vibro acoustic behavior of an unbaffled rotating disk. Jeyraj [18] used an isotropic plate with arbitrarily varying thickness to determine its vibro-acoustic behavior using the finite element method (FEM). Ranjan and Ghosh [19] studied the forced response of a thin plate of uniform thickness with attached dynamic absorbers. Bipin et al. [20] analyzed an isotropic plate with attached discrete patches and point masses with different thickness variation with different taper ratios to determine its vibro acoustic response. Lee and Singh [21] investigated the annular disk acoustic radiation using structural modes through analytical formulations. Rdzanek et al. [22] investigated the sound radiation and sound power of a planar annular membrane for axially-symmetric free vibrations. Doganli [23] determined the sound power radiation from clamped annular plates of uniform thickness. Nakayama et al. [24] investigated the acoustic radiation of a circular plate for a single sound pulse. Hasegawa and Yosioka [25] determined the acoustic radiation force used on the solid elastic sphere. Lee and Singh [26] used a simplified disk brake rotor to investigate the acoustic radiation through a semi-analytical method. Thompson et al. [27,28] analyzed the modal approach for different boundary conditions to calculate the average radiation efficiency of a rectangular plate. Rayleigh [29] determined the sound radiation from flat finite structures. Maidanik [30] analyzed the total radiation resistance for ribbed and simple plates using a simplified asymptotic formulation. Heckl [31] used the wave number domain and Fourier transform to analyses the acoustic power. Williams [32] determined the wave number as a series in ascending power to estimate the sound radiation from a planar source. Keltie and Peng [33] analyzed the sound radiation using the cross- modal coupling from a plane. Snyder and Tanaka [34] demonstrated the importance of cross-modal contributions for a pair of modes through total sound power output using modal radiation efficiency. Martini et al. [35] investigated the structural and elastodynamic analysis of rotary transfer machines by a finite element model. Croccolo et al. [36] determined the lightweight design of modern transfer machine tools using the finite element model. Martini and Troncossi [37] determined the upgrade of an automated line for plastic cap manufacture based on experimental vibration analysis. Pavlovic et al. [38] investigated the modal analysis and stiffness optimization: the case of a tool machine for ceramic tile surface finishing using FEM.

While reviewing the literature, this comes to a discussion at a common point that has inspired the present paper based on a comparison of vibroacoustic behavior of a parabolic tapered annular circular plate with attached rectangular and concentric patches at different positions. The paper is significant for the analysis of the comparison of vibroacoustic behavior of such clamped free tapered plate, which is done by keeping the mass of the plate and patch constant. Therefore, this paper is based on the vibroacoustic analysis of a clamped free parabolic tapered annular circular plate with different attachments of rectangular and concentric stiffener patches for different positions with different taper ratios under time-varying harmonic excitations.

2. Mathematical Modeling and Analysis

2.1. Plate Free Vibration

Let us considered a plate with outer radius 'a' and inner radius 'b' as shown in Figure 1. In this paper, the modal analysis is performed to estimate the natural frequency and modes shape of the plate is given by the following equation:

$$\left([K] - \omega^2[M]\right)\psi_{mn} = 0 \tag{1}$$

where $[M]$ is the mass matrix and $[K]$ is the stiffness matrix where as ψ_{mn} is the mode shape and ω is the respective natural frequency of the plate in rad/sec. The non-dimensional frequency parameter λ^2 is given by the following equation:

$$\lambda^2 = \omega a^2 \sqrt{\frac{\rho h}{D}} \tag{2}$$

where D, the flexure rigidity $= \frac{Eh^3}{12(1-v^2)}$, a = outer radius, E = Young's modulus of elasticity, v = Poisson's ratio, h = thickness of the plate and ρ = density of plate.

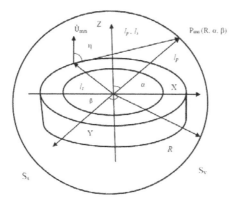

Figure 1. Sound radiation investigated for thick annular circular plate in Z direction enclosed in a sphere.

2.2. Analytical and Numerical Formulation for Acoustic Radiation from Tapered Annular Circular Plate

It is considered that an annular circular plate of inner radius 'b' and outer radius 'a' in flexural vibration is set on flat rigid baffle having infinite extent as reported in Figure 1. Acoustic scattering of the edges of a vibrating structure is neglected in this study. Let P be the sound pressure amplitude, S_s be the surface of the sound source, q be the Green methods function in free field, l_s and l_p be the position vectors of source and receiver and the surface normal vector at l_s be f; then structure sound radiation can be obtained by the Rayleigh integral [10] as given by Equation (3):

$$P(l_p) = \int_{S_s} \left(\frac{\partial q}{\partial f}(l_p, l_s) P(l_p) - \frac{\partial P}{\partial f}(l_s) q(l_p, l_s) \right) ds(l_s) \tag{3}$$

The sound pressure, radiated from non-planar source in far and free field environment based on plane wave approximation can be expressed by Equation (4):

$$P(l_p) = \frac{\rho_0 c_0 B}{4\pi} \int_{S_s} \frac{e^{iB|l_p - l_s|} U(l_s)}{|l_p - l_s|} (1 + \cos \eta) dS \tag{4}$$

Let ρ_0 be the mass density of air, c_0 be the speed of sound in air, B be the corresponding acoustic wave number, and $\dot{U}$ and $\dot{u}$ be the corresponding vibratory velocity amplitude and spatial dependent vibratory velocity amplitude in the z direction at l_s, then from a normal plane [10], the modal sound pressure P_{mn} for an annular plate with $(m, n)^{th}$ mode is obtained from simplifying Equation (4) with Hankel transform and is expressed by Equations (5) and (6):

$$P_{mn}(R, \alpha, \beta) = \frac{\rho_0 c_0 B e^{iB_{mn} R_d}}{2R_d} \cos n\beta (-i)^{n+1} A_n \left[\dot{u}(l) \right] (1 + \cos \eta) \tag{5}$$

$$A_f \left[\dot{u}(l) \right] = \int_0^\infty \dot{u}(l) J_n(B_l l) l \, dl, \quad Bl = B \sin \theta; \quad R_d = |l_p - l_s| \tag{6}$$

where X_n is Bessel function of order n, (α, β) are the cone and azimuthal angles of the observation positions, respectively, η is the angle between the surface normal vector and the vector from source position to receiver position, and A is the Hankel transform. According to the far field condition, R_d in the denominator is approximated by R where $R = |l_p|$ is considered to be radius of the sphere. Consider that on a sphere S_v the observation positions are represented by some points having equal angular increments $(\Delta\varphi, \Delta\alpha)$. If '$\Delta\varphi$' represents the small increment in the circumferential direction of the plate, at all of the observation positions, the sound pressures is given by Equations (4)–(6). The modal sound power S_{mn} for the $(m, n)^{th}$ mode [10,16] from the far-field is given by Equation (7):

$$S_{mn} = (D_{mn} S_v)_s = \frac{1}{2} \int_0^{2\pi} \int_0^\pi \frac{P_{mn}^2}{\rho_0 c_0} R^2 \sin \alpha \, d\alpha \, d\beta \tag{7}$$

where the acoustic intensity is represented by D_{mn} and area of the control surface is represented by S_v. The radiation efficiency σ_{mn} of the plate [10] is given by Equation (8):

$$\sigma_{mn} = \frac{S_{mn}}{\left| \dot{u}_{mn}^2 \right|_{ts}}, \quad \left| \dot{u}_{mn}^2 \right|_{ts} = \frac{1}{2\pi(a^2 - b^2)} \int_b^a \int_0^{2\pi} u^2 \, d\varphi \, dl \tag{8}$$

where, for the two normal surfaces of the plate the spatially average r.m.s velocity is represented as $\left| \dot{u}_{mn}^2 \right|_{ts}$. Considering the plate thickness (h) effect, the sum of sound radiations [16] from two normal surfaces of the plate at (Z = 0.5 h and −0.5 h) will represent the modal sound power, which can be given by Equations (9)–(11):

$$P_{mn}(R, \alpha, \beta) = (1 + \cos \alpha) P_{mn}^s(R, \alpha, \beta) + (1 - \cos \alpha) P_{mn}^0(R, \alpha, \beta) \tag{9}$$

$$P_{mn}^s(R, \alpha, \beta) = \frac{\rho_0 c_0 B_{mn} e^{iB_{mn} R}}{2R} e^{-iB_{mn}\left(\frac{h}{2}\right) \cos \alpha} \cos n\beta (-i)^{n+1} A_n \left[\dot{U}(l) \right] \tag{10}$$

$$P_{mn}^0(R, \alpha, \beta) = \frac{\rho_0 c_0 B_{mn} e^{iB_{mn} R}}{2R} e^{-iB_{mn}\left(\frac{h}{2}\right) \cos \alpha} \cos n(\beta + \phi)(-i)^{n+1} A_n \left[\dot{U}(l) \right] \tag{11}$$

where, the corresponding acoustic wave number of the $(m, n)^{th}$ mode is represented by $B_{m,n}$, s and o in Equations (10) and (11) represent the source side and the opposite to the source side.

For numerical analysis, we have used ANSYS (ANSYS, Inc., Canonsburg, PA, USA) as a tool. The plates with rectangular and concentric patches are modeled in ANSYS with Plane 185 with 8 brick nodes and having three degrees of freedom at each node. The mesh is not exactly equal to all the cases of different thickness variation and stiffener. The number of element and nodes for uniform unloaded plates ends up being 5883 and 1664, respectively. For plates with different cases of thickness variation with different stiffener we tried to keep the mesh as close to the mesh of the unloaded plate. For vibration analysis and for a Case I plates with 1 rectangular stiffener, the modal structure consists of 5685 elements with 1638 nodes whereas for Case I plates with 1 concentric stiffener, it has 5524 elements and 1618 nodes. With other combinations of rectangular and concentric stiffener with different parabolic thickness, a variation of 5% of the mesh from that of the unloaded plate is considered. The numerical results obtained using FEM are compared with the existing literature. The structure is modeled as such that the total volume of the plate plus stiffeners is equal to the total volume of the uniform unloaded plate. As a result the whole mass of the plate plus stiffener is equal to the whole mass of the uniform unloaded plate. So for all the cases of plate, the mass will be constant. FLUID 30 and FLUID130 elements are used to create the acoustic medium environment around the plate. For fluid-structure interaction FLUID 30 is used. For the surface on outer sphere, FLUID 130 elements are created by imposing a condition of infinite space around the source and to prevent the back reflection of sound waves to the source. For acoustic analysis, the number of element and nodes for a uniform unloaded plate ends up being 14,680 and 3639, respectively. For a Case I plate with 1 rectangular stiffener, and after proper convergence of modeling, the numbers of elements and nodes ends up being 14,124 and 3465 respectively while for Case I plate with 1 concentric stiffener, the numbers of elements and nodes found to be 13,934 and 3345 respectively. Again, for other combination of rectangular and concentric stiffeners with different parabolic thickness, a variation of 5% of mesh from that of the unloaded plate is taken. Consider the air medium where the plate is vibrating with air density $\rho_0 = 1.21$ kg/m^3. At 20 °C, the speed of sound c_0 of air is taken as 343 m/s. The structural damping coefficient of the plate is assumed as 0.01.

2.3. Thickness Variation of the Plate

In this study, three different parabolic thickness variations of plates is considered for analysis and is reported in Figure 2. The radial direction is considered for thickness variation by keeping the total mass of the plate plus patch constant. In the radial direction the plate thickness is given by $h_x = h [1 - T_x \{f(x)\}^n]$, where '$h$' is the maximum thickness of the plate where,

$$f(x) = \{^{0,x=b}_{1,x=a} \text{ and } f(x) = \frac{x-b}{a-b} \text{ where } b < x < a \qquad (12)$$

The taper parameter or taper ratio (T_x) is given by the equation:

$$T_x = \left(1 - \frac{h_{min}}{h}\right) \qquad (13)$$

The Case I plate of (Figure 2 with parabolically decreasing thickness variation is given by the equation:

$$h_x = h\left\{1 - T_x\left(\frac{x-b}{a-b}\right)^n\right\} \qquad (14)$$

The Case II plate (parabolically decreasing-increasing) and Case III plate (parabolically increasing-decreasing) thickness variation of (Figure 2) are given by the equations:

$$h_x = h\left\{1 - T_x\left(1 - abs\left(1 - 2\frac{(x-b)}{(a-b)}\right)\right)^n\right\} \qquad (15)$$

$$h_x = h\left\{1 - T_x abs\left(1 - 2\frac{(x-b)}{(a-b)}\right)^n\right\} \tag{16}$$

where, n = 2 for parabolic thickness variation. The total volume of the plate plus patches as well as the unloaded plate is kept constant and is given by the equation:

$$\text{Volume} = \pi(a^2 - b^2)h = \int_b^a (a^2 - b^2)h_x dx \tag{17}$$

In this paper, a comparison for the effect of frequency parameters, effect of sound power levels, average radiation efficiency and peak sound power level is obtained for parabolic tapered plates. The out of plane (m, n)th modes in Z direction for the plate with different attachment of rectangular and concentric stiffener patches at different positions with different parabolically tapered varying thickness is considered. The plate is made tapered with different taper ratios of 0.25, 0.50 and 0.75. The mass of the plate plus rectangular or concentric patches are kept constant for this analysis. The inner clamped and outer free boundary condition is taken. Three arrangements of plates with different combinations of rectangular or concentric stiffener patches are considered as shown in Figure 3. The selection of different combinations of rectangular or concentric stiffener patches are such that the mass of the unloaded plate is equal to mass of the rectangular or concentric stiffener patches plus plate and in all three cases mass of the plate with stiffener patches remains constant. The specifications and the material properties of an annular circular plate with attached rectangular and concentric stiffener patches are reported in Table 1. Rayleigh integral has been used for sound power calculation and ANSYS has been used as a tool for computation.

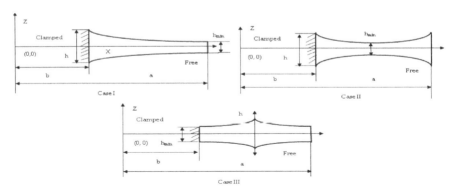

Figure 2. Plate with different parabolic varying thickness variations.

Table 1. The specifications and the material properties of an annular circular plate with different attachment of rectangular and concentric stiffener patches.

Dimension of the Plate with Different Stiffener Patches	Uniform Unloaded Plate	Plate with 1 Rectangular/Concentric Stiffener Patch	Plate with 2 Rectangular/Concentric Stiffeners Patches	Plate with 4 Rectangular/Concentric Stiffeners Patches
Outer radius (a) m	0.1515	0.1515	0.1515	0.1515
Inner radius (b) m	0.0825	0.0825	0.0825	0.0825
Radii ratio, (b/a)	0.54	0.54	0.54	0.54
Length of rectangular stiffener (m)	-	0.069	0.069	0.069
Width of rectangular stiffener (m)	-	0.0145	0.00828	0.00483
Thickness (h) of plate with rectangular stiffener (m)	0.0315	0.040	0.035	0.030
Density, ρ (kg/m³)	7905.9	7905.9	7905.9	7905.9
Young's modulus, E (GPa)	218 × 10⁹	218 × 10⁹	218 × 10⁹	218 × 10⁹
Poisson's ratio, υ	0.305	0.305	0.305	0.305
Width of concentric stiffener (m)	-	0.0069	0.0069	0.0069
Thickness (h) of plate with concentric stiffener (m)	0.03150	0.007875	0.0039375	0.00196875

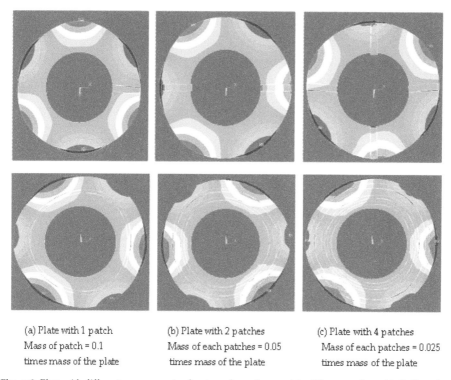

(a) Plate with 1 patch
Mass of patch = 0.1
times mass of the plate

(b) Plate with 2 patches
Mass of each patches = 0.05
times mass of the plate

(c) Plate with 4 patches
Mass of each patches = 0.025
times mass of the plate

Figure 3. Plate with different arrangements of rectangular and concentric stiffener patches with (0, 3) modes.

3. Results and Discussion

3.1. Validation of Natural Frequency Parameter and Acoustic Power Calculation

In this paper, the natural frequency parameter of a uniform unloaded annular circular plate is validated with the published result of Lee et al. [10] and is reported in Table 2. In reference [10], Lee et al. provide the solution for the natural frequency parameter of a uniform annular circular plate by Thick and thin plate theories. In our study we have calculated our result using FEM by taking the same dimension of plate as that of Lee et al. From Table 2, it is clearly understand that in this paper the results obtained are almost equal to the published results [10]. For the acoustic power calculation, the computed analytical, numerical and published experimental results [10] are considered as reported in Figure 4. From Figure 4, a good agreement of computed acoustic results is seen to be obtained analytically and numerically in line with published experimentally results [10].

Table 2. Validation and comparison of natural frequency parameter λ^2 of uniform clamped-free annular circular plate obtained in the present work with that of the published result of Lee et al. [10].

Plate	Mode	Non Dimensional Frequency Parameter, λ^2	
		H. Lee et al. [10]	Present Work
Uniform plate b/a = 0.54 h/a = 0.21	(0,0)	11.96	13.4929
	(0,1)	13.43	13.4946
	(0,2)	15.28	14.1185
	(0,3)	18.75	16.6681

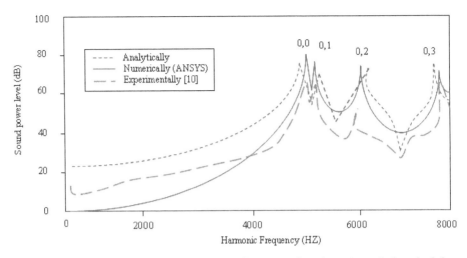

Figure 4. Comparison of sound power level analytically, numerically and experimentally for unloaded plate with uniform thickness for taper ratio $T_x = 0.00$.

3.2. Effect of Natural Frequency Parameter (λ^2) of Plate with Different Combinations of Rectangular and Concentric Stiffener Patches with Different Taper Ratios

In this paper, the effect of the natural frequency parameter (λ^2) is investigated for annular plates with different attachment of rectangular and concentric stiffener patches for different positions. The analysis is made for annular plates with different cases of parabolic thickness variations keeping the mass of the plate plus patches constant. Table 3 compares the first four natural frequency parameters numerically of a uniform unloaded plate for taper ratio, $T_x = 0.00$, with different attachments of rectangular and concentric stiffener patches along with the percentage variation of λ^2. It is clear from Table 3 that the plate with different arrangements of rectangular stiffener patches has the same effect of natural frequency parameters as that of the unloaded plate for taper ratio, $T_x = 0.00$. However, for concentric stiffener patches, the natural frequency parameter decreases with addition of patches and minimum for 4 concentric stiffener patches. In Table 3, the negative variation of frequency parameter is calculated by $\frac{l^2(stiffener) - l^2(original)}{l^2(original)} * 100$. Figures 5 and 6 show the comparison of negative % variation of natural frequency parameter with different modes for a plate with taper ratio, $T_x = 0.00$ and with different attachment of rectangular and concentric stiffener patches. From Figure 5 it is shown that due to less stiffness associated with these modes, the (0, 2) mode of plate with 4 rectangular stiffener patches and (0, 0) mode of plate with 1 rectangular stiffener patch show the lowest percentage variation of λ^2. However, due to greater stiffness associated with the (0, 1) mode of plate with 1 rectangular stiffener patch, it showed the highest percentage variation of λ^2. Furthermore, from Figure 6 it is observed that for concentric stiffener patches the (0, 1) mode of plate with all combinations of patches shows the highest value of percentage variation of λ^2 due to greater stiffness associated with this mode. However, for all the remaining modes (0, 0), (0, 2) and (0, 3) of plates with concentric stiffener patches the stiffness decreases and as a result the percentage variation of λ^2 decreases associated with these modes. Figure 7 shows the numerical comparison of natural frequency parameters λ^2 with modes for an unloaded plate and for a plate with 4 rectangular and 4 concentric stiffener patches for taper ratio, $T_x = 0.00$. It is clear from Figure 7 that the effect of natural frequency parameter due to 4 rectangular stiffener patches is almost same as that of the unloaded plate. However, a plate with 4 concentric stiffener patches shows little decrease in the frequency parameter due to greater stiffness associated with this plate with concentric patch. Tables 4–6 numerically compare the first four natural frequency parameter λ^2 of a plate with different combinations of rectangular and concentric stiffener patches for different cases of thickness variation with different taper ratios. It is observed from Tables 4–6 that the natural frequency parameter for a plate with concentric patches for all thickness variations reduces more in comparison to rectangular patches with increasing taper ratios. This may be due to the lower stiffness of the plate associated with concentric patches. Furthermore, it is observed that the frequency parameter for a Case II plate (parabolically decreasing–increasing thickness variation) for all cases of thickness variation with different combinations of rectangular and concentric stiffener patches reduces more in comparison to a Case I plate (parabolic decreasing thickness variation). This is due to the lower stiffness associated with the Case II plate than that of the Case I plate. It is further investigated that the effect of the frequency parameter for a Case III plate with different attachment of rectangular and concentric stiffener patches (parabolic increasing–decreasing thickness variation) is almost same as that of uniform unloaded plate due to more stiffness associated with the Case III plate. However, for all cases of different rectangular and concentric stiffener patches, plate with different parabolically thickness variations alters its modes at higher taper ratios.

Table 3. Numerical comparison of first four natural frequency parameter λ^2 of uniform unloaded plate with different attachment of rectangular and concentric stiffener patches for taper ratio, $T_x = 0.00$.

Type of Stiffener Patches	Mode	Uniform Unloaded Plate	Plate with 1 Stiffener Patch	% of Negative Variation in λ^2	Plate with 2 Stiffener Patches	% of Negative Variation in λ^2	Plate with 4 Stiffener Patches	% of Negative Variation in λ^2
Rectangular Stiffener	(0,0)	13.4929	13.4819	−0.0815	13.4862	−0.04960	13.4877	−0.0385
	(0,1)	13.4946	13.4942	−0.00296	13.4937	−0.00669	13.4914	−0.0237
	(0,2)	14.1185	14.1148	−0.0262	14.1099	−0.06090	14.1053	−0.0934
	(0,3)	16.6681	16.6638	−0.0258	16.6638	−0.02580	16.6658	−0.0258
Concentric Stiffener	(0,0)	13.4915	13.4223	−0.518	13.3615	−0.964	13.3242	−1.260
	(0,1)	13.5023	13.4842	−0.148	13.4427	−0.444	13.3614	−1.037
	(0,2)	14.1214	14.0815	−0.283	14.0235	−0.708	13.9212	−1.416
	(0,3)	16.6762	16.6014	−0.419	16.5056	−1.019	16.3843	−1.740

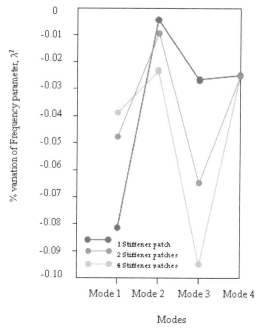

Figure 5. Comparison of % variation of natural frequency parameter with different modes for plate with taper ratio, $T_x = 0.00$ and with different attachments of rectangular stiffener patches.

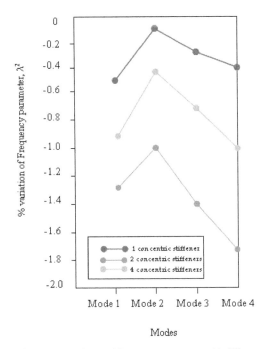

Figure 6. Comparison of % variation of natural frequency parameter with different modes for a uniform plate with taper ratio, $T_x = 0.00$ and with different attachment of concentric stiffener patches.

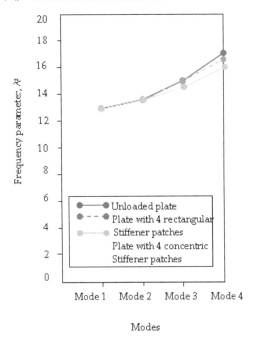

Figure 7. Comparison of variation of different frequency parameter with different modes for an unloaded plate and for a plate with 4 rectangular and 4 concentric stiffener patches.

Table 4. Numerical comparison of natural frequency parameter λ^2 of a plate with 1 rectangular and 1 concentric stiffener patch for different thickness variations and for different taper ratios T_x.

| Case | Mode | Natural Frequency Parameter, λ^2 | | | | | | | | | | | |
| | | $T_x = 0.00$ | | $T_x = 0.25$ | | $T_x = 0.50$ | | $T_x = 0.75$ | | | | | |
		Rectangular Stiffener	Concentric Stiffener	Rectangular Stiffener	Concentric Stiffener	Rectangular Stiffener	Concentric Stiffener	Rectangular Stiffener	Concentric Stiffener
I	(0,0)	13.4819	13.4223	12.9912	12.9212	12.4711	12.2552	11.9311	11.8025
	(0,1)	13.4957	13.4842	12.9759	13.0540	12.4449	12.7821	11.8929	12.1634
	(0,2)	14.1148	14.0815	13.6148	13.8225	13.0942	13.3244	12.5513	13.0512
	(0,3)	16.6638	16.6014	16.0741	16.1024	15.4619	15.6012	14.8262	15.1020
II	(0,0)	13.4819	13.4223	12.8960	12.8021	12.2729	12.3230	11.6210	11.6254
	(0,1)	13.4957	13.4842	12.8799	12.9523	12.2424	12.5422	11.5764	11.7651
	(0,2)	14.1148	14.0815	13.5199	13.5535	12.8951	12.8641	12.2392	12.6821
	(0,3)	16.6638	16.6014	15.9622	16.0202	15.2286	15.4021	14.4611	14.9248
III	(0,0)	13.4819	13.4223	13.4891	13.3844	13.4917	13.4012	13.4914	13.4030
	(0,1)	13.4957	13.4842	13.4791	13.4214	13.4816	13.4414	13.4814	13.4342
	(0,2)	14.1148	14.0815	14.1116	14.0025	14.1142	14.0521	14.1139	14.0614
	(0,3)	16.6638	16.6014	16.6601	16.5254	16.6632	16.5528	16.6629	16.5403

Table 5. Numerical comparison of natural frequency parameter λ^2 of plate with 2 rectangular and 2 concentric stiffener patches for different thickness variations and for different taper ratios T_x.

Case	Mode	Natural Frequency Parameter, λ^2							
		$T_x = 0.00$		$T_x = 0.25$		$T_x = 0.50$		$T_x = 0.75$	
		Rectangular Stiffener	Concentric Stiffener	Rectangular Stiffener	Concentric Stiffener	Rectangular Stiffener	Concentric Stiffener	Rectangular Stiffener	Concentric Stiffener
I	(0,0)	13.4862	13.3625	12.9886	12.8938	12.4699	12.4829	11.9583	12.0021
	(0,1)	13.4937	13.4421	12.9782	12.9436	12.4463	12.5335	11.9217	12.0554
	(0,2)	14.1099	14.0275	13.6108	13.9637	13.0904	13.2724	12.5774	13.0068
	(0,3)	16.6638	16.5028	16.0744	15.9829	15.4625	15.5421	14.8610	15.0227
II	(0,0)	13.4862	13.3665	12.8937	12.7717	12.2720	12.2929	11.6478	11.7631
	(0,1)	13.4937	13.4485	12.8810	12.8342	12.2435	12.3227	11.6042	11.9929
	(0,2)	14.1099	14.0267	13.5156	13.8732	12.9216	12.7811	12.2648	12.6456
	(0,3)	16.6638	16.5057	15.9625	15.9018	15.2646	15.2882	14.4954	14.7824
III	(0,0)	13.4862	13.3619	13.4831	13.3228	13.5170	13.3415	13.5245	13.3238
	(0,1)	13.4937	13.4467	13.4860	13.4039	13.5248	13.4262	13.5168	13.402472
	(0,2)	14.1099	14.0228	14.1070	13.9442	14.1421	13.9594	14.1421	13.9684
	(0,3)	16.6638	16.5025	16.6603	16.4311	16.7020	16.4435	16.7017	16.4224

Appl. Sci. **2018**, *8*, 2542

Table 6. Numerical comparison of natural frequency parameter λ^2 of plate with 4 rectangular and 4 concentric stiffener patches for different thickness variations and for different taper ratios T_X.

| Case | Mode | Natural Frequency Parameter, λ^2 | | | | | | | |
| | | $T_x = 0.00$ | | $T_x = 0.25$ | | $T_x = 0.50$ | | $T_x = 0.75$ | |
		Rectangular Stiffener	Concentric Stiffener	Rectangular Stiffener	Concentric Stiffener	Rectangular Stiffener	Concentric Stiffener	Rectangular Stiffener	Concentric Stiffener
I	(0,0)	13.4877	13.3214	12.9897	12.8342	12.4711	11.9235	11.9314	11.4862
	(0,1)	13.4914	13.3624	12.9788	12.8547	12.4463	11.9567	11.8937	11.5324
	(0,2)	14.1053	13.9232	13.6062	13.8615	13.0858	13.6681	12.5439	12.4683
	(0,3)	16.6658	16.3814	16.0761	15.8823	15.4639	14.9884	14.8279	14.3616
II	(0,0)	13.4877	13.3227	12.8948	12.7175	12.2732	11.7225	11.6216	11.1449
	(0,1)	13.4914	13.3612	12.8813	12.7619	12.2432	11.8034	11.5770	11.3425
	(0,2)	14.1053	13.9224	13.5110	13.7721	12.8876	12.1841	12.2320	12.0434
	(0,3)	16.6658	16.3825	15.9642	15.8015	15.2314	14.6248	14.4629	14.1828
III	(0,0)	13.4877	13.3234	13.4845	13.2837	13.4871	13.3016	13.4868	13.2926
	(0,1)	13.4914	13.3624	13.4883	13.3227	13.4908	13.3326	13.4908	13.3223
	(0,2)	14.1053	13.9285	14.1024	13.8212	14.1047	13.8418	14.1044	13.8434
	(0,3)	16.6658	16.3824	16.6624	16.2865	16.6652	16.2712	16.6649	16.2824

3.3. Acoustic Response Solution of Tapered Annular Circular Plate with Different Combination of Rectangular and Concentric Stiffener Patches with Different Taper Ratios

In this paper, the sound power level (dB, reference = 10^{-12} watts) of an annular circular plate with a different attachment of rectangular and concentric stiffener patches is estimated. The plate for the sound power level is analyzed for all cases of different parabolic thickness variation due to transverse vibration. The taper ratio is maintained from a range (0.00–0.75). The sound power level is investigated by applying 1 N concentrated load under time-varying harmonic excitations at different excitation locations at different nodes, and a harmonic frequency range of (0–8000) HZ is taken to determine the sound radiation characteristic. The Case I plate with parabolic decreasing thickness variation is taken as a convergence study. Figures 8 and 9 compare the sound power level for a Case I plate obtained analytically and numerically for the taper ratio $T_x = 0.75$ for 4 rectangular stiffener patches and 4 concentric stiffener patches, respectively, for different modes. A good agreement of computed results is seen in the comparison of sound power as depicted from Figures 8 and 9. Figures 10–12 shows the numerical comparison of sound power level for Case I plate with different combinations of rectangular stiffener patches for different taper ratios and for different modes under forced excitation. From Figure 10, a sound power level up to 30 dB is seen, and we do not get any broad range of frequencies for different taper ratios for plate with 1 rectangular stiffener patch. However, for a sound power level up to 40 dB, we get all taper ratios, $T_x = 0.00, 0.25, 0.50$ and 0.75, with a broad range of frequencies in frequency band A only with 1 rectangular stiffener.

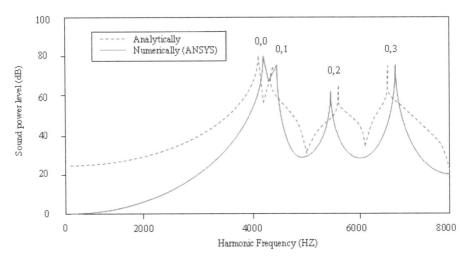

Figure 8. Comparison of sound power level analytically and numerically for annular plate attached with 4 rectangular stiffener patches and having parabolic decreasing thickness variations (Case I) for taper ratio $T_x = 0.75$.

Appl. Sci. **2018**, *8*, 2542

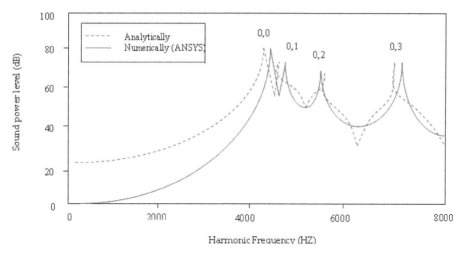

Figure 9. Comparison of sound power level analytically and numerically for annular plate attached with 4 concentric stiffener patches and having parabolic decreasing thickness variations (Case I) for taper ratio $T_X = 0.75$.

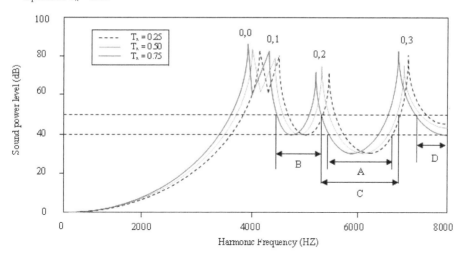

Figure 10. Numerical comparison of sound power level for annular plate attached with 1 rectangular stiffener patch and having parabolic decreasing thickness variations (Case I) for different taper ratios T_X.

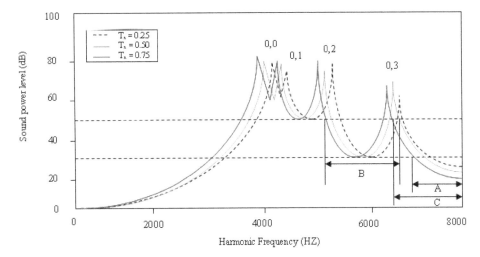

Figure 11. Numerical comparison of sound power level for annular plate attached with 2 rectangular stiffener patches and having parabolic decreasing thickness variations (Case I) for different taper ratios T_x.

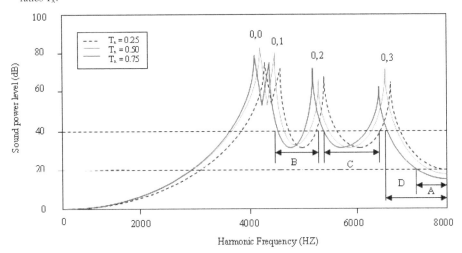

Figure 12. Numerical comparison of sound power level for annular plate attached with 4 rectangular stiffener patches and having parabolic decreasing thickness variations (Case I) for different taper ratios T_x.

Patches are as reported in Figure 10. It is noteworthy that for sound power level up to 50 dB, we get a broader range of frequencies for sound power level in different frequency bands, i.e., B, C and D, as reported in Figure 10. From Figure 11, it is apparent that for a sound power level up to 20 dB, we do not get any broad range of frequencies for plate with 2 rectangular stiffener patches. However for a sound power level up to 30 dB, we get the broad range of frequencies in frequency band A only with taper ratio, $T_x = 0.00$, 0.25, 0.50 and 0.75 as available design alternative. For a sound power level up to 50 dB, we get more broad range of frequencies for the sound power level in different frequency bands, i.e., B and C, as reported in Figure 11. From Figure 12, it is shown that for a sound power level up to 10 dB, we do not get any broad range of frequencies for plate with 4 rectangular stiffener patches. But for a sound power level up to 20 dB, we get the broad range of frequencies in frequency bands A only with all taper ratios, $T_x = 0.00$, 0.25, 0.50 and 0.75 and, therefore, this is the available design alternative. However, for a sound power level up to 40 dB, we get more design options for the sound power level in different frequency bands, i.e., B, C and D as reported in Figure 12. Furthermore, Figures 13–15 show the numerical comparison of the sound power level for a Case I plate with different combinations of concentric stiffener patches for different taper ratios. From Figure 13, it is seen that for a sound power level up to 30 dB, we do not get any design options for different taper ratios for the plate with both 1 concentric stiffener patch and 4 concentric stiffener patches; and for 2 concentric stiffener patches, we do not find any sound power level upto 10 dB. However, for a sound power level up to 40 dB, we get all taper ratios, $T_x = 0.00$, 0.25, 0.50 and 0.75 as design options in frequency bands A and B for plate with 1 concentric stiffener patch combinationas reported in Figure 13. It is noteworthy that for a sound power level up to 50 dB, we get more design options for the sound power level in different frequency bands, i.e., C, D and E as reported in Figure 13. From Figure 14, it is apparent that for a sound power level up to 20 dB, then in frequency band A only taper ratio $T_x = 0.00$, 0.25, 0.50 and 0.75 are available design alternatives for a plate with the 2 concentric stiffener patches combination. But for sound power level up to 30 dB, we get wider frequency bands, B, C and D for different taper ratios as reported in Figure 14. From Figure 15, it is seen that for a sound power level up to 40 dB is possible only in frequency bands A only with all taper ratios, $T_x = 0.00$, 0.25, 0.50 and 0.75 and, therefore, this is the available design alternative for plate with 4 concentric stiffener patches combination. However, for a sound power level up to 60 dB, we get a broader range of frequency denoted as B and C for all taper ratios as reported in Figure 15.

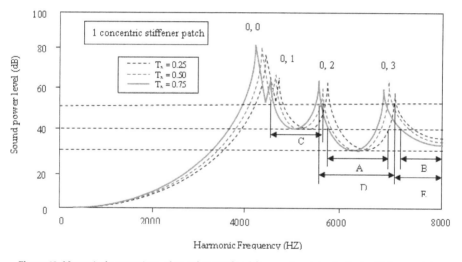

Figure 13. Numerical comparison of sound power level for annular plate attached with 1 concentric stiffener patch and having parabolic decreasing thickness variations (Case I) for different taper ratios T_x.

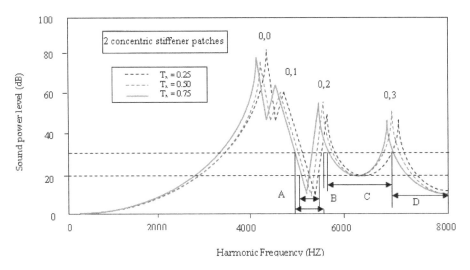

Figure 14. Numerical comparison of sound power level for annular plate attached with 2 concentric stiffener patches and having parabolic decreasing thickness variations (Case I) for different taper ratios T_x.

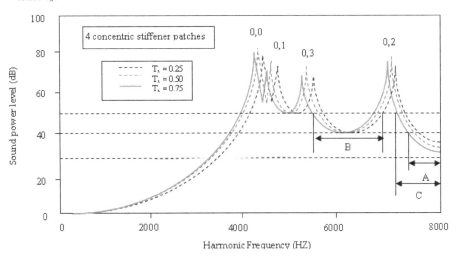

Figure 15. Numerical comparison of sound power level for annular plate attached with 4 concentric stiffener patches and having parabolic decreasing thickness variations (Case I) for different taper ratios T_x.

It may be inferred from Figures 10–15 that a plate with different combinations of rectangular and concentric stiffener patches plays a significant role in sound power reduction in different frequency bands. A plate with 4 rectangular stiffener patches combination causes maximum sound power level reduction in comparison to 1 rectangular stiffener patch and 2 rectangular stiffener patch combinations for a Case I plate; whereas, for a plate with 4 concentric stiffener patches the lowest sound power is observed in comparison to other combinations. However, the stiffness contribution due to various taper ratios has a very limited impact on sound power level reduction in comparison to that of modes and excitation locations of plate with different combinations of rectangular and concentric stiffener patches. Furthermore, from Figures 10–15, it is observed that for an excitation frequency up to 2000 HZ, the effect of different combinations of rectangular and concentric stiffener patches and stiffness variation due to different taper ratios do not have a significant effect on sound power radiation for clamped-free boundary condition. However, when the excitation frequency increases beyond 2000 HZ and up to the first peak, the sound power level is higher for only higher taper ratios for a Case I plate with both 1 rectangular stiffener and 1 concentric stiffener patch, and variation of sound power level due to variation of peaks for different taper ratios is observed for a plate with both 2 rectangular stiffener and 2 concentric stiffener patches and for 4 rectangular stiffener and 4 concentric stiffener patch combinations. For a Case II plate, beyond a forcing frequency 2000 HZ, the highest sound power level is associated with plate for 2 rectangular stiffener patches combinations; while for a Case III plate, the sound power level is seen to decrease for all combinations of rectangular stiffener patches with increasing taper ratio. Similar effect is observed for plate with concentric stiffener patches where plate with 2 concentric stiffener patches is seen to have highest radiation power. Again for case III plate the sound power is seen to be decreased for all combination of concentric stiffener patches. Furthermore, different modes do influence the sound power peaks as evident from Figures 10–15. Sound power level peak obtained for different modes (0, 0) and (0, 1) almost remain same for different taper ratios. However, no such sound power similarity of modes (0, 0) and (0, 1) is observed for plates with concentric stiffener patches. The sound power level does shift towards a lower frequency with increasing taper ratio for all combinations of rectangular and concentric stiffener patches. For a higher frequency beyond 4000 HZ, it is observed that different taper ratios alter its stiffness at higher forcing frequency for different cases of thickness variation. It is noteworthy that for a higher frequency beyond 4000 HZ and up to 8000 HZ, a plate with different combination of rectangular and concentric stiffener patches alters its stiffness at higher forcing frequency and the acoustic power curve tends to intersect each other at this high forcing region. Table 7 compares the peak sound power level of a plate having different parabolically varying thickness with different combinations of rectangular and concentric stiffener patches for a taper ratio $T_x = 0.75$. It is interesting to note that a plate with 4 rectangular stiffener patches combination shows the lowest peak sound power level among all cases of thickness variations, and the lowest peak sound of 77 dB is obtained for Case III plate whereas the highest peak sound power level of 84 dB is obtained for a Case II plate with 2 rectangular stiffener patches combination. Similar effect is again observed for plate with concentric stiffener patch combination. The lowest sound power of 76 dB is observed for the plate with the 4 concentric stiffener patches combination, and the highest power of 83 dB is observed for plate with 1 concentric stiffener patch combination. Figures 16–21 shows the numerical comparison of sound power levels for Case I, Case II and Case III plates for different combinations of rectangular and concentric stiffener patches for taper ratio $T_x = 0.75$. From Figures 16–21, it is observed that for all cases of thickness variation and for excitation frequency up to 2000 HZ, different parabolic thickness variation does not have any significant effect on sound power radiation. Furthermore, from Figures 16–18 it is seen that beyond excitation frequency of 2000 HZ and up to the first peak, a Case II plate with 2 rectangular stiffener patches shows the highest radiation power of 84 dB in comparison to a radiation power of 82 dB for a Case I plate with 1 rectangular stiffener patch combination. However, at this forcing frequency of 2000 HZ case III plate remains unaffected and shows the lowest peak sound level for all cases of thickness variations and so it is suggested that Case III plate is the lowest sound power radiator among all cases of thickness variation with different combinations of rectangular stiffener patches. Again, a similar effect is observed for plate with the concentric stiffener patches combination. Beyond 200 HZ, Case II plate with 2 concentric stiffener patches is a very good sound radiator of sound power 83 dB in comparison to 82 dB of plate with 1 stiffener patch combination. A Case III plate with all combination of concentric stiffener patches is found to be a poor sound radiator.

Table 7. Numerical comparison of peak sound power level and radiation efficiency for annular plate having different parabolically thickness variations with different attachments of rectangular and concentric stiffener patches for T$_x$ = 0.75.

Type of Stiffener Patches	Plate Thickness Variation for Taper Ratio T$_x$ = 0.75	Unloaded Plate Thickness Variation for Taper Ratio T$_x$ = 0.75		Plate with 1 Rectangular/Concentric Stiffener Patch		Plate with 2 Rectangular/Concentric Stiffener Patches		Plate with 4 Rectangular/Concentric Stiffener Patches	
		Sound Power Level (dB)	Radiation Efficiency (σ_{mn})	Sound Power Level (dB)	Radiation Efficiency (σ_{mn})	Sound Power Level (dB)	Radiation Efficiency (σ_{mn})	Sound Power Level (dB)	Radiation Efficiency (σ_{mn})
Rectangular Stiffener	Case I	83	1.079	82	1.058	81	1.048	79	1.015
	Case II	85	1.135	81	1.045	84	1.094	78	1.002
	Case III	79	1.020	79	1.020	78	1.007	77	0.994
Concentric Stiffener	Case I	83	1.079	82	1.058	81	1.048	77	0.994
	Case II	85	1.135	81	1.045	83	1.079	79	1.020
	Case III	79	1.020	78	1.020	78	1.007	76	0.935

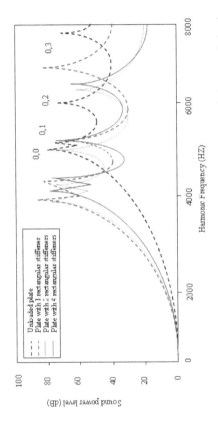

Figure 16. Numerical comparison of sound power level for annular plate having parabolic decreasing thickness variation (case I) for different attachments of rectangular stiffener patches for taper ratio T$_x$ = 0.75.

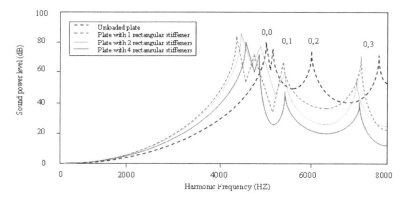

Figure 17. Numerical comparison of sound power level for annular plate having parabolic decreasing increasing thickness variation (Case II) for different attachments of rectangular stiffener patches for taper ratio $T_x = 0.75$.

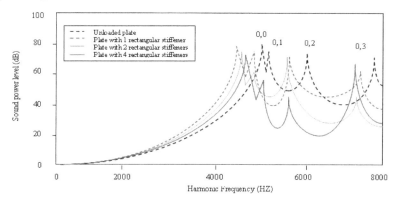

Figure 18. Numerical comparison of sound power level for annular plate having parabolic increasing decreasing thickness variations (Case III) for different combinations of rectangular stiffener patches for taper ratio $T_x = 0.75$.

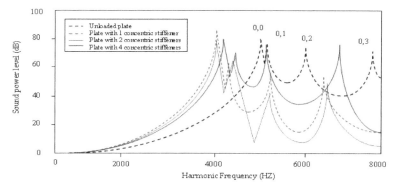

Figure 19. Numerical comparison of sound power level for annular plate having parabolic decreasing thickness variations (Case I) for different combinations of concentric stiffener patches for taper ratio $T_x = 0.75$.

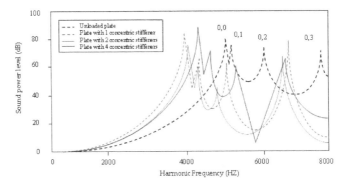

Figure 20. Numerical comparison of sound power level for annular plate having parabolic decreasing increasing thickness variations (Case II) for different combinations of concentric stiffener patches for taper ratio $T_x = 0.75$.

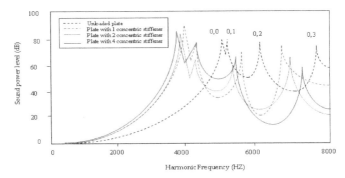

Figure 21. Numerical comparison of sound power level for annular plate having parabolic increasing decreasing thickness variations (Case III) for different combinations of concentric stiffener patches for taper ratio $T_x = 0.75$.

Figures 22 and 23 compare the analytical and numerical comparison of radiation efficiency (σ_{mn}) for a Case I plate with 4 rectangular stiffener patches and 4 concentric stiffener patches respectively having parabolically decreasing thickness variation for taper ratio $T_x = 0.75$. A good agreement of results is seen in the comparison of radiation efficiency as reported in Figures 22 and 23. Figures 24 and 25 show the variation of radiation efficiency with different taper ratios T_x for different combination of rectangular and concentric stiffener patches for a Case I plate with parabolically decreasing thickness variation. It is seen that for all combinations of rectangular and concentric stiffener patches, the effect of radiation efficiency due to different taper ratios is independent of exciting frequency up to 1000 HZ, but at a given forcing frequency a higher taper ratio causes higher radiation efficiency beyond 1000 HZ. However, sound power level peaks do shift towards a lower frequency as taper ratio increases. For higher frequency beyond 2000 HZ, different taper ratios alter its stiffness at higher forcing frequency and the radiation efficiency curves tend to intersect each other at this high forcing region. It is interesting to note that the radiation curve tends to unity in the frequency band 6800–7200 HZ and a clear peak is seen at this frequency band for all combination of rectangular and concentric stiffener patches. Furthermore, from Figures 24 and 25 it is seen that the radiation efficiency increases with the taper ratio for all combinations of rectangular and concentric stiffener patches. Out of these combinations, the Case II plate with 2 rectangular stiffener patches and 2 concentric stiffener patches delivers the highest radiation efficiency whereas Case I plate with both 1 rectangular and 1 concentric stiffener patch and 4 rectangular and 4 concentric stiffener patches is seen to be a moderate radiator as depicted in Table 7. However, at higher forcing frequency, it is seen that both a plate with 4 rectangular and 4 concentric stiffener patches with all cases of thickness variation (Cases I, II and III) shows the least radiation efficiency for all combinations of rectangular stiffener patches, as evident from Table 7.Therefore, it is interesting to mention that a Case III plate shows the lowest radiation efficiency (σ_{mn}) for all cases of thickness variations and is a poor radiation emitter among all the thickness variations with different combinations of rectangular and concentric stiffener patches. Figure 26 shows the numerical comparison of radiation efficiency for a plate with both 4 rectangular and 4 concentric stiffener patches for taper ratio $T_x = 0.75$. It is found that the plate shows almost the same radiation efficiency as depicted from Figure 26. Figure 27 shows the numerical comparison of the sound power level for a plate with 4 rectangular and 4 concentric stiffener patches for taper ratio $T_x = 0.75$. It is observed that both the plates show almost the same peak for taper ratio $T_x = 0.75$. Hence, the effect of stiffness variation along with the modes has negligible effect for both the combinations.

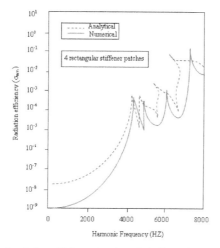

Figure 22. Comparison of radiation efficiency (σ_{mn}) analytically and numerically for annular plate attached with 4 rectangular stiffener patches and having parabolic decreasing thickness variations (Case I) for taper ratio $T_x = 0.75$.

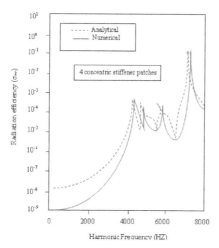

Figure 23. Comparison of radiation efficiency (σ_{mn}) analytically and numerically for annular plate attached with 4 concentric stiffener patches and having parabolic decreasing thickness variations (Case I) for taper ratio $T_x = 0.75$.

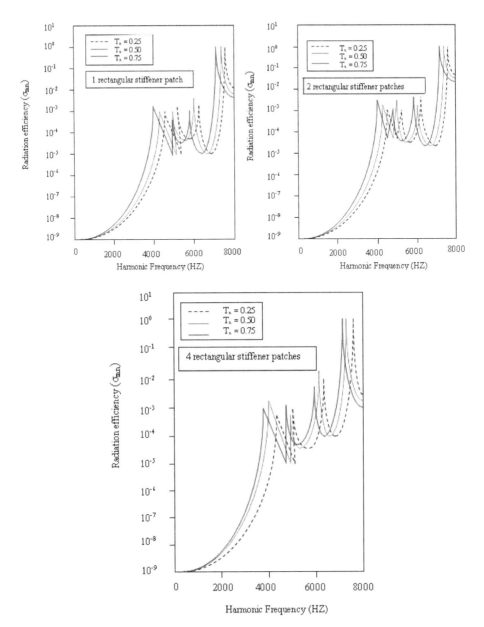

Figure 24. Numerical comparison of radiation efficiency (σ_{mn}) for annular plate having parabolically decreasing thickness variations (Case I) with different attachment of (a) 1 rectangular stiffener patch (b) 2 rectangular stiffener patches (c) 4 rectangular stiffener patches for taper ratio T_x =0.75.

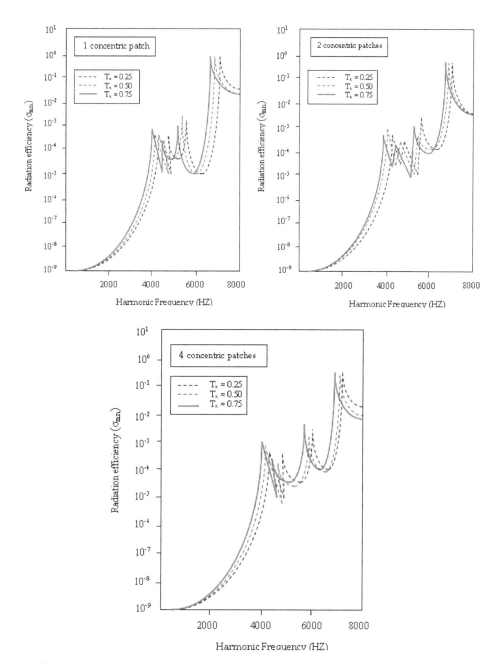

Figure 25. Numerical comparison of radiation efficiency (σ_{mn}) for annular plate having parabolically decreasing thickness variations (Case I) with different attachment of (a) 1 concentric patch (b) 2 concentric patches and (c) 4 concentric patches for taper ratio $T_X = 0.75$.

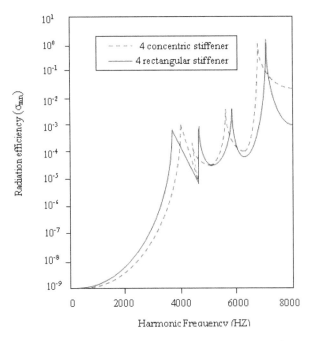

Figure 26. Numerical comparison of radiation efficiency (σ_{mn}) for annular plate attached with 4 rectangular and 4 concentric stiffener patches and having parabolic decreasing thickness variations (Case I) for taper ratio $T_x = 0.75$.

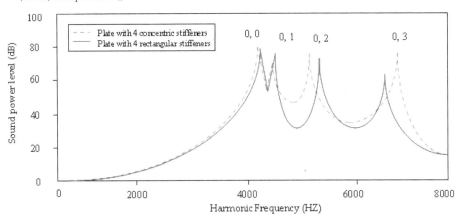

Figure 27. Numerical comparison of sound power level (dB) for annular plate attached with 4 rectangular and 4 concentric stiffener patches having parabolic decreasing thickness variations (Case I) for taper ratio $T_x = 0.75$.

3.4. Peak Sound Power Level Variation with Different Taper Ratios for All Combinations of Rectangular and Concentric Stiffener Patches Attached to a Plate

Peak sound power level for a plate was estimated for annular plates with different attachment of rectangular and concentric stiffener patches. The peak sound was considered for plates with different parabolically varying thickness. The different taper ratios were taken as reported in Figures 28 and 29, respectively. Furthermore, peak sound power level for different attachments of rectangular and concentric stiffener patches attached to a plate is reported at the first peak which corresponds to (0, 0) mode of the plate. From Figure 28 it is seen that for a Case I plate with 1 rectangular stiffener patch combination, peak sound power level increases for increasing value of taper ratio whereas for 2 rectangular stiffener patches and 4 rectangular stiffener patches combinations, variations of peak sound power levels are observed for an increasing value of taper ratio. For a Case I plate, the maximum peak sound power level is obtained for taper ratio $T_x = 0.75$ for plate with 1 rectangular stiffener patch combination. Furthermore, it is seen that peak is minimum for taper ratio, $T_x = 0.25$ and maximum for taper ratio, $T_x = 0.50$ for plate with 4 rectangular stiffener patches combination, whereas for a plate with 2 rectangular stiffener patches combination, the peak is at a minimum for taper ratio $T_x = 0.25$ and maximum for taper ratio $T_x = 0.75$. Similarly, for a Case II plate, the maximum peak sound power level is obtained for taper ratio $T_x = 0.75$ and minimum peak is seen for taper ratio $T_x = 0.50$ for a plate with 2 rectangular stiffener patches combination. Also, for a Case II plate, it is interesting to note that peak sound power level increases for increasing value of taper ratio for plate with 1 rectangular stiffener patch combination and the peak is seen to be maximum for a taper ratio $T_x = 0.75$. However, for a plate with 4 rectangular stiffeners patches combination, the peak is seen to be at a minimum for taper ratio $T_x = 0.75$ and maximum for taper ratio $T_x = 0.50$. Furthermore, it is investigated that for case III plate, peak sound power level decreases for increasing value of taper ratio for all combinations of rectangular stiffener patches attached to a plate. For a Case III plate, the maximum peak of the sound power level is obtained for taper ratio $T_x = 0.25$ for a plate with 2 rectangular stiffener patches combination and minimum peak is observed for taper ratio $T_x = 0.75$ for a plate with 4 rectangular stiffener patches combination. From Figure 29, it is observed that for a Case I plate peak is maximum for taper ratio, $T_x = 0.75$ for plate with 1 concentric stiffener patches and minimum for taper ratio $T_x = 0.50$ for 2 concentric stiffener patches. For a Case II plate the highest peak is seen for taper ratio $T_x = 0.75$ for 2 concentric stiffener patches. Similarly, for case III plate lowest peak is observed for taper ratio, $T_x = 0.75$ for plate with 4 concentric stiffener patches. However, from Figures 28 and 29, it is necessary to mention that for a Case II unloaded tapered plate the highest peak sound power level is seen for taper ratio $T_x = 0.75$. Furthermore, it is also observed that the peak sound power level increases for case I plate for taper ratio $T_x = 0.75$ and peak sound power level decreases for a Case III plate for taper ratio, $T_x = 0.75$.

It is thus quite obvious that different combinations of rectangular and concentric stiffener patches have a significant impact on peak sound power level corresponding to the (0, 0) mode. Furthermore, different combinations of rectangular and concentric stiffener patches with different taper ratios provide us design options for peak sound power level. For example, for peak sound power reduction, taper ratio $T_x = 0.75$ with 4 rectangular stiffener patches and 4 concentric stiffener patches combination, as well as taper ratio $T_x = 0.50$ with 2 rectangular stiffener patches combination, for a Case III plate may be the options. Similarly, for sound power actuation, taper ratio $T_x = 0.75$ with 1 rectangular stiffener patch and 1 concentric stiffener patch combination for a Case I plate and 2 rectangular stiffener patches and 4 concentric stiffener patches combination for a Case II plate may be the alternative solution. However, for an unloaded tapered plate, it can be added that for taper ratio $T_x = 0.75$ for a Case III plate may be considered as a poor sound emitter and a taper ratio $T_x = 0.75$ for a Case I and Case II plate may be considered as the highest sound emitter.

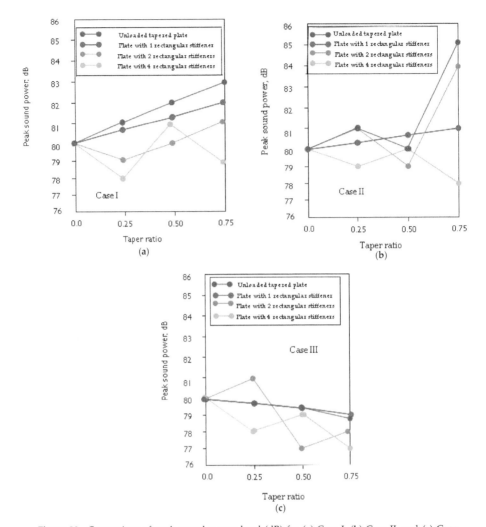

Figure 28. Comparison of peak sound power level (dB) for (**a**) Case I, (**b**) Case II, and (**c**) Case III plates having different parabolic thickness variations with different attachments of rectangular stiffener patches.

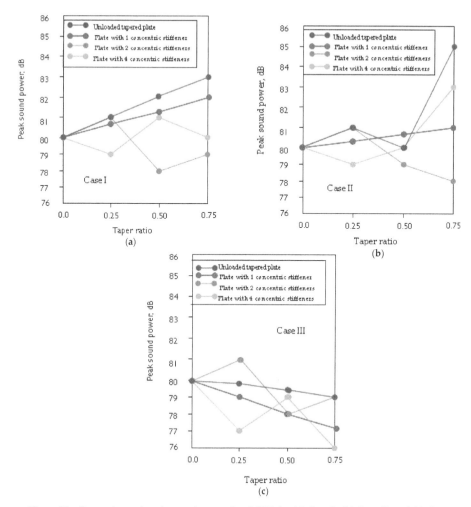

Figure 29. Comparison of peak sound power level (dB) for (**a**) Case I, (**b**) Case II, and (**c**) Case III plates having different parabolic thickness variations with different combinations of concentric stiffener patches.

4. Conclusions

A comparison is made of vibroacoustic behavior of tapered annular circular plates having different parabolically varying thickness with different combinations of rectangular and concentric stiffeners patches keeping the mass of the plate plus patch constant for a clamped-free boundary condition. It is observed that due to lower stiffness associated with the Case II plate, the non-dimensional frequency parameter of a Case II plate reduces more in comparison to a Case I plate. However, for a Case III plate, the non-dimensional frequency parameter is same as that of unloaded plate. In response to acoustic behavior, it is observed that different combinations of rectangular and concentric stiffener patches and modes variation have significant impacts on the sound power level in comparison to the stiffness variation due to the taper ratio. It is observed that for a sound power level up to 50 dB, and for a plate with different parabolically varying thickness, we get all taper ratios, $T_x = 0.00, 0.25, 0.50$ and 0.75, with a broad range of frequencies as design options in different frequency bands for different combinations of rectangular and concentric stiffener patches. It is further shown that a plate with 4 rectangular and 4 concentric stiffener patches combination shows the minimum sound power level for all cases of thickness variation, whereas the highest power is obtained for a Case II plate with 2 rectangular and concentric stiffener patches combination. It is interesting to note that a Case III plate has the lowest sound power level among all variations and is seen to be the lowest sound radiator. Further different combinations of rectangular and concentric stiffener patches with different taper ratios provide us design options for peak sound power level. For example, for peak sound power reduction, taper ratios, $T_x = 0.75$ with a 4 rectangular stiffener patches 4 concentric stiffener patches combination, and taper ratio $T_x = 0.50$ with a 2 rectangular stiffener patches combination for a Case III plate may be the options. Similarly, for sound power actuation, a taper ratio $T_x = 0.75$ with 1 rectangular stiffener patch and 1 concentric stiffener patch combination for a Case I plate and 2 rectangular stiffener patches and 4 concentric stiffener patches combination for a Case II plate may be an alternative solution. Furthermore, for unloaded tapered plates, it can be added that taper ratio $T_x = 0.75$ for a Case III plate may be considered as a poor sound emitter and a taper ratio $T_x = 0.75$ for Case I and Case II plates may be considered as the highest sound emitter.

Author Contributions: V.R. and M.R. supervised the research. A.C. and M.S.A. developed the research concept, developed the theory and performed the analysis. M.S.A. collected the data. A.C. wrote the paper. V.R. and M.R. revised the manuscript and made important suggestions technically and grammatically. A.C. provided the APC funding.

Funding: The work was carried out in the Indian Institute of Technology (ISM) Dhanbad, India. The APC will be funded by the corresponding author only.

Conflicts of Interest: The authors declare no conflicts of interest.

References

1. Wang, C.M.; Hong, G.M.; Tan, T.J. Elasting buckling of tapered circular plates. *Comput. Struct.* **1995**, *55*, 1055–1061. [CrossRef]
2. Gupta, A.P.; Goyal, N. Forced asymmetric response of linearly tapered circular plates. *J. Sound Vib.* **1999**, *220*, 641–657. [CrossRef]
3. Vivio, F.; Vullo, V. Closed form solutions of axisymmetric bending of circular plates having non-linear variable thickness. *Int. J. Mech. Sci.* **2010**, *52*, 1234–1252. [CrossRef]
4. Sharma, S.; Lal, R.; Neelam, N. Free transverse vibrations of non-homogeneous circular plates of linearly varying thickness. *J. Int. Acad. Phys. Sci.* **2011**, *15*, 187–200
5. Wang, C.Y. The vibration modes of concentrically supported free circular plates. *J. Sound Vib.* **2014**, *333*, 835–847. [CrossRef]
6. Liu, T.; Kitipornchai, S.; Wang, C.M. Bending of linearly tapered annular Mindlin plates. *Int. J. Mech. Sci.* **2001**, *43*, 265–278. [CrossRef]
7. Duana, W.H.; Wang, C.M.; Wang, C.Y. Modification of fundamental vibration modes of circular plates with free edges. *J. Sound Vib.* **2008**, *317*, 709–715. [CrossRef]
8. Gupta, U.S.; Lal, R.; Sharma, S. Vibration of non-homogeneous circular Mindlin plates with variable thickness. *J. Sound Vib.* **2007**, *302*, 1–17. [CrossRef]
9. Kang, J.H. Three-dimensional vibration analysis of thick circular and annular plates with nonlinear thickness variation. *Comput. Struct.* **2003**, *81*, 1663–1675. [CrossRef]
10. Lee, H.; Singh, R. Acoustic radiation from out-of-plane modes of an annular disk using thin and thick plate theories. *J. Sound Vib.* **2005**, *282*, 313–339. [CrossRef]
11. Thompson Jr, W. The computation of self- and mutual-radiation impedances for annular and elliptical pistons using Bouwkamp integral. *J. Sound Vib.* **1971**, *17*, 221–233. [CrossRef]
12. Levine, H.; Leppington, F.G. A note on the acoustic power output of a circular plate. *J. Sound Vib.* **1988**, *21*, 269–275. [CrossRef]
13. Rdzanek, W.P., Jr.; Engel, W. Asymptotic formula for the acoustic power output of a clamped annular plate. *Appl. Acoust.* **2000**, *60*, 29–43. [CrossRef]
14. Wodtke, H.W.; Lamancusa, J.S. Sound power minimization of circular plates through damping layer placement. *J. Sound Vib.* **1998**, *215*, 1145–1163. [CrossRef]

15. Wanyama, W. Analytical Investigation of the acoustic radiation from linearly-varying circular plates. Doctoral Dissertation, Texas Tech University, Lubbock, TX, USA, 2000.

16. Lee, H.; Singh, R. Self and mutual radiation from flexural and radial modes of a thick annular disk. *J. Sound Vib.* **2005**, *286*, 1032–1040. [CrossRef]

17. Cote, A.F.; Attala, N.; Guyader, J.L. Vibro acoustic analysis of an unbaffled rotating disk. *J. Acoust. Soc. Am.* **1998**, *103*, 1483–1492. [CrossRef]

18. Jeyraj, P. Vibro-acoustic behavior of an isotropic plate with arbitrarily varying thickness. *Eur. J. Mech. A/Solids* **2010**, *29*, 1088–1094. [CrossRef]

19. Ranjan, V.; Ghosh, M.K. Forced vibration response of thin plate with attached discrete dynamic absorbers. *Thin Walled Struct.* **2005**, *43*, 1513–1533. [CrossRef]

20. Kumar, B.; Ranjan, V.; Azam, M.S.; Singh, P.P.; Mishra, P.; PriyaAjit, K.; Kumar, P. A comparison of vibro acoustic response of isotropic plate with attached discrete patches and point masses having different thickness variation with different taper ratios. *Shock Vib.* **2016**, *2016*, 8431431.

21. Lee, M.R.; Singh, R. Analytical formulations for annular disk sound radiation using structural modes. *J. Acoust. Soc. Am.* **1994**, *95*, 3311–3323. [CrossRef]

22. Rdzanek, W.J.; Rdzanek, W.P. The real acoustic power of a planar annular membrane radiation for axially-symmetric free vibrations. *Arch. Acoust.* **1997**, *4*, 455–462.

23. Doganli, M. Sound Power Radiation from Clamped-Clamped Annular Plates. Master's Thesis, Texas Tech University, Lubbock, TX, USA, 2000.

24. Nakayama, I.; Nakamura, A.; Takeuchi, R. Sound Radiation of a circular plate for a single sound pulse. *Acta Acust. United Acust.* **1980**, *46*, 330–340.

25. Hasegawa, T.; Yosioka, K. Acoustic radiation force on a solid elastic sphere. *J. Acoust. Soc. Am.* **1969**, *46*, 1139–1143. [CrossRef]

26. Lee, H.; Singh, R. Determination of sound radiation from a simplified disk brake rotor using a semi-analytical method. *Noise Control Eng. J.* **2000**, *52*. [CrossRef]

27. Squicciarini, G.; Thompson, D.J.; Corradi, R. The effect of different combinations of boundary conditions on the average radiation efficiency of rectangular plates. *J. Sound Vib.* **2014**, *333*, 3931–3948. [CrossRef]

28. Xie, G.; Thompson, D.J.; Jones, C.J.C. The radiation efficiency of baffled plates and strips. *J. Sound Vib.* **2005**, *280*, 181–209. [CrossRef]

29. Rayleigh, J.W. *The Theory of Sound*, 2nd ed.; Dover: New York, NY, USA, 1945.

30. Maidanik, G. Response of ribbed panels to reverberant acoustic fields. *J. Acoust. Soc. Am.* **1962**, *34*, 809–826. [CrossRef]

31. Heckl, M. Radiation from plane sound sources. *Acust.* **1977**, *37*, 155–166.

32. Williams, E.G. A series expansion of the acoustic power radiated from planar sources. *J. Acoust. Soc. Am.* **1983**, *73*, 1520–1524. [CrossRef]

33. Keltie, R.F.; Peng, H. The effects of modal coupling on the acoustic power radiation from panels. *J. Vib. Acoust. Stress Reliab. Des.* **1987**, *109*, 48–55. [CrossRef]

34. Snyder, S.D.; Tanaka, N. Calculating total acoustic power output using modal radiation efficiencies. *J. Acoust. Soc. Am.* **1995**, *97*, 1702–1709. [CrossRef]

35. Martini, A.; Troncossi, M.; Vincenzi, N. Structural and elastodynamic analysis of rotary transfer machines by Finite Element model. *J. Serb. Soc. Comput. Mech.* **2017**, *11*, 1–16. [CrossRef]

36. Croccolo, D.; Cavalli, O.; De Agostinis, M.; Fini, S.; Olmi, G.; Robusto, F.; Vincenzi, N. A Methodology for the Lightweight Design of Modern Transfer Machine Tools. *Machines* **2018**, *6*, 2. [CrossRef]

37. Martini, A.; Troncossi, M. Upgrade of an automated line for plastic cap manufacture based on experimental vibration analysis. *Case Stud. Mech. Syst. Signal Process.* **2016**, *3*, 28–33. [CrossRef]

38. Pavlovic, A.; Fragassa, C.; Ubertini, F.; Martini, A. Modal analysis and stiffness optimization: The case of a tool machine for ceramic tile surface finishing. *J. Serb. Soc. Comput. Mech.* **2016**, *10*, 30–44. [CrossRef]

Article

Time-Domain Hydro-Elastic Analysis of a SFT (Submerged Floating Tunnel) with Mooring Lines under Extreme Wave and Seismic Excitations

Chungkuk Jin * and Moo-Hyun Kim

Department of Ocean Engineering, Texas A&M University, Haynes Engineering Building, 727 Ross Street, College Station, TX 77843, USA; m-kim3@tamu.edu
* Correspondence: kenjin0519@gmail.com; Tel.: +1-979-204-3454

Received: 19 October 2018; Accepted: 22 November 2018; Published: 26 November 2018

Abstract: Global dynamic analysis of a 700-m-long SFT section considered in the South Sea of Korea is carried out for survival random wave and seismic excitations. To solve the tunnel-mooring coupled hydro-elastic responses, in-house time-domain-simulation computer program is developed. The hydro-elastic equation of motion for the tunnel and mooring is based on rod-theory-based finite element formulation with Galerkin method with fully coupled full matrix. The dummy-connection-mass method is devised to conveniently connect objects and mooring lines with linear and rotational springs. Hydrodynamic forces on a submerged floating tunnel (SFT) are evaluated by the modified Morison equation for a moving object so that the hydrodynamic forces by wave or seismic excitations can be computed at its instantaneous positions at every time step. In the case of seabed earthquake, both the dynamic effect transferred through mooring lines and the seawater-fluctuation-induced seaquake effect are considered. For validation purposes, the hydro-elastic analysis results by the developed numerical simulation code is compared with those by a commercial program, OrcaFlex, which shows excellent agreement between them. For the given design condition, extreme storm waves cause higher hydro-elastic responses and mooring tensions than those of the severe seismic case.

Keywords: submerged floating tunnel (SFT); mooring line; coupled dynamics; hydro-elastic responses; wet natural frequencies; mooring tension; seismic excitation; wave excitation; seaquake

1. Introduction

The submerged floating tunnel (SFT) is an innovative solution used to cross deep waterways [1,2]. The SFT consists mainly of a tunnel for vehicle transportation and mooring lines for station-keeping. The tunnel is usually positioned at a certain submergence depth, typically greater than 20 m, with positive net buoyancy that is balanced by mooring lines anchored in the seabed [3,4].

Considering that wave/current/wind effects are greatly reduced, the cost is almost constant along the length [5], and ship passage is not obstructed by the structure, the SFT has been regarded as a competitive alternative to floating bridges and immersed tunnels. In this regard, since Norway's first patent in 1923 [6], many proposals and case studies have been published worldwide, which includes Høgsfjord/Bjørnafjord in Norway [7–9], the Strait of Messina in Italy [10], Funka Bay in Japan [11,12], Qiandao Lake in China [13,14], and the Mokpo-Jeju SFT in Korea [15]. Even though there is no actually installed structure in the world despite extensive research [16], the first construction of the SFT is being considered by Norwegian Public Road Administration (NPRA) with global interest [17].

To provide sufficient confidence for the concept, feasibility studies under diverse catastrophic environmental conditions, such as extreme waves and earthquakes, must be extensively studied in advance. Along this line, numerous researches have been carried out to verify structural safety in wave and seismic excitations on the SFT. Regarding wave-excitation effects, Kunisu et al. [18] evaluated the effect of mooring-line configurations on SFT dynamic responses including possible snap loading. Lu et al. [11] and Hong et al. [19] focused on slack mooring phenomena at various buoyancy-weight ratios (BWRs) of the SFT and inclination angles of mooring lines. Long et al. [3] conducted parametric studies to investigate the effects of the BWR and mooring-line stiffness. Dynamic motions at varying BWRs and the corresponding comfort index were investigated by Long et al. [20]. Seo et al. [21] compared experimental results with simplified numerical approach for a segment of the SFT. Chen et al. [22] evaluated the influence of VIV (vortex induced vibration) of mooring lines on the SFT dynamic responses using a simplified numerical model. In addition, with regard to seismic-excitation effects, Di Pilato et al. [4] carried out a coupled dynamic analysis to investigate the effect of wave and seismic excitations. Martinelli et al. [13] suggested detailed procedures to generate artificial seismic excitations and performed the corresponding structural analysis. Dynamic responses at various shore connections under transverse earthquake were investigated by Xiao and Huang [23]. Martinelli et al. [24] and Wu et al. [25] focused on hydrodynamic fluid-structure interaction induced by vertical fluid fluctuations known as the seaquake. Mirzapour et al. [26] derived simplified analytical solutions for 2D and 3D cases and computed SFT dynamic responses in diverse stiffness conditions. Muhammad et al. [6] compared the dynamic effects induced by wave and seismic excitations.

During the past decade, various SFT-related studies have been carried out in the second author's research lab. Cifuentes et al. [27] compared the dynamics of a moored-SFT segment in regular waves for various BWRs and mooring types between experimental results and numerical simulations. For the numerical simulations, both commercial program (OrcaFlex) and in-house program CHARM3D (Coupled Hull And Riser Mooring 3D) were used for cross-checking. Lee et al. [16] further investigated the dynamics of the short tunnel segment under irregular waves and random seabed earthquakes. Then, the initial studies of hydro-elastic responses of a long SFT with many mooring lines by random waves and seabed earthquakes were conducted by Jin and Kim [28] and Jin et al. [29] by using commercial software, OrcaFlex. However, when using OrcaFlex for seismic excitations, an indirect modeling with many seabed dummy masses has to be introduced instead of direct inputs of dynamic boundary conditions at those anchor points.

In this research, to add the capability of hydro-elastic analyses of a long SFT with many mooring lines in the in-house coupled dynamic-analysis program, a new approach called 'dummy-connection-mass method' is developed. The equation of motion for the line element is derived by rod theory, and finite element modelling is implemented by using Galerkin method. Linear and rotational springs are employed to conveniently connect several objects with given connection conditions. The Adams–Moulton implicit integration method combined with the Adam–Bashforth explicit scheme, is used for the time-domain-integration method so that stable and time-efficient numerical integration can be done without iteration. The newly developed program is applied to calculate the hydro-elastic responses of a 700-m-long SFT (with both ends fixed) with many mooring lines by extreme random waves or severe random earthquakes. The results from the newly developed program are cross-checked against those from OrcaFlex program. In the case of seabed earthquake, the seabed motions are transferred to the SFT through mooring lines and through seawater fluctuations called seaquake, which is extensively discussed in Section 4 based on the produced numerical results. In the present study, the effect of seismic-induced acoustic pressure is not considered since the resulting frequency range is much higher [30], and thus it is of little importance for the mooring design.

2. Configuration of the System

Figure 1 shows 2D and 3D views of the entire structure, and Table 1 summarizes major design parameters of the tunnel and mooring lines. The tunnel, which has a diameter of 23 m and a length of 700 m, is made of high-density concrete. Since the structure in this study is a section of the 30-km-long SFT, the fixed-fixed boundary condition at both ends are applied assuming that strong fixtures (or towers) will be built at 700-m intervals, as shown in the Figure 1. Considering that the water depth of the planned site is 100 m, the submergence depth, a vertical distance between free surface and the tunnel centerline, is set to be 61.5 m. The BWR is fixed at 1.3, and the tunnel thickness is 2.3 m. The tunnel thickness is actually greater than the real value to have the equivalent tunnel bending stiffness including inner compartment structures. The axial and bending stiffnesses are calculated based on the given data of Table 1.

Chain mooring lines with a nominal diameter of 180 mm are used. High static and dynamic mooring tensions are expected based on the given BWR and wave condition [28]. In addition, the maximum mooring tension should be smaller than the MBL (minimum breaking load) divided by safety factor (SF). Thus, chain might be the best choice considering high MBL of 30,689 kN for Grade R5. As shown in Figure 1, four 60-degree-inclined mooring lines are installed for every 25 m interval toward the center locations. The lengths of mooring lines are 51.1 m for line #1 and #2 and 37.8 m for line #3 and #4. The wet natural frequencies of the tunnel hydro-elastic responses coupled with mooring lines are calculated and presented in Table 2.

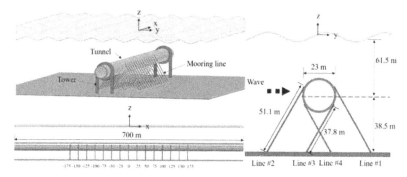

Figure 1. 2D and 3D views of the entire structure.

Table 1. Major parameters of the tunnel and mooring lines.

Component	Parameter	Value	Unit
Tunnel	Length	700	m
	Outer diameter	23	m
	End boundary condition	Fixed-fixed condition	-
	Material	High-density concrete	-
	Young's modulus	30	GPa
	Bending stiffness (EI)	2.34×10^{11}	kN·m²
	Axial stiffness (EA)	4.27×10^9	kN
	Buoyancy-weight ratio (BWR)	1.3	-
	Added mass coefficient	1.0	-
	Drag coefficient	0.55 [31]	-
Mooring lines (Chain, Stud-less type)	Length	51.1 (Line # 1 and 2), 37.8 (Line # 3 and 4)	m
	Mass/unit length	644.7	kg/m
	Nominal diameter (d) for wave drag force calculation	0.18	m
	Equivalent outer diameter (D_E) for wave inertia force calculation	0.324 ($D_E = 1.8d$)	m
	Bending stiffness (EI)	0	kN·m²
	Axial stiffness (EA)	2.77×10^6	kN
	Added mass coefficient	1.0	
	Drag coefficient	2.4 [32]	
	Minimum breaking load (MBL)	30,689 (Grade R5) [33]	kN

Table 2. Wet natural frequencies of the tunnel hydro-elastic responses coupled with mooring lines.

Component	Wet Natural Frequency (rad/s)	Mode Number
Tunnel (Horizontal direction)	1.92	1st mode
	2.70	2nd mode
	4.53	3rd mode
Tunnel (Vertical direction)	3.12	1st mode
	3.45	2nd mode
	4.89	3rd mode
Mooring lines #1 and #2 (Center)	5.78	1st mode
Mooring lines #3 and #4 (Center)	9.04	1st mode

3. Numerical Model

Tunnel-mooring coupled dynamic analysis was conducted by using the in-house program, CHARM3D. This in-house code has been developed by second author's research lab for the coupled dynamic simulations of complex offshore structures with mooring lines and risers during the past two decades [34,35]. In addition, the capability has further been expanded for various applications including multiple bodies connected by lines, wind turbines [36], dynamic positioning [37], and ice-structure interactions [38]. The program is further extended in this paper to study the SFT hydro-elastic dynamics for seismic excitations. In addition, some of the computed results are compared with those by widely-used commercial program OrcaFlex for cross-checking. In the following equations, bold variables represent vectors or matrices.

3.1. Governing Equations of Dynamic Simulation

The entire structure is modelled by rod elements and the rod theory suggested by Garrett [39] is used. The behavior of the rod element is determined by the position of the rod centerline. The equation of motion is solved in general coordinate whose tangential direction follows the line profile; therefore, coordinate transformations, which increase computation time, are not required. In addition, geometric non-linearity is considered without specific assumptions associated with the shape or orientation of lines [34]. The equation of motion and the extensible condition are presented in Equations (1) and (2).

$$- (EI\mathbf{r}'')'' + (\lambda \mathbf{r}')' + \mathbf{q} = m\ddot{\mathbf{r}} \tag{1}$$

$$\frac{1}{2}(\mathbf{r}' \cdot \mathbf{r}' - 1) = \frac{T}{A_I E} \approx \frac{\lambda}{A_I E} \tag{2}$$

where $\mathbf{r}(s,t)$ is a position vector, which is a function of arc length s and time t in order to define space curve, E is Young's modulus, I is second moment of sectional area, λ is Lagrange multiplier, $\mathbf{q}$ and m are the distributed load and mass per unit length, T is the tension, and A_I is the cross sectional area filled with the material. In addition, dot and apostrophe denote time and spatial derivatives, respectively. The distributed load includes the weight of the rod and hydrostatic and hydrodynamic loads induced by the surrounding fluid. The hydrostatic load is subdivided into buoyancy and force induced by hydrostatic pressure. The hydrodynamic force is estimated by Morison equation for moving objects, which consists of linear wave inertia and nonlinear wave drag forces. Thus, Morison equation, which is given by Equation (3), enables to compute wave force per unit length at instantaneous rod-element positions at each time step.

$$\mathbf{F_d} = -C_A \rho A_E \ddot{\mathbf{r}}^n + C_M \rho A_E \dot{\mathbf{V}}^n + \frac{1}{2} C_D \rho D \left| \mathbf{V}^n - \dot{\mathbf{r}}^n \right| (\mathbf{V}^n - \dot{\mathbf{r}}^n) \tag{3}$$

where C_M, C_A, and C_D are the inertia, added mass, and drag coefficients, ρ is density of water, and A_E is the cross-sectional area for the element, D is the outer diameter, and $\mathbf{V}^n$ and $\dot{\mathbf{V}}^n$ represent velocity and acceleration of a fluid particle normal to the rod centerline. The inertia coefficient of the tunnel and mooring lines is 2.0 considering that the added mass is the same as displaced mass [40]. The drag coefficient of the tunnel is a function of Reynolds number, KC (Keulegan-Carpenter) number, and relative surface roughness, and the representative value of 0.55 is used here based on the experimental results (e.g., [31]). The drag coefficient of mooring lines is 2.4 for stud-less chain [32]. It was shown in Cifuentes et al. [27] that the use of Morison equation for SFT dynamics is good enough compared to the case by using 3D diffraction/radiation panel program. Here, the Morison equation is further modified to include hydrodynamic force induced by vertical pressure variations during earthquake excitations i.e., the seaquake effect, as supported by Islam and Ahrnad [41], Martinelli et al. [24], Mousavi et al. [42], and Wu et al. [25]. In the equation, inertia and drag force terms are modified by introducing seismic velocity $\mathbf{v}_g^n$ and acceleration $\dot{\mathbf{v}}_g^n$ as shown in Equation (4). The vertical component of seismic velocity and acceleration is considered only for the seaquake simulations.

$$\mathbf{F_d} = -C_A\rho A_E \ddot{\mathbf{r}}^n + C_M\rho A_E(\dot{\mathbf{V}}^n + \dot{\mathbf{v}}_g^n) + \frac{1}{2}C_D\rho D\left|\mathbf{V}^n + \mathbf{v}_g^n - \dot{\mathbf{r}}^n\right|(\mathbf{V}^n + \mathbf{v}_g^n - \dot{\mathbf{r}}^n) \tag{4}$$

Therefore, the final form of the equation of motion is given by Equations (5)–(9):

$$m\ddot{\mathbf{r}} + C_A\rho A_E \ddot{\mathbf{r}}^n + (EI\mathbf{r}'')'' - (\tilde{\lambda}\mathbf{r}')' = \tilde{\mathbf{w}} + \tilde{\mathbf{F}}_d \tag{5}$$

$$\tilde{\mathbf{F}}_d = C_M\rho A_E(\dot{\mathbf{V}}^n + \dot{\mathbf{v}}_g^n) + \frac{1}{2}C_D\rho D\left|\mathbf{V}^n + \mathbf{v}_g^n - \dot{\mathbf{r}}^n\right|(\mathbf{V}^n + \mathbf{v}_g^n - \dot{\mathbf{r}}^n) \tag{6}$$

$$\tilde{\lambda} = \tilde{T} - EI\kappa^2 \tag{7}$$

$$\tilde{\mathbf{w}} = \mathbf{w} + \mathbf{B} \tag{8}$$

$$\tilde{T} = T + P \tag{9}$$

where κ is local curvature, $\tilde{\mathbf{w}}$ is wet weight of the rod per unit length, which is comprised of weight $\mathbf{w}$ and buoyancy $\mathbf{B}$, $\tilde{T}$ is effective tension in the rod, and P is the hydrostatic pressure, which is a scalar, at the position $\mathbf{r}$ on the rod. Therefore, Equation (5) combined with the stretching condition given in Equation (2) are the governing equations for dynamic simulations.

The governing equations are further formulated by Galerkin finite element method [39,43]. The position vector and Lagrange multiplier for a single element of the length L are expressed as follows:

$$\mathbf{r}(s,t) = \sum_m A_m(s)\mathbf{U}_m(t) \tag{10}$$

$$\lambda(s,t) = \sum_n P_n(s)\lambda_n(t) \tag{11}$$

where A_m and P_n are shape functions defined on the interval $0 \leq s \leq L$. The weak form of the governing equation is generated by using the Galerkin method and integration by part:

$$\int_0^L \left[A_m(m\ddot{\mathbf{r}} + C_A\rho A_E \ddot{\mathbf{r}}^n) + EIA''_m\mathbf{r}'' + \tilde{\lambda}A'_m\mathbf{r}' - A_m(\tilde{\mathbf{w}} + \tilde{\mathbf{F}}^d)\right]ds = EI\mathbf{r}''A'_m\Big|_0^L + \left[\tilde{\lambda}\mathbf{r}' - (EI\mathbf{r}'')'\right]A_m\Big|_0^L \tag{12}$$

$$\int_0^L P_n\left\{\frac{1}{2}(\mathbf{r}' \cdot \mathbf{r}' - 1) - \frac{\lambda}{A_IE}\right\}ds = 0 \tag{13}$$

where first and second terms of the right-hand side in Equation (12) are related to moment and force at the boundary. Cubic and quadratic shape functions, which are continuous on the element, are defined for the position vector and Lagrange multiplier, respectively:

$$A_1 = 1 - 3\xi^2 + 2\xi^3, \quad A_2 = L(\xi - 2\xi^2 + \xi^3),$$
$$A_3 = 3\xi^2 - 2\xi^3, \quad A_4 = L(-\xi^2 + \xi^3),$$
$$P_1 = 1 - 3\xi + 2\xi^2, \quad P_2 = 4\xi(1 - \xi),$$
$$P_3 = \xi(2\xi - 1)$$

$$(14)$$

where $\xi = s/L$. The position vector, tangent of the position vector, and Lagrange multiplier are chosen to be continuous at the node between the neighboring elements. Therefore, the parameters $\mathbf{U}_m$ and λ_n can be written as:

$$\mathbf{U}_1 = \mathbf{r}(0,t), \quad \mathbf{U}_2 = \mathbf{r}'(0,t),$$
$$\mathbf{U}_3 = \mathbf{r}(L,t), \quad \mathbf{U}_4 = \mathbf{r}'(L,t),$$
$$\lambda_1 = \lambda(0,t), \quad \lambda_2 = \lambda(L/2,t), \quad \lambda_3 = \lambda(L,t)$$

$$(15)$$

The position and its tangent vectors are obtained at both ends of the element while the Lagrange multiplier are computed at both ends and the middle point of the element. The final finite element formulation of the governing equation for the 3 dimensional problem are presented in Equation (16).

$$(M_{ijlk} + M^a_{ijlk})\ddot{\mathbf{U}}_{jk} + (K^1_{ijlk} + \lambda_n K^2_{nijlk})\mathbf{U}_{jk} = \mathbf{F}_{il}$$

$$(16)$$

For $\mathbf{U}_{jk}$, subscript j is dimension, which is 1–3 for the 3 dimensional problem, and subscript k is for 1–4 given in Equation (15). In Equations (17)–(21), the general mass, the added mass, the general stiffness from the bending stiffness and rod tension, and external force matrices are defined with Kronecker Delta function δ_{ij}:

$$M_{ijlk} = \int_0^L m A_l A_k \delta_{ij} ds$$

$$(17)$$

$$M^a_{ijlk} = C_A \rho A_E \left[\int_0^L A_l A_k \delta_{ij} ds - (\int_0^L A_l A_k A'_s A'_t ds) \mathbf{U}_{it} \mathbf{U}_{js} \right]$$

$$(18)$$

$$K^1_{ijlk} = \int_0^L EI A''_l A''_k \delta_{ij} ds$$

$$(19)$$

$$K^2_{nijlk} = \int_0^L P_n A'_l A'_k \delta_{ij} ds$$

$$(20)$$

$$\mathbf{F}_{il} = \int_0^L (\tilde{\mathbf{w}}_i + \tilde{\mathbf{F}}^d_i) A_l ds$$

$$(21)$$

In addition, the stretching condition can be formulated as given in Equation (22):

$$G_m = A_{mil} \mathbf{U}_{kl} \mathbf{U}_{ki} - B_m - C_{mt} \lambda_t$$

$$(22)$$

where

$$A_{mil} = \frac{1}{2} \int_0^L P_m A'_i A'_l ds$$

$$(23)$$

$$B_m = \frac{1}{2} \int_0^L P_m ds$$

$$(24)$$

$$C_{mt} = \frac{1}{A_l E} \int_0^L P_m P_t ds$$

$$(25)$$

A dummy 6 DOF rigid body, which is equipped with negligible properties, is introduced to conveniently connect the tunnel and mooring lines. The dummy mass means negligible mass (1 kg in proto type) of dummy rigid body used only for connection purpose. Therefore, force and moment are transferred by using both linear and rotational springs of very large stiffness from the tunnel and mooring lines through the rigid body. Force and moment transmitted from the mooring line to the rigid body are computed as follows [43]:

$$\mathbf{F}_P = \tilde{K}(\tilde{T}_P\tilde{\mathbf{u}}_P - \tilde{\mathbf{u}}_I) + \tilde{C}(\tilde{T}_P\dot{\tilde{\mathbf{u}}}_P - \dot{\tilde{\mathbf{u}}}_I) \qquad (26)$$

where $\tilde{K}$ and $\tilde{C}$ represent coupling stiffness and damping matrices, $\tilde{T}_P$ denotes a transformation matrix between the rigid body origin and the connection location, $\tilde{\mathbf{u}}_P$ and $\tilde{\mathbf{u}}_I$ are the displacements of the rigid body and the connecting location. Infinite stiffness values are used in the coupling stiffness matrix to tightly connect lines, and damping matrix is not utilized in the simulations. Therefore, the entire stiffness matrix that couples tunnel elements with mooring lines is created as shown in Figure 2.

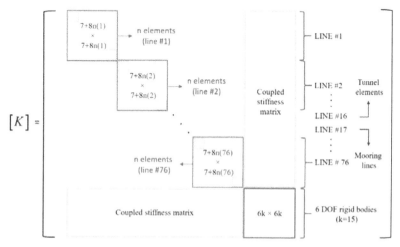

Figure 2. Stiffness matrix for the simulated SFT (line #1~#16 are for tunnel and line #17~#N are for mooring lines, n(1) means number of sub-elements of line #1, k is the number of the 6 DOF rigid body).

Newton's iteration method is used in static analysis of the SFT. The Adams–Moulton implicit integration method, which has 2nd-degree of accuracy, is used for the time-domain-integration method. Since instantaneous velocity and acceleration are required to calculate hydrodynamic force from Morison equation, the Adam–Bashforth explicit scheme is combined with the Adams–Moulton implicit scheme to avoid iteration.

3.2. Theory of OrcaFlex

A similar approach is used to model the whole structure in OrcaFlex, a well-known commercial program. The tunnel and mooring lines are modelled by line elements, and the line-element theory is based on the lumped mass method. The line element consists of a series of nodes and segments. Force properties are lumped in the node, which includes weight, buoyancy, and drag etc. Stiffness components i.e., axial, bending, and torsional stiffness, are represented by massless springs [44]. The equation of motion is expressed in Equation (27).

$$\mathbf{M}(\mathbf{p},\mathbf{a}) + \mathbf{C}(\mathbf{p},\mathbf{v}) + \mathbf{K}(\mathbf{p}) = \mathbf{F}(\mathbf{p},\mathbf{v},t) \qquad (27)$$

where $\mathbf{M}(\mathbf{p},\mathbf{a})$, $\mathbf{C}(\mathbf{p},\mathbf{v})$, and $\mathbf{K}(\mathbf{p})$ are mass, damping, and stiffness matrices. $\mathbf{F}(\mathbf{p},\mathbf{v},t)$ is external force vector, which is hydrodynamic force in this case. Symbols, $\mathbf{p}$, $\mathbf{v}$, $\mathbf{a}$, and t denote position, velocity, acceleration vectors, and time, respectively. Hydrodynamic force is also computed by the same Morison equation for a moving object with consideration for relative velocity and acceleration. The advantage of the developed program compared to OrcaFlex for the present application can be summarized as follows: (i) In OrcaFlex, hydrodynamic force generated from the seaquake effects is not included. (ii) In CHARM3D, higher-order rod FE elements are used compared to lumped-mass-based OrcaFlex. (iii) The seabed movements can be directly imputed in the developed program.

3.3. Environmental Conditions

Simultaneous random-wave and seismic excitations are considered for global performance analysis. The same wave and seismic time histories are inputted in both programs for cross-checking. JONSWAP wave spectrum is used to generate time histories of random waves. Significant wave height and peak period for the 100-year-storm condition are 11.7 m and 13.0 s. Enhancement parameter is 2.14 that is the average value in Korea [45]. Random waves are generated by superposing 100 component waves with randomly perturbed frequency intervals to avoid signal repetition. The lowest and highest cut-off frequencies of input spectrum is 0.3 rad/s and 2.3 rad/s, respectively. The wave direction is perpendicular to a longitudinal direction of the tunnel. A 3-h simulation is carried out to analyze the statistics of dynamic behaviors and mooring tensions under the storm condition. Figure 3 shows theoretical JONSWAP wave spectrum and the reproduced spectrum from the time histories of wave elevation. It also shows the time histories of wave elevation produced by the JONSWAP wave spectrum.

Regular (sinusoidal) and recorded irregular seismic excitations data are also employed. The amplitude of regular seismic motion in the vertical direction is 0.01 m at diverse frequencies from 0.781 rad/s to 7.805 rad/s. Figure 4 shows the time histories of seismic displacements and corresponding spectra for recorded irregular seismic excitations in three directions, which are obtained by USGS [46]. The earthquake occurred in 78 km WNW of Ferndale, California, USA in 2014, and the magnitude of this earthquake is 6.8 in Richter scale. Seismic displacements in three directions are inputted for each anchor point of mooring lines and two ends of the tunnel fixture at every time step. Hydrodynamic force from the seaquake effect is also computed for the tunnel and mooring lines.

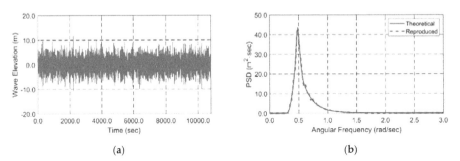

Figure 3. Wave time histories produced by JONSWAP wave spectrum (**a**) and theoretical JONSWAP wave spectrum and reproduced spectrum from wave time histories using FFT (fast Fourier transform) for validation (**b**).

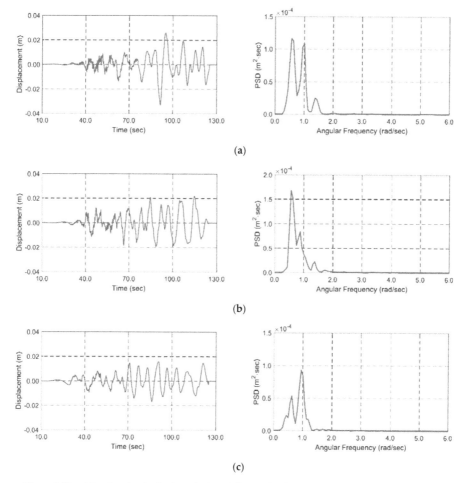

Figure 4. Time histories of real seismic excitations in longitudinal = x (**a**), transverse = y (**b**), and vertical = z (**c**) directions and its corresponding spectra.

4. Results and Discussions

4.1. Static Analysis

The developed code is first cross-checked with OrcaFlex in the static condition before dynamic simulations. Because static displacements of the tunnel are only affected by weight, buoyancy, and stiffness components of tunnel and mooring lines, direct comparison can be made after initial modeling of the entire SFT system. Figure 5 shows the vertical displacements of tunnel and mooring tension in the static condition. The results produced by the developed program coincide well with OrcaFelx's results. The reference dashed line in the tension figure indicates the allowable tension (minimum break load divided by safety factor).

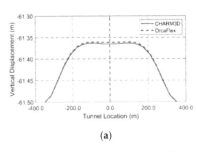
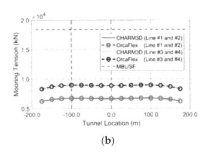

(a)   (b)

Figure 5. Submerged floating tunnel (SFT) vertical displacement (**a**) and mooring tension (**b**) in the static condition.

4.2. Dynamic Behaviors under Extreme Wave Excitations

Dynamic simulations under the 100-year-strom condition (Hs = 11.7 m and Tp = 13.0 s) are performed for three hours. As mentioned before, the same wave time histories are inputted to both programs to directly compare the dynamics results. Both computer programs produce almost identical results. Figure 6 shows the envelopes of maximum and minimum for SFT displacements and mooring tension. The maximum horizontal and vertical responses and mooring tension occur in the middle location. The horizontal responses are larger than the vertical responses since the 1st natural frequency of horizontal motion is closer to the input wave spectrum than that of vertical motion. Mooring-tension results show that shorter mooring lines (Line #3) have higher mooring tension than longer mooring lines (Line #1). The maximum mooring tension at the middle section is smaller than the MBL (minimum breaking load) divided by the SF (safety factor), which is presented in Figure 6b as a pink line. Recall that the MBL is 30,689 kN for Grade R5, which is obtained by DNV regulation [33]. The SF 1.67 is used as recommended by API RP 2SK [47]. Even if the extreme 100-year-storm condition is considered, the maximum mooring tension is still smaller than the allowable tension.

Figures 7–9 show the time histories and corresponding spectra of horizontal/vertical responses of the tunnel and mooring tension at the middle section. The spectra of responses indicate that wave-induced motions are dominant since the lowest natural frequencies in both directional motions (1.92 and 3.12 rad/s for horizontal and vertical directions) are away from the dominant input-wave spectral range of Figure 3. It means that there is negligible contribution from the structural elastic resonances. In case of mooring tension, under the given BWR = 1.3, snap-loadings characterized by extraordinary high peaks do not occur, as shown in the time series. However, it should be noted that the snap-loadings tend to occur at lower BWRs [28]. Obviously, smaller dynamic motions and mooring tensions can be obtained by further increasing submergence depth [28]. The relevant statistics obtained from the time series are summarized in Table 3.

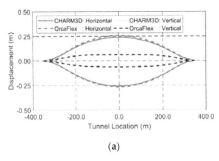
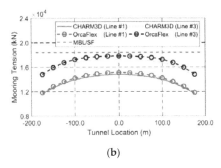

(a)   (b)

Figure 6. Envelopes of the maximum and minimum displacements of the tunnel (**a**) and mooring tension (**b**) in the 100-year-strom condition.

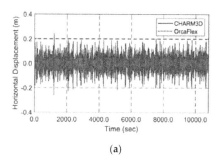
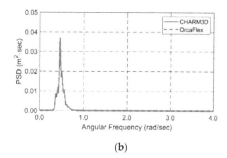

(a)   (b)

Figure 7. Time histories (**a**) and spectrum (**b**) of horizontal displacement of the tunnel in the middle location under the 100-year-strom waves.

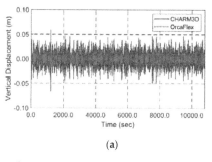
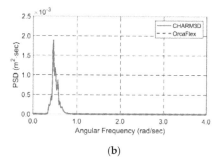

(a)   (b)

Figure 8. Time histories (**a**) and spectrum (**b**) of vertical displacement of the tunnel in the middle location under the 100-year-strom waves.

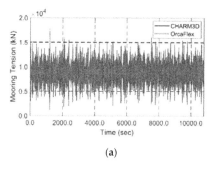
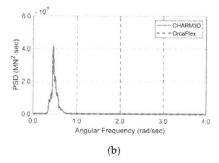

(a)   (b)

Figure 9. Time histories (**a**) and spectrum (**b**) of mooring tension (#3) in the middle location under the 100-year-strom waves.

Table 3. Statistics of the SFT motions and mooring tensions at the middle location under 100-yr irregular wave excitations (from the time series of Figures 7–9).

Parameter	Maximum	Minimum	Standard Deviation	Unit
Horizontal displacement	0.243	−0.261	0.059	
Vertical displacement	0.058	−0.066	0.014	m
Mooring tension (line #1)	14,765.75	885.56	1917.55	
Mooring tension (line #2)	15,276.12	902.94	1919.01	
Mooring tension (line #3)	17,334.93	1206.64	2015.53	kN
Mooring tension (line #4)	16,542.11	953.24	2014.32	

4.3. Dynamic Behaviors under Severe Seismic Excitations

Regular and irregular seismic excitations are utilized for SFT dynamic analysis. Since the fixed–fixed boundary condition is applied at both ends of the tunnel, both ends as well as all anchoring points are assumed to move together with seismic motions. As a result, seismic time histories are inputted to every anchor location of mooring lines and both ends of the tunnel. The hydrodynamic forces generated by sea-water fluctuations under vertical seismic motions are computed by using modified Morison equation (e.g., Islam and Ahrnad [41], Martinelli et al. [24], Mousavi et al. [42], and Wu et al. [25]). The effect is well known and called seaquake. As a result, there are two mechanisms causing SFT dynamics under seabed seismic motions. First, the seismic motions are transferred through mooring lines. Second, sea-water fluctuations in the vertical direction. In this paper, the former will be called earthquake effect and the latter will be called seaquake effect. To investigate the seaquake effect, regular seismic cases only in the vertical direction are simulated and the resulting SFT dynamics are analyzed. Subsequently, strong real seismic displacements are applied to the SFT system to check the global performance and structural robustness.

Figure 10 shows tunnel's vertical motion amplitudes at the mid-section and the corresponding vertical responses of mooring line #1 at its center under regular (sinusoidal) seismic excitations. Vertical motions of tunnel are largely amplified at 3.12 rad/s and 4.89 rad/s, the 1st and 3rd natural frequencies. The amplified tunnel motions at those frequencies directly influence high mooring dynamics, as shown in Figure 10b. A small peak can also be observed at 5.78 rad/s, the lowest natural frequency of mooring lines #1 itself.

The hydrodynamic force by seaquake directly acts on the tunnel with earthquake frequencies. Whereas, the seismic excitations are delivered to the tunnel through mooring lines, as discussed earlier. Then, the resulting tunnel response also causes hydrodynamic force on the tunnel. Therefore, there exist phase effects between the two components. We can see that the tunnel dynamics are significantly reduced after including the seaquake effect when compared to the earthquake-only case. The reason can be found from Figure 11 by plotting the contribution of each constituent component separately. In the figure, the phase of the tunnel response induced by earthquake is opposite to that induced by seaquake at the tunnel's natural frequencies, 3.12 rad/s and 4.89 rad/s. Therefore, there is cancellation effect between the two components so that the total vertical response amplitude can be reduced compared to the earthquake-only case. On the other hand, when earthquake frequency is greater than 5.7 rad/s, the two components become in phase, so the tunnel vertical responses are increased compared to the earthquake-only case although the resulting increment is small. The seaquake effects are not generated by the horizontal seismic motions if the seabed is flat since the horizontal seabed motions do not influence seawater fluctuating motions.

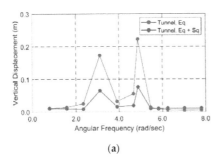
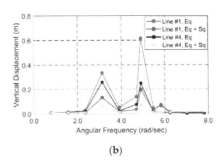

(a)   (b)

Figure 10. Amplitudes of vertical displacements of the tunnel (a) and mooring line #1 (b) at the middle location under regular seismic excitations of various frequencies (Eq: earthquake only considered; Eq + Sq: both earthquake and seaquake considered).

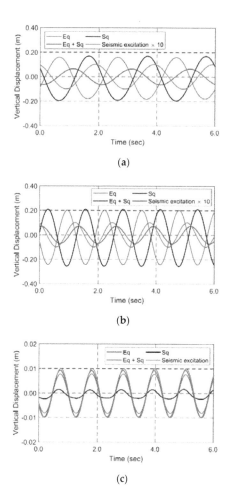

(a)

(b)

(c)

Figure 11. Time histories of vertical displacements of the tunnel at the middle section by respective force components under regular seismic excitations of 3.12 rad/s (**a**), 4.89 rad/s (**b**), and 5.78 rad/s (**c**) (Eq: earthquake only considered; Eq + Sq: both earthquake and seaquake considered; Sq: seaquake only considered; time histories of seismic excitations are multiplied by 10 for better visualization).

Figures 12–14 show the time histories of horizontal/vertical responses of the tunnel and the corresponding mooring tensions at the tunnel's middle section under the real seismic excitations, as given in Figure 4. The case of earthquake effect only is compared with that of earthquake plus seaquake. Firstly, in the earthquake-only case, the tunnel responses are greater than the input seismic motions, horizontally about 3 times and vertically about twice larger. The horizontal responses are more amplified because its lowest natural frequency is closer to the dominant frequency range of seismic excitations than that of vertical response. The corresponding tunnel-response spectra show that they have the first small peak at the seismic frequency, the next highest peak at the lowest natural frequency, and the next small peak at the third-lowest natural frequency. Mooring tensions are mostly influenced by the SFT horizontal and vertical motions at their lowest natural frequencies, while there is virtually little contribution near seismic frequencies. The maximum tensions for this earthquake case are much smaller than those caused by extreme wave excitations, as previously considered. However, the earthquake-induced tunnel dynamics can be significantly more amplified when the lowest natural frequencies of the tunnel's elastic responses are closer to dominant seismic frequencies. In the figure, the same dynamic simulation results by OrcaFlex are also given for cross-checking. The two independent computer programs produced almost identical results.

In the spectral plots of Figures 12–14, the spectra of tunnel responses and mooring tensions after adding seaquake effects are also given. In Figure 12, there is little change in the case of SFT horizontal motions since the seaquake mainly influences only the vertical responses, as was pointed out earlier. In Figure 13, there is a big reduction in the vertical-response spectrum at its lowest natural frequency (3.12 rad/s) after including the seaquake effect. It is due to the phase-cancellation effects, as discussed in the previous regular-earthquake case of Figure 11a,b. So, this reduction effect directly reflects the reduction in tension i.e., in Figure 14, the tension spectral amplitude is greatly reduced near 3.12 rad/s but remains the same at the lowest natural frequency of horizontal response, 1.92 rad/s. This trend can also be seen in the corresponding time-series comparisons (Figure 15) for the two cases (with and without considering the seaquake effect) regarding vertical tunnel responses and mooring tensions. The relevant statistics obtained from the time series are summarized in Table 4. It is seen that the inclusion of seaquake effect reduces both vertical SFT responses and mooring tensions, as discussed earlier.

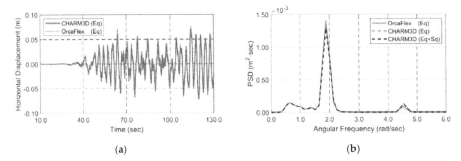

Figure 12. Time histories (without seaquake) (**a**) and spectra (**b**) of horizontal tunnel responses at the middle location under seismic excitations

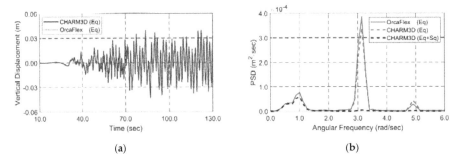

Figure 13. Time histories (without seaquake) (**a**) and spectra (**b**) of vertical tunnel responses at the middle location under seismic excitations.

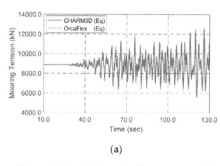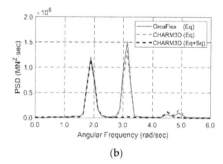

(a)   (b)

Figure 14. Time histories (without seaquake) (a) and spectra (b) of mooring tension #4 at the middle location under seismic excitations

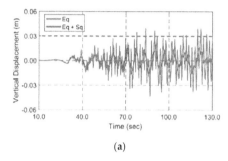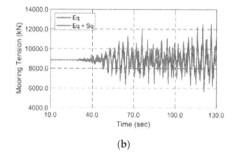

(a)   (b)

Figure 15. Time histories of vertical responses of the tunnel (a) and mooring tension #4 (b) in the middle location under seismic excitations with and without seaquake effect.

Table 4. Statistics of the SFT motions and mooring tensions at the middle location under irregular seismic excitations (Eq: earthquake, Sq: seaquake).

Parameter	Numerical Model	Maximum	Minimum	Standard Deviation	Unit
Horizontal displacement	Eq	0.070	−0.073	0.023	
	Eq + Sq	0.070	−0.072	0.023	m
Vertical displacement	Eq	0.039	−0.042	0.013	
	Eq + Sq	0.019	−0.019	0.006	
Mooring tension (line #1)	Eq	9872.22	3783.56	801.58	
	Eq + Sq	8728.01	4631.84	649.80	
Mooring tension (line #2)	Eq	9722.09	3291.79	829.07	
	Eq + Sq	8778.38	4491.79	649.33	kN
Mooring tension (line #3)	Eq	12,295.18	4918.66	958.46	
	Eq + Sq	11,009.40	6691.91	649.58	
Mooring tension (line #4)	Eq	12,512.53	5633.35	925.67	
	Eq + Sq	11,001.66	6772.99	652.55	

5. Conclusions

Global performance analysis of the SFT was carried out for survival random wave and seismic excitations. To solve tunnel-mooring coupled hydro-elastic responses, an in-house time-domain-simulation computer program was developed. The hydro-elastic equation of motion for the tunnel and mooring was based on rod-theory-based finite element formulation with Galerkin method. The dummy-connection-mass method was devised to conveniently connect multiple segmented objects and mooring lines with linear and rotational springs. Considering the slender shape of the structure, hydrodynamic forces were computed by the modified Morison equation. The numerical results produced by the developed program were in good agreement with those by the commercial program

OrcaFlex based on lumped-mass method. The extreme wave excitations caused the maximum SFT dynamic motions of 24 cm and 6 cm in the horizontal and vertical directions and the corresponding mooring tensions below allowable level. Snap motions and loadings of mooring lines were not observed. Under regular seismic excitations, large resonant responses of the tunnel were observed at 1st and 3rd natural frequencies. In the case of seabed earthquake, the seabed motions are transferred to SFT through mooring lines and through seawater fluctuations called seaquake. When the latter is further considered, horizontal responses were not affected but vertical responses become significantly reduced especially at its lowest natural frequency. After analyzing the behaviors of the two contributions, it was found that the reduction was caused by the phase-cancellation effect. However, in other cases, the phases could enhance each other to increase the total responses of the SFT. Under extreme irregular seismic excitations, the maximum SFT dynamic motions of 7 cm and 2 cm were generated and the corresponding mooring tensions were about 30% smaller compared to the extreme wave case. However, when the frequencies of seismic excitations are closer to SFT natural frequencies, larger dynamic amplifications are expected.

Author Contributions: All authors have equally contributed to publish this article related to design of target model, validation of numerical modeling, simulations, analysis, and writing.

Funding: This work was supported by the National Research Foundation of Korea (NRF) grant funded by the Korea government (MSIT) (No. 2017R1A5A1014883).

Conflicts of Interest: The authors declare no conflict of interest.

References

1. Ge, F.; Lu, W.; Wu, X.; Hong, Y. Fluid-structure interaction of submerged floating tunnel in wave field. *Procedia Eng.* **2010**, *4*, 263–271. [CrossRef]

2. Paik, I.Y.; Oh, C.K.; Kwon, J.S.; Chang, S.P. Analysis of wave force induced dynamic response of submerged floating tunnel. *KSCE J. Civ. Eng.* **2004**, *8*, 543–550. [CrossRef]

3. Long, X.; Ge, F.; Wang, L.; Hong, Y. Effects of fundamental structure parameters on dynamic responses of submerged floating tunnel under hydrodynamic loads. *Acta Mech. Sin.* **2009**, *25*, 335–344. [CrossRef]

4. Di Pilato, M.; Perotti, F.; Fogazzi, P. 3D dynamic response of submerged floating tunnels under seismic and hydrodynamic excitation. *Eng. Struct.* **2008**, *30*, 268–281. [CrossRef]

5. Faggiano, B.; Landolfo, R.; Mazzolani, F. *The sft: An innovative Solution for waterway Strait Crossings*; IABSE Symposium Report; International Association for Bridge and Structural Engineering. Lisbon, Portugal, 2005; pp. 36–42.

6. Muhammad, N.; Ullah, Z.; Choi, D.-H. Performance evaluation of submerged floating tunnel subjected to hydrodynamic and seismic excitations. *Applied Sci.* **2017**, *7*, 1122. [CrossRef]

7. Skorpa, L. Innovative norwegian fjord crossing. How to cross the høgsjord, alternative methods. In Proceedings of the 2nd Congress AIOM (Marine and Offshore Engineering Association), Naples, Italy, 15–17 November 1989; pp. 15–17.

8. Remseth, S.; Leira, B.J.; Okstad, K.M.; Mathisen, K.M.; Haukås, T. Dynamic response and fluid/structure interaction of submerged floating tunnels. *Comput. Struct.* **1999**, *72*, 659–685. [CrossRef]

9. Engebretsen, K.B.; Jakobsen, K.K.; Haugerud, S.A.; Minoretti, A. A submerged floating tube bridge concept for the bjørnafjord crossing: Marine operations. In Proceedings of the ASME 2017 36th International Conference on Ocean, Offshore and Arctic Engineering, Trondheim, Norway, 25–30 June 2017; American Society of Mechanical Engineers: New York, NY, USA, 2017; p. V07BT06A027.

10. Faggiano, B.; Landolfo, R.; Mazzolani, F. Design and modelling aspects concerning the submerged floating tunnels: An application to the messina strait crossing. *Krobeborg. Strait Crossing* **2001**, 511–519.

11. Lu, W.; Ge, F.; Wang, L.; Wu, X.; Hong, Y. On the slack phenomena and snap force in tethers of submerged floating tunnels under wave conditions. *Mar. Struct.* **2011**, *24*, 358–376. [CrossRef]

12. Fujii, T. Submerged floating tunnels project in funka bay design and execution. In Proceedings of the International Conference on Submerged Floating Tunnel, Sandnes, Norway, 29–30 May 1996.

13. Martinelli, L.; Barbella, G.; Feriani, A. A numerical procedure for simulating the multi-support seismic response of submerged floating tunnels anchored by cables. *Eng. Struct.* **2011**, *33*, 2850–2860. [CrossRef]

14. Mazzolani, F.; Landolfo, R.; Faggiano, B.; Esposto, M.; Perotti, F.; Barbella, G. Structural analyses of the submerged floating tunnel prototype in qiandao lake (pr of china). *Adv. Struct. Eng.* **2008**, *11*, 439–454. [CrossRef]

15. Han, J.S.; Won, B.; Park, W.-S.; Ko, J.H. Transient response analysis by model order reduction of a mokpo-jeju submerged floating tunnel under seismic excitations. *Struct. Eng. Mech.* **2016**, *57*, 921–936. [CrossRef]

16. Lee, J.; Jin, C.; Kim, M. Dynamic response analysis of submerged floating tunnels by wave and seismic excitations. *Ocean Syst. Eng. Int. J.* **2017**, *7*, 1–19. [CrossRef]

17. Ghimire, A.; Prakash, O. Intangible study for the design and construction of submerged floating tunnel. *Imp. J. Interdiscip. Res.* **2017**, *3*, 721–724.

18. Kunisu, H.; Mizuno, S.; Mizuno, Y.; Saeki, H. Study on submerged floating tunnel characteristics under the wave condition. In Proceedings of the The Fourth International Offshore and Polar Engineering Conference, Osaka, Japan, 10–15 April 1994; International Society of Offshore and Polar Engineers: Osaka, Japan, 1994; pp. 27–32.

19. Hong, Y.; Ge, F.; Lu, W. On the two essential concepts for sft: Synergetic buoyancy-weight ratio and slack-taut map. *Procedia Eng.* **2016**, *166*, 221–228. [CrossRef]

20. Long, X.; Ge, F.; Hong, Y. Feasibility study on buoyancy–weight ratios of a submerged floating tunnel prototype subjected to hydrodynamic loads. *Acta Mech. Sin.* **2015**, *31*, 750–761. [CrossRef]

21. Seo, S.-I.; Mun, H.-S.; Lee, J.-H.; Kim, J.-H. Simplified analysis for estimation of the behavior of a submerged floating tunnel in waves and experimental verification. *Mar. Struct.* **2015**, *44*, 142–158. [CrossRef]

22. Chen, Z.; Xiang, Y.; Lin, H.; Yang, Y. Coupled vibration analysis of submerged floating tunnel system in wave and current. *Appl. Sci.* **2018**, *8*, 1311. [CrossRef]

23. Xiao, J.; Huang, G. Transverse earthquake response and design analysis of submerged floating tunnels with various shore connections. *Procedia Eng.* **2010**, *4*, 233–242. [CrossRef]

24. Martinelli, L.; Domaneschi, M.; Shi, C. Submerged floating tunnels under seismic motion: Vibration mitigation and seaquake effects. *Procedia Eng.* **2016**, *166*, 229–246. [CrossRef]

25. Wu, Z.; Ni, P.; Mei, G. Vibration response of cable for submerged floating tunnel under simultaneous hydrodynamic force and earthquake excitations. *Adv. Struct. Eng.* **2018**, *21*, 1761–1773. [CrossRef]

26. Mirzapour, J.; Shahmardani, M.; Tariverdilo, S. Seismic response of submerged floating tunnel under support excitation. *Ships Offshore Struct.* **2017**, *12*, 404–411. [CrossRef]

27. Cifuentes, C.; Kim, S.; Kim, M.; Park, W. Numerical simulation of the coupled dynamic response of a submerged floating tunnel with mooring lines in regular waves. *Ocean Syst. Eng.* **2015**, *5*, 109–123. [CrossRef]

28. Jin, C.; Kim, M. Dynamic and structural responses of a submerged floating tunnel under extreme wave conditions. *Ocean Syst. Eng. Int. J.* **2017**, *7*, 413–433.

29. Jin, C.; Lee, J.; Kim, H.; Kim, M. Dynamic responses of a submerged floating tunnel in survival wave and seismic excitations. In Proceedings of the The 27th International Ocean and Polar Engineering Conference, San Francisco, CA, USA, 25–30 June 2017; International Society of Offshore and Polar Engineers: San Francisco, CA, USA, 2017; pp. 547–551.

30. Lee, J.H.; Seo, S.I.; Mun, H.S. Seismic behaviors of a floating submerged tunnel with a rectangular cross-section. *Ocean Eng.* **2016**, *127*, 32–47. [CrossRef]

31. Thompson, N. Mean forces, pressure and flow field velocities for circular cylindrical structures: Single cylinder with two-dimensional flow. *EDU Data Item* **1980**, 80025.

32. Veritas, D.N. *Offshore Standard dnv-os-e301: Offshore Standard-Position Mooring*; Det Norske Veritas (DNV) Oslo: Oslo, Norway, 2010.

33. Veritas, D.N. *Offshore Standard dnv-os-e302: Offshore Mooring Chain*; Det Norske Veritas (DNV) Oslo: Oslo, Norway, 2009.

34. Kim, M.; Koo, B.; Mercier, R.; Ward, E. Vessel/mooring/riser coupled dynamic analysis of a turret-moored fpso compared with otrc experiment. *Ocean Eng.* **2005**, *32*, 1780–1802. [CrossRef]

35. Koo, B.; Kim, M. Hydrodynamic interactions and relative motions of two floating platforms with mooring lines in side-by-side offloading operation. *Appl. Ocean Res.* **2005**, *27*, 292–310. [CrossRef]

36. Bae, Y.; Kim, M. Coupled dynamic analysis of multiple wind turbines on a large single floater. *Ocean Eng.* **2014**, *92*, 175–187. [CrossRef]

37. Kim, S.; Kim, M.; Kang, H. Turret location impact on global performance of a thruster-assisted turret-moored fpso. *Ocean Syst. Eng. Int. J.* **2016**, *6*, 265–287. [CrossRef]

38. Jang, H.; Kang, H.; Kim, M. Numerical simulation of dynamic interactions of an arctic spar with drifting level ice. *Ocean Syst. Eng. Int. J.* **2016**, *6*, 345–362. [CrossRef]

39. Garrett, D. Dynamic analysis of slender rods. *J. Energy Resour. Technol.* **1982**, *104*, 302–306. [CrossRef]

40. Faltinsen, O. *Sea Loads on Ships and Offshore Structures*; Cambridge University Press: London, UK, 1993.

41. Islam, N.; Ahrnad, S. Nonlinear seismic response of articulated offshore tower. *Def. Sci. J.* **2003**, *53*, 105–113. [CrossRef]

42. Mousavi, S.A.; Bargi, K.; Zahrai, S.M. Optimum parameters of tuned liquid column–gas damper for mitigation of seismic-induced vibrations of offshore jacket platforms. *Struct. Control Health Monit.* **2013**, *20*, 422–444. [CrossRef]

43. Ran, Z.; Kim, M.; Zheng, W. Coupled dynamic analysis of a moored spar in random waves and currents (time-domain versus frequency-domain analysis). *J. Offshore Mech. Arct. Eng.* **1999**, *121*, 194–200. [CrossRef]

44. Orcina. Orcaflex Manual. Available online: https://www.orcina.com/SoftwareProducts/OrcaFlex/index.php (accessed on 17 October 2018).

45. Suh, K.-D.; Kwon, H.-D.; Lee, D.-Y. Some statistical characteristics of large deepwater waves around the korean peninsula. *Coast. Eng.* **2010**, *57*, 375–384. [CrossRef]

46. USGS. National Strong-Motion Project Earthquake Data Sets. Available online: https://escweb.wr.usgs.gov/nsmp-data/nsmn_eqdata.html (accessed on 27 August 2018).

47. API. *Recommended Practice for Design and Analysis of Stationkeeping Systems for Floating Structures: Exploration and Production Department. Api Recommended Practice 2sk (rp 2sk): Effective Date: March 1, 1997*; American Petroleum Institute: Washington, DC, USA, 1996.

Article

The Influence of Dynamic Tissue Properties on HIFU Hyperthermia: A Numerical Simulation Study

Qiaolai Tan [1,2], Xiao Zou [1], Yajun Ding [3], Xinmin Zhao [1] and Shengyou Qian [1,*]

[1]   School of Physics and Electronics, Hunan Normal University, Changsha 410081, China;
tanql1981@smail.hunnu.edu.cn (Q.T.); shawner@hunnu.edu.cn (X.Z.); 470241651@hunnu.edu.cn (X.Z.)
[2]   School of Electronic Information and Electrical Engineering, Xiangnan University, Chenzhou 423000, China
[3]   College of Information Science and Engineering, Hunan Normal University, Changsha 410081, China;
yajunding@hunnu.edu.cn
*   Correspondence: shyqian@hunnu.edu.cn

Received: 2 September 2018; Accepted: 10 October 2018; Published: 16 October 2018

Abstract: Accurate temperature and thermal dose prediction are crucial to high-intensity focused ultrasound (HIFU) hyperthermia, which has been used successfully for the non-invasive treatment of solid tumors. For the conventional method of prediction, the tissue properties are usually set as constants. However, the temperature rise induced by HIFU irradiation in tissues will cause changes in the tissue properties that in turn affect the acoustic and temperature field. Herein, an acoustic–thermal coupling model is presented to predict the temperature and thermal damage zone in tissue in terms of the Westervelt equation and Pennes bioheat transfer equation, and the individual influence of each dynamic tissue property and the joint effect of all of the dynamic tissue properties are studied. The simulation results show that the dynamic acoustic absorption coefficient has the greatest influence on the temperature and thermal damage zone among all of the individual dynamic tissue properties. In addition, compared with the conventional method, the dynamic acoustic absorption coefficient leads to a higher focal temperature and a larger thermal damage zone; on the contrary, the dynamic blood perfusion leads to a lower focal temperature and a smaller thermal damage zone. Moreover, the conventional method underestimates the focal temperature and the thermal damage zone, compared with the simulation that was performed using all of the dynamic tissue properties. The results of this study will be helpful to guide the doctors to develop more accurate clinical protocols for HIFU treatment planning.

Keywords: HIFU; dynamic tissue property; Westervelt equation; thermal damage zone

---

1. Introduction

Cancer is one of the serious diseases that threatens the life and health of humans. According to cancer statistics released by the National Cancer Center of China, 3.804 million new cancer cases were diagnosed and 2.296 million cancer deaths were reported in 2014 [1]. Traditional therapies for cancer include surgery resection, chemotherapy, and radiotherapy. In recent years, other alternative therapies such as microwave ablation, laser ablation, cryoablation, and high-intensity focused ultrasound (HIFU) hyperthermia also have developed rapidly [2,3]. HIFU therapy is a non-invasive technology in which an ultrasound beam carries sufficient energy, and the energy is focused onto the target area to cause a local temperature rise, which is sufficiently high to make the lesion tissue undergo coagulative necrosis without causing damage to the overlaying or surrounding tissue [4,5]. It has many advantages such as non-invasive, non-contact, non-ionization, and low cost [6,7], and has been successfully used in clinics to treat solid malignant tumors, including cancers of the prostate, liver, kidney, breast, and pancreas [8]. The clinical success of HIFU hyperthermia depends on the accurate thermal dose at the lesion location. Unfortunately, it is difficult to accurately measure the thermal dose at a depth of the tissue in most clinical situations. Instead, a numerical simulation method is usually used to predict the transient temperature profiles and thermal dose to assess the thermal damage that will occur in tissue during HIFU ablation [9].

In the conventional method, the numerical simulation of HIFU hyperthermia is usually based on the acoustic model Westervelt equation and the thermal model Pennes bioheat transfer equation, and the tissue properties are set as constants. However, that the tissue properties varied with temperature had been observed in several experimental studies [10–13]. Moreover, the temperature-dependent tissue properties in turn affect the acoustic field and temperature field. Several researchers have considered some temperature-dependent tissue properties to perform the numerical study of HIFU hyperthermia [9,14,15]. For example, Hallaj [14] studied the effect of dynamic sound speed in the liver with and without a fat layer undergoing HIFU surgery. Christopher [15] examined the importance of the thermal lens effect with a phased array transducer in the liver with the fat layer when considering dynamic sound speed and a dynamic acoustic absorption coefficient in the HIFU hyperthermia study using three-dimensional model. Guntur [9] studied the influence of temperature-dependent thermal parameters on temperature during HIFU irradiation by comparing the conventional prediction of temperature and the thermal damage zone with that for different thermal parameters (i.e., specific heat capacity and thermal conductivity) at the given temperatures. However, only one or two dynamic tissue properties were considered in the above studies; to our knowledge, other dynamic tissue properties such as density and blood perfusion have never been considered. Furthermore, the joint effect of two dynamic tissue properties was investigated in the above studies, but the individual influence of each tissue property on HIFU hyperthermia is still unclear. Therefore, we first study the evolutions of the acoustic and temperature fields with each dynamic tissue property independently, and clarify the physical significance of each tissue property. In addition, we develop an acoustic–thermal model to evaluate the joint effect of all of the dynamic tissue properties on temperature distribution and thermal damage, including sound speed, acoustic absorption coefficient, non-linearity parameter, specific heat capacity, thermal conductivity, density, and blood perfusion. The results provide a more accurate prediction of temperature distribution and thermal damage, gaining insight into the complex dynamic processes during HIFU hyperthermia, which are useful for doctors making treatment planning.

2. Theory

2.1. Acoustic Model for Ultrasound Wave Propagation

Generally, the Westervelt equation [16,17] is used to model the ultrasound wave propagation in the thermoviscous medium:

$$\left(\nabla^2 - \frac{1}{c^2}\frac{\partial^2}{\partial t^2}\right)p + \frac{\delta}{c^4}\frac{\partial^3 p}{\partial t^3} + \frac{\beta}{\rho c^4}\frac{\partial^2 p^2}{\partial t^2} = 0 \tag{1}$$

where ∇^2, p, c, t are the Laplace operator, acoustic pressure, sound speed, and time, respectively; the non-linearity coefficient β is related to the non-linearity parameter B/A by $\beta = 1 + (B/2A)$; and $\delta = 2\alpha c^3/\omega^2$ is the acoustic diffusivity accounting for the thermoviscous effect in the fluid [18], where ω is the acoustic angular frequency and α is the acoustic absorption coefficient.

The acoustic field is computed by Westervelt equation in two-dimensional (2D) cylindrical coordinate using the finite-difference time-domain (FDTD). The z-axis is the acoustic axis of the ultrasonic transducer, and r is the radial coordinate measured from the z-axis. The excitation of the ultrasonic transducer is:

$$p(t) = p_0 \sin(\omega t) \tag{2}$$

where p_0 is the amplitude of acoustic pressure on the ultrasonic transducer.

An absorbing boundary condition (ABC) is imposed at the edge of the computation domain to prevent or minimize the reflection from the edges of the domain, and a first-order Mur's absorption boundary condition is employed [19]:

$$\frac{\partial p}{\partial x} - \frac{1}{c}\frac{\partial p}{\partial t} = 0 \tag{3}$$

where x denotes z or r in their own ultrasonic wave propagation direction.

2.2. Thermal Energy Model for Tissue Heating

The transfer of heat in the tissue under HIFU irradiation is modeled using the Pennes bioheat transfer equation [20]:

$$\rho_t C_t \frac{\partial T}{\partial t} = k\nabla^2 T - W_b C_b (T - T_a) + Q_{ext} \tag{4}$$

where C_t and ρ_t are the specific heat and density of tissue, respectively; C_b, W_b, and T_a are the specific heat, perfusion rate, and ambient temperature of blood, respectively. Q_{ext} is the ultrasound heat deposition term, which can be calculated by employing the time averaged over one acoustic period by numerical integration [14]:

$$Q_{ext} = \frac{2\alpha}{\rho c \omega^2}\left\langle \left(\frac{\partial p}{\partial t}\right)^2\right\rangle \tag{5}$$

To evaluate the performance of the HIFU treatment, thermal dose is usually used to estimate the tissue damage. The thermal dose depends on the final time t_f and temperature level T, which was developed by Sapareto and Dewey [21]:

$$t_{43} = \int_0^{t_f} R^{(T-43)}\,dt \approx \sum_0^{t_f} R^{(T-43)}\Delta t \tag{6}$$

where t_{43} is the thermal dose equivalent time at 43 °C. $R = 2$ if $T \geq 43$ °C, and $R = 4$ if 37 °C $< T <$ 43 °C. The threshold value of an isothermal dose value of 240 min at 43 °C was usually selected to predict the size of the thermal lesion region.

Another way to quantify the tissue thermal damage is to use the Arrhenius equation [22]:

$$\Omega = \int_0^{t_f} A\exp\left(\frac{-E_a}{R_a T}\right) dt \tag{7}$$

where A, E_a, and R_a are the frequency factor, activation energy, and universal gas constant, respectively. For liver thermal damage, $A = 9.4 \times 10^{104}$ s^{-1}, $E_a = 6.68 \times 10^5$ J mol^{-1}, and $R_a = 8.31$ J mol^{-1} K^{-1} [23]. The undamaged fraction of the tissue and the damaged fraction can be estimated by $f_u = \exp(-\Omega)$ and $f_d = 1 - f_u$, respectively [24].

2.3. Dynamic Tissue Propertiesc

The acoustic and thermal parameters of tissue were strongly dependent on tissue temperature, and many experimental data had been obtained [10–13]. The data for acoustic absorption coefficient and sound speed were derived from measurements in liver tissue by Damianou [10] and Bamber [11], respectively. The polynomials to fit the acoustic absorption coefficient and sound speed in liver tissue to experimental data are [15]:

$$\begin{aligned}\alpha_{liver} = &\ 5.5367 - 2.9950 \times 10^{-1}T + 3.3357 \times 10^{-2}T^2 - 1.6058 \times 10^{-3}T^3 + 3.4382 \times 10^{-5}T^4 \\ &- 3.2486 \times 10^{-7}T^5 + 1.1181 \times 10^{-9}T^6 \qquad 30\ °C \leq T \leq 90\ °C.\end{aligned} \tag{8}$$

$$c_{liver} = 1529.3 + 1.6856T + 6.1131 \times 10^{-2}T^2 - 2.2967 \times 10^{-3}T^3$$
$$+2.2657 \times 10^{-5}T^4 - 7.1795 \times 10^{-8}T^5 \qquad 30\,^\circ\text{C} \le T \le 90\,^\circ\text{C} \tag{9}$$

In this study, the experimental data for the change in the non-linearity parameter with temperature in the liver tissue are derived from measurements by Choi [12], and the experimental data for the changes in the specific heat capacity, thermal conductivity, and density with temperature in liver tissue are derived from measurements by Guntur [13]. We obtain the expressions of the temperature-dependent non-linearity parameter, specific heat capacity, thermal conductivity, and density respectively by the least squares polynomial fitting their experimental data in liver tissue:

$$\left(\frac{B}{A}\right)_{liver} = 6.68 - 0.41448\,T + 0.03364\,T^2 - 0.00101\,T^3 + 1.34407 \times 10^{-5}T^4$$
$$-6.35346 \times 10^{-8}T^5 \qquad 30\,^\circ\text{C} \le T \le 75\,^\circ\text{C} \tag{10}$$

$$C_{liver} = 3600 + 53.55552T - 3.96009T^2 + 0.10084T^3 - 0.00106T^4$$
$$+4.01666 \times 10^{-6}T^5 \qquad 20\,^\circ\text{C} \le T \le 90\,^\circ\text{C} \tag{11}$$

$$K_{liver} = 0.84691 - 0.02094T + 3.89971 \times 10^{-4}T^2 - 5.47451 \times 10^{-7}T^3$$
$$-4.14455 \times 10^{-8}T^4 + 2.97188 \times 10^{-10}T^5 \qquad 20\,^\circ\text{C} \le T \le 90\,^\circ\text{C} \tag{12}$$

$$\rho_{liver} = 1084.09352 - 2.97434T + 0.0042T^2 + 0.00293T^3 - 6.14447 \times 10^{-5}T^4$$
$$+3.33019 \times 10^{-7}T^5 \qquad 20\,^\circ\text{C} \le T \le 90\,^\circ\text{C} \tag{13}$$

The polynomials above are shown in Figure 1, which have the validity in their own temperature range. In this study, their use is also strictly restricted to their respective temperature ranges.

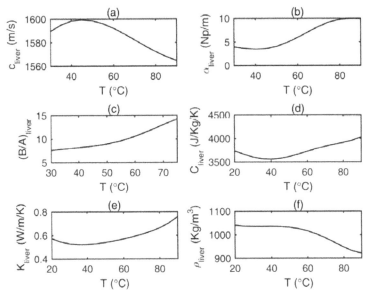

Figure 1. Temperature-dependent tissue properties in liver tissue.

The variation of the blood perfusion rate with temperature and thermal damage can be described by:

$$W_{b,liver}(T,\Omega) = W_{b,0}f_T f_u \tag{14}$$

where $W_{b,0}$ is the constitutive blood perfusion rate, 18.2 Kg m^{-3} s^{-1} for liver, and f_T is a dimensionless function that accounts for vessel dilation at slightly elevated temperatures, which can be approximated as [23,24]:

$$f_T = \begin{cases} 4 + 0.6\,(T - 42)\ 37\,°C \leq T \leq 42\,°C \\ 4T \geq 42\,°C \end{cases} \tag{15}$$

The blood perfusion rate increases as the temperature rises, but as tissue coagulation develops, it is decreased to zero due to the factor of thermal damage [25].

In order to study the effects of dynamic tissue properties on HIFU hyperthermia, we compare the simulation using dynamic tissue properties with the conventional method using tissue properties with constant values. In this study, the constant values of tissue properties were obtained from the values of the above fitting formula at 37 °C. The values of the acoustic and thermal parameters are listed in Tables 1 and 2, respectively.

Table 1. Values of acoustic parameters in this study (37 °C).

Material	ρ(Kg m^{-3})	c(m s^{-1})	α(Np m^{-1} MHz^{-1})	β
Water	1000	1500	2.88×10^{-4}	3.5
Liver	1036	1596	3.5	5

Table 2. Values of thermal parameters in this study (37 °C).

Material	K(W m^{-1}°C^{-1})	C(J Kg^{-1} °C^{-1})	W_b(Kg m^{-3} s^{-1})
Water	0.6	4180	0
Liver	0.5	3560	18.2

2.4. Description of the Simulation

The HIFU transducer is a spherical cap with an aperture radius a of 35 mm, a focal length F of 62.64 mm, and a center frequency f of 1 MHz. The transducer and liver tissue are placed in the water at 37 °C, and a geometric configuration of the physical model is shown in Figure 2.

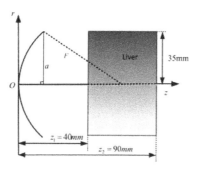

Figure 2. Geometric configuration of the physical model. The liver tissue is a cylinder with a radius of 35 mm and a length of 50 mm, and is placed at $z_1 = 40$ mm.

As the tissue temperature rises, the tissue properties also change dynamically. These tissue properties need to be updated in real time according to the temperature, and the updated tissue properties are fed back into the calculation of the acoustic field and temperature field. The flowchart in Figure 3 shows how to carry out the coupling calculation of the acoustic and temperature field under such dynamic conditions. The acoustic field and temperature field are coupled by the heat deposition term Q_{ext}, which is computed from the acoustic pressure. In the practical simulation, the temperature field is calculated periodically, and the resulting temperature data is used to renew the tissue properties using the function above at each spatial point on the tissue domain. The updated tissue properties are then used as an input to recalculate the acoustic field. Therefore, the acoustic field, temperature field, and tissue properties are mutually influenced. In this study, the acoustic parameters and acoustic field are updated for the simulations here every 1 s unless otherwise noted, and the thermal parameters are updated in real time. This coupling method is based on the time rates of change of the tissue properties being slow enough in the given period interval.

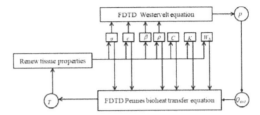

Figure 3. Flowchart of the iterative method for coupling acoustic pressure and temperature calculation.

For this study, the acoustic field and temperature field are calculated on a polar cylindrical grid using the explicit finite-difference time-domain (FDTD) method as described by Hallaj [26]. The spatial grids for the simulation are: $\Delta z = \Delta r = 10^{-4}$ m. The time step for the acoustic field and temperature field simulation are 10^{-8} s and 0.01 s, respectively [26]. All of the simulations are performed with MATLAB programming based on the FDTD method.

3. Result and Discussion

In this manuscript, we focus on the effect of each dynamic tissue property independent from each other, and compared these effects with the conventional method of keeping the tissue properties as constant. When the effect of one dynamic tissue property is studied, the other tissue properties remain constant unless otherwise noted. In the following study, the amplitude of acoustic pressure on the sound source face p_0 is 1.4×10^5 Pa unless otherwise noted.

3.1. Dynamic Acoustic Absorption Coefficient

Simulations are carried out that only consider the change of the acoustic absorption coefficient with temperature independently. To get more accurate simulation results, the acoustic absorption coefficient and acoustic field are updated here every 0.2 s. Figure 4a depicts the axial profile of the acoustic absorption coefficient during 3 s of HIFU irradiation. The acoustic absorption coefficient near ultrasonic focus increases with the time of HIFU irradiation. At time t = 3 s, Figure 4b illustrates the axial distribution of peak acoustic pressure. Clearly, the peak acoustic pressures are almost the same between dynamic α_{liver} and constant α_{liver}. It can be explained that the temperature only has a great effect on the acoustic absorption coefficient near the ultrasonic focus, as shown in Figure 4a. In Figure 4c, the maximum value of Q_{ext} is 8.938×10^7 W/m² for constant α_{liver}, and 2.179×10^8 W/m² for dynamic α_{liver}. The features can be explained that Q_{ext} is proportional to the acoustic absorption coefficient, according to Formula (5). Figure 4d contrasts the evolution of the focal temperature with time for dynamic and constant α_{liver}. Before t = 1 s, the rate of the focal temperature rise is almost the same for simulations using dynamic α_{liver} and constant α_{liver}. This may be due to the small change of the acoustic absorption coefficient during the early HIFU irradiation stage, as shown in Figure 4a. After t = 1 s, the focal temperature for dynamic α_{liver} rises much faster than that for constant α_{liver}. When t = 3 s, the focal temperature is 65.94 °C for constant α_{liver} and 85.53°C for dynamic α_{liver}. Figure 4e plots the shape of the thermal damage zone, representing the heated region for more than 240 min equivalent time at 43 °C. The thermal damage zone is an ellipse of 0.51 cm × 0.12 cm size for constant α_{liver}, and an ellipse of 0.6 cm × 0.16 cm size for dynamic α_{liver}. These phenomena indicate that dynamic α_{liver} has a greater effect on the focal temperature and thermal damage zone as the

HIFU irradiation time increases, compared with constant α_{liver}, which can be explained by the greater acoustic absorption coefficient being related to a greater the value of Q_{ext}, higher focal temperature, and larger thermal damage zone. Figure 4f describes the axial profile of the thermal dose, and the black dotted line denotes the value $\log_{10}(240)$ min. The axial length AB of the thermal damage zone is 0.51 cm for constant α_{liver}, and the axial length CD of the thermal damage zone is 0.6 cm for dynamic α_{liver}, which are consistent with the axial length of thermal damage zone in Figure 4e. Meanwhile, the thermal dose of ultrasonic focus for dynamic α_{liver} is much greater than that for constant α_{liver}.

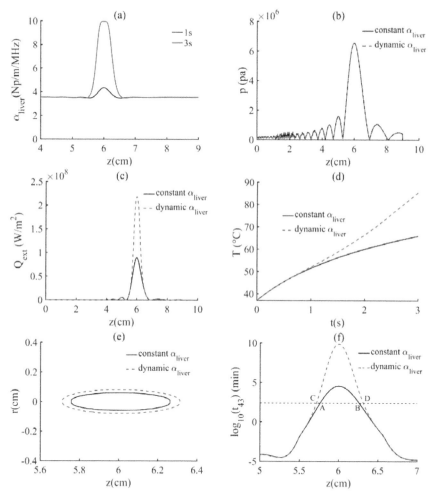

Figure 4. (a) The evolution of α_{liver} at t = 1 s and 3 s. The effects of the dynamic acoustic absorption coefficient on: (b) p (c) Q_{ext} (d) T at the ultrasonic focus (e) thermal damage zone (f) t_{43} at t = 3 s.

3.2. Dynamic Non-linearity Parameter

Simulation is carried out considering the change of the non-linearity parameter with temperature independently. The HIFU irradiation time is set to 5 s to ensure the validity of the dynamic non-linearity parameter used in the range of 30 °C to 75 °C, and the non-linearity parameter and acoustic field are updated here every 0.5 s. In Figure 5b, the axial profile of peak acoustic pressure at 5 s is almost identical for dynamic and constant $(B/A)_{liver}$. This phenomenon can be attributed to the local increase of the non-linearity parameter near the ultrasonic focus with the increase of HIFU irradiation time, as shown in Figure 5a. According to Formula (5), the value of Q_{ext} for dynamic $(B/A)_{liver}$ is almost the same as that for constant $(B/A)_{liver}$. Consequently, the change of the focal temperature with time and the thermal damage zone are almost identical for dynamic and constant $(B/A)_{liver}$, as shown in Figure 5c,d. It can be concluded that the dynamic acoustic non-linear parameter has little effect on the HIFU hyperthermia.

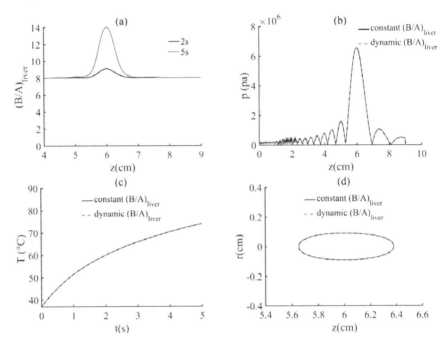

Figure 5. (a) The evolution of $(B/A)_{liver}$ at t = 2 s and 5 s. The effects of the dynamic non-linearity parameter on: (b) p and (c) T at the ultrasonic focus, and (d) the thermal damage zone at t = 5 s.

3.3. Dynamic Sound Speed, Specific Heat Capacity, Thermal Conductivity, and Density

Simulations are carried out using the dynamic sound speed, dynamic specific heat capacity, dynamic thermal conductivity, and dynamic density, respectively, and the HIFU irradiation time is 10 s. Figure 6 describes the axial profiles of dynamic sound speed, dynamic specific heat capacity, dynamic thermal conductivity, and dynamic density, which are affected only by the temperature in the vicinity of the ultrasonic focus.

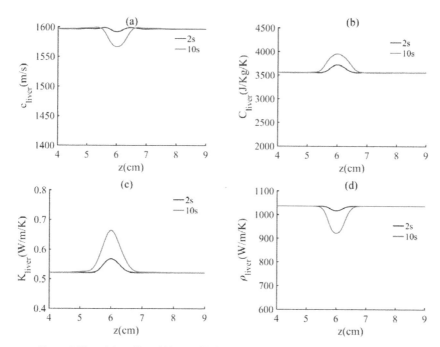

Figure 6. The axial profiles of (a) c_{liver} (b) C_{liver} (c) K_{liver}, and (d) ρ_{liver} at t = 2 s and 10 s.

Figure 7a shows that the axial profiles of the peak acoustic pressure for simulations with dynamic C_{liver}, dynamic K_{liver}, and dynamic ρ_{liver} are almost the same as that for simulation with constant tissue properties. At ultrasonic focus, the peak acoustic pressure with dynamic c_{liver} is a little greater than that with constant tissue properties, which is consistent with previously reported results [14]. Figure 7b demonstrates the evolutions of focal temperature with time for simulations using dynamic c_{liver}, dynamic C_{liver}, dynamic K_{liver}, dynamic ρ_{liver}, and constant tissue properties, respectively. Before t = 2 s, the rate of focal temperature rise is almost the same for simulations using dynamic c_{liver}, dynamic C_{liver}, dynamic K_{liver}, dynamic ρ_{liver}, and constant tissue properties. This may be due to the very small change in the tissue properties during the early HIFU irradiation stage, as shown in Figure 6. After t = 2 s, Figure 6 shows that the sound speed and density decrease with the increase of HIFU irradiation time, but the specific heat capacity and thermal conductivity have the opposite trend. According to Formula (5), the values of focal Q_{ext} for dynamic c_{liver} and dynamic ρ_{liver} are both greater than that for the constant tissue properties. Consequently, the rate of focal temperature rise for dynamic c_{liver} and dynamic ρ_{liver} is faster than that for constant tissue properties, as shown in Figure 7b. The rate of focal temperature rise for dynamic C_{liver} is slower than that for constant tissue properties. This feature can be explained by the physical significance of specific heat capacity, which is defined as the amount of energy that is required to increase the temperature of a unit mass of tissue by 1 °C [27]. In other words, for the same amount of heat energy and mass, the larger the specific heat capacity, the smaller the temperature rise. The focal temperature for dynamic K_{liver} rises slower than that for constant K_{liver}, and the focal temperature for dynamic K_{liver} is 5.33 °C lower than that for the constant tissue properties at 10 s. This feature can be explained by the greater thermal conductivity meaning that more thermal energy is lost from the treated areas because of thermal diffusion [28]. Therefore, it can be concluded that greater thermal conductivity leads to a slower focal temperature rise, which is similar to Guntur's result [9]. The maximum focal temperatures for simulations using dynamic c_{liver}, dynamic C_{liver}, dynamic ρ_{liver}, and constant tissue properties are 86.66 °C, 84.58 °C, 88.62 °C, and 85.47 °C, respectively, indicating that dynamic c_{liver}, dynamic C_{liver}, and dynamic ρ_{liver} have little effect on the temperature during HIFU hyperthermia. This is mainly due to the local variations of tissue property near the ultrasonic focus. Figure 7c shows that the thermal damage zones for dynamic c_{liver}, dynamic C_{liver}, dynamic K_{liver}, dynamic ρ_{liver}, and constant tissue properties are almost the same, which can also be confirmed from Figure 7d. It's interesting to note that the maximum focal temperature for dynamic K_{liver} is lower than that for constant K_{liver}, but the thermal damage zone is almost the same for dynamic K_{liver} and constant K_{liver}. It is mainly because the size of the thermal damage zone depends on the thermal dose above 240 min at 43 °C, rather than the maximum focal temperature.

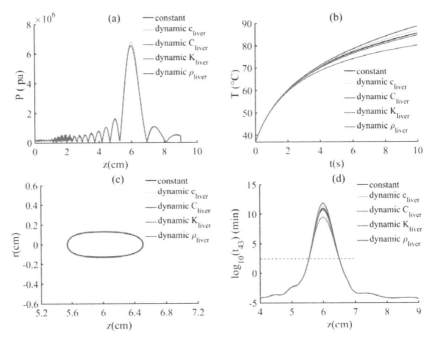

Figure 7. The effects of dynamic sound speed, dynamic specific heat capacity, dynamic thermal conductivity, and dynamic density on: (**a**) p and (**b**) T at the ultrasonic focus, (**c**) the thermal damage zone, and (**d**) t_{43} at t = 10 s.

3.4. Dynamic Blood Perfusion

The simulation is performed only considering the dynamic change of blood perfusion, and the HIFU irradiation time is 10 s. The blood perfusion firstly increased, then remained unchanged, and finally decreased to zero as shown in Figure 8a. This is because the initial increase in temperature causes the blood vessels to inflate to increase blood perfusion. As the temperature continues to increase, the tissue damage fraction increases so that the blood perfusion decreases. When the tissue undergoes coagulation necrosis, the blood perfusion decreases to zero. Figure 8b describes the axial profile of dynamic blood perfusion at t = 2 s and 10 s. Compared with other tissue properties, the temperature has a greater impact on the blood perfusion of the surrounding tissue around the central axis. At time t = 10 s, the axial profile of the peak acoustic pressure for dynamic $W_{b,liver}$ is almost the same as that for the constant $W_{b,liver}$, as shown in Figure 8c. In Figure 8d, the temperature rise for dynamic $W_{b,liver}$ is slower at first; then, it is faster than that for the constant $W_{b,liver}$. It is because blood perfusion is first increased to four times the constant blood perfusion, and then remains unchanged, and is finally reduced to zero, as shown in Figure 8a. Figure 8e shows that the thermal damage zone is an ellipse of 0.97 cm × 0.26 cm size for constant $W_{b,liver}$ and an ellipse of 0.89 cm × 0.24 cm size for dynamic $W_{b,liver}$, respectively. Thus, it can be seen that the thermal damage zone for dynamic $W_{b,liver}$ is smaller than that for constant $W_{b,liver}$, which can also be verified by the axial thermal dose distribution of Figure 8f.

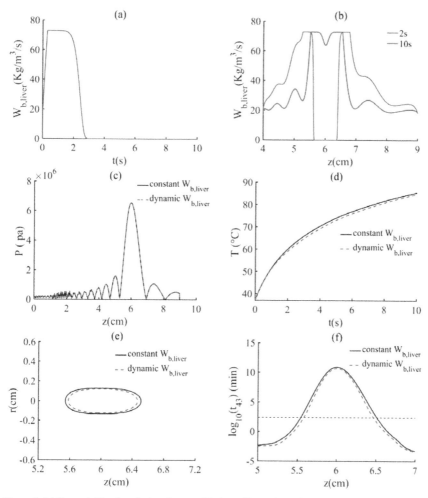

Figure 8. (a) Dynamic blood perfusion change with time. (b) Axial profile of dynamic blood perfusion at t = 2 s and 10 s. The effects of dynamic blood perfusion on: (c) p and (d) T at the ultrasonic focus and (e) thermal damage zone; (f) Q_{43} at t = 10 s.

3.5. Considering All Dynamic Tissue Properties

In the above research results, the individual influence of each tissue property on HIFU hyperthermia was studied independently. Therefore, in this section, it is necessary to perform the simulation using all of the dynamic tissue properties to explore the joint influence on HIFU hyperthermia by comparing them with simulations using dynamic α_{liver} and constant tissue properties. Note that we assume that the non-linearity parameter above 75 °C is replaced by that at 75 °C to simplify the physical model owing to (i) the dynamic non-linear parameter being found to have little effect on the acoustic pressure, temperature, and thermal damage zone in our calculation (Figure 5b–d); and (ii) above 75 °C, biological tissue having been coagulated. To ensure that all of the dynamic tissue properties are valid within their respective temperature ranges, the HIFU irradiation time is set to

3 s. At time t = 3 s, Figure 9a shows that the peak acoustic pressure for simulation using all of the dynamic tissue properties is larger than that using dynamic α_{liver} and constant tissue properties due to the influence of dynamic sound velocity, and the peak acoustic pressures are almost the same between the dynamic α_{liver} and constant tissue properties. In Figure 9b, the peak value of Q_{ext} is the greatest for simulation using all of the dynamic tissue properties, followed by that for simulation using dynamic α_{liver}, and the smallest for simulation using constant tissue properties. In Figure 9c, before t = 1 s, the focal temperature is almost the same for simulation using dynamic α_{liver} across all of the dynamic tissue properties and constant tissue properties; after t = 1 s, the rate of focal temperature rise is fastest for simulation using dynamic α_{liver}, followed by that for simulation using all of the dynamic tissue properties, and slowest for simulation using constant tissue properties. The maximum focal temperature for all of the dynamic tissue properties, dynamic α_{liver}, and constant tissue properties are 81.56 °C, 85.53 °C, and 65.94 °C, respectively, indicating that the maximum focal temperature for all of the dynamic tissue properties is lower than that for dynamic α_{liver}, although the peak value of Q_{ext} for all of the dynamic tissue properties is greater than that for dynamic α_{liver}. Based on the above research, this is mainly due to the influence of the comprehensive factors such as dynamic C_{liver}, dynamic K_{liver}, and dynamic $W_{b,liver}$ on focal temperature, especially the influence of dynamic K_{liver}. In Figure 9d, the thermal damage zone is an ellipse of 0.57 cm × 0.16 cm size for all of the dynamic tissue properties, an ellipse of 0.6 cm × 0.16 cm size for dynamic α_{liver}, and an ellipse of 0.51 cm × 0.12 cm size for constant tissue properties, respectively. Consequently, it is can be concluded that the simulation using constant tissue properties significantly underestimates the focal temperature and thermal damage zone compared with the simulation using all of the dynamic tissue properties or dynamic α_{liver}. Meanwhile, although the dynamic acoustic absorption coefficient plays the most important role in relation to the focal temperature and thermal damage zone, other dynamic tissue properties ought to be considered.

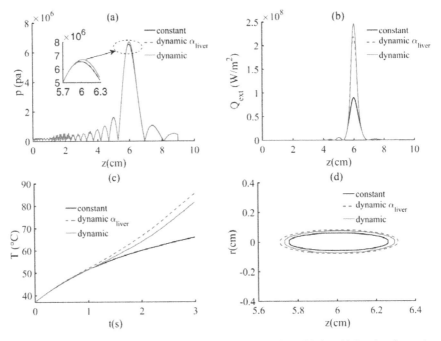

Figure 9. The effects of all of the dynamic tissue properties on: (**a**) *p*, (**b**) Q_{ext}, (**c**) *T* at the ultrasonic focus, and (**d**) thermal damage zone.

4. Conclusions

The influence of each dynamic tissue property on HIFU hyperthermia is studied independently based on the reported experimental data of dynamic tissue properties. The findings in the present study suggest that the acoustic pressure is insensitive to the dynamic tissue properties. The numerical results also show that the dynamic acoustic absorption coefficient significantly affects the temperature and thermal damage zone; on the contrary, the dynamic non-linearity parameter has almost no effect on the temperature and thermal damage zone. It is found that the thermal damage zone for dynamic $W_{b,liver}$ is smaller than that for constant $W_{b,liver}$, and the influence of a dynamic sound speed, dynamic specific heat capacity, and dynamic density on the thermal damage zone is slight. It is also worth mentioning that the maximum focal temperature for dynamic K_{liver} is lower than that for constant K_{liver}, but the thermal damage zone is almost the same for dynamic K_{liver} and constant K_{liver}. Among all of the individual dynamic tissue properties, the dynamic acoustic absorption coefficient has the greatest influence on the temperature and thermal damage zone. Knowing the influence of each dynamic tissue property is beneficial to our deep understanding of the principle of HIFU therapy. Besides studying the influence of each individual dynamic tissue property, the simulation considering all of the dynamic tissue properties to explore the comprehensive influence on HIFU hyperthermia is performed. The numerical results show that the maximum focal temperature and thermal damage zone for simulation using all of the dynamic tissue properties increase, compared with those for simulation using constant tissue properties, implying that the simulation using constant tissue properties underestimates the focal temperature and thermal damage zone compared with the simulation using all of the dynamic tissue properties. Moreover, it is interesting to point out that the thermal energy absorbed by the tissue for simulation using all of the dynamic tissue properties is greater than that for simulation using dynamic α_{liver}, but the maximum focal temperature and thermal damage zone for simulation using all of the dynamic tissue properties decrease, compared with those for simulation using dynamic α_{liver}. Consequently, when doctors develop a more accurate clinical protocol for HIFU treatment planning, it is necessary to consider all of the dynamic tissue properties to assess the size of thermal damage zone, so as not to damage normal tissue.

Author Contributions: S.Q. conceived and designed the research idea and the framework; Q.T. and X.Z. (Xiao Zou) performed the simulations; Q.T. and X.Z. (Xiao Zou) wrote the paper; S.Q., Q.T. and Y.D. analyzed the data, S.Q. and X.Z. (Xinmin Zhao) modified the paper.

Acknowledgments: This work is partially supported by the National Nature Science Foundation of China (No. 11474090, 11774088, 11174077, 61502164), Hunan Provincial Natural Science Foundation of China (No. 2016JJ3090), Scientific Research Fund of Hunan Provincial Education Department (No. 16B155), Aid program for Science and Technology Innovative Research Team in Higher Educational Institutions of Hunan Province, Science and Technology Research Program of Chenzhou City (No. CZ2014039) and Research Program of Xiangnan University (No. 2014XJ63).

Conflicts of Interest: The authors declare no conflict of interest.

References

1. Chen, W.; Sun, K.; Zheng, R.; Zeng, H.; Zhang, S.; Xia, C.; Yang, Z.; Li, H.; Zou, X.; He, J. Cancer incidence and mortality in China, 2014. *Chin. J. Cancer Res.* **2018**, *30*, 1–12. [CrossRef] [PubMed]
2. Cranston, D. A review of high intensity focused ultrasound in relation to the treatment of renal tumours and other malignancies. *Ultrason. Sonochem.* **2015**, *27*, 654–658. [CrossRef] [PubMed]
3. Manthe, R.L.; Foy, S.P.; Krishnamurthy, N.; Sharma, B.; Labhasetwar, V. Tumor Ablation and Nanotechnology. *Mol. Pharm.* **2010**, *7*, 1880–1898. [CrossRef] [PubMed]
4. Zhou, Y. Acoustic power measurement of high-intensity focused ultrasound transducer using a pressure sensor. *Med. Eng. Phys.* **2015**, *37*, 335–340. [CrossRef] [PubMed]
5. Xiao, J.; Sun, T.; Zhang, S.; Ma, M.; Yang, X.; Zhou, J.; Zhu, J.; Wang, F. HIFU, a noninvasive and effective treatment for chyluria: 15 years of experience. *Surg. Endosc.* **2018**, *32*, 3064–3069. [CrossRef] [PubMed]

6. Nover, A.B.; Hou, G.Y.; Han, Y.; Wang, S.; O'Connell, G.D.; Ateshian, G.A.; Konofagou, E.E.; Hung, C.T. High intensity focused ultrasound as a tool for tissue engineering: Application to cartilage. *Med. Eng. Phys.* **2016**, *38*, 192–198. [CrossRef] [PubMed]

7. Huang, C.W.; Sun, M.K.; Chen, B.T.; Shieh, J.; Chen, C.S.; Chen, W.S. Simulation of thermal ablation by high-intensity focused ultrasound with temperature-dependent properties. *Ultrason. Sonochem.* **2015**, *27*, 456–465. [CrossRef] [PubMed]

8. Qian, K.; Li, C.; Ni, Z.; Tu, J.; Guo, X.; Zhang, D. Uniform tissue lesion formation induced by high-intensity focused ultrasound along a spiral pathway. *Ultrasonics* **2017**, *77*, 38–46. [CrossRef] [PubMed]

9. Guntur, S.R.; Choi, M.J. Influence of Temperature-Dependent Thermal Parameters on Temperature Elevation of Tissue Exposed to High-Intensity Focused Ultrasound: Numerical Simulation. *Ultrasound Med. Biol.* **2015**, *41*, 806–813. [CrossRef] [PubMed]

10. Damianou, C.A.; Sanghvi, N.T.; Fry, F.J.; Maass-Moreno, R. Dependence of ultrasonic attenuation and absorption in dog soft tissues on temperature and thermal dose. *J. Acoust. Soc. Am.* **1997**, *102*, 628–634. [CrossRef] [PubMed]

11. Bamber, J.C.; Hill, C.R. Ultrasonic attenuation and propagation speed in mammalian tissues as a function of temperature. *Ultrasound Med. Biol.* **1979**, *5*, 149–157. [CrossRef]

12. Choi, M.J.; Guntur, S.R.; Lee, J.M.; Paeng, D.G.; Lee, K.I.L.; Coleman, A. Changes in Ultrasonic Properties of Liver Tissue In Vitro During Heating-Cooling Cycle Concomitant with Thermal Coagulation. *Ultrasound Med. Biol.* **2011**, *37*, 2000–2012. [CrossRef] [PubMed]

13. Guntur, S.R.; Lee, K.I.; Paeng, D.G.; Coleman, A.J.; Choi, M.J. Temperature-Dependent Thermal Properties of Ex Vivo Liver Undergoing Thermal Ablation. *Ultrasound Med. Biol.* **2013**, *39*, 1771–1784. [CrossRef] [PubMed]

14. Hallaj, I.M.; Cleveland, R.O.; Hynynen, K. Simulations of the thermo-acoustic lens effect during focused ultrasound surgery. *J. Acoust. Soc. Am.* **2001**, *109*, 2245–2253. [CrossRef] [PubMed]

15. Christopher, W.C.; Kullervo, H. Bio-acoustic thermal lensing and nonlinear propagation in focused ultrasound surgery using large focal spots: A parametric study. *Phys. Med. Biol.* **2002**, *47*, 1911–1928.

16. Doinikov, A.A.; Novell, A.; Calmon, P.; Bouakaz, A. Simulations and measurements of 3-D ultrasonic fields radiated by phased-array transducers using the westervelt equation. *IEEE Trans. Ultrason. Ferroelectr. Freq. Control* **2014**, *61*, 1470–1477. [CrossRef] [PubMed]

17. Solovchuk, M.A.; Sheu, T.W.H.; Thiriet, M.; Lin, W.L. On a computational study for investigating acoustic streaming and heating during focused ultrasound ablation of liver tumor. *Appl. Therm. Eng.* **2013**, *56*, 62–76. [CrossRef]

18. Liu, C.; Jayathilake, P.G.; Khoo, B.C. Perturbation method for the second-order nonlinear effect of focused acoustic field around a scatterer in an ideal fluid. *Ultrasonics* **2014**, *54*, 576–585. [CrossRef] [PubMed]

19. Mur, G. Absorbing Boundary Conditions for the Finite-Difference Approximation of the Time-Domain Electromagnetic-Field Equations. *IEEE Trans.* **1981**, *-23*. [CrossRef]

20. Pennes, H.H. Analysis of tissue and arterial blood temperatures in the resting human forearm. *J. Appl. Physiol.* **1948**, *1*, 93–122. [CrossRef] [PubMed]

21. Sapareto, S.A.; Dewey, W.C. Thermal dose determination in cancer therapy. *Int. J. Radiat. Oncol.* **1984**, *10*, 787–800. [CrossRef]

22. Jiang, S.; Zhang, X. Effects of dynamic changes of tissue properties during laser-induced interstitial thermotherapy (LITT). *Laser Med. Sci.* **2005**, *19*, 197–202. [CrossRef] [PubMed]

23. Shibib, K.S.; Munshid, M.A.; Lateef, H.A. The effect of laser power, blood perfusion, thermal and optical properties of human liver tissue on thermal damage in LITT. *Laser Med. Sci.* **2017**, *32*, 2039–2046. [CrossRef] [PubMed]

24. Zhou, J.; Chen, J.K.; Zhang, Y. Simulation of Laser-Induced Thermotherapy Using a Dual-Reciprocity Boundary Element Model with Dynamic Tissue Properties. *IEEE Trans. Biomed. Eng.* **2010**, *57*, 238–245. [CrossRef] [PubMed]

25. London, R.A.; Glinsky, M.E.; Zimmerman, G.B.; Bailey, D.S.; Eder, D.C.; Jacques, S.L. Laser–tissue interaction modeling with LATIS. *Appl. Opt.* **1997**, *36*, 9068–9074. [CrossRef] [PubMed]

26. Hallaj, I.M.; Cleveland, R.O. FDTD simulation of finite-amplitude pressure and temperature fields for biomedical ultrasound. *J. Acoust. Soc. Am.* **1999**, *105*, L7–L12. [CrossRef] [PubMed]

27. Yang, D.; Converse, M.C.; Mahvi, D.M.; Webster, J.G. Expanding the Bioheat Equation to Include Tissue Internal Water Evaporation During Heating. *IEEE Trans. Biomed. Eng.* **2007**, *54*, 1382–1388. [CrossRef] [PubMed]

28. Roggan, A.; Mueller, G.J. Two-dimensional computer simulations for real-time irradiation planning of laser-induced interstitial thermotherapy (LITT). In *Medical Applications of Lasers II*; International Society for Optics and Photonics: Lille, France, 1994; pp. 242–253.

Article

Fingerprinting Acoustic Localization Indoor Based on Cluster Analysis and Iterative Interpolation

Shuopeng Wang [1,*], Peng Yang [1,2] and Hao Sun [1,2]

[1] School of Artificial Intelligence, Hebei University of Technology, Tianjin 300130, China; yphebut@163.com (P.Y.); sunhao@hebut.edu.cn (H.S.)
[2] Engineering Research Center of Intelligent Rehabilitation, Ministry of Education, Tianjin 300130, China
* Correspondence: wangsp87921@hotmail.com; Tel.: +86-183-2260-1354

Received: 6 August 2018; Accepted: 4 October 2018; Published: 10 October 2018

Abstract: Fingerprinting acoustic localization usually requires tremendous time and effort for database construction in sampling phase and reference points (RPs) matching in positioning phase. To improve the efficiency of this acoustic localization process, an iterative interpolation method is proposed to reduce the initial RPs needed for the required positioning accuracy by generating virtual RPs in positioning phase. Meanwhile, a two-stage matching method based on cluster analysis is proposed for computation reduction of RPs matching. Results reported show that, on the premise of ensuring positioning accuracy, two-stage matching method based on feature clustering partition can reduce the average RPs matching amount to 30.14% of the global linear matching method taken. Meanwhile, the iterative interpolation method can guarantee the positioning accuracy with only 27.77% initial RPs of the traditional method needed.

Keywords: fingerprinting acoustic localization; iterative interpolation; K-Means clustering; Two-stage matching; Adjacent RPs

1. Introduction

With the development of signal processing technology and artificial intelligence technology, voice interaction has been gaining extensive attention in the smart device field [1–3]. Nowadays, the autonomous robot, as the representative of intelligent equipments, is expected to interact with people in a human-like way [4], and voice interaction can effectively improve its intelligence level. During human–robot interaction (HRI) process, acoustic localization technology can provide necessary reference for robot's pose adjustment to enhance the HRI reliability [5,6]. In recent years, great advancement of theory and application has been made in acoustic localization field. Most existing acoustic localization methods are parametric positioning methods, which are based on the space geometrical propagation models of acoustical signal [6–11]. Usually, these models are simplified with the assumptions of the sound source and the transmission channel listed as follow:

(1) The sound source is a particle without size and shape.
(2) The signal propagates in homogeneous space.
(3) The sound signal is omnidirectional.

The geometry model acoustic localization methods can achieve acceptable results outdoors, where the actual signal propagation model is similar to the ideal assumptions mentioned above. However, for indoor circumstances, the signal propagation model may be altered by the multipath effect, shadowing effect, fading effect, and delay distortion from walls, floors, furniture, or ceilings [12,13]. Meanwhile, it is difficult to provide compensation for model distortion analytically [14,15].

Different from the acoustic localization methods based on geometry model, the fingerprinting acoustic localization method simply adopted in our previous work [16], as a nonparametric location approach, can effectively accomplish sound positioning task according to the idea of environment perception. Compared with the precondition of the parametric localization method mentioned above, avoiding dramatic environment changes in target area, as the necessary requirement of the fingerprinting localization method, is easier in the practical application [17,18].

Many studies indicate that the positioning accuracy of non-parametric positioning method largely depends on the sampling density [19,20]. Therefore, for high resolution positioning indoor, considerable amount of sampling work for the database construction is needed in the offline phase. Additionally, during the online phase, the involved algorithms need considerable number of data, and large amounts of memory and computation resources to carry out the target position estimation in real time.

Interpolation is a mathematical tool to estimate the value of a function at a certain point using available values at other arguments. Interpolation methods for scattered data are widely implemented in mathematical, industrial and manufacturing applications. Radial basis function (RBF) [21], Linear [22], Inverse Distance Weighting (IDW) [23], and kriging [24] are well-known interpolation methods for positioning database expansion. Even though the numerous methods are effective to reduce sampling task quantity under the premise of ensuring the positioning accuracy [25], there are still many problems in the existing interpolation methods to be solved. In our previous work [26], the interpolation methods were executed in global interpolation way, which resulted in the rapid expansion of virtual RPs quantity and increase the calculation amount for RP matching [27]. Meanwhile, the conventional interpolation methods usually rely on the experience of the implementer and cannot accurately reckon the quantity of virtual RPs needed. In this paper, we propose an iterative interpolation method to refine the interpolation scope and, at the same time, the interpolation process can be monitored to avoid unnecessary virtual RPs. The estimation result of each iteration process can be compared with the one of the previous iteration process, and the interpolation will end when the different of the estimation results between the two adjacent process is less then the given threshold value.

The Selective matching combination of the target point (TP) and the RPs can reduce the matching task, thus improve the positioning efficiency of fingerprinting acoustic localization. Therefore, the positioning database is considered to be divided into a certain number of sub-databases in the offline phase, and then the matching scope can be shrunk to a smaller one through the search of adjacent sub-database [28]. For the database division method, the coordinate space partition is investigated firstly, which is easy to implement and can reduce the influence of outliers. However, this method has the defects of uncertain partitioning results and large positioning error. That is mainly because the division result is greatly influenced by the subjective judgment in the coordinate space partition. The adjacent RPs of the TP may be divided into different sub-databases, which will cause the RPs matching error. In machine learning technology, cluster analysis, as a precursor process of other learning tasks, is often used for classification of unlabeled samples. The RPs with similar features can be assigned to the same sub-database automatically by cluster analysis technique to accomplish the purpose of the database partition [29].

The rest of the paper is organized as follows: In Section 2, the general process of the fingerprinting acoustic localization is briefly introduced. In Section 3, the positioning database partition by cluster analysis and the adjacent RPs searching based on the two-stage matching method are stated, and then the iterative interpolation method is proposed to generate the virtual RPs for ensuring the target position estimation accuracy with few initial RPs. Section 4 presents the implementation details and evaluates the performance of the novel methods from the results obtained. Finally, some conclusions are drawn in Section 5.

2. The Fingerprinting Localization Model

Fingerprinting localization method is a database matching approach. As Figure 1 shows, the fingerprinting localization method uses the position information and the related features measured in the target region to establish the positioning database. In the actual positioning, the signal captured by the positioning system will match with the samples in the positioning database, and the samples most similar to the target signal are selected to accomplish the position estimation.

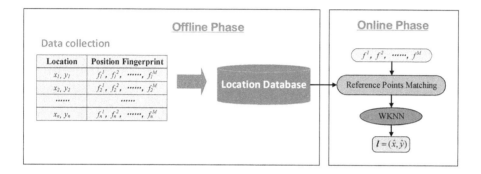

Figure 1. Illustration of the fingerprinting localization process.

As what has been introduced in our previous work [16], fingerprinting acoustic localization method requires an offline phase to construct the positioning database and an online phase for acoustic target location [30].

In the offline phase, the positioning database can be constructed by the coordinate of position marks and the corresponding features. The coordinates of the position marks are usually determined according to the site environment of the positioning area and the location accuracy requirements of the task. As the location-related feature in this work, time difference of arrival (TDOA) is widely used in real-time acoustic positioning applications for its low computational complexity and small data size [31–35]. Finally, samples, also known as position fingerprints, are formed by the coordinate of position marks and their corresponding features. The fingerprints in each position mark are collected and the location fingerprint database is established.

In the online phase, the feature vector of the observed target signal is matched with each sample of the positioning database. A specific number of samples are selected as the adjacent RPs according to the similarity with the target. Finally, the position of the target can be calculated by the specific position estimation algorithm based on the adjacent RPs according to the matching result. Exactly the same estimation algorithm used in the RADAR system, weighted-nearest neighbor (WKNN) [36] is usually used for the fingerprinting localization process.

3. The Proposed Fingerprinting Acoustic Localization Approach

In the traditional fingerprinting localization approach, the target signal needs to match with all samples in the database to select its adjacent RPs for location estimation. Therefore, the large scale positioning database that the fingerprinting localization accuracy depends on means the complexity of matching operation improvement and the efficiency of positioning reduction. Aiming at solving the contradictory in traditional fingerprinting localization approach, this paper makes some improvements. As Figure 2 shows, after the database construction in the offline sampling phase, the entire positioning database is divide into sub-databases. Then, in the online positioning phase, the matching scope can be narrowed by the adjacent sub-database matching stage, and the adjacent RPs searching can be accomplished with a small amount of matching computation.

Meanwhile, the offline phase of the fingerprinting localization approach requires a great sampling effort as the mobile sensor has to be placed at every position marks in the location area. To reduce the initial sampling effort, the database can be constituted using the sparse samples in the target area and extended afterwards by interpolation functions. As Figure 2 shows, an iterative interpolation method is presented to further refine the interpolation scope to avoid unnecessary virtual RPs. At the same time, the estimation result of each iteration process is compared with the one of the previous iterations, and the interpolation ends when the difference of the estimation results between the two adjacent process is less than the given threshold value.

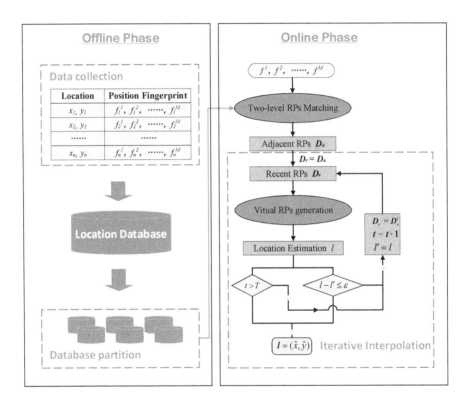

Figure 2. Fingerprinting acoustic localization process based on cluster analysis and iterative interpolation.

The novel acoustic localization approach consists of three main stages:

(1) Divide the positioning database into a certain number of sub-databases by K-Means clustering algorithm after database construction in the offline sampling phase.
(2) Select the adjacent RPs of target point by the two-stage RPs matching method.
(3) Conduct the position estimation of the target point by the iterative interpolation method.

3.1. Database Partition by Clustering Method

Clustering is an elements grouping process according to some specific features, which is called the cluster key, such as the TDOA value we choose in this paper. The prototype-based clustering method has the advantages of simple, fast and efficient process for big datasets classification. As one of the prototype-based clustering algorithms, K-Means clustering algorithm is a classical and efficient algorithm for cluster analysis [37].

The database partition process by K-Means clustering algorithm includes three steps. Suppose there is a database D with N samples that needs to be partitioned to K ($K < N$) groups. The clustering algorithm for the sound-position fingerprint database can be described as Algorithm 1.

Algorithm 1: The K-Means clustering algorithm for positioning database partition.

Input: database $D = [S_1, S_2, \cdots, S_N]^T$; cluster class number K.
Output: $D = \{C_1, C_2, \cdots, C_K\}$
1 Randomly selected K samples from D as the initial cluster centers $:\{\mu_1, \mu_2, \cdots, \mu_K\}$;
2 **while** $flag > 0$ **do**
3   cluster centers update flag: $flag = 0$;
4      $C_i = \emptyset (i = 1, 2, \ldots, K)$;
5   **for** $j = 1, 2, \ldots, N$ **do**
6   calculate the distance between S_j and each cluster center μ_i: $d_{ji} = \left\| S_j - \mu_i \right\|_2$;
7   definite the cluster mark of S_j by the nearest cluster center: $\lambda_j = \arg\min_{i \in \{1,2,\cdots,K\}} d_{ji}$;
8   classify the sample S_j into the corresponding cluster: $C_{\lambda_j} = C_{\lambda_j} \cup S_j$;
9   **for** $i = 1, 2, \ldots, K$ **do**
10   calculate the new cluster center $\mu'_i = \frac{1}{|C_i|} \sum_{S \in C_i} S$;
11   **if** $\mu'_i \neq \mu_i$ **then**
12   update the current value of μ_i to μ'_i;
13           $flag = flag + 1$;
14   **else**
15   the current value μ_i remain the same;

Firstly, K samples are randomly selected from the positioning database D as the initial cluster center $[\mu_1, \mu_2 \cdots \mu_K]$. Then, the remaining samples are assigned to the most similar clusters according to the similarity with each cluster center in feature space. Then, the cluster center is updated by $\mu_i = \frac{1}{|C_i|} \sum_{S \in C_i} S$, where S means the samples clustered to C_i. The clustering process is repeated until the cluster centers stop updating, and finally $D = \{C_1, C_2 \cdots C_K\}$.

3.2. Two-Stage RPs Matching

The vocal target can be located in the positioning area after the RPs sampling and the database construction. In the positioning phase, a two-stage matching algorithm is proposed to compare the feature vector of vocal target $F = [f^1, f^2, \cdots, f^M]$ with each simple in database D to find the adjacent RPs with the minimum matching error.

The Euclidean distance between target point F and cluster center μ_i of cluster C_i can be defined by:

$$Dis_i = \|F - \mu_i\|_2, \quad i = 1, 2, \ldots, K. \tag{1}$$

The adjacent cluster can be chosen through:

$$C_a = C_{\arg\min_{i \in \{1,2,\ldots,K\}} Dis_i} \tag{2}$$

Then, as shown in Figure 3b, the adjacent RPs can be searched according to the Euclidean distance distance dis_j between the target point and each sample of the adjacent cluster in feature space. The distance can be defined as:

$$dis_j = \left\| F - F_a^j \right\|_2, \quad j = 1, 2, \ldots, n_c. \tag{3}$$

where F_a^j is the feature vector of the jth RP in adjacent clustering C_a, and n_c denotes the total number of samples in C_a. Adjacent RPs set D_a can be gathered by:

$$D_a = D_a \cup S_{\arg\min_{j \in \{1,2,...,n_c\}} dis_j} \qquad (4)$$

Because the complexity of matching process is far greater than the other parts of the location process, and the computation complexity of the other parts in the positioning process is almost the same, the computational complexity of matching operations is investigated in this paper. N denotes the total number of RPs in the database and K is the number of clusters. The complexity of linear matching process is $O(N)$, and the average complexity of matching process based on cluster analysis is $O(N/K)$. Comparing with the conventional matching method, the proposed approach can reduce the complexity of matching process to its $1/K$.

Firstly, as Figure 3a shows, adjacent cluster is determined based on the Euclidean distance between the target point and each cluster center in feature space.

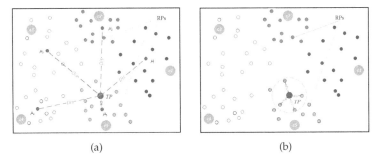

(a)   (b)

Figure 3. Two-stage RPs matching process: (**a**) the adjacent cluster matching process; and (**b**) the adjacent RPs matching process. The purple, orange, yellow, blue and green dots are the RPs clustered to different clusters marked as $c1$ to $c5$; the dark blue dots are the cluster centers marked as μ_1 to μ_5; the red dot denotes the TP; DIS means the Euclidean distance between TP and each cluster center; and dis means the Euclidean distance between TP and each RP in adjacent cluster.

3.3. Location Estimation Based on Iterative Interpolation

To reduce the sampling effort, global interpolation methods are usually used to improve the positioning accuracy under the sparse sample points collection. However, the global interpolation method usually results in the rapid expansion of virtual RPs quantity, and cannot accurately reckon the quantity of virtual RPs required for the satisfactory positioning accuracy. In this paper, we propose an iterative interpolation method to avoid the unnecessary virtual RPs and further improve the location efficiency.

In the online positioning process, the virtual RPs can be generated by the iterative interpolation method, as Figure 4 shows, where the iteration interpolation process is based on four adjacent RPs. In the interpolation process, the first generation virtual RPs are defined as the adjacent RPs selected before as $D_v^1 = D_a$. During the iteration interpolation process, the elements of D_v^t are refreshed by:

$$\begin{cases} S_n^{t+1} = \omega_n^t S_n^t + \omega_{n+1}^t S_{n+1}^t, & n < N \\ S_n^{t+1} = \omega_n^t S_n^t + \omega_1^t S_1^t, & n = N \end{cases} \qquad (5)$$

where S_n^t is the nth element of tth generation D_v^t, and ω_n^t is the according weight that can be calculated through IDW method by the feature space Euclidean distance between recent generation virtual RPs and the target point that needs to be located.

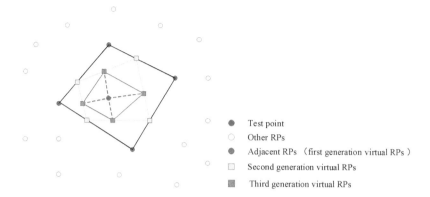

Figure 4. Virtual RPs generation process.

As Algorithm 2 shows, online positioning process based on the iteration process continues until the iteration time beyond the maximum iteration value T or the difference between two-estimation process results is satisfied with $|l - l'| \leq \varepsilon$, where ε is the iterative process end threshold value.

Algorithm 2: The location estimation process based on iterative interpolation.

    Input: adjacent RPs D_a

    Output: test point location estimation results l

1 iteration time $t = 1$;

2 first generation virtual RPs $D_v^1 = D_a$;

3 **while** $t < T \,\|\, |l - l'| \leq \varepsilon$ **do**

4      $t = t + 1$;

5      $l' = l$;

6   new generation virtual RPs D_v^{t+1} can be generated by Equation (5);

7   location estimation l can be calculated by the WKNN algorithm ;

---

4. Experimental Validation

To demonstrate the performance of the proposed acoustic localization approach based on the cluster analysis and iterative interpolation, real-world experiments have been carried out in a practical room. The room is w $9.64 \times 7.04 \times 2.95$ m^3, where the noise is about 40 dB and the walls are not insulated. The scene and equipment of the experiments are shown in Figure 5. The target area is a rectangular plane with the length of about 6 m and width of about 5 m. The four-channel microphone array is composed of the MPA201 microphones produced by the BSWA Technology Co., Ltd., Beijing, China. The microphones are installed at the four vertices of the positioning area with the height about 1.35 m above the floor. The type of the acquisition card is known as NI9215A from NI company, Austin, USA. The sampling frequency is set as 100 kHz, and the sampling period is 1 s. The sound source is a Bluetooth speaker with the same height as microphone array. A system-provided text tone called "Popcom" in iPhone 6 is selected for localization sound signal.

Figure 5. The fingerprint-based acoustic localization system and experiment scene.

In the sampling process, the coordinates of the samples are uniformly distributed in the location area by grid division, and the distance between each samples is 0.593 m. The total number of the samples prepared for database construction is 72, and 13 test points are used for target point estimation.

4.1. Analysis of the Two-Level RPs Matching Method

According to Section 3.3, more subsets in the positioning database partitioned means more online positioning efficient improvement by two-level RPs matching method. However, the same as the coordinate space partition method, when the subsets reaches a certain number, distinguishing between sub-databases partitioned by feature clustering partition method is no longer obvious. Then, the adjacent RPs may be divided into different sub-databases, which will cause RP matching error.

To investigate the effect of the division number on location accuracy, we explored the localization results with division number from 1 to 6, where 1 means the the matching process without partition, that is, global linear matching localization.

As shown in Figure 6, when division number increased from 2 to 4, positioning accuracy slightly improved compared with global linear matching. That is mainly because, according to the clustering results, the outlier points with large measurement errors in the sub-database can be eliminated by outlier test method. However, when the number of sub-databases increased to 5, the localization result began to deteriorate significantly. Moreover, when the division number increased to 6, the average error exceeded 0.18 m, while the maximum error reached 0.2780 m and 61.5% of the test point positioning accuracy could not meet the 0.20 m positioning requirements.

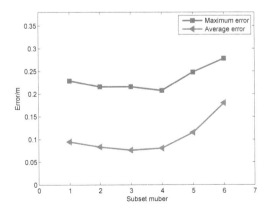

Figure 6. The effect of sub-database number on positioning.

To compare the positioning effect between the coordinate space partition method and feature clustering partition method, the average matching amount, matching time and the average positioning error were considered when the division number is 4. As shown in Table 1, the average matching amount and average matching time of the two different partition methods in the RPs matching process are basically the same, and their online positioning efficiencies greatly improved compared with the global linear matching method. Among them, the feature clustering partition method can reduce the average matching amount and the average matching time to 30.13% and 29.89% of the global linear matching method, respectively, while the coordinate space partition method was 30.97% and 30.13%. In comparison to the positioning accuracy, the positioning error of 0.0813 m based on feature clustering partition method is significantly superior to the 0.1214 m based on the coordinate partitioning partition method, and the positioning accuracy is improved by 13.97% compared with the traditional linear matching method.

Table 1. Comparison of the influence on positioning effect between database partition methods. A-amount, average matching amount; A-time, average matching time; A-error, average positioning error.

Cases	A-Amount	A-Time (s)	A-Error (m)
No partition	72	0.0271	0.0945
Coordinate partition	22.3	0.0084	0.1214
Cluster analysis partition	21.7	0.0081	0.0813

4.2. Analysis of the Iterative Interpolation Method

In this work, four adjacent RPs selected by the global linear matching method are used for target location estimation. In the process of virtual RPs generation, the maximum number of iterations is set as $T_{max} = 10$. In addition, when the difference between the results of two adjacent iterations interpolation positioning process is less than $\varepsilon = 0.0001$ m, the iteration process will end.

The global positioning results by examining the average error and maximum error of the positioning results of 13 test points has been evaluated. As shown in Figure 7, the iterative interpolation method can reduce the average error from 0.0945 m to 0.0406 m, and the maximum error from 0.2290 m to 0.0818 m. In the process of iterative interpolation, the effect of location accuracy improvement is obvious in the first six iterations. However, along with the iterative process and the improvement of positioning accuracy, the effect is gradually weakened. The same phenomenon also occurred at the maximum error.

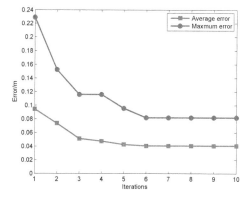

Figure 7. The changes of mean error and maximum error of location estimation according with the iterate interpolation process.

As Table 2 shows, 6 cases were considered to compare the positioning effect of different interpolation methods. The maximum errors and average errors were selected as the evaluation indicators. For iterative interpolation methods, the V-RPs is defined as the average value of the virtual RPs generated during the localization process of the test points.

Table 2. The location results comparison of different fingerprinting acoustic localization methods. I-RPs, initial RPs; V-RPs, virtual RPs; M-error, maximum positioning error; A-error, average positioning error.

Cases	Total RPs	I-RPs	V-RPs	M-Error (m)	A-Error (m)
No interpolation	20	20	0	0.4505	0.2385
	72	72	0	0.2290	0.0945
Global interpolation	255	20	235	0.1656	0.0772
	255	72	183	0.1330	0.0534
Iterative interpolation	48.9	20	28.9	0.1773	0.0791
	91.2	72	19.2	0.1542	0.0652

According to Table 2, in the fingerprinting acoustic localization process without interpolation, 72 RPs can provide apparently higher accuracy than the one with 20 RPs. The results confirmed the viewpoint that improving RPs density can directly improve the positioning accuracy.

In the cases of global interpolation method, the interpolation method can make further improvement for the positioning accuracy. On the other side, the initial RPs ratio can also affect the location results. That is, when the total RPs of the acoustic localization process based on global interpolation method are the same, more initial RPs means better positioning accuracy, but the influence of initial RPs ratio is weaker than the number of the total RPs.

In the case of iterative interpolation method, it is easy to see that iterative interpolation method needs only 12.3% virtual RPs of the global interpolation method for similar precise location results when the number of initial RPs is 20. When the number of initial RPs is 72, iterative interpolation method needs only 10.5% virtual RPs of the global interpolation method for a slightly less precise positioning results.

4.3. Analysis of the Novel Method

The fingerprinting acoustic localization approach based on iterative interpolation and cluster analysis is presented in this work. The positioning database consisting of 72 initial RPs is divided into four sub-databases by K-Means clustering algorithm, and four adjacent RPs selected by the two-stage matching method are used for 13 test point's location estimation based on iterative interpolation.

As Figure 8 shows, all of the estimated positions of the 13 target points obtained good concordance with the true positions. Meanwhile, the interpolation process at most target points ended in five iterations. Take Test Point 3, for instance: the location error decreased during iterative interpolation process and ended at the seventh iteration.

To analyze the influence at different test points, the position accuracy comparison of the novel method and the original method were taken on each test point. As Figure 9 shows, the novel method brought significant improvement of positioning accuracy for 11 of the 13 test points. In the novel method, the errors of Test Points 3, 7, 9, 11 and 12 decreased more than 50% from the original location method. However, Test Points 2, 4, 5, 6, 10 and 13 were not sensitive to interpolation process because they already had relatively high positioning accuracy. It must to be pointed out that the location results of Test Points 1 and 8 got worse and result in no apparent improvement in maximum error. That is because these points were located at the boundary of two sub-databases, and their adjacent RPs were assigned to different clusters by feature clustering partition method. The causes of location error are complex and varied; to further decrease the location error, improvements of other links in fingerprinting acoustic localization process are also needed.

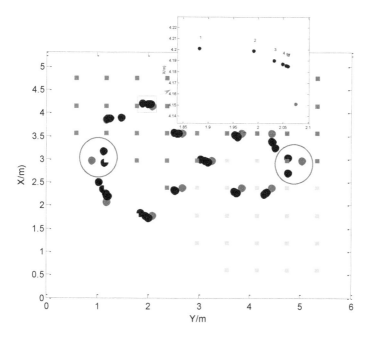

Figure 8. The positioning results of the fingerprinting acoustic localization based on iterative interpolation method. The green, pink, blue and yellow dots are the RPs clustered to different clusters, red dots denote the test points, and the black dots are the estimation results of each interpolation.

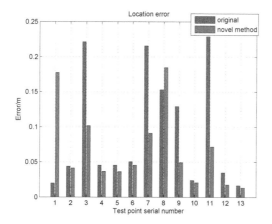

Figure 9. The positioning results comparison of the acoustic localization without interpolation process and acoustic localization base on iterative interpolation process.

5. Conclusions

In this paper, the iterative interpolation method and cluster analysis method has been presented for improving the positioning efficiency of indoor fingerprinting acoustic localization. In the fingerprinting acoustic localization process, the calibration efforts in offline phase can be reduced due to the sparse sampling treatment, and the satisfactory positioning accuracy can be guaranteed by virtual RPs generated by the iterative interpolation method. Meanwhile, the K-Means cluster analysis method was adopted for database partition, and a two-level RPs matching method was used to speeding up the online positioning phase. The results show that the fingerprinting acoustic localization method can achieve satisfactory accuracy with few initial RPs sampling in offline phase and a more rapid RPs matching process in online phase by iterative interpolation and cluster analysis. As future works, an extension of the clustering method to reduce the location results deterioration of the frontier points and various types of complex tasks for further verification of the novel method are being considered .

Author Contributions: S.W. conceived and designed the experiments and wrote the paper. P.Y. and H.S. contributed to project research scheme formulation. All authors contributed to the final version.

Funding: This research was funded by the National Natural Science Foundation of China (No. 61373017), Natural Science Foundation of Hebei Province (No. F2014202121), and Graduate Student Innovation Funding Project of Hebei Province (No. 220056).

Acknowledgments: The authors thank all the reviewers and editors for their valuable comments and work.

Conflicts of Interest: The authors declare no conflict of interest.

References

1. Zhen, J.G.; Aoki, H.; Sato-Shimokawara, E.; Yamaguchi, T. Interactive system for sharing objects information by gesture and voice recognition between human and robot with facial expression. In Proceedings of the IEEE/SICE International Symposium on System Integration, Fukuoka, Japan, 16–18 December 2012; pp. 293–298.
2. Li, X.F.; Liu, H. A survey of sound source localization for robot audition. *CAAI Trans. Intell. Syst.* **2012**, *1*, 9–20.
3. Yu, Q.; Yuan, C.; Fu, Z. Research of the Localization of Restaurant Service Robot. *Int. J. Adv. Robot. Syst.* **2010**, *7*, 227–238.
4. Breazeal, C. Social interactions in HRI: The robot view. *IEEE Trans. Syst. Man Cybern. Part C* **2004**, *34*, 181–186. [CrossRef]
5. Li, X.F.; Liu, H. Sound Source Localization for HRI Using FOC-Based Time Difference Feature and Spatial Grid Matching. *IEEE Trans. Cybern.* **2013**, *4*, 1199–1212. [CrossRef] [PubMed]
6. Park, J.S.; Kim, J.H.; Oh, Y.H. Feature vector classification based speech emotion recognition for service robots. *IEEE Trans. Consum. Electron.* **2009**, *3*, 1590–1596. [CrossRef]
7. Wang, H.; Kaveh, M. Coherent signal subspace processing for the detection and estimation of angles of arrival of mutiple wide-band sources. *IEEE Trans. Acoust. Speech Signal Process.* **1985**, *4*, 823–831. [CrossRef]
8. Wax, M.; Kailath, T. Optimum localization of multiple sourcesby passive arrays. *IEEE Trans. Acoust. Speech Signal Process.* **1983**, *5*, 1210–1217. [CrossRef]
9. Cater, G.C. Variance bounds for passively locating an acoustic source with a symmetric line array. *J. Acoust. Soc. Am.* **1977**, *4*, 922–926. [CrossRef]
10. Knapp, C.H.; Carter, G.C. The generalized correlation method for estimation of time delay. *IEEE Trans. Acoust. Speech Signal Process.* **1976**, *4*, 320–327. [CrossRef]
11. Carter, G.C. Special issue on time delay estimation. *IEEE Trans. Acoust. Speech Signal Process.* **1981**, *3*, 461–624. [CrossRef]
12. Liu, H.; Darabi, H.; Banerjee, P.; Jing, L. Survey of wireless indoor positioning techniques and systems. *IEEE Trans. Syst. Man Cybern. Part C* **2007**, *6*, 1067–1080. [CrossRef]
13. Dehkordi, M.B.; Abutalebi, H.R.; Taban, M.R. Sound source localization using compressive sensing-based feature extraction and spatial sparsity. *Digit. Signal Process.* **2013**, *4*, 1239–1246. [CrossRef]
14. Song, Z.; Jiang, G.; Huang, C. A Survey on Indoor Positioning Technologies. In Proceedings of the Second International Conference on International Conference on Theoretical and Mathematical Foundations of Computer Science, Singapore, 5–6 May 2011; pp. 198–206
15. Gu, Y.; Lo, A.; Niemegeers, I. A survey of indoor positioning systems for wireless personal networks. *IEEE Commun. Surv. Tutor.* **2009**, *11*, 13–32. [CrossRef]
16. Wang, S.P.; Sun, H.; Yang, P. Indoor sound-position fingerprint method based on scenario analysis. *J. Beijing Univ. Technol.* **2017**, *2*, 224–229.

Appl. Sci. **2018**, *8*, 1862

17. Wan, Q. From parametric localization to non-parametric localization. In *Indoor Positioning Theory, Method and Application*; Electronic Industry Press: Beijing, China, 2012; pp. 8–14.

18. Chen, Z.; Li, Z.; Wang, S.W.; Yin, F.L. A microphone position calibration method based on combination of acoustic energy decay model and TDOA for distributed microphone array. *Appl. Acoust.* **2015**, *95*, 13–19. [CrossRef]

19. He, S.; Ji, B.; Chan, S.H.G. Chameleon: Survey-free updating of a fingerprint database for indoor localization. *IEEE Pervasive Comput.* **2016**, *15*, 66–75. [CrossRef]

20. Chen, L.N.; Li, B.H.; Zhao, K.; Rizos, C.; Zheng, Z.Q. An Improved Algorithm to Generate a Wi-Fi Fingerprint Database for Indoor Positioning. *Sensors* **2013**, *8*, 11085–11096. [CrossRef] [PubMed]

21. Krumm, J.; Platt, J.C. *Minimizing Calibration Effort for an Indoor 802.11 Device Location Measurement System*; Technical Report for MSRTR-2003-82; Microsoft Research Microsoft Corporation One Microsoft Way: Washington, DC, USA, 2003.

22. Li, B.; Wang, Y.; Lee, H.K.; Dempster, A.; Rizos, C. Method for yielding a database of location fingerprints in WLAN. *IEE Proc.-Commun.* **2005**, *152*, 580–586. [CrossRef]

23. Ouyang, R.W.; Wong, K.S.; Lea, C.T.; Chiang, M. Indoor Location Estimation with Reduced Calibration Exploiting Unlabeled Data via Hybrid. *IEEE Trans. Mob. Comput.* **2012**, *11*, 1613–1626. [CrossRef]

24. Kuo, S.P.; Tseng, Y.C. Discriminant minimization search for largescale RF-based localization systems. *IEEE Trans. Mob. Comput.* **2011**, *10*, 291–304.

25. Lee, M.; Han, D. Voronoi Tessellation Based Interpolation Method for Wi-Fi Radio Map Construction. *IEEE Commun. Lett.* **2012**, *16*, 404–407. [CrossRef]

26. Yang, P.; Xu, J.; Wang, S. Position fingerprint localization method based on linear interpolation in robot auditory system. In Proceedings of the Chinese Automation Congress, Jinan, China, 20–22 October 2017; pp. 2766–2771.

27. Atia, M.M.; Noureldin, A.; Korenberg, M.J. Dynamic online-calibrated radio maps for indoor positioning in wireless local area networks. *IEEE Trans. Mob. Comput.* **2013**, *9*, 1774–1787. [CrossRef]

28. Yook, D.; Lee, T.; Cho, Y. Fast sound source localization using two-level search space clustering. *IEEE Trans. Cybern.* **2016**, *46*, 20–26. [CrossRef] [PubMed]

29. Abusara, A.; Hassan, M.S.; Ismail, M.H. Reduced-complexity fingerprinting in WLAN-based indoor positioning. *Telecommun. Syst.* **2017**, *65*, 407–417. [CrossRef]

30. Au, A.W.S.; Chen, F.; Shahrokh, V.; Sophia, R.; Sameh, S.; Samuel, N.M.; Deborah, G.; Keith, G.; Moshe, E. Indoor tracking and navigation using received signal strength and compressive sensing on a mobile device. *IEEE Trans. Mob. Comput.* **2013**, *10*, 2050–2062. [CrossRef]

31. Steen, K.A.; Mcclellan, J.H.; Green, O.; Karstoft, H. Acoustic source tracking in long baseline microphone arrays. *Appl. Acoust.* **2015**, *87*, 38–45. [CrossRef]

32. Wang, G.; Li, Y.; Ansari, N.A. Semidefinite Relaxation Method for Source Localization Using TDOA and FDOA Measurements. *IEEE Trans. Veh. Technol.* **2013**, *2*, 853–862. [CrossRef]

33. Kim, U.H.; Nakadai, K.; Okuno, H.G. Improved sound source localization in horizontal plane for binaural robot audition. *Appl. Intell.* **2015**, *1*, 63–74. [CrossRef]

34. Kwak, K.C.; Kim, S.S. Sound source localization with the aid of excitation source information in home robot environments. *IEEE Trans. Consum. Electron.* **2008**, *2*, 852–856. [CrossRef]

35. Tian, Y.; Chen, Z.; Yin, F.L. Distributed Kalman filter-based speaker tracking in microphone array networks. *Appl. Acoust.* **2015**, *89*, 71–77. [CrossRef]

36. Bahl, P.; Padmanabhan, V.N. RADAR: An In-Building RF-based User Location and Tracking System. In Proceedings of the Nineteenth Joint Conference of the IEEE Computer and Communications Societies, Tel Aviv, Israel, 26–30 March 2000; pp. 775–784.

37. Zhou, Z.H. Prototype-based clusterings. In *Machine Learning*; Tsinghua University Press: Beijing, China, 2016; pp. 202–211.

Article

Spatial Information on Voice Generation from a Multi-Channel Electroglottograph

Lamberto Tronchin [1,*], Malte Kob [2] and Claudio Guarnaccia [3]

[1]   Department of Architecture, University of Bologna, 47521 Cesena (FC), Italy
[2]   Hochschule für Musik, University of Detmold, 32756 Detmold, Germany; kob@hfm-detmold.de
[3]   Department of Civil Engineering, University of Salerno, 84084 Fisciano (SA) Italy; cguarnaccia@unisa.it
*   Correspondence: lamberto.tronchin@unibo.it; Tel.: +39-0547-338-356

Received: 13 August 2018; Accepted: 3 September 2018; Published: 5 September 2018

Abstract: In the acoustics of human voice, an important role is reserved for the study of larynx movements. One of the most important aspects of the physical behavior of the larynx is the proper description and simulation of swallowing and singing register changes, which require complex laryngeal *manoeuvres*. In order to describe (and solve, in some cases) these actions, it is fundamental to analyze the accurate synchronization of vocal fold adduction/abduction and the change of the larynx position. In the case of dysfunction, which often occurs for professional singers, this synchronization can be disturbed. The simultaneous assessment of glottal dynamics (typically electroglottograph, EGG signal) and larynx position might be useful for the diagnosis of disordered voice and swallowing. Currently, it is very difficult to instantaneously gather this information because of technology problems. In this work, we implemented a time-multiplex measurement approach of space-resolved transfer impedances through the larynx (Multi-Channel electroglottograph MC-EGG). For this purpose, we developed specific software (Labview code) for the visualization of the main waveforms in the study of the EGG signals. Moreover, the data acquired by the Labview code have been used to create a theoretical algorithm for deriving the position of the larynx inside the neck. Finally, we verified the results of the algorithm for the 3D larynx movement by comparing the data acquired with the values described in the literature. The paths of the larynx and the displacement on the sagittal and transverse plans matched the ones known for the emission of low/high notes and for swallowing. Besides, we have introduced the possibility to study the movement on the coronal (x) plan (so far, unexplored), which might be a starting point for further analysis.

Keywords: voice generation; multichannel electroglottograph; larynx acoustics

---

1. Introduction

The study of musical acoustics includes several aspects about the physics of musical instruments, and the main purpose consists of describing their sound [1], including the development of new physical parameters [2]. One of the most important applications of these studies is to emulate their sound by means of the proper description of their behavior, by means of convolution between the music piece played by the musician and impulse responses of the instrument [3]. However, sound production in humans is a complex process depending on different singing styles, which involves several anatomic structures [4]. This process is responsible for the generation of formant frequencies [5]. For these reasons, it is necessary to properly describe their movements, also including nonlinear aspects, in order to emulate nonlinearities using novel approaches [6,7]. Considering these aspects, it would be feasible to obtain a proper reconstruction of the diffuseness of musical signals for subjective evaluations [8].

Nevertheless, the interest in the description and modelling of the phonetic act includes researchers working in medicine and singing teaching. This interest has grown in the last few years and is continuing to grow even more.

Scientific studies of the human voice started with Helmholtz, who gave a detailed explanation of this phenomenon in 1863, describing that the voice is produced by a steady flow of air from the lungs, segmented at the laryngeal level into a series of air puffs at a fundamental frequency (f_0) that generates higher harmonics in the cavity of the upper airway. The supra-laryngeal cavity plays the role of a resonator, only filtering some frequencies, and finally the mouth and nose cavities modify the air flux, generating sound [9].

Mechanically, the phenomenon can be compared with the pression provoked by a piston. The air pressure forces the vocal folds to open. As the suction produced by the drop in pressure in the region of the folds plus static tissue forces begins to counterbalance the subglottic pressure in the region of the lungs, the folds begin to move inward, and the narrowing channel causes an increase in suction until the folds snap shut. Once the vocal fold cycle is completed, the folds return to the starting position.

Complex laryngeal manoeuvres occur during swallowing and singing register changes. These actions require an accurate synchronization of vocal fold adduction or abduction and the change of the larynx position. The simultaneous assessment of glottal dynamics and larynx position could be beneficial for several reasons: it might be an important instrument for the diagnosis of disordered voice or speech production and swallowing, it might be useful in the research of effective correlations between the control of the speech frequency f_0 and the position of the larynx, and it can also be an instrument for the mechanic evaluation of singing techniques. Currently, the existing tools normally available do not allow this simultaneous assessment because of their features (e.g., the incompatibility between MRI and other electric devices) or low resolution (e.g., CT).

For the aforementioned reasons, there is interest in a device which might be capable of making both the measurements at the same time. This is the reason why a prototype of MC-EGG (Multi-Channel Electroglottograph) was realized. This new device differs from a standard EGG in that more electrodes are rapidly switched to give information about the larynx position inside the neck [10].

2. Multi-Channel Electroglottograph

There are several different devices that might be used for the evaluation of the glottal dynamic. One of the most important is the EGG, which was utilized in this research [11].

This device evaluates the TEC (Transverse Electrical Conductance) between two electrodes placed on the sides of the neck. The first electrode sends a low intensity-high frequency current stimulus that is received by the second electrode.

The typical EGG signal appears as in Figure 1; the maximum conductance is at the maximum contact point of vocal folds and the minimum is at the maximum opening point. A standard EGG has two electrodes (one sender and one receiver), while the MC-EGG uses two six-electrode arrays (Figures 2 and 3).

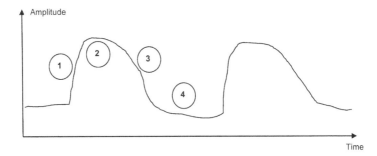

Figure 1. Phases of the idealized EGG waveform related to the vibratory cycle of the folds: 1: closing phase; 2: maximum contact; 3 opening phase; 4: open, no contact.

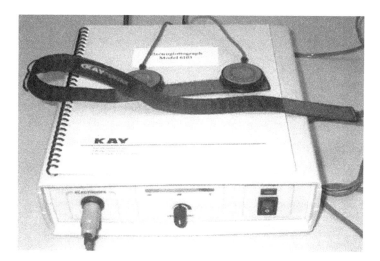

Figure 2. A standard EGG.

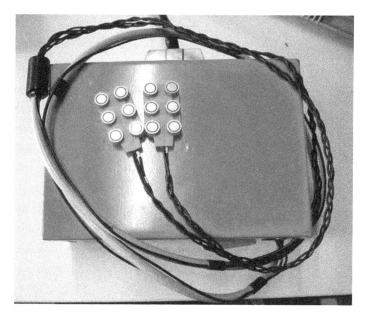

Figure 3. An example of MC-EGG, used for the experiments.

For each electrode's switch in the transmitter array, we have a fast switch of all the electrodes on the receiving array (every 25 ms). In this way, we could obtain all the 36 possible paths of current inside the neck [10]. In other words, by using an MC-EGG, it is possible to simultaneously obtain much more information if compared with a normal EGG. In this way, the resolution of the possible movements of the larynges increases. Further information about the behavior of the MC-EGG could be found in [10].

3. Methodologies

In order to describe the laryngeal *manoeuvres*, it was very important to focus on acquiring, visualizing, and saving data on a computer from the MC-EGG. Moreover, an algorithm for the evaluation of the larynx position inside the neck has been developed.

For the acquisition, we used a DAQ 6035E, 38.5 kHz, and we developed a Labview (National Instrument) tool to interface the device with a laptop. This tool included a user interface (Front Panel) and a code interface (Block Diagram), following the numerical description of the phenomenon [12].

The Front Panel tool consisted of a macro-box with three folders (Figure 4): the first was used for the electrodes' positioning; the second, called the EGG, was developed to acquire data and to evaluate the larynx position inside the neck. The same box also includes a graph that shows the real-time dynamic of one channel (user defined), which represents the TEC variation in time (the typical EGG signal that evaluates the glottal dynamic).

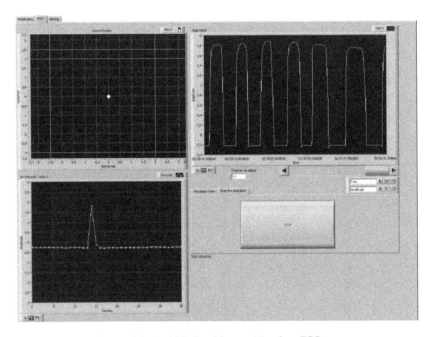

Figure 4. Labview Main Panel for acquiring data: EGG page.

The second folder included another box that allowed the user to set the simulation time or to manually stop it.

The third (last) folder, called "Setting", enclosed all the settable parameters. The acquired-data matrix is also visualized in that folder. This might be saved as a text file, which is useful for a Matlab post-processing, in a spreadsheet (.xls) file, or both [13].

The output matrix contained 36 columns, with each one representing a possible current path inside the neck between a sender and a receiver electrode: the number of rows depends on the simulation time. The algorithm for the evaluation of the larynx position has been developed in a "light" version, in terms of computational cost, in order to work online with Labview. In Matlab, the algorithm is more complex and more precise because it might work offline. The EGG signal consists of an AM (Amplitude Modulated) signal; its value is bigger when the current flows through the vocal folds' plane. On the other hand, it becomes smaller if that plane is partially crossed or not crossed at all. We approximated the field between two electrodes as a cylindrical shape. Therefore, we used the information given by the EGG signal to obtain information about the distance between the axis of each cylinder and the vocal folds' plane.

When the distance between electrodes is known, we can calculate all the 36 possible cylinders representing the 36 current paths (Figure 5).

The mathematical equation that should be solved for each cylinder is:

$$(X - X_{i0})^2 + (Y - Y_{i0})^2 + (Z - Z_{i0})^2 - \left[(X - X_{i0})v_{ijy} + (Y - Y_{i0})v_{ijy} + (Z - Z_{i0})v_{ijy}\right]^2 - R_{ij}^2 = C_{ij} \quad (1)$$

where $i = 1, \ldots, 6$ is the index for the sending electrodes; $j = 1, \ldots, 6$ is for the receiver ones; and X, Y, Z represents the larynx position coordinates, which start from position (X_{i0}, Y_{i0}, Z_{i0}). They should potentially assume any value inside the volume mapped by the two electrodes' arrays.

In Formula (1), C represents the "cost function". Therefore, for each set of X, Y, Z, we could obtain 36 possible cost functions. The sum of the cost function for all 36 paths will give the global cost function. The most probable point where the larynx is located is obtained by minimizing the global cost function, by varying the X, Y, Z sets.

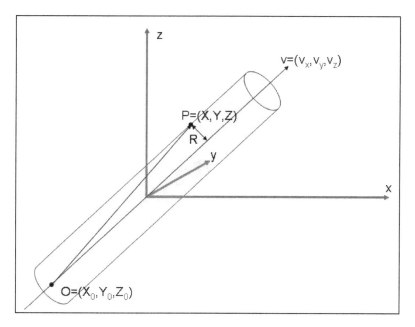

Figure 5. Cylindrical field between two electrodes.

Since in a volume we could localize infinite values for X, Y, and Z, it was necessary to divide them into finite elements in each direction, otherwise the problem could not be solved in a continuous medium.

In order to obtain a finite number of values for X, Y, and Z in a specific volume, we have divided the global volume through three grids on the main axis, evaluating only the intersection points. Nevertheless, even in this case, we would have had an excessive number of points, aiming to reach a good spatial resolution (i.e., millimeters). Figure 6 reports the position of MC-EGG for humans.

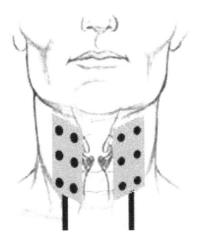

Figure 6. Application of MC-EGG on humans.

This problem has been solved using the EGG values, acquired at the beginning of each measure cycle, to restrict the number of possible values.

In order to reduce the number of tested points, we introduced two logic trees, one for the Y coordinates and the other one for the Z coordinates. These trees exclude, by logical operation of the 36 EGG signal, the zones of the mapped volume that could not be interested by the position of the larynx; in this way, we reduced the number of tested points. The full algorithm has been implemented in Matlab to obtain an accurate solution.

The lateral displacement of the larynx (on the X axis) has never been studied and there is no literature material about it; nevertheless, this algorithm gives the user the possibility to also set a displacement range on the X axis.

The Labview code worked online and the implementation of the whole algorithm was not possible; for this reason, we built another, lighter algorithm, that uses just the two logic trees to define a range of possible values on Y and Z. This algorithm considered the midpoint as the most probable point for the larynx position.

4. Comparison between Software and Experiments

The new developed software has been tested by studying the larynx displacement during two well-known vocal acts: the alternate emission of the vowel/**a**, first with a low note and then with a high note; and swallowing.

In order to guarantee a correct synchronization between the physical (measured) signal and the acquisition (samples acquired), the acquisition chain in Labview should be set to read data in Finite Mode. The sampling rate that allows the best synchronization was estimated to be around 38.5 kHz: using this sampling rate, we could acquire 36 samples every 3.5 ms. It is also important to remember that the EGG signals are normally studied during the emission of a low note characterized by a fundamental frequency of $f_0 \approx 100$ Hz. It is also important to note that during the alternate emission of a low and high note, the larynx has a marked displacement, in the range between 18 and 22 mm [14].

Moreover, the high level of background noise caused some difficulties of accuracy during the acquisition of the experimental data, since the EGG signal has a magnitude of around 1 mV, which is comparable with the background noise. However, this background noise, which represents the main issue during these experiments with an EGG device, is often discussed in scientific literature [14,15].

In order to reduce the background noise, the first attempt consisted of using a proper contact gel which could increase the ECC signal, improving the contact between the electrode and the skin surface. The second attempt consisted of using a higher voltage range, which was in the order of tens of mV.

The two codes were initially tested using the Labview code, and then the Matlab code. As expected, Labview allowed us to visualize the larynx in a downward position during the emission of the low note and an upward displacement during the emission of the same note at a higher frequency. There is no back-forward displacement of the larynx during this phonetic act. The range of displacement was 0 mm on the sagittal plane and 18.2 mm on the vertical plane, as described in scientific literature [15].

The data acquired by the Labview code was processed with the Matlab algorithm. In this way, the larynx movement was graphically visualized. The resulting movement was similar to the Labview one, and the resulting displacement was 0 mm for the sagittal plane and 18.4 mm for the vertical one. Besides the graphs, a video of the displacement was also obtained. Figure 7 reports some frames of the video.

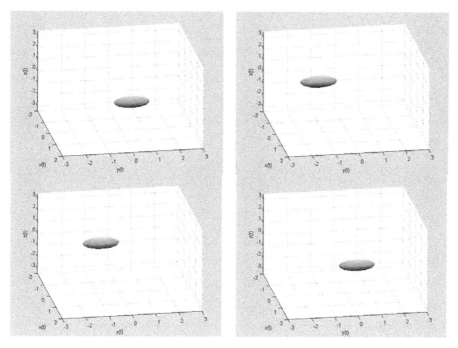

Figure 7. Frames from the created Matlab movie for the larynx movement.

During swallowing, the path of the larynx inside the neck is more complex; the larynx responds to this act by rising up in the first moment to push down the bolus, the epiglottis then moves backward to avoid the bolus penetration into the respiratory airways and, when the bolus is passed, the larynx returns to the original position.

The results obtained from the evaluation made through Labview and Matlab confirmed this path [13]. We recorded a 19.7 mm vertical displacement (both in Labview and Matlab), while the sagittal movement was 16.65 mm with Labview and 16.75 mm in Matlab. All these values are inside the range described by scientific literature [15]. Figure 8 reports the swallowing displacement as elaborated by Matlab.

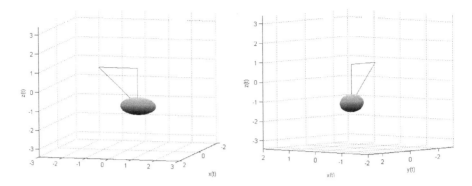

Figure 8. Trajectory of the larynx evaluated by the Matlab code for swallowing.

5. Conclusions

The purpose of this work was to develop a tool able to visualize the glottal dynamic and the displacement of the larynx inside the neck during phonetic acts. This task was possible to achieve by means of fast EGG data acquisition, properly designed and configured, and by means of the development of a specific algorithm to process the data acquired.

The Matlab code also allowed us to study the larynx displacement on the coronal plane. Currently, there is not enough knowledge about this kind of movement and this research could be a starting point for further analysis.

Moreover, there is the prospective to extend the potentiality of the numerical code for exploiting the number of electrodes. In this way, it might be possible to study the behavior of the ventricular (or false) vocal folds. These are not exactly vocal folds, because they have different tissues and do not display muscular activity. The false vocal folds are not usually used in the normal phonation, but could be used for some kinds of singing styles, and they take the place of the true vocal folds in some voice diseases. So far, the ventricular (false) vocal folds have been less investigated, but the interest in them is growing. This research could contribute to, for example, detecting voice disorders in a non-invasive way.

Author Contributions: L.T., M.K. and C.G. contributed equally for writing original draft, control, review and editing, for setting up the experiments and the codes, for formal analysis and funding.

Funding: This research received no external funding.

Acknowledgments: The Authors wish to thank Andrea Casadei for having collaborated with the measurements.

Conflicts of Interest: The authors declare no conflict of interest.

References

1. Campbell, D.M. Evaluating musical instruments. *Phys. Today* **2014**, *67*, 35–40. [CrossRef]
2. Tronchin, L. Modal analysis and Intensity of Acoustic Radiation of the kettledrum. *J. Acoust. Soc. Am.* **2005**, *117*, 926–933. [CrossRef] [PubMed]
3. Farina, A.; Tronchin, L. On the "Virtual" reconstruction of sound quality of trumpets. *Acta Acust. Acust.* **2000**, *86*, 737–745.
4. Kato, K.; Fujii, K.; Hirawa, T.; Kawai, K.; Yano, T.; Ando, Y. Investigation of the relation between minimum effective duration of running autocorrelation function and operatic singing with different interpretation styles. *Acta Acust. Acust.* **2007**, *93*, 421–434.
5. Nakayama, M.; Kato, K.; Matsunaga, M. Statistical analysis and modeling of formant frequencies of vowels phonated by traditional Japanese *Shigin* singers. *Int. J. Innov. Comput. Inf. Control* **2017**, *13*, 1441–1452.

6.   Tronchin, L. The emulation of nonlinear time-invariant audio systems with memory by means of Volterra series. *J. Audio Eng. Soc.* **2012**, *60*, 984–996.

7.   Tronchin, L.; Coli, V.L. Further investigations in the emulation of nonlinear systems with Volterra series. *J. Audio Eng. Soc.* **2015**, *63*, 671–683. [CrossRef]

8.   Shimokura, R.; Tronchin, L.; Cocchi, A.; Soeta, Y. Subjective diffuseness of music signals convolved with binaural impulse responses. *J. Sound Vib.* **2011**, *330*, 3526–3537. [CrossRef]

9.   Titze, I. *The Myoelastic Theory of Phonation*; National Center for Voice and Speech: Iowa City, IA, USA, 2006.

10.   Kob, M.; Frauenrath, T. A system for parallel measurement of glottis opening and larynx position. *Biomed. Signal Process. Control* **2009**, *4*, 221–228. [CrossRef]

11.   Maresek, K. Description of the EGG Waveform. Available online: www.ims.uni-stuttgart.de/phonetik/EGG/pagee2.htm (accessed on 4 September 2018).

12.   Deller, J.R.; Hansen, J.H.; Proakis, J.G. *Discrete-Time Processing of Speech Signals*; Macmillan: New York, NY, USA, 1993.

13.   Casadei, A.; Tronchin, L.; Kob, M. Derivation of Spatial Information from a Multi-Channel Electroglottograph. In Proceedings of the Forum Acusticum 2011, Aalborg, Denmark, 27 June–1 July 2011; pp. 415–419.

14.   Lindqvist, J.; Sawashima, M.; Hirose, H. *An Investigation of Vertical Movement of the Larynx in a Swedish Speaker*; Annual bulletin, No. 7; Research Institute of Logopedics and Phoniatrics: Tokyo, Japan, 1973.

15.   Abe, S.; Kaneko, H.; Nakamura, Y.; Watanabe, Y.; Shintani, M.; Hashimoto, M.; Yamane, G.; Ide, Y.; Shimono, M.; Ishikawa, T.; et al. Experimental device for detecting laryngeal movement during swallowing. *Bull. Tokyo Dent. Coll.* **2002**, *43*, 99–203. [CrossRef]

Article

Enhancing Target Speech Based on Nonlinear Soft Masking Using a Single Acoustic Vector Sensor

Yuexian Zou [1,*], Zhaoyi Liu [1] and Christian H. Ritz [2]

[1]   ADSPLAB, School of Electronic Computer Engineering, Peking University Shenzhen Graduate School,
      Shenzhen 518055, China; 1701213615@sz.pku.edu.cn
[2]   School of Electrical, Computer, and Telecommunications Engineering, University of Wollongong,
      Wollongong, NSW 2500, Australia; critz@uow.edu.au
*   Correspondence: zouyx@pkusz.edu.cn; Tel.: +86-75526032016

Received: 15 June 2018; Accepted: 25 July 2018; Published: 23 August 2018

Abstract: Enhancing speech captured by distant microphones is a challenging task. In this study, we investigate the multichannel signal properties of the single acoustic vector sensor (AVS) to obtain the inter-sensor data ratio (ISDR) model in the time-frequency (TF) domain. Then, the monotone functions describing the relationship between the ISDRs and the direction of arrival (DOA) of the target speaker are derived. For the target speech enhancement (SE) task, the DOA of the target speaker is given, and the ISDRs are calculated. Hence, the TF components dominated by the target speech are extracted with high probability using the established monotone functions, and then, a nonlinear soft mask of the target speech is generated. As a result, a masking-based speech enhancement method is developed, which is termed the AVS-SMASK method. Extensive experiments with simulated data and recorded data have been carried out to validate the effectiveness of our proposed AVS-SMASK method in terms of suppressing spatial speech interferences and reducing the adverse impact of the additive background noise while maintaining less speech distortion. Moreover, our AVS-SMASK method is computationally inexpensive, and the AVS is of a small physical size. These merits are favorable to many applications, such as robot auditory systems.

Keywords: Direction of Arrival (DOA); time-frequency (TF) mask; speech sparsity; speech enhancement (SE); acoustic vector sensor (AVS); intelligent service robot

1. Introduction

With the development of information technology, intelligent service robots will play an important role in smart home systems. Auditory perception is one of the key technologies of intelligent service robots [1]. Research has shown that special attention is currently being given to human–robot interaction [2], and especially speech interaction in particular [3,4]. It is clear that service robots are always working in noisy environments, and there are possible directional spatial interferences such as the competing speakers located in different locations, air conditioners, and so on. As a result, additive background noise and spatial interferences significantly deteriorate the quality and intelligibility of the target speech, and speech enhancement (SE) is considered the most important preprocessing technique for speech applications such as automatic speech recognition [5].

Single-channel SE and two-channel SE techniques have been studied for a long time, while practical applications have a number of constraints, such as limited physical space for installing large-sized microphones. The well-known single channel SE methods, including spectral subtraction, Wiener filtering, and their variations, are successful for suppressing additive background noise, but they are not able to suppress spatial interferences effectively [6]. Besides, mask-based SE methods have predominantly been applied in many SE and speech separation applications [7]. The key idea behind mask-based SE methods is to estimate a spectrographic binary or soft mask to suppress the unwanted spectrogram components [7–11]. For binary mask-based SE methods, the spectrographic masks are "hard binary masks" where a spectral component is either set to 1 for the target speech component or set to 0 for the non-target speech component. Experimental results have shown that the performance of binary mask SE methods degrades with the decrease of the signal-to-noise ratio (SNR) and the masked spectral may cause the loss of speech components due to the harsh black or white binary conditions [7,8]. To overcome this disadvantage, the soft mask-based SE methods have been developed [8]. In soft mask-based SE methods, each time-frequency component is assigned a probability linked to the target speech. Compared to the binary mask SE methods, the soft-mask SE methods have shown better capability to suppress the noise with the aid of some priori information. However, the priori information may vary with time, and obtaining the priori information is not an easy task.

By further analyzing the mask-based SE algorithms, we have the following observations. (1) It is a challenging task to estimate a good binary spectrographic mask. When noise and competing speakers (speech interferences) exist, the speech enhanced by the estimated mask often suffers from the phenomenon of "musical noise". (2) The direction of arrival (DOA) of the target speech is considered as a known parameter for the target SE task. (3) A binaural microphone and an acoustic vector sensor (AVS) are considered as the most attractive front ends for speech applications due to their small physical size. For the AVS, its physical size is about 1–2 cm^3 and AVS also has the merits such as signal time alignment and a trigonometric relationship of signal amplitudes [12–16]. A high-resolution DOA estimation algorithm with a single AVS has been proposed by our team [12–16]. Some effort has also been made for the target SE task with one or two AVS sensors [17–21]. For example, with the minimum variance distortionless response (MVDR) criterion, Lockwood et al. developed a beamforming method using the AVS [17]. Their experimental results showed that their proposed algorithm achieves good performance for suppressing noise, but brings certain distortion of the target speech.

As discussed above, in this study, we focus on developing the target speech enhancement algorithm with a single AVS from a new technical perspective in which both the ambient noise and non-target spatial speech interferences can be suppressed effectively and simultaneously. The problem formulation is presented in Section 2. Section 3 shows the derivation of the proposed SE algorithm. The experimental results are given in Section 4, and conclusions are drawn in Section 5.

2. Problem Formulation

In this section, the sparsity of speech in the time-frequency (TF) domain is discussed first. Then, the AVS data model and the corresponding inter-sensor data ratio (ISDR) models are presented for completeness, which was developed by our team in a previous work [13]. After that, the derivation of monotone functions between ISDRs and the DOA is given. Finally, the nonlinear soft TF mask estimation algorithm is derived specifically.

2.1. Time-Frequency Sparsity of Speech

In the research of speech signal processing, the TF sparsity of speech is a widely accepted assumption. More specifically, when there is more than one speaker in the same spatial space, the speech TF sparsity implies the following [5]. (1) It is likely that only one speaker is active during certain time slots. (2) For the same time slot, if more than one speaker is active, it is probable that the different TF points are dominated by different speakers. Hence, the TF sparsity of speech can be modeled as:

$$S_m(\tau, \omega) S_n(\tau, \omega) = 0, m \neq n \qquad (1)$$

where $S_m(\tau, \omega)$ and $S_n(\tau, \omega)$ are the speech spectral at (τ, ω) for the mth speaker and nth speaker, respectively. (3) In practice, at a specific TF point (τ, ω), it is most probably true that only one speech source with the highest energy dominates, and the contributions from the other sources can be negligible.

2.2. AVS Data Model

An AVS unit generally consists of J co-located constituent sensors, including one omnidirectional sensor (denoted as o-sensor) and J-1 orthogonally oriented directional sensors. Figure 1 shows the data capture setup with a single AVS. It is noted that the left bottom plot in Figure 1 shows a 3D-AVS unit implemented by our team, which consists of one o-sensor with three orthogonally oriented directional sensors depicted as the u-sensor, v-sensor, and w-sensor, respectively. In theory, the directional response of the oriented directional sensor has dipole characteristics, as shown in Figure 2a, while the omnidirectional sensor has the same response in all of the directions, as shown in Figure 2b. In this study, one target speaker is considered. As shown in Figure 1, the target speech $S(t)$ is impinging from (θ_s, ϕ_s) meanwhile, interference $S_i(t)$ are impinging from (θ_j, ϕ_j), where ϕ_s, $\phi_i \in (0°, 360°)$ are the azimuth angles, and θ_s, $\theta_i \in (0°, 180°)$ are the elevation angles.

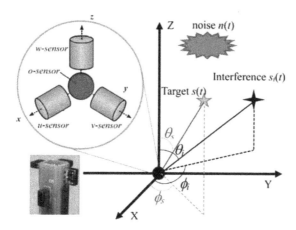

Figure 1. Illustration of a single acoustic vector sensor (AVS) for data capturing.

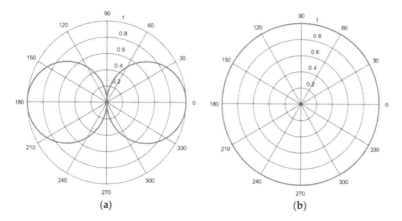

Figure 2. (a) The directional response of oriented directional sensor; (b) The directional response of omnidirectional sensor.

For simplifying the derivation, without considering room reverberation, the received data of the AVS can be modeled as [13]:

$$x_{avs}(t) = a(\theta_s, \phi_s)s(t) + \sum_{i=1}^{M_i} a(\theta_i, \phi_i)s_i(t) + n_{avs}(t) \tag{2}$$

where $x_{avs}(t)$, $n_{avs}(t)$ and $a(\theta_s, \phi_s)$ are defined respectively as:

$$x_{avs}(t) = [x_u(t), x_v(t), x_w(t), x_o(t)] \tag{3}$$

$$n_{avs}(t) = [n_u(t), n_v(t), n_w(t), n_o(t)] \tag{4}$$

$$a(\theta_s, \phi_s) = [u_s, v_s, w_s, 1]^T = [\sin\theta_s \cos\phi_s, \sin\theta_s \sin\phi_s, \cos\theta_s, 1]^T \tag{5}$$

$$a(\theta_i, \phi_i) = [u_i, v_i, w_i, 1]^T = [\sin\theta_i \cos\phi_i, \sin\theta_i \sin\phi_i, \cos\theta_i, 1]^T \tag{6}$$

In Equation (3), $x_u(t)$, $x_v(t)$, $x_w(t)$, $x_o(t)$ are the received data of the u-sensor, v-sensor, w-sensor, and o-sensor, respectively; $n_u(t)$, $n_v(t)$, $n_w(t)$, $n_o(t)$ are assumed as the additive zero-mean white Gaussian noise captured at the u-sensor, v-sensor, w-sensor, and o-sensor, respectively; $s(t)$ is the target speech; $s_i(t)$ are the ith interfering speech; the number of interferences is M_i; $a(\theta_s, \phi_s)$ and $a(\theta_j, \phi_j)$ are the steering vectors of $s(t)$ and $s_i(t)$, respectively. $[.]^T$ denotes the vector/matrix transposition.

From the AVS data model given in Equation (2), taking the short-time Fourier transform (STFT), for a specific TF point (τ, ω), we have:

$$X_{avs}(\tau, \omega) = a(\theta_s, \phi_s)S(\tau, \omega) + \sum_{i=1}^{M_i} a(\theta_i, \phi_i)S_i(\tau, \omega) + N_{avs}(\tau, \omega) \tag{7}$$

where $X_{avs}(\tau, \omega) = [X_u(\tau, \omega), X_v(\tau, \omega), X_w(\tau, \omega), X_o(\tau, \omega)]^T$; $X_u(\tau, \omega)$, $X_v(\tau, \omega)$, $X_w(\tau, \omega)$, and $X_o(\tau, \omega)$ are the STFT of $x_u(t)$, $x_v(t)$, $x_w(t)$, and $x_o(t)$, respectively. Meanwhile, $N_{avs}(\tau, \omega) = [N_u(\tau, \omega), N_v(\tau, \omega), N_w(\tau, \omega), N_o(\tau, \omega)]^T$; $N_u(\tau, \omega)$, $N_v(\tau, \omega)$, $N_w(\tau, \omega)$, and $N_o(\tau, \omega)$ are the STFT of $n_u(t)$, $n_v(t)$, $n_w(t)$, and $n_o(t)$, respectively. Since the target speech spectral is $S(\tau, \omega)$, let us define a quantity as follows:

$$N_{total}(\tau, \omega) = \sum_{i=1}^{M_i} a(\theta_i, \phi_i)S_i(\tau, \omega) + N_{avs}(\tau, \omega) \tag{8}$$

where we define $N_{total}(\tau, \omega) = [N_{tu}(\tau, \omega), N_{tv}(\tau, \omega), N_{tw}(\tau, \omega), N_{to}(\tau, \omega)]^T$ to represent the mixture of the interferences and additive noise. Therefore, from Equations (7) and (8), we have the following expressions:

$$X_u(\tau, \omega) = u_s S(\tau, \omega) + N_{tu}(\tau, \omega) \tag{9}$$

$$X_v(\tau, \omega) = v_s S(\tau, \omega) + N_{tv}(\tau, \omega) \tag{10}$$

$$X_w(\tau, \omega) = w_s S(\tau, \omega) + N_{tw}(\tau, \omega) \tag{11}$$

$$X_o(\tau, \omega) = S(\tau, \omega) + N_{to}(\tau, \omega) \tag{12}$$

In this study, we make the following assumptions. (1) $s(t)$ and $s_i(t)$ are uncorrelated and are considered as far-field speech sources; (2) $n_u(t)$, $n_v(t)$, $n_w(t)$ and $n_o(t)$ are uncorrelated. (3) The DOA of the target speaker is given as (θ_s, ϕ_s); the task of target speech enhancement is essentially to estimate $S(\tau, \omega)$ from $X_{avs}(\tau, \omega)$.

2.3. Monotone Functions between ISDRs and the DOA

Definition and some discussions on the inter-sensor data ratio (ISDR) of the AVS are presented in our previous work [13]. In this subsection, we briefly introduce the definition of ISDR first, and then present the derivation of the monotone functions between the ISDRs and the DOA of the target speaker.

The ISDRs between each channel of the AVS are defined as:

$$I_{ij}(\tau, \omega) = X_i(\tau, \omega)/X_j(\tau, \omega) \text{ where } (i \neq j) \tag{13}$$

where i and j are the channel index, which refers to u, v, w, and o, respectively. Obviously, there are 12 different computable ISDRs, which are shown in Table 1. In the following context, we carefully evaluate I_{ij}, and it is clear that only three ISDRs (I_{uv}, I_{vu} and I_{wo}) hold the approximate monotone function between ISDR and the DOA of the target speaker.

Table 1. Twelve computable inter-sensor data ratios (ISDRs).

Sensor	u	v	w	o
u	NULL	I_{vu}	I_{wu}	I_{ou}
v	I_{uv}	NULL	I_{wv}	I_{ov}
w	I_{uw}	I_{vw}	NULL	I_{ow}
o	I_{uo}	I_{vo}	I_{wo}	NULL

According to the definition of ISDRs given in Equation (13), we look at I_{uv}, I_{vu} and I_{wo} first. Specifically, we have:

$$I_{uv}(\tau, \omega) = X_u(\tau, \omega)/X_v(\tau, \omega) \tag{14}$$

$$I_{vu}(\tau, \omega) = X_v(\tau, \omega)/X_u(\tau, \omega) \tag{15}$$

$$I_{wo}(\tau, \omega) = X_w(\tau, \omega)/X_o(\tau, \omega) \tag{16}$$

Substituting Equations (9) and (10) into Equation (14) gives:

$$I_{uv}(\tau, \omega) = \frac{u_s S(\tau, \omega) + N_{tu}(\tau, \omega)}{v_s S(\tau, \omega) + N_{tv}(\tau, \omega)} = \frac{u_s + N_{tu}(\tau, \omega)/S(\tau, \omega)}{v_s + N_{tv}(\tau, \omega)/S(\tau, \omega)} = \frac{u_s + \varepsilon_{tus}(\tau, \omega)}{v_s + \varepsilon_{tvs}(\tau, \omega)} \tag{17}$$

where $\varepsilon_{tus}(\tau,\omega) = N_{tu}(\tau,\omega)/S(\tau,\omega)$, and $\varepsilon_{tvs}(\tau,\omega) = N_{tv}(\tau,\omega)/S(\tau,\omega)$.

Similarly, we get I_{uw} and I_{wo}:

$$I_{vu}(\tau, \omega) = \frac{v_s S(\tau, \omega) + N_{tv}(\tau, \omega)}{u_s S(\tau, \omega) + N_{tu}(\tau, \omega)} = \frac{v_s + N_{tv}(\tau, \omega)/S(\tau, \omega)}{u_s + N_{tu}(\tau, \omega)/S(\tau, \omega)} = \frac{v_s + \varepsilon_{tvs}(\tau, \omega)}{u_s + \varepsilon_{tus}(\tau, \omega)} \tag{18}$$

$$I_{wo}(\tau, \omega) = \frac{w_s S(\tau, \omega) + N_{tw}(\tau, \omega)}{S(\tau, \omega) + N_{to}(\tau, \omega)} = \frac{w_s + N_{tw}(\tau, \omega)/S(\tau, \omega)}{1 + N_{to}(\tau, \omega)/S(\tau, \omega)} = \frac{w_s + \varepsilon_{tws}(\tau, \omega)}{1 + \varepsilon_{tos}(\tau, \omega)} \tag{19}$$

In Equation (19), $\varepsilon_{tws}(\tau,\omega) = N_{tw}(\tau,\omega)/S(\tau,\omega)$ and $\varepsilon_{tos}(\tau,\omega) = N_{to}(\tau,\omega)/S(\tau,\omega)$.

Based on the assumption of TF sparsity of speech shown in Section 2.1, we can see that if the TF points (τ,ω) are dominated by the target speech from (θ_s,ϕ_s), the energy of the target speech is high, and the value of $\varepsilon_{tus}(\tau,\omega)$, $\varepsilon_{tvs}(\tau,\omega)$, $\varepsilon_{tws}(\tau,\omega)$ and $\varepsilon_{tos}(\tau,\omega)$ tends to be small. Then, Equations (17)–(19) can be accordingly approximated as:

$$I_{uv}(\tau, \omega) \approx u_s/v_s + \varepsilon_1(\tau, \omega) \tag{20}$$

$$I_{vu}(\tau, \omega) \approx v_s/u_s + \varepsilon_2(\tau, \omega) \tag{21}$$

$$I_{wo}(\tau, \omega) \approx w_s + \varepsilon_3(\tau, \omega) \tag{22}$$

where ε_1, ε_2, and ε_3 can be viewed as the ISDR modeling error with zero-mean introduced by interferences and background noise. Moreover, $\varepsilon_i(\tau,\omega)$ (i = 1, 2, 3) is inversely proportion to the local SNR at (τ,ω).

Furthermore, from Equation (5), we have $u_s = \sin\theta_s\cdot\cos\phi_s$, $v_s = \sin\theta_s\cdot\sin\phi_s$ and $w_s = \cos\theta_s$. Then, substituting Equation (5) into Equations (20)–(22), we obtain the following equations:

$$I_{uv}(\tau,\omega) \approx \frac{\sin\theta_s\cos\phi_s}{\sin\theta_s\sin\phi_s} + \varepsilon_1(\tau,\omega) = \cot\phi_s + \varepsilon_1(\tau,\omega) \tag{23}$$

$$I_{vu}(\tau,\omega) \approx \frac{\sin\theta_s\sin\phi_s}{\sin\theta_s\cos\phi_s} + \varepsilon_2(\tau,\omega) = \tan(\phi_s) + \varepsilon_2(\tau,\omega) \tag{24}$$

$$I_{wo}(\tau,\omega) \approx w_s + \varepsilon_3(\tau,\omega) = \cos(\theta_s) + \varepsilon_3(\tau,\omega) \tag{25}$$

From Equations (23)–(25), it is desired to see that the approximate monotone functions between I_{uv}, I_{vu}, and I_{wo} and the DOA (θ_s or ϕ_s) of the target speaker have been obtained since arccot, arctan, and arccos functions are all monotone functions.

However, except for I_{uv}, I_{vu}, and I_{wo}, other ISDRs do not hold such a property. Let's take I_{uw} as an example. From the definition in Equation (13), we can get:

$$I_{uw}(\tau,\omega) = \frac{u_s S(\tau,\omega) + N_{tu}(\tau,\omega)}{w_s S(\tau,\omega) + N_{tw}(\tau,\omega)} = \frac{u_s + N_{tu}(\tau,\omega)/S(\tau,\omega)}{w_s + N_{tw}(\tau,\omega)/S(\tau,\omega)} = \frac{u_s + \varepsilon_{tus}(\tau,\omega)}{w_s + \varepsilon_{tws}(\tau,\omega)} = \frac{u_s}{w_s} + \varepsilon_4(\tau,\omega) \tag{26}$$

where ε_4 can be viewed as the ISDR modeling error with zero-mean introduced by unwanted noise. Obviously, Equation (26) is valid when w_s is not equal to zero. Substituting Equation (5) into Equation (26) yields:

$$I_{uw}(\tau,\omega) \approx \frac{\sin\theta_s\cos\phi_s}{\cos\theta_s} + \varepsilon_4(\tau,\omega) = \tan\theta_s\cos\phi_s + \varepsilon_4(\tau,\omega) \tag{27}$$

From Equation (27), we can see that I_{uw} is a function of both θ_s and ϕ_s.

In summary, after analyzing all of the ISDRs, we find that the desired monotone functions between ISDRs and θ_s or ϕ_s, which are given in Equations (23)–(25), respectively. It is noted that Equations (23)–(25) are derived conditioned by assuming v_s, u_s, and w_s are not equal to zero. Therefore, we need to find out where v_s, u_s, and w_s are equal to zero. For presentation clarity, let's define an ISDR vector $\mathbf{I}_{isdr} = [I_{uv}, I_{vu}, I_{wo}]$.

From Equation (5), it is clear that when the target speaker is at angles of $0°$, $90°$, $180°$, and $270°$, one of v_s, u_s, and w_s becomes zero, and it means that $\mathbf{I}_{isdr}$ is not fully available. Specifically, we need to consider the following cases:

Case 1: the elevation angle θ_s is about $0°$ or $180°$. In this case, $u_s = \sin\theta_s\cdot\cos\phi_s$ and $v_s = \sin\theta_s\cdot\sin\phi_s$ are close to zero. Then, the denominator in Equations (20) and (21) is equal to zero, and we cannot obtain I_{uv} and I_{vu}, but we can get I_{wo}.

Case 2: θ_s is away from $0°$ or $180°$. In this condition, we need to look at ϕ_s carefully.

(1) ϕ_s is about $0°$ or $180°$. Then, $v_s = \sin\theta_s\sin\phi_s$ is close to zero, and the denominator in Equation (20) is equal to zero, which leads to I_{uv} being invalid. In this case, we can compute I_{vu} and I_{wo} properly.

(2) ϕ_s is about $90°$ or $270°$. Then, $u_s = \sin\theta_s\cdot\cos\phi_s$ is close to zero, and the denominator in Equation (21) is equal to zero, which leads to I_{vu} being invalid. In this case, we can obtain I_{uv} and I_{wo} properly.

(3) ϕ_s is away from $0°$, $90°$, $180°$, and $270°$, we can obtain all of the I_{uv}, I_{vu} and I_{wo} values properly.

To visualize the discussions above, a decision tree of handling the special angles in computing $\mathbf{I}_{isdr}$ is plotted in Figure 3.

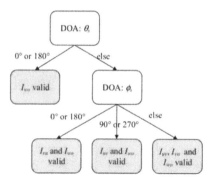

Figure 3. The decision tree of handling the special angles in computing I_{isdr}.

When $I_{isdr} = [I_{uv}, I_{vu}, I_{wo}]$ has been computed properly, with simple manipulation from Equations (23)–(25), we get:

$$\phi_s(\tau, \omega) = \text{arccot}(I_{uv}(\tau, \omega) - \varepsilon_1(\tau, \omega)) \tag{28}$$

$$\phi_s(\tau, \omega) = \arctan(I_{vu}(\tau, \omega) - \varepsilon_2(\tau, \omega)) \tag{29}$$

$$\theta_s(\tau, \omega) = \arccos(I_{wo}(\tau, \omega) - \varepsilon_3(\tau, \omega)) \tag{30}$$

From Equations (28)–(30), we can see that arccot, arctan, and arccos functions are all monotone functions, which are what we expected. Besides, we also note that (θ_s, ϕ_s) is given, and I_{uv}, I_{vu} and I_{wo} can be computed by Equations (14)–(16). However, ε_1, ε_2, and ε_3 are unknown, which reflect the impact of noise and interferences. According to the assumptions made in Section 2.1, if we are able to select the TF components (θ_s, ϕ_s) dominated by the target speech, and the local SNR at this (τ, ω) is high, then ε_1, ε_2, and ε_3 can be ignored, since they will have values approaching zero at these (τ, ω) points. In such conditions, we obtain the desired formulas to compute (θ_s, ϕ_s):

$$\phi_s(\tau, \omega) \approx \text{arccot}(I_{uv}(\tau, \omega)), \phi_s(\tau, \omega) \approx \arctan(I_{vu}(\tau, \omega)) \text{and} \theta_s(\tau, \omega) \approx \arccos(I_{wo}(\tau, \omega)) \tag{31}$$

2.4. Nonlinear Soft Time-Frequency (TF) Mask Estimation

As discussed above, Equation (31) is valid when the (τ, ω) points are dominated by target speech with high local SNR. Besides, we have three equations to solve two variables, θ_s and ϕ_s. In this study, from Equation (31), we estimate θ_s and ϕ_s in the following way:

$$\hat{\phi}_{s1}(\tau, \omega) = \text{arccot} I_{uv}(\tau, \omega) + \Delta \eta_1 \tag{32}$$

$$\hat{\phi}_{s2}(\tau, \omega) = \arctan I_{vu}(\tau, \omega) + \Delta \eta_2 \tag{33}$$

$$\hat{\phi}_s(\tau, \omega) = mean(\hat{\phi}_{s1}, \hat{\phi}_{s2}) \tag{34}$$

$$\hat{\theta}_s(\tau, \omega) = \arccos I_{wo}(\tau, \omega) + \Delta \eta_3 \tag{35}$$

where $\Delta \eta_1$ and $\Delta \eta_2$ are estimation errors. Comparing Equation (31) and Equations (32)–(35), we can see that if the estimated DOA values $(\hat{\phi}_s(\tau, \omega), \hat{\theta}_s(\tau, \omega))$ approximate to the real DOA values (θ_s, ϕ_s), then $\Delta \eta_1$ and $\Delta \eta_2$ should be small. Therefore, for the TF points (τ, ω) dominated by the target speech, we can derive the following inequality:

$$|\hat{\phi}_s(\tau, \omega) - \phi_s| < \delta_1 \tag{36}$$

$$|\hat{\theta}_s(\tau, \omega) - \theta_s| < \delta_2 \tag{37}$$

where $\hat{\phi}_s(\tau,\omega)$ and $\hat{\theta}_s(\tau,\omega)$ are the target speaker's DOA estimated by Equations (34) and (35), respectively. θ_s and ϕ_s are given the DOA of the target speech for the SE task. The parameters δ_1 and δ_2 can be set as the predefined permissible parameters (referring to an angle value). Following the derivation up to now, if Equations (36) and (37) are met at (τ,ω) points, we can infer that these (τ,ω) points are dominated by the target speech with high probability. Therefore, using Equations (36) and (37), the TF points (τ,ω) can be extracted, and a mask associated with these (τ,ω) points dominated by the target speech can be designed accordingly. In addition, we need to take the following facts into account. (1) The value of ϕ_s belongs to $(0,2\pi]$. (2) The principal value interval of the arccot function is $(0,\pi)$, and the arctan function is $(-\pi/2,\pi/2)$. (3) The value range of θ_s is $(0,2\pi]$. (4) The principal value interval of the arccos function is $[0,\pi]$. (5) To make the principal value of the anti-trigonometric function match the value of θ_s and ϕ_s, we need to add $L\pi$ to avoid ambiguity. As a result, a binary TF mask for preserving the target speech is designed as follows:

$$mask(\tau,\omega) = \begin{cases} 1, \text{ if } \begin{cases} \Delta\phi(\tau,\omega) = \left|\hat{\phi}_s(\tau,\omega) - \phi_s + L\pi\right| < \delta_1 \\ \Delta\theta(\tau,\omega) = \left|\hat{\theta}_s(\tau,\omega) - \theta_s + L\pi\right| < \delta_2 \end{cases} \\ 0, \text{ else} \end{cases} \tag{38}$$

where $L = 0, \pm 1$. $(\Delta\phi(\tau,\omega), \Delta\theta(\tau,\omega))$ is the estimation difference between the estimated DOA and the real DOA of the target speaker at TF point (τ,ω). Obviously, the smaller the value of $(\Delta\phi(\tau,\omega), \Delta\theta(\tau,\omega))$, the more probable it is that the TF point (τ,ω) is dominated by the target speech. To further improve the estimation accuracy and suppress the impact of the outliers, we propose a nonlinear soft TF mask as:

$$mask(\tau,\omega) = \begin{cases} \frac{1}{1+e^{-\xi(1-(\Delta\phi(\tau,\omega)/\delta_1 + \Delta\theta(\tau,\omega)/\delta_2)/2)}} & \Delta\phi < \delta_1 \,\&\, \Delta\theta < \delta_2 \\ \rho & else \end{cases} \tag{39}$$

where ξ is a positive parameter and ρ $(0 \le \rho < 1)$ is a small positive parameter tending to be zero, which reflects the noise suppression effect. The parameters Δ_1 and Δ_2 control the degree of the estimation difference $(\Delta\phi(\tau,\omega), \Delta\theta(\tau,\omega)$. When parameters Δ_1, Δ_2, and ρ become larger, the capability of suppressing noise and interferences degrades, and the possibility of the (τ,ω) being dominated by the target speech also degrades. Hence, selecting the values of ρ, Δ_1, and Δ_2 is important. In our study, these parameters are determined through experiments. Future work could focus on selecting these parameters based on models of human auditory perception. In the end, we need to emphasize that the mask designed in Equation (39) has the ability to suppress the adverse effects of the interferences and background noise, and preserve the target speech simultaneously.

3. Proposed Target Speech Enhancement Method

The diagram of the proposed speech enhancement method (termed as AVS-SMASK) is shown in Figure 4, which is processed in the time-frequency domain. The details of each block in Figure 4 will be addressed in the following context.

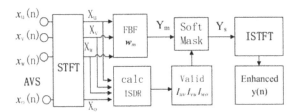

Figure 4. Block diagram of our proposed AVS-SMASK algorithm (STFT: Short-Time Fourier Transform; FBF: a fixed beamformer; ISTFT: inverse STFT; y(n): enhanced target speech).

3.1. The FBF Spatial Filter

As shown in Figure 4, the input signals to the FBF spatial filter are the data captured by the u, v, and w-sensor of the AVS. With the given DOA (θ_s, ϕ_s), the spatial matched filter (SMF) is employed as the FBF spatial filter, and its output can be described as:

$$Y_m(\tau, \omega) = w_m^H X_{avs}(\tau, \omega) \tag{40}$$

where $w_m^H = a^H(\theta_s, \phi_s) / \|a(\theta_s, \phi_s)\|^2$ is the weight vector of the SMF, and $a(\theta_s, \phi_s)$ is given in Equation (5). $[.]^H$ denotes the vector/matrix conjugate transposition. Substituting the expressions in Equations (5), (3), and (9)–(11) in Equation (40) yields:

$$
\begin{aligned}
Y_m(\tau, \omega) &= u_s X_u(\tau, \omega) + v_s X_v(\tau, \omega) + w_s X_w(\tau, \omega) \\
&= u_s^2 S(\tau, \omega) + u_s N_{tu}(\tau, \omega) + v_s^2 S(\tau, \omega) + v_s N_{tv}(\tau, \omega) + w_s^2 S(\tau, \omega) + w_s N_{tw}(\tau, \omega) \\
&= (u_s^2 + v_s^2 + w_s^2) S(\tau, \omega) + N_{tuvw}(\tau, \omega) \\
&= S(\tau, \omega) + N_{tuvw}(\tau, \omega)
\end{aligned}
\tag{41}
$$

where $N_{tuvw}(\tau, \omega)$ is the total noise component given as:

$$
\begin{aligned}
N_{tuvw}(\tau, \omega) &= u_s N_{tu}(\tau, \omega) + v_s N_{tv}(\tau, \omega) + w_s N_{tw}(\tau, \omega) \\
&= u_s(u_i S_i(\tau, \omega) + N_u(\tau, \omega)) + v_s(v_i S_i(\tau, \omega) + N_v(\tau, \omega)) \\
&\quad + w_s(w_i S_i(\tau, \omega) + N_w(\tau, \omega)) \\
&= (u_s u_i + v_s v_i + w_s w_i) S_i(\tau, \omega) + u_s N_u(\tau, \omega) + v_s N_v(\tau, \omega) + w_s N_w(\tau, \omega)
\end{aligned}
\tag{42}
$$

It can been seen that $N_{tuvw}(\tau, \omega)$ in Equation (42) consists of the interferences and background noise captured by directional sensors, while $Y_m(\tau, \omega)$ in Equation (41) is the mix of the desired speech source $S(\tau, \omega)$ and unwanted component $N_{tuvw}(\tau, \omega)$.

3.2. Enhancing Target Speech Using Estimated Mask

With the estimated mask in Equation (39) and the output of the FBF spatial filter $Y_m(\tau, \omega)$ in Equation (42), it is straightforward to compute the enhanced target speech as follows:

$$Y_s(\tau, \omega) = Y_m(\tau, \omega) \times mask(\tau, \omega) \tag{43}$$

where $Y_s(\tau, \omega)$ is then the spectra of the enhanced speech or an approximation of the target speech.

For presentation completeness, our proposed speech enhancement algorithm is termed as an AVS-SMASK algorithm, which is summarized in Table 2.

Table 2. The pseudo-code of our proposed AVS-SMASK algorithm.

(1)	Segment the output data captured by the u-sensor, v-sensor, w-sensor, and o-sensor of the AVS unit by the N-length Hamming window;
(2)	Calculate the STFT of the segments: $X_u(\tau, \omega)$, $X_v(\tau, \omega)$, $X_w(\tau, \omega)$ and $X_o(\tau, \omega)$;
(3)	Calculate the ISDR vector $I_{isdr} = [I_{uv}, I_{vu}, I_{wo}]$ by Equations (14)–(16);
(4)	Obtain the valid I_{isdr} according to the known direction of arrival (DOA) (θ_s, ϕ_s) and the summary of Section 2.3;
(5)	Utilize the valid I_{isdr} to estimate the DOA $(\hat{\theta}_s, \hat{\phi}_s)$ of the target speech for each time-frequency (TF) point;
(6)	Determine TF points belong to the target speech by Equations (36) and (37);
(7)	Calculate the nonlinear soft TF mask: $mask(\tau, \omega)$ by Equation (39);
(8)	Calculate the output of the FBF $Y_m(\tau, \omega)$ by Equation (40);
(9)	Compute the enhanced speech spectrogram by Equation (43);
(10)	Get the enhanced speech signal $y(n)$ by ISTFT.

4. Experiments and Results

The performance evaluation of our proposed AVS-SMASK algorithm has been carried out with simulated data and recorded data. Five commonly used performance measurement metrics—SNR, the signal-to-interference ratio (SIR), the signal-to-interference plus noise ratio (SINR), log spectral division (LSD), and the perceptual evaluation of speech quality (PESQ)—have been adopted. The definitions are given as follows for presentation completeness.

(1) Signal-to-Noise Ratio (SNR):

$$SNR = 10\log\left(\|s(t)\|^2 / \|n(t)\|^2\right) \qquad (44)$$

(2) Signal-to-Interference Ratio (SIR)

$$SIR = 10\log\left(\|s(t)\|^2 / \|s_i(t)\|^2\right) \qquad (45)$$

(3) Signal-to-Interference plus Noise Ratio (SINR):

$$SINR = 10\log\left(\|s(t)\|^2 / \|x(t) - s(t)\|^2\right) \qquad (46)$$

where $s(t)$ is the target speech, $n(t)$ is the additive noise, $s_i(t)$ is the ith interference, and $x(t) = s(t) + s_i(t) + n(t)$ is the received signal of the o-sensor. The metrics are calculated by averaging over frames to get more accurate measurement [22].

(4) Log Spectral Deviation (LSD), which is used to measure the speech distortion [22]:

$$LSD = \|\ln\left(\psi_{ss}(f) / \psi_{yy}(f)\right)\| \qquad (47)$$

where $\psi_{ss}(f)$ is the power spectral density (PSD) of the target speech, and $\psi_{yy}(f)$ is the PSD of the enhanced speech. It is clear that smaller LSD values indicate less speech distortion.

(5) Perceptual Evaluation of Speech Quality (PESQ). To evaluate the perceptual enhancement performance of the speech enhancement algorithms, the ITU-PESQ software [23] is utilized.

In this study, the performance comparison is carried out with the comparison algorithm AVS-FMV [17] under the same conditions. We do not take other SE methods into account since they use different transducers for signal acquisition. One set of waveform examples that is used in our experiments is shown in Figure 5, where $s(t)$ is the target speech, $s_i(t)$ is the i-th interference speech, $n(t)$ is the additive noise, and $y(t)$ is the enhanced speech.

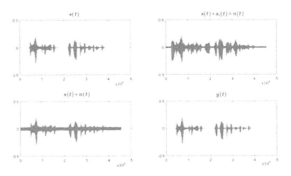

Figure 5. Waveform examples: $s(t)$ is the target speech, $s_i(t)$ is the interference speech, $n(t)$ is the additive noise, and $y(t)$ is the enhanced speech signal.

4.1. Experiments on Simulated Data

In this section, three experiments have been carried out. The simulated data of about five seconds duration is generated, where the target speech $s(t)$ is male speech, and two speech interferences $s_i(t)$ are male and female speech, respectively. Moreover, the AURORA2 database [24] was used, which includes subway, babble, car, exhibition noise, etc. Without loss of generality, all of the speech sources are placed one meter away from the AVS.

4.1.1. Experiment 1: The Output SINR Performance under Different Noise Conditions

In this experiment, we have carried out 12 trials (numbered as trial 1 to trial 12) to evaluate the performance of the algorithms under different spatial and additive noise conditions following the experimental protocols in Ref. [25]. The details are given below:

(1) The DOAs of target speech, the first speech interference (male speech) and the second speech interference (female speech) are at $(\theta_s, \phi_s) = (45°, 45°)$, $(\theta_1, \phi_1) = (90°, 135°)$, and $(\theta_2, \phi_2) = (45°, 120°)$, respectively. The background noise is chosen as babble noise $n(t)$;

(2) We evaluate the performance under three different conditions: (a) there exists only additive background noise: $n(t) \neq 0$ and $s_i(t) = 0$; (b) there exists only speech interferences: $n(t) = 0$ and $s_i(t) \neq 0$; (c) there exists both background noise and speech interferences: $n(t) \neq 0$ and $s_i(t) \neq 0$;

(3) The input SINR (denoted as SINR-input) is set as -5 dB, 0 dB, 5 dB, and 10 dB, respectively. Following the setting above, 12 different datasets are generated for this experiment.

In addition, the parameters of algorithms are set as follows. (1) The sampling rate is 16 kHz, 1024-point FFT (Fast Fourier Fransform), and 1024-point Hamming window with 50% overlapping are used. (2) For our proposed AVS-SMASK algorithm, we set $\delta_1 = \delta_2 = 25°$, $\rho = 0.07$, and $\xi = 3$. (3) For comparing algorithm AVS-FMV: F = 32, M = 1.001 followed Ref. [17]. The experimental results are given in Table 3.

Table 3. Output signal-to-interference plus noise ratio (SINR) under different noise conditions.

Noise Conditions	SINR-Input (dB)	AVS-FMV [17] (dB)		AVS-SMASK (dB)	
		SINR-Out	Average	SINR-Out	Average
Trial 1 ($n(t) = 0$ and $s_i(t) \neq 0$)	-5	4.96		7.32	
Trial 2 ($n(t) = 0$ and $s_i(t) \neq 0$)	0	5.60		9.38	
Trial 3 ($n(t) = 0$ and $s_i(t) \neq 0$)	5	7.81	4.88	11.53	8.14
Trial 4 ($n(t) = 0$ and $s_i(t) \neq 0$)	10	11.15		14.31	
Trial 5 ($n(t) \neq 0$ and $s_i(t) = 0$)	-5	4.77		6.70	
Trial 6 ($n(t) \neq 0$ and $s_i(t) = 0$)	0	5.51		10.17	
Trial 7 ($n(t) \neq 0$ and $s_i(t) = 0$)	5	6.76	4.97	13.03	9.11
Trial 8 ($n(t) \neq 0$ and $s_i(t) = 0$)	10	12.83		16.55	
Trial 9 ($n(t) \neq 0$ and $s_i(t) \neq 0$)	-5	3.66		4.70	
Trial 10 ($n(t) \neq 0$ and $s_i(t) \neq 0$)	0	5.70		7.22	
Trial 11 ($n(t) \neq 0$ and $s_i(t) \neq 0$)	5	7.10	4.42	10.46	6.66
Trial 12 ($n(t) \neq 0$ and $s_i(t) \neq 0$)	10	11.20		14.27	

As shown in Table 3, for all of the noise conditions (Trial 1 to Trial 12), our proposed AVS-SMASK algorithm outperforms AVS-FMV [17]. From Table 3, we can see that our proposed AVS-SMASK algorithm gives about 3.26 dB, 4.14 dB, and 2.25 dB improvement compared with that of AVS-FMV under three different experimental settings, respectively. We can conclude that our proposed AVS-SMASK is effective in suppressing the spatial interferences and background noise.

4.1.2. Experiment 2: The Performance versus Angle Difference

This experiment evaluates the performance of SE methods versus the angle difference between the target and interference speakers. Let's define the angle difference as $\Delta\phi = \phi_s - \phi_I$ and $\Delta\theta = \theta - \theta_i$

(here, the subscripts s and i refer to the target speaker and the interference speaker, respectively). Obviously, the closer the interference speaker is to the target speaker, the speech enhancement is more limited. The experimental settings are as follows. (1) PESQ and LSD are used as metrics. (2) The parameters of algorithms are set as the same as those used in *Experiment 1*. (3) Without loss of generality, the SIR-input is set 0 dB, while SNR-input is set 10 dB. (4) We consider two cases.

- Case 1: $\Delta\theta$ is fixed and $\Delta\phi$ is varied, $(\theta_1,\phi_1) = (45°,0°)$, the DOA of the target speaker moves from $(\theta_s,\phi_s) = (45°,0°)$ to (θ_s,ϕ_s) $(45°,180°)$ with $20°$ increments. Hence, the angle difference $\Delta\phi$ changes from $0°$ to $180°$ with $20°$ increments. Figure 6 shows the results of Case 1. From Figure 6, it is clear to see that when $\Delta\phi \to 0°$ (the target speaker moves closer to the interference speaker), for both algorithms, the PESQ drops significantly, and the LSD values are also big. These results indicate that the speech enhancement is very much limited if $\Delta\phi \to 0°$. However, when $\Delta\phi > 20°$, the PESQ gradually increases, and LSD drops. It is quite encouraging to see that the performance of PESQ and LSD of our proposed AVS-SMASK algorithm is superior to that of the AVS-FMV algorithm for all of the angles. Moreover, our proposed AVS-SMASK algorithm has the absolute advantage when $\Delta\phi \geq 40°$.
- Case 2: $\Delta\phi$ is fixed and $\Delta\theta$ is varied, $(\theta_1,\phi_1) = (10°,75°)$, the DOA of the target speaker moves from $(\theta_s,\phi_s) = (10°,75°)$ to $(\theta_s,\phi_s) = (170°,75°)$ with $20°$ increments. Then, the angle difference $\Delta\theta$ changes from $0°$ to $160°$ with $20°$ increments. Figure 7 shows the results of Case 2. From Figure 7, we can see that when $\Delta\theta \to 0°$ (the target speaker moves closer to the interference speaker), for both algorithms, the performance of PESQ and LSD are also poor. This means that the speech enhancement is very much limited if $\Delta\theta \to 0°$. However, when $\Delta\theta > 20°$, it is quite encouraging to see that the performance of PESQ and LSD of our proposed AVS-SMASK algorithm outperforms that of the AVS-FMV algorithm for all of the angles. In addition, it is noted that the performance of two algorithms drops again when the $\Delta\theta > 140°$ (the target speaker moves closer to the interference speaker around a cone). However, from Figure 6, this phenomenon does not exit.

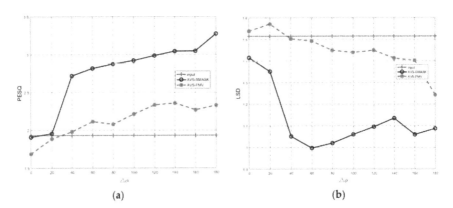

(a)   (b)

Figure 6. (Experiment 2) The perfomance versus $\Delta\phi$. (a) Perceptual evaluation of speech quality (PESQ) results and (b) Log spectral division (LSD) results (Case 1: ϕ_s of the target speaker changes from $0°$ to $180°$) (Case 1).

In summary, from the experimental results, it is clear that our proposed AVS-SMASK algorithm is able to enhance the target speech and suppress the interferences when the angle difference between the target speaker and the interference is larger than $20°$.

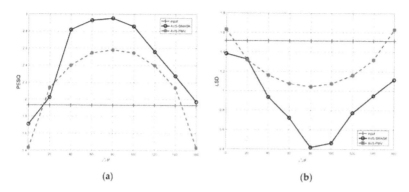

(a)   (b)

Figure 7. (Experiment 2) The performance versus $\Delta\theta$. (a) PESQ results and (b) LSD results (Case 2: θ_s of the target speaker changes from $0°$ to $160°$).

4.1.3. Experiment 3: The Performance versus DOA Mismatch

In practice, the DOA estimation of the target speaker may be inaccurate or the target speaker may make a small movement that causes the DOA mismatch problem. Hence, this experiment evaluates the impact of the DOA mismatch on the performance of our proposed speech enhancement algorithm. The experimental settings are as follows. (1) The parameters of algorithms are set as same as the *Experiment 1*. (2) $(\theta_s,\phi_s) = (45°,45°)$ and $(\theta_1,\phi_1) = (90°,135°)$. (3) The SIR-input is set to 0 dB, while the SNR-input is set to 10 dB; the performance measurement metrics are chosen as SINR and LSD. (4) We consider two cases:

Case 1: Only ϕ_s is mismatched, and the mismatch $(\partial\phi_s)$ ranges from $0°$ to $30°$ with $5°$ increments.
Case 2: Only θ_s is mismatched, and the mismatch $(\partial\theta_s)$ ranges from $0°$ to $30°$ with $5°$ increments.

Experimental results are given in Figures 8 and 9 for Case 1 and Case 2, respectively. From these results, we can clearly see that when the DOA mismatch is less than $20°$, our proposed AVS-SMASK algorithm is not sensitive to DOA mismatch. Besides, our AVS-SMASK algorithm outperforms the AVS-FMV algorithm under all of the conditions. However, when the DOA mismatch is larger than $20°$, the performance of our proposed AVS-SMASK algorithm drops significantly. Fortunately, it is easy to achieve $20°$ DOA estimation accuracy.

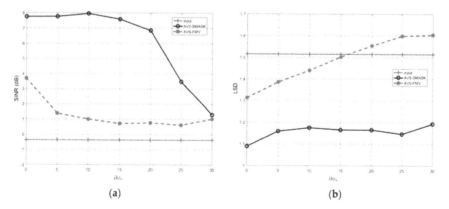

(a)   (b)

Figure 8. (Experiment 3) The performance versus the $\partial\phi_s$. (a) SINR results and (b) LSD results (Case 1).

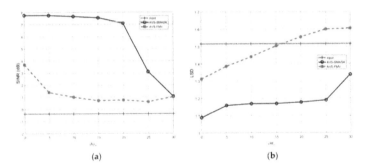

Figure 9. (Experiment 3, Case 2) The performance versus the $\partial\theta_s$. (a) SINR results and (b) LSD results (Case 2).

4.2. Experiments on Recorded Data in an Anechoic Chamber

In this section, two experiments have been carried out with the recorded data captured by an AVS in an anechoic chamber [25]. Every set of recordings lasts about six seconds, which is made by the situation that the target speech source and the interference source are broadcasting at the same time along with the background noise, as shown in Figure 1. The speech sources taken from the Institute of Electrical and Electronic Engineers (IEEE) speech corpus [26] are placed in the front of the AVS at a distance of one meter, and the SIR-input is set to 0 dB, while the SNR-input is set to 10 dB, and the sampling rate was 48 kHz, and then down-sampled to 16 kHz for processing.

4.2.1. Experiment 4: The Performance versus Angle Difference with Recorded Data

In this experiment, the performance of our proposed method has been evaluated versus the angle difference between the target and interference speakers ($\Delta\phi = \phi_s - \phi_i$ and $\Delta\theta = \theta_s - \theta_i$). The experimental settings are as follows. (1) PESQ is taken as the performance measurement metric. (2) The parameters of algorithms are set as the same as that of *Experiment 1*. (3) Considering page limitation, here, we only consider the changing of azimuth angle ϕ_s while $\theta_s = 90°$. The interfering speaker $s_1(t)$ is at $(\theta_1, \phi_1) = (90°, 45°)$. ϕ_s varies from $0°$ to $180°$ with $20°$ increments. Then, there are 13 recorded datasets. The experimental results are shown in Figure 10. It is noted that the x-axis represents the azimuth angle ϕ_s. It is clear to see that the overall performance of our proposed AVS-SMASK algorithm is superior to that of the comparing algorithm. Specifically, when ϕ_s approaches $\phi_1 = 45°$, the PESQ degrades quickly for both algorithms. When the angle difference $\Delta\phi$ is larger than $30°$ (ϕ_s is smaller than $15°$ or larger than $75°$), the PESQ of our proposed AVS-SMASK algorithm goes up quickly, and is not sensitive to the angle difference.

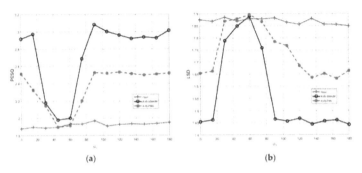

Figure 10. (Experiment 4) The performance versus ϕ_s. (a) PESQ results and (b) LSD results.

4.2.2. Experiment 5: Performance versus DOA Mismatch with Recorded Data

This experiment is carried out to evaluate the performance of speech enhancement algorithms when there are DOA mismatches. The experimental settings are as follows. (1) PESQ and LSD are taken as the performance measurement metric. (2) The parameters of algorithms are set the same as those of *Experiment 1*. (3) The target speaker is at $(\theta_s, \phi_s) = (45°, 45°)$, and the interference speaker is at $(\theta_1, \phi_1) = (90°, 135°)$. The azimuth angle ϕ_s is assumed to be mismatched. We consider the mismatch of ϕ_s (denoted as ϕ_s'') varying from $0°$ to $30°$ with $5°$ increments. The experimental results are shown in Figure 11, where the x-axis is the mismatch of the azimuth angle ϕ_s (ϕ_s''). It is noted that our proposed AVS-SMASK is superior to the compared algorithm under all conditions. It is clear to see that our proposed algorithm is not sensitive to DOA mismatch when the DOA mismatch is smaller than $23°$.

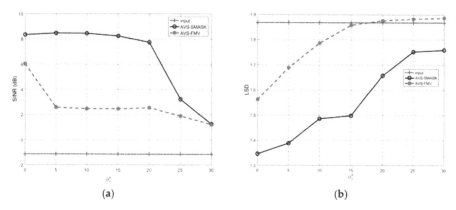

(a) (b)

Figure 11. (Experiment 5) The performance versus the ϕ_s mismatch ϕ_s''. (a) PESQ results and (b) LSD results.

We are encouraged to conclude that our proposed algorithm will offer a good speech enhancement performance in practical applications when the DOA may not be accurately estimated.

5. Conclusions

In this paper, aiming at the hearing technology of service robots, a novel target speech enhancement method has been proposed systematically with a single AVS to suppress spatial multiple interferences and additive background noise simultaneously. By exploiting the AVS signal model and its inter-sensor data ratio (ISDR) model, the desired monotone functions between ISDR and the DOA of the target speaker is derived. Accordingly, a nonlinear soft mask has been designed by making use of speech time-frequency (TF) sparsity with the known DOA of the target speaker. As a result, a single AVS-based speech enhancement method (named as AVS-SMASK) has been formulated and evaluated. Comparing with the existing AVS-FMV algorithm, extensive experimental results using simulated data and recorded data validate the effectiveness of our AVS-SMASK algorithm in suppressing spatial interferences and the additive background noise. It is encouraging to see that our AVS-SMASK algorithm is able to maintain less speech distortion. Due to page limitations, we did not show the derivation of the algorithm under reverberation. The signal model and ISDR model under reverberant conditions will be presented in our paper [27]. Our preliminary experimental results show that the PESQ of our proposed AVS-SMASK degrades gradually when the room reverberation becomes stronger (RT60 > 400 ms), but LSD is not sensitive to the room reverberation. Besides, there is an argument that learning-based SE methods achieve the state-of-art. In our opinion, in terms of SNR, PESQ, and LSD, this is true. However, learning-based SE methods ask for large amounts of training data, and require much larger memory size and a high computational cost. In contrast, the application scenarios of this research are different to learning-based SE methods, and our solution is more suitable for low-cost embedded systems. A real demo system was established in our lab, and the conducted trials further confirmed the effectiveness of our method where room reverberation is moderate (RT60 < 400 ms). We are confident that with only four-channel sensors and without any additional training data collected, the subjective and objective performance of our proposed AVS-SMASK is impressive. Our future study will investigate the deep learning-based SE method with a single AVS to improve its generalization and capability to handle different noise and interference conditions.

Author Contributions: Original draft preparation and writing, Y.Z. and Z.L.; Review & Editing, C.H.R., Y.Z. and Z.L. carried out the studies of the DOA estimation and speech enhancement with Acoustic Vector Sensor (AVS), participated in algorithm development, carried out experiments as well as drafted the manuscript. C.H.R. contributed to the design of the experiments, analyzed the experimental results and helped to review and edit the manuscript. All authors read and approved the final manuscript.

Funding: This research was funded by National Natural Science Foundation of China (No: 61271309), Shenzhen Key Lab for Intelligent MM and VR (ZDSYS201703031405467) and the Shenzhen Science & Technology Fundamental Research Program (JCYJ20170817160058246).

Conflicts of Interest: The authors declare no conflict of interest.

References

1. Yang, Y.; Song, H.; Liu, J. Service robot speech enhancement method using acoustic micro-sensor array. In Proceedings of the International Conference on Advanced Intelligence and Awareness Internet (IET), Beijing, China, 23–25 October 2010; pp. 412–415.
2. Gomez, R.; Ivanchuk, L.; Nakamura, K.; Mizumoto, T.; Nakadai, K. Utilizing visual cues in robot audition for sound source discrimination in speech-based human-robot communication. In Proceedings of the IEEE International Conference on Intelligent Robots and Systems, Hamburg, Germany, 28 September–2 October 2015; pp. 4216–4222.
3. Atrash, A.; Kaplow, R.; Villemure, J.; West, R.; Yamani, H.; Pineau, J. Development and validation of a robust speech interface for improved human-robot interaction. *Int. J. Soc. Rob.* **2009**, *1*, 345–356. [CrossRef]
4. Chen, M.; Wang, L.; Xu, C.; Li, R. A novel approach of system design for dialect speech interaction with NAO robot. In Proceedings of the International Conference on Advanced Robotics (ICAR), Hong Kong, China, 10–12 July 2017; pp. 476–481.
5. Philipos, C.; Loizou, P.C. *Speech Enhancement: Theory and Practice*, 2nd ed.; CRC Press: Boca Raton, FL, USA, 2017.
6. Chen, J.; Benecty, J.; Huang, Y. New insights into the noise reduction Wiener filter. *IEEE Trans. Audio Speech Lang. Process.* **2006**, *14*, 1218–1234. [CrossRef]
7. Reddy, A.M.; Raj, B. Soft mask methods for single-channel speaker separation. *IEEE Trans. Audio Speech Lang. Process.* **2007**, *15*, 1766–1776. [CrossRef]
8. Lightburn, L.; De Sena, E.; Moore, A.; Naylor, P.A.; Brookes, M. Improving the perceptual quality of ideal binary masked speech. In Proceedings of the IEEE International Conference on Acoustics, Speech and Signal Processing (ICASSP), New Orleans, LA, USA, 5–9 March 2017; pp. 661–665.
9. Wang, Z.; Wang, D. Mask Weighted Stft Ratios for Relative Transfer Function Estimation and Its Application to Robust ASR. In Proceedings of the IEEE International Conference on Acoustics, Speech and Signal Processing (ICASSP), Calgary, AB, Canada, 15–20 April 2018.
10. Xiao, X.; Zhao, S.; Jones, D.L.; Chng, E.S.; Li, H. On time-frequency mask estimation for MVDR beamforming with application in robust speech recognition. In Proceedings of the IEEE International Conference on Acoustics, Speech and Signal Processing (ICASSP), New Orleans, LA, USA, 5–9 March 2017; pp. 3246–3250.
11. Heymann, J.; Drude, L.; Haeb-Umbach, R. Neural network based spectral mask estimation for acoustic beamforming. In Proceedings of the IEEE International Conference on Acoustics, Speech and Signal Processing (ICASSP), Shanghai, China, 20–25 March 2016; pp. 196–200.
12. Li, B.; Zou, Y.X. Improved DOA Estimation with Acoustic Vector Sensor Arrays Using Spatial Sparsity and Subarray Manifold. In Proceedings of the IEEE International Conference on Acoustics, Speech and Signal Processing (ICASSP), Kyoto, Japan, 25–30 March 2012; pp. 2557–2560.

13. Zou, Y.X.; Shi, W.; Li, B.; Ritz, C.H.; Shujau, M.; Xi, J. Multisource DOA Estimation Based On Time-Frequency Sparsity and Joint Inter-Sensor Data Ratio with Single Acoustic Vector Sensor. In Proceedings of the IEEE International Conference on Acoustics, Speech and Signal Processing (ICASSP), Vancouver, BC, Canada, 26–31 May 2013; pp. 4011–4015.

14. Zou, Y.X.; Guo, Y.; Zheng, W.; Ritz, C.H.; Xi, J. An effective DOA estimation by exploring the spatial sparse representation of the inter-sensor data ratio model. In Proceedings of the IEEE China Summit & International Conference on Signal and Information Processing (ChinaSIP), Xi'an, China, 9–13 July 2014; pp. 42–46.

15. Zou, Y.X.; Guo, Y.; Wang, Y.Q. A robust high-resolution speaker DOA estimation under reverberant environment. In Proceedings of the International Symposium on Chinese Spoken Language Processing (ISCSLP), Singapore, 13–16 December 2014; p. 400.

16. Zou, Y.; Gu, R.; Wang, D.; Jiang, A.; Ritz, C.H. Learning a Robust DOA Estimation Model with Acoustic Vector Sensor Cues, In Proceedings of the Asia-Pacific Signal and Information Processing Association (APSIPA). Kuala Lumpur, Malaysia, 12–15 December 2017.

17. Lockwood, M.E.; Jones, D.L.; Bilger, R.C.; Lansing, C.R.; O'Brien, W.D.; Wheeler, B.C.; Feng, A.S. Performance of Time- and Frequency-domain Binaural Beamformers Based on Recorded Signals from Real Rooms. *J. Acoust. Soc. Am.* **2004**, *115*, 379–391. [CrossRef] [PubMed]

18. Lockwood, M.E.; Jones, D.L. Beamformer Performance with Acoustic Vector Sensors in Air. *J. Acoust. Soc. Am.* **2006**, *119*, 608–619. [CrossRef] [PubMed]

19. Shujau, M.; Ritz, C.H.; Burnett, I.S. Speech Enhancement via Separation of Sources from Co-located Microphone Recordings. In Proceedings of the IEEE International Conference on Acoustics, Speech and Signal Processing (ICASSAP), Dallas, TX, USA, 14–19 March 2010; pp. 137–140.

20. Wu, P.K.T.; Jin, C.; Kan, A. A Multi-Microphone SPE Algorithm Tested Using Acoustic Vector Sensors. In Proceedings of the International Workshop on Acoustic Echo and Noise Control, Tel-Aviv, Israel, 30 August–2 September 2010.

21. Zou, Y.X.; Wang, P.; Wang, Y.Q.; Ritz, C.H.; Xi, J. Speech enhancement with an acoustic vector sensor: An effective adaptive beamforming and post-filtering approach. *EURASIP J. Audio Speech Music Process* **2014**, *17*. [CrossRef]

22. Zou, Y.X.; Wang, P.; Wang, Y.Q.; Ritz, C.H.; Xi, J. An effective target speech enhancement with single acoustic vector sensor based on the speech time-frequency sparsity. In Proceedings of the 19th International Conference on Digital Signal Processing (DSP), Hong Kong, China, 20–23 August 2014; pp. 547–551.

23. Gray, R.; Buzo, A.; Gray, A.; Matsuyama, Y. Distortion measures for speech processing. *IEEE Trans. Acoust. Speech Signal Process.* **1980**, *28*, 367–376. [CrossRef]

24. ITU-T. *862-Perceptual Evaluation of Speech Quality (PESQ): An Objective Method for End-to-End Speech Quality Assessment of Narrow-Band Telephone Networks and Speech Codecs*; International Telecommunication Union-Telecommunication Standardization Sector (ITU-T): Geneva, Switzerland, 2001.

25. Hirsch, H.G.; Pearce, D. The AURORA experimental framework for the performance evaluations of speech recognition systems under noisy conditions. In Proceedings of the Automatic Speech Recognition: Challenges for the Next Millennium, Paris, France, 18–20 September 2000; pp. 29–32.

26. Shujau, M.; Ritz, C.H.; Burnett, I.S. Separation of speech sources using an Acoustic Vector Sensor. In Proceedings of the IEEE International Workshop on Multimedia Signal Processing, Hangzhou, China, 19–22 October 2011; pp. 1–6.

27. Rothauser, E.H. IEEE Recommended Practice for Speech Quality Measurements. *IEEE Trans. Audio Electroacoust.* **1969**, *17*, 225–246.

Article

The Accuracy of Predicted Acoustical Parameters in Ancient Open-Air Theatres: A Case Study in Syracusae

Elena Bo [1,*], Louena Shtrepi [1], David Pelegrín Garcia [2], Giulio Barbato [3], Francesco Aletta [4] and Arianna Astolfi [1]

1   Department of Energy, Politecnico di Torino, 10129 Turin, Italy; louena.shtrepi@polito.it (L.S.); arianna.astolfi@polito.it (A.A.)
2   Laboratory for Soft Matter and Biophysics, Department of Physics & Astronomy, KU Leuven, 3001 Leuven, Belgium; david.pelegringarcia@kuleuven.be
3   Department of Management and Production Engineering, Politecnico di Torino, 10129 Turin, Italy; giulio.barbato@polito.it
4   UCL Institute for Environmental Design and Engineering, The Bartlett, University College London (UCL), London WC1H 0NN, UK; f.aletta@ucl.ac.uk
*   Correspondence: elena.bo@polito.it; Tel.: +39-011-090-4496

Received: 20 June 2018; Accepted: 10 August 2018; Published: 17 August 2018

Featured Application: The work aims to give more insights into the relation between the sensitivity of the simulated objective parameters and the software input parameters for open-air ancient theatres. It is meant to raise awareness on the use of predictive acoustic software for unconventional outdoor environments in order to validate the possibility of re-using them as performance spaces.

Abstract: Nowadays, ancient open-air theatres are often re-adapted as performance spaces for the additional historical value they can offer to the spectators' experience. Therefore, there has been an increasing interest in the modelling and simulation of the acoustics of such spaces. These open-air performance facilities pose several methodological challenges to researchers and practitioners when it comes to precisely measure and predict acoustical parameters. Therefore this work investigates the accuracy of predicted acoustical parameters, that is, the Reverberation Time (T_{20}), Clarity (C_{80}) and Sound Strength (G), taking the ancient Syracusae open-air theatre in Italy as a case study. These parameters were derived from both measured and simulated Impulse Responses (IR). The accuracy of the acoustic parameters predicted with two different types of acoustic software, due to the input variability of the absorption and scattering coefficients, was assessed. All simulated and measured parameters were in good agreement, within the range of one "just noticeable difference" (JND), for the tested coefficient combinations.

Keywords: open-air theatres; acoustical measurements; prediction models; historical acoustics

---

1. Introduction

The recent interest in the design of ancient theatres and in their acoustical characteristics has drawn attention to the lack of methodologies in metrology for historical acoustics [1]. The ISO 3382-1 standard [2] was used in the European ERATO project [3] to evaluate the acoustical apparatus of ancient theatres through room acoustic parameters, such as the Early Decay Time (EDT), Reverberation Time (RT), Clarity (C_{80}), and Sound Strength (G). However, ISO 3382-1 basically refers to indoor environments and temporal decay parameters seem to be less suitable for open-air conditions [4–8].

Farnetani et al. [4] reported that EDT is not a robust predictor of the acoustic quality of open-air theatres. The lack of robustness in EDT is due to a marked and intrinsic variability of this parameter, according to the source position, which defines the delay and incidence direction of the first reflections to the receivers. The same study asserted that RT behaviour in an open-air theatre is clearly different from that dealt with in the classical reverberation theory, which refers to a reference room volume. However, this parameter showed a limited variability. Chourmouziadou et al. [5] also suggested the use of RT when comparative studies are performed. However, it should be utilised with caution since it is usually used to evaluate enclosed spaces. Mo et al. [6] conducted a listening test with monaural and binaural auralisations of an open-air space. They stated that the perceived reverberance in an unroofed space is not only affected by the temporal characteristics during the decay process, but also by the spatial characteristics, due to the distribution of the reflections. The results showed that the conventional RT described in ISO 3382-1, which only deals with the sound energy decay rate, is not suitable for evaluating the reverberance of an unroofed space. Thus, more insight is needed into the adoption of an indoor acoustic measurements standard for the investigation of the acoustic conditions of open-air theatres. These sites represent particular environments that have their own specific sound field, which is rather different from the ideal diffuse field.

Besides the doubts about the applicability of the aforementioned indoor standard to outdoor case studies, other specific problems could arise when conducting measurements in ancient theatres. In fact, archaeological field measurements are also clearly influenced by the current conditions of the architecture of the theatres. Most ancient theatres have undergone damage of anthropologic and atmospheric nature. It was attested in Farnetani et al. [4] that the measured values of RT, G, and C_{80} in ancient theatres are affected to a great extent by the state of conservation of the theatres themselves, with particular reference to the completeness of the architectural elements. Therefore, it is currently difficult to design acoustical correction guidelines for their contemporary reuse as performance spaces. Moreover, particular attention should be paid to the outdoor environmental conditions, such as temperature (t), relative humidity (RH), and air velocity, which could affect the variability of the measurement results, in the same way as for indoor measurements [9,10].

The topic of acoustical characterisation has already been examined in detail for indoor spaces, through statistical analysis, in order to investigate the reproducibility of measurements, the accuracy of the parameter calculation, the influence of source-receiver position displacement, and the measurement chains of different systems [9,11,12].

An alternative to the experimental acoustical characterisation is the virtual reconstruction of the theatre, using room acoustics simulation software. Since they were introduced, geometrical acoustic (GA) software applications have been used as the standard room acoustics models [13]. In order to enable a better acoustic design of existing buildings, the simulations first need to replicate the real acoustical conditions of the examined environment through three important steps: (1) appropriate geometry modelling; (2) material properties; and (3) simulation settings. This procedure, namely, the calibration of the model, is even more complicated for open-air theatres as the acoustic scattering and diffraction phenomena are more relevant than in closed theatres [14]. An appropriate calculation method and a geometrically detailed model are of fundamental importance to achieve accurate predicted results [15].

The reliability of simulations is an on-going matter of discussion and interest, as testified by the Round Robin comparisons of room acoustic modelling tools [16–18], and the more recent overview on the uncertainties of input data in simulations [13]. In the latter overview, it was reported that the specific uncertainties that characterise the absorption coefficient (α_w) and scattering coefficient (s) of materials [19,20] could affect the estimation accuracy of room acoustic parameters in the end. Such parameters are derived from simulated Impulse Responses (IR) or from energy reflectograms, depending on which analysis algorithm of the room acoustics software is being used. In situ and scale measurements [4] have revealed that the IRs of ancient theatres are composed of the direct sound and of two major reflections, which come from the orchestra floor and the scaenae frons (the ancient stage building), respectively, when these parts of the theatres still exist. Therefore, in the case of open-air theatres, the IR should be modelled with a limited number of specular reflections and a high number of scattered reflections, because of the irregularities in the steps of the cavea [21]. This configuration is difficult to handle using geometrical acoustic-based software (GA), such as Odeon (Version) and CATT-Acoustic [22,23]. Yet, most of researchers still rely on such tools also for open-air theatres in everyday practice; thus, special attention should be given to properly controlling the boundary conditions. In fact, open-air theatres represent a special case, which creates a challenge for these prediction algorithms. The absence of a roof, and therefore of a reverberant field, urges to have a high reliability in the prediction of the early reflections. Moreover, the concave shape of these theatres is responsible for the creation of "shadow zones" of the mirroring surfaces in great lateral areas of the cavea [14]. This affects the deterministic method of the Image Source, which is used by the GA software to build the early part of the IRs.

The aim of this work is to assess the performance of predictive software in calculating a set of acoustic parameters for ancient theatres, a particular type of open-air spaces, taking the case study of the ancient theatre of Syracusae (SR). The objective is to give more insight about the relation of the sensitivity of the simulated results to the input parameters. It is mainly referred to raise awareness on the use of this kind of software for outdoor unconventional environments. The theatre is located in Sicily, an island in the South of Italy, a region where ancient Greek culture had historically a lot of influence. The simulation accuracy of two kinds of software, Odeon and CATT-Acoustic, is considered. This theatre was selected because it was relatively easy to model due to the lack of contemporary additional elements. In this manner, the virtual model of SR could be considered as a valid archetype model. The paper is organised as follows:

- Section 2 (Case Study) includes a brief description of the state of conservation of the theatre chosen for this research.
- Section 3 (In Situ Measurements) includes a description of the acoustical measurement campaign carried out in the investigated theatre.
- Section 4 (Uncertainty Expression of the Acoustic Prediction Models) comprises the assessment of the uncertainty contribution related to the absorption (α_w) and scattering (s) input data assigned to the materials, predicted with Odeon (v. 13.02) (Odeon A/S, Lyngby, Denmark), and with CATT-Acoustic (v. 9) (CATT, Gothenburg, Sweden) software.
- Section 5 (Discussion) is focused on analysing the differences between measured (in situ) and predicted (through software) acoustic parameters, and it includes a discussion on the overall limitations of the study.

2. Case Study

The theatre of Syracusae (SR) was chosen as case study for a measurement campaign carried out by the Department of Energy at the Politecnico di Torino, from the 5th to 7th September 2015. SR (Figure 1) has Greek origins, dating back to the 5th century BC, but it was later modified by the Romans. Apart from a few ruins, nothing is visible of the original scaenae frons, but the surviving part of the rock-cut cavea has a diameter of 105 m.

Several studies that refer to the acoustics of SR have been retrieved from literature. These studies refer to measurements on a scale model of the ancient theatre and its contemporary use [4,24], to acoustic and lighting simulations [25], and to in situ acoustical characterisations with temporary scenery [26]. Measurements had only been carried out in empty conditions at one point of the orchestra area, as a pilot study in which different techniques were used [27].

This ancient open-air theatre is intensively used during cyclic summer season festivals in its current (deteriorated) condition, and acoustic measurements are made also for conservation purposes. Therefore, this study concerns the "historical acoustics" research field, which is the study of the auditory and acoustic environment of historic sites and monuments [1], with a valorisation purpose. The empty condition has been chosen for obvious practical reasons, as with the public present it is very difficult to carry out reliable measurements due to high background noise levels and unsteady boundary conditions [28]. Moreover, in order to simulate correctly the presence of public or the placement of an acoustic shell for renovation purposes, the reliability of simulated data must be verified, starting from the calibration of the acoustic model.

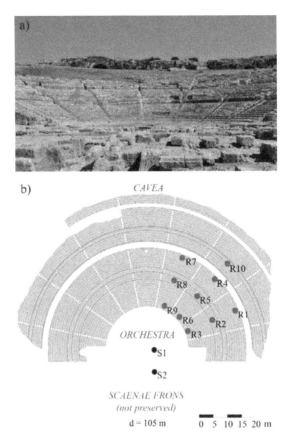

Figure 1. Present conditions of the ancient theatre of Syracusae (**a**) and measurement set-up (**b**). S1 and S2 represent the source positions. R1 to R10 indicate the receiver positions.

3. In Situ Measurement Methods

Standard measurements have been performed in unoccupied conditions, with omnidirectional sound sources and receivers, as stated in the ISO 3381-1 [2]. Different considerations on ancient theatre measurements, defined during the European ERATO Project [3], were taken into account. The measurement results for SR have been used in Section 4 for the calibration of the simulation model and as references for the acoustic parameters predicted through computer simulations. The source and receiver positions for the theatre are shown in Figure 1b.

Receivers were positioned on three radial axes of the cavea in the theatre, 1.2 m above the floor resembling the height of the ear of a sitting person. An omnidirectional microphone (Shoeps CMC 5-U, Durlach, Germany) was used to record the IRs. Ten receiver positions were considered. There was only a single microphone, meaning that all position measurements were carried out sequentially.

Measurements were repeated two or three times for each source position for most of the receivers, in order to evaluate the repeatability of the results. Two source positions were investigated: S1 was shifted horizontally by 1 m from the centre of the orchestra, in order to avoid any acoustical focus [29]; S2 was located behind S1, closer to the ancient scaenae frons position. The S1–S2 distance was equal to 7.6 m. Firecrackers were used as impulsive sources to measure the IRs ("Raudo Manna New Ma1b" Napoli, Italy and "Perfetto C00015 Raudo New", Napoli, Italy). The IRs were measured directly by recording the impulse produced by the firecracker blast. Firecrackers were used in S1 and S2, in order to overcome the problem of the low Signal-to-Noise Ratio (SNR): they maximise the SNR and this constitutes a significant advantage in outdoor measurements, but on the other hand, caution should be used as they are also more likely to be influenced by random effects (e.g., atmospheric conditions and random directivity). According to San Martin et al. [30], in the case of firecrackers the generated impulse is nearly omnidirectional. Its directivity index is, on average, around 1 dB for the octave bands between 125 Hz and 16 kHz. In addition, both its time curve and spectral power are highly repetitive, resulting in levels above 115 dB (reference 1 pW) within the aforementioned range.

The Background Noise Level (BNL) was measured as an equivalent continuous A-weighted sound pressure level (L_{Aeq}) over a period of 10 min, before the measurement sessions. The measured BNL was 45 dB (A), in unoccupied conditions.

The sound source was positioned at a height of 1.5 m from the floor, and a custom-made tripod was used to hold the firecrackers in a fixed position. Aurora (version 4.4, Parma, Italy) was used as acquisition software.

The air temperature and relative humidity were monitored during the whole measurement campaign, using a thermometer/hygrometer, Testo 608-H1 (Croydon South, VIC, Australia). The wind speed was measured by means of an anemometer, Testo 450-V1 (Croydon South, VIC, Australia). The environmental parameters acquired during the measurements campaign were t = 33 °C, RH = 65%, wind speed = 0.30 m/s. These did not change significantly during the measurement campaign.

In order to characterise the acoustical conditions of a performance space, the ISO 3382-1 standard lists a series of parameters that can be obtained from the IRs measured at each receiver position. Although open-air theatres cannot be considered typical performance spaces, like closed theatres or concert-halls, the ISO 3382-1 standard was used as the reference for the acoustical characterisation. In particular, the following room acoustic parameters were measured, as these are considered the most relevant parameters for the acoustical characterisation of open-air theatres [4]:

- Reverberation time, RT, (s): duration required for the space-averaged sound energy density in an enclosure to decrease by 60 dB after the source emission has stopped. The integrated impulse response method was applied to obtain the RT from the IR [2]. RT can be evaluated on a smaller dynamic range than 60 dB and extrapolated to a decay time of 60 dB. It is then labelled accordingly. The RT in SR was derived from decay values of 5 dB to 25 dB below the initial level, and it was therefore labelled T_{20}.

- Clarity, C_{80}, (dB): the balance between early- and late-arriving energy. This was calculated for an 80 ms early time limit, as the results were intended to relate to music conditions, using equation:

$$C_{80} = 10 \, \log \frac{\int_0^{80} p^2(t)dt}{\int_{80}^{\infty} p^2(t)dt} \tag{1}$$

where p(t) is the instantaneous sound pressure of the impulse response measured at the measurement point.

- Sound Strength, G, (dB): the logarithmic ratio of the measured sound energy (i.e., the squared and integrated sound pressure) to the sound energy that would arise in a free field at a distance of 10 m from a calibrated omnidirectional sound source, as expressed in the following equations:

$$G = 10 \, \log \frac{\int_0^{\infty} p^2(t)dt}{\int_0^{\infty} p_{10}^2(t)dt} = L_{pE} - L_{pE,10} \tag{2}$$

in which

$$L_{pE} = 10 \log \left[\frac{1}{T_0} \int_0^\infty \frac{p^2(t)dt}{p_0^2} \right] \qquad (3)$$

and

$$L_{pE,\,10} = 10 \log \left[\frac{1}{T_0} \int_0^\infty \frac{p_{10}^2(t)dt}{p_0^2} \right] \qquad (4)$$

where:

$p(t)$ is the instantaneous sound pressure of the impulse response measured at the measurement point;

$p_{10}(t)$ is the instantaneous sound pressure of the impulse response measured at a distance of 10 m in a free field;

L_{pE} (dB) is the sound exposure level of $p(t)$;

$L_{pE,10}$ (dB) is the sound exposure level of $p_{10}(t)$;

p_0 is the reference sound pressure of 20 µPa;

T_0 is the reference time interval of 1 s.

In the above equations, $t = 0$ corresponds to the start of the direct sound, i.e., which corresponds to the arrival of the direct sound at the receiver, and ∞ should correspond to a time that is greater than or equal to the point at which the decay curve has decreased by 30 dB [2].

G requires a calibration procedure for the sound power of the source. Different procedures have been described previously [2]. $L_{pE,10}$ can be calculated from the sound pressure $p_d(t)$ measured at a source-to-receiver distance d ($\geq$3 m) according to the following equation:

$$L_{pE,10} = L_{pE,d} + 20 \log \left(\frac{d}{10} \right) \qquad (5)$$

where:

$L_{pE,d}$ (dB) is the sound exposure level of $p_d(t)$, obtained from (3) (using p_d instead of p_{10}).

The Aurora plugin was used for the calculation of G with the firecrackers [31]. According to this procedure, the anechoic segment (direct sound) of each IR was used for calibration, providing the distance between the source and the receiver that allows for the estimation of $L_{pE,10}$, and it is recommended to keep a length of the IR of at least 1 s and to silence the signal just after the end of the direct sound. In this way, the smearing out in time caused by the octave filtering does not push the energy outside the time window, even at a low frequency, and the correct value of the signal level can be computed. A calibration file was obtained in situ from each analysed IR and was used to calculate the G value for that measurement path, with the knowledge of the exact source-to-receiver distance.

The resulting dataset is composed of the octave-band values from 125 Hz to 8 kHz of the acoustic parameters calculated by the Aurora software (v. 4.4) [31] from the measured IRs.

Measurements Results

The measurement results at receiver positions R1–R10 are reported in Table 1, expressed as T_{20}, C_{80}, and G acoustical parameters obtained with firecrackers at source positions S1 and S2. All the values are the averages of two or three repetitions at each receiver position and of the central 500 Hz and 1 kHz octave band frequencies, as indicated in ISO 3382-1 [2]. In accordance with the ISO 3382-1, spatial averages for each row were also reported in Table 1. It was assumed that each row can be considered as a homogeneous area, as in open-air theatres the direct sound and the distance from the source play a predominant role in the acoustic response. The Impulse Response-to-Noise Ratio, INR, (dB) is also reported as a parameter for judging the validity of the measurement, in order to establish the reliability of the outdoor acoustical measurements [31]. According to ISO 3382-1, the source level should be at least 35 dB above the background noise level in the corresponding frequency band for the case of T_{20}. All the measurements considered in this study had INR values well above 35 dB and up to 60 dB for the octave bands from 250 Hz to 8 kHz. It is important to underline that in the case of the T_{20} values, the larger standard deviation is due to the presence of only one strong reflection from the orchestra after the direct sound (as shown in Figure 2), which determines an irregular course of the decay curve and a greater variability in the slope of the decay curve.

Table 1. Mean values and standard deviations of the measurements for the T_{20}, C_{80} and G acoustical parameters for the firecrackers, in positions S1 and S2. The data refer to the averages of the 500 Hz and 1 kHz octave bands and to the repetitions for the same receiver position. The rows spatial means are also reported. The standard deviations of the spatial means outside the JND range (for a definition see Section 4) are shown in bold. INR is also reported to help assess the quality of the measurements.

Row	Receiver	No. of Repetitions		Distance from Source (m)		Acoustical Parameters T_{20} (s) (St. Dev.)		C_{80} (dB) (St. Dev.)		G (dB) (St. Dev.)		INR (dB)	
		S1	S2	S1	S2	S1	S2	S1	S2	S1	S2	S1	S2
First row	R3	2	2	13.8	18.7	0.58 (0.09)	0.80 (0.02)	20.8 (3.0)	13.3 (0.6)	−0.3 (0.3)	−3.0 (0.1)	58	60
	R6	3	2	14.6	21.2	0.70 (0.04)	0.84 (0.01)	16.9 (2.3)	15.2 (0.6)	−0.9 (0.4)	−5.2 (0.1)	51	57
	R9	2	2	15.5	23.0	0.31 (0.05)	0.60 (0.14)	22.1 (0.5)	16.9 (0.8)	−2.4 (0.1)	−9.7 (0.0)	52	55
Sp. mean				14.6	20.9	0.53 (**0.20**)	0.75 (**0.13**)	19.9 (2.7)	15.1 (**1.8**)	−1.2 (**1.1**)	−6.0 (**3.4**)	-	-
Second row	R2	2	2	23.3	27.5	0.81 (0.07)	0.83 (0.01)	16.5 (0.7)	11.0 (0.8)	−4.8 (0.2)	−6.4 (0.1)	57	55
	R5	2	2	24.0	30.2	0.66 (0.19)	0.85 (0.03)	19.2 (3.8)	13.3 (0.2)	−5.4 (1.0)	−7.2 (0.1)	50	52
	R8	2	2	24.9	32.3	0.71 (0.16)	0.91 (0.11)	15.9 (2.8)	13.0 (1.6)	−6.0 (0.0)	−8.5 (0.4)	50	54
Spatial mean				24.1	30.0	0.73 (**0.08**)	0.87 (**0.04**)	17.2 (**1.8**)	12.4 (**1.3**)	−5.4 (0.6)	−7.4 (**1.1**)	-	-
Third row	R1	2	2	31.7	35.6	0.98 (0.05)	0.96 (0.01)	15.4 (2.0)	11.2 (0.1)	−7.8 (0.5)	−8.3 (0.1)	52	53
	R4	3	2	32.4	38.4	0.94 (0.15)	0.98 (0.03)	16.9 (1.8)	12.7 (1.1)	−8.0 (0.7)	−9.6 (0.3)	51	51
	R7	2	2	33.3	40.6	1.04 (0.09)	1.05 (0.01)	15.7 (0.2)	13.9 (0.0)	−9.3 (0.2)	−10.5 (0.0)	51	50
Spatial mean				32.5	38.2	0.99 (**0.05**)	1.00 (**0.05**)	16.0 (0.8)	12.6 (**1.3**)	−8.4 (0.8)	−9.4 (**1.1**)	-	-
	R10	2	2	39.2	45.6	1.31 (0.03)	1.91 (0.05)	14.3 (0.2)	12.3 (0.1)	−10.2 (0.3)	−11.2 (0.4)	52	49

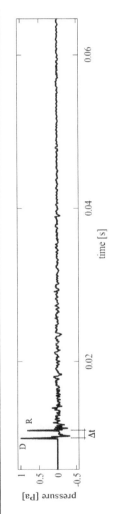

Figure 2. Measured Impulse Response (IR) in Syracusae (SR) for the S1-R6 measurement path, for the firecracker source. Δt is the time interval between the direct sound (D) and the first reflection (R) from the orchestra floor.

4. Uncertainty of the Geometrical Acoustic Prediction Models

In the acoustic domain, it is important to recall that the parameters have the aim of evaluating the perception of the acoustic signal, namely the average capability of a "conventional" listener to notice sound variations. An important factor that correlates the subjective field to objective measures has been defined as the Just Noticeable Difference (JND), that is, the smallest perceivable change in a given acoustical parameter, which is specified for information in Annex A of ISO 3382-1 [2] for central frequencies (500 Hz and 1 kHz), but which is also acceptable for lower and higher frequencies [32–35]. This issue will be further discussed, when analysing the accuracy of the acoustic prediction models.

The uncertainty contribution of the input data, propagated to the results obtained from two different types of room acoustic software, Odeon version 13.02 and CATT-Acoustic version 9, was assessed and compared with the measurements values.

Odeon version 13.02 [22] is based on a hybrid calculation method. Early reflections are calculated through a mixture of the Image Source Method and the Ray-Tracing Method (RTM), by means of a stochastic scattering process that uses secondary sources. Late reflections are calculated by means of a special RTM, where the secondary sources radiate energy locally from the surfaces and are assigned a frequency-dependant directionality, namely the reflection-based scattering coefficient. The secondary sources may have a Lambert, Lambert oblique, or Uniform directivity: this directivity depends on the properties of the reflections as well as on the calculation settings.

CATT-Acoustic version 9 [23] is made up of two modules: CATT-A is the main programme, and it handles the modelling, surface properties, and directivity libraries, and TUCT (The Universal Cone Tracer), which is the main prediction and auralisation programme. TUCT can use three alternative cone-tracing algorithms: the first algorithm is based on stochastic diffuse rays, while the second and third algorithms are based on the split-up of the actual diffuse rays. The difference between these modules is that the latter handles two orders of diffuse split-up reflections in a deterministic way, thus resulting in lower random run-to-run variations.

Both CATT-Acoustic and Odeon base their scattering algorithms on two main implementations, which are described in detail in a previous paper [36]. These two methods are the Hybrid Reflectance Model (HRM) and vector mixing (VM). The HRM method complies with the definition of the scattering coefficient based on ISO 17497-1 [20] which defines it in a quantitative way as the fraction of the non-symmetrically reflected energy. In the HRM method, a random number between 0 and 1 is used to determine whether the reflection is specular or scattered. This number is compared with the surface scattering coefficient (s) assigned to the surface. In case it exceeds the value of s, the scattered energy is assumed to be distributed according to Lambert's Law, i.e., the intensity of the reflected ray is independent on the angle of incidence but proportional to the cosine of the angle of reflection. This is the basic concept implemented in CATT-Acoustic [23] and in Odeon for the uniform and Lambert directivity scattering [22]. On the contrary, the VM is based on the linear interpolation of the specular and diffuse reflection [37]. In this way the direction of a reflection vector is calculated by adding the specular vector scaled by a factor (1-s) to a scattered vector following a certain direction that has been scaled by a factor s. This is the basic concept implemented in Odeon [22,38,39], named "vector-based scattering", where the scattered vector follows a random direction, generated according to the Lambert distribution named oblique Lambert directivity.

4.1. General Procedure for the Implementation of the Models

In order to compare the two software packages and to obtain the best match with the measurement results, it was necessary to perform simulations with the same geometric model and source/receiver positions as in the measurements. To the best of the authors' knowledge, this preliminary benchmark procedure has never been performed before on ancient open-air theatres, although many studies on indoor environments have been conducted [13–15]. Both types of software used for the simulation, that is, Odeon and CATT-Acoustic, have been validated in Round Robin tests. One of the main findings of these tests was that precise knowledge of the characteristics of the surface material is an important prerequisite for a reliable room simulation. Thus, a more detailed analysis on absorption and scattering coefficient changes was proposed.

A preliminary benchmark test study was carried out on SR, whose model had previously been used in different investigations, e.g., simulations concerning its ancient conditions, during the European ERATO project, [3] and in investigations on its contemporary use [25]. Figure 3a shows the 3D model configuration of SR.

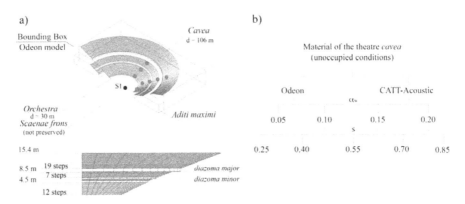

Figure 3. 3D model and source-receiver simulation set-up of SR (**a**) and scheme of the characteristics of the material chosen for the cavea (**b**).

The procedure applied for the comparison of the simulation tools was focused on solving the following issues:

1. Geometrical model: MATLAB software, version R2015b, was used to create a parametric open-air theatre script. Two 3D cavea model script outputs were created, one suitable for Odeon (dxf file) and the other for CATT-Acoustic (.geo file). In order to reduce the simulation time, the theatre geometry was simplified and designed as symmetric. A few geometrical simplifications have been performed in both Odeon and CATT-A models. The number of surfaces was 1357 in CATT-Acoustic and 1362 in Odeon. As recommended previously [14], the steps were modelled. The higher number of surfaces in Odeon corresponds to an additional boundary box with totally absorbing walls and top which is required in Odeon to simulate open-air conditions [22]. CATT-A algorithms are implemented in order to detect lost rays, i.e., rays that escape from the geometrical model. In open cases, such as an open-air ancient theatre, rays disappear whenever they do not hit any surface during the calculation time. This principle is similar to the one used in Odeon, where the escaping rays disappear since they are totally absorbed by the boundary box. The circular geometry was modelled with 20 segments, as recommended by Charmouziadou [40], who showed that a number between 12 and 24 segments is optimal with respect to the influence on the objective acoustic parameters.

2. Source-receivers: The source-receiver path was defined as in the measurement set-up, considering the theatre as unoccupied. For an easier comparison, only the source in position S1 was considered, as shown in Figure 3a.

3. Surface material properties: The main surface considered in the model was the cavea, the stone and steps of which are not well-conserved. In both types of software, 20 material alternatives were assigned to the cavea stone, that is, from the most reflective one, with $\alpha_w = 0.05$ and s = 0.25, to the most porous one, with $\alpha_w = 0.2$ and s = 0.85, and all the intermediate combinations of α_w were tested in steps of 0.05, while s was tested in steps of 0.15, as explained in Figure 4b [14]. Other elements were then added: the remains of the ancient entrances to the orchestra area (aditi

maximi), which was considered as an aperture ($\alpha_w = 0.9$; s = 0), and the floor, which includes the ruins of the scaenae frons ($\alpha_w = 0.8$; s = 0.8) and the better conserved orchestra area ($\alpha_w = 0.1$; s = 0.2). Odeon and CATT-Acoustic software allow for frequency dependent absorption coefficients. The same absorption coefficients have been used for both software. The Odeon software allows giving as input value for the scattering coefficient the value as an average between 500 and 1000 Hz, and considers a frequency dependent scattering by using default interpolation curves as shown in the Manual. These curves have been used in CATT-Acoustic, i.e., a frequency dependent scattering coefficient, by inserting each value for each octave-band. The values given in Figure 4b refer to the mean values at 500 and 1000 Hz.

4. General settings: The following settings were considered for all the simulations: a 100 dB source sound power level, 1500 ms as the impulse response length, and 4 million rays. The Transition Order (TO) in Odeon was limited to 1, which better resembles the impulse response characteristics in the real condition with only one specular reflection from the stage floor. The third calculation algorithm in CATT-Acoustic, described above, was chosen as it is the most suitable for the simulation of open-air spaces. Scattering and diffraction settings were defined as in Table 2 in order to allow a more coherent comparison between the two software. The diffraction phenomenon occurs when a sound wave hits edges, i.e., intersections between surfaces, or when the surface dimensions are limited. Both these events are taken into account by the software Odeon and CATT-A when the Reflection-Based Scattering and Diffuse reflection method are enabled. In Odeon, the Lambert and the Oblique Lambert functions for scattering were disabled, as suggested in a previous paper [39]. The uniform scattering distribution was considered more suitable for the cavea which is made of steps that can be reassembled as periodic triangular section [39–41], as shown in Figure 3a. In CATT-Acoustic, the diffraction after 1st order option was deactivated, even though it is usually suggested for ancient theatres [42], in order to take into account the current large amount of damage to the cavea steps in SR. In this way, it was possible to avoid the typical "chirp" echo due to diffraction phenomenon which has been attested to come from the regular stone steps in ancient theatres in empty conditions [43]. This phenomenon was not encountered during in situ measurements or recordings in SR. However, the first order diffraction has been taken into account since it occurs in coherence to the scattering phenomena. Moreover, based on the literature [44], it was found that higher orders and combinations of edge diffraction components were not usually as significant as first-order diffraction components when the receiver was visible to the source. The environmental data considered in both of the prediction tools were those obtained during the in situ measurements (t = 33 °C, RH = 65%).

5. Data analysis: the analysis algorithm has been taken into account. Odeon conducts an energy based analysis, while the CATT-Acoustic software conducts both energy and pressure-based analyses. The variation of different types of analysis algorithms can lead to different results, as pointed out in Katz [45]. Thus, in order to avoid further uncertainty in the results, the simulated IRs have been exported and analysed by means of Aurora, version 4.4, in the same way as the measurements.

Table 2. Scattering and diffraction set-up in Odeon and CATT-Acoustic.

Phenomenon	Model	Odeon	CATT-Acoustic
Scattering	Lambert	Disabled	Late part of the IR (not manageable by the user)
	Oblique Lambert	Disabled	Not managed by the software
	Uniform	Enabled, for early and late part of the IR	Early part of the IR (not manageable by the user)
Edge + surface diffraction		Enabled (i.e., Reflection-Based Scattering)	Enabled (in CATT-A, i.e., Diffuse reflection)
Diffraction after 1st order		Not managed by the software	Disabled (in TUCT)

As reported in Vorländer [13], the level of detail in the model, besides the curved surfaces, is considered a systematic source of uncertainty. Besides the number of rays employed in the simulation, the absorption and scattering coefficients are defined as random sources of uncertainty. Both kinds of software use a ray-tracing method to build the late part of the IR. Since this method is based on stochastic calculation, which depends on the input general set-up data, it could affect the uncertainty of the resulting parameters when a run-to-run analysis is considered. All the aforementioned random sources of uncertainty were subjected to analysis, considering both the Odeon and CATT-Acoustic software, which, for the sake of an easier presentation of the results, will hereafter be referred to as O and C, respectively; the results are shown in the following sub-sections.

4.2. Run-to-Run Variation

The run-to-run variations of the applied algorithms are due to the stochastic implementation of the ray-tracing algorithm in the GA software. In order to test this effect, ten repeated simulations were performed with the GA model of SR, using both kinds of software. An analysis based on the assessment of the Normalized Error [46] was performed on the T_{20}, C_{80}, and G results, considering a confidence level of 95%. The results for each receiver position and octave-band frequency were all within the upper and lower limits of the respective limit range. This confirms the results obtained in analogous analyses conducted on an enclosed space [47].

4.3. Number of Rays

GA software usually distinguishes between deterministic and stochastic ray-tracing, depending on which algorithm is applied: The first algorithm is used to detect the image sources, while the second is used to estimate the reverberant tail. It is possible to select separately the number of early and late rays in O. Early rays are used in the deterministic ray-tracing, while late rays determine the ray density in the late part of the IR. The number of rays/cones in C only refers to the stochastic ray-tracing; that is the construction of the late rays. It becomes important to investigate the variation in results due to stochastic ray-tracing, which is a random source of uncertainty in GA.

Stochastic ray-tracing was here investigated by comparing simulations with different numbers of rays (4000–40,000–400,000–4 million). A Normalized Error analysis revealed that the results for each receiver position and octave-band frequency were all within the upper and lower limits of the respective limit range. This investigation was performed in order to verify the stochastic fluctuation, which may result as numerical errors in the results due to the low number of rays. This has been extensively studied and validated in systematic experiments [48]. The number of rays is strictly related to the systematic uncertainty in the final results of the parameters, and independently on the used method of the ray tracing, the fluctuations can be reduced by increasing the number of rays or by averaging repeated simulations. The choice of the number of rays becomes important in cases where large environments with uneven distribution of the absorption are considered. Therefore, a compromise should be reached between a very large number of rays and a smaller one since it may affect significantly the computation time. In fact, the reverberant field in a simulated open-air theatre is spatially uneven. The absorbing area is concentrated on the ceiling of the boundary box (in the case of O), while the theatre itself is mostly reflective. Thus, despite the prolongation of the computation time, a number of rays above 1 million would be preferable for the correct estimation of the reverberation tails at different receiver positions [22]. It is assumed that at least one ray is received at the longest source-to-receiver distance, which in this case is about 40 m (R10). The receiving area is considered as a spherical receiver with a radius r_d of about 0.06 m, thus the area of the visibility cone per ray A(ray) was 0.01 m^2. Considering that the total surface covered by the emitted rays is a sphere of radius 40 m, whose surface A(sph) is equal to 20,096 m^2, it is possible to calculate the minimum required number of rays N_{min}(rays) by means of Equation (6), which was also indicated in Vorländer [13]:

$$N_{min}(\text{rays}) = \frac{A(\text{sph})}{A(\text{ray})} = \frac{4(ct)^2}{r_d^2} \qquad (6)$$

where c and t are the speed of sound in air and the max arrival time counted from source excitation, respectively.

N_{min}(rays) is equal to 2 million rays. Thus, 4 million rays are necessary to ensure that at least two rays (instead of one) arrive at the receiver at a distance of 40 m from the source.

4.4. Absorption and Scattering Coefficients

The predictive software considers α_w and s as input variables that have to be assigned to the surfaces of the model. Thus, it is important to evaluate the uncertainty (U) of the calculated values, due to the uncertainty of the absorption ($U\alpha_w$) and scattering (Us) variables. These uncertainties were estimated to be higher than 0.05 and 0.15, respectively, as was found on the basis of the user's experience in Vorländer [13] and Shtrepi et al. [49]. This case study considered only a few materials: stones and grass in particular. This allowed variations due to different α_w and s combinations regarding the cavea stone, which is the main surface considered in the model, to be investigated. To this aim, as shown in Figure 4b, twenty alternative materials were considered in both kinds of software, with α_w equal to 0.05, 0.10, 0.15, and 0.20, and with s equal to 0.25, 0.40, 0.55, 0.70, and 0.85. These values considered the possibility of having different degrees of damage on the steps of the cavea. In the case of the scattering coefficients of 0.85 [41], a perfectly preserved periodic triangular section with an angle of 45° has been considered, whereas in the case of scattering coefficient of 0.25, a heavily damaged cavea was represented.

As suggested previously [50], the sensitivity coefficients were calculated in order to evaluate the uncertainty propagation. This evaluation was conducted considering the average simulation results of the 500 Hz and 1 kHz octave bands [2]. The variability of each simulated receiver was calculated, and no systematic effects were detected. Thus, the sensitivity coefficients were calculated considering the normalized values, with respect to the relevant average value. An appropriate mathematical model, based on linear regression, was defined so as to relate the simulated values of each acoustical parameter to the absorption and scattering coefficients [50,51]. The expanded uncertainty was obtained as 2σ, where σ is the standard deviation of the model [50].

The expanded uncertainties for the O and C simulation software (U_O and U_C) are shown in Table 3. The uncertainty, due to the input variability of α_w and s, is lower than the JND for all the parameters, except for T_{20} and C_{80} when the C software is used. The lower uncertainty values are due to the software algorithm, which is less sensitive to variations in α_w and s.

Table 3. Just Noticeable Difference (JND) of the T_{20}, C_{80}, and G acoustical parameters, the expanded uncertainty due to the variability of the input values of α_w and s for the simulation software O and C (U_O and U_C). Values higher than the JNDs are reported in bold.

Acoustical Parameter	JND	U_O	U_C
T_{20} (s)	5% $\approx$ 0.03	0.01	**0.05**
C_{80} (dB)	1	0.50	**1.20**
G (dB)	1	0.01	0.30

5. Discussion

This work aimed at providing an overview of the many methodological challenges that should be faced when dealing with the acoustics of open-air ancient theatres, both in the case of measured (i.e., for the acoustical characterisation of the current state) and predicted (i.e., for the simulation of a no longer/not yet existing state) room acoustics parameters. Measurement and simulation are strictly interconnected, also considering that the former is often required to validate the latter; the rationale for addressing both these aspects within the framework of this paper is that this is particularly true for open-air ancient theatres. Indeed, measurements of such unroofed spaces have been shown to be problematic with the application of current standards. Achieving reliable acoustical measurements is important in order to provide calibration data for the simulation software. In the context of cultural heritage research, and specifically for archaeological or historical acoustics, simulation becomes crucial because of the need to investigate (in most of cases) physical conditions which no longer exist (acoustics of the past), due to, among other aspects, the deterioration of the architectural elements. For these reasons, while measurements and simulations are concerned with different uncertainty issues, it was decided to compare the measured and calculated parameters (Section 5.1), as well as discussing the overall limitations of the considered protocols (Section 5.2).

5.1. Comparison of the Measured and Simulated Results

The aim of acoustical simulations is to obtain predictions that would closely match measured data. A well-calibrated model should minimise the perceivable differences between simulation and measurements for any considered acoustic parameter.

The subsequent considerations were also based on the α_w and s values of the cavea surface and its variations. The differences between the measured and simulated results are shown in Figure 4, which reports the acoustical behaviour during the calibration of both kinds of software, considering the variations due to the 20 alternative combinations (5 scattering coefficients × 4 absorption coefficients), for all the receivers, and the average between the 500 Hz and 1 kHz octave-bands. The isolevel curves shown in Figure 4 have been obtained by a two-dimensional data interpolation using the MATLAB function "interp2" with the "spline" method active. This method was chosen in order to have smooth first and second derivatives throughout the curves. Figure 4a,b, which pertain to O and C, respectively, refer to parameter T_{20}, while Figure 4c,d refer to C_{80} and Figure 4e,f to G. The light yellow colour in the graphs shows the α_w and s combinations for which the simulated values were closest to the measured ones. These isolevel curves were based on SAD, i.e., the Sum of the Absolute Differences between the simulated values, s_n, and the measured ones, m_n, for each receiver position, expressed as follows by Equation (7) [52]:

$$SAD = \sum_{1}^{n} |s_n - m_n| \tag{7}$$

The results show that, depending on which parameter is considered, the best agreement between the simulated and measured values could not be obtained for the same combination of α_w and s. From the isolevel curves layout it is observable that, apart from T_{20}, Odeon software is more sensitive to variations of α_w than of s, while the opposite occurs for CATT-Acoustic. For T_{20}, lower differences between the simulated and measured values are detectable for both high and low absorption and scattering values in the case of Odeon software, while mainly for high scattering values over the whole range of absorption values in the case of CATT-Acoustic. For C_{80}, a good matching between measured and simulated values occurs with high absorption values over the whole range of scattering coefficients in the case of Odeon software, while it occurs with low scattering coefficients over the whole range of absorption coefficients in the case of CATT-Acoustic. For G, the best matching between measured and simulated values occurs with low absorption values over the whole range of scattering coefficients in the case of Odeon software, while it occurs with a medium scattering coefficient over the whole range of absorption coefficients in the case of CATT-Acoustic. Only in the case of G do both kinds of software show an agreement that is obtained in a range around the values of $\alpha_w = 0.10$ and s = 0.55. Thus, this combination was considered for the calibration of the model.

Table 4 shows all the simulation results of the calibrated model of SR, expressed as T_{20}, C_{80}, and the G acoustical parameters, considering both O and C. All the values are averaged over the central 500 Hz and 1 kHz octave-band frequencies and spatial values have been added for each row. In this way, the results can be compared directly with those of the corresponding measurements. A good agreement has been shown between the results obtained with the two different types of software, as can be also seen from the graph in Figure 5, where the average G for each row is represented along the average distance from the source, in the cases of measurements and simulations with Odeon and CATT-Acoustic.

In particular, the average values for each row obtained from the two software are always within or at the limit of the JND for each parameter, except C_{80} in the first row. The differences between the simulated and measured results, in terms of average values per each row, are within two to seven times the JND for T_{20}, without any systematic behaviour related to the row. In the case of C_{80}, the differences from simulated and measured values are higher for the first row, with average simulated values that are three and five times the JND with Odeon and Catt-Acoustic, respectively, within 2 JND for the second and third rows, and within the JND for the last row, for both the software. For G, the average simulated values for each row are always within or quite close the JND compared to measured values for both the software, with a slightly worse behaviour for Odeon. Figure 5 shows as both the software correctly simulated the reduction of G with the distance from the source, with slopes in dB per distance doubling (dB/dd) that are 6.6 dB/dd and 6.3 dB/dd, for Odeon and Catt-Acoustic, respectively, compared to 6.3 dB/dd for the measurements.

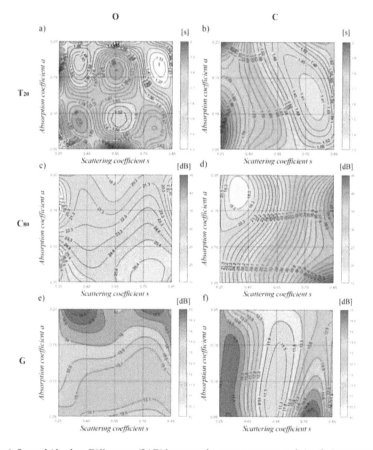

Figure 4. Sum of Absolute Differences (SAD) between the measurements and simulations overall the receivers, for T_{20}, C_{80}, and G in Odeon (**a,c,e**) and for CATT (**b,d,f**). Light yellow refers to very similar values between simulation and measurements.

Table 4. Mean values of the measurements and simulations for the T_{20}, C_{80}, and G acoustical parameters for the source in position S1. The data refer to the averages of the 500 Hz and 1 kHz octave bands and to the repetitions for the same receiver position. Spatial means refer to receivers on the same row.

Row	Receiver	Acoustical Parameters								
		T_{20} (s)			C_{80} (dB)			G (dB)		
		Measured	Pred. (Odeon)	Pred. (Catt)	Measured	Pred. (Odeon)	Pred. (Catt)	Measured	Pred. (Odeon)	Pred. (Catt)
First row	R3	0.58	0.66	0.68	20.8	17.0	15.1	−0.3	−1.7	−1.0
	R6	0.70	0.66	0.62	16.9	17.0	15.2	−0.9	−2.3	−2.6
	R9	0.31	0.70	0.64	22.1	15.4	15.3	−2.4	−1.8	−1.4
spatial mean		0.53	0.67	0.65	19.9	16.5	15.2	−1.2	−1.9	−1.7
Second row	R2	0.81	0.92	0.97	16.5	14.6	14.5	−4.8	−6.7	−6.4
	R5	0.66	0.93	0.92	19.2	15.3	14.6	−5.4	−7.1	−6.7
	R8	0.71	0.90	0.77	15.9	14.5	15.4	−6.0	−7.0	−7.0
spatial mean		0.73	0.92	0.89	17.2	14.8	14.8	−5.4	−6.9	−6.7
Third row	R1	0.98	1.10	1.00	15.4	14.0	13.0	−7.8	−9.0	−9.0
	R4	0.94	1.03	1.05	16.9	13.9	12.5	−8.0	−9.8	−9.6
	R7	1.04	0.99	0.90	15.7	13.8	14.2	−9.3	−9.8	−9.7
spatial mean		0.99	1.04	0.98	16.0	13.9	13.2	−8.4	−9.5	−9.4
	R10	1.31	1.11	0.89	14.3	14.5	15.2	−10.2	−11.4	−10.6

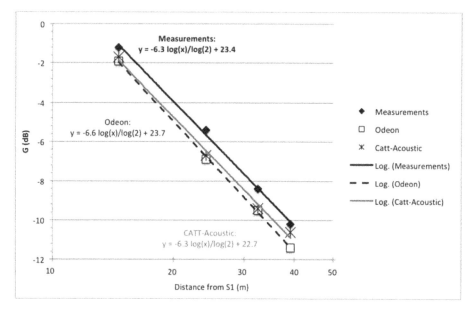

Figure 5. G values averaged over the central 500 Hz and 1 kHz octave-band frequencies, and for each row, represented along the average distance from the source, derived from the measurements and from the simulations with Odeon and CATT-Acoustic.

5.2. Limitations of the Study

Given the complexity of the task, there are, of course, a number of limitations in the methodological approach implemented in the present study. Most of such shortcomings are related, as previously mentioned, to the actual applicability of the ISO 3382-1, intended for roofed performance spaces, to open-air environments.

Certainly, Section 7 of the ISO 3382-1 deals with the "Measurement uncertainty" and specifies that for practical evaluation of the measurement uncertainty of reverberation time using the integrated impulse response method, it can be considered as being of the same order of magnitude as that using an average of n = 10 measurements in each position with the interrupted noise method. No additional averaging is necessary to increase the statistical measurement accuracy for each position. However, considering the variability due to the atmospheric conditions, more than one repetition is needed. On the other side, anyone who has performed measurements in ancient open-air theatres knows that a large number of repetitions is rarely feasible, for a number of practical reasons due to the stability of the boundary conditions; thus, the scope of this study was to assess the protocols' reliability with fewer measurements.

Table 5 summarises the most salient aspects and recommendations provided in the different sections of the ISO 3382-1, confirms on whether such requirements were met and reports briefly on each circumstance ("notes" column).

Moreover, another limitation of the work derives from the use of GA software. The differences between simulations and measurements are mainly related to the approximations of GA with respect to the real wave effects, which result to be important for an open-air environment where the number of surfaces is limited and the generation of a diffuse field becomes critical. The GA principals are valid above the Schroeder frequency, which is not easy to estimate for an ancient theatre. The limits of GA are related to large rooms, low absorption coefficients, and broadband signals [48]. Furthermore, they neglect phase. As shown in different Round Robin tests [16,17], the GA based software differ between each other even when the same input data of absorption coefficients are given to the surfaces. Therefore, the major drawback, for the state of the art modelling software, is that the different simulation tools require different input data [53]. In practice, the absorption and scattering coefficient values are calibrated, i.e., varied within the range of their measurement uncertainty, in order to match the simulation results to the measured values. This may result in different values of these coefficients for the different software.

Table 5. ISO 3382-1 recommendations and their applications in the measurement campaign (X).

ISO 3382-1 Section	Recommendation	Implemented	Notes
4. Measurement conditions	Temperature and Relative Humidity: these quantities should be measured with an accuracy of ±1 °C and 5%, respectively.	X	
	Equipment: omnidirectional sources and receivers. Maximum deviations of directivity for an omnidirectional source are indicated.	X	The deviation of directivity of the used sound source respected the maximum values indicated by the standards [30,54,55].
	Number of source positions: minimum 2, located where the natural sound source would take position. Height of sources: 1.5 m.	X	
	Number of microphone positions: Microphone positions should be at positions representative of positions where listeners would normally be located. For reverberation time measurements, it is important that the measurement positions sample the entire space; for the room acoustic parameters, they should also be selected to provide information on possible systematic variations with position in the room. Height of the receivers: 1.2 m.	X	
5. Measurement procedures	Integrated Impulse Response method: any source is allowed provided that its spectrum is broad enough to cover from 125 Hz to 4 kHz. The peak sound pressure level has to ensure a decay curve starting at least 35 dB above the BNL.	X	In some receiving positions, the 125 Hz frequency band did not guarantee the required 35 dB over the BNL, with the firecrackers.
	Time averaging: it is necessary to verify that the averaging process does not alter the measured impulse responses.		
6. Decay curves	Regression analysis: a least-squares fit line shall be computed for the decay curve. If the curves are wavy or bent, this may indicate a mixture of modes with different reverberation times and thus the result may be unreliable.		The open-air condition is characterised by a cliff-decay curve [54] linked to a few strong reflections, but this case is not considered by the standard.

Appl. Sci. **2018**, *8*, 1393

6. Concluding Remarks

This work deals with the accuracy of acoustical measurements and prediction models related to the ancient open-air theatre of Syracusae. Measurements based on ISO 3382-1 were conducted in unoccupied conditions. Firecrackers were used, because of the relatively high background noise level. The acoustical parameters described in the ISO 3382-1 standard, that is, Reverberation Time (T_{20}), Clarity (C_{80}), and Sound Strength (G), were obtained from the IRs measured at each receiver position. The uncertainty contributions due to the input values of sound absorption and scattering coefficients, α_w and s, have also been calculated with two simulation tools, that is, Odeon, version 13.02, and CATT-Acoustic, version 9. The models have been calibrated on the basis of the best match between the simulated and measured parameter values. Other sources of uncertainties, that is, the run-to-run variations and number of rays, have also been analysed and the obtained results have all been found to be under or at commonly accepted limit values of the Just Noticeable Differences (JNDs). The variability of the results is related to the algorithms used to approximate the acoustic phenomenon of the absorption and scattering. This kind of software are based on geometric acoustic principles, which rely on a statistical approach used to include diffuse sound scattering and predict the reverberant tail of an impulse response [22,23].

The following main results have been found from the uncertainty analysis that was conducted on the simulations of the Syracusae theatre:

- The uncertainty, due to the input variability of α_w and s, is lower than the JND for T_{20} and C_{80}, when the Odeon software is considered, and for G when both types of software are considered;
- Apart from T_{20}, Odeon software is more sensitive to variation of sound absorption than of sound scattering, while the opposite occurs for CATT-Acoustics;
- Comparable behaviour of the simulated values of G has been shown for both types of software; G has been found to be the most suitable parameter for the calibration of the open-air theatre model;
- A good agreement with the measured values has been found, at the limit of the JNDs, in the calibrated model for all the parameters, in spite of the limitation of the GA software that has emerged in this case study, for both types of software.

Future studies will be conducted on a larger number of case studies, considering the influence of the architectural state of conservation, completeness, and dimensions on the acoustic field. Moreover, more suitable parameters for the acoustical characterisation of the open-air theatres than those described in ISO 3382-1 standard are the subject of continuous research [49].

Author Contributions: E.B. and A.A. conceived and designed the data collection campaigns and simulations; E.B. collected data on site and performed the simulations with Odeon together with L.S., D.P.G. performed the simulations with CATT. L.S. and G.B. collaborated for the statistical analysis of the uncertainties. F.A. offered support on the applicability of the ISO standard. All authors wrote and revised the paper.

Funding: This research was funded through a Ph.D. scholarship awarded to the first author by the Politecnico di Torino (Turin, Italy).

Acknowledgments: The authors are grateful to Fabrizio Bronuzzi, Rocco Costantino, and Maurizio Bressan from the Acoustics Laboratory of Politecnico di Torino for their technical contribution to this project, as well as to George Koutsouris and Claus Lynge Christensen from ODEON support team. The authors would like to thank also Soprintendenza BB.CC.AA di Siracusa, Istituto Nazionale Dramma Antico and Andrea Tanasi for their support during the measurements in the theatre. Finally, the authors are grateful to Nicola Prodi and Andrea Farnetani from the Univeristy of Ferrara, Angelo Farina from the University of Parma, and Monika Rychtarikova, from KU Leuven, for their suggestions and advice.

Conflicts of Interest: The authors declare no conflicts of interest.

References

1. Scarre, C.; Lawson, G. *Archeoacoustics*; McDonald Institute for Archaeological Research, University of Cambridge: Cambridge, UK, 2006.
2. ISO 3382-1:2009. *Measurement of Room Acoustic Parameters—Part 1: Performance Spaces*; International Organization for Standardization: Geneva, Switzerland, 2009.
3. Rindel, J.H. *ERATO*; Final Report, INCO-MED Project ICA3-CT-2002-10031; ERATO Project: Lyngby, Denmark, 2006.
4. Farnetani, A.; Prodi, N.; Pompoli, R. On the acoustic of ancient Greek and Roman Theatres. *J. Acoust. Soc. Am.* **2008**, *124*, 157–167. [CrossRef] [PubMed]
5. Chourmouziadou, K.; Kang, J. Acoustic evolution of ancient Greek and Roman theatres. *Appl. Acoust.* **2008**, *69*, 514–529. [CrossRef]
6. Mo, F.; Wang, J. The Conventional RT is Not Applicable for Testing the Acoustical Quality of Unroofed Theatres. *Build. Acoust.* **2013**, *20*, 81–86. [CrossRef]
7. Iannace, G.; Trematerra, A.; Masullo, M. The large theatre of Pompeii: Acoustic evolution. *Build. Acoust.* **2013**, *20*, 215–227. [CrossRef]
8. Iannace, G.; Trematerra, A. The rediscovery of Benevento Roman theatre acoustics. *J. Cult. Herit.* **2014**, *15*, 698–703. [CrossRef]
9. Guski, M. Influences of External Error Sources on Measurements of Room Acoustic Parameters. Ph.D. Thesis, RWTH Aachen University, Aachen, Germany, 2015.
10. Akama, T.; Suzuki, H.; Omoto, A. Distribution of selected monaural acoustical parameters in concert halls. *Appl. Acoust.* **2010**, *71*, 564–577. [CrossRef]
11. Pelorson, X.; Vian, J.P.; Polack, J.D. On the variability of room acoustical parameters: Reproducibility and statistical validity. *Appl. Acoust.* **1992**, *37*, 175–198. [CrossRef]
12. Malecki, P.; Zastawnik, M.; Wiciak, J.; Kamisinski, T. The influence of the measurement chain on the impulse response of a reverberation room and its application listening tests. *Acta Phys. Pol. A* **2011**, *119*, 1027–1030. [CrossRef]
13. Vorländer, M. Computer simulations in room acoustics: Concepts and uncertainties. *J. Acoust. Soc. Am.* **2013**, *133*, 1203–1213. [CrossRef] [PubMed]
14. Lisa, M.; Rindel, J.H.; Christensen, C.L. Predicting the acoustics of open-air theatres: The importance of calculation methods and geometrical details. In Proceedings of the Baltic-Nordic Acoustics Meeting, Mariehamn, Aland, 8–10 June 2004.
15. Gade, A.C.; Lynge, C.; Lisa, M.; Rindel, J.H. Matching simulations with measured acoustic data from Roman Theatres using the Odeon program. In Proceedings of the Forum Acusticum, Budapest, Hungary, 29 August–2 September 2005.
16. Vorländer, M. International round robin on room acoustical computer simulations. In Proceedings of the 15th International Congress on Acoustics, Trondheim, Norway, 26–30 June 1995.
17. Bork, I. A comparison of room simulation software—the 2nd Round Robin on Room Acoustical Computer Simulations. *Acta Acust. United Acust.* **2000**, *86*, 943–946.
18. Bork, I. Report on the 3rd Round Robin on Room Acoustical Computer Simulation—Part II: Calculations. *Acta Acust. United Acust.* **2005**, *91*, 753–763.
19. ISO 354:2003. *Acoustics—Measurement of Sound Absorption in a Reverberation Room*; International Organization for Standardization: Geneva, Switzerland, 2003.
20. ISO 17497-1:2004. *Acoustics—Sound-Scattering Properties of Surface—Part 1: Measurement of the Random-Incidence Scattering Coefficient in a Reverberation Room*; International Organization for Standardization: Geneva, Switzerland, 2004.
21. Farnetani, A.; Prodi, N.; Roberto, P. Measurements of the sound scattering of the steps of the *cavea* in ancient open air theatres. In Proceedings of the International Symposium of Room Acoustics, Seville, Spain, 10–12 September 2007.
22. Christensen, C.L.; Koutsouris, G. Odeon Room Acoustics Software. Version 13. Full User's Manual, Odeon A/S, Lyngby, Denmark. 2015. Available online: https://www.odeon.dk/ (accessed on 10 January 2015).
23. Dalenback, B.I.L. CATT-A v9.0, User's Manual, CATT-Acoustic v9, CATT, Sweden. 2011. Available online: https://www.catt.se/ (accessed on 10 January 2015).

24. Prodi, N.; Farnetani, A.; Fausti, P.; Pompoli, R. On the use of ancient open-air theatres for modern unamplified performances: A scale model approach. *Acta Acust. United Acust.* **2013**, *99*, 58–63. [CrossRef]

25. Bo, E.; Astolfi, A.; Pellegrino, A.; Pelegrín Garcia, D.; Puglisi, G.E.; Shtrepi, L.; Rychtarikova, M. The modern use of ancient theatres related to acoustic and lighting requirements: Stage design guidelines for the Greek theatre of Syracuse. *Energy Build.* **2015**, *95*, 106–115. [CrossRef]

26. Gullo, M.; La Pica, A.; Rodonò, G.; Vinci, V. Acoustic characterization of the ancient theatre at Syracuse. In Proceedings of the Acoustics Conference, Paris, France, 29 June–4 July 2008.

27. Farina, A. Personal Communications, Syracuse Measurements Data (Realised in 2003). 2013. Available online: http://www.angelofarina.it/Siracusa/ (accessed on 30 September 2016).

28. Bo, E.; Bergoglio, M.; Astolfi, A.; Pellegrino, A. Between the Archaeological Site and the Contemporary Stage: An Example of Acoustic and Lighting Retrofit with Multifunctional Purpose in the Ancient Theatre of Syracuse. *Energy Procedia* **2015**, *78*, 913–918. [CrossRef]

29. Rindel, J.H. Echo problems in ancient theatres and a comment to the sounding vessels described by Vitruvius. In Proceedings of the Acoustics of Ancient Theatres Conference, Patras, Greece, 18–21 September 2011.

30. San Martin, R.; Arana, M.; Machin, J.; Arregui, A. Impulse source versus dodecahedral loudspeaker for measuring parameters derived from the impulse response in room acoustics. *J. Acoust. Soc. Am.* **2013**, *134*, 275–284. [CrossRef] [PubMed]

31. Angelo Farina's personal Home Page. Available online: http://pcfarina.eng.unipr.it/Aurora_XP/index.htm (accessed on 4 July 2016).

32. Martellotta, F. The just noticeable difference of center time and clarity index in large reverberant spaces. *J. Acoust. Soc. Am.* **2010**, *128*, 654–663. [CrossRef] [PubMed]

33. Blevins, M.G.; Buck, A.T.; Peng, Z.; Wang, L.M. Quantifying the just noticeable difference of reverberation time with band-limited noise centered around 1000 Hz using a transformed up-down adaptive method. In Proceedings of the International Symposium on Room Acoustics, Toronto, ON, Canada, 9–11 June 2013.

34. Cox, T.J.; Davies, W.J.; Lam, Y.W. The sensitivity of listeners to early sound field changes in auditoria. *Acta Acust. United Acust.* **1993**, *79*, 27–41.

35. Bradley, J.S.; Reich, R.; Norcross, S.G. A just noticeable difference in C50 for speech. *Appl. Acoust.* **1999**, *58*, 99–108. [CrossRef]

36. Schröder, D.; Pohl, A. Modeling (non-)uniform scattering distributions in geometrical acoustics. In Proceedings of the International Congress on Acoustics, ICA 2013, Montreal, QC, Canada, 2–7 June 2013.

37. Stephenson, U.M. Eine Schallteilchen-Computer-Simulation zur Berechnung der für die Hörsamkeit in Konzertsälen maßgebenden Parameter. *Acta Acust. United Acust.* **1985**, *59*, 1–20.

38. Rindel, J.H. A new scattering method that combines roughness and diffraction effects. In Proceedings of the Forum Acusticum, Budapest, Hungary, 29 August–2 September 2005.

39. Shtrepi, L.; Astolfi, A.; Puglisi, G.E.; Masoero, M.C. Effects of the Distance from a Diffusive Surface on the Objective and Perceptual Evaluation of the Sound Field in a Small Simulated Variable-Acoustics Hall. *Appl. Sci.* **2017**, *7*, 224. [CrossRef]

40. Charmouziadou, K. Ancient and Contemporary Use of the Open-Air Theatres: Evolution and Acoustic Effects of Scenery Design. Ph.D. Thesis, School of Architecture, The University of Sheffield, Sheffield, UK, 2007.

41. Cox, T.J.; D'Antonio, P. *Acoustic Absorbers and Diffusers: Theory, Design and Application*; Spon: New York, NY, USA, 2004; pp. 1–476.

42. Economou, P.; Charalampous, P. The significance of sound diffraction effects in predicting acoustics in ancient theatres. *Acta Acust. United Acust.* **2013**, *99*, 48–57. [CrossRef]

43. Declercq, N.F.; Degrick, J.; Briers, R.; Leroy, O. A theoretical study of special acoustic effects caused by the staircase of the El Castillo pyramid at the Maya ruins of Chichen-Itza in Mexico. *J. Acoust. Soc. Am.* **2004**, *116*, 3328–3335. [CrossRef] [PubMed]

44. Torres, R.R.; Svensson, U.P.; Kleiner, M. Computation of edge diffraction for more accurate room acoustics auralization. *J. Acoust. Soc. Am.* **2001**, *109*, 600–610. [CrossRef] [PubMed]

45. Katz, B.F.G. International round robin on room acoustical response analysis software. *Acoust. Res. Lett.* **2004**, *5*, 158–164. [CrossRef]

46. ISO/IEC Guide 43-1. *Proficiency Testing by Interlaboratory Comparisons. Part 1: Development and Operation of Proficiency Testing Schemes*; International Organization for Standardization: Geneva, Switzerland, 1997.

47. Postma, B.N.J.; Katz, B.F.G. Creation and calibration method of acoustical models for historic virtual reality auralizations. *Virtual Real.* **2015**, *19*, 161–180. [CrossRef]
48. Vorländer, M. *Auralization: Fundamentals of Acoustics, Modeling, Simulation, Algorithms and Acoustic Virtual Reality*; Springer: Berlin, Germany, 2008.
49. Shtrepi, L.; Astolfi, A.; Pelzer, S.; Vitale, R.; Rychtarikova, M. Objective and perceptual assessment of the scattered sound field in a simulated concert hall. *J. Acoust. Soc. Am.* **2015**, *138*, 1485–1497. [CrossRef] [PubMed]
50. JCGM 100:2008. *Expression of Measurement Data—Guide to the Expression of Uncertainty in Measurement*; Bureau International des Poids et Mesures: Sèvres, France, 2008.
51. Barbato, G.; Germak, A.; Genta, G. *Measurements for Decision Making. Measurements and Basic Statistics*; Esculapio: Bologna, Italy, 2013.
52. Li, Z.; Ding, Q.; Zhang, W. A Comparative Study of Different Distances for Similarity Estimation. In *Intelligent Computing and Information Science*; Chen, R., Ed.; Communications in Computer and Information Science; Springer: Berlin/Heidelberg, Germany, 2011; Volume 134.
53. Lam, Y.W. A comparison of three diffuse reflection modeling methods used in room acoustics computer models. *J. Acoust. Soc. Am.* **1996**, *100*, 2181–2192. [CrossRef]
54. Barron, M. Interpretation of Early Decay Time in concert auditoria. *Acta Acust. United Acust.* **1995**, *81*, 320–331.
55. Sumarac-Pavlovic, D.; Mijic, M.; Kurtovic, H. A simple impulse sound source for measurements in room acoustics. *Appl. Acoust.* **2008**, *69*, 378–383. [CrossRef]

Article

Acoustic Localization for a Moving Source Based on Cross Array Azimuth

Junhui Yin [1], Chao Xiong [1] and Wenjie Wang [2],*

[1] Department of Artillery Engineering, Army Engineering University, Shijiazhuang 050003, China;
    yuanzhidao@163.com (J.Y.); xiongchao@tsinghua.org.cn (C.X.)
[2] Fluid and Acoustic Engineering Laboratory, Beihang University, Beijing 100191, China
*   Correspondence: wangwenjie@buaa.edu.cn; Tel.: +86-108-231-7407

Received: 5 June 2018; Accepted: 28 July 2018; Published: 1 August 2018

Featured Application: the work introduced in this paper can be used for wildlife conservation, health protection, and other engineering applications.

Abstract: Acoustic localization for a moving source plays a key role in engineering applications, such as wildlife conservation and health protection. Acoustic detection methods provide an alternative to traditional radar and infrared detection methods. Here, an acoustic locating method of array signal processing based on intersecting azimuth lines of two arrays is introduced. The locating algorithm and the precision simulation of a single array shows that such a single array has good azimuth precision and bad range estimation. Once another array of the same type is added, the moving acoustic source can be located precisely by intersecting azimuth lines. A low-speed vehicle is used as the simulated moving source for the locating experiments. The length selection of short correlation and moving path compensation are studied in the experiments. All results show that the proposed novel method locates the moving sound source with high precision (<5%), while requiring fewer instruments than current methods.

Keywords: acoustic localization; cross array; moving sound source; discrete sampling; error analysis

---

1. Introduction

The localization of moving sources represents a major issue in engineering applications. Similar to other detection technologies, acoustic-localization methods have been developed rapidly over the years. Meanwhile, the noise generated by low-speed vehicles (LPVs) is a key issue, especially in connection with acoustics mitigation, where noise pollution continues to be a major health problem, with a whole host of health effects, such as: sleep disorders with awakenings [1], learning impairment [2,3], hypertension ischemic heart disease [4], and especially annoyance [5], a widely used indicator to study the effect of different noise sources on wellbeing. In this context, the main effort has been done to mitigate the main sources of noise: road traffic [6–8], railway traffic [9,10], airport [11,12], and industrial [13]. Specifically about road noise, the most important interaction producing noise, more than just the engine noise used for the LPV, was also road/tire interaction [14,15] and aerodynamic noise for high-speed vehicles. Furthermore, a relatively new noise source is impacting modern society in areas where background noise is low. Wind farms are being installed continuously every year to supply energy demand, but people are being affected by its noise, which is more disturbing than other sources [16,17] and the scientific community is moving towards its assessment [18].

In this paper, the LPV is the research object. As for all moving vehicles, exhaust systems and chain tracks are the main noise sources of LPVs, with exhaust systems representing the dominant factor. Therefore, exhaust systems could be chosen as the moving noise source. The most common localization methods for noise sources are Nearfield Acoustic Holography (NAH), beamforming, and array signal processing [19]. The sound field of a moving vehicle is effectively measured based on NAH with a moving acoustic plane [20] and coordinate compensation [21,22]. Far-field measurements of a moving source can be achieved by the short-time beamforming method, but these require extensive computational resources for processing the acquired data for the acoustic plane frame at every moment. The false noise source (ghost image) would also be easily generated [23,24]. For array signal processing, the required computations are fast and can be performed to high precision [25,26]. This is so since the necessary calculations to be performed on the signals are only one-dimensional and, therefore, substantially less demanding than those for a whole acoustic plane.

The localization method of a moving sound source for the new method described here is achieved by intersecting the azimuth lines of cross arrays. Initial testing of the localization algorithm and the data analysis were performed for a single array and revealed a good performance. Therefore, a second array was added to cross the azimuth lines. The locating experiments were conducted with the engine noise of an LPV as a moving noise source. The data length determined by short-time correlation and path compensation were also introduced. The new method succeeded in effective localization of moving vehicles, requiring less expensive instrumentation than existing methods. Moreover, it was found that it continues to perform properly even under adverse ambient conditions, such as bad weather or at low light levels at night.

2. Localization Analysis of Single Array

2.1. Localization Algorithm

The LPV used for the current study travelled on level ground such that its height remained constant relative to the array sensors. The height of the vehicle was about 2.0 m, which was approximately 1.5 m higher than the arrays themselves. Compared to the range of about 100 m or more, the constant height difference between the vehicle and the arrays had little influence on the localization performance and accuracy. Therefore, the localization was operated in the *x-y* coordinate system while height difference was ignored. The five-element cross array was taken as a basic array pattern, as illustrated in Figure 1.

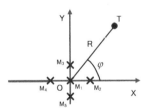

Figure 1. Model of single array.

The coordinates are defined within the plane of the array. The center acoustic sensor is located at *O* (0, 0), while the remaining four were M_2 (D, 0), M_3(0, D), M_4 (−D, 0), and M_5 (0, −D), where D represents the distance from M_i to *O*. The noise source is assumed to be located at *T* (*x, y*), with an angle φ between *OT* and *x* axis as indicated in Figure 1. The time delays between the arrival time of noise at the center sensor and the other four sensors are referred to as τ_{1i}. Similarly, d_{1i} (*i* = 2, 3, 4, 5) represents the distance between the center sensor and the neighboring sensors, such that $d_{1i} = c \times \tau_{1i}$ (*c* is current sound velocity). R is the distance from *O* to *T*.

From the simple geometry in Figure 1, the distances can be expressed.

$$\begin{cases} x^2 + y^2 = R^2 \\ (x-D)^2 + y^2 = (R+d_{12})^2 \\ x^2 + (y+D)^2 = (R+d_{13})^2 \\ (x+D)^2 + y^2 = (R+d_{14})^2 \\ x^2 + (y-D)^2 = (R+d_{15})^2 \end{cases} \tag{1}$$

The solution of Equation (1) is

$$\begin{cases} x = \frac{2R(d_{14}-d_{12})+d_{14}^2-d_{12}^2}{2D} \\ y = \frac{2R(d_{13}-d_{15})+d_{13}^2-d_{15}^2}{2D} \end{cases} \tag{2}$$

$$\begin{cases} \tan\varphi = \frac{y}{x} = \frac{(\tau_{15}-\tau_{13})[2R-c(\tau_{15}-\tau_{13})]}{(\tau_{14}-\tau_{12})[2R-c(\tau_{14}-\tau_{12})]} \\ R = \sqrt{x^2+y^2} = \frac{4D^2-d_{12}^2-d_{13}^2-d_{14}^2-d_{15}^2}{2(d_{12}+d_{13}+d_{14}+d_{15})} \end{cases} \tag{3}$$

when $R \gg c \times \tau_{1i}$,

$$\tan\varphi \approx \frac{(\tau_{15}-\tau_{13})}{(\tau_{14}-\tau_{12})} \tag{4}$$

then Equation (3) can be simplified:

$$\begin{cases} \varphi = \arctan\frac{(\tau_{15}-\tau_{13})}{(\tau_{14}-\tau_{12})} \\ R = (4D^2 - c^2\sum\limits_{i=2}^{5}\tau_{1i}^2)/2c\sum\limits_{i=2}^{5}\tau_{1i} \end{cases} \tag{5}$$

The location of the noise source is given by Equations (2) and (5) and, with reference to their derivation, it is evident that the localization algorithm is based on the time delays between the arrival times of noise at the sensors in the array.

2.2. Precision Analysis for Localization

The algorithm for localizing the noise source, as described by Equations (2) and (5) in Section 2.1, and the associated accuracy depend on sound velocity c, array size D and, in particular, the error involved in estimating the time delay σ_τ. Since D and c remain constant, for any particular array and measurement environment, the dominant factor affecting the precision of the proposed method is associated with the error involved in measuring σ_τ. Due to the symmetric arrangement of the sensors with regard to the central sensor, the standard errors for the time delay of all sensors were assumed to be equal, such that $\sigma_\tau = \sigma_{\tau 1i}$.

In Equation (5), quadratic function was included in the expression of coordinates (x, y), which makes it different to calculate the transmission. Then, after the precision calculation of coordinates was transferred into angular coordinates, the localization was described with azimuth φ and range R as illustrated in Equation (5).

2.3. Azimuth Precision

According to Equation (5), azimuth Φ was a function of time delay τ.

$$\Phi = F(\tau) = F(\tau_{12}, \tau_{13}, \tau_{14}, \tau_{15}) \tag{6}$$

The transmission form of azimuth error σ_φ can be expressed.

$$\sigma_\varphi^2 = \left(\frac{\partial\varphi}{\partial\tau_{12}}\sigma\tau\right)^2 + \left(\frac{\partial\varphi}{\partial\tau_{13}}\sigma\tau\right)^2 + \left(\frac{\partial\varphi}{\partial\tau_{14}}\sigma\tau\right)^2 + \left(\frac{\partial\varphi}{\partial\tau_{15}}\sigma\tau\right)^2 \tag{7}$$

Taking derivative of τ in Equation (5):

$$\begin{cases} \frac{\partial \varphi}{\partial \tau_{12}} = -\frac{\partial \varphi}{\partial \tau_{14}} = \frac{1}{1+\tan^2 \varphi} \cdot \frac{\tau_{15}-\tau_{13}}{(\tau_{14}-\tau_{12})^2} \\ \frac{\partial \varphi}{\partial \tau_{13}} = -\frac{\partial \varphi}{\partial \tau_{15}} = -\frac{1}{1+\tan^2 \varphi} \cdot \frac{1}{\tau_{14}-\tau_{12}} \end{cases} \tag{8}$$

So the expression of azimuth error σ_φ is

$$\sigma_\varphi = \frac{\sigma\tau}{1+\tan^2 \varphi} \sqrt{\frac{2(\tau_{14}-\tau_{12})^2 + 2(\tau_{15}-\tau_{13})^2}{(\tau_{14}-\tau_{12})^4}} \tag{9}$$

Solving Equations (2) and (5):

$$\begin{cases} (\tau_{14}-\tau_{12})^2 + (\tau_{15}-\tau_{13})^2 = \frac{D^2}{v^2} \\ (\tau_{14}-\tau_{12})^2 = \frac{D^2}{v^2(1+\tan^2 \varphi)} \end{cases} \tag{10}$$

Substituting Equation (10) into Equation (9):

$$\sigma_\varphi = \frac{\partial \varphi}{\partial \tau_i} = \frac{\sqrt{2}c}{D}\sigma_\tau \tag{11}$$

Thus, the azimuth error is determined by c, D, and σ_τ. We assume a value of $c = 343$ m/s for the sound velocity and employ a sampling rate of 5000 Hz. The sampling interval is 200 μs and the distribution of azimuth error is shown in Figure 2.

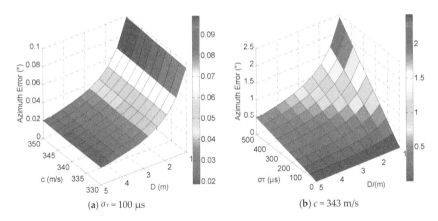

(a) $\sigma_\tau = 100$ μs   (b) $c = 343$ m/s

Figure 2. Distribution of azimuth error of c and σ_τ.

Figure 2 shows that the relationship of σ_φ and c was linear, as well as σ_τ. However, the one between σ_φ and D was inverse. In the condition of $D \geq 2$ m, σ_φ stays at an optimal level as $0.03°$ in Figure 2a; when $\sigma_\tau = 100$ μs in Figure 2b it stays $0.1°$ when c was set as 343 m/s.

2.4. Range Precision

The range is also a function of time delay τ, and the transmission error is:

$$\sigma_R{}^2 = \left(\frac{\partial R}{\partial \tau_{12}}\sigma\tau\right)^2 + \left(\frac{\partial R}{\partial \tau_{13}}\sigma\tau\right)^2 + \left(\frac{\partial R}{\partial \tau_{14}}\sigma\tau\right)^2 + \left(\frac{\partial R}{\partial \tau_{15}}\sigma\tau\right)^2 \tag{12}$$

Evaluating the partial derivatives of τ in Equation (5) gives:

$$\frac{\partial R}{\partial \tau_{1i}} = [2c^2\tau_{1i}\sum_{j=2}^{5}\tau_{1j} - (c^2\sum_{j=2}^{5}\tau_{1j}^{2} - 4D^2)]/2c(\sum_{j=2}^{5}\tau_{1j})^2 \tag{13}$$

Substituting Equation (5) into Equation (13) yields:

$$\frac{\partial R}{\partial \tau_{1i}} = (c\tau_{1i} - R)/\sum_{j=2}^{5}\tau_{1j} \tag{14}$$

According to the geometric relation of array and target:

$$\begin{aligned}\tau_{1i} &= \tfrac{1}{c}\{R - \sqrt{R^2+D^2 - 2RD\cos[\varphi - (i-1)\tfrac{\pi}{2}]}\} \\ &= \tfrac{R}{c} - \tfrac{R}{c}\sqrt{1+[\tfrac{D}{R}]^2 - 2[\tfrac{D}{R}]\cos[\varphi - (i-1)\tfrac{\pi}{2}]}\end{aligned} \tag{15}$$

then the Taylor expansion of Equation (15) is:

$$\begin{aligned}\tau_{1i} &= \tfrac{R}{v} - \tfrac{R}{v}\{1+\tfrac{1}{2}[(\tfrac{D}{R})^2 - 2(\tfrac{D}{R})\cos[\varphi - (i-1)\tfrac{\pi}{2}]\} \\ &\quad -\tfrac{1}{8}\{(\tfrac{D}{R})^2 - 2(\tfrac{D}{R})\cos^2[\varphi - (i-1)\tfrac{\pi}{2}]\} \\ &\approx \tfrac{R}{v}\{\tfrac{1}{2}[(\tfrac{D}{R})^2 - (\tfrac{D}{R})\cos[\varphi - (i-1)\tfrac{\pi}{2}] \\ &\quad -\tfrac{1}{2}(\tfrac{D}{R})^2\cos^2[\varphi - (i-1)\tfrac{\pi}{2}]\}\end{aligned} \tag{16}$$

$$\begin{aligned}\sum_{i=2}^{5}\tau_{1i} &= -\tfrac{2D^2}{Rc} + \tfrac{D}{c}\sum_{i=2}^{5}\cos[\varphi - (i-1)\tfrac{\pi}{2}] \\ &\quad +\tfrac{D^2}{2Rc}\sum_{i=2}^{5}\cos^2[\varphi - (i-1)\tfrac{\pi}{2}]\end{aligned} \tag{17}$$

Substituting $\begin{cases}\sum_{i=2}^{5}\cos[\varphi - (i-1)\tfrac{\pi}{2}] = 0 \\ \sum_{i=2}^{5}\cos^2[\varphi - (i-1)\tfrac{\pi}{2}] = 1\end{cases}$ into Equation (17):

$$\sum_{i=2}^{5}\tau_{1i} \approx -\frac{2D^2}{Rc} \tag{18}$$

and then substituting Equation (18) into Equation (14) yields:

$$\frac{\partial R}{\partial \tau_{1i}} \approx -\frac{2Rc(c\tau_{1i} - R)}{3D^2} \tag{19}$$

Therefore, Equation (14) then becomes

$$\sigma_R = \frac{4Rc\sqrt{D^2 + R^2}}{3D^2}\sigma_\tau \tag{20}$$

Equation (20) reveals that the range error σ_R is determined by range R, array size D, sound velocity c, and the delay error. Compared to the azimuth error, the influence of R here is an additional effect on the error. Assuming values of 5 kHz for the sampling rate, $R = 100$ m, and array size changed from 1 m to 5 m, the distribution of range error is shown in Figure 3.

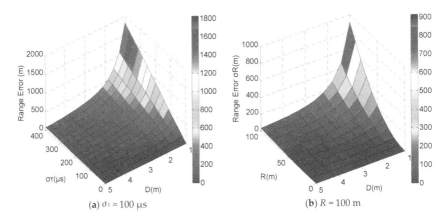

(a) $\sigma_\tau = 100\ \mu s$ (b) $R = 100$ m

Figure 3. Distributions of the range error of R and σ_τ.

In Figure 3, the range error was quite big. In both cases, the error reduces with increasing array size, and it increases with both the range R and the time delay σ_τ. In general, however, the error is overall at a fairly high level. In Figure 3a, the relative error was almost 40% under the optimal condition, and the biggest error was 950%. The error distribution in Figure 3b is the same as in Figure 3a, with higher level, the optimal error was 50%, and the biggest one is twenty times.

In summary, a single five-element cross array has good directional ability. The azimuth error can stay below 0.1° under reasonable conditions. However, the range ability is rather bad. The error is nearly 40% even under best conditions, which makes it impossible to achieve satisfactory sound source localization.

3. Localization Analysis of Double Arrays

3.1. Localization Principle

Although the single array has poor range-detection ability, its good directional ability ensures that the direction of the sound source is accurately determined. In order to improve the range-detection ability, a second array was added to the setup by means of intersecting the azimuth lines.

The array in Figure 1 remained positioned as shown in the figure and is referred to as Array 1. The second array, with identical characteristics, was added to the X-Y plane as Array 2. The centre of Array 2 is located at O_1 $(L, 0)$. The angle between the line OT (sound source T to origin O) and the axis-X is referred to as φ_1, while the angle between O_1T and X is φ_2. The time delay when the sound signal reaches the sensors in Array 2 is $\tau_{1i}{}'$. The range differences are $d_{1i}{}'$ $(i = 2, 3, 4, 5)$, so $d_{1i}{}' = c \times \tau_{1i}{}'$. The geometry of the double-array setup is shown in Figure 4.

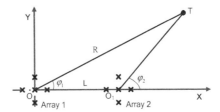

Figure 4. Model of double array.

Since the structure of both arrays is the same, the form of the azimuth formula is the same and the relevant expressions follow from Equation (3) as:

$$
\begin{cases}
tan\varphi_1 \approx \dfrac{(\tau_{15}-\tau_{13})}{(\tau_{14}-\tau_{12})} \\
tan\varphi_2 \approx \dfrac{(\tau_{15}'-\tau_{13}')}{(\tau_{14}'-\tau_{12}')}
\end{cases}
\tag{21}
$$

From the geometric relationship,

$$
\begin{cases}
k_1 = \dfrac{y}{x} = tan\varphi_1 \\
k_2 = \dfrac{y}{x-L} = tan\varphi_2
\end{cases}
\tag{22}
$$

The simplification of Equation (22) is:

$$
\begin{cases}
x = \dfrac{Lk_2}{k_2-k_1} \\
y = \dfrac{Lk_1k_2}{k_2-k_1}
\end{cases}
\tag{23}
$$

$$
\begin{cases}
\varphi = arctan(k_1) \\
R = \dfrac{Lk_2\sqrt{1+k^2_1}}{k_2-k_1}
\end{cases}
\tag{24}
$$

Equations (23) and (24) represent two alternative expressions for the localization of sound source. In these two expressions, the variables are array distance L and slopes k_1 and k_2. The slopes can be inferred from the time delay at each one of the two sensors by means of Equation (21). In the experiments, localization was obtained from time delay $\tau_{1i}^{(')}$ and array distance L.

3.2. Precision Analysis for Localization

Compared to the single array, the variable L has been added to the localization expression for double arrays. However, the time delay remains the key variable. As the structure and sensors of the two arrays are identical, the standard time delay errors of both are equal ($\sigma_\tau = \sigma_{\tau 12} = \sigma_{\tau 13} = \sigma_{\tau 14} = \sigma_{\tau 15} = \sigma'_{\tau 12} = \sigma'_{\tau 13} = \sigma'_{\tau 14} = \sigma'_{\tau 15}$).

As the direction was determined by Array 1, the azimuth error was analyzed according to Equation (24). Meanwhile, range error σ_R is influenced by azimuth error σ_φ and array distance L. Range precision is obviously determined by the azimuth precision according to Equation (24). Range precision can be expressed with error transmission as

$$
\sigma_R = \frac{\partial R}{\partial \tau_i} = \frac{\partial R}{\partial \varphi}\frac{\partial \varphi}{\partial \tau_i} = \left(\frac{\partial R}{\partial \varphi_1} + \frac{\partial R}{\partial \varphi_2}\right)\frac{\partial \varphi}{\partial \tau_i}
\tag{25}
$$

$$
\sigma_R = \frac{\sqrt{2}cr(sin(\varphi_1 + \varphi_2) + sec\,\varphi_1\,cos\,\varphi_2)}{sin(\varphi_1 - \varphi_2)}\sigma_\tau
\tag{26}
$$

Applying the sin theorem on ΔTO_1O_2 gives:

$$
\frac{sin\angle TO_2O_1}{R} = \frac{sin\angle O_2TO_1}{L}
\tag{27}
$$

$$
\frac{sin\,\varphi_2}{R} = \frac{sin(\varphi_2 - \varphi_1)}{L}
\tag{28}
$$

$$
R = \frac{L\,sin\,\varphi_2}{sin(\varphi_2 - \varphi_1)}
\tag{29}
$$

Taking partial derivative in Equation (29) yields:

$$
\sigma_R = \frac{\sqrt{2}cL\,cos\,\varphi_2}{D\,sin(\varphi_2 - \varphi_1)}\sigma_\tau
\tag{30}
$$

Substituting in Equation (29) gives

$$\sigma_R = \frac{\sqrt{2}cR}{D\tan\varphi_2}\sigma_\tau \tag{31}$$

In Equation (31), range precision is determined by sound velocity c, array size D, azimuth φ_2 of Array 2, error of time delay σ_τ, and range R. The distribution of the range error is shown in Figure 5. In Figure 5a, $R = 100$ m, $c = 343$ m/s, $\sigma_\tau = 100$ μs. In Figure 5b, $D = 3$ m, $R = 100$ m, $c = 343$ m/s.

Figure 5 reveals that φ_2 affected range precision substantially. In $(1°, 20°)$ and $(160°, 179°)$, the range error remains very high. In $(20°, 160°)$, the error was much lower and acceptable. In Figure 5a, array size D significantly affected precision when $D < 3$ m. Error was 10.78% when φ_2 was 8.12°. The error reduced as angle φ_2 is increased. When φ_2 was 15.24°, range error was 5.65% and it reduced to 3.78% as φ_2 was increased to 22.36. In Figure 5b, distribution was the same to Figure 5a, and the error stayed below 5% when φ_2 is above 20°.

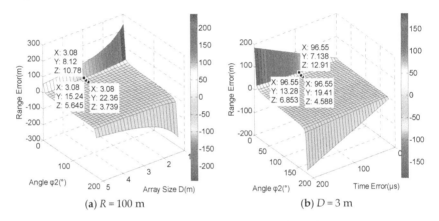

(a) $R = 100$ m   (b) $D = 3$ m

Figure 5. Distributions of the range error of R and D.

Since array distance L is independent to time t, the error expression of L is:

$$\sigma_R^L = \frac{\partial R}{\partial L} \tag{32}$$

Take partial derivatives of L in Equation (24):

$$\sigma_R^L = \frac{k_2\sqrt{1+k_1^2}}{2(k_2-k_1)} = \frac{k_2k_1\sqrt{1+\frac{1}{k1^2}}}{2(k_2-k_1)} \tag{33}$$

Substituting $\begin{cases} y = R\sin\varphi_1 \\ k_1 = \tan\varphi_1 \end{cases}$ into Equation (33) gives

$$\sigma_R^L = \frac{R}{L} \tag{34}$$

Equation (34) was the expression of the range error with factor L. The error was affected by range R and array distance L. The relative error is $1/L$; therefore, the relative error theoretically stays constant when L was designated. The error is below 6.67% when $L \geq 15$ m. The distribution is shown in Figure 6.

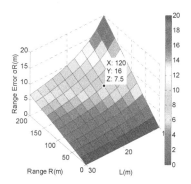

Figure 6. Distributions of the range error of *L*.

In summary, the locating ability of the double array is good. The range error stays below 5 m under a range condition of 100 m in most areas, and the azimuth error remains below 0.2° for all conditions. Considering the environmental factor (*c*) and calculated factor (σ_τ), it is advisable to choose large array sizes to improve localization precision. However, larger array sizes result in higher costs and increased complexity of the system.

4. Experiments

4.1. Experiment Setting

The locating experiments were conducted in the natural environment. The test area was open with a size of 150 × 150 m², and there were no tall reflectors along the boundary of the measurement domain. According to the empirical sound speed formula, the velocity of sound propagation was 343 m/s for an air temperature of 21 °C. During the experiments, wind speed was very low and the localization range was about 100 m, such that the influence of wind can be assumed negligible. Since it's difficult to keep an LPV going straight and travel at a constant speed, a simulated sound source with a smaller size was used to replace the vehicle noise.

The simulated source consisted of a 0.1 kW loudspeaker and a power amplifier. The biggest noise sources of LPV were the exhaust system and track system. The track noise was random and nonstatistical. So, the actual measurement of periodic exhaust noise was the sound signal that was collected during the running of the vehicle engine with rotating speed *r* = 1200 rpm. The sound signal is shown in Figure 7.

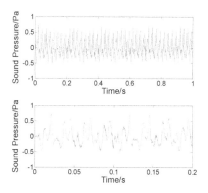

Figure 7. Sound signal for moving source.

The moving sound source travelling at a constant speed was achieved by dragging the loudspeaker with a fixed pulley at a constant rotating speed. The source was traveling along a straight line. By monitoring the distance travelled as a function of time, the speed of motion of the source was obtained.

In Figure 8, two five-element cross arrays were set according to Figure 4. The distance L between two central sensors was 10 m, and 2 m for array size D. A NI-PXI system with 10 channels and a sample rate of 248 kS/s was the testing instrument. The array microphones used G.R.A.S with the sensitivity of 40 mv/pa.

Figure 8. Microphone array.

The sound source started moving from point A ($-28.8, 97.92$) to B ($28.8, 97.92$) in the X-Y plane, and then returned back to point A.

4.2. Data Length for Correlation

Since the signal collected was not even during the interval while the source moved, it is necessary to extract part of the whole signal periodically for short-time correlation. To ensure efficiency, the extraction must cover one whole period in each short-time correlation.

As the sound source was being actuated by a pulley, the speed of motion was relatively slow, which should be less than 5 m/s.

$$f' = (\frac{v_0 \pm v_t}{v_0 \mp v_s})f \tag{35}$$

According to the Doppler effect formula, the difference in value between Doppler frequency f' and original frequency f was about $0.01 f$. Meanwhile, sound velocity v_0 was 343 m/s, velocity v_t of the observer was zero and velocity of source v_s took the maximum velocity 5 m/s. Therefore, data bias resulting from the Doppler effect can be ignored.

The longest distance that one acoustic wave travels in the single array is $2 D$, such that the maximum travel time is $2 D/c$. The sampling interval is $T_N = 1/F$ when the sampling frequency is assumed to be F. The data length n for short-time correlation describes the theoretical length of each correlation in Equation (36):

$$n \geq \frac{1/f + 2D/c}{1/F} \tag{36}$$

4.3. Time Compensation during Signal Transmission

Since the acoustic signal travels a long distance before it is detected by the test system, there exists a time delay between this signal and the instant when it was emitted by the noise source. Point $x(t)$ is the position of the moving source at the instant t. The signal as used for the data analysis has been generated at the source at point $x(t_0)$, which is located at a distance r from $x(t)$, as was illustrated in

Figure 9. Thus, the identified location at the instant time *t* is in fact the position of noise source at the moment t_0. It is essential to compensate for the difference.

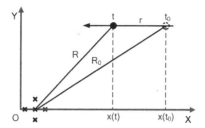

Figure 9. Path of motion of the sound source.

This constellation, as graphically illustrated in Figure 8, can be expressed as:

$$\begin{cases} R_0 = \sqrt{x(t_0)^2 + y(t_0)^2} \\ R = \sqrt{x(t)^2 + y(t)^2} \\ r = |x(t_0) - x(t)| = vR_0/c \end{cases} \tag{37}$$

To locate the noise source in actual conditions, point $x(t_0)$ moves with velocity v and sound velocity c can be calculated. Therefore, compensation r is available and needs be taken into consideration in the actual localization procedure.

4.4. Experiment Results

Additional environmental noise cannot be avoided either. This superposed additional noise will negatively affect the correlation of the signals. Therefore, preprocessing of the measured signals is required before obtaining the correlations. Since there were no other obvious sound sources and since the superposed additional noise is of high frequency, the wavelet-filtering method was chosen to remove unwanted noise, whereby the wavelet basis was "db10"(No. 10 of Daubechies Series Basis [27]). As the signal was relatively simple, it was decomposed into three layers. Then, the lower part in frequency was taken to perform short-time correlation. The signal-filtering process of one channel is shown in Figure 10.

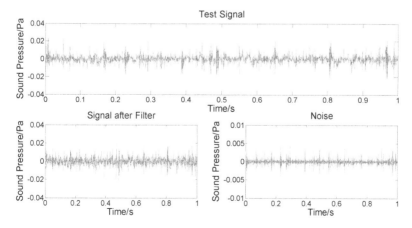

Figure 10. Filter process of signal.

Reference to Figure 10 reveals the noise contained in the test signal. There were two kinds of noise. One is the high-frequency noise for which the amplitude is about 1/20 of test signal. The other is the impulse with about 1/3 of the amplitude. Both of these noise contributions detrimentally affect the correlations of the array signal. Hence, the signal form without noise, as shown in the lower left part of Figure 9, was used to operate correlations.

During the path A→B, traveling distance was S = 57.6 m, with associated travel time t = 45.6 s and sampling length N = 228,800. Distance S was divided into 11 elements, while travel time was the same. The localization of the moving source was achieved by locating the central point of the 11 elements. The results obtained are shown in Table 1.

Table 1. Point information of path A→B.

No.	Coordinates/m	Time/s	Central Point
1	(−23.56, 97.92)	4.15	20,727
2	(−18.33, 97.92)	8.30	41,454
3	(−13.09, 97.92)	12.45	62,181
4	(−7.85, 97.92)	16.60	82,908
5	(−2.62, 97.92)	20.75	103,635
6	(2.62, 97.92)	24.90	124,362
7	(7.85, 97.92)	29.05	145,089
8	(13.09, 97.92)	33.20	165,816
9	(18.33, 97.92)	37.35	186,543
10	(23.56, 97.92)	41.50	207,270

In Figure 11, it's obvious that basic frequency was 600 Hz with some doubling frequency component. From Section 3.2, it is known that the period of noise T was $1/f$ = 1666 µs. Maximum traveling time of a single wave between array sensors is $2\,D/c \approx 11.66$ ms. The length of signal extracted for one correlation must be bigger than 11.66 ms. As the sampling interval was 200 µs, the minimum length of signal extraction was 58.

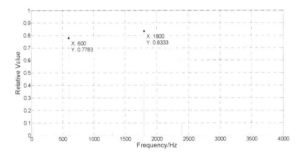

Figure 11. Frequency spectrum of the signal.

The localization was carried out at the first central point to study the choosing principle of extraction length that participated in one short-time correlation. Based on the signals of Sensor 1 and Sensor 2 in Array 1, the first central was set at point 20,727, and the length of extraction was assigned as 30, 60, 65, 70, 100, 200, and 300. After extraction of the signal and 100 times of interpolations, the correlations of different length were calculated, as shown in Figure 12a.

The sampling interval decreased to 2 µs after 100 times of interpolation. Maximum value is located at N = 3233, while the central point of correlation was N = 3000 when extraction n = 30. Then, time delay is $\tau_{n=30}$ = (3223 − 3000) × 2 = 446 µs and delays of other lengths can be calculated in the same way, as shown in Figure 12b.

Figure 12 illustrated that it is not possible to obtain correct time delays as time delay varies randomly when $n \leq 60$. When $n \geq 65$, delay value only varies marginally and remains steady at a level of about 520 μs, which represents the correct value of time delay.

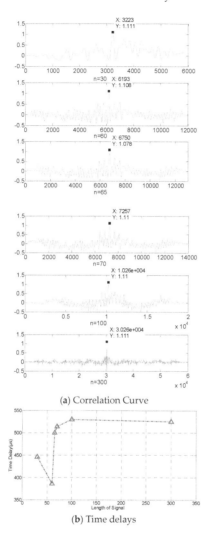

(a) Correlation Curve

(b) Time delays

Figure 12. Correlation performance of different lengths.

The locating results of different signal lengths are shown in Table 2.

Table 2. Locating result of different lengths.

Length	Actual Coordinates	Test Coordinates
30		(18.24, 52.31)
60		(6.79, 7.34)
65	$(-23.56, 97.92)$	(−24.36, 92.95)
70		(−24.35, 92.70)
100		(−24.20, 92.80)
300		(−24.07, 92.17)

Table 2 illustrates that localization has failed with a rather high error when the length of the signal involved in the correlation was $n \leq 60$. When $n \geq 65$, the test point was located nearly around the actual point. Therefore, the length of the signal extraction was assigned as $n = 65$, participating the calculation after interpolation.

All the locating results are summarized in Table 3, and the moving paths are shown in Figure 13.

Table 3. Locating result of moving source.

No.	Actual Coordinates	Test Coordinates/m	
		A→B	B→A
1	(−23.56, 97.92)	(−24.01, 92.82)	(−23.65, 98.29)
2	(−18.33, 97.92)	(−19.44, 94.33)	(−18.24, 96.86)
3	(−13.09, 97.92)	(−14.29, 94.62)	(−15.85, 99.70)
4	(−7.85, 97.92)	(−9.75, 98.12)	(−9.55, 98.38)
5	(−2.62, 97.92)	(−4.58, 101.67)	(−3.38, 98.15)
6	(2.62, 97.92)	(1.33, 98.53)	(2.81, 96.97)
7	(7.85, 97.92)	(6.39, 96.68)	(8.70, 97.28)
8	(13.09, 97.92)	(11.06, 93.80)	(13.68, 95.85)
9	(18.33, 97.92)	(16.61, 95.15)	(19.27, 96.00)
10	(23.56, 97.92)	(21.58, 94.49)	(25.04, 95.39)

Both Table 3 and Figure 13 depict that the discrete points along the moving path were obtained accurately in the localization experiment with only small associated error.

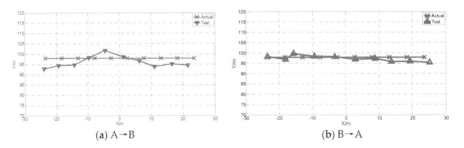

(a) A→B   (b) B→A

Figure 13. Actual and tested path of source.

According to Section 3.2, after compensation was considered, the relative locating error distribution is shown in Figure 14 before and after fixing.

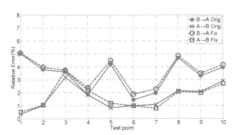

Figure 14. Distribution of relative error.

The lines in Figure 14 represent the original error of the localization, while dotted lines for error after fixing. To some extent, the locating result improved. In summary, the accurate localization

Appl. Sci. **2018**, *8*, 1281

of a moving acoustic source was achieved and the error stayed below 5%. In further application, the complete moving path could be obtained by increasing the number of parts divided.

5. Conclusions

The precision analysis of a locating method of a moving sound source based on intersecting azimuth lines was studied in this paper. Simulations showed that, after another single array was added, it had better precision and lower error. The experiments were conducted outdoors after choosing principle of signal length in correlation. Accurate localization of a moving source was achieved with the associated error for locating the source staying below 5%.

The work in this paper indicates applications for low-speed noise sources, such as wildlife conservation, health protection, wind turbine noise, and other engineering applications in the wild. However, the property change of the acoustic signal was ignored with low velocity and assumption of point source during the simulated source localization. Further research is required in the actual application. Further developments should focus on improvements of array size and shape. Meanwhile, the localization of high-speed moving sources and long-distance sources is not just an extension of this research.

Author Contributions: Conceptualization, J.Y. and C.X.; Methodology, C.X. and W.W.; Writing-Review & Editing, W.W.; Supervision, Project Administration and Funding Acquisition, C.X. and W.W.

Funding: This paper is supported by the China Scholarship Council (No. 201809110025).

Conflicts of Interest: The authors declare no conflict of interest.

References

1. Muzet, A. Environmental noise, sleep and health. *Sleep Med. Rev.* **2007**, *11*, 135–142. [CrossRef] [PubMed]
2. Zacarías, F.F.; Molina, R.H.; Ancela, J.L.C.; López, S.L.; Ojembarrena, A.A. Noise exposure in preterm infants treated with respiratory support using neonatal helmets. *Acta Acust. United Acust.* **2013**, *99*, 590–597. [CrossRef]
3. Babisch, W.; Beule, B.; Schust, M.; Kersten, N.; Ising, H. Traffic noise and risk of myocardial infarction. *Epidemiology* **2005**, *16*, 33–40. [CrossRef] [PubMed]
4. Miedema, H.; Oudshoorn, C. Annoyance from transportation noise: Relationships with exposure metrics DNL and DENL and their confidence intervals. *Environ. Health Perspect.* **2001**, *109*, 409–416. [CrossRef] [PubMed]
5. Cueto, J.L.; Petrovici, A.M.; Hernández, R.; Fernández, F. Analysis of the Impact of Bus Signal Priority on Urban Noise. *Acta Acust. United Acust.* **2017**, *103*, 561–573. [CrossRef]
6. Morley, D.W.; Hoogh, K.; Fecht, D.; Fabbri, F.; Bell, M.; Goodman, P.S.; Elliott, P.; Hodgson, S.; Hansell, A.L.; Gulliver, G. International scale implementation of the CNOSSOS-EU road traffic noise prediction model for epidemiological studies. *Environ. Pollut.* **2015**, *206*, 332–341. [CrossRef] [PubMed]
7. Ruiz-Padillo, A.; Ruiz, D.P.; Torija, A.J.; Ramos-Ridaob, Á. Selection of suitable alternatives to reduce the environmental impact of road traffic noise using a fuzzy multi-criteria decision model. *Environ. Impact Assess. Rev.* **2016**, *61*, 8–18. [CrossRef]
8. Licitra, G.; Fredianelli, L.; Petri, D.; Vigotti, M.A. Annoyance evaluation due to overall railway noise and vibration in Pisa urban areas. *Sci. Total Environ.* **2016**, *568*, 1315–1325. [CrossRef] [PubMed]
9. Bunn, F.; Trombetta, Z.P.H. Assessment of railway noise in an urban setting. *Appl. Acoust.* **2016**, *104*, 16–23. [CrossRef]
10. Paolo, G.; Luca, F.; Duccio, S.; Gaetano, L. ADS-B System as a Useful Tool for Testing and Redrawing Noise Management Strategies at Pisa Airport. *Acta Acust. United Acust.* **2017**, *103*, 543–551.
11. Iglesias-Merchan, C.; Luis, D.; Mario, S. Transportation planning and quiet natural areas preservation: Aircraft overflights noise assessment in a National Park. *Transp. Res. Part Transp. Environ.* **2015**, *41*, 1–12. [CrossRef]

12. Kephalopoulos, S.; Paviotti, K.; Maercke, D.V.; Shilton, S.; Jones, N. Advances in the development of common noise assessment methods in Europe: The CNOSSOS-EU framework for strategic environmental noise mapping. *Sci. Total Environ.* **2014**, *482*, 400–410. [CrossRef] [PubMed]

13. Morel, J.; Marquis-Favre, C.; Gille, L.A. Noise annoyance assessment of various urban road vehicle pass-by noises in isolation and combined with industrial noise: A laboratory study. *Appl. Acoust.* **2016**, *101*, 47–57. [CrossRef]

14. Sakhaeifar, M.; Banihashemrad, A.; Liao, G.Y.; Waller, B. Tyre-pavement interaction noise levels related to pavement surface characteristics. *Road Mater. Pavement Des.* **2018**, *5*, 1044–1056. [CrossRef]

15. Donavan, P.R. The effect of pavement type on low speed light vehicle noise emission. *SAE Tech. Pap.* **2005**, *1*, 2416.

16. Gille, L.A.; Marquis-Favre, C.; Morel, J. Testing of the European Union exposure-response relationships and annoyance equivalents model for annoyance due to transportation noises: The need of revised exposure-response relationships and annoyance equivalents model. *Environ. Int.* **2016**, *94*, 83–94. [CrossRef] [PubMed]

17. Janssen, S.A.; Vos, H.; Eisses, A.R.; Pedersen, E. A comparison between exposure-response relationships for wind turbine annoyance and annoyance due to other noise sources. *J. Acoust. Soc. Am.* **2011**, *130*, 3743–3756. [CrossRef] [PubMed]

18. Gallo, P.; Fredianelli, L.; Palazzuoli, D.; Licitra, G.; Fidecaro, F. A procedure for the assessment of wind turbine noise. *Appl. Acoust.* **2016**, *114*, 213–217. [CrossRef]

19. Schulte-Werning, B.; Jäger, K.; Strube, R.; Willenbrink, L. Recent developments in noise research at Deutsche Bahn. *J. Sound Vib.* **2003**, *267*, 689–699. [CrossRef]

20. Park, S.H.; Kim, Y.H. An improved moving frame acoustic holography for coherent band limited noise. *Acoust. Soc. Am.* **1998**, *104*, 3179–3189. [CrossRef]

21. Buckingham, M.J. On the sound field from a moving source in a viscous medium. *Acoust. Soc. Am.* **2003**, *11*, 3112–3118. [CrossRef]

22. Park, S.H.; Kim, Y.H. Visualization of pass by noise by means of moving frame acoustic holography. *J. Acoust. Soc. Am.* **2001**, *109*, 2326–2339. [CrossRef]

23. Park, C.S.; Kim, Y.H. Time domain visualization using acoustic holography implement by temporal and spatial complex envelope. *J. Acoust. Soc. Am.* **2009**, *126*, 1659–1662. [CrossRef] [PubMed]

24. Boone, M.M.; Kinneging, N.; Dool, T.V. Two-dimensional Noise Source Imaging with T-shaped Microphone Cross Array. *J. Acoust. Soc. Am.* **2000**, *108*, 2884–2890. [CrossRef]

25. Si, C.D.; Chen, E.L.; Yang, S.P.; Wang, C.Y. Experimental study on noise sources identification of vehicle based on microphone array technology. *J. Vib. Shock* **2009**, *28*, 173–175.

26. Wang, W.J.; Thomas, P.J. Low-frequency active noise control of an underwater large-scale structure with distributed giant magnetostrictive actuators. *Sens. Actuators A Phys.* **2017**, *263*, 113–121. [CrossRef]

27. Ingrid, D. *Ten Lectures on Wavelets*; Society for Industrial and Applied Mathematics: Philadelphia, PA, USA, 1992.

Article

Automatic Bowel Motility Evaluation Technique for Noncontact Sound Recordings

Ryunosuke Sato [1,*], Takahiro Emoto [2,*] , Yuki Gojima [1] and Masatake Akutagawa [2]

[1] Graduate School of Advanced Technology and Science, Tokushima University, Tokushima 770-8506, Japan; ygojima1029@gmail.com

[2] Graduate School of Technology, Industrial and Social Sciences, Tokushima University, Tokushima 770-8506, Japan; makutaga@ee.tokushima-u.ac.jp

* Correspondence: e.crhyv.t@gmail.com (R.S.); emoto@ee.tokushima-u.ac.jp (T.E.); Tel.: +81-88-656-7476 (T.E.)

Received: 9 May 2018; Accepted: 13 June 2018; Published: 19 June 2018

Abstract: Information on bowel motility can be obtained via magnetic resonance imaging (MRI)s and X-ray imaging. However, these approaches require expensive medical instruments and are unsuitable for frequent monitoring. Bowel sounds (BS) can be conveniently obtained using electronic stethoscopes and have recently been employed for the evaluation of bowel motility. More recently, our group proposed a novel method to evaluate bowel motility on the basis of BS acquired using a noncontact microphone. However, the method required manually detecting BS in the sound recordings, and manual segmentation is inconvenient and time consuming. To address this issue, herein, we propose a new method to automatically evaluate bowel motility for noncontact sound recordings. Using simulations for the sound recordings obtained from 20 human participants, we showed that the proposed method achieves an accuracy of approximately 90% in automatic bowel sound detection when acoustic feature power-normalized cepstral coefficients are used as inputs to artificial neural networks. Furthermore, we showed that bowel motility can be evaluated based on the three acoustic features in the time domain extracted by our method: BS per minute, signal-to-noise ratio, and sound-to-sound interval. The proposed method has the potential to contribute towards the development of noncontact evaluation methods for bowel motility.

Keywords: bowel sound; bowel motility; automatic detection/evaluation; power-normalized cepstral coefficients; noncontact instrumentation

---

1. Introduction

The decrease in or loss of bowel motility is a problem that seriously affects quality of life (QOL) and daily eating habits of patients; examples of this include functional gastrointestinal disorders (FGID), in which patients experience bloating and pain when bowel motility is impaired due to stress or other factors. Such bowel disorders are diagnosed by evaluating the bowel motility. Bowel motility is currently measured using X-ray imaging or endoscopy techniques; however, these methods require complex testing equipment and place immense mental, physical, and financial burdens on patients, which make these methods unsuitable for repeated monitoring.

In recent years, the acoustic features obtained from bowel sounds (BS) have been used to evaluate bowel motility. BS are created when transportation of gas and digestive contents through the digestive tract occurs due to peristaltic movement [1]. BS can be easily recorded by applying an electronic stethoscope to the surface of the body. In recent years, a method has been developed for evaluating bowel motility by automatically extracting BS from the audio data recorded using electronic stethoscopes [2–7]. In quiet conditions, BS can be perceived at a slight distance without the use of an electronic stethoscope. As such, our recent research has demonstrated that even when data is acquired using a noncontact microphone, bowel motility can be evaluated based on BS in a manner the same as that when an electronic stethoscope is used [8]. However, in this study, BS were required to be manually extracted from the audio data that was recorded using noncontact microphones, and a large amount of time was spent on carefully labeling the sounds. The sound pressure of BS recorded using noncontact microphones was lower than that of BS recorded with electronic stethoscopes placed directly on the surface of the body. Furthermore, compared to recordings from electronic stethoscopes, there may have been sounds other than BS mixed in at higher volumes. As such, a BS extraction system that is robust against extraneous noise must be developed to reduce the time- and labor-intensive work of BS labeling.

To resolve these issues, this study proposes a new system for evaluating bowel motility on the basis of results obtained by automatically extracting BS from the audio data recorded with a noncontact microphone. The proposed method is primarily made up of the following four steps: (1) segment detection using the short-term energy (STE) method; (2) automatic extraction of two acoustic features—mel-frequency cepstral coefficients (MFCC) [9,10] and power-normalized cepstral coefficients (PNCC) [11–14]—from segments; (3) automatic classification of segments as BS/non-BS based on an artificial neural network (ANN); and (4) evaluation of bowel motility on the basis of the acoustic features in the time domain of the BS that were automatically extracted. On the basis of audio data recorded from 20 human participants before and after they consumed carbonated water, we verified (i) the validity of automatic BS extraction by the proposed method and (ii) the validity of bowel motility evaluation based on acoustic features in the time domain.

2. Materials and Methods

2.1. Subject Database

This study was conducted with the approval of the research ethics committee of the Institute of Technology and Science at Tokushima University in Japan. A carbonated water tolerance test was performed using 20 male participants (age: 22.9 ± 3.4, body mass index (BMI): 22.7 ± 3.8) who had provided their consent to the research content and their participation. The test was conducted after 12 or more hours of fasting by the participants, over a 25-min period (comprised of a 10-min period of rest before consuming carbonated water and a 15-min period of rest after consuming carbonated water). During the test, sound data was recorded using a noncontact microphone (NT55 manufactured by RODE), an electronic stethoscope (E-Scope2 manufactured by Cardionics), and a multitrack recorder (R16 manufactured by ZOOM). The primary frequency components of BS have generally been reported to be present between 100 Hz and 500 Hz [15]. Based on these reports, sound data was stored at a sampling frequency of 4000 Hz and digital resolution of 16 bits. Furthermore, sound data was filtered by a third-order Butterworth bandpass filter with a cutoff frequency of 100–1500 Hz. The participants were in a supine position during testing, with an electronic stethoscope positioned 9 cm to the right of the navel and a microphone 20 cm above the navel [8].

BS present in the sound data obtained using the noncontact microphone were also present in the sound data obtained using the electronic stethoscope. Based on this, as in our previous studies, we used audio playback software to listen carefully to both types of sound recordings, and classified as a BS episode any episode that was 20 ms or more in duration and could be distinguished by the ear at the same time position in both recordings [7].

For the analysis, we divided the sound data into sub-segments with a window range of 256 samples and a shift range of 64 samples. The STE method was used to calculate the power of each window range, making it possible to detect sub-segments above a certain signal-to-noise ratio (SNR). SNR, as used in this study, is defined as follows:

$$SNR = 10 log_{10} \frac{P_S}{P_N} \tag{1}$$

Here, P_S represents the signal power and P_N represents the noise power. P_N can be calculated based on a one-second interval of silence determined by conducting the abovementioned listening process, and it is a time-averaged value. Sub-segments detected successively using the STE method are treated as a single segment (also called sound episode (SE)). If a detected segment corresponds to a BS episode, then it is defined as a BS segment; otherwise, it is defined as a non-BS segment.

2.2. Automatic BS Extraction on the Basis of Acoustic Features

The acoustic feature presented to the ANN is either MFCC or PNCC. MFCC is widely used in fields such as speech recognition and analysis of biological sounds such as lung or heart sounds [9,16–18]. MFCC is calculated by performing a discrete cosine transformation on the output from triangular filter banks evenly spaced along a logarithmic axis; this is referred to as a mel scale, and it approximates the human auditory frequency response. PNCC is a feature value developed to improve the robustness of voice recognition systems in noisy environments [11–14]. Because BS captured using noncontact microphones are generally low in volume and have degraded SNR, PNCC can be expected to be effective; it improves the process of calculating MFCC to make it more similar to certain physiological aspects of humans. Moreover, PNCC differs from MFCC primarily in the following three ways: First, instead of the triangular filter banks used in MFCC, PNCC uses gamma-tone filter banks based on an equivalent rectangular bandwidth to imitate the workings of the cochlea. Second, it uses bias subtraction based on the ratio of the arithmetic mean to the geometric mean (AM-to-GM ratio) for the sound that undergoes intermediate processing, which is not done in the MFCC calculation process. Third, it replaces the logarithmic nonlinearity (used in MFCC) with power nonlinearity. Owing to these differences, PNCC is expected to provide sound processing with excellent resistance to noise. For BS extraction in this work, a SE is divided into frames with a frame size of 200 samples and a shift size of 100 samples. Considering the number of dimensions often used in the field of voice recognition, we use 13-dimension MFCC and PNCC obtained from 24-channel filter banks, averaged over all the frames in each episode.

On the basis of these acoustic features, an artificial neural network (ANN) is used as a classifier to categorize segments detected with the STE method into BS segments and non-BS segments. The ANN is structured as a hierarchical neural network made up of three layers: namely, the input, intermediate, and output layers. The number of units in the input, intermediate, and output layers are, respectively, 13, 25 and 1. The output function of the intermediate layer units is a hyperbolic tangent function, and the transfer function of the output layer units is a linear function. As a target signal, the value of 1 is assigned to analysis sections in which sound is present if the sound is BS, whereas 0 is assigned if it is non-BS. The ANN learns from this categorization using an error back-propagation algorithm based on the Levenberg–Marquardt method [19,20]. To calculate sensitivity and specificity based on the post-training ANN output, a receiver operating characteristic (ROC) curve can be drawn. Through the analysis of the ROC curve, an optimum threshold (T_h) is estimated for use when classifying testing data sets. The optimum threshold used at this point is the threshold that is the shortest Euclidean distance from the positions at which sensitivity = 1 and specificity = 1 on the ROC curve [21]. Using this threshold for the ANN test output $\hat{b}$, it is possible to calculate the classification accuracy using sensitivity (Sen), specificity (Spe), positive predictive value (PPV), negative predictive value (NPV), and accuracy (Acc).

As shown in Figure 1, automatic BS extraction performance in this ANN-based method is evaluated by dividing the BS and non-BS segments obtained from the 20-person sound database at a ratio of 3:1, and using them respectively as training and testing data. This study calculated the average classification accuracy by performing multiple trials of ANN training and testing, in which (1) initial values of combined load were randomly assigned or (2) test data was randomly assigned.

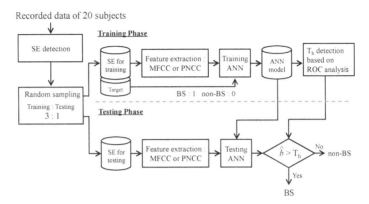

Figure 1. Block diagram showing the proposed method for automatic BS extraction based on acoustic features. SE: sound episode; MFCC: Mel Frequency Cepstral Coefficients; PNCC: Power Normalized Cepstral Coefficients; ANN: artificial neural network; ROC: receiver operating characteristic; BS: bowel sound; b: ANN test output; T_h: threshold obtained via ROC analysis.

2.3. Evaluation of Bowel Motility Based on Automatically Extracted BS

Our past research demonstrated significant differences in the following time domain acoustic features extracted before and after consumption of carbonated water by the participants: BS detected per minute, SNR, length of BS, and interval between BS (sound to sound (SS) interval). These differences suggest that bowel motility can be evaluated on the basis of these acoustic features [8]. As such, this study examines whether bowel motility can be automatically evaluated based on these acoustic features, as investigated in the previous study. To evaluate bowel motility from the data of one participant, the acoustic features of time domains were extracted based on multiple BS automatically extracted by performing leave-one-out cross validation for the proposed method. As in past studies, the differences between the previously mentioned acoustic features before and after the participant consumed carbonated water was evaluated using a Wilcoxon signed-rank test. The block diagram in Figure 2 shows the process leading up to the evaluation of bowel motility.

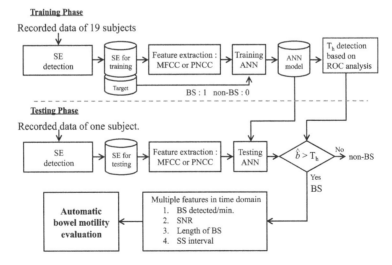

Figure 2. Block diagram showing the proposed method for automatic evaluation of bowel motility.

3. Results

To investigate the effect of SNR thresholds used in the STE method on the automatic evaluation performance and evaluation of bowel motility by the method, experiments were performed in which the SNR thresholds used in the STE method were 0, 0.5, 1 and 2 dB.

3.1. Automatic Bowel Sound Detection

Table 1 lists the number and length of BS and non-BS segments obtained at each SNR threshold used in the STE method.

Table 1. Number and length of BS and non-BS segments obtained at each SNR threshold used in the STE method.

Threshold of SNR (dB)	Before Soda Intake				After Soda Intake			
	No. of Segments		Length of Segments(s)		No. of Segments		Length of Segments(s)	
	BS	Non-BS	BS	Non-BS	BS	Non-BS	BS	Non-BS
2	396	4840	0.42 ± 0.59	0.34 ± 1.49	1538	6372	0.59 ± 1.82	0.30 ± 1.15
1	439	10202	0.59 ± 0.91	0.36 ± 1.28.	1463	15614	0.85 ± 2.47	0.29 ± 0.82
0.5	441	13444	0.90 ± 2.27	0.39 ± 1.35	1378	21202	1.23 ± 3.61	0.32 ± 0.53
0	409	14904	1.51 ± 3.83	0.47 ± 1.72	1264	23522	1.86 ± 4.59	0.39 ± 0.65

Table 1 reveals the following pattern for both cases (before and after consumption of carbonated water by participants): As the SNR threshold decreases, the numbers of both BS and non-BS segments increase until a certain threshold, after which the numbers of segments decrease. Additionally, the values in the table confirm that the lengths of both segments also increase with decrease in SNR. The values of length and number of both segments were larger after consumption of carbonated water than those before consumption, and BS segments were longer than non-BS segments.

To evaluate the automatic extraction performance of the proposed method, the respective segments were divided in a ratio of 3:1 for training data and testing data. Tables 2 and 3, respectively, present the results of 100 ANN-based approach trials that used MFCC and PNCC as acoustic features to derive the average classification accuracy.

Table 2 reveals that for the case before consumption of carbonated water, accuracy slightly degraded with decrease in the SNR threshold, whereas the accuracy increased with decrease in SNR threshold in the case after consumption. Table 3 demonstrates that when PNCC is used, classification accuracy increases as SNR threshold decreases, for cases both before and after consumption of carbonated water. Furthermore, we can see that the highest accuracy is obtained when the SNR threshold is 0 dB. Figure 3 shows the results of a comparative analysis of extraction accuracy before and after consumption of carbonated water when using MFCC and PNCC, respectively. Table 3 shows that PNCC is more accurate than MFCC for all SNR thresholds. When the SNR threshold is 0 dB before the consumption of carbonated water, the average of PNCC becomes sufficiently larger compared to that of MFCC. In general, a BS with lower sound-pressure occurs before consumption of carbonated water than after consumption. This suggests that PNCC is effective in classifying such sounds. On the basis of the abovementioned observation, a subsequent automatic evaluation of bowel motility was conducted using PNCC with an ANN-based approach.

Table 2. Results of automatic BS extraction using an ANN-based approach based on MFCC (using performance evaluation through random sampling). Sne: sensitivity; Spe: specificity; PPV: positive predictive value; NPV: negative predictive value; Acc: Accuracy.

Threshold of SNR (dB)	MFCC Before Soda Intake				
	Sen (%)	Spe (%)	PPV (%)	NPV (%)	Acc (%)
2	78.8 ± 4.7	83.9 ± 2.1	28.8 ± 2.3	98.0 ± 0.4	83.5 ± 1.8
1	79.7 ± 4.9	83.3 ± 2.1	17.1 ± 1.4	99.0 ± 0.2	83.2 ± 1.9
0.5	78.9 ± 4.1	83.2 ± 2.0	13.5 ± 1.4	99.2 ± 0.2	83.1 ± 1.9
0	78.0 ± 4.8	81.1 ± 2.0	10.2 ± 1.0	99.3 ± 0.2	81.0 ± 1.9

Threshold of SNR (dB)	MFCC After Soda Intake				
	Sen (%)	Spe (%)	PPV (%)	NPV (%)	Acc (%)
2	77.6 ± 2.6	79.9 ± 1.6	48.2 ± 1.6	93.7 ± 0.6	79.4 ± 1.0
1	81.9 ± 2.4	83.8 ± 1.1	32.2 ± 1.3	98.0 ± 0.2	83.6 ± 0.9
0.5	82.4 ± 2.4	84.7 ± 1.2	26.0 ± 1.3	98.7 ± 0.2	84.6 ± 1.0
0	82.6 ± 2.8	83.6 ± 1.4	21.4 ± 1.2	98.9 ± 0.2	83.6 ± 1.2

Table 3. Results of automatic BS extraction using an ANN-based approach based on PNCC (using performance evaluation through random sampling).

Threshold of SNR (dB)	PNCC Before Soda Intake				
	Sen (%)	Spe (%)	PPV (%)	NPV (%)	Acc (%)
2	79.4 ± 4.5	85.3 ± 2.0	30.9 ± 2.6	98.1 ± 0.4	84.9 ± 1.7
1	81.9 ± 3.8	86.6 ± 1.7	20.9 ± 1.8	99.1 ± 0.2	86.4 ± 1.6
0.5	82.2 ± 4.0	87.4 ± 1.4	17.8 ± 1.4	99.3 ± 0.1	87.3 ± 1.3
0	85.6 ± 4.3	87.8 ± 1.6	16.3 ± 1.6	99.6 ± 0.1	87.8 ± 1.5

Threshold of SNR (dB)	PNCC After Soda Intake				
	Sen (%)	Spe (%)	PPV (%)	NPV (%)	Acc (%)
2	80.6 ± 2.3	83.5 ± 1.6	54.1 ± 2.1	94.7 ± 0.6	82.9 ± 1.1
1	84.4 ± 1.9	87.6 ± 1.1	38.9 ± 1.9	98.4 ± 0.2	87.3 ± 0.9
0.5	87.1 ± 1.8	88.2 ± 0.9	32.4 ± 1.4	99.1 ± 0.1	88.1 ± 0.8
0	87.0 ± 2.1	88.3 ± 0.9	28.7 ± 1.4	99.2 ± 0.1	88.2 ± 0.8

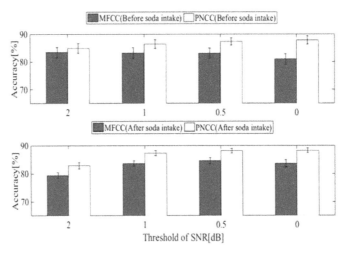

Figure 3. Comparison of accuracies of ANN-based approaches based on MFCC and PNCC, respectively.

3.2. Bowel Motility Evaluation

In this study, leave-one-out cross validation was performed for each participant, and the classification accuracy of an ANN-based approach using PNCC was verified. Table 4 presents the average classification accuracies for which the corresponding accuracy was the highest for each participant after leave-one-out cross validation was performed 50 times.

Table 4. Results of automatic BS extraction using an ANN-based approach based on PNCC (using performance evaluation through leave-one-out cross validation).

Threshold of SNR (dB)	PNCC Before Soda Intake				
	Sen (%)	Spe (%)	PPV (%)	NPV (%)	Acc (%)
2	71.5 ± 23.1	85.0 ± 12.0	25.5 ± 21.5	96.5 ± 6.6	85.5 ± 9.6
1	75.2 ± 23.8	88.8 ± 8.0	17.6 ± 15.1	98.8 ± 2.0	88.7 ± 7.3
0.5	74.1 ± 23.5	90.2 ± 5.8	15.4 ± 13.7	99.2 ± 1.0	90.0 ± 5.5
0	72.4 ± 23.0	90.4 ± 6.4	15.2 ± 13.5	99.4 ± 0.8	90.2 ± 6.3

Threshold of SNR (dB)	PNCC After Soda Intake				
	Sen (%)	Spe (%)	PPV (%)	NPV (%)	Acc (%)
2	78.6 ± 8.1	85.2 ± 6.0	54.7 ± 12.8	94.4 ± 2.8	84.4 ± 4.1
1	82.9 ± 8.7	88.6 ± 5.3	40.0 ± 12.9	98.2 ± 1.1	88.3 ± 4.7
0.5	85.4 ± 7.2	90.1 ± 3.4	34.5 ± 12.9	98.9 ± 0.7	89.9 ± 3.1
0	84.4 ± 8.7	90.0 ± 3.6	30.7 ± 12.8	99.1 ± 0.7	89.8 ± 3.2

As was noted in a prior study [8], Table 5 shows that the acoustic features—BS detected per minute, SNR, and SS interval—can capture the differences in bowel motility before and after a participant consumes carbonated water, up to a point at which the SNR threshold decreases to nearly 0 dB. Note that these results are related to the accuracy of automatic BS extraction. However, unlike in the prior study [8], no significant difference in BS length before and after consumption of carbonated water was found. This suggests that when the SNR threshold reduces to 0 dB, the acoustic features of BS detected per minute, SNR, and SS interval can evaluate the bowel motility without being affected by the reduction in SNR threshold.

Table 5. Results of automatic bowel motility evaluation using acoustic features in four time domains: BS detected/min, SNR (dB), length of BS (s), and SS interval (s).

Threshold of SNR (dB)	BS Detected/min.			SNR (dB)		
	Before Soda Intake	After Sofa Intake	P Value	Before Soda Intake	After Sofa Intake	P Value
2	3.97 ± 3.62	6.97 ± 3.83	>0.001 *	5.38 ± 0.88	6.20 ± 0.64	0.007 *
1	6.59 ± 4.70	9.54 ± 4.28	0.015 *	3.83 ± 0.77	4.65 ± 0.56	0.002 *
0.5	7.79 ± 5.48	10.69 ± 3.82	0.008 *	3.05 ± 0.82	3.86 ± 0.56	0.002 *
0	7.90 ± 5.35	11.18 ± 3.70	0.014 *	2.44 ± 0.94	3.06 ± 0.54	0.036 *

Threshold of SNR (dB)	Length of BS(s)			SS Interval(s)		
	Before Soda Intake	After Sofa Intake	P Value	Before Soda Intake	After Sofa Intake	P Value
2	0.56 ± 0.46	0.52 ± 0.15	0.126 *	23.32 ± 17.38	10.26 ± 5.66	0.001 *
1	0.85 ± 0.80	0.67 ± 0.20	0.457 *	12.59 ± 10.11	6.84 ± 3.55	0.013 *
0.5	1.20 ± 1.30	0.98 ± 0.45	0.323	9.83 ± 8.25	5.37 ± 2.51	0.014 *
0	1.89 ± 2.55	1.30 ± 0.61	0.379	9.54 ± 7.94	4.74 ± 2.57	0.021 *

4. Discussion and Conclusions

This study proposes a system for automatic evaluation of bowel motility on the basis of acoustic features in BS time domains obtained by automatically extracting BS from sound data recorded using a noncontact microphone. Although studies related to bowel motility using BS have been conducted previously [2–7], those studies used electronic stethoscopes that were applied to the surface of the body. Our recent research has demonstrated that bowel motility can be evaluated from sound data recorded using a noncontact microphone the same way as it can be evaluated using data recorded with a stethoscope [8]. However, the extraction of BS from sound data performed in this study was based on manual labeling. The sound pressure of BS recorded using noncontact microphones is lower than that of BS recorded using electronic stethoscopes applied to the surface of the human body, and there are fewer perceptible BS. As such, using sound data recorded without contact requires an automatic BS extraction method that is resistant to extraneous noise. Even so, the results suggest that the system proposed herein—which uses PNCC and has excellent noise resistance—is able to automatically extract BS with approximately 90% accuracy if the SNR threshold is 0 dB. Furthermore, even when the SNR threshold drops to 0 dB, results suggest that bowel motility can be evaluated using the acoustic features other than those from the BS length time domain, such as BS detected per minute, SNR, and SS interval.

The proposed method could extract more sound by decreasing the SNR threshold used in the STE method, further extending segment length to increase the information provided to the system for BS/non-BS differentiation. We believe that as a result of this extension, we could improve the performance of automatic BS extraction. However, this also suggests that proper BS length cannot be obtained because of the extension in BS segment length caused by the decrease in the SNR threshold used in the STE method.

Compared to the results of the performance evaluation based on random sampling, the results based on leave-one-out cross validation tended to have a larger standard deviation and decreased sensitivity in the proposed method, particularly before the consumption of carbonated water by participants. The cause of this was thought to be the small number of participants, meaning that sufficient BS segments were not available for use in leave-one-out cross validation. As such, we expect an improvement with increase in the number of subjects. To further improve system performance, a combination of the following two measures would likely be useful: (1) replacing the STE method with another method for detecting segments having sound; and (2) selecting acoustic features with excellent resistance to extraneous noise.

In this study, we have provided new knowledge for noncontact automatic evaluation of bowel motility. It is hoped that the foundations of the system developed in this study can assist in the further development of the evaluation of bowel motility using noncontact microphones and research related to diagnostic support for bowel disorders.

Author Contributions: T.E., R.S., and Y.G. conceived and designed the experiments; R.S. and Y.G. performed the experiments; R.S. analyzed the data; R.S. and M.A. contributed materials/analysis tools; T.E. and R.S. wrote the paper.

Acknowledgments: This study was partly supported by the Ono Charitable Trust for acoustics.

Conflicts of Interest: The authors declare no conflict of interest.

References

1. Zaloga, G.P. Blind bedside placement of enteric feeding tubes. *Tech. Gastrointest. Endosc.* **2001**, *3*, 9–15. [CrossRef]
2. Shono, K.; Emoto, T.; Abeyratne, T.O.U.R.; Yano, H.; Akutagawa, M.; Konaka, S.; Kinouchi, Y. Automatic evaluation of gastrointestinal motor activity through the analysis of bowel sounds. In Proceedings of the 10th IASTED International Conference on Biomedical Engineering, BioMed 2013, Innsbruck, Austria, 11–13 February 2013; ACTA Press: Calgary, AB, Canada, 2013; pp. 136–140.
3. Ulusar, U.D. Recovery of gastrointestinal tract motility detection using Naive Bayesian and minimum statistics. *Comput. Boil. Med.* **2014**, *51*, 223–228. [CrossRef] [PubMed]
4. Craine, B.L.; Silpa, M.L.; O'toole, C.J. Two-dimensional positional mapping of gastrointestinal sounds in control and functional bowel syndrome patients. *Dig. Dis. Sci.* **2002**, *47*, 1290–1296. [CrossRef] [PubMed]
5. Goto, J.; Matsuda, K.; Harii, N.; Moriguchi, T.; Yanagisawa, M.; Sakata, O. Usefulness of a real-time bowel sound analysis system in patients with severe sepsis (pilot study). *J. Artif. Organs* **2015**, *18*, 86–91. [CrossRef] [PubMed]

6.   Dimoulas, C.; Kalliris, G.; Papanikolaou, G.; Petridis, V.; Kalampakas, A. Bowel-sound pattern analysis using wavelets and neural networks with application to long-term, unsupervised, gastrointestinal motility monitoring. *Expert Syst. Appl.* **2008**, *34*, 26–41. [CrossRef]

7.   Ranta, R.; Louis-Dorr, V.; Heinrich, C.; Wolf, D.; Guillemin, F. Digestive activity evaluation by multichannel abdominal sounds analysis. *IEEE Trans. Biomed. Eng.* **2010**, *57*, 1507–1519. [CrossRef] [PubMed]

8.   Emoto, T.; Abeyratne, U.R.; Gojima, Y.; Nanba, K.; Sogabe, M.; Okahisa, T.; Kinouchi, Y. Evaluation of human bowel motility using non-contact microphones. *Biomed. Phys. Eng. Express.* **2016**, *2*, 45012. [CrossRef]

9.   Lu, X.; Dang, J. An investigation of dependencies between frequency components and speaker characteristics for text-independent speaker identification. *Speech Commun.* **2008**, *50*, 312–322. [CrossRef]

10.   Karunajeewa, A.S.; Abeyratne, U.R.; Hukins, C. Multi-feature snore sound analysis in obstructive sleep apnea–hypopnea syndrome. *Physiol. Meas.* **2010**, *32*, 83–97. [CrossRef] [PubMed]

11.   Kim, C.; Stern, R.M. Power-normalized cepstral coefficients (PNCC) for robust speech recognition. In Proceedings of the 2012 IEEE International Conference on Acoustics, Speech and Signal Processing (ICASSP), Kyoto, Japan, 25–30 March 2012; pp. 4101–4104.

12.   Chenchah, F.; Lachiri, Z. A bio-inspired emotion recognition system under real-life conditions. *Appl. Acoust.* **2017**, *115*, 6–14. [CrossRef]

13.   Kim, C.; Stern, R.M. Feature extraction for robust speech recognition based on maximizing the sharpness of the power distribution and on power flooring. In Proceedings of the 2010 IEEE International Conference on Acoustics Speech and Signal Processing (ICASSP), Dallas, TX, USA, 14–19 March 2010; pp. 4574–4577.

14.   Kim, C.; Stern, R.M. Feature extraction for robust speech recognition using a power-law nonlinearity and power-bias subtraction. In Proceedings of the Interspeech 2009, Tenth Annual Conference of the International Speech Communication Association, Brighton, UK, 6–10 September 2009; pp. 28–31.

15.   Cannon, W.B. Auscultation of the rhythmic sounds produced by the stomach and intestines. *Am. J. Physiol. Legacy Content* **1905**, *14*, 339–353. [CrossRef]

16.   Chauhan, S.; Wang, P.; Lim, C.S.; Anantharaman, V. A computer-aided MFCC-based HMM system for automatic auscultation. *Comput. Boil. Med.* **2008**, *38*, 221–233. [CrossRef] [PubMed]

17.   Rubin, J.; Abreu, R.; Ganguli, A.; Nelaturi, S.; Matei, I.; Sricharan, K. Classifying heart sound recordings using deep convolutional neural networks and mel-frequency cepstral coefficients. In Proceedings of the 2016 Computing in Cardiology Conference (CinC), Vancouver, BC, Canada, 11–14 September 2016; pp. 813–816.

18.   Duckitt, W.D.; Tuomi, S.K.; Niesler, T.R. Automatic detection, segmentation and assessment of snoring from ambient acoustic data. *Physiol. Meas.* **2006**, *27*, 1047. [CrossRef] [PubMed]

19.   Levenverg, K. A Method for the Solution of Certain Problems in Least Squares. *Q. Appl. Math.* **1944**, *2*, 164–168. [CrossRef]

20.   More, J.J. The Levenberg–Marquardt Algorithm: Implementation and theory. In *Numerical Analysis*; Lecture Notes in Mathematics 630; Watson, G.A., Ed.; Springer: Berlin/Heidelberg, Germany, 1970; pp. 105–116.

21.   Akobeng, A.K. Understanding diagnostic tests 3: Receiver operating characteristic curves. *Acta Paediatr.* **2007**, *96*, 644–647. [CrossRef] [PubMed]

Article

A Multi-Frame PCA-Based Stereo Audio Coding Method

Jing Wang *, Xiaohan Zhao, Xiang Xie and Jingming Kuang

School of Information and Electronics, Beijing Institute of Technology, 100081 Beijing, China; jonestorrons@gmail.com (X.Z.); xiexiang@bit.edu.cn (X.X.); jmkuang@bit.edu.cn (J.K.)
* Correspondence: wangjing@bit.edu.cn; Tel.: +86-138-1015-0086

Received: 18 April 2018; Accepted: 9 June 2018; Published: 12 June 2018

Abstract: With the increasing demand for high quality audio, stereo audio coding has become more and more important. In this paper, a multi-frame coding method based on Principal Component Analysis (PCA) is proposed for the compression of audio signals, including both mono and stereo signals. The PCA-based method makes the input audio spectral coefficients into eigenvectors of covariance matrices and reduces coding bitrate by grouping such eigenvectors into fewer number of vectors. The multi-frame joint technique makes the PCA-based method more efficient and feasible. This paper also proposes a quantization method that utilizes Pyramid Vector Quantization (PVQ) to quantize the PCA matrices proposed in this paper with few bits. Parametric coding algorithms are also employed with PCA to ensure the high efficiency of the proposed audio codec. Subjective listening tests with Multiple Stimuli with Hidden Reference and Anchor (MUSHRA) have shown that the proposed PCA-based coding method is efficient at processing stereo audio.

Keywords: stereo audio coding; Principal Component Analysis (PCA); multi-frame; Pyramid Vector Quantization (PVQ)

1. Introduction

The goal of audio coding is to represent audio in digital form with as few bits as possible while maintaining the intelligibility and quality required for particular applications [1]. In audio coding, it is very important to deal with the stereo signal efficiently, which can offer better experiences of using applications like mobile communication and live audio broadcasting. Over these years, a variety of techniques for stereo signal processing have been proposed [2,3], including M/S stereo, intensity stereo, joint stereo, and parametric stereo.

M/S stereo coding transforms the left and right channels into a mid-channel and a side channel. Intensity stereo works on the principle of sound localization [4]: humans have a less keen sense of perceiving the direction of certain audio frequencies. By exploiting this characteristic, intensity stereo coding can reduce the bitrate with little or no perceived change in apparent quality. Therefore, at very low bitrate, this type of coding usually yields a gain in perceived audio quality. Intensity stereo is supported by many audio compression formats such as Advanced Audio Coding (AAC) [5,6], which is used for the transfer of relatively low bit rate, acceptable-quality audio with modest internet access speed. Encoders with joint stereo such as Moving Picture Experts Group (MPEG) Audio Layer III (MP3) and Ogg Vorbis [7] use different algorithms to determine when to switch and how much space should be allocated to each channel (the quality can suffer if the switching is too frequent or if the side channel does not get enough bits). Based on the principle of human hearing [8,9], Parametric Stereo (PS) performs sparse coding in the spatial domain. The idea behind parametric stereo coding is to maximize the compression of a stereo signal by transmitting parameters describing the spatial image. For stereo input signals, the compression process basically follows one idea: synthesizing one signal from the two input channels and extracting parameters to be encoded and transmitted in order to add spatial cues for synthesized stereo at the receiver's end. The parameter estimation is made in the frequency domain [10,11]. AAC with Spectral Band Replication (SBR) and parametric stereo is defined as High-Efficiency Advanced Audio Coding version 2 (HE-AACv2). On the basis of several stereo algorithms mentioned above, other improved algorithms have been proposed [12], which causes Max Coherent Rotation (MCR) to enhance the correlation between the left channel and the right channel, and uses MCR angle to substitute the spatial parameters. This kind of method with MCR reduces the bitrate of spatial parameters and increases the performance of some spatial audio coding, but has not been widely used.

Audio codec usually uses subspace-based methods such as Discrete Cosine Transform (DCT) [13], Fast Fourier Transform (FFT) [14], and Wavelet Transform [15] to transfer audio signal from time domain to frequency domain in suitably windowed time frames. Modified Discrete Cosine Transform (MDCT) is a lapped transform based on the type-IV Discrete Cosine Transform (DCT-IV), with the additional property of being lapped. Compared to other Fourier-related transforms, it has half as many outputs as inputs, and it has been widely used in audio coding. These transforms are general transformations; therefore, the energy aggregation can be further enhanced through an additional transformation like PCA [16,17], which is one of the optimal orthogonal transformations based on statistical properties. The orthogonal transformation can be understood as a coordinate one. That is, fewer new bases can be selected to construct a low dimensional space to describe the data in the original high dimensional space by PCA, which means the compressibility is higher. Some work was done on the audio coding method combined with PCA from different views. Paper [18] proposed a novel method to match different subbands of the left channel and the right channel based on PCA, through which the redundancy of two channels can be reduced further. Paper [19] mainly focused on the multichannel procession and the application of PCA in the subband, and it discussed several details of PCA, such as the energy of each eigenvector and the signal waveform after PCA. This paper introduced the rotation angle with Karhunen-Loève Transform (KLT) instead of the rotation matrix and the reduced-dimensional matrix compared to our paper. The paper [20] mainly focused on the localization of multichannel based on PCA, with which the original audio is separated into primary and ambient components. Then, these different components are used to analyze spatial perception, respectively, in order to improve the robustness of multichannel audio coding.

In this paper, a multi-frame, PCA-based coding method for audio compression is proposed, which makes use of the properties of the orthogonal transformation and explores the feasibility of increasing the compression rate further after time-frequency transition. Compared to the previous work, this paper proposes a different method of applying PCA in audio coding. The main contributions of this paper include a new matrix construction method, a matrix quantization method based on PVQ, a combination method of PCA and parametric stereo, and a multi frame technique combined with PCA. In this method, the encoders transfer the matrices generated by PCA instead of the coefficients of the frequency spectrum. The proposed PCA-based coding method can hold both a mono signal and a stereo signal combined with parametric stereo. With the application of the multi-frame technique, the bitrate can be further reduced with a small impact on quality. To reduce the bitrate of the matrices, a method of matrix quantization based on PVQ [21] is put forward in this paper.

The rest of the paper is organized as follows: Section 2 describes the multi-frame, PCA-based coding method for mono signals. Section 3 presents the proposed design of the matrix quantization. In Section 4, the PCA-based coding method for the mono signal is extended to stereo signals combined with improved parametric stereo. The experimental results, discussion, and conclusion are presented in Sections 5–7, respectively.

2. Multi-Frame PCA-Based Coding Method

2.1. Framework of PCA-Based Coding Method

The encoding process can be described as follows: after time-frequency transformation such as MDCT, the frequency coefficients are used in the module of PCA, which includes the multi-frame technique. Several matrices are generated after PCA is quantized and encoded to bitstream. The decoder is the mirror image of the encoder, after decoding and de-quantizing, matrices are used to generate frequency domain signals by inverse PCA (iPCA). Finally, after frequency-time transformation, the encoder can export audio. Flowcharts of encoder and decoder for mono signals are shown in Figures 1 and 2. The part of MDCT is used to concentrate energy of signal on low band in frequency domain, which is good for the process of matrix construction (details are shown in Section 2.4). Some informal listening experiments have been carried out on the performance applying PCA without MDCT. The experimental results show that without MDCT, the performance of PCA has slight reduction, which means more bits are needed by the scheme without MDCT in order to achieve the same output quality of the scheme with MDCT. Thus, in this paper MDCT is assumed to enhance the performance of the PCA, although it will bring more computational complexity.

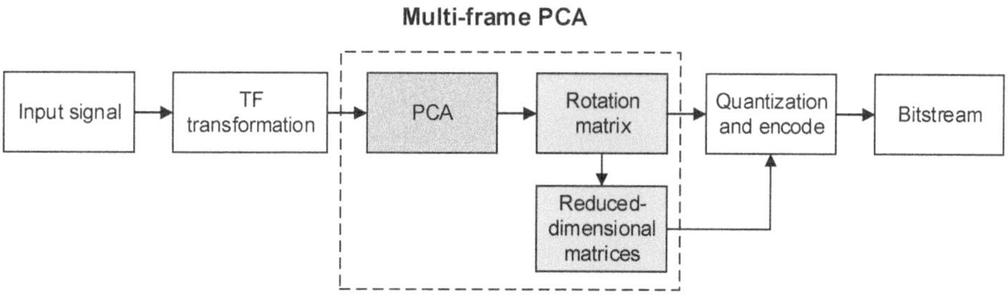

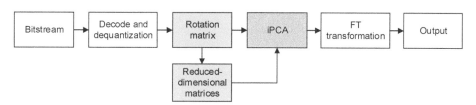

Figure 1. Flowchart of mono encoder. (TF, Time-to-Frequency; PCA, Principle Component Analysis).

Figure 2. Flowchart of mono decoder. (iPCA, inverse Principle Component Analysis; FT, Frequency-to-Time).

2.2. Principle of PCA

The PCA's mathematical principle is as follows: after coordinate transformation, the original high-dimensional samples with certain relevance can be transferred to a new set of low-dimensional samples that are unrelated to each other. These new samples carry most information of the original data and can replace the original samples for follow-up analysis.

There are several criteria for choosing new samples or selecting new bases in PCA. The typical method is to use the variance of new sample F_1 (i.e., the variance of the original sample mapping on the new coordinates). The larger Var (Fi) is, the more information Fi contains. So, the first principal component should have the largest variance F_1. If the first principal component F_1 is not qualified to replace the original sample, then the second principal component F_2 should be considered. F_2 is the principal component with the largest variance except F_1, and F_2 is uncorrelated to F_1, that is,

$\text{Cov}(F_1, F_2) = 0$. This means that the base of F_1 and the base of F_2 are orthogonal to each other, which can reduce the data redundancy between new samples (or principal components) effectively. The third, fourth, and p-th principal component can be constructed similarly. The variance of these principal components is in descending order, and the corresponding base in new space is uncorrelated to other new base. If there are m n-dimensional data, the procession of PCA is shown in Table 1.

Table 1. PCA ALGORITHM.

Algorithm: PCA (Principle Component Analysis)
(i) Obtain matrix by columns: $X_{[\ m\ \ n\ ]}$
(ii) Zero-mean columns in $X_{[\ m\ \ n\ ]}$ to get matrix X
(iii) Calculate $C = \frac{1}{m} X^T X$
(iv) Calculate eigenvalues $\lambda_1, \lambda_2, \lambda_3 \ldots \lambda_n$ and eigenvectors $a_1, a_2, a_3 \ldots a_n$ of C
(v) Use eigenvector to construct $P_{[\ n\ \ n\ ]}$ according to the eigenvalue
(vi) Select the first k columns of $P_{[\ n\ \ n\ ]}$ to construct the rotation matrix $P_{[\ n\ \ k\ ]}$
(vii) Complete dimensionality reduction by $Y_{[\ m\ \ k\ ]} = X_{[\ m\ \ n\ ]} \times P_{[\ n\ \ k\ ]}$

The contribution rate of the principal component reflects the proportion that each principal component accounts for the total amount of data after coordinate transformation, which can effectively solve the problem of dimension selection after dimensionality reduction. In PCA application, people often use the cumulative contribution rate as the basis for principal components selection. The cumulative contribution rate M_k of the first k principal components is

$$M_k = \frac{\sum_{i=1}^{k} \lambda_i}{\sum_{i=1}^{n} \lambda_i} \tag{1}$$

If the contribution rate of the first k principal components meets the specific requirements (the contribution rates are different according to different requirements), the first k principal components can be used to describe the original data to achieve the purpose of dimensionality reduction.

PCA is a good transformation due to its properties, as follows:

(i) Each new base is orthogonal to the other new base;
(ii) Mean squared error of the data is the minimum after transformation;
(iii) Energy is more concentrated and more convenient for data processing.

It is worthwhile noting that PCA does not simply delete the data of little importance. After PCA transformation, the dimension-reduced data can be transformed to restore most of the high-dimensional original data, which is a good character for data compression. In this paper, as is shown in Figure 3, the spectrum coefficients of the input signal are divided into multiple samples according to specific rules; then, these samples will be constructed to the original matrix X. After the principal component analysis, matrix X is decomposed into reduced-dimensional matrix Y and rotation matrix P; the process of calculating matrix Y and P is shown in Table 1. The matrix Y and P are transmitted to the decoder after quantization and coding. In decoder, the original matrix can be restored by multiplying reduced-dimensional matrix and transposed rotation matrix. There is some data loss during dimension reduction, but the loss is much less, so we can ignore it. For example, we can recover 99.97% information through a 6-dimension matrix, when the autocorrelation matrix has the 15th dimension. Ideally the original matrix X can be restored by reduced-dimensional matrix Y and rotation matrix P with

$$X \approx X_{restore} = Y \times P^T \tag{2}$$

in which $X_{restore}$ is the matrix restored in decoder and P^T is the transposition rank of matrix P. Then, $X_{restore}$ is reconstructed to spectral coefficients.

414

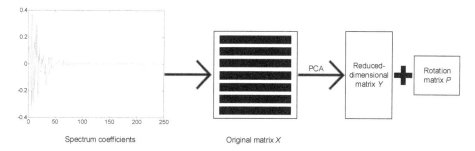

Figure 3. Scheme of PCA-based coding method. (PCA, Principle Component Analysis).

2.3. Format of Each Matrix

In encoder, when the sampling rate is 48 kHz, the frame has 240 spectral coefficients after MDCT (in this paper, the MDCT frame size is 5 ms with 50% overlap). There are many forms of matrices like 6 × 40, 12 × 20, 20 × 12, and so on; each format of matrix brings different compression rates. In a simple test, several formats of original matrix were constructed. Then, a subjective test was devised using those different dimensional rotation matrices. 10 listeners recorded the number of dimensions when the restored audio had acceptable quality. Then, the compression rate was calculated by the number of dimensions. As is shown in Figure 4, the matrix has the largest compression rate when it has 16 rows. So, the matrix $X_{[\ 16\ \ 15\ ]}$ with 16 rows and 15 columns is selected for transient frame in this paper. That means a 240-coefficient-long frequency domain signal is divided into 16 samples, each sample having 15 dimensions.

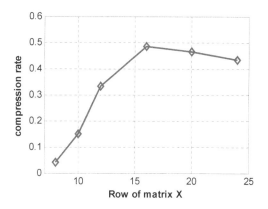

Figure 4. Compression rate for different format of matrix.

2.4. Way of Matrix Construction

An appropriate way to obtain the 16 samples from frequency domain coefficients is necessary. This paper proposes one method as follows: suppose the coefficients of one frame in frequency domain are $a_1, a_2 \ldots a_{240}$. a_1 is filled in the first column and the first row $X_{[\ 1\ \ 1\ ]}$, a_2 is filled in the first column and the second row $X_{[\ 2\ \ 1\ ]}$, and a_{16} is filled in the first column and the 16th row $X_{[\ 16\ \ 1\ ]}$. Then, a_{17} is filled in the first row and second column $X_{[\ 1\ \ 2\ ]}$, a_{18} is filled in the second row and second column $X_{[\ 2\ \ 2\ ]}$, and so on, until all the coefficients have been filled in the original matrix $X_{[\ 16\ \ 15\ ]}$; that is,

$$X_{[\ 16\quad 15\ ]} = \begin{bmatrix} a_{1,} & \cdots & a_{225} \\ \vdots & \ddots & \vdots \\ a_{16,} & \cdots & a_{240} \end{bmatrix} \tag{3}$$

This method has two obvious advantages, which can be find in Figure 5:

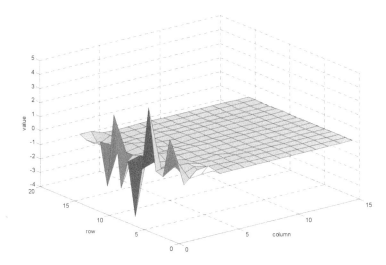

Figure 5. Example for matrix construction ("value" means the value of cells in original matrix, "column" means the column of original matrix, and "row" means the row of original matrix).

(i) This method takes advantage of the short-time stationary characteristic of signals in the frequency domain. Therefore, the difference between different rows in the same column of the matrix constructed by this sampling method is small. In other words, the difference between the same dimensions of different samples in the matrix is small, and different dimensions have similar linear relationships, which is very conducive to dimensionality reduction.

(ii) This method allows signal energy to gather still in the low-dimensional region of the new space. The energy of the frequency domain signal is concentrated in the low frequency region; after PCA, the advanced column of reduced-dimensional matrix still has the most signal energy. Thus, after dimensionality reduction, we can still focus on the low-dimensional region.

2.5. Multi-Frame Joint PCA

In the experiment, a phenomenon was observed that the rotation matrices of adjacent frames are greatly similar. Therefore, it is possible to do joint PCA with multiple frames to generate one rotation matrix, that is, multiple frames use the same rotation matrix. Therefore, the codec can transmit fewer rotation matrices, and bitrate can be reduced.

Below is one way to do joint PCA with least error. First, frequency domain coefficients of n sub-frames are constructed as n original matrices $X1_{[\ 16\quad 15\ ]}$, $X2_{[\ 16\quad 15\ ]} \cdots Xn_{[\ 16\quad 15\ ]}$, respectively; then, the original matrices of each sub-frame are used to form one original matrix $X_{[\ 16n\quad 15\ ]}$. This matrix is used to obtain one rotation matrix and n reduced-dimensional matrices.

If too many matrices are analyzed at the same time, the codec delay will be high, which is unbearable for real-time communication. Besides, the average quality of restored audio signal decreases with the increase in the number of frames. Therefore, the need to reduce bitrate and real-time communication should be comprehensively considered. A subjective listening test was designed to find the relationship between the number of frames and the quality of restored signal. 10 audio materials from European Broadcasting Union (EBU) test materials were coded with multi-frame PCA with different numbers of frames. The Mean Opinion Score (MOS) [22] of the restored music was recorded by 10 listeners. The statistical results are shown in Figure 6.

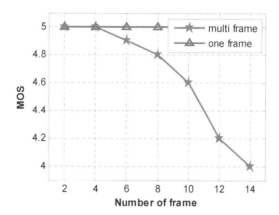

Figure 6. Subjective test results for different number of frames.

As is shown in Figure 6, when the number of frames is less than 6 or 8, the decrease of audio quality is not obvious. A suitable number of frames is then subjected to joint PCA. Taken together, when 8 sub-frames are analyzed at the same time, the bitrate and the delay of encoder is acceptable, that is, for every 40 ms signal, 8 sub-frame reduced-dimensional (Rd) matrices and one rotation matrix are transferred. Main functions of the mono encoder and decoder combined with multi-frame joint PCA are shown in Figures 7 and 8. In encoder, 40 ms signal is used to produce 8 Rd matrices and 1 rotation matrix. In decoder, after receiving 8 Rd matrices and 1 rotation matrix, 8 frames are restored to generate 40 ms signal.

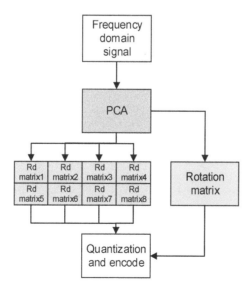

Figure 7. Multi-frame in encoder. (PCA, Principle Component Analysis; Rd, reduced-dimensional).

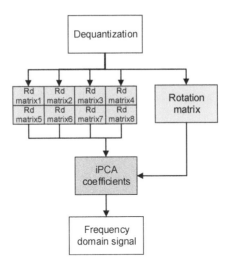

Figure 8. Multi-frame in decoder. (iPCA, inverse Principle Component Analysis; Rd, reduced-dimensional).

3. Quantization Design On PVQ

According to the properties of matrix multiplication, if the error of one point in matrix Y or P is large, the restored signal may have a large error. Therefore, uniform quantization cannot limit the error of every point in the matrix in the acceptable range with bitrate limitation. So, it is necessary to set a series of new quantization rules based on the properties of the dimensionality matrix and the rotation matrix. It is assumed that the audio signal obeys the distribution of Laplace [23], and both PCA and MDCT in the paper are orthogonal transformations. Thus, the distribution of matrix coefficients is maintained in Laplace distribution. Meanwhile, we have observed the values in reduced-dimensional matrix and rotation matrix. It is shown that most values of cells in matrix are close to 0, and the bigger the absolute value, the smaller the probability is. Based on the above two statements, the distribution of coefficients in reduced-dimensional matrix and rotation matrix can be regarded as Laplace distribution. Lattice vector quantization (LVQ) is widely used in the codec because of its low computational complexity. PVQ is one method of LVQ that is suitable for Laplace distribution. Thus, this section presents a design of quantization for reduced-dimensional matrix and rotation matrix combined with PVQ.

3.1. Quantization Design of the Reduced-Dimensional Matrix

In the reduced-dimensional matrix, the first column is the first principal component, the second column is the second principal component, etc. According to the property of PCA, the first principal component has the most important information of the original signal, and information carried by other principal components becomes less and less important. In fact, more than 95% of the original signal energy, which can be also called information, is restored only by the first principal component. That means if the quantization error of the first principal component is large, compared with the original signal, the restored signal also has a large error. Therefore, the first principal component needs to be allocated more bits, and the bits for other principal components should be sequentially reduced. For some kinds of audio, 4 principal components are enough to obtain acceptable quality, while for other kinds of audio 5 principal components may be needed. We choose 6 principal components, because they can satisfy almost all kinds of audio. In fact, the fifth and sixth principal components play a small role in the restored spectral; therefore, little quantization accuracy is needed for the last two principal components.

Based on the above conclusion, the reduced dimensional matrix can be divided into certain regions, as is shown in Figure 9. Different regions have different bit allocations: the darker color means more bits needed.

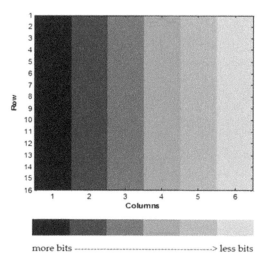

Figure 9. Bits allocation for reduced-dimensional matrix (darker color means more bits needed).

A PVQ quantizer was used to quantify the distribution of different bits in each principal component of the reduced-dimensional matrix. Several subjective listening tests have been carried out, and the bits assignments policy is determined according to the quality of the restored audio under different bit assignments. Finally, the bits that need to be allocated for each principal component are determined. Table 2 gives the number of bits required for each principal component of non-zero reduced-dimensional matrix under the PVQ quantizer.

Table 2. Quantization bit for reduced-dimensional matrix.

Principal Component	Bits Needed (bit/per Point)
First principal component	3
Second principal component	3
Third principal component	2.5
Fourth principal component	1.5
Fifth principal component	0.45
Sixth principal component	0.45

3.2. Quantization Design of the Rotation Matrix

According to $Y = XP$ in encoder and $X_{restore} = YP^T$ in decoder, some properties of the rotation matrix can be found:

(i) The higher row in matrix P is used to restore the region of higher frequencies in the restored signal.
(ii) The first column in matrix P corresponds to the first principal component in the reduced-dimensional matrix. That means that the first column of the rotation matrix only multiplies with the first column (first principal component) of the reduced-dimensional matrix when calculating the restored signal in the decoder. The second column of the rotation matrix only

multiplies with the second column (second principal component) of the reduced-dimensional matrix, and so on. According to the above properties of the rotation matrix, the quantization distribution of the rotation matrix has been made clearer, that is, the larger the row number is, and the larger the column number is, the fewer allocation bits there are.

In addition to the above two properties of the rotation matrix, there is another important property. Generally, the data in the first four rows around the diagonal are bigger than others. The thinking of this characteristic in this paper is as follows: common audio focuses more energy on low-band in frequency domain, and the method of matrix construction described in Section 2.4 can keep the coefficients of low-band stay in low-column. Thus, the first diagonal value that is calculated from the first column must be the largest one of overall values in rotation matrix or autocorrelation matrix. The second diagonal value could quite possibly be the second-largest value, and so on. That means these data are more important for decoder, so the quantization accuracy of these regions with larger absolute values can determine the error between the restored signal and the original signal. Therefore, the data around the diagonal need to be allocated with more bits. Figure 10 shows the "average value" rotation matrix of a piece of audio as an example to show this property more clearly.

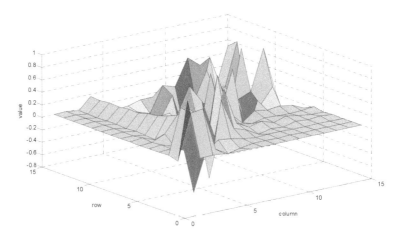

Figure 10. An example rotation matrix ("value" means the average value of cells in rotation matrices, "column" means the column of rotation matrix, and "row" means the row of rotation matrix).

The rotation matrix also has the following quantization criterion:

(i)   The first column of the rotation matrix needs to be precisely quantized, because the first principal component of the reduced-dimensional signal is only multiplied by the first column of P in decoder to restore signal.

(ii) Data in columns 2–6 in row 1 have little effect on the restored signal, so that few bits can be allocated for this region.

(iii) The higher row in matrix P is used to restore the region of higher frequencies in the restored signal. The data in lines 13, 14, and 15 correspond to the frequency that exceeds the range of frequencies perceptible to the human ear, so these data do not need to be quantized.

According to the above quantization criteria, the rotation matrix that is divided into the following regions according to bit allocation is shown in Figure 11. The darker the color is, the more bits should be allocated.

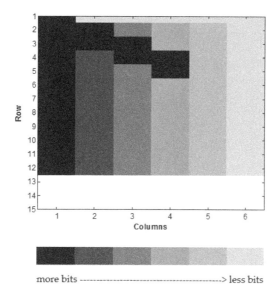

more bits --> less bits

Figure 11. Bit allocation for rotation matrix (darker color means more bits needed; white color means no bits).

The same test method as the one for reduced-dimensional matrix was used to determine the number of bits needed in each region in rotation matrix.

In Table 3, the first region corresponds to the region with the darkest color in Figure 11; the second corresponds to the area with the second-darkest color, and so on. The white color means there are no bits allocated to that area.

Table 3. Quantization bits for rotation matrix.

Region	Bits Needed (bit/per Point)
The first region	4
The second region	3
The third region	2
The fourth region	2
The fifth region	0.5
The sixth region	0.5
The seventh region	0

3.3. Design of the Low-Pass Filter

The noise generated from quantization and matrix calculation is white noise. There are two ways to reduce it. The first way is introducing noise shaping to make noise more comfortable for human hearing, and the second way is introducing a filter in decoder.

For most signals, the energy concentrates on low frequency domain, therefore the noise in low frequency domain does not sound obvious because of simultaneous masking. While in the high frequency part, if the original signal does not have high frequency components, the noise signal will not be masked and can be heard. So, a low-pass filter can be set to mask the high frequency noise signal, without affecting the original signal. The key point of the filter design is to determine the cut-off frequency.

Given the original matrix $X_{[\ 16\quad 15\ ]} = \begin{bmatrix} a_1 & \cdots & a_{225} \\ \vdots & \ddots & \vdots \\ a_{16} & \cdots & a_{240} \end{bmatrix}$, there are 15 subbands in

X, in which the first subband is the first row, the second subband is the second row, and so on. When $C = \frac{1}{m}X^T X$ is calculated in PCA, the first value e_1 on the diagonal line $e_1, e_2 \ldots e_{15}$ is calculated by

$$
\begin{aligned}
e_1 &= ((a_1 - \bar{a}) * (a_1 - \bar{a}) + (a_2 - \bar{a}) * (a_2 - \bar{a}) + \ldots (a_{16} - \bar{a}) * (a_{16} - \bar{a}))/16 \\
&= (a_1^2 + a_2^2 + \ldots a_{16}^2 + 16\bar{a}^2 - 2\bar{a}(a_1 + a_2 + \ldots a_{16}))/16 \\
&= (a_1^2 + a_2^2 + \ldots a_{16}^2 - 16\bar{a}^2)/16
\end{aligned}
\tag{4}
$$

in which $a_1^2 + a_2^2 + \ldots a_{16}^2$ is equal to the energy of the first subband E_1, and $\bar{a}$ is the average value of the first subband. Therefore, the relationship between E_1 and e_1 is

$$
E_1 = 16(e_1 + \bar{a}^2)
\tag{5}
$$

Actually, the value of $\bar{a}^2$ is far less than e_1, so E_1 is equal to $16e_1$, and the relationships between $E_2 \ldots E_{15}$ and $e_2 \ldots e_{15}$ can be gotten by analogy. Therefore, through PCA, the energy of each subband is calculated, and the filter can be determined by the energy of each band. Considering the proportion of energy accumulation, A_k is

$$
A_k = \frac{\sum_{i=1}^{k} e_i}{\sum_{i=1}^{15} e_i}
\tag{6}
$$

According to some experiments, when $A_k = 99.6\%$, k is the proper cut-off band. When the signal passes through the filter, the noise signal will be filtered out, and the signal itself will not be too much damaged.

Considering the frequency characteristics of the audio signal, the stop band setting is not low, and the signal with more than 20,000 Hz is often ignored by default, so each band of the above 15 bands will not be transmitted. Taken together, $e_1, e_2, e_3, e_{12}, e_{13}, e_{14}, e_{15}$ will not be transmitted, and the index of the left 8 bands are quantized by 3 bits, so the bitrate for cut-off band is 75 bps.

4. PCA-Based Parametric Stereo

The stereo coding method proposed in this paper, as the extension of mono coding method mentioned before, is shown in Figures 12 and 13. The encoder and decoder for stereo audio use the same module of PCA and quantization as mono audio. The differences between mono coding and stereo coding are elaborated in the following sections. In encoder, the two channels' signal carries out MDCT and the two channels' coefficients gather to generate an original matrix to do PCA; then, an improved parametric stereo module is used to downmix and calculate parameters of the high-band. Finally, a module based on PVQ is used for quantizing coefficients of matrix, and so on. In decoder, coefficients of mid downmix matrix and rotation matrix are used to generate mid channel; then, spatial parameters and other information are introduced to restore stereo signals. After inverse MDCT (iMDCT) and filtering, the signal can be regarded as the output signal.

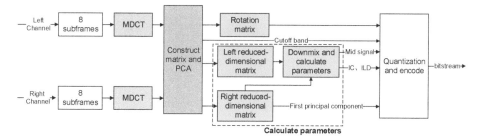

Figure 12. Flowchart of stereo encoder. (MDCT, Modified Discrete Cosine Transform; PCA, Principle Component Analysis; IC, Interaural Coherence; ILD, Interaural Level Difference).

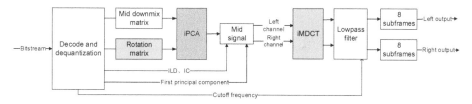

Figure 13. Flowchart of stereo decoder. (MDCT, Modified Discrete Cosine Transform; iPCA, inverse Principle Component Analysis; IC, Interaural Coherence; ILD, Interaural Level Difference).

4.1. Procession of Stereo Signal

Since the signals in two channels of the stereo tend to have high correlation. The signal of the left and right channels can be constructed into one original matrix. Firstly, the coefficients from left channel and right channel construct original matrices $X_{l[m\ n]}$ and $X_{l[m\ n]}$ respectively. Then, matrices $X_{l[m\ n]}$ and $X_{r[m\ n]}$ are used to form a new matrix X, in which $= \begin{bmatrix} X_{l[m\ n]} \\ X_{r[m\ n]} \end{bmatrix}$. Matrix X is used to obtain one rotation matrix $P_{[n\ k]}$ by PCA, and $P_{[n\ k]}$ can handle both left and right channel signals. That is,

$$Y_{l[m\ k]} = X_{l[m\ n]} \times P_{[n\ k]} \tag{7}$$

$$Y_{r[m\ k]} = X_{r[m\ n]} \times P_{[n\ k]} \tag{8}$$

If the first six principal components are preserved, most mono audio signals can be well restored. At this time, we keep the first six bases in principal component matrix and obtain rotation matrix $P_{[\ 15\ \ 6\ ]}$. The reduced-dimensional matrices of each sub-frame are $Y_{1[156]}, Y_{2[156]}, \cdots, Y_{8[156]}$. Experiments were done to verify the design for stereo signals: 10 normal audio files and 5 artificial synthesized audio files (the left channel and right channel have less correlation) were chosen as the test materials. Results of the subjective listening experiments are shown in Figures 14 and 15. We can consider that for most stereo signals, in which two channels have high relevance with each other, the proposed method for stereo signals perform as well as for mono signals.

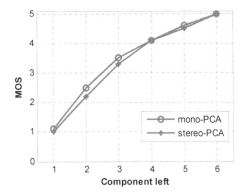

Figure 14. Subjective MOS of high-relation stereo signal. (MOS, Mean Opinion Score).

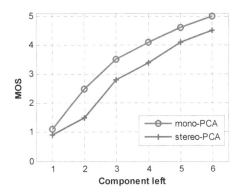

Figure 15. Subjective MOS of low-relation stereo signal.

4.2. Parameters in Parametric Stereo

In parametric stereo, Interaural Level Difference (ILD), Interaural Time Difference (ITD), and Interaural Coherence (IC) are used to describe the difference between two channels' signals. In MDCT domain, the above parameters in subband b are calculated by:

$$\text{ILD}[b] = 10log_{10}\frac{\sum_{k=A_{b-1}}^{A_b-1} X_l(k)X_l(k)}{\sum_{k=A_{b-1}}^{A_b-1} X_r(k)X_r(k)} \tag{9}$$

$$\text{IC}[b] = \text{R}(X_{bl}(k), X_{br}(k)) = \frac{\langle X_{bl}(k), X_{br}(k)\rangle}{|X_{bl}(k)||X_{br}(k)|} \tag{10}$$

While in MDCT domain, calculating ITD must introduce Modified Discrete Sine Transform (MDST) to calculate Interaural Phase Difference (IPD) instead of ITD, in which MDST is:

$$Y(k) = \sum_{n=0}^{N-1} x(n)w(n)\sin\left[\frac{2\pi}{N}\left(n + \frac{1}{2} + \frac{N}{4}\right)\left(k + \frac{1}{2}\right)\right], \ k = 0,1,2\ldots, \frac{N}{2} - 1 \tag{11}$$

in which $Y(k)$ is the spectrum coefficients, $x(n)$ is the input signal in time domain, and $w(n)$ is the window function. Then, a new transform MDFT is introduced, $Z(k) = X(k) + jY(k)$, in which $X(k)$ is the MDCT spectral coefficients, $Y(k)$ is the MDST spectral coefficients, and IPD can be calculated by

$$\text{IPD}[b] = \angle \left(\sum_{k=A_{b-1}}^{A_b-1} Z_l(k) Z_r{}^*(k) \right) \tag{12}$$

Traditional decoder uses these parameters and a downmix signal to restore left channel's signal and right channel's signal. Compared with formula (4, 9, 10), when the method described in Section 4.1 is used to deal with stereo signals, $\sum_{k=A_{b-1}}^{A_b-1} X_l(k)X_l(k)$ and $\sum_{k=A_{b-1}}^{A_b-1} X_r(k)X_r(k)$ can be calculated in the processing of PCA; therefore, parametric stereo and PCA have high associativity. After PCA, we can get ILD and IC only by calculating $\langle X_{bl}(k), X_{br}(k) \rangle$. In addition, we also need to calculate IPD by Formula (12); however, introducing MDST will bring computational complexity, and ITD or IPD mainly works on signals below 1.6 kHz that play smaller roles in high frequency domain. Thus, some improvements can be made to the parametric stereo according to the nature of the PCA.

4.3. PCA-Based Parametric Stereo

Given that the original matrix is $X = \begin{bmatrix} a_1 & \cdots & a_{225} \\ \vdots & \ddots & \vdots \\ a_{16} & \cdots & a_{240} \end{bmatrix}$, and the rotation matrix is $P = \begin{bmatrix} p_1 & \cdots & p_{76} \\ \vdots & \ddots & \vdots \\ p_{15} & \cdots & p_{90} \end{bmatrix}$, the reduced-dimensional matrix is $Y = XP = \begin{bmatrix} b_1 & \cdots & b_{49} \\ \vdots & \ddots & \vdots \\ b_{16} & \cdots & b_{64} \end{bmatrix}$. For the coefficients in the reduced-dimensional matrix Y,

$$b_1 = a_1 p_1 + a_{17} p_2 + \ldots a_{225} p_{15} \tag{13}$$

$$b_2 = a_2 p_1 + a_{18} p_2 + \ldots a_{226} p_{15} \tag{14}$$

$$b_{16} = a_{16} p_1 + a_{33} p_2 + \ldots a_{240} p_{15} \tag{15}$$

The first column is only related to the first column of P (the first base). As Figure 9 shows, main energy of the first base in the rotation matrix is entirely concentrated on the data in the first column of the first row. Therefore, the matrix Y can be approximated as

$$b_1 = a_1 p_1 \tag{16}$$

$$b_2 = a_2 p_1 \tag{17}$$

$$b_{16} = a_{16} p_1 \tag{18}$$

While p_1 in the matrix P is approximately equal to 1. Therefore the first column in the matrix Y is equal to the first column originally in matrix X. When the sampling rate is 48 kHz, the first column in X indicates the coefficients from 0 to 1.6 kHz, which means that when calculating the restored signal, the points below 1.6 kHz in the frequency domain happen to be the first principal component. So, the first principal component can be used to restore signals below 1.6 kHz in frequency domain instead of introducing MDST and estimating binaural cues. In decoder, the spectrum of the left and right channels above 1.6 kHz can be restored according to the downmix reduced-dimensional matrix, rotation matrix, and spatial parameters. The spectrum of the left and right channels below 1.6 kHz can be restored according to the first principal component and the downmix reduced-dimensional matrix.

4.4. Subbands and Bitrate

The spectrums of signal are divided into several segments based on Equivalent Rectangular Bands (ERB) model. The subbands are shown in Table 4.

Table 4. Subband division.

index	0	1	2	3	4	5	6	7
start	0	100	200	300	400	510	630	760
index	8	9	10	11	12	13	14	15
start	900	1040	1200	1380	1600	1860	2160	2560
index	16	17	18	19	20	21	22	23
start	3040	3680	4400	5300	6400	7700	9500	12,000
index	24	25						
start	15,500	19,880						

The quantization of space parameters uses ordinary vector quantization. The codebook with different parameters is designed based on the sensitivity of the human ear and the range of the parameter fluctuation of the experimental corpus. The codebooks of ILD and IC are shown in Tables 5 and 6, respectively.

Table 5. Codebook for ILD. (ILD, Interaural Level Difference).

index	0	1	2	3	4	5	6	7
ILD	−20	−15	−11	−8	−5	−3	−1	0
index	8	9	10	11	12	13	14	15
ILD	1	3	5	8	11	13	15	20

Table 6. Codebook for IC. (IC, Interaural Coherence).

index	0	1	2	3	4	5	6	7
IC	1	0.94	0.84	0.6	0.36	0	−0.56	−1

According to the above codebooks, the ILD parameters of each subband are quantized using 4 bits, and the IC parameters of each subband are quantized using 3 bits. According to the above sub-band division, the number of sub-bands higher than 1.6 kHz accounts for half of the total number of sub-bands in the whole frequency domain, which is 13, so the number of bits needed for each frame's spatial parameter is $13 \times 7 = 91$. For frequencies above 1.6 kHz, the rate of quantitative parameters is about 4.5 kbps. In the frequency domain less than 1.6 kHz, the first principal component is used to describe the signal directly. The rate of transmission of the first principal component is around 10 kbps, so the parameter rate of PCA-based parametric stereo is around 15 kbps. In traditional parametric stereo [24], IPD of each subband is quantized by 3 bits, so the parameter rate of the traditional parameter stereo is about $(4 + 3 + 3 + 3) \times 25 \times 50 = 16.25$ kbps. Therefore, compared with traditional parametric stereo, the rate of PCA-based parametric stereo is slightly reduced.

Figure 16 shows the results of a 0–1 test for spatial sense. In this test, 12 stereo music from EBU test materials is chosen. Score 0 means the sound localization is stable, and score 1 means there are some unstable sound localization in test materials. The ratio in Figure 16 is calculated from the times of unstable localization, and lower ratio means better performance in the quality of spatial sense. Experiments show that compared with the traditional parametric stereo encoding method, the spatial sense of the audio source has been obviously improved through the PCA-based parametric stereo. Through the use of PCA, almost half of the amount of parameter estimation can be reduced, while the computational complexity still rises because of the increasing complexity of PCA.

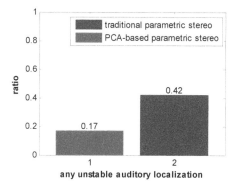

Figure 16. Test results for spatial sense.

5. Test and Results

The method proposed in this paper performs significantly better with stereo signals compared to mono signals. Thus, this section only presents the results for stereo signals. In order to verify the encoding and decoding performance of the PCA-based stereo coding method, some optimized modules such as DTX, noise shaping, and other efficient coding tools in the codec were not used in testing

5.1. Design of Test Based on MUSHRA

The key points of the MUSHRA [25] test are as follows:

5.1.1. Test Material

(i) Several typical EBU test sequences were selected: piano, trombone, percussion, vocals, song of rock, multi sound source background and mixed voice, and so on.

(ii) Contrast test objects: PCA-based codec signal that transmits two channels separately, PCA-based codec signal with traditional parametric stereo, PCA-based codec signal with improved parametric stereo, G719 codec signal with traditional parametric stereo [24], HE-AACv2 codec signal, anchor signal, and original signal. In the algorithm proposed in this paper, the relationship between the quality of the restored signal and bitrate is not linear, as Figure 17 shows, which uses a simple subjective test with different bitrate allocation; therefore, the test chooses a case in which the qualities of restored signal and bitrate are both acceptable.

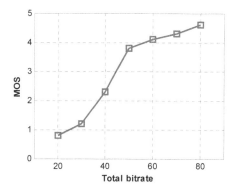

Figure 17. Relationship between quality and bitrate.

The bits allocations of each module in PCA–based codec for stereo signal are shown in Table 7.

Table 7. Bitrate allocation in encoder.

Module	Bitrate
Reduced-dimensional matrix	35 kbps
Rotation matrix	5 kbps
First principal component	10 kbps
Spatial parameters and side information	5 kbps

(iii) In order to eliminate psychological effects, the order and the name of each test material in each group are random. The listener needs to select the original signal from the test signals and score 100 points, and the rest of the signals are scored by 0–100 according to overall quality, including sound quality and the spatial reduction degree.

5.1.2. Listeners

10 people with certain listening experiences were selected for the listening test, of which 5 were male, 5 were female, and each listener has normal hearing.

5.1.3. Auditory Environment

All 10 listeners use headphones connected to a laptop in quiet environments.

5.2. Test Results

After the test is finished, we calculated average value and the 95% confidence interval based on the listeners' scores. The average confidence interval of each test codec is [77.2, 87.0], [74.4, 84.2], [70.5, 80.7], [65.8, 76.6], [56.1, 66.9], and [78.6, 86.2]. After removing three outlier data (data beyond confidence interval), the test results of MUSHRA are shown in Figures 18 and 19.

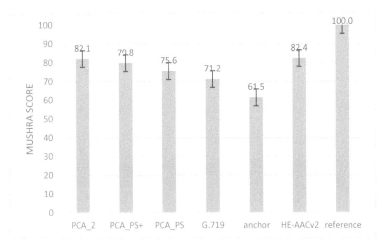

Figure 18. Results of MUSHRA test. (PCA_2 represents the PCA-based codec signal that is transmitted over two channels separately (75 kbps), PCA_PS+ represents PCA-based codec signal with improved parametric stereo (55 kbps), PCA_PS represents PCA-based codec signal with traditional parametric stereo (56 kbps), G.719 represents G.719 codec signal with traditional parametric stereo (56 kbps), anchor represents anchor signal, HE_AACv2 represents HE-AACv2 signal (55 kbps), and reference represents hidden reference signal).

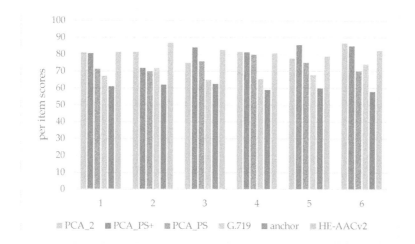

Figure 19. MUSHRA score of per item test. (PCA_2 represents the PCA-based codec signal that is transmitted over two channels separately (75 kbps), PCA_PS+ represents PCA-based codec signal with improved parametric stereo (55 kbps), PCA_PS represents PCA-based codec signal with traditional parametric stereo (56 kbps), G.719 represents G.719 codec signal with traditional parametric stereo (56 kbps), anchor represents anchor signal; HE-AACv2 represents HE-AACv2 signal (55 kbps), and hidden reference material has been removed. 1–6 represents different test materials).

Compared with traditional parametric stereo, the PCA-based parametric stereo has less bitrate, higher quality, and better spatial sense. Compared with G719 with traditional parametric stereo with the same bitrate, PCA-based codec signal has better quality. Compared with HE-AACv2 signal, the average score of the PCA-based parametric stereo is slightly less than HE-AACv2. HE-AACv2 is a mature codec that uses several techniques to improve the quality, including Quadrature Mirror Filter (QMF), Spectral Band Replication (SBR), noise shaping and so on. The complexity of PCA is less than the part of the 32-band QMF in HE-AACv2. Considering the high complexity and maturity of HE-AACv2, the test results are optimistic. Conclusions can be drawn that the PCA-based codec method possesses good performance, especially for stereo signal in which the audio quality and spatial sense can be recovered well.

5.3. Complexity Analysis

The module of principal component analysis can be regarded as a part of the singular value decomposition (SVD): the calculate procession of the right singular matrix and the singular value of original matrix $X_{[m \ n]}$, therefore the algorithm complexity of principal component analysis module is O(n^3). According to the properties of SVD, when $n < m$, the computation complexity of the right singular matrix is half of the computation complexity of SVD for $X_{[m \ n]}$. Therefore, the algorithm complexity and delay of PCA are far less than those of SVD. In the Intel i5-5200U processor, 4 GB memory, 2.2 GHz work memory, it takes 20 ms to finish one part of PCA. Given the time reduction of parametric stereo, the delay of PCA-based codec algorithm is in the acceptable range. In the part of multi-frame joint PCA, the forming of the original matrix takes 40 ms. When the first frame finishes MDCT, the process of forming original matrix will begin. Besides, the thread of PCA is different from matrix construction, and MDCT windowing also belongs to the calculating thread. Suppose the time for MDCT of first frame is t_1; the whole delay can be regarded as around $40 + t_1$ ms, which is around 50 ms. The delay of the algorithm proposed in this paper still has space to be improved, and we can make the balance of delay and bitrate better by adjusting the number of multi frames using a more intelligent strategy in the future.

6. Discussion

This paper just presents a preliminary algorithm. There is still much space for improvement in real applications. One question worth further study is how to eliminate the noise. In the experiment, when the number of bits or the number of principal components is too small, the noise spectrum has special nature, as Figures 20–22 show. Signal in Figure 20 is restored by three components; compared with signal in Figures 21 and 22, the spectrum of noise in high-frequency domain has obvious repeatability, which occurs once every 1.6 kHz. Therefore, low pass filter mentioned in Section 3.3 is not the best way to get rid of this noise: the damage of original signal is unavoidable. Ideally, an adaptive notch filter can filter the spectrum of noise clearly and not damage original signal. However, the design of such an adaptive notch filter needs to be studied more in the future.

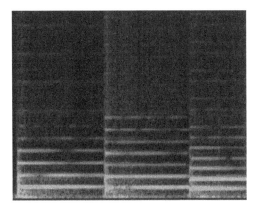

Figure 20. The spectrogram of the signal restored by three components.

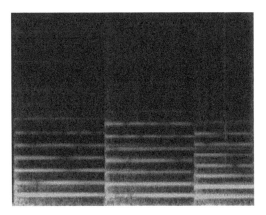

Figure 21. The spectrogram of the signal restored by four components.

Appl. Sci. **2018**, *8*, 967

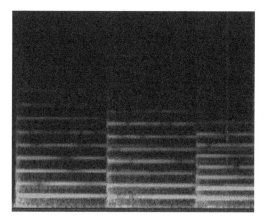

Figure 22. The spectrogram of the signal restored by five components.

7. Conclusions

The framework of proposed multi-frame PCA-based audio coding method has several differences compared to other codecs; therefore, there are lots of barriers to the design of an optimal algorithm. This paper proposed several ways to remove those barriers. For mono signal, the design of PCA-based coding method in this paper, including multi-frame signal processing, matrix design, and quantization design can hold it efficiently. As to stereo signal, PCA has high associativity with parametric stereo, which makes PCA-based parametric stereo certainly feasible and significant. Experimental results show satisfactory performance of the multi-frame PCA-based stereo audio coding method compared with the traditional audio codec.

In summary, research on the multi-frame PCA-based codec, both for mono and stereo, has certain significance and needs further improvement. This kind of stereo audio coding method has good performance in processing different kinds of audio signals, but further studies are still needed before it can be widely applied.

Author Contributions: J.W. conceived the method and modified the paper, X.Z. performed the experiments and wrote the paper, X.X. and J.K. contributed suggestions, and J.W. supervised all aspects of the research.

Funding: National Natural Science Foundation of China (No. 61571044).

Acknowledgments: The authors would like to thank the reviewers for their helpful suggestions. The work in this paper is supported by the cooperation between BIT and Ericsson.

References

1. Bosi, M.; Goldberg, R.E. *Introduction to Digital Audio Coding and Standards*; Kluwer Academic Publishers: Dordrecht, The Netherlands, 2003; pp. 399–400, ISBN 1402073577.
2. Fatus, B. Parametric Coding for Spatial Audio. Master's Thesis, KTH, Stockholm, Sweden, 2015.
3. Faller, C. Parametric joint-coding of audio sources. In Proceedings of the AES 120th Convention, Paris, France, 20–23 May 2016.
4. Blauert, J. *Spatial Hearing: The Psychophysics of Human Sound Localization*; MIT Press: Cambridge, MA, USA, 1983; pp. 926–927, ISBN 0262021900.
5. ISO/IEC. *13818-7: Information Technology—Generic Coding of Moving Pictures and Associated Audio Information—Art 7: Advanced Audio Coding (AAC)*; ISO/IEC JTC 1/SC 29: Klagenfurt, Austria, 2006.

6. Herre, J. From Joint Stereo to Spatial Audio Coding—Recent Progress and Standardization. In Proceedings of the 7th International Conference of Digital Audio Effects (DAFx), Naples, Italy, 5–8 October 2004.

7. Jannesari, A.; Huda, Z.U.; Atr, R.; Li, Z.; Wolf, F. Parallelizing Audio Analysis Applications—A Case Study. In Proceedings of the 39th International Conference on Software Engineering: Software Engineering Education and Training Track (ICSE-SEET), Buenos Aires, Argentina, 20–28 May 2017; pp. 57–66.

8. ISO/IEC. *14496-3:2001/Amd 2: Parametric Coding for High-Quality Audio*; ISO/IEC JTC 1/SC 29: Redmond, WA, USA, 2004.

9. Breebaart, J.; Par, S.V.D.; Kohlrausch, A.; Schuijers, E. Parametric coding of stereo audio. *EURASIP J. Appl. Signal Process.* **2005**, *9*, 1305–1322. [CrossRef]

10. Faller, C.; Baumgarte, F.D. Binaural cue coding—Part I: Psychoacoustic fundamentals and design principles. *IEEE Trans. Speech Audio Process.* **2003**, *11*, 509–519. [CrossRef]

11. Faller, C.; Baumgarte, F.D. Binaural cue coding—Part II: Schemes and applications. *IEEE Trans. Speech Audio Process.* **2003**, *11*, 520–531. [CrossRef]

12. Zhang, S.H.; Dou, W.B; Lu, M. Maximal Coherence Rotation for Stereo Coding. In Proceedings of the 2010 IEEE International conference on multimedia & Expo (ICME), Suntec City, Singapore, 19–23 July 2010; pp. 1097–1101.

13. Ahmed, N.; Natarajan, T.; Rao, K.R. Discrete cosine transform. *IEEE Trans. Comput.* **1974**, *1*, 90–93. [CrossRef]

14. Smith, S.W. *The Scientist and Engineer's Guide to Digital Signal Processing*, 2nd ed.; California Technical Publishing: San Diego, CA, USA, 1997; ISBN 0-9660176.

15. Skodras, A.; Christopoulos, C.; Ebrahimi, T. The jpeg 2000 still image compression standard. *IEEE Signal Process. Mag.* **2001**, *18*, 36–58. [CrossRef]

16. Pearson, K. On Lines and Planes of Closest Fit to Systems of Points in Space. *Philos. Mag.* **1901**, *2*, 559–572. [CrossRef]

17. Jia, M.S.; Bao, C.C.; Liu, X.; Li, R. A novel super-wideband embedded speech and audio codec based on ITU-T Recommendation G.729.1. In Proceedings of the 2009 Annual Summit and Conference of Asia-Pacific Signal and Information Processing Association (APSIPA ASC), Sapporo, Japan, 4–7 October 2009; pp. 522–525.

18. Jia, M.S.; Bao, C.C.; Liu, X.; Li, X.M.; Li, R.W. An embedded stereo speech and audio coding method based on principal component analysis. In Proceedings of the International Symposium on Signal Processing and Information Technology (ISSPIT), Bilbao, Spain, 14–17 December 2011; Volume 42, pp. 321–325.

19. Briand, M.; Virette, D.; Martin, N. Parametric representation of multichannel audio based on Principal Component Analysis. In Proceedings of the 120th Convention Audio Engineering Society Convention, Paris, France, 1–4 May 2006.

20. Goodwin, M. Primary-Ambient Signal Decomposition and Vector-Based Localization for Spatial Audio Coding and Enhancement. In Proceedings of the IEEE Conference on Acoustics, Speech, and Signal Processing (ICASSP), Honolulu, HI, USA, 15–20 April 2007; pp. I:9–I:12.

21. Chen, S.X.; Xiong, N.; Park, J.H.; Chen, M.; Hu, R. Spatial parameters for audio coding: MDCT domain analysis and synthesis. *Multimed. Tools Appl.* **2010**, *48*, 225–246. [CrossRef]

22. Robert, C.S.; Stefan, W.; David, S.H. Mean opinion score (MOS) revisited: Methods and applications, limitations and alternatives. *Multimed. Syst.* **2016**, *22*, 213–227. [CrossRef]

23. Hou, Y.; Wang, J. Mixture Laplace distribution speech model research. *Comput. Eng. Appl.* **2014**, *50*, 202–205. [CrossRef]

24. Jiang, W.J.; Wang, J.; Zhao, Y.; Liu, B.G.; Ji, X. Multi-channel audio compression method based on ITU-T G.719 codec. In Proceedings of the Ninth International Conference on Intelligent Information Hiding and Multimedia Signal Processing (IINMSP), Beijing, China, 8 July 2014.

25. ITU-R. *BS.1534: Method for the Subjective Assessment of Intermediate Quality Level of Coding Systems*; International Telecommunication Union: Geneva, Switzerland, 2001.

Article

Prediction of HIFU Propagation in a Dispersive Medium via Khokhlov–Zabolotskaya–Kuznetsov Model Combined with a Fractional Order Derivative

Shilei Liu [1], Yanye Yang [1], Chenghai Li [1], Xiasheng Guo [1,*], Juan Tu [1] and Dong Zhang [1,2,*]

[1] Key Laboratory of Modern Acoustics (MOE), Department of Physics,
    Collaborative Innovation Centre of Advanced Microstructure, Nanjing University, Nanjing 210093, China;
    lsl666@foxmail.com (S.L.); dz1622045@smail.nju.edu.cn (Y.Y.); lichh-82008@163.com (C.L.);
    juantu@nju.edu.cn (J.T.)
[2] The State Key Laboratory of Acoustics, Chinese Academy of Science, Beijing 10080, China
*   Correspondence: guoxs@nju.edu.cn (X.G.); dzhang@nju.edu.cn (D.Z.)

Received: 17 March 2018; Accepted: 10 April 2018; Published: 12 April 2018

Abstract: High intensity focused ultrasound (HIFU) has been proven to be promising in non-invasive therapies, in which precise prediction of the focused ultrasound field is crucial for its accurate and safe application. Although the Khokhlov–Zabolotskaya–Kuznetsov (KZK) equation has been widely used in the calculation of the nonlinear acoustic field of HIFU, some deviations still exist when it comes to dispersive medium. This problem also exists as an obstacle to the Westervelt model and the Spherical Beam Equation. Considering that the KZK equation is the most prevalent model in HIFU applications due to its accurate and simple simulation algorithms, there is an urgent need to improve its performance in dispersive medium. In this work, a modified KZK (mKZK) equation derived from a fractional order derivative is proposed to calculate the nonlinear acoustic field in a dispersive medium. By correcting the power index in the attenuation term, this model is capable of providing improved prediction accuracy, especially in the axial position of the focal area. Simulation results using the obtained model were further compared with the experimental results from a gel phantom. Good agreements were found, indicating the applicability of the proposed model. The findings of this work will be helpful in making more accurate treatment plans for HIFU therapies, as well as facilitating the application of ultrasound in acoustic hyperthermia therapy.

Keywords: KZK equation; fractional order derivative; ultrasound hyperthermia; HIFU; acoustic simulation; Kramers–Kronig relation

1. Introduction

Although pioneering clinical studies of focused ultrasound were carried out as early as the 1940s [1,2], it did not attract intensive research interest until the end of the 20th century and the beginning of 21st century, during which several theoretical models were developed, improved and then broadly accepted [3–9]. In the past decades, high intensity focused ultrasound (HIFU) has played an increasingly significant role in the study of non-invasive therapies by demonstrating unique advantages in safety, effectiveness and high efficiency [10–14]. However, the applications of HIFU are still limited, and clinical treatments are only available for limited sites [11,14–16].

One of the challenges confronting HIFU treatments is the spatial precision of tissue ablation. Several techniques such as ultrasound B-Scan and Magnetic Resonance Imaging (MRI) have been combined with HIFU to achieve real-time monitoring of focal areas [17–20]. With these methods, the actual focal profiles were usually found to deviate from those predicted through theoretical models [21,22]. As was indicated by Petrusca et al., the shift of the focal point away from the prescribed position caused by acoustic aberrations and non-linear wave propagating effects make it mandatory to evaluate the spatial accuracy of HIFU ablation [21]. Li et al. observed a focal shift of 1–2 mm and ascribed it to the layered distribution of tissues. Several different mechanisms might contribute to these deviations, such as the thermos-lensing effect [23,24], bubble formation [25–27], acoustic radiation force [27] and the nonlinear nature of acoustic waves [21,22]. Connor et al. pointed out that the positioning error of the focal spots could be mainly related to the thermos-lensing effect and nonlinear propagation of ultrasonic waves [23]. When accounting for the complexity of wave propagation, e.g., ribs, abdomen tissues, blood vessels and other celiac organs between the transducers and the targeted area, these could constitute sources of acoustic scattering, diffraction, attenuation and dispersion etc. [28,29], and negatively affect the precision of HIFU treatments. Therefore, it is very important to take into account the complexities of the acoustic paths by developing theoretical models for higher accuracy.

Using the state of the art methods, the spatial distribution of HIFU field can be simulated with the well-known Khokhlov–Zabolotskaya–Kuznetsov (KZK) equation [5,6], the Westervelt model [22] or the Spherical Beam Equation (SBE) [9]. Because of its reasonable parabolic approximation, the KZK equation has been widely used for describing the propagation of finite amplitude acoustic beams emitted by focused transducers, with the only restriction being that the half angle of divergence of the transducer does not exceed $16°$ [7]. Existing algorithms to solve the KZK equation are usually balanced between accuracy and simplicity, making it possible to calculate and adjust HIFU fields in real time during treatments. In contrast, the heavy calculation burden of the Westervelt equation and the SBE limits their application. Therefore, it is not unexpected that the KZK equation has been preferred against the other two in both clinical situations and industry. In the KZK equation, the effects of diffraction, nonlinearity and attenuation have been taken into consideration and are described by separate terms. Therefore, it has been recognized to be more effective than the Rayleigh integral in simulating HIFU propagation [8]. Meanwhile, the parabolic approximation is also accurate enough in practice, by which the near-axis HIFU field can be calculated relatively accurately [7,30,31]. However as Meaney et al. have pointed out, it is hard to explain the shifts of focal positions with the traditional KZK equation [24]. Therefore, several questions are still open to discussion. Firstly, the Kramers–Kronig dispersion relations indicate a power law between the attenuation and the working frequency, i.e., $\alpha \propto \omega^y$, $y = 1 - 2$. For most tissues, the attenuation factor y sits in the range from 1 to 1.7 [32,33]. However, in the KZK model, the propagation loss caused by viscosity and thermal conduction is considered to be proportional to the square of the working frequency ($y = 2$), which is actually only valid for fresh water [32,33]. Furthermore, $y = 2$ for fresh water causes a third derivative term to appear in the KZK equation. The third derivative as well as the derivative of acceleration have not yet been well clarified in the physical overview. In the context of Newton's law of motion, acceleration is directly affected by the force, and the derivative of acceleration has no obvious physical meaning. Secondly, sound velocity, which is treated as a constant in the KZK equation, usually changes with frequency in biological tissues. In HIFU fields, propagation nonlinearity could be non-negligible due to the high acoustic pressure, and the orders of harmonic waves could be rather high in many cases [5–7,9,29]. Therefore, the descriptions of both attenuation and sound velocity should be modified when biological tissues are present on the wave path.

To overcome these shortcomings in the existing models, efforts need to be made to account for the case of a non-integer power index in the Kramers–Kronig relationship. In the frequency domain, with the help of the classic Laplace Transform, the acoustic field could be easily accessible, but only valid for linear models. The Rayleigh model proposed by Wojcik et al. in 1995 was also not applicable in cases using wide-bandwidth acoustic pulses [34]. The fractional order derivative method described by Makris and Constantinou [35] was believed to be an effective approach to solve this problem. However, the complexity in mathematics has made numerical analysis hard to achieve, hence restrained its application in developing HIFU theories. After that, Szabo et al. utilized a convolution integral and further developed the theory of the fractional order derivative, in which Fourier Transform was adopted to define a derivative of non-integer order, while the basic principles of the Kramers–Kronig relationship was still preserved [36–38]. Some other calculation algorithms have also been proposed. Treeby et al. developed a popular, open source nonlinear simulation tool named the k-wave toolbox to simulate nonlinear wave propagation [39]. Prieur et al. proposed time-fractional acoustic wave equations [40]. Inspired by these algorithms, simulation of the KZK equation has made definite progress in the past few years.

In this paper, a fractional order derivative was introduced to modify the KZK model to address the power-law relationship (non-integer power index), and a frequency-dependent sound velocity was employed to account for the dispersive behaviors of media. Numerical and experimental verifications were carried out to demonstrate the different behaviors between the modified model and the original KZK model. Results showed that the obtained modified KZK (mKZK) equation is in better agreement with the experimental results. The findings in this paper will promote not only the prediction and design of focused sound fields and acoustic transducers, but also the therapeutic applications of ultrasonic treatments.

2. Theory and Experiments

2.1. The KZK Equation

The KZK equation is an extended form of Burgers model, and an approximation of the Westervelt equation, written as [41]

$$\frac{\partial^2 p}{\partial z \partial \tau} = \frac{c_0}{2}\Delta_\perp p + \frac{\delta}{2c_0{}^3}\frac{\partial^3 p}{\partial \tau^3} + \frac{\beta}{2\rho_0 c_0{}^3}\frac{\partial^2 p^2}{\partial \tau^2}, \tag{1}$$

in which p is the acoustic pressure, c_0 is the sound velocity, δ is the sound diffusivity, β is the nonlinearity coefficient, ρ_0 is the ambient density of the medium, $\Delta_\perp$ is the transverse Laplace operator (defined as $\Delta_\perp = \frac{1}{r}\frac{\partial}{\partial r}\left(r\frac{\partial}{\partial r}\right) + \frac{1}{r^2}\frac{\partial^2}{\partial \theta^2}$ in cylindrical coordinates), and $\tau = t - z/c_0$ is the time delay at the axial distance of z with t being the time. The three terms on the right side of Equation (1) represent the diffraction, attenuation and nonlinearity, respectively. The attenuation caused by viscosity and thermal conduction is hence proportional to the square of the angular frequency ω, i.e., $\alpha(\omega) = \omega^2\delta/(2c_0^3)$. However, in biological tissues, the attenuation factor is usually a non-integer less than 2, the attenuation term is hence only valid for describing wave propagation in fresh water (attenuation factor $y = 2$). If the attenuation parameter is directly set for fresh water rather than considering the actual properties of the media, the accuracy of calculation would certainly be undermined. Furthermore, the sound velocity c_0 here is regarded as a constant for all harmonic components, which is in conflict with the inherent Kramers–Kronig dispersion relations where the phase velocity varies with increasing frequency.

2.2. The Modified KZK Model

By introducing Fourier Transform, the nth time derivative of time-dependent acoustic pressure $p(t)$ could be written as

$$\frac{d^n p(t)}{dt^n} = F^-\left\{(i\omega)^n F^+[p(t)]\right\}. \tag{2}$$

Here F^+ and F^- represents operators of the Fourier transform and its inversion, respectively. As the order number y is a non-integer, the fractional order derivative is then defined as a convolution [38],

$$\frac{d^y p(t)}{dt^y} = \frac{1}{\Gamma(-y)}\int_{-\infty}^{t}\frac{p(t')}{(t-t')^{(1+y)}}dt', \tag{3}$$

where $\Gamma(\cdot)$ is the Gamma Function, i.e., $\Gamma(x) = \int_0^\infty \xi^{x-1}e^{-\xi}d\xi$. The definition in Equation (3) is hence the Riemann–Liouville fractional derivative [42]. Therefore, the fractional order derivative is not only determined by the pressure value at the time point t, but is also related to its history range from $-\infty$ to t.

Since the attenuation factor y is a non-integer in lossy media, the wavenumber $\widetilde{k}$ could then be considered as a complex, while its squared form in the low-frequency approximation is

$$\widetilde{k}^2 \approx \frac{\omega^2}{c_0^2} + 2i\frac{\omega}{c_0}\alpha_0\omega^y = \frac{\omega^2}{c_0^2} - 2\frac{\alpha_0}{c_0}\frac{(-i\omega)^{y+1}}{(-i)^y}. \tag{4}$$

in which i is the imaginary unit, and $(-i)^y$ can be expressed with trigonometric functions as $(-i)^y = \cos(y\pi/2) - i\sin(y\pi/2)$. With the wave number $\widetilde{k} = \omega/c_0 + i\alpha_0(-i\omega)^y/[\cos(y\pi/2)]$, the phase velocity can be calculated as

$$\frac{1}{c(\omega)} = \frac{\mathrm{Re}\left(\widetilde{k}\right)}{\omega} = \frac{1}{c_0} + \alpha_0\tan(y\pi/2)|\omega|^{y-1}, \tag{5}$$

where $\mathrm{Re}(\cdot)$ represents the real part of the complex value.

Considering the non-integer attenuation factor y and the dispersion of phase velocity, the classical KZK equation could be modified as

$$\frac{\partial^2 p}{\partial z\partial\tau} = \frac{c(\omega)}{2}\nabla_\perp^2 p + \frac{\delta}{2c(\omega)^3}\frac{\partial^{y+1}p}{\partial\tau^{y+1}} + \frac{\beta}{2\rho_0 c(\omega)^3}\frac{\partial^2 p^2}{\partial\tau^2}, \tag{6}$$

where the attenuation term and the phase velocity $c(\omega)$ could be calculated according to Equations (4) and (5), respectively. Consistent conclusions can be found between Equation (6) and the work of Zhao et al. [43], which is an extension of an earlier model for ultrasound propagation in power-law media proposed by Kelly et al. [44]. In addition, Equation (6) will not hold if $y = 1$. However, in the framework of discussing HIFU propagation problems, the $y = 1$ case is of no importance [37,38] and is not of interest here.

2.3. The Numerical Algorithm

In the numerical analysis, both the KZK model and its modified form were solved with the finite difference time domain (FDTD) method for the acoustic field emitted from a single-element self-focusing transducer. For the traditional model, coordinate transformation $Z = z/F$, $R = r/a$, $T = \omega t$, $P = p/P_0$ were introduced, with F, r, a, P_0 being the geometrical focal length, the radial coordinate, the aperture radius of the transducer, and the surficial acoustic pressure, respectively. The following assumptions were then made,

$$G = \frac{ka^2}{2F}, \tag{7}$$

$$A = \frac{\omega^2\delta}{2\rho_0 c_0^3}, \quad F = \alpha'F, \tag{8}$$

$$N = \frac{F}{\rho_0 c_0^3/(P_0\beta\omega)} = \frac{F}{l_d'}, \tag{9}$$

in order that the KZK equation could be normalized to the following form,

$$\frac{\partial^2 P}{\partial T\partial Z} = \frac{1}{4G}\triangle_\perp P + A\frac{\partial^3 P}{\partial T^3} + \frac{N}{2}\frac{\partial^2 P^2}{\partial T^2}. \tag{10}$$

The values of the physical constants used for acoustic modeling were $\rho_0 = 1000$ kg/m^3, $c_0 = 1486$ m/s, $\beta = 3.5$, $\alpha = 0.025$ Np/m at 1 MHz, and $\mu = 2$ for fresh water [45]. The FDTD algorithm adopted here was generally the same as that used in a previous study [45,46], in which Equation (10) was decomposed into three independent equations, accounting for the diffraction, attenuation and nonlinearity, respectively. Based on an orthogonal spatial grid, the discretized forms of these equations were expressed as,

$$\frac{P_{i,j+1}^n - P_{i,j}^n}{dZ} = \frac{1}{4G} \int_{T_{min}}^{T} \left(\frac{P_{i+1,j+1}^n - 2P_{i,j+1}^n + P_{i-1,j+1}^n}{dR^2} + \frac{P_{i+1,j+1}^n - P_{i-1,j+1}^n}{2kdR^2} \right) dt, \tag{11}$$

$$\frac{P_{i,j+1}^n - P_{i,j}^n}{dZ} = A \frac{P_{i,j+1}^{n+1} - 2P_{i,j+1}^n + P_{i,j+1}^{n-1}}{dT^2}, \tag{12}$$

And

$$P_{j+1}^n = \begin{cases} P_j^n \left(1 - N \dfrac{P_j^{n+1} - P_j^n}{dT} dZ \right)^{-1} & P_j^n \geq 0 \\[4mm] P_j^n \left(1 - N \dfrac{P_j^n - P_j^{n-1}}{dT} dZ \right)^{-1} & P_j^n < 0 \end{cases}. \tag{13}$$

Here i and j were the spatial coordinate indexes in the radial and axial directions, respectively, and n was the time step, i.e., $P_{i,j}^n = P(i \cdot dr, j \cdot dz; n \cdot dt)$.

In the simulations for the mKZK equation, the major difference was the differential form of the fractional derivative in Equation (6),

$$\frac{\partial^{y+1} p}{\partial \tau^{y+1}} = \frac{A}{\Delta \tau^y} \times \left[\Delta \tau^{-1} \sum_{r=0}^{n} \omega_r^{(y)} \left(P_{i,j}^{n-r+1} + P_{i,j}^{n-r} \right) + \omega_{n+1}^{(y)} P'(0)_{i,j}^k \right], \tag{14}$$

where

$$A = 2\Gamma(-y)\Gamma(y+1)\cos[(y+1)\pi/2]/\pi, \tag{15}$$

$$\omega_r^{(y)} = (-1)^r y(y-1)\Gamma(y-r+1)/r!. \tag{16}$$

The FDTD simulations were then accomplished through a self-developed FORTRAN code package running on a ×64 PC platform. In the calculation, the boundary condition was symmetrical and had equal amplitude acoustic pressure driving conditions, which was also the boundary condition generally used to calculate HIFU [29,41].

2.4. Experimental Methods

2.4.1. Phantom Preparation

A tissue phantom was prepared based on the recipe of polyacrylamide electrophoresis gel [47], in which micron-sized polystyrene microspheres were added to adjust its attenuation and phase velocity dispersion. The formula of phantom contained 100 mL degassed water, 10 g acrylamide (A9099, Sigma-Aldrich, St Louis, MO, USA), 0.05 g ammonium persulfate (A9164, Sigma-Aldrich), 0.3 g methylene double acrylamide (146072, Sigma-Aldrich), 0.2 mL TEMED (411019, Sigma-Aldrich), and 4 mL 10-micron microsphere solution (P107798, Aladdin, Shanghai, China, original concentration 5% w/v).

2.4.2. Experimental Setup

The thickness of each phantom sample was L_s = 42.3 mm, with a density of ρ_1 = 1000 kg/m^3 so that it could stably suspend in the water. The distance between transducer and the phantom was 37.7 mm. Following the same protocol used in our previous work [45], the nonlinearity parameter was measured as β = 4.2. Since the only difference in gel recipe between the two works is the introduction of amino polystyrene microspheres in this paper, the same β value indicates that microspheres did not influence the nonlinear propagation of waves. To confirm this, we measured the ratio between the second harmonic and fundamental components, and found it was identical to that in [45] under the same sonication conditions. The attenuation coefficient and sound velocity of phantom, as functions of frequency, were measured through a broadband spectrum method [48,49]. In the measurement, two planar piston transducers (Immersion, Unfocused, Panametrics, Waltham, MA, USA) calibrated with a needle hydrophone (HNC-1000, ONDA Corp., Sunnyvale, CA, USA), were placed on the opposite sides of the phantom, with one of them driven by a broadband pulse generator (5900PR, Panametrics). The reflected and transmitted acoustic signals were then acquired by the transducers and digitalized with a digital oscilloscope (54830B, Agilent, Santa Clara, CA, USA). (Device connection was similar to that in [49]). In the measurements, 8 continuous pulse sequences were acquired and averaged to reduce the signal-to-noise ratio (SNR).

Prior to the HIFU experiments, a low-level driving voltage was used to drive a customized HIFU transducer (Chongqing Haifu Med. Tech. Co., Ltd., Chongqing, China) to emit a linear sound field. Simultaneously, effective parameters of the transducer, such as effective radius, radiation profile, and angle of divergence were obtained by adjusting the transducer parameters in the KZK calculations, such that the linear field predicted via KZK was consistent with that measured [45]. The effective parameters were then used in the simulations of both the KZK and mKZK models. As a result, the HIFU transducer (working frequency 1.12 MHz) had an effective aperture radius of 48.6 mm and a geometrical focal length of 101.5 mm. As illustrated in Figure 1, the transducer was immersed in water and driven with signals from a signal generator (33250A, Agilent, Santa Clara, CA, USA) amplified by a broadband power amplifier (2200L, E&I, Rochester, NY, USA). The input voltage was set as 465 mV (20 cycles; burst period, 10 ms; duty cycle, 0.18%). Another needle hydrophone (HNA-0400, ONDA Corp., Sunnyvale, CA, USA) was mounted on a customized three-dimensional (3D) scanning system (Controller Model: XPS-C8, Newport, CA, USA) to scan the HIFU field. To suppress possible acoustic cavitation in surrounding liquid, the water was processed with a self-developed water degassing and deionizing system. The temporal and spatial scanning procedure was controlled via the GPIB interface (National Instruments, Austin, TX, USA).

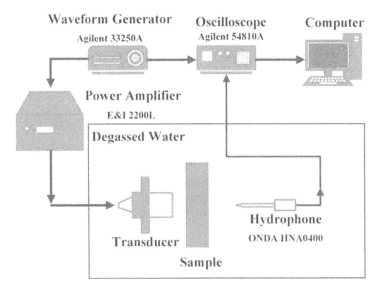

Figure 1. The experimental setup.

3. Results and Discussions

3.1. Non-Dispersive Water

In order to verify the validity of the modified model, the mKZK equation was used to predict the sound field distributions generated from the transducer. In this case, the media in the direction of propagation was degassed water, and the surface pressure of the transducer was set to be 0.4 MPa. Then, the results were compared with those obtained from experimental measurements as well as from the original KZK model. The results are presented in Figure 2 for comparison. Due to the axial-symmetry of the sound field, only the sound field distribution in the axial direction was studied. Also, since the major concern of this paper is to investigate how the focal-shift could be accurately predicted, the pressure profiles are all presented in a normalized way, so that the focal-shift effect is more intuitive and easy to observe. For the total pressure distributions presented in Figure 2a, the axial distribution of acoustic pressure seems identical for the modified and original KZK models, which provides reliable proof that the current theoretical modification did not compromise the accuracy of the sound field prediction in non-dispersive media. With the help of fast Fourier transform (FFT) algorithm, further analysis was then carried out by decomposing the total sound pressure into the superposition of linear and nonlinear components. In Figure 2b–d, all the components exhibited good agreement between the results calculated from the modified and the original KZK models, showing that both models are applicable for sound field prediction under the experimental conditions mentioned above. It should also be noted that, in Figure 2 the locations of the pressure peaks from the two models were exactly the same for all components, although the measured axial beam-widths seem a bit narrower than both theoretical predictions, especially for harmonic components. Since the linear fields were found to be almost identical for the measured and predicted results, we speculate that some far-field attenuation factors such as dissolved oxygen in water, or bubbles might exist. However, this does not affect the conclusion on the location of the maximum pressure.

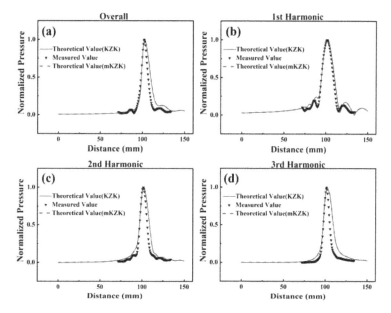

Figure 2. The normalized acoustic pressure distributions along the axis of the HIFU transducer without the phantom: (**a**) the overall pressure; (**b**) the fundamental component; (**c**) the second harmonic; (**d**) the third harmonic.

3.2. Dispersive Phantom

To incorporate the phantom model into the study, the acoustic parameters of the phantom was firstly characterized according to procedures described in earlier studies [48,49]. In brief, the acoustic phase velocity $c(f)$ inside the phantom material was determined by [48]

$$c(f) = c_w \left[1 + 2 \frac{\theta_w(f) - \theta_s(f)}{\theta_2(f) - \theta_1(f)} \right], \tag{17}$$

where the sound velocity in water c_w was considered as 1500 m/s. When the sound velocity inside the phantom was measured, the acoustic signals p_s and p_f were the acoustic pressure acquired by a transducer before and after the phantom was inserted into the acoustic path. While p_1 and p_2 were the pressure of the reflected signals from the first and second water/phantom interfaces, respectively, $\theta_w(f)$, $\theta_s(f)$, $\theta_1(f)$ and $\theta_2(f)$ were the corresponding phase spectra. The frequency dependence of the attenuation coefficient $\alpha(f)$ was calculated according to [48,49]

$$\alpha(f) = \frac{1}{L_s} \left[\ln\left(\frac{A_1}{A_2}\right) - \ln\left(\frac{A_w}{A_s}\right) \right], \tag{18}$$

where A_w, A_s, A_1 and A_2 were the amplitude spectra corresponding to the above-mentioned phase responses. Figure 3 plots the measured attenuation coefficient and acoustic velocity as a function of frequency. In the frequency range 1.5–3.1 MHz, the sound velocity increased by about 26 m/s. The acoustic attenuation coefficient increased from 0.59 dB/cm at 1.5 MHz to 1.79 dB/cm at 3.1 MHz, indicating the attenuation factor being $y = 1.83$ for the phantom. The attenuation factor was obtained from the exponential fitting using a curve fitting toolbox in Matlab. It should be noted that sound velocity could fluctuate due to the thermal effect of focused ultrasound. However, in this work the duty cycle was as low as 0.18%, and a medium-level surface pressure of up to 0.4 MPa was chosen for the transducer. Thus, no significant temperature elevation was observed during the experiments, and the thermal-induced change in sound velocity could be neglected [50]. It should also be mentioned, that to account for the thermal-effect, an appropriate bio-heat transferring equation should be incorporated with the current model. The main concern then lies in the dispersion due to microsphere scattering. In clinical applications, the major dispersion originates from tissues like fat, whose attenuation coefficient is generally larger than that of body tissue. Therefore, in comparison with other research in which the parameters of the phantom were nearly the same as body tissue (e.g., liver and spleen), the choice of phantoms with larger attenuation coefficients in the present experiments might provide results in better agreement with the actual situation. Meanwhile, since the measured physical parameters are close to those of dense fat, this setup could be regarded as a simplified mimic of the abdomen in HIFU therapies.

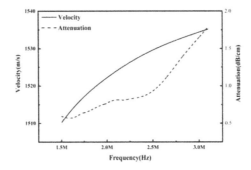

Figure 3. Measured frequency-dependent acoustic velocity and attenuation coefficient of the phantom sample.

As is illustrated in Figure 4, the acoustic field distribution along the axial direction is examined by sitting the phantom on the acoustic path of the transducer. The results show the comparison between measured data and simulated results obtained from both the modified and traditional KZK models. Figure 4a describes the total pressure and Figure 4b–d shows the fundamental, second harmonic and third harmonic components, respectively. For the total pressure distribution, it is clearly observed that the peak-pressure location predicted through the mKZK model agrees well with that acquired in experiments, while the data calculated from the traditional KZK model show a deviation from the previous two groups. Note that the axial peak-pressure deviation is quite small for the fundamental components, but it gradually becomes significant when more harmonic components appear. Thus, the overall peak-pressure deviation observed in Figure 4a is mainly caused by higher-order harmonic components, indicating that the significance of the theoretical modification relies on high nonlinearity, such as in HIFU. This phenomenon could be addressed with the results shown in Figure 3, where higher-order harmonic waves that occupy higher frequency bands exhibit more notable sound velocity/attenuation dispersions, thus play a more dominant role in the modification of the KZK model. Meanwhile, it can be seen from Figure 4 that the modified equation has larger variation than the KZK equation. The effect of the mKZK equation on the simulation results can be thus clarified as providing more precise prediction for experimental results. The results in Figure 4 give persuasive proof that theoretical modifications made previously are necessary and valid. The beam narrowing effect caused by data normalization still exists. However, in actual treatment more attention is paid to the location of the focus point, because the location determines the heat distribution area, which significantly alters the biological properties of the treatment area.

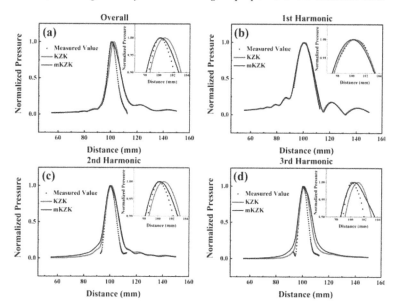

Figure 4. The normalized acoustic pressure distributions along the axis of the HIFU transducer, with the phantom placed at 37.7 mm away from the transducer. (**a**) the overall pressure; (**b**) the fundamental component; (**c**) the second harmonic; (**d**) the third harmonic.

Among existing studies, most researchers have focused on the thermos-lensing effect [23] and bubble formation induced focal region distortion [25,26], while some also mentioned the acoustic radiation force induced tissue displacement [27]. With the results presenting the dispersion of sound velocity and attenuation in tissues, the deviation of focal spots can be well explained in combination with the strong nonlinearity nature of HIFU.

3.3. Dispersion-Induced Focus Shift

It is of great importance to evaluate how the mKZK model demonstrates its significance in HIFU applications. In Figure 5, comparisons are carried out to display how the wave distribution behaves differently before and after inserting the phantom sample into the wave propagation path. It can be observed that, for either data from experiments or from the mKZK model, although only a slight difference is seen in the axial wave profile when examining the fundamental components, an axial focus shift is more evident for the overall acoustic pressure since it is highly affected by the harmonic components. It is explained here that, due to the dispersive nature of the phantom, higher-order harmonic components require larger values of both sound velocity and attenuation coefficient in the KZK model, urging the overall wave profile to move forward to the transducer.

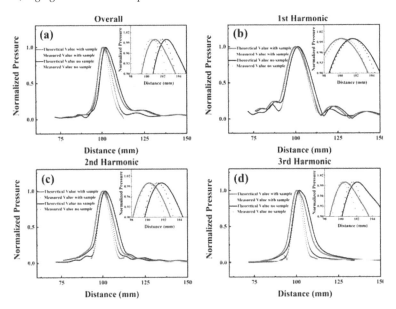

Figure 5. Comparison of the axial distributed acoustic pressure between the cases of with and without the phantom sample: (**a**) the overall pressure; (**b**) the fundamental component; (**c**) the second harmonic; (**d**) the third harmonic.

In the present study, the focus shift distance was quantified for the overall axial acoustic pressure distribution and its decomposed components, and this is listed in Table 1.

Table 1. Focus shift induced by acoustic-dispersive phantom sample (in mm).

Method	Overall	Fundamental	2nd Harmonic	3rd Harmonic
mKZK	1.47	1.41	1.51	1.62
Experiment	1.42 ± 0.04	1.40 ± 0.03	1.45 ± 0.04	1.51 ± 0.04

The focus shift, which is found to be of millimeter magnitude for the studied condition, cannot be ignored, especially for clinical HIFU studies. On the one hand, in typical HIFU applications, the much higher surface pressure would induce even stronger acoustic nonlinearity in the focus area, and up to tens of orders of harmonic components might contribute to the overall acoustic responses. In that case, dispersion in acoustic velocity as well as the attenuation coefficient could cause larger focus shifts. On the other hand, even for the millimeter-level focus shift exhibited with the current setup, an impressive amount of acoustic energy would be deposited outside the designated focus area. As demonstrated in Figure 6, about 25% of the −3 dB focal region would fall outside of the one predicted by the traditional KZK equation, when the phantom sample is introduced into the transducer axis. In addition, the absence of shockwaves in the experiment should be noted, and this indicates that the influence of shockwaves could be ignored in the theoretical model under such experimental parameters.

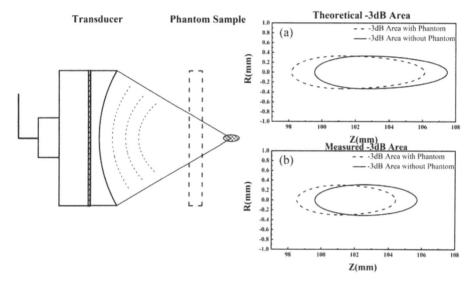

Figure 6. Illustration of the focus shift induced by the presence of phantom sample: (**a**) the theoretical −3 dB area; (**b**) the measured −3 dB area.

3.4. Discussion

In previous studies, shifts between the predicted and observed focal regions have been frequently reported in HIFU-related studies. Usually, the shift was measured to be around 1–3 mm for 1-MHz HIFU transducers [24], 4–5 mm for ~2.2-MHz excitation [51], and an empirical formula was also used to calculate the focus shift [22]. Although different possible mechanisms have been proposed and a series of correlation studies have been carried out [21–27], detailed theoretical proof that could quantitatively explain the inherent mechanisms of the observed focal shifts is still lacking. The proposed model modification here clarifies how the acoustic dispersion played a role in the complicated physics of this problem.

The experiments carried out here have demonstrated that the observed focal shift should come from the dispersive behavior of sound velocity and attenuation in media. For different harmonic components in the HIFU beam, their sound velocities could be different at the gel/water interface. Speculating from Snell's law, these different components could actually propagate along slightly different paths in the phantom, causing the focal point to shift its location. That is also why the main difference between the two models is observed for the harmonics in Figures 4 and 5—because the sound velocity of the harmonics deviated further away as their frequencies were higher. During this process, the attenuation should also have contributed in a dispersive way. However, the influence of dispersive attenuation could not be separated from that of sound velocity.

Despite the general recognition that the Westervelt equation and SBE provide higher precision in non-dispersive sound field prediction than KZK, difficulties in predicting accurate sound field distribution in strong dispersion media still remains a great challenge [22]. Moreover, due to the existing complexity of the Westervelt equation and SBE, adding more modifications to these two equations could be over-whelming and/or time-consuming for clinical applications that require real-time monitoring. Considering that KZK shows a balance between accuracy and computation burden, this paper chose to further modify the KZK equation as a straightforward way to achieve improved simulation of HIFU propagation, especially to work out how the focal-shift could be predicted accurately. In this work, for consistency of the experiment, the mKZK and KZK in ordinary media was first confirmed. Then, when measuring the parameters in the strong dispersion medium, mKZK gave a more accurate prediction of the focal shift. Thus, by using the mKZK equation we can quickly and accurately predict the HIFU field in complex media.

However, further theoretical studies are still needed since the current model is unable to eradicate all the possible unfavorable factors in predicting the characteristics of HIFU. For example, defocusing effects or shifts in focal position might be caused by the layered tissue effect, where sound speed/attenuation may vary in different tissues layers. The thermal lesion effect, in which the tissue properties change due to the heating of HIFU, could be more difficult to include in the modeling. A full solution could be even more challenging if other possible mechanisms, including acoustic radiation force, acoustic cavitation and inconsistent thermal deposition [27] are also considered.

The focal shifts observed in phased-array-based HIFU devices are also notable. For instance, a focal shift of about 2-mm was observed along the transducer axis in multiple-layered soft tissues sonicated with a 65-element phased array transducer [22]. To overcome this problem in phased-array HIFU, possible solutions could be obtained by drawing lessons from the underlying mechanisms discussed above.

4. Conclusions

Although HIFU technology based on the KZK equation calculation has been widely accepted and used in the clinical setting and transducer designs, the absence of an accurate theory to predict the sound field inevitably limits the application of the ultrasound focusing. In this work, a mKZK equation is proposed to predict the HIFU field established with a spherical focusing transducer. Meanwhile, the accuracy of the methods is verified through experiment. This method could improve the computational accuracy of the KZK equation in dispersive media, which is similar to human tissue. Simulation and experimental results show that the focus area will shift towards the transducer and the offset increases as the nonlinearity becomes higher. Therefore, in the process of HIFU transducer design, the impact of dispersion on the results need to be taken into account, in order that accurate sonication can be achieved. By modifying the KZK equation, the findings will help with transducer design and the application of the HIFU. This will also help to ensure the stability and safety of HIFU and further accelerate its clinical applications.

Acknowledgments: This work was partially supported by the National Natural Science Foundation of China (Grant Nos. 81627802, 11774166, 11474161, 11474001 and 11674173), QingLan Project and Nanjing University Innovation and Creative Program for Ph.D. candidate, No. CXCY17-13.

Author Contributions: Xiasheng Guo and Dong Zhang conceived and designed the experiments. Shilei Liu and Yanye Yang performed the experiments. Shilei Liu, Xiasheng Guo and Juan Tu analyzed the data. Shilei Liu and Chenghai Li performed the derivation theory. Shilei Liu and Xiasheng Guo contributed reagents/materials/analysis tools. Shilei Liu, Xiasheng Guo and Dong Zhang wrote the manuscript.

Conflicts of Interest: The authors declare no conflict of interest.

References

1. Lynn, J.G.; Zwemer, R.L.; Chick, A.J.; Miller, A.E. A new method for the generation and use of focused ultrasound in experimental biology. *J. Gen. Physiol.* **1942**, *26*, 179. [CrossRef] [PubMed]

2. Fry, W.J.; Fry, F.; Barnard, J.; Krumins, R.; Brennan, J. Ultrasonic lesions in mammalian central nervous system. *Science* **1955**, *122*, 1091. [CrossRef]
3. Westervelt, P.J. Parametric acoustic array. *J. Acoust. Soc. Am.* **1963**, *35*, 535–537. [CrossRef]
4. Hallaj, I.M.; Cleveland, R.O. FDTD simulation of finite-amplitude pressure and temperature fields for biomedical ultrasound. *J. Acoust. Soc. Am.* **1999**, *105*, L7–L12. [CrossRef] [PubMed]
5. Zabolotskaya, E. Quasi-plane waves in the nonlinear acoustics of confined beams. *Sov. Phys. Acoust.* **1969**, *15*, 35–40.
6. Kuznetsov, V. Equation of nonlinear acoustics. *Sov. Phys. Acoust.* **1971**, *16*, 467–470.
7. Tjotta, J.N.; Tjotta, S.; Vefring, E.H. Effects of focusing on the nonlinear interaction between two collinear finite amplitude sound beams. *J. Acoust. Soc. Am.* **1991**, *89*, 1017–1027. [CrossRef]
8. Kamakura, T.; Ishiwata, T.; Matsuda, K. Model equation for strongly focused finite-amplitude sound beams. *J. Acoust. Soc. Am.* **2000**, *107*, 3035–3046. [CrossRef] [PubMed]
9. Kamakura, T.; Ishiwata, T.; Matsuda, K. A new theoretical approach to the analysis of nonlinear sound beams using the oblate spheroidal coordinate system. *J. Acoust. Soc. Am.* **1999**, *105*, 3083–3086. [CrossRef]
10. ter Haar, G. Biological effects of ultrasound in clinical applications. In *Ultrasound: Its Chemical, Physical and Biological Effects*; VCH Publishers: New York, NY, USA, 1988.
11. ter Haar, G. Turning up the power: High intensity focused ultrasound (HIFU) for the treatment of cancer. *Ultrasound* **2007**, *15*, 73–77. [CrossRef]
12. Wu, F.; Chen, W.-Z.; Bai, J.; Zou, J.-Z.; Wang, Z.-L.; Zhu, H.; Wang, Z.-B. Pathological changes in human malignant carcinoma treated with high-intensity focused ultrasound. *Ultrasound Med. Biol.* **2001**, *27*, 1099–1106. [CrossRef]
13. Gelet, A.; Chapelon, J.; Bouvier, R.; Rouviere, O.; Lasne, Y.; Lyonnet, D.; Dubernard, J. Transrectal high-intensity focused ultrasound: Minimally invasive therapy of localized prostate cancer. *J. Endourol.* **2000**, *14*, 519–528. [CrossRef] [PubMed]
14. Orsi, F.; Arnone, P.; Chen, W.; Zhang, L. High intensity focused ultrasound ablation: A new therapeutic option for solid tumors. *J. Cancer Res. Ther.* **2010**, *6*, 414. [CrossRef] [PubMed]
15. Illing, R.; Kennedy, J.; Wu, F.; ter Haar, G.; Protheroe, A.; Friend, P.; Gleeson, F.; Cranston, D.; Phillips, R.; Middleton, M. The safety and feasibility of extracorporeal High-Intensity Focused Ultrasound (HIFU) for the treatment of liver and kidney tumours in a Western population. *Br. J. Cancer* **2005**, *93*, 890. [CrossRef] [PubMed]
16. Zhou, Y.-F. High intensity focused ultrasound in clinical tumor ablation. *World J. Clin. Oncol.* **2011**, *2*, 8. [CrossRef] [PubMed]
17. Gudur, M.S.R.; Kumon, R.E.; Zhou, Y.; Deng, C.X. High-frequency rapid B-mode ultrasound imaging for real-time monitoring of lesion formation and gas body activity during high-intensity focused ultrasound ablation. *IEEE Trans. Ultrason. Ferroelectr. Freq. Control* **2012**, *59*, 1687–1699. [CrossRef] [PubMed]
18. Kemmerer, J.; Ghoshal, G.; Oelze, M. Quantitative ultrasound assessment of HIFU induced lesions in rodent liver. In Proceedings of the 2010 IEEE International Ultrasonics Symposium, San Diego, CA, USA, 11–14 October 2010; pp. 1396–1399.
19. Wijlemans, J.; Bartels, L.; Deckers, R.; Ries, M.; Mali, W.T.M.; Moonen, C.; Van Den Bosch, M. Magnetic resonance-guided high-intensity focused ultrasound (MR-HIFU) ablation of liver tumours. *Cancer Imaging* **2012**, *12*, 387. [CrossRef] [PubMed]
20. Zhang, L.; Chen, W.; Liu, Y.; Hu, X.; Zhou, K.; Chen, L.; Peng, S.; Zhu, H.; Zou, H.; Bai, J. Feasibility of magnetic resonance imaging-guided high intensity focused ultrasound therapy for ablating uterine fibroids in patients with bowel lies anterior to uterus. *Eur. J. Radiol.* **2010**, *73*, 396–403. [CrossRef] [PubMed]
21. Petrusca, L.; Viallon, M.; Breguet, R.; Terraz, S.; Manasseh, G.; Auboiroux, V.; Goget, T.; Baboi, L.; Gross, P.; Sekins, K.M. An experimental model to investigate the targeting accuracy of MR-guided focused ultrasound ablation in liver. *J. Transl. Med.* **2014**, *12*, 12. [CrossRef] [PubMed]
22. Li, D.; Shen, G.; Bai, J.; Chen, Y. Focus shift and phase correction in soft tissues during focused ultrasound surgery. *IEEE Trans. Biomed. Eng.* **2011**, *58*, 1621–1628. [CrossRef] [PubMed]
23. Connor, C.W.; Hynynen, K. Bio-acoustic thermal lensing and nonlinear propagation in focused ultrasound surgery using large focal spots: A parametric study. *Phys. Med. Biol.* **2002**, *47*, 1911. [CrossRef] [PubMed]
24. Meaney, P.M.; Cahill, M.D.; ter Haar, G. The intensity dependence of lesion position shift during focused ultrasound surgery. *Ultrasound Med. Biol.* **2000**, *26*, 441–450. [CrossRef]

25. Zderic, V.; Foley, J.; Luo, W.; Vaezy, S. Prevention of post-focal thermal damage by formation of bubbles at the focus during high intensity focused ultrasound therapy. *Med. Phys.* **2008**, *35*, 4292–4299. [CrossRef] [PubMed]

26. Zhou, Y.; Wilson Gao, X. Variations of bubble cavitation and temperature elevation during lesion formation by high-intensity focused ultrasound. *J. Acoust. Soc. Am.* **2013**, *134*, 1683–1694. [CrossRef] [PubMed]

27. Laughner, J.I.; Sulkin, M.S.; Wu, Z.; Deng, C.X.; Efimov, I.R. Three potential mechanisms for failure of high intensity focused ultrasound ablation in cardiac tissue. *Circulation* **2012**, *5*, 409–416. [CrossRef] [PubMed]

28. Bobkova, S.; Gavrilov, L.; Khokhlova, V.; Shaw, A.; Hand, J. Focusing of high-intensity ultrasound through the rib cage using a therapeutic random phased array. *Ultrasound Med. Biol.* **2010**, *36*, 888–906. [CrossRef] [PubMed]

29. Liu, Z.; Fan, T.; Zhang, D.; Gong, X. Influence of the abdominal wall on the nonlinear propagation of focused therapeutic ultrasound. *Chin. Phys. B* **2009**, *18*, 4932–4937.

30. Rosnitskiy, P.B.; Yuldashev, P.V.; Sapozhnikov, O.A.; Maxwell, A.D.; Kreider, W.; Bailey, M.R.; Khokhlova, V.A. Design of HIFU Transducers for Generating Specified Nonlinear Ultrasound Fields. *IEEE Trans. Ultrason. Ferroelectr. Freq. Control* **2017**, *64*, 374–390. [CrossRef] [PubMed]

31. Soneson, J.E. A parametric study of error in the parabolic approximation of focused axisymmetric ultrasound beams. *J. Acoust. Soc. Am.* **2012**, *131*, EL481–EL486. [CrossRef] [PubMed]

32. O'Donnell, M.; Jaynes, E.; Miller, J. Kramers-Kronig relationship between ultrasonic attenuation and phase velocity. *J. Acoust. Soc. Am.* **1981**, *69*, 696–701. [CrossRef]

33. Kudo, N.; Kamataki, T.; Yamamoto, K.; Onozuka, H.; Mikami, T.; Kitabatake, A.; Ito, Y.; Kanda, H. Ultrasound attenuation measurement of tissue in frequency range 2.5–40 MHz using a multi-resonance transducer. In Proceedings of the Ultrasonics Symposium, Toronto, ON, Canada, 5–8 October 1997; pp. 1181–1184.

34. Wojcik, G.; Mould, J.; Abboud, N.; Ostromogilsky, M.; Vaughan, D. Nonlinear modeling of therapeutic ultrasound. In Proceedings of the Ultrasonics Symposium, Seattle, WA, USA, 7–10 November 1995; pp. 1617–1622.

35. Makris, N.; Constantinou, M. Fractional-derivative Maxwell model for viscous dampers. *J. Struct. Eng.* **1991**, *117*, 2708–2724. [CrossRef]

36. Szabo, T.L. Time domain wave equations for lossy media obeying a frequency power law. *J. Acoust. Soc. Am.* **1994**, *96*, 491–500. [CrossRef]

37. Szabo, T.L. Causal theories and data for acoustic attenuation obeying a frequency power law. *J. Acoust. Soc. Am.* **1995**, *97*, 14–24. [CrossRef]

38. Szabo, T.L.; Wu, J. A model for longitudinal and shear wave propagation in viscoelastic media. *J. Acoust. Soc. Am.* **2000**, *107*, 2437–2446. [CrossRef] [PubMed]

39. Treeby, B.E.; Jaros, J.; Rendell, A.P.; Cox, B.T. Modeling nonlinear ultrasound propagation in heterogeneous media with power law absorption using a *k*-space pseudospectral method. *J. Acoust. Soc. Am.* **2012**, *131*, 4324–4336. [CrossRef] [PubMed]

40. Prieur, F.; Holm, S. Nonlinear acoustic wave equations with fractional loss operators. *J. Acoust. Soc. Am.* **2011**, *130*, 1125–1132. [CrossRef] [PubMed]

41. Rosnitskiy, P.B.; Yuldashev, P.V.; Vysokanov, B.A.; Khokhlova, V.A. Setting boundary conditions on the Khokhlov-Zabolotskaya equation for modeling ultrasound fields generated by strongly focused transducers. *Acoust. Phys.* **2016**, *62*, 151–159. [CrossRef]

42. Heymans, N.; Podlubny, I. Physical interpretation of initial conditions for fractional differential equations with Riemann-Liouville fractional derivatives. *Rheol. Acta* **2006**, *45*, 765–771. [CrossRef]

43. Zhao, X.; McGough, R.J. The Khokhlov-Zabolotskaya-Kuznetsov (KZK) equation with power law attenuation. In Proceedings of the IEEE International Ultrasonics Symposium, Chicago, IL, USA, 3–6 September 2014; pp. 2225–2228.

44. Kelly, J.F.; McGough, R.J.; Meerschaert, M.M. Analytical time-domain Green's functions for power-law media. *J. Acoust. Soc. Am.* **2008**, *124*, 2861–2872. [CrossRef] [PubMed]

45. Fan, T.; Liu, Z.; Zhang, D.; Tang, M. Comparative study of lesions created by high-intensity focused ultrasound using sequential discrete and continuous scanning strategies. *IEEE Trans. Biomed. Eng.* **2013**, *60*, 763–769. [CrossRef] [PubMed]

46. Fan, T.; Zhang, D.; Gong, X. Estimation of the tissue lesion induced by a transmitter with aluminium lens. *J. Phys.* **2011**, *279*, 012020. [CrossRef]

47. Lafon, C.; Zderic, V.; Noble, M.L.; Yuen, J.C.; Kaczkowski, P.J.; Sapozhnikov, O.A.; Chavrier, F.; Crum, L.A.; Vaezy, S. Gel phantom for use in high-intensity focused ultrasound dosimetry. *Ultrasound Med. Biol.* **2005**, *31*, 1383–1389. [CrossRef] [PubMed]

48. He, P. Measurement of acoustic dispersion using both transmitted and reflected pulses. *J. Acoust. Soc. Am.* **2000**, *107*, 801–807. [CrossRef] [PubMed]

49. He, P.; Zheng, J. Acoustic dispersion and attenuation measurement using both transmitted and reflected pulses. *Ultrasonics* **2001**, *39*, 27–32. [CrossRef]

50. Fan, T.; Zhang, D.; Zhang, Z.; Ma, Y.; Gong, X. Effects of vapour bubbles on acoustic and temperature distributions of therapeutic ultrasound. *Chin. Phys. B* **2008**, *17*, 3372–3377. [CrossRef]

51. Camarena, F.; Adrián-Martínez, S.; Jiménez, N.; Sánchez-Morcillo, V. Nonlinear focal shift beyond the geometrical focus in moderately focused acoustic beams. *J. Acoust. Soc. Am.* **2013**, *134*, 1463–1472. [CrossRef] [PubMed]

Article

Application of Elastic Wave Velocity for Estimation of Soil Depth

Hyunwook Choo [1], Hwandon Jun [2] and Hyung-Koo Yoon [3,*]

[1] Department of Civil Engineering, Kyung Hee University, Yongin 17-104, Korea; choohw@gmail.com
[2] Department of Civil Engineering, Seoul National University of Science and Technology,
    Seoul 139-743, Korea; hwjun@seoultech.ac.kr
[3] Department of Construction and Disaster Prevention Engineering, Daejeon University,
    Daejeon 300-716, Korea
* Correspondence: hyungkoo@dju.ac.kr

Received: 21 March 2018; Accepted: 5 April 2018; Published: 11 April 2018

Abstract: Because soil depth is a crucial factor for predicting the stability at landslide and debris flow sites, various techniques have been developed to determine soil depth. The objective of this study is to suggest the graphical bilinear method to estimate soil depth through seismic wave velocity. Seismic wave velocity rapidly changes at the interface of two different layers due to the change in material type, packing type, and contact force of particles and thus, it is possible to pick the soil depth based on seismic wave velocity. An area, which is susceptible to debris flow, was selected, and an aerial survey was performed to obtain a topographic map and digital elevation model. In addition, a seismic survey and a dynamic cone penetration test were performed in this study. The comparison between the soil depth based on dynamic cone tests and the graphical bilinear method shows good agreement, indicating that the newly suggested soil depth estimating method may be usefully applied to predict soil depth.

Keywords: graphical bilinear method; seismic survey; dynamic cone penetration test; soil depth; time-distance curve

1. Introduction

In assessing the stability of landslide or debris flow areas, both hydrological and geotechnical properties are the key parameters [1–6] among the various geotechnical properties such as soil strength, hydraulic conductivity, and friction angle, the soil depth is the most important parameter because the capacity for inflow and outflow of water is related to the soil thickness. Note that soil depth can be defined as the thickness of the soil from ground surface to consolidated medium [7,8].

The most reliable method to estimate soil depth at a given location is the test pit method, which involves direct excavation of a testing site in a square shape. However, excavation of multiple pits for the estimation of soil depth is very expensive and time-consuming. To overcome these limitations, Ref. [9] used a cone-tipped metal probe with a diameter of 18 mm for estimating local soil depth based on probe penetration resistance. Note that the methods mentioned above can only be performed at the selected local points; therefore, the soil depth of unmeasured areas is assumed to be the same as the measured depth of nearby points. However, soil depth shows substantial spatial variation; thus, the reliability of the above methods may be low, leading to the development of several models that consider the spatial distribution of soil depth. Ref. [10] developed the regolith-mantle slope method to predict the thickness of original parent material on a slope. Ref. [11] estimated soil depth based on the mass balance between soil production from underlying bedrock and divergence of diffusive soil transport. Even though these models are advantageous for obtaining soil depth across the whole area, the use of these models may not be easy because they require detailed information regarding hydrological, geotechnical, and geochemical parameters. Therefore, the geophysical method has been widely applied to estimate soil depth because it can quickly and cost-effectively provide various soil characteristics across an entire area. Electromagnetic waves have been used to measure electrical resistivity [12] and electrical conductivity [2], and the measured values can be converted into soil depths. Refs. [13,14] used seismic survey to predict soil depth and however, the studies merely showed the distributions of results without methodological content for picking soil depth. In addition, the applied previous geophysical methods require special reference values to determine soil depth and additional invasive experiments to enhance the reliability of the estimated soil depth.

This study proposes a graphical bilinear method to estimate soil depth based only on seismic survey. A seismic survey was performed, and the dynamic cone penetration test (DCPT) was also used to verify the soil depths deduced by the suggested technique. The measured information, including geological map, location, particle distribution, and distribution of seismic wave velocity were introduced first. Then the graphical bilinear method for determining soil depth, is demonstrated. The estimated soil depths using the suggested method are compared with values deduced by the dynamic cone penetration tests and their reliability is assessed.

2. Testing Site

2.1. Site Description

The selected testing site experienced debris flow a few years ago and is still susceptible to additional debris flow because of several geological characteristics of the site, including steep slope angle (over 32°) and saturated soil condition on slope. According to the geological map issued by the Korea Institute of Geology, Analysis, Mining (KIGAM), the testing area mainly consists of gneissose granite terrane. The testing site belongs to Mt. Geohwa, South Korea, where the altitude and area are approximately 200 m and 1.2 km^2, respectively, and the latitude and longitude of the top of the mountain are N 36°29′15.7313″ and E 127°18′40.5986″. A drone aerial survey was performed to obtain a topographic map and digital elevation model. Figure 1 shows the topographic map: the area consists of one main stream and several branch streams. The vertical length and area of the main stream are approximately 206 m and 14,675 m^2, respectively. Figure 1 also indicates the presence of the debris barrier and check dam at the site to prevent additional debris flow (N 36°29′04.3147″, E 127°18′44.7017″). The main stream was divided at four points, selected by considering slope characteristics: point A (N 36°29′10.8530″, E 127°18′44.1891″), point B (N 36°29′12.045″, E 127°18′39.5912″), point C (N 36°29′14.8329″, E 127°18′30.7079″), and point D (N 36°29′18.2046″, E 127°18′23.6158″). Point A, which is located at the bottom of the main stream, shows various fallen trees and weeds undergoing decay and rot, suggesting that considerable time has passed since the occurrence of the last debris flow. The slope of point A was measured to be approximately 10°, which is low compared to the slopes at points B and C. Thus, point A is covered with various flowed materials from the streams above due to debris flow. Point B shows a dramatically steeper slope of approximately 27° and the area mainly consists of sedimentary basin. Point C has the largest rapid slope (~32°) and width (maximum 105 m) among the selected points. Note that point C is located at a large catchment basin. Finally, point D indicates the initial zone where the debris flow occurred as ascertained by the presence of upright trees and a stable subsurface without collapsed conditions.

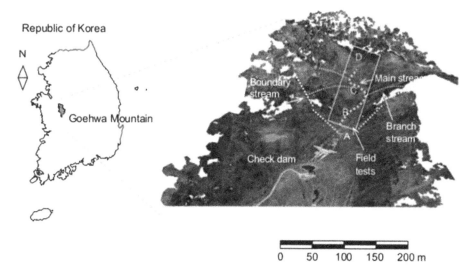

Figure 1. Aerial photography of the testing area. Points A and D denote the bottom and top streams, respectively. Points B and C indicate the middle stream. The field tests were performed from points A to D.

2.2. Soil Classification

A hand auger with an outer diameter of 20 cm was used to excavate the subsurface and a soil sampler was used to gather the specimen at the wall of a borehole. The length of the hand auger was limited to 2 m due to a shortage of engine output, and thus the maximum depth of extracted soil in this study was 2 m. Even though the hand auger can potentially excavate soil down to a maximum of 2 m, the actual excavated depth was reduced due to the presence of gravel and weathered rock. Therefore, the actual maximum extracted depths were 1, 1, and 2 m at locations near points A, B, and C, respectively. The results from the excavations at points A and B show various deposited materials, including gravel and boulder (Figure 2). At point C near the initial zone, a larger extraction depth (~2 m) could be achieved due to the presence of a relatively deep soil layer (Figure 2). Soil samples were collected at 4–5 different depths: 10, 40, 60, and 100 cm for points A and B; and 10, 40, 90, 140, and 200 cm for point C. Sieve tests were performed using the extracted specimens according to [15].

The results of sieve tests are shown in Figure 2. The diameters at passing percentages of 10, 30 and 60% are calculated for every specimen, to classify the specimen based on the unified soil classification system. The calculated coefficients of uniformity are 10, 5.92, and 9.33 for extracted soils at points A, B, and C, respectively. The coefficients of curvature at points A, B, and C are determined to be 1.34, 1.01, and 1.51, respectively. The measured coefficients of uniformity show values almost greater than 6 and the coefficients of curvature are in the range of 1~3; hence, the specimens at testing sites are classified as SW (well-graded sand). Additionally, Figure 2 indicates that the grain size distributions of extracted samples at different depths are very similar.

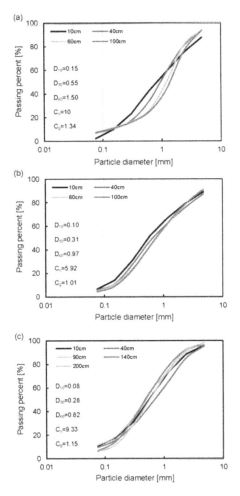

Figure 2. Sieve test results: (**a**) bottom stream (point A in Figure 1); (**b**) middle stream (point B); (**c**) top stream (point C). D_{10}, D_{30}, and D_{60} denote the diameters at passing percentages of 10, 30, and 60%, respectively. C_u and C_g are coefficients of uniformity and curvature, respectively.

3. Methodology

A seismic survey and dynamic cone penetrometer test were performed to obtain primary wave velocity and dynamic cone penetration index (DPI), respectively, and the detailed descriptions are as follows.

3.1. Seismic Survey

A seismic survey has the advantage of assessing a whole area without altering the fabric state; thus, the measured values can greatly reflect the state and behavior of soils. A seismic wave propagates in solid media by travelling through particle connections, and thus the wave velocity increases with an increase in the particle contact area (or applied confining stress or soil depth). The seismic refraction method was applied in this study to obtain a profile of compressional (or primary) wave velocity. Four transects were determined with consideration for spatial variability in the geological characterization of each stream. Figure 3 shows schematic drawings of the seismic transections.

The bottom-top stream (BTS) line reflects the main stream from south to north (from points A to D in Figure 1), whereas the bottom stream (BS), middle stream (MS), and top stream (TS) lines are horizontally set to reflect bottom, middle, and top valleys, respectively. The center of the MS line is located between points B and C in Figure 1. The lengths of the transection lines are 90, 20, 20, and 20 m for BTS, BS, MS, and TS, respectively. A geophone was installed as a sensor for gathering seismic waves every 2 m; therefore, the installed geophone numbers in the profiles of BTS, BS, MS and TS are 45, 10, 10, and 10, respectively. A sledgehammer was used to generate vibrations, and the impactions were performed at start, middle, and end positions to enhance signal-to-noise ratios. Impactions were performed five times in the same position for each test to minimize random noise.

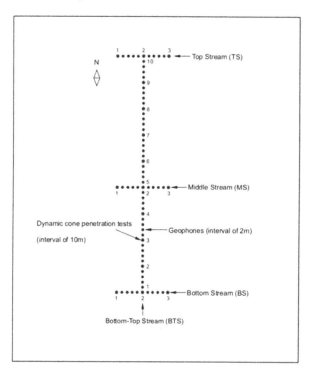

Figure 3. Locations of the field investigations, including seismic survey and dynamic cone penetrometer (DCP). Note: the numbers with red circles denote the DCP testing sites (BS: 3 holes, MS: 3 holes, TS: 3 holes, and BTS: 10 holes); and the distance between each number in the figure is 10 m.

3.2. Dynamic Cone Penetrometer Test

A dynamic cone penetrometer (DCP) test was selected as an invasive method to verify the soil depths estimated by seismic waves. Note that the DCP method can readily detect the thickness of soil layer because the continuous penetration of DCP enables the easy recognition of the presence of different layers [16]: the penetration depth through stiff material is relatively small. The DCP technique has been widely applied to investigate soil properties in geotechnical engineering, especially for railways and roads [16–18]. The test records the penetration depth of DCP when a hammer drops from a fixed height according to [19]. The hammer weight and drop height were fixed to 78.8 N and 575 mm, respectively. The tip diameter of the DCP was 20 mm with an apex angle of 60°, and driving energy was 45 J. The penetration depth was converted into dynamic penetration index (DPI), which represents (mm/blow) as follows.

$$DPI = \frac{P_{i+1} - P_i}{B_{i+1} - B_i}$$ (1)

where P = penetration depth (mm); B = blow count; and i and $i + 1$ = experimental number.

4. Results and Analysis

The first arrival time of the measured seismic wave signals was determined through PickWin software (OYO corporation, Tokyo, Japan), and the time-distance curve was obtained by using the SeisImager-Poltrefa program (OYO corporation, Tokyo, Japan). An interactive process was performed to increase resolution when carrying out the inversion process. Two groups, which are based on the picked first-arrival times and the results analyzed by the simultaneous iterative method, are plotted on a travel-time curve in Figure 4. Figure 4 shows that the two groups are nearly identical, reflecting that the quality of the measured data is high. The distribution of seismic wave velocity is plotted in Figure 5, and the soil layers are divided into various different layers according to the reference P-wave velocities suggested by [20]: 0–0.7 km/s (landfill and alluvial soil), 0.7–1.2 km/s(weathered soil), 1.2–1.9 km/s (weathered rock), and over 1.9 km/s (soft rock). The BTS, BS, MS, and TS lines were found to comprise four, four, three, and three layers, respectively based on Reynolds (2003). Note that 0.7 km/s of primary wave velocity is an accepted criterion in South Korea for distinguishing soil layers from other geomaterials [21,22].

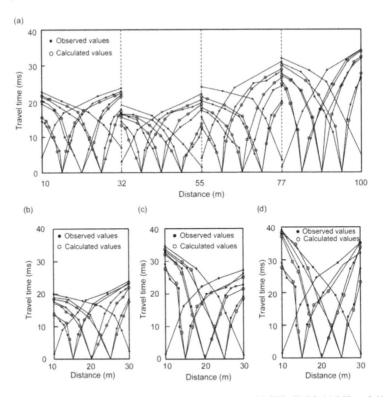

Figure 4. Travel time-distance curves through seismic survey: (a) BTS; (b) BS; (c) MS; and (d) TS. Note 10, 20, 30, and 100 m distances indicate points 1, 2, 3, and 10, respectively, in Figure 3.

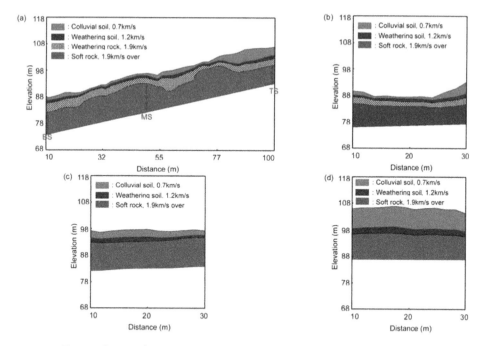

Figure 5. Converted primary wave velocity profiles: (**a**) BTS; (**b**) BS; (**c**) MS; and (**d**) TS.

The calculated DPI values from dynamic cone penetration tests are shown in Figure 6 for each stream. The BTS line shows high variation of DPI values according to the testing locations, which are described in Figure 3, because the line covers the bottom, middle, and top streams. Note that the high variation of DPI values means that the soil depth changes along the stream in the BTS line. Number 4, 8 and 9 holes in Figure 6a show high DPI values at shallow depth, reflecting the presence of weak soil layers. Being different from other testing holes, the DPI was measured at deeper depths at No. 7 and 9 holes, demonstrating that the soil depth near the top stream is deeper than that of other locations. Note that Figure 5a and the maximum extracted soil depth using a hand auger also support the presence of deep soil layer near the top stream in the BTS line. In the case of the BS line, high DPI values at the initial stage and high penetrated depth were observed at the right side of the stream. A similar DPI trend was observed in the MS line, and the distributions of soil depth in the No. 1, 2, and 3 holes look alike, reflecting similar soil depths along the MS line. Even though the DPI of the TS line shows a similar penetrated depth near the surface, the DPI was continuously measured at the only No. 2 hole. Therefore, it is predicted that the soil depth at the No. 2 hole, which is the center of the TS line, is high. The depth where the DPI value is close to zero is referred to as the soil depth of the testing site in this study, and the average soil depths based on DPI are calculated to be approximately 0.92, 0.75, 0.86 and 0.85 m for the BTS, BS, MS, and TS lines, respectively.

Comparison between Figures 5 and 6 reveals that the soil depth of the testing site based on dynamic cone penetration test is generally shallower than 1 m; while, the depth based on seismic wave velocity (0.7 km/s) is around 2 m or more (Table 1), reflecting that the subsurface classification based on previous reference values of seismic wave velocity cannot precisely estimate the soil depth. Therefore, in this study, the graphical bilinear method is newly suggested for determining soil depth based on seismic wave velocity.

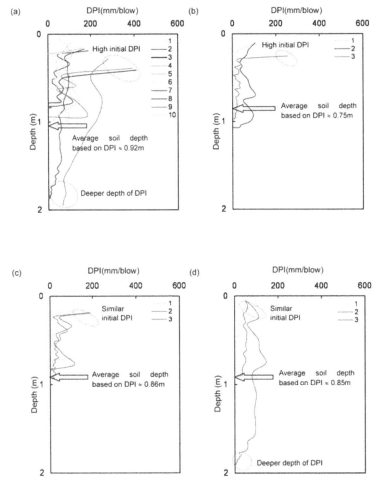

Figure 6. Measured DPI values of each stream: (**a**) BTS; (**b**) BS; (**c**) MS; and (**d**) TS. Note the numbers in each figure indicate the location of dynamic cone penetration test described in Figure 3.

Table 1. Comparison between Measured and Estimated Soil Depths.

Position		Soil Depth (m)			Error Ratio (%)	
		Measured	Estimated			
		Dynamic Cone Test	Reference P-Wave Velocity	Graphical Bilinear Method	Reference P-Wave Velocity	Graphical Bilinear Method
	1	1.1	2.3	1.1	109.1	0.0
	2	0.8	2.0	0.8	150.0	0.0
	3	1.0	1.8	1.0	80.0	0.0
	4	0.4	1.9	0.5	375.0	25.0
	5	0.5	1.4	0.6	180.0	20.0
BTS	6	0.2	1.8	0.5	800.0	150.0
	7	0.6	1.4	0.5	133.3	16.7
	8	0.7	2.4	1.0	242.9	42.9
	9	1.9	5.6	1.9	194.7	0.0
	10	Over 2.0	3.5	3.0	-	-
			Average		251.7	28.2

Table 1. *Cont.*

Position		Soil Depth (m)			Error Ratio (%)	
		Measured	Estimated			
		Dynamic Cone Test	Reference P-Wave Velocity	Graphical Bilinear Method	Reference P-Wave Velocity	Graphical Bilinear Method
BS	1	0.2	1.7	0.2	750.0	0.0
	2	0.9	1.4	0.9	55.6	0.0
	3	1.0	3.4	1.2	240.0	20.0
			Average		348.5	6.6
MS	1	1.1	2.4	1.0	118.2	9.1
	2	0.8	2.6	0.9	225.0	12.5
	3	1.0	2.4	1.1	140.0	10.0
			Average		161.1	10.5
TS	1	0.4	4.4	0.6	1000.0	50.0
	2	1.9	5.2	2.2	173.7	15.8
	3	0.9	4.2	0.9	366.7	0.0
			Average		513.5	21.9

Note: error ratio = (estimated value − measured value)/measured value; the numbers in Position column indicate the testing locations in Figure 3.

5. Discussion

The seismic refraction method measures the travel time of the waves refracted at the interface between different sublayers with different wave velocities or impedances. Thus, the thickness of each layer can be calculated using the wave velocities of two consecutive layers and the time intercept in the travel time-distance curve as:

$$Z_i = \frac{t_i \cdot V_i \cdot V_{i+1}}{2\sqrt{(V_{i+1}^2 - V_i^2)}} \qquad (2)$$

where Z = thickness of layer; t = time intercept; V = wave velocity; and i and $i + 1$ = different layers. The thickness of the soil layer (or the first layer) can be determined by the velocities of the first layer and the second layer (V_1 and V_2), and the time intercept (t_1). However, the clear determination of V_1, V_2, and t_1 is not easy because wave velocity in a given layer is not constant, resulting from the natural soil deposits rarely being homogeneous and wave velocity slightly increasing with an increase in depth at a given layer. Note that seismic waves propagate in a medium through connected particles, and seismic wave velocity depends on soil structure and stress condition. Seismic wave propagation is primarily affected by the stiffness of the fabric in a particular material, and thus seismic wave velocity and effective stress (σ') have a certain relationship with experimentally determined coefficients (α and β).

$$\text{Seismic wave velocity} = \alpha(\sigma')^\beta \qquad (3)$$

The α and β coefficients are dependent on the packing type (porosity and coordination number) and the contact force (Hertzian contact and Coulombic force).

The seismic wave velocity slightly increases with depth even in one layer because of the increase in effective stress (Equation (3)), which is the function of mass density and depth. However, the wave velocity changes rapidly at layer boundaries due to the change in material type, mass density, particle size, and others. Hence, the relation between seismic wave velocity and depth can produce different slopes along the depth. Figure 7 shows the seismic wave velocity with the depth at which the geophones are installed. It can be observed in Figure 7 that there are four different lines "a", "b", "c", and "d": Line "a" indicates the initial slope between wave velocity and depth; thus, it is related to the first soil layer; Line "b" indicates the second slope between wave velocity and depth; thus, it is related to the second layer. Therefore, the intersection point between lines "a" and "b" may correspond to the thickness of the first soil layer, which is point "e". Figure 8 shows the calculated soil depths based on the suggested graphical bilinear method at the selected testing sites. It is shown in Figure 8 that various soil depths can be estimated by the proposed method. The detailed soil depths based on the suggested graphical bilinear method are summarized in Table 1.

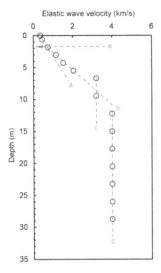

Figure 7. Method for selecting soil depth through measured elastic wave velocity. Line "a" denotes the first slope between wave velocity and depth. Lines "b", "c", and "d" represent the second, third and fourth slopes, respectively. Point "e" is the intersection point between lines "a" and "b". Note the figure is the relation between wave velocity and depth of point 1 (or 10 m distance) in the BTS line (Figure 3).

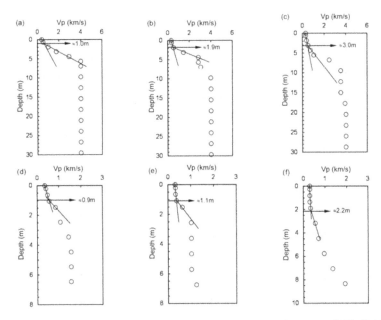

Figure 8. Estimated soil depth profile based on graphical bilinear method: (**a**) 30 m of BTS; (**b**) 80 m of BTS; (**c**) 100 m of BTS; (**d**) 20 m of BS; (**e**) 30 m of MS; and (**f**) 20 m of TS.

The measured DPI values at the selected testing sites are plotted in Figure 9 to determine soil depth based on the impaction study. Soil depth is estimated to the maximum penetrated depth at which the DPI value is nearly zero. Even though the DPI shows variation with depth within a borehole, the final penetrated depth can be easily calculated based on the depth where the value of DPI is close to zero. Figure 9c shows that DPI does not converge close to zero until a depth of 2 m. The maximum possible experiment depth is fixed to 2 m due to the rod length and energy transfer during impaction, thus the soil depth is expected to be greater than 2 m. This deeper expected soil depth can be anticipated because the soil thickness was estimated to be approximately 3.0 m based on the graphical bilinear method shown in Figure 8c for the same position as in Figure 9c. Table 1 summarizes the soil depth calculated by dynamic cone penetration tests.

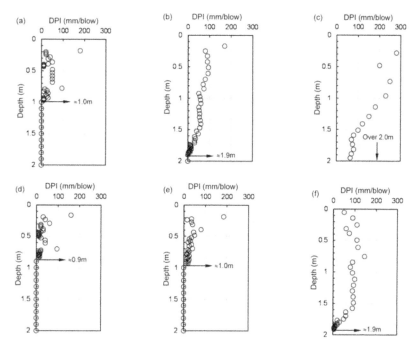

Figure 9. Soil depth based on dynamic cone penetration test: (**a**) 30 m of BTS; (**b**) 80 m of BTS; (**c**) 100 m of BTS; (**d**) 20 m of BS; (**e**) 30 m of MS; and (**f**) 20 m of TS.

Figure 10 shows the comparison of soil depths predicted by the reference P-wave velocity (0.7 km/s) and the graphical bilinear method with the measured depths using dynamic cone penetration test. Note that the depth based on the results of the dynamic cone penetration test (DPI values) has relatively high reliability because the data is gathered through direct penetration of the probe into soil; therefore, the soil depth measured by DPI can be regarded as the real soil depth. The soil depth predicted based on the reference P-wave velocity shows high variation and high soil depth compared with those based on DPI values. In contrast, the soil depth based on the graphical bilinear method shows small variation and the estimated values are comparable with the measured values by dynamic cone test, reflecting the enhanced reliability of the estimated soil depth by using the suggested method. Furthermore, the deduced soil depths based on DPI, P-wave velocity and graphical bilinear method are compared through box and whisker plot as shown in Figure 11. The median values of DPI and graphical bilinear method show 0.35 m and 0.3 m, and however the median soil depth based on P-wave velocity is highly deduced to 0.5 m. The first and third quartiles of DPI and graphical bilinear method show also similar ranges of 0.15–0.55 m and 0.2–0.6 m, respectively and it shows the difference is just 0.5 m. However, the soil depth estimated by existing method shows unreasonable ranges of 1.05–1.8 m. The ranges of minimum and maximum values based on DPI (0.2–2 m) and suggested method (0.2–3 m) are also demonstrated to similar trends while those deduced by P-wave velocity exhibited huge variations (1.4–5.6 m). And thus, the Figure 11 shows that the suggested method can provide the reliable soil depth under ≈3 m with consideration of 2 m intervals of geophone.

Table 1 shows the calculated error ratios (error ratio = (estimated value − measured value)/ measured value). The estimated soil depths based on reference P-wave velocity show high error ratios, ranging from 55% to 1000%, and the average value shows approximately over 100%, reflecting the estimation of soil depth using the classification of soil layers based on reference wave velocity is not reliable. In contrast, the graphical bilinear method shows relatively small average error ratios of 28.2, 6.6, 10.5, and 21.9% for BTS, BS, MS, and TS, respectively, demonstrating that the graphical bilinear method suggested in this study provides reliable estimation of soil depth.

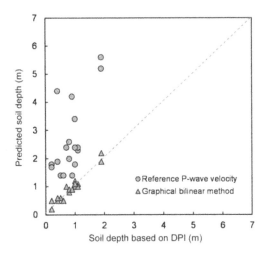

Figure 10. Comparison of soil depths estimated by seismic wave velocity and measured by dynamic cone penetration test.

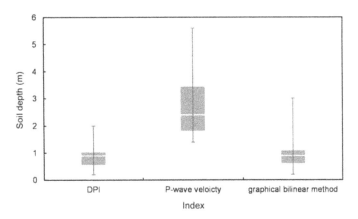

Figure 11. Comparison of soil depths through box plot.

Appl. Sci. **2018**, *8*, 600

6. Conclusions

This paper suggests methods for the determination of soil depth through seismic wave velocity measurements. The graphical bilinear method is newly introduced, and the soil depth estimated with the suggested method is compared with the result of dynamic cone penetration tests. The results of this study demonstrate the following:

- The estimated soil depth using the soil classification based on the reference P-wave velocity shows high variation and high soil depth compared with that based on dynamic cone penetration test, reflecting the estimation of soil depth based on the reference wave velocity is not reliable.
- The seismic wave velocity slightly increases with depth even in one layer and it changes rapidly at layer boundaries. Because the graphical bilinear method newly suggested in this study is based on the change in the slope between wave velocity and depth, the estimated soil depths using the suggested method are comparable with the measured values by dynamic cone test, reflecting the enhanced reliability in estimating soil depth by seismic survey.

Acknowledgments: This work was supported by the National Research Foundation of Korea (NRF) grant funded by the Korea government (MSIP) (NRF-2017R1A2B4008157).

Author Contributions: Hyunwook Choo, Hwandon Jun and Hyung-Koo Yoon performed field experiments and wrote paper together.

Conflicts of Interest: The authors declare no conflict of interest.

References

1. Kamatchi, P.; Rajasankar, J.; Iyer, N.R.; Lakshmanan, N.; Ramana, G.V.; Nagpal, A.K. Effect of depth of soil stratum on performance of buildings for site-specific earthquakes. *Soil Dyn. Earthq. Eng.* **2010**, *30*, 647–661. [CrossRef]
2. Francés, A.P.; Lubczynski, M.W. Topsoil thickness prediction at the catchment scale by integration of invasive sampling, surface geophysics, remote sensing and statistical modeling. *J. Hydrol.* **2011**, *405*, 31–47. [CrossRef]
3. Ho, J.Y.; Lee, K.T.; Chang, T.C.; Wang, Z.Y.; Liao, Y.H. Influences of spatial distribution of soil thickness on shallow landslide prediction. *Eng. Geol.* **2012**, *124*, 38–46. [CrossRef]
4. Adhikary, S.; Singh, Y.; Paul, D.K. Effect of soil depth on inelastic seismic response of structures. *Soil Dyn. Earthq. Eng.* **2014**, *61*, 13–28. [CrossRef]
5. Bao, X.; Liao, W.; Dong, Z.; Wang, S.; Tang, W. Development of Vegetation-Pervious Concrete in Grid Beam System for Soil Slope Protection. *Materials* **2017**, *10*, 96. [CrossRef] [PubMed]
6. Han, Z.; Wang, Y.; Qing, X. Characteristics Study of In-Situ Capacitive Sensor for Monitoring Lubrication Oil Debris. *Sensors* **2017**, *17*, 2851. [CrossRef] [PubMed]
7. Heimsath, A.M.; Dietrich, W.E.; Nishiizumi, K.; Finkel, R.C. Cosmogenic nuclides, topography, and the spatial variation of soil depth. *Geomorphology* **1999**, *27*, 151–172. [CrossRef]
8. Kuriakose, S.L.; Devkota, S.; Rossiter, D.G.; Jetten, V.G. Prediction of soil depth using environmental variables in an anthropogenic landscape, a case study in the Western Ghats of Kerala, India. *Catena* **2009**, *79*, 27–38. [CrossRef]
9. Trustrum, N.A.; De Rose, R.C. Soil depth-age relationship of landslides on deforested hillslopes, Taranaki, New Zealand. *Geomorphology* **1988**, *1*, 143–160. [CrossRef]
10. Kirkby, M.J. A model for the evolution of regolith-mantled slopes. In *Models Geomorphology*; Allen and Unwin: London, UK, 1985; pp. 213–237.
11. Dietrich, W.E.; Reiss, R.; Hsu, M.L.; Montgomery, D.R. A process-based model for colluvial soil depth and shallow landsliding using digital elevation data. *Hydrol. Process.* **1995**, *9*, 383–400. [CrossRef]
12. Gallipoli, M.; Lapenna, V.; Lorenzo, P.; Mucciarelli, M.; Perrone, A.; Piscitelli, S.; Sdao, F. Comparison of geological and geophysical prospecting techniques in the study of a landslide in southern Italy. *Eur. J. Environ. Eng. Geophys.* **2000**, *4*, 117–128.
13. De Vita, P.; Agrello, D.; Ambrosino, F. Landslide susceptibility assessment in ash-fall pyroclastic deposits surrounding Mount Somma-Vesuvius: Application of geophysical surveys for soil thickness mapping. *J. Appl. Geophys.* **2006**, *59*, 126–139. [CrossRef]

Appl. Sci. **2018**, *8*, 600

14. Min, D.H.; Park, C.H.; Lee, J.S.; Yoon, H.K. Estimating Soil Thickness in a Debris Flow using Elastic Wave Velocity. *J. Eng. Geol.* **2016**, *26*, 143–152. [CrossRef]

15. *Standard Test Method for Sieve Analysis of Fine and Coarse Aggregates*; ASTM, C136; ASTM: West Conshohocken, PA, USA, 1984.

16. Mohammadi, S.D.; Nikoudel, M.R.; Rahimi, H.; Khamehchiyan, M. Application of the Dynamic Cone Penetrometer (DCP) for determination of the engineering parameters of sandy soils. *Eng. Geol.* **2008**, *101*, 195–203. [CrossRef]

17. Brough, M.; Stirling, A.; Ghataora, G.; Madelin, K. Evaluation of railway trackbed and formation: A case study. *NDT & E Int.* **2003**, *36*, 145–156.

18. Salgado, R.; Yoon, S. Dynamic cone penetration test (DCPT) for subgrade assessment. *Jt. Transp. Res. Program* **2003**, *73*, 20–28.

19. *Standard Test Method for Use of the Dynamic Cone Penetrometer in Shallow Pavement Applications*; ASTM D6951/D6951M-09; ASTM: West Conshohocken, PA, USA, 2015.

20. Reynolds, J.M. *An Introduction to Applied and Environmental Geophysics*; Wiley: New York, NY, USA, 2003.

21. Lee, K.M.; Kim, H.; Lee, J.H.; Seo, Y.S.; Kim, J.S. Analysis on the Influence of Groundwater Level Changes on Slope Stability using a Seismic Refraction Survey in a Landslide Area. *J. Eng. Geol.* **2007**, *17*, 545–554.

22. Hong, W.P.; Kim, J.H.; Ro, B.D.; Jeong, G.C. Case Study on Application of Geophysical Survey in the Weathered Slope including Core Stones. *J. Eng. Geol.* **2009**, *19*, 89–98.

applied
sciences

Article

A CFD Results-Based Approach to Investigating Acoustic Attenuation Performance and Pressure Loss of Car Perforated Tube Silencers

Hao Zhang, Wei Fan and Li-Xin Guo *

School of Mechanical Engineering and Automation, Northeastern University, Shenyang 110819, China; peterat2011@foxmail.com (H.Z.); 1410110@stu.neu.edu.cn (W.F.)
* Correspondence: lxguo@mail.neu.edu.cn

Received: 1 March 2018; Accepted: 29 March 2018; Published: 2 April 2018

Abstract: This paper proposes an approach to investigating the effect of different temperatures and flow velocities on the acoustic performance of silencers in a more accurate and meticulous fashion, based on steady computational results of the flow field inside the silencer using computational fluid dynamics (CFD). This approach can transfer the CFD results—including temperature and flow velocity distribution—to acoustic meshes by mesh mapping. A numerical simulation on the sound field inside the silencer is then performed, using the CFD results as a boundary condition. This approach facilitates the analysis of complex silencer designs such as perforated tube silencers, and the numerical predictions are verified by a comparison with available experimental data. In the case of the three-pass perforated tube silencer of a car, the proposed approach is implemented to calculate the transmission loss (*TL*) of the silencer at different temperatures and flow velocities. We found that increasing the air temperature shifts the TL curve to a higher frequency and reduces the acoustic attenuation at most frequencies. As the air flow increases, the curve moves to a slightly lower frequency and the acoustic attenuation increases slightly. Additionally, the pressure loss of perforated tube silencers could be calculated according to the total pressure distribution of their inlet and outlet from the steady computational results using CFD.

Keywords: perforate tube silencer; transmission loss (*TL*); pressure loss; computational fluid dynamics (CFD); temperature; air flow velocity

1. Introduction

The silencer is the main device used for suppressing automobile noise. The most important goal for a high-performance silencer is to simultaneously have good aerodynamic and acoustic attenuation characteristics. However, these often contradict each other. Perforated tube silencers are well-balanced in terms of both characteristics. Thus, they are applied extensively to the exhaust systems of automobiles.

Pressure loss is the main indicator used for evaluating the aerodynamic performance of a silencer. It has a negative influence on the efficiency of an engine. If the loss exceeds the backpressure limit, it will result in reduced engine power and an increase in fuel consumption [1]. Currently, the pressure loss is predicted by performing a 3D steady computation using computational fluid dynamics (CFD). This method has been widely accepted among researchers because of its high level of accuracy and adaptability [2–4]. Middelberg et al. [5] computed the pressure loss of a simple expansion chamber muffler using a CFD simulation. Lee et al. [6] predicted the pressure loss of concentric tube silencers with five different patterns of perforated elements using CFD analysis. Ren et al. [7] employed the CFD approach to predict the pressure loss of an exhaust muffler, which was influenced by the insert duct, the position of the baffle, and the inlet air velocity.

Transmission loss (*TL*) is one of the most important indicators used for evaluating the acoustic attenuation performance of a silencer. The gas from an automotive engine has a high temperature and speed, which will have a strong influence on the acoustic attenuation performance. High temperature changes, gas density, sound velocity and acoustic impedance [8], and the effects of flow velocity embody two sides: one is to affect the sound wave propagation in a medium, and the other is to reduce aerodynamic noise resulting from turbulence [9]. Various theoretical and experimental studies were conducted in the past in order to investigate the acoustic attenuation performance of a silencer under the influence of different temperatures and air flow [10–12]. Kim et al. [13] investigated the acoustic characteristics of an expansion chamber with a constant mass flow and a steady temperature gradient. Tsuji et al. [14] applied finite element and boundary element methods to evaluate the acoustic wave transmission characteristics in a medium with uniform and steady mean flow. Kirby [15] compared two numerical methods and two analytical methods of modelling automotive dissipative silencers with a uniform mean gas flow of Mach number (M). The comparison indicated a close similarity to the *TL* predictions that were obtained for the silencers that were examined. With the rapid development of high-performance computers, the 3D numerical simulation method plays an increasingly significant role in sound field analysis. Broatch et al. [16] proposed a 3D time-domain technique based on the CFD approach to calculate the *TL* of exhaust mufflers with different chambers. Sánchez-Orgaz et al. [17] proposed a hybrid finite element approach—combining an acoustic velocity potential formulation in the central airway with a pressure-based wave equation in the outer chamber—to study the *TL* of perforated dissipative silencers with heterogeneous properties in the presence of mean flow. Dong et al. [18] employed the CFD approach to perform a 3D steady computation and obtain a temperature distribution inside a two-pass perforated tube silencer, and then defined the elements with a 5 °C difference in temperature as a collection. The air parameters of the corresponding temperatures were assigned to the defined collections using SYSNOISE software, in order to calculate the *TL*.

However, most of the present works associated with predicting the *TL* of perforated tube silencers assume the mean flow to be uniform. Similarly, the temperature fields are simplified as constant or linear fields, which may result in inaccurate analytical results—especially in the case of silencers with complicated temperature and flow velocity distributions. In light of the aforementioned disadvantages, this paper proposes an approach based on CFD results to investigate the acoustic attenuation performance of silencers. This approach comprises the following steps: (1) the flow field inside the silencer is calculated by performing a 3D steady computation using CFD; (2) the CFD results—including temperature and air flow velocity—are transferred to acoustic meshes by mesh mapping, so that the results can be used as a boundary condition of sound field analysis; (3) the *TL* of the silencer is calculated at a high temperature and air flow. Compared to previous works, the proposed approach avoids the use of simplified temperature and flow velocity distributions. Consequently, the effects of temperature variation and air flow on the acoustic attenuation performance of silencers can be calculated and observed in greater detail.

This paper proceeds as follows: Following the introduction, Section 2 describes the proposed computational approach and verifies its accuracy. Section 3 builds an internal fluid model of a perforated tube silencer of a car, and generates CFD meshes and acoustic meshes of the model, respectively. Following this, a 3D steady computation using CFD is performed in order to calculate the pressure loss of the silencer and obtain its temperature and flow velocity distributions. Section 4 employs the proposed approach to investigate the acoustic attenuation performance of the silencer under the influence of different temperatures and air flow. Section 5 concludes the study.

2. Methods

2.1. Mesh Mapping

The most critical step in the proposed approach to calculating the acoustic attenuation of silencers is to set a mesh mapping for transferring data. The mesh mapping is used to transfer the node data of the CFD mesh (source mesh) into the acoustic mesh (target mesh). However, there is not usually a one-to-one correspondence between the nodes of different meshes. Thus, an appropriate mapping algorithm should be employed.

A maximum distance algorithm is often applied in Virtual.lab acoustics software to set a mesh mapping between different meshes with the same geometrical shape, but a different density of nodes. The algorithm includes the following two necessary parameters [19]:

1. Number of nodes (N): The maximum number of nodes from the source mesh that are considered for mapping with one node of the target mesh.
2. Maximum distance (R): Only the nodes of the source mesh that lie inside a sphere with a radius R—centered at the node of the target mesh—are taken into account.

The N closest nodes to a given source node are used to transfer data to the target node. The data value assigned to the target node is a weighted average of the values at the N source nodes. The weights are:

$$W_i = \frac{1}{d_i} \bigg/ \sum_{i=1}^{N} \frac{1}{d_i}. \tag{1}$$

The transferred value of the target node is then:

$$P_{\text{Target}} = \sum_{i=1}^{N} \frac{P_i^{\text{Source}}}{d_i} \bigg/ \sum_{i=1}^{N} \frac{1}{d_i}, \tag{2}$$

where d_i is the distance between the source node and the target node, and P_i^{Source} is the value of the source node.

For example, when the target node is defined as a center, there are three source nodes ($N = 3$) lying inside a sphere with $R = 100$ mm. Thus, the mapping data transfer relation is depicted in Figure 1. The value at A is given by:

$$P_{\text{Target}} = \frac{\frac{1}{d_1} P_1^{\text{Source}} + \frac{1}{d_2} P_2^{\text{Source}} + \frac{1}{d_3} P_3^{\text{Source}}}{\frac{1}{d_1} + \frac{1}{d_2} + \frac{1}{d_3}}. \tag{3}$$

After transferring data by the mesh mapping method, Virtual.lab software is used to conduct a numerical simulation on the internal sound field of the silencer, and the TL is then determined by

$$TL = 10 \log_{10} \left(\frac{W_1}{W_2} \right), \tag{4}$$

with

$$W_1 = |p_1|^2 A_1 / Z_1, W_2 = |p_2|^2 A_2 / Z_2. \tag{5}$$

Substituting (5) into (4) yields

$$TL = 10 \log_{10} \left(\frac{p_1 \overline{p_1} Z_2 A_1}{p_2 \overline{p_2} Z_1 A_2} \right), \tag{6}$$

where W_1 and W_2 are the acoustic power of the inlet and outlet of the silencer, respectively; p_1 and p_2 are the sound pressure of the incident and the transmitted waves, respectively; $\overline{p_1}$ and $\overline{p_2}$ are conjugate complex numbers of p_1 and p_2, respectively; Z_1 and Z_2 are the acoustic impedance of the inlet and outlet of the silencer, respectively; and A_1 and A_2 are the cross-sectional areas of the inlet and outlet of the silencer, respectively.

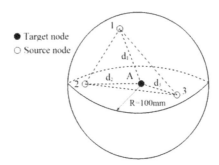

Figure 1. Sketch of data transfer (Number of nodes $N = 3$, Maximum distance $R = 100$ mm).

2.2. Method Validation

In order to verify the accuracy of the proposed approach, a straight-through perforated tube silencer with two different perforated patterns that was proposed by the authors of [20,21] and a cross-flow perforated tube silencer that was proposed by the authors of [10] was considered. Previously, Liu et al. [22] employed the 3D time-domain CFD approach to calculate the *TL* of two perforated tube silencers at different flow velocities and temperatures, and their predictions were verified by experimental data. Moreover, they found that the distribution of flow velocity inside the silencer was anisotropic and nonhomogeneous. Therefore, it is not sufficiently accurate to consider the actual gas flow as the mean flow. The approach proposed in this paper avoids this problem effectively.

Figure 2 presents the straight-through perforated tube silencer. The diameters of the inner and outer cavities are $d = 32$ mm and $D = 110$ mm, respectively; the length of the silencer is $l = 200$ mm; the thickness of the wall is 2 mm; and the diameter of the hole and the porosityare $dh = 4$ mm and $\sigma = 4.7\%$, respectively, for Pattern 1, and $dh = 8$ mm and $\sigma = 14.7\%$, respectively, for Pattern 2. The cross-flow perforated silencer is illustrated in Figure 3. The diameters of the inner and outer cavities are $d = 49.3$ mm and $D = 101.6$ mm, respectively; the lengths of tubes on both sides of the baffle are $l_1 = l_2 = 128.6$ mm; each tube is perforated with 160 orifices, with a porosity of 3.9%, a diameter of 2.49 mm; and a wall thickness of 0.81 mm.

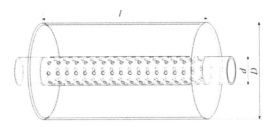

Figure 2. Straight-through perforated tube silencer.

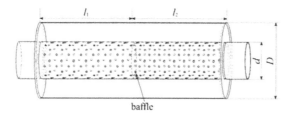

Figure 3. Cross-flow perforated tube silencer.

The proposed approach is applied to calculate the *TL* for the silencers under different boundary conditions, and the predictions were compared with the available measurement results. Figure 4a and 4b compare the *TL* curves of the straight-through perforated tube silencer for Pattern 1 and Pattern 2, respectively. Considering the predictions and measurements at an air flow of $M = 0.1$ and a temperature of $T = 288$ K, it is evident that the predictions are in line with the experimental data. The predicted and measured TL curves for Pattern 1, with $M = 0.2$ and $T = 288$ K, are presented in Figure 5, and it is evident that these predictions are also in line with the experimental data. Figure 6 presents a comparison of the numerical results and the measurements for the cross-flow silencer with an air flow velocity of $v = 17$ m/s and $T = 347$ K, which again presents an excellent agreement.

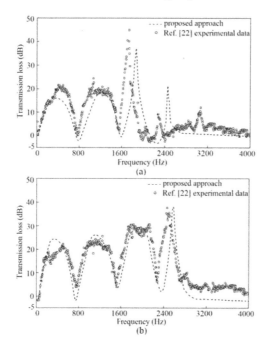

Figure 4. Comparison of the predicted and measured transmission loss (*TL*) for the straight-through perforated tube silencer (Mach number $M = 0.1$, Temperature $T = 288$ K): (**a**) Pattern 1; (**b**) Pattern 2.

Based on the above comparisons, it is evident that the proposed approach displayed a high degree of accuracy in investigating the acoustic attenuation performance of perforated tube silencers. This high degree of accuracy should be attributed to the avoidance of using simplified temperature fields and air flow in the calculation process. In doing so, the actual working condition of the silencer can be better reflected. Additionally, it must be stated that the computations may be highly time-consuming because there are too many orifices in the silencers, which increases the amount of mesh.

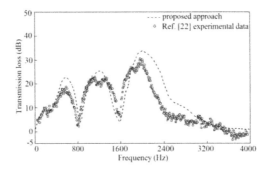

Figure 5. Comparison of the predicted and measured *TL* for Pattern 1 ($M = 0.2$, $T = 288$ K).

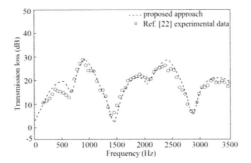

Figure 6. Comparison of the calculated and measured *TL* for the cross-flow silencer ($v = 17$ m/s, $T = 374$ K).

3. Modeling and Steady Computation Using CFD

3.1. Modeling

The three-pass perforated tube silencer of a car that is considered in this paper is presented in Figure 7. The silencer is divided into three chambers by two baffles. The inlet tube and the middle buffer tube are respectively perforated by 40 and 48 orifices with 4 mm diameter, and the porosity of the tubes are 2.7% and 5.8%, respectively. There is no perforation in the outlet tube. Additionally, the axes of the three tubes are not on the same plane. Consequently, it is difficult to build a fluid model directly inside the silencer because of its complicated structure. Therefore, firstly CATIA (version V5R20, Dassault Systemes, Paris, France, 2010) software was used to build its structural model, and then it was filled using ANSYS Workbench (version 14.5, ANSYS Inc., Canonsburg, PA, USA, 2012) software to obtain the fluid model. Following this, the fluid model was split into several parts to generate mesh individually. Tetrahedral mesh was applied to the perforation area and the transition tube, and hexahedral mesh was applied to the remaining parts.

Figure 8a,b illustrate the CFD mesh model and the acoustic mesh model of the fluid model, which will be used for CFD steady computation and acoustic computation, respectively. There is a different node density between the two models. The CFD mesh is refined in the perforation area in order to obtain more accurate results. However, the accuracy of the acoustic computation is determined by the overall mesh. Local mesh refinement can not improve its accuracy, and thus the size of the acoustic mesh element should be kept as uniform as possible.

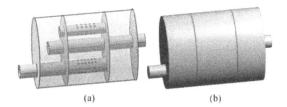

Figure 7. 3D geometrical model for the three-pass perforated tube silencer: (**a**) structural model; (**b**) flow field model.

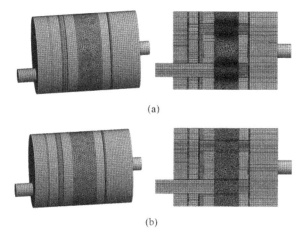

Figure 8. Mesh model for the three-pass perforated tube silencer: (**a**) computational fluid dynamics (CFD) mesh model; (**b**) acoustic mesh model.

The steady flow computation is carried out using the CFD mesh model. The governing equations for pressure velocity coupling—based on a finite volume method—are solved by the Semi-Implicit Method for Pressure Linked Equations (SIMPLE) algorithm. Turbulence is examined using the standard k-epsilon model [23]. The fluid material is air—with the density conforming to the ideal gas law—and is considered as incompressible. The boundary conditions are concretely set as follows: (1) the velocity inlet boundary condition is defined at the inlet of the silencer. According to the information provided by the manufacturer, when the engine runs at 5500 r/min, the velocity and temperature of the inlet are 55 m/s and 760 K, respectively; (2) the pressure outlet boundary condition is defined at the outlet of the silencer, and the gauge pressure is 0 Pa—which is relative to one standard atmospheric pressure; and, (3) the walls are assumed to be stationary—with no slip condition—and adiabatic.

3.2. The CFD Results

Figure 9 shows the temperature distribution in the axial cross sections of the three tubes of the silencer. Overall, the temperature value decreases along the direction of the air flow. There is little difference in temperature between the first and second chambers, where the value is approximately 574 K. However, in the third chamber, the value increases to approximately 670 K. The highest temperature—which is approximately 751 K—occurs in the inlet tube. The temperature in the outlet tube is close to the value of the first chamber, and it is distributed evenly. A greater temperature gradient exists in the perforation area. Additionally, the temperature of the edge chamber is the lowest, which results from the large difference in temperature between the inner and outer walls.

Figure 10 depicts the velocity distribution in the axial cross sections of the three tubes of the silencer. When air flows through the perforation area of the inlet tube and the middle buffer tube, some of the air will enter into the second chamber through the orifices. At this moment, an eddy forms in the perforation area (as shown in Figure 11) because of the disturbance of the airflows with different velocities and flow directions, which may induce a greater velocity gradient in the perforation area. Additionally, there exists a greater velocity gradient next to the nozzle of each tube, resulting from the sudden change of the flow sections.

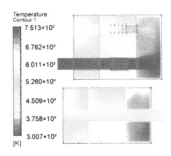

Figure 9. Contour of temperature in the cross section of the silencer.

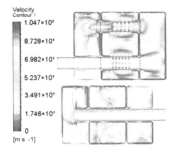

Figure 10. Contour of velocity in the cross section of the silencer.

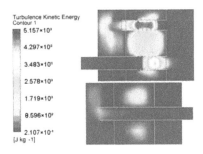

Figure 11. Contour of turbulence kinetic energy in the cross section of the silencer.

Figure 12 illustrates the total pressure distribution in the axial cross sections of the three tubes of the silencer. The highest pressure occurs in the inlet tube—especially in the perforation area—where the value is approximately 23,660 Pa. In the second and third chambers, the pressure is reduced to approximately 19,560 Pa. The middle buffer tube plays the role of the transition region with a greater pressure gradient. Gradually, however, the pressure distribution tends to be uniform in the first chamber. The pressure is distributed more evenly in the outlet tube than in the other regions, and the value is approximately −2313 Pa (the pressure, which is lower than the atmospheric pressure, may be a result of the Venturi effect).

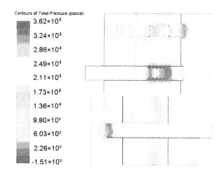

Figure 12. Contour of total pressure in the cross section of the silencer.

Based on the CFD results, it is evident that the distributions of temperature and flow velocity are non-uniform. Therefore, it is not accurate enough to consider them as uniform distributions.

3.3. Pressure Loss Calculation

The pressure loss equals the total pressure difference between the inlet and the outlet of a silencer. Following the method of calculating the pressure loss that was proposed by the authors of [24]—whose accuracy has been validated by experiments—nine evenly distributed points were selected in the cross sections of the inlet and the outlet, respectively. The average total pressure of the nine points was considered to be the total pressure of the corresponding cross section. Therefore, the pressure loss could be expressed as

$$\Delta p = \overline{p_{t_1}} - \overline{p_{t_2}}, \tag{7}$$

where $\overline{p_{t_1}}$ is the total pressure of the inlet, and $\overline{p_{t_2}}$ is the total pressure of the outlet.

Figure 13 presents the cross section of the silencer in question for the pressure loss calculation. The total pressure distribution—predicted by the steady computation using CFD—is applied to obtain the pressure loss of the silencer. The value of the pressure loss is approximately 37,710 Pa. The following explanation is a plausible explanation for the pressure loss that occurred: When air flow moves from the chamber to the tube—or vice versa—the mechanical energy of the air flow is reduced dramatically because of the sudden change of the flow sections. Moreover, there are large-scale eddy zones present in the perforation area that restrict air flow and lead to energy dissipation, which results from the increased turbulence kinetic energy (as shown in Figure 11).

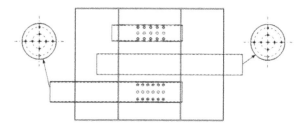

Figure 13. The selected points for the total pressure calculation in the cross sections of the inlet and outlet of the silencer.

4. Sound Field

4.1. Check the Maximum Frequency Value of the Acoustic Mesh

When there is mean flow, the maximum frequency value that is allowed by the acoustic mesh is defined by

$$f_{max} = \frac{c(1 - M)}{6L}, \tag{8}$$

where c is the local sound velocity, M is the mean flow Mach number, and L is the mesh element size [19]. Therefore, the calculated frequency must satisfy the condition $f \leq f_{max}$ in order to guarantee accuracy during the acoustic calculation. This means that the maximum size of the acoustic mesh should be small enough to allow at least six elements to fit into the wavelength of the calculated frequency.

The CFD results that were acquired in Section 3.2—including the temperature and flow velocity—are transferred to the acoustic mesh (presented in Figure 8b) by mesh mapping. Following the data transfer, the maximum frequency value of the investigated acoustic mesh was obtained through checking the maximum frequency report in Virtual.Lab. The observed value is 3676.1 Hz. The noise frequency that needs to be calculated for the silencer is usually below 3000 Hz. Therefore, the investigated acoustic mesh could guarantee computational accuracy.

4.2. Effects of Temperature Changes on the TL of the Silencer

Prior to performing the acoustic computation, the inlet of the silencer is supplied with a unit vibration velocity source. Additionally, an anechoic end duct property is defined at the end of the outlet to simulate the anechoic termination condition. It is known that the existing acoustic mesh is valid up to a frequency of 3676.1 Hz. Therefore, the analysis is conducted from 20 Hz to 2000 Hz, in linear steps of 10 Hz. Additionally, an input point and an output point are chosen from the inlet and the outlet of the silencer, respectively. Figure 14 presents the sound pressure level of the inlet and outlet of the silencer, calculated by the proposed approach. Figure 14a–c present the sound pressure level of the inlet and outlet of the silencer with the same flow velocity but different temperatures.

Based on the sound pressure of the inlet and the outlet, the TL of the silencer is calculated using Equation (6). Figure 15 indicates the TL predictions with the same flow velocity (55 m/s) and different temperatures. It may be noted that, as the temperature rises, the TL curves move to a higher frequency and the acoustic attenuation is reduced at most frequencies. Considering the TL curves from Figure 15a as an example, a valley at 850 Hz is moved to 1160 Hz with an increase in temperature from 560 K to 760 K. Similarly, a peak moves from 1240 Hz to 1560 Hz, while their corresponding TL values are reduced by 10 dB and 20 dB, respectively.

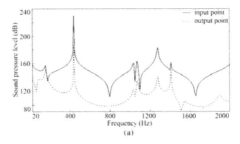

Figure 14. *Cont.*

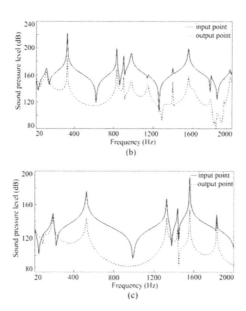

Figure 14. The sound pressure level of the inlet and outlet of the silencer (v = 55 m/s): (**a**) T = 760 K; (**b**) T = 560 K; (**c**) T = 960 K.

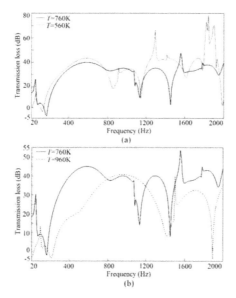

Figure 15. Effects of temperature on the TL of the silencer (v = 55 m/s) : (**a**) 760 K versus 560 K; (**b**) 760 K versus 960 K.

4.3. Effects of Flow Velocity Changes on the TL of the Silencer

It is evident from the flow velocity distribution in Figure 10 that the velocity inside the silencer is lower. Therefore, it is assumed that the air flow mainly influences the sound wave propagation, and the turbulence noise is neglected [25]. Following the proposed approach, the sound pressure level curves of the inlet and outlet of the silencer—with the same temperature but different flow velocities—are presented in Figure 16.

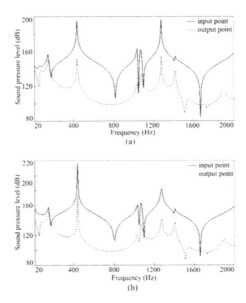

Figure 16. The sound pressure level of the inlet and outlet of the silencer (T = 760 K): (**a**) v = 20 m/s; (**b**) v = 35 m/s.

Figure 17 indicates the effects of increasing the flow velocity on the *TL* predictions at the same temperature (760 K). It can be observed that, as the velocity goes up, the *TL* curves move to a slightly lower frequency and the acoustic attenuation is increased slightly at most frequencies. The variation of the TL curve may be attributed to the fact that the higher velocity increases the acoustic resistance of the perforations and reduces the effective flow area of the orifices [26]. This causes the frequency response of the silencer to differ from a static state. Generally, however, the propagation of the sound wave is only changed dramatically if the air flow Mach number is greater than 0.3 [27]. Therefore, the variation of the curves in Figure 17 is small.

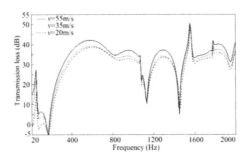

Figure 17. Effects of flow velocity on the *TL* of the silencer (T = 760 K).

Appl. Sci. **2018**, *8*, 545

5. Conclusions

This paper proposes a new approach—based on CFD results—to investigating the effects of temperature and flow velocity on the acoustic attenuation performance of a perforated tube silencer of a car. The validation results suggest that the proposed approach can make accurate predictions. In actual fact, the proposed approach is distinguished from the existing approaches by defining the fluid properties of every acoustic mesh element under the CFD results (including temperature and flow velocity). In doing so, the simplification of the temperature and flow velocity distribution is avoided, thus the effect can be predicted in greater detail. The results indicate that air flow velocity and temperature variation may have an effect on the investigated three-pass perforated tube silencer. An increase in temperature will shift the TL curve to a higher frequency and reduce the acoustic attenuation at most frequencies. As flow velocity increases and the curve is moved to a slightly lower frequency, the acoustic attenuation is enhanced, although the tendency is not particularly obvious. Additionally, the pressure loss of the perforated tube silencer can be calculated through the total pressure distribution from the CFD results. When the engine runs at 5500 r/min, the calculated value is approximately 37,710 Pa.

Acknowledgments: This work was supported by the National Natural Science Foundation of China (51275082, 11272273).

Author Contributions: Li-Xin Guo conceived and designed the research idea and the framework; Wei Fan performed the simulations; Hao Zhang analyzed the data and wrote the paper.

Conflicts of Interest: The authors declare no conflict of interest.

References

1. Shrivastava, A.K.; Tewari, V.K.; Santosh, K. Effect of Exhaust Back Pressure on Noise Characteristic of Tractor Mufflers. *AMA Agric. Mech. Asia* **2014**, *45*, 79–83.
2. Panigrahi, S.N.; Munjal, M.L. Backpressure Considerations in Designing of Cross Flow Perforated-element Reactive Silencers. *Noise Control Eng. J.* **2007**, *55*, 504–515. [CrossRef]
3. Bouldin, B.; Vunnam, K.; Hernanz-Manrique, J.A.; Ambit-Marin, L. CFD Analysis and Full Scale Testing of a Complex Auxiliary Power Unit Intake System. In Proceedings of the ASME Turbo Exposition, Vancouver, BC, Canada, 6–10 June 2011.
4. Chen, W.X.; Chong, D.T.; Yan, J.J.; Dong, S.C.; Liu, J.P. Numerical Investigation of Two-Phase Flow in Natural Gas Ejector. *Heat Transf. Eng.* **2014**, *35*, 738–745. [CrossRef]
5. Middelberg, J.M.; Barber, T.J.; Byme, K.P.; Leong, S.S.; Leonadi, E. *CFD Analysis of the Acoustic and Mean Flow Performance of Simple Expansion Chamber Mufflers*; ASME Paper No. IMECE2004-61371; ASME: New York, NY, USA, 2004.
6. Lee, S.H.; Ih, J.G. Effect of Non-Uniform Perforation in the Long Concentric Resonator on Transmission Loss and Back Pressure. *J. Sound Vib.* **2008**, *311*, 280–296. [CrossRef]
7. Ren, J.W.; Jiang, Q.Y.; Wang, Z. CFD Simulation and Computation of Pressure Loss of Resistance Muffler. In Proceedings of the 4th International Conference on Manufacturing Science and Engineering, Dalian, China, 30–31 March 2013.
8. Wylen, G.J.V.; Sonntag, R.E.; Borgnakke, C. *Fundamentals of Classical Thermodynamics*; Wiley: New York, NY, USA, 1985.
9. Wang, C.N.; Tse, C.C.; Chen, S.C. Flow Induced Aerodynamic Noise Analysis of Perforated Tube Mufflers. *J. Mech.* **2013**, *29*, 224–231. [CrossRef]
10. Sullivan, J.W. A Method for Modeling Perforated Tube Muffler Components: I. Theory; II. Applications. *J. Acoust. Soc. Am.* **1979**, *66*, 772–788. [CrossRef]
11. Sujith, R.I. Transfer Matrix of a Uniform Duct with an Axial Mean Temperature Gradient. *J. Acoust. Soc. Am.* **1996**, *100*, 2540–2542. [CrossRef]
12. Selamet, A.; Ji, Z.L. Acoustic Attenuation Performance of Circular Chambers with Single-inlet and double-outlet. *J. Sound Vib.* **2000**, *299*, 3–19. [CrossRef]

Appl. Sci. **2018**, *8*, 545

13. Kim, Y.H.; Choi, J.W.; Lim, B.D. Acoustic characteristics of an expansion chamber with constant mass flow and steady temperature gradient (theory and numerical simulation). *ASME J. Vib. Acoust.* **1990**, *112*, 460–467. [CrossRef]

14. Tsugi, T.; Tsuchiya, T.; Kagawa, Y. Finite Element and Boundary Element Modeling for the Acoustic Wave Transmission in Mean Flow Medium. *J. Sound Vib.* **2002**, *255*, 849–866. [CrossRef]

15. Kirby, R. A Comparison between Analytic and Numerical Methods for Modelling Automotive Dissipative Silencers with Mean Flow. *J. Sound Vib.* **2009**, *325*, 565–582. [CrossRef]

16. Broath, A.; Margot, X.; Gila, A. A CFD Approach to the Computation of the Acoustic Response of Exhaust Mufflers. *J. Comput. Acoust.* **2005**, *13*, 301–316. [CrossRef]

17. Sanchez-Orgaz, E.M.; Denia, F.D.; Martinez-Casas, J.; Baeza, L. 3D Acoustic Modelling of Dissipative Silencers with Nonhomogeneous Properties and Mean Flow. *Adv. Mech. Eng.* **2014**, *6*, 537935. [CrossRef]

18. Dong, H.L.; Deng, Z.X.; Lai, F. Analysis and Improvement of Muffler Acoustic Performance Considering Temperature Influence. *J. Vib. Eng.* **2009**, *22*, 70–75.

19. LMS Virtual. Lab. *LMS Virtual. Lab Online Help*; LMS International N.V.: Leuven, Belgium, 2013; Volume 11.

20. Guo, L.X.; Fan, W. A comparison between various numerical simulation methods for predicting the transmission loss in silencers. *J. Eng. Res.* **2017**, *5*, 163–180.

21. Lee, S.H.; Ih, J.G. Empirical Model of the Acoustic Impedance of a Circular Orifice in Grazing Mean Flow. *J. Acoust. Soc. Am.* **2003**, *114*, 98–113. [CrossRef] [PubMed]

22. Liu, C.; Ji, Z.L. Computational Fluid Dynamics-Based Numerical Analysis of Acoustic Attenuation and Flow Resistance Characteristics of Perforated Tube Silencers. *ASME J. Vib. Acoust.* **2014**, *136*, 21006. [CrossRef]

23. *ANSYS Fluent 12.1 in Workbench User's Guide*; ANSYS Worbench 12.1; ANSYS Inc.: New York, NY, USA, 2009.

24. Fang, J.H.; Zhou, Y.Q.; Hu, X.D.; Lin, Z.S. Study on Aerodynamic Quality and Fluid Simulation of Expansion Mufflers. *J. Syst. Simul.* **2009**, *21*, 6399–6404.

25. Pan, D.Y.; Sun, J.Q.; Fang, D.Q.; Dong, J.Y. Research and Application on Flow Characteristics of Muffler. *Noise Vib. Control* **1984**, *5*, 9–14.

26. Karlsson, M.; Abomb, M. Aeroacoustics of T-Junctions—An Experimental Investigation. *J. Sound Vib.* **2010**, *329*, 1793–1808. [CrossRef]

27. Zhan, F.L.; Xu, J.W. *Virtual. Lab Acoustics Simulation from Entry to Master*; Northwestern Polytechnical University Press: Xi'an, China, 2013.

MDPI

Article

A Pseudo-3D Model for Electromagnetic Acoustic Transducers (EMATs)

Wuliang Yin [1,*], Yuedong Xie [2,*], Zhigang Qu [1] and Zenghua Liu [3]

[1]   College of Electronic Information and Automation, Tianjin University of Science & Technology, 1038 DaguNan Road, Hexi District, Tianjin 300222, China; zhigangqu@tust.edu.cn

[2]   School of Electrical and Electronic Engineering, University of Manchester, 60 Sackville Street, Manchester M13 9PL, UK

[3]   College of Mechanical Engineering and Applied Electronics Technology, Beijing University of Technology, 100 Pingleyuan, Beijing 100124, China; liuzenghua@bjut.edu.cn

*   Correspondence: wuliang.yin@gmail.com (W.Y.); yuedong.xie@manchester.ac.uk (Y.X.); Tel.: +86-15002257471 (W.Y.); +44-759-844-9793 (Y.X.)

Received: 31 January 2018; Accepted: 9 March 2018; Published: 15 March 2018

Abstract: Previous methods for modelling Rayleigh waves produced by a meander-line-coil electromagnetic acoustic transducer (EMAT) consisted mostly of two-dimensional (2D) simulations that focussed on the vertical plane of the material. This paper presents a pseudo-three-dimensional (3D) model that extends the simulation space to both vertical and horizontal planes. For the vertical plane, we combines analytical and finite-difference time-domain (FDTD) methods to model Rayleigh waves' propagation within an aluminium plate and their scattering behaviours by cracks. For the horizontal surface plane, we employ an analytical method to investigate the radiation pattern of Rayleigh waves at various depths. The experimental results suggest that the models and the modelling techniques are valid.

Keywords: Analytical solutions; FDTD; EMATs; beam directivity

---

1. Introduction

A wide group of non-destructive testing (NDT) techniques are commonly used in biomedical industries, such as ultrasonic techniques, electromagnetic techniques, and laser testing [1–4]. Due to the non-contact nature, more and more attention has been paid to the NDT technique with electromagnetic acoustic transducers (EMATs), and EMATs have gradually been used in industrial applications, such as thickness gauging and defect detection [5–8].

A classic EMAT sensor is made of a meander-line-coil and a permanent magnet (Figure 1). There are two major coupling principles for EMATs: magnetostriction is for ferromagnetic metallic materials, and the Lorentz force mechanism is for conductive and ferromagnetic materials [9]. This work focussed on only the Lorentz force mechanism performing on an aluminium plate. The Lorentz force mechanism is: the meander-line-coil placed above the sample generates eddy currents **J** within the sample. A permanent magnet placed above the coil generates a static magnetic field **B** to the sample. The interaction between **J** and **B** produces Lorentz force density **F**, as shown in Equation (1):

$$\mathbf{F} = \mathbf{J} \times \mathbf{B} \tag{1}$$

Substantial works have been reported on EMATs modelling, which comprises an electromagnetic (EM) model and an ultrasonic (US) model [10–12]. The EM model was accomplished by the finite element method (FEM) and the analytical method, while the US model was accomplished with FEM, the finite-difference time-domain method (FDTD), and the analytical solutions. Some of the previous work modelled EMATs by combining FEM and analytical solutions, i.e., FEM for EM modelling, and analytical solutions for US modelling [10–12]. On the other hand, some of the previous work utilised FEM for both EM and US modelling, i.e., COMSOL (a commercial EM simulation package) for EM modelling, and Abaqus for US modelling [13,14]. Authors have proposed several methods to model EMATs, including a method combining FEM and FDTD, a method combining analytical solutions and FDTD, and a wholly analytical method [15–18].

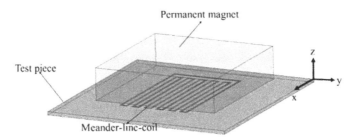

Figure 1. The configuration of a typical meander-line-coil electromagnetic acoustic transducer (EMAT). Reproduced with permission from [15], IEEE, 2016.

Most of EMATs' simulations were two-dimensional (2D), and can only focus on the specific plane. This article is attempting to build a pseudo-three-dimensional (3D) model in order to further study EMATs by combining a surface plane 2D model (the *x-y* cross-section shown in Figure 1) and a vertical plane 2D model (the *y-z* cross-section shown in Figure 1) together. More specifically, the Lorentz force density obtained from the vertical plane of the sample is imported to the surface plane of the sample as the driving source to generate Rayleigh waves. Previously, only the beam directivity at the surface of the sample ($z = 0$ in Figure 1) was investigated [18,19]. However, Rayleigh waves not only distribute along the surface ($z = 0$), they also distribute within a depth of the Rayleigh waves' wavelength. Some industrial defects are within the sample instead of on the surface of the sample; in order to use Rayleigh waves to detect such defects, the beam directivity of the Rayleigh waves at various depths is worth investigating. In this article, the equations to study the beam directivity are more complete compared to the approximate equations presented in [18] by Xie et al., and the beam directivity at various depths are investigated. This study lays a solid industrial foundation for near-surface defects detection using Rayleigh waves, and can be a starting point to build an advanced 3D EMAT model in the future. Except for near-surface detection, the proposed strategy can be used to perform body detection, i.e., to model bulk waves, including longitudinal waves and shear waves. In addition, the proposed 3D EMAT model can be used to characterise other EMAT structures to generate surface waves, such as unidirectional Rayleigh waves EMATs and multiple directional Rayleigh waves EMATs. The 2D simulation on the vertical plane utilizes the analytical method and FDTD; the 2D simulation on the vertical plane and experimental validations are introduced in Section 2. The pseudo-3D model is presented in Section 3 to investigate the radiation pattern of Rayleigh waves at various depths by utilizing a wholly analytical solution. This work is an extension of the work published in [15,18] by Xie et al.

2. Vertical Plane Modelling

Previously, the authors have conducted EMAT modelling for Rayleigh waves focussing on the vertical plane of the material. This model combines the analytical method and FDTD to model EMATs [15]. The dimension and material of the test piece, the coil, and the permanent magnet are the same as the ones used in [15] by Xie et al. Based on such design, the working frequency used to form the interference of Rayleigh waves is 483 kHz.

2.1. EMAT-EM Model

This section introduces the EMAT-EM model to analyse the distribution of **F** (Lorentz force density). Firstly, the classic Dodd and Deeds solution [20] to the vector potential is described, and the strategy of adapting the circular analytical solutions for a straight wire is introduced (Section 2.1.1). The FEM is employed to validate the adapted solution (Section 2.1.2). Finally, based on the adapted analytical solutions, the distribution of Lorentz force density is presented (Section 2.1.3).

2.1.1. Adapted Analytical Solutions to the Vector Potential for a Straight Wire

The governing equations for calculating eddy currents are described in Equations (2)–(4):

$$\nabla^2 \mathbf{A} = -\mu \mathbf{I} + \mu\sigma\frac{\partial \mathbf{A}}{\partial t} + \mu\varepsilon\frac{\partial^2 \mathbf{A}}{\partial t^2} + \mu\nabla\left(\frac{1}{\mu}\right) \times (\nabla \times \mathbf{A}) \tag{2}$$

$$\mathbf{E} = -j\omega\mathbf{A} \tag{3}$$

$$\mathbf{J} = \sigma\mathbf{E} \tag{4}$$

where **A** is the vector potential generated by **I**, ω and **I** are the angular frequency and the density of the applied alternating current (AC), respectively, ε, μ and σ are the permittivity, permeability and conductivity of the test piece respectively, and **E** and **J** are the induced electric field and eddy current density, respectively [20].

As described in [15] by Xie et al., for a small radius circular coil, the distribution of the vector potential **A** at $z = 0$ (surface of the sample) is not symmetrical with the radius due to the bent wire.

The coil used in this work consists of straight wires; thus, the analytical solution for a straight wire is needed. The adapted solution for a straight wire has been presented in [15] by Xie et al. Here is a brief introduction. A hypothesis is proposed: if the radius of the circular coil, compared with its width, is very large, a bent wire can be viewed as a straight wire, and the distribution of **A** should be symmetrical. To validate such a hypothesis, a model is built with a large-radius circular coil above the aluminium plate. The aluminium sample used has a dimension of 80 mm × 30 mm, and the inner radius and the outer radius of the circular coil are set to 5.0395 m and 5.0405 m, respectively. At kHz, the current density applied to the circular coil is 1 A/m², and the lift-off of the coil is 1 mm. The permeability and the conductivity of the aluminium plate are 1.257 × 10⁻⁶ H/m and 3.8 × 10⁷ Siemens/m, respectively.

The distribution of the magnitude of **A** based on the adapted solution is shown in Figure 2. The red marker denotes the maximum magnitude of the vector potential. The distribution of the magnitude of **A** is symmetrical with the radius of 5.04 m, where the coil is located. The result verifies the hypothesis that, when the radius of the circular coil is very large, a bent wire serves as a straight wire.

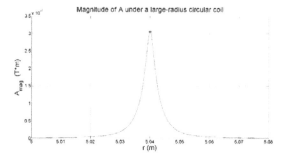

Figure 2. The magnitude distribution of A under a large-radius circular coil. Reproduced with permission from [15], IEEE, 2016.

2.1.2. Comparison between the Adapted Solution and FEM

In order to compare the adapted solutions to FEM, Maxwell Ansoft, which is a FE solver, is utilised. The FEM model has a rectangular cross-sectional coil located above the cross-sectional aluminium plate, and is surrounded by a vacuum region that is four times larger than the sample. The FEM subdivides the large model to smaller elements, and this FEM solves the calculation by minimising the energy error. In this work, when the elements number is beyond 20,000, the energy error is as low as 0.05%, which is sufficiently accurate for the FEM computation. In this work, the mesh number used is 20,395, and the boundary used is a balloon boundary to simulate an infinite space. The distribution of **A** at the sample's surface (z = 0) is presented in Figure 3. The analytical solution and FEM present a good agreement at an operational frequency of 1 kHz. However, at a working frequency of 1 MHz, the distribution of **A** from the FEM is not smooth compared to that from the analytical solution; the reason is that the FEM is affected by the elements density and numerical approximation is unavoidable, etc. Therefore, the adapted analytical method presents a more accurate result compared to FEM, especially for a high working frequency.

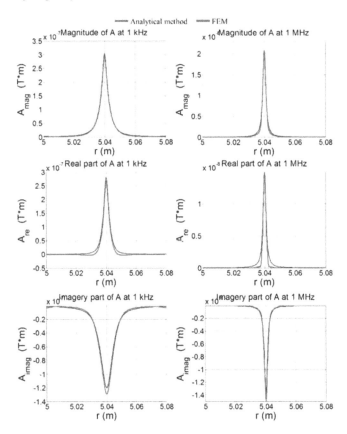

Figure 3. The distribution of A from the adapted analytical solution and the finite element method (FEM). The left curves are the results at 1 kHz, while the right curves are the results at 1 MHz. The red curve is the result from the FEM, and the blue curve is the result from the analytical solutions. Reproduced with permission from [15], IEEE, 2016.

2.1.3. EMAT-Lorentz Force Calculation

The analytical solution to a straight wire has been described in Section 2.1.2. **A** (vector potential) generated by a meander-line-coil is the addition of **A** generated by every single straight wire. The zone where the meander-line-coil mainly operates on is selected to model EM simulation to increase the modelling effectiveness. The distribution of **A** and **F** at z = 0 are shown in Figures 4 and 5. The generated periodic fields have different directions for any neighbouring wires, since their applied ACs are opposite, and therefore, the periodic distribution of **A** and **F** has six positive values and six negative values. The outermost **A** and outermost **F** are largest, because **A** is under the outermost wires, and thus is only determined by the fields on one side.

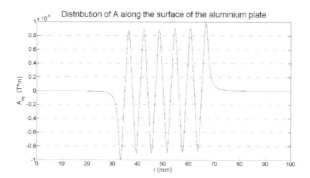

Figure 4. For a meander-line-coil, the distribution of the vector potential **A** at z = 0. Reproduced with permission from [15], IEEE, 2016.

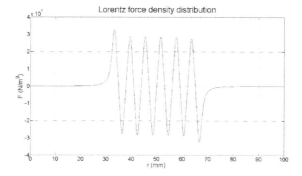

Figure 5. For a meander-line-coil, the distribution of the Lorentz force density **F** at z = 0. Reproduced with permission from [15], IEEE, 2016.

2.2. EMAT-US Simulation

2.2.1. Elastodynamic Equations

Elastodynamic equations (Equations (5) and (6)) are a group of partial differential equations to model the wave propagation:

$$\rho(x)\frac{\partial v_i}{\partial t}(x,t) = \sum_{j=1}^{d}\frac{\partial T_{ij}}{\partial x_j}(x,t) + f_i(x,t) \tag{5}$$

$$\frac{\partial T_i}{\partial t} - \sum_{j=1}^{d}\sum_{i=1}^{d} c_{ijkl}(x)\frac{\partial v_k}{\partial x_l} + \theta_{ij}(x,t) \tag{6}$$

where ρ and c_{ijkl} are the density and the fourth stiffness tensor of the material, and f_i and θ_{ij} are the force source and strain tensor rate source, respectively.

2.2.2. Combination of EMAT-EM and EMAT-US Models

As described in Section 2.1.3, **F** is obtained from the EMAT-EM calculation. In this section, **F**, which is used as the force source, is imported to the EMAT-US model to produce ultrasound (Figure 6). Since **F** is calculated in the frequency domain and FDTD is a time-domain solver, the excitation signal for the EMAT-US model is a time sequence signal with the peak equalling the peak values of **F**. The excitation signal used is a Gaussian-modulated sinusoidal with a fractional width of 0.18. A crack and a receiver R are located within the sample, as shown in Figure 6. Regarding the FDTD setup in the ultrasonic model, there are two main parameters: the spatial step, and the time step. The spatial step used is 0.2 mm, which approximately equals to 1/30th of the wavelength. The time step is 0.0222 μs, which is calculated based on the Courant–Friedrichs–Lewy (CFL) condition. Free surface conditions are utilised on the surface of the sample. Perfectly-matched layers (PML) with a thickness of 16 mm are utilised to absorb ultrasound.

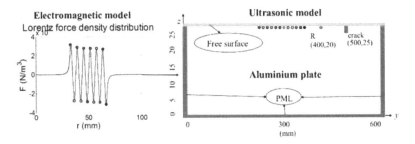

Figure 6. On the vertical plane of the material, the combination of the EMAT-electromagnetic (EM) and EMAT-ultrasonic (US) models. Reproduced with permission from [15], IEEE, 2016.

2.2.3. Wave Propagations

Based on FDTD calculations, the velocity fields of ultrasound waves are obtained. The velocity fields at 27 μs and 83 μs are shown in Figure 7, which describes the Rayleigh waves' propagation and their scattering behaviours, respectively. The white arrows denote the propagation path.

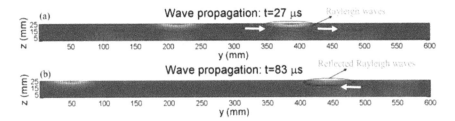

Figure 7. Wave propagations. (**a,b**) denote the velocity fields at 27 μs and 83 μs, respectively. Reproduced with permission from [15], IEEE, 2016.

2.3. EMAT-Reception Simulation

The EMAT reception process has been reported quite a lot [21]. The received signal from simulations is shown in Figure 8, where DRW and RRW denote the directly transmitted Rayleigh waves and the reflected Rayleigh waves, respectively. The propagation distance of DRW and RRW is 100 mm and 300 mm, respectively, and the velocity of Rayleigh waves is 2.93 mm/μs; hence, the theoretically arrival time of DRW and RRW is 34 μs and 102.4 μs, respectively. Figure 8 shows the numerically arrival time of DRW and RRW; these numerical and experimental results present a good agreement.

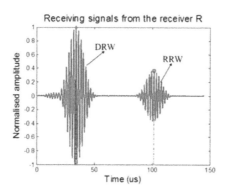

Figure 8. The received signal from simulations. Reproduced with permission from [15], IEEE, 2016.

2.4. Experimental Validations

Experiments were carried out to validate the proposed modelling method. The setup of the experiments is the same as that of our previous work [15]. The received signals from experiments are shown in Figure 9, where three signals are observed. The "Main bang" is the interference signal due to a high power excitation, arriving before DRW and RRW. The red curve and the blue curve are the envelope and the time series signal, respectively.

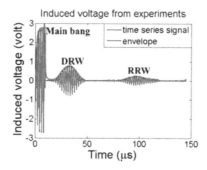

Figure 9. The received signal from experiments. The blue curve denotes the induced voltage in the received coil, and the red curve denotes the envelope of the received signal. Reproduced with permission from [15], IEEE, 2016.

Figure 10 shows the envelope of DRW and RRW from experiments and simulations. The experimental arrival times of DRW is 34 μs, which is consistent with the numerical results. However, the experimental arrival time of RRW is slightly different from the simulations. This is because of the approximated model used and the inevitable experimental noise.

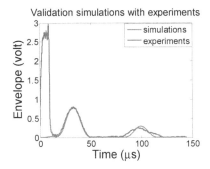

Figure 10. The envelope of the received signals. The blue curve and the red curve are the envelope of the received signal from simulations and experiments, respectively.

3. Horizontal Surface Plane Modelling—Directivity Analysis of Rayleigh Waves

The main mode of propagation on the horizontal surface of the material is the Rayleigh wave. The radiation pattern and the beam directivity of Rayleigh waves at z = 0 (Figure 1) were reported in [18] by Xie et al. Rayleigh waves not only distribute along the surface (z = 0), they also distribute within a depth of the Rayleigh waves' wavelengths. However, Rayleigh waves not only concentrate on the surface, they also concentrate within a depth of one wavelength. In this section, the beam directivity of Rayleigh waves at various depths are investigated utilising an analytical method.

3.1. The Analytical Solution to the Displacement of Rayleigh Waves

N. A. Haskell proposed an analytical solution to Rayleigh waves [22,23]. Ref. [24] by Love introduced these solutions in detail, and investigated the beam directivity of Rayleigh waves at z = 0 with approximated equations. However, Rayleigh waves not only concentrate on the surface; they also concentrate within a depth of one wavelength. The complete equations of the Rayleigh waves' displacement at various depths are:

$$u_r = A(\kappa, r, z) e^{-\frac{\pi i}{4}} \frac{2\kappa(\gamma - 1)}{v_\beta} F(\frac{2}{\gamma} - 1 + \frac{\gamma - 1}{\gamma} e^{-z(v_\alpha - v_\beta)}) \tag{7}$$

$$u_z = \frac{i\gamma v_\alpha u_r}{\kappa(\gamma - 1)} \tag{8}$$

where:

$$A(\kappa, r, z) = \frac{\kappa^2 \gamma v_\beta}{4\rho(\frac{2\gamma^2 v_\alpha v_\beta}{\kappa^3})} \sqrt{\frac{2}{\pi \kappa r}} e^{(-i\kappa r - z v_\beta)} \tag{9}$$

$$\gamma = \cos(\theta) \tag{10}$$

$$v_\alpha = \begin{cases} \sqrt{\kappa^2 - (\omega/c_L)^2} & \kappa > \omega/c_L \\ i\sqrt{(\omega/c_L)^2 - \kappa^2} & \kappa < \omega/c_L \end{cases} \tag{11}$$

$$v_\beta = \begin{cases} \sqrt{\kappa^2 - (\omega/c_S)^2} & \kappa > \omega/c_S \\ i\sqrt{(\omega/c_S)^2 - \kappa^2} & \kappa < \omega/c_S \end{cases} \tag{12}$$

$$\kappa = \frac{\omega}{c_R} \tag{13}$$

where u_r and u_z are the in-plane and the out-of-plane displacement to be calculated, respectively; r is the distance between the source point and the field point (Figure 11); F is the excitation source; ρ is the density of the material; θ is the angle between the force vector and the in-plane displacement vector; ω is the operational angular frequency; z is the depth; and c_L, c_S, c_R, and κ are the velocity of the longitudinal waves, shear waves, Rayleigh waves, and the wave number, respectively.

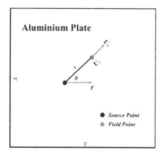

Figure 11. The model used to simulate Rayleigh waves. Reproduced with permission from [18], Elsevier, 2017.

3.2. Linking EMAT-EM and EMAT-US Models

On the surface plane of the material, the combination between the EMAT-EM model and the EMAT-US model is described in Figure 12. The calculated *F* values, acting as the excitation source, are imported to each surface layer at various depths. The Rayleigh waves' distribution is the addition of the Rayleigh waves generated by each point source. Table 1 illustrates the parameters used for the EMAT-US model.

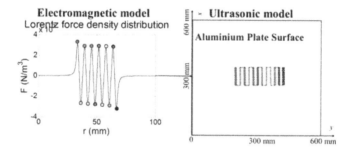

Figure 12. On the surface plane of the material, the transformation between the EM and US models. Reproduced with permission from [18], Elsevier, 2017.

Table 1. Detailed parameters used for the EMAT-US modelling.

Description	Symbol	Value
Length of the aluminium plate	Y	600 mm
Width of the aluminium plate	X	600 mm
Field spatial step	Δx_f	1 mm
Length of the meander-line-coil	L	50 mm
Source spatial step for each wire	Δx_s	0.2 mm
Density of the aluminium plate	ρ	2700 kg/m^3
Frequency	f	483 kHz
Longitudinal waves' velocity	C_l	6.375 mm/µs
Shear waves' velocity	C_s	3.14 mm/µs
Rayleigh waves' velocity	C_r	2.93 mm/µs

3.3. Analysis of the Beam Directivity of Rayleigh Waves

Figure 13 shows the calculated Rayleigh waves' radiation pattern at $z = 0$, which is symmetrical with the center of the coil. Rayleigh waves are mainly generated along the y direction (main lobe) and some undeniable directions (side lobe). The characteristics of Rayleigh waves are quantitatively investigated by means of beam directivity, as shown in the red arc in Figure 13 ($r = 250$ mm, $\theta_1 = -70°$, and $\theta_2 = 70°$).

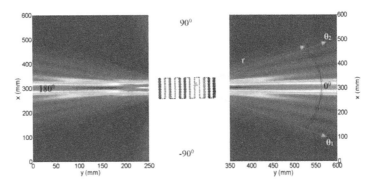

Figure 13. The radiation pattern of Rayleigh waves generated by the meander-line-coil EMAT. Reproduced with permission from [18], Elsevier, 2017.

The beam directivity of Rayleigh waves at $z = 0$ is shown in Figure 14. A main lobe contains a larger magnitude compared to the side lobes, which contain a smaller magnitude. The main lobe is centered at $0°$, and the side lobes are roughly centered at $-25.5°$, $-18°$, $-10°$, $10°$, $18°$, and $25.5°$, respectively. The largest magnitude of the side lobes is 25.87% that of the main lobe. In most applications, side lobes are usually undesirable.

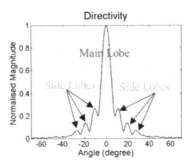

Figure 14. The simulated beam directivity of Rayleigh waves at $z = 0$. Reproduced with permission from [15], IEEE, 2016.

The beam directivity of the Rayleigh waves at various depths is shown in Figure 15. The magnitude of the beam directivity is normalised. From this image, the magnitude of the Rayleigh waves decays with the depth, especially for the depth larger than one Rayleigh wavelength. At a depth equalling to one Rayleigh wave's wavelength, $z = 6$ mm, the magnitude of the Rayleigh waves is 34.9% of that at $z = 0$. At a depth of 7 mm, the magnitude of the Rayleigh waves decays to 22.8% of that at the surface of the test piece ($z = 0$). This observation confirms that Rayleigh waves mainly distribute within a depth equal to one Rayleigh wavelength.

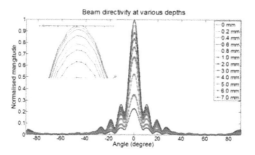

Figure 15. The simulated beam directivity of Rayleigh waves at various depths.

3.4. Experimental Validations

In Figure 15, based on the analytical simulations, the distribution of the Rayleigh waves at various depths is presented. In this part, the measured results at $z = 0$ are picked to compare with the simulation results. The experimental setup was the same as the one in Section 2.4. The measured beam directivity at $z = 0$ is obtained by placing the receiver along the scanning path, as shown in Figure 16. Figure 17 shows the beam directivity results at $z = 0$ from simulations and experiments. Thirty-three measuring points on the scanning path were picked with a moving step of 2.5°. The measured beam directivity at $z = 0$ agrees well with the simulated beam directivity, which validates the proposed method. There are some non-overlapping points between these two curves due to several factors, which include the experimental noise and the tolerance of the receiver's position, etc.

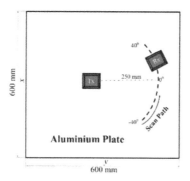

Figure 16. The scanning path of the receiver.

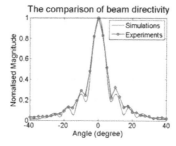

Figure 17. The comparison between the simulated beam directivity and the experimental beam directivity at $z = 0$. The red curve is the beam directivity from simulations, while the blue curve is the beam directivity from experiments.

4. Conclusions

A pseudo-3D model for simulating meander-line-coil EMATs was proposed. A method combining the analytical method for the EM model and FDTD for the US model was utilised to simulate the Rayleigh waves' propagation. On the other hand, a wholly analytical method was utilised to simulate the radiation pattern of the Rayleigh waves. For both cases, analytical solutions to the EM model were adapted from the classic Dodd and Deeds solution in order to calculate eddy currents under straight wires. By comparing with the FEM, the analytical solutions are more accurate. Experiments were conducted in order to validate the proposed method, and these showed a good consistency. The beam directivity of Rayleigh waves at various depths were investigated, and results confirmed that Rayleigh waves mainly distribute within a depth of one wavelength of Rayleigh waves. Overall, this pseudo-3D model combines both the vertical plane and surface plane of EMAT models, and provides the beam directivity of Rayleigh waves at various depths, which have not been reported previously. Therefore, this work can be a starting point to build an advanced EMAT 3D model in the future. There are some limitations of the proposed 3D EMAT model. Firstly, it can only be applied in a homogeneous medium. It is worth investigating the 3D EMAT model for a multiple-layer medium, since multiple-layer samples are widely used in applicable industries. Secondly, the EMAT-US model on the vertical plane of the sample uses an approximated model with only point sources (Lorentz force density) to generate Rayleigh waves. A more detailed model with volume sources within the skin depth is worth considering in the future. In addition, it is worth investigating the scattering behaviours of defects in other orientations, as it is a variable in practical application.

Acknowledgments: This work was financially supported by Engineering and Physical Sciences Research Council (Grant No. EP/M020835/1).

Author Contributions: Wuliang Yin conceived and reviewed the manuscript, did some simulations and designed the experiments. Yuedong Xie performed the experiments and wrote this manuscript. Zhigang Qu contributed some simulation results. Zenghua Liu provided related instruments for carrying out experiments.

Conflicts of Interest: The authors declare no conflict of interest.

References

1. Buchner, S.P.; Miller, F.; Pouget, V.; McMorrow, D.P. Pulsed-laser testing for single-event effects investigations. *IEEE Trans. Nucl. Sci.* **2013**, *60*, 1852–1875. [CrossRef]
2. Taheri, H.; Delfanian, F.; Du, J. Ultrasonic phased array techniques for composite material evaluation. *J. Acoust. Soc. Am.* **2013**, *134*, 4013. [CrossRef]
3. Yin, W.; Dickinson, S.; Peyton, A. A multi-frequency impedance analysing instrument for eddy current testing. *Meas. Sci. Technol.* **2006**, *17*, 393. [CrossRef]
4. Yin, W.; Peyton, A.J.; Dickinson, S.J. Simultaneous measurement of distance and thickness of a thin metal plate with an electromagnetic sensor using a simplified model. *IEEE Trans. Instrum. Meas.* **2004**, *53*, 1335–1338. [CrossRef]
5. Dixon, S.; Edwards, C.; Palmer, S. High accuracy non-contact ultrasonic thickness gauging of aluminium sheet using electromagnetic acoustic transducers. *Ultrasonics* **2001**, *39*, 445–453. [CrossRef]
6. Edwards, R.S.; Dixon, S.; Jian, X. Non-contact ultrasonic characterization of defects using EMATs. *AIP Conf. Proc.* **2005**, *760*, 1568.
7. Edwards, R.S.; Sophian, A.; Dixon, S.; Tian, G.-Y.; Jian, X. Dual EMAT and PEC non-contact probe: Applications to defect testing. *NDT E Int.* **2006**, *39*, 45–52. [CrossRef]
8. Hutchins, D.; Schindel, D. Advances in non-contact and air-coupled transducers (US materials inspection). In Proceedings of the Ultrasonics Symposium, Cannes, France, 31 October–3 November 1994; IEEE: Piscataway, NJ, USA, 1994.
9. Hirao, M.; Ogi, H. *EMATs for Science and Industry: Noncontacting Ultrasonic Measurements*, 1st ed.; Springer Science & Business Media: New York, NY, USA, 2003; 372p.
10. Jian, X.; Dixon, S.; Grattan, K.T.V.; Edwards, R.S. A model for pulsed Rayleigh wave and optimal EMAT design. *Sens. Actuators A Phys.* **2006**, *128*, 296–304. [CrossRef]

11. Wang, S.; Kang, L.; Li, Z.; Zhai, G.; Zhang, L. A novel method for modeling and analysis of meander-line-coil surface wave EMATs. In *Life System Modeling and Intelligent Computing*; Kang, L., Fei, M., Jia, L., Irwin, G.W., Eds.; Springer: Berlin/Heidelberg, Germany, 2010; pp. 467–474.

12. Kang, L.; Dixon, S.; Wang, K.; Dai, J. Enhancement of signal amplitude of surface wave EMATs based on 3-D simulation analysis and orthogonal test method. *NDT E Int.* **2013**, *59*, 11–17. [CrossRef]

13. Dhayalan, R.; Balasubramaniam, K. A hybrid finite element model for simulation of electromagnetic acoustic transducer (EMAT) based plate waves. *NDT E Int.* **2010**, *43*, 519–526. [CrossRef]

14. Dhayalan, R.; Balasubramaniam, K. A two-stage finite element model of a meander coil electromagnetic acoustic transducer transmitter. *Nondestruct. Test. Eval.* **2011**, *26*, 101–118. [CrossRef]

15. Xie, Y.; Rodriguez, D.S.; Yin, W.; Peyton, A.; Liu, Z.; Hao, J.; Zhao, Q.; Wang, B. Simulation and experimental verification of a meander-line-coil electromagnetic acoustic transducers (EMATs). In Proceedings of the Instrumentation and Measurement Technology Conference (I2MTC), Taipei, Taiwan, 23–26 May 2016; IEEE International: Piscataway, NJ, USA, 2016.

16. Xie, Y.; Yin, L.; Rodriguez, S.G.; Yang, T.; Liu, Z.; Yin, W. A wholly analytical method for the simulation of an electromagnetic acoustic transducer array. *Int. J. Appl. Electromagn. Mech.* **2016**, *51*, 1–15. (Preprint) [CrossRef]

17. Xie, Y.; Yin, W.; Liu, Z.; Peyton, A. Simulation of ultrasonic and EMAT arrays using FEM and FDTD. *Ultrasonics* **2016**, *66*, 154–165. [CrossRef] [PubMed]

18. Xie, Y.; Liu, Z.; Yin, L.; Wu, J.; Deng, P.; Yin, W. Directivity analysis of Meander-Line-Coil EMATs with a wholly analytical method. *Ultrasonics* **2017**, *73*, 262–270. [CrossRef] [PubMed]

19. Wang, S.; Kang, L.; Li, Z.; Zhai, G.; Zhang, L. 3-D modeling and analysis of meander-line-coil surface wave EMATs. *Mechatronics* **2012**, *22*, 653–660. [CrossRef]

20. Dodd, C.; Deeds, W. Analytical Solutions to Eddy-Current Probe-Coil Problems. *J. Appl. Phys.* **1968**, *39*, 2829–2838. [CrossRef]

21. Jian, X.; Dixon, S.; Quirk, K.; Grattan, K.T.V. Electromagnetic acoustic transducers for in-and out-of plane ultrasonic wave detection. *Sens. Actuators A Phys.* **2008**, *148*, 51–56. [CrossRef]

22. Haskell, N. Radiation pattern of Rayleigh waves from a fault of arbitrary dip and direction of motion in a homogeneous medium. *Bull. Seismol. Soc. Am.* **1963**, *53*, 619–642.

23. Haskell, N. Radiation pattern of surface waves from point sources in a multi-layered medium. *Bull. Seismol. Soc. Am.* **1964**, *54*, 377–393.

24. Love, A.E.H. *A Treatise on the Mathematical Theory of Elasticity*; Dover Publication: New York, NY, USA, 1944; 643p.

Article

Calculation of Noise Barrier Insertion Loss Based on Varied Vehicle Frequencies

Haibo Wang [1,2], Peng Luo [3] and Ming Cai [1,2,*]

1 School of Engineering, Sun Yat-sen University, Guangzhou 510275, China; wanghb9@mail.sysu.edu.cn
2 Guangdong Provincial Key Laboratory of Intelligent Transportation System, Guangzhou 510275, China
3 Dongguan Geographic Information and Urban Planning Research Center, Dongguan 523129, China; liangy98@mail2.sysu.edu.cn
* Correspondence: caiming@mail.sysu.edu.cn; Tel.: +86-20-393-3227-2803

Received: 14 December 2017; Accepted: 5 January 2018; Published: 11 January 2018

Abstract: A single frequency of 500 Hz is used as the equivalent frequency for traffic noise to calculate the approximate diffraction in current road barrier designs. However, the noise frequency changes according to the different types of vehicles moving at various speeds. The primary objective of this study is the development of a method of calculating the insertion loss based on frequencies. First, the noise emissions of a large number of vehicles classified by speed and type were measured to obtain data the noise spectrum. The corresponding relation between vehicle type, speed, and noise frequency was obtained. Next, the impact of different frequencies on the insertion loss was analyzed and was verified to be reasonable in experiments with different propagation distances compared to the analysis of a pure 500 Hz sound. In addition, calculations were applied in a case with different traffic flows, and the effect of a road noise barrier with different types of constituents and flow speeds were analyzed. The results show that sound pressure levels behind a barrier of a heavy vehicle flow or with a high speed are notably elevated.

Keywords: noise barrier; insertion loss; vehicle frequencies; diffraction; flow speed

1. Introduction

The rapidly increasing number of vehicles is in the world, especially in many developing countries, has raised the serious problem of traffic noise [1–3]. Traffic noise is disturbing to the daily routine of residents along the roads [4]. Many issues, physiological [5] and psychological [6] health problems, are demonstrated to be related to a noisy environment. To reduce traffic noise, one of the most effective methods is to set noise barriers along the roads [7].

Current research on noise barriers primarily focuses on noise calculation [8,9], the effects of barrier shapes [10,11], and the design of barriers [12–15]. As one of the important aspects, prediction of the insertion loss by the noise barrier has been attracting the attention of scholars, and both mathematical and experimental approaches have been developed. The mathematical formula was first developed by Sommerfeld in 1896 [16] and has been subsequently developed by many other scholars. As such, rigorous diffraction results can be calculated in mathematical methods, e.g., the Boundary Element Method [17], the Finite Element Method (FEM) [18], and the Finite Difference Method (FDM) [19]. In many cases, the typical procedure for the mathematical computation is too complicated to be applicable to the engineering application. Alternatively, comprehensive sets of data have been measured to plot the experimental attenuation of noise barrier since 1940 [20]. The most famous data set is the Maekawa chart [21], and many experimental formulae were developed based on Maekawa's original chart [22,23], including Kurze and Anderson's derived formula [24]. Kurze and Anderson's formula is widely used and is regarded as the simplest experimental practice for application in China and many other countries. Fresnel number, which is the single parameter in Kurze and

Anderson's diffraction formulae, is the ratio between the path length difference and half of the sound wavelength [25,26].

In practical application, the path length difference of a noise barrier is defined by a particular source-barrier-receiver geometry and easily measured. However, because of the multiple frequency components of traffic noise, the computation of diffraction is always time consuming. The noise frequency changes according to the different types of vehicles travelling at various speeds. Hence, the noise protection performance of the barrier depends on the traffic flow states. For simplicity, an equivalent frequency of 550 Hz can be set to represent the total frequency band in the calculation of diffraction suggested in the FHWA traffic noise model [27]. In the 1/3rd-octave-band spectrum, the frequency of 500 Hz is used in the approximate calculation of diffraction and noise reduction effect of a sound barrier [9,28–31]. However, the data used was collected from 1993 and 1995, and the characteristics of vehicle noise have changed since then because of the variety of vehicle types and speeds [32]. Moreover, in the micro traffic noise prediction model, the reduction effect of a noise barrier should consider every vehicle to be point source. Therefore, a single 500 Hz frequency is not adequate in the noise reduction calculation of a sound barrier.

In this paper, data of the noise spectrum of different vehicles in a variety of speed ranges were measured to calculate the equivalent frequency. The corresponding relationships among vehicle type, speed, and noise frequency were obtained. Several path length differences were preset to calculate the 1/3rd-octave-band diffraction and the total diffraction. Next, verification experiments were established to ensure the calculation accuracy of barrier diffraction based on different equivalent frequencies. Last, an application of the approach to different traffic flows is implemented, and the effects of a barrier to traffic noise for different speeds and different types of constituents are analyzed.

2. Methods

2.1. Measurement Method of Vehicle Noise Spectrum Data

The measurements were conducted on the roadside of seven typical dry roads in Guangzhou, China, with a surface constructed of asphalt concrete. The parameters of surface could refer to the Chinese standard JTG F40-2004 (Technical Specification for Construction of Asphalt Pavements) [33]. To avoid possible disturbances, the samples were collected in places far away from intersections, rivers, and populated areas. The chosen urban roads have the following speed limits: 2 roads with 70 km/h, 3 roads with 60 km/h, 1 road with 50 km/h, and 1 road with 40 km/h. The spectrum was captured when the A-weighted sound pressure level during a vehicle pass-by was at its maximum, and the sampling frequency of sound level meter was 23 Hz, i.e., 23 spectra per second. Only one vehicle was measured at a time. In Figure 1, the locations of the measurement sites are shown. Both sites were 1.2 m high and 6 m to the roadside with a distance of 50 m to each other. In the experiment, the subject vehicles only moved from left to right alone. As the single vehicle passed by measurement site 1, the noise spectrum data were recorded by a digital noise recorder. The vehicle speed had little variation between points 1 and 2 while traveling on a long-straight road with a very sparse traffic flow. The vehicle speed was detected by a laser speedometer at measurement site 2.

In this paper, the vehicles are classified into three categories: heavy cars, middle cars, and light cars. The 1/3rd-octave-band sound pressure level (SPL) and the total SPL are included in the noise spectral data. All the noise data are A-weighted.

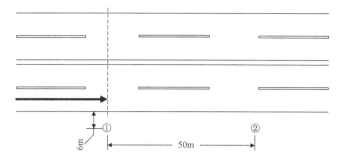

Figure 1. Locations of the Measurement Spots.

In this paper, 1351 sets of valid data were collected, among which 973 sets were light car data, 166 sets were middle car data, and 212 sets were heavy car data. The distribution of the samples is shown in Table 1.

Table 1. Distribution of the samples of vehicle noise spectra.

Speed	Light Car	Middle Car	Heavy Car
[0, 40) km/h	39	23	41
[40, 50) km/h	186	48	80
[50, 60) km/h	317	57	55
[60, 70) km/h	222	38	36
[70, 80) km/h	134		
≥80 km/h	75		

2.2. Noise Spectra of Different Vehicle Types and Speeds

To analyze the frequency characteristics of traffic noise of different types of vehicle with various speed, the 1/3rd octave band spectrum data of vehicles were recorded, including 34 frequencies from 10 Hz to 20,000 Hz. To facilitate the analysis of the frequency characteristics, the following ranges are defined: frequencies from 10 Hz to 315 Hz are low frequency; 400 Hz to 800 Hz are medium-low frequency; 1000 Hz to 2500 Hz are medium-high frequency; 3150 Hz to 20,000 Hz are high frequency. To analyses the frequency characteristics of traffic noise and exclude the effect of SPL, the noise energy percentage of the 1/3rd octave band spectrum was calculated [34], as given by the following formulas (Equations (1)–(3)):

$$E(i,j) = 10^{0.1*L(i,j)} \tag{1}$$

$$E'(i,j) = \frac{E(i,j)}{\sum_j E(i,j)} \tag{2}$$

$$E_S(j) = \frac{\sum_i E'(i,j)}{i} \tag{3}$$

where i is the number of vehicles, j is the number of frequencies; $L(i,j)$ is the SPL (dB), $E(i,j)$ is the relative value of noise energy (dimensionless quantities), $E'(i,j)$ is the noise energy percentage (%), and $E_S(j)$ is the mean percentage of noise energy (%).

From the equations, the SPL of each frequency of each vehicle is transformed to the noise energy percentage of each frequency. Figure 2 depicts the average noise energy percentage versus frequency for each type of vehicle with different speeds.

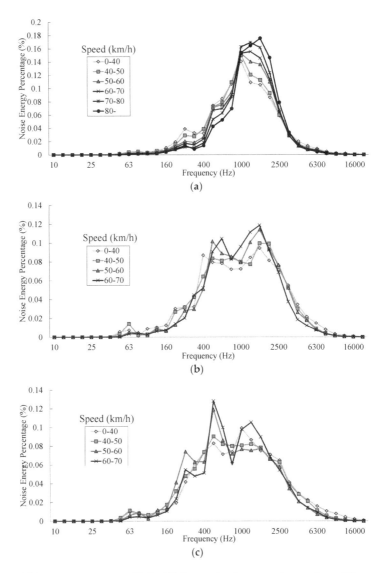

Figure 2. Noise energy percentage with the 1/3rd octave band spectrum frequencies at different speeds of three types of vehicles. (**a**) light vehicles; (**b**) middle vehicles; (**c**) heavy vehicles.

As shown in Figure 2, the noise energy percentage of the different speeds of three types of vehicles were performed with the 1/3rd octave band spectrum frequencies in the range of 10–20,000 Hz. The following points can be made:

A. For a light vehicle, the noise energy is concentrated in the range of 500–2500 Hz, with peak frequency at approximately 1250 Hz. For a middle vehicle, the noise energy is concentrated in the range of 315–2500 Hz, with peak frequency at approximately 1600 Hz. For a heavy vehicle, the noise energy is concentrated in the range of 315–2500 Hz, with peak frequency at approximately 500 Hz.

B. Speed will affect the SPL of a vehicle and the frequency characteristics of a vehicle. Noise spectra exhibit a trend of concentrating as the speed increase. Take the light vehicle case as an example, the noise energy percentage at the primary frequency range of 1000 Hz to 2500 Hz increases with speed, whereas the percentage of insignificant frequencies, such as low and medium-low frequencies, decrease with speeds. In addition, the main frequencies increase as the speed grows, especially for a light vehicle.

C. The noise frequencies of the heavy vehicles are mainly medium-low frequency and medium-high frequency, which is quite different from the noise frequencies of light and middle vehicles, as their frequencies are mainly medium-high frequency.

D. The emission noise of a vehicle can be divided into engine noise and tire/asphalt noise. Most of the light vehicles have gasoline engines. The contribution of engine noise is far less than the tire/asphalt noise. The proportion of diesel engines is high when for middle vehicles and heavy vehicles. The noise caused by diesel engine represents a large proportion, leading to a bi-modal trend in the noise spectra, which is in agreement with the rules presented by previous studies [34,35].

2.3. The Calculation Method of Insertion Loss

The center frequency with the minimum mean deviation between the 1/3rd-octave-band diffraction and the total diffraction is the equivalent frequency [36]. The equivalent frequency could be used in the approximate calculation of barrier attenuation. The results of the approximate calculation are close to the direct calculation using the whole spectrum as parameters, and the total error is always less than 1 dB. The procedures for calculating the insertion loss in this paper are presented below. Generally, several path length differences were preset first to calculate the 1/3rd-octave-band diffraction and the total diffraction. The detailed calculations are shown as follows:

(1) Seven path length differences from 0.01 m to 10 m were preset first: 0.01 m, 0.1 m, 0.5 m, 1 m, 2.5 m, 5 m, 10 m;

(2) The data were collected at the noise measurement site without the noise reduction effect of the barrier, and contained the 1/3rd-octave-band SPL L_i and the total SPL L, where i is the sequence number of the 1/3rd-octave-band center frequency;

(3) Through the diffraction formula, the diffraction ΔL_{di} of every 1/3rd-octave-band center frequency could be calculated as follows (Equation (4)):

$$\Delta L_{di} = 20\lg\frac{\sqrt{2\pi N}}{\tanh\sqrt{2\pi N}} + 5\text{dB}, N = \frac{\delta \cdot f}{170} \tag{4}$$

where δ is the path length difference and f is the 1/3rd-octave-band center frequency;

(4) For a certain path length difference, the total SPL L' with the reduction effect of noise barrier can be calculated as (Equations (5) and (6))

$$L_i' = L_i - \Delta L_{di} \tag{5}$$

$$L' = 10\lg\left(\sum_i 10^{L_i'/10}\right) \tag{6}$$

where L_i' is the 1/3rd-octave-band SPL with the reduction effect of barrier;

(5) The following equation is used to yield the total diffraction ΔL_d(Equation (7)):

$$\Delta L_d = L - L' \tag{7}$$

(6) After calculating the total diffraction $\Delta L_d(\delta_j)$ and the 1/3rd-octave-band diffraction $\Delta L_{di}(\delta_j)$ with seven preset path length differences, the mean deviation Δd_i between $\Delta L_d(\delta_j)$ and $\Delta L_{di}(\delta_j)$ can be obtained (Equation (8)).

$$\Delta d_i = \frac{1}{7}\sum_{j=1}^{7}|\Delta L_d(\delta_j) - \Delta L_{di}(\delta_j)| \qquad (8)$$

3. Experimental Verification

Two verifying experiments were conducted: an insertion loss experiment and a distance attenuation experiment. In the experiments, vehicle noise was recorded next to the roads and subsequently played in the audio amplifier to emulate a point sound source that was used to measure actual insertion loss of noise barrier. For every type of car at different speed range, the synthesized point source consisted of five recorded audio samples. The length of each audio sample is 8 s.

The insertion loss experiment was conducted to collect sound level pressure data in front of and behind the noise barrier. A schematic of the experiment is shown in Figure 3.

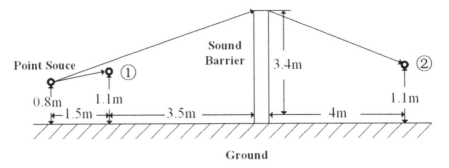

Figure 3. Insertion loss experiment.

Two noise recorders were installed, one at measurement site 1 and the other at measurement site 2, to collect noise data and compute the noise levels L_1 and L_2, respectively. The insertion loss of barrier is the difference between L_1 and L_2. The computed equivalent noise levels are shown in Table 2.

Table 2. Noise levels computed in the insertion loss experiment (dB).

Speed (km/h)	Light Car		Middle Car		Heavy Car	
	L_1	L_2	L_1	L_2	L_1	L_2
[0, 40)	75.21	46.31	84.26	52.11	83.91	54.38
[40, 50)	80.69	48.98	83.85	52.42	80.12	48.84
[50, 60)	80.09	49.03	86.59	53.76	85.95	53.39
[60, 70)	82.17	50.14	85.53	53.99	86.68	54.17
[70, 80)	82.74	50.65				
≥80	85.30	52.48				

To obtain the actual value of diffraction, noise attenuation as a function of distance between the two measurement sites should be considered. The noise data were measured in an open space near the sound barrier to avoid possible errors introduced by the change of the experimental locations. Because noise attenuation with distance is independent of the frequency of the sound source [20], the sound source audios were merged into three categories in the experiment: heavy car, middle car, and light car. The geometric configuration of measurement is shown in Figure 4.

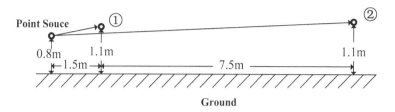

Figure 4. Noise attenuation experiment with distance.

Noise recorders were installed in measurement site 1 and measurement site 2. The data are summarized in Table 3. Noise attenuation with distance for heavy, middle, and light vehicles are 11.16 dB, 11.03 dB, and 11.13 dB, respectively. The average of the three is defined as ΔL. In this instance, $\Delta L = 11.11$ dB.

Table 3. Data collected in the noise attenuation experiment with distance (dB).

Vehicle Type	L_1'	L_2'	ΔL
Light Car	82.02	70.86	11.16
Middle Car	84.13	73.10	11.03
Heavy Car	84.02	72.89	11.13

From the experimental verification, the actual value of diffraction by barrier was determined as follows (Equation (9)):

$$\Delta L_d = (L_1 - L_2) - \Delta L \tag{9}$$

The theoretical value of diffraction is defined as (Equation (10))

$$\Delta L'_d = \begin{cases} 20\lg\frac{\sqrt{2\pi N}}{\tanh\sqrt{2\pi N}} + 5dB & ,N > 0 \\ 5dB & ,N = 0 \end{cases} \tag{10}$$

where N is the Fresnel number, $N = \frac{\delta \cdot f}{170}$.

Equations (9) and (10) are mainly applied to near-field conditions. According to reference [20], Equation (11) should be satisfied (Equation (11)).

$$D < 2d^2/\lambda \tag{11}$$

where D is the distance from the sound source to the receivers' center, d is the step of the receivers, and λ is the wave length. The speed of sound is 340 m/s.

In the experiment, D is calculated as approximately 5.26 m, d is set as 0.5 m in environmental acoustics calculation, and the wave length λ is in a range of [2.125, 0.054] with a primary frequency range of 160 Hz to 6300 Hz. Thus, Equation (11) is satisfied. The experimental setup can be considered with the theoretical formulation of the Fresnel number N.

Two types of equivalent frequencies were used to yield the theoretical value of diffraction: one was the 500 Hz frequency, and the other was the varied equivalent frequencies computed in this paper. The actual value and the theoretical values are shown in Figure 5. The figure reveals that the error of diffraction calculated with the variety of noise equivalent frequencies is less than that calculated with 500 Hz frequency alone. Under the same experimental conditions, the average error using the 500 Hz frequency is approximately 2.3 dB, whereas the average error with the variety of noise equivalent frequencies is approximately 0.9 dB. With a difference of approximately 1.4 dB, the accuracy of the proposed method in calculating the noise barrier diffraction is undoubtedly more justified.

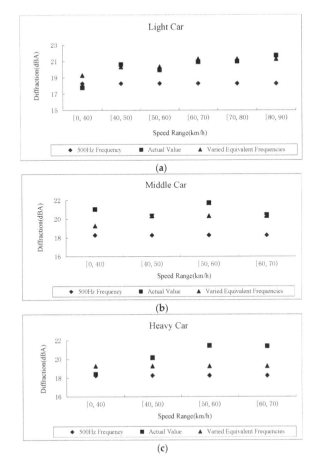

Figure 5. Comparisons between 500 Hz frequency and the variety of equivalent frequencies for diffraction by barrier. (**a**) light vehicle; (**b**) middle vehicle; (**c**) heavy vehicle.

4. Effects of a Road Noise Barrier with Different Flow States

As shown in Figure 6, there is a two-dimensional sound field involving a sufficiently long enough sound barrier and a line road traffic noise source that is parallel to both the sound barrier and the ground. Three receivers (R_1, R_2, and R_3) are chosen to analyze the effect of a road noise barrier with different type constituents and flow speed; all the receivers are sheltered by the barrier. The insertion losses of all receivers are calculated with different flow speed and the proportion of heavy vehicles. For each point, the SPL is calculated. The insertion loss of barrier can be defined as (Equation (12))

$$L_D = SPL_0 - SPL_b \tag{12}$$

where SPL_0 is the SPL at the receiver when the barrier is absent and SPL_b is the SPL at the same receiver when the barrier is present.

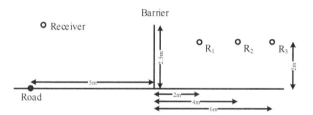

Figure 6. A case of noise calculation behind a road sound barrier.

4.1. Effect of Vehicle Type

Traffic noise caused by heavy vehicles is quite different in frequencies compared to other vehicles based on the results of Section 2.2. Hence, the effects of the proportion of heavy vehicles on insertion loss of barrier were analyzed. The speed of the vehicles in this part is set as 50 km/h, and the flow of 1000 vehicles per hour consists of light and heavy vehicles with 11 different proportions. To ensure that the results are not affected by the different vehicles' sound source strong, the emission intensity of each vehicle source is set as 85 dB. Traffic noise for different heavy vehicle proportions (0, 10%, 20%, 30%, 40%, 50%, 60%, 70%, 80%, 90%, and 100%) on each point was calculated.

With different proportions of heavy vehicles, the insertion losses at three points are shown in Figure 7. L_D presents a distinct pattern regarding points R_1, R_2, and R_3, and the insertion loss at all of the receivers have relationships with the proportion of heavy vehicles. For point 1, L_D and heavy vehicles a (%) follow the linear relationship of $L_D = -0.0293a + 24.661$, i.e., 10 percent of heavy vehicles can cause a 0.29 dB decline on insertion loss in this scene. The results show that the SPL behind a barrier of a heavy vehicle flow is approximately 2.3 dB higher than that a light vehicle flow with the same source emission intensity. The same rule is also applicable to the overall shadow area. The heavy vehicles have more acoustical constituents at low frequencies compared to light vehicles, which are more significantly shaded by barriers. The insertion loss of heavy vehicle flow is the highest, followed by the insertion loss of mixed traffic flow, and the insertion loss of light vehicle flow is the lowest; this result indicates that the lower-frequency sound can bypass the barrier more easily. Thus, flow control of heavy vehicles whose sound is concentrated in low frequencies is an effective measure to improve the acoustic environment near roadways.

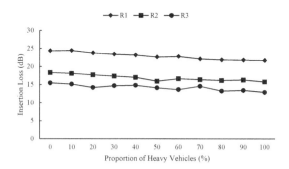

Figure 7. Insertion loss of the road barrier with different proportions of heavy vehicles.

4.2. Effect of Flow Speed

Vehicle noise characteristics of various speeds are different in frequencies according to the results of Section 2.2. Hence, the effects of different traffic flow speeds on the insertion loss of a noise barrier were analyzed. The flow of 1000 vehicles per hour consists of light and heavy vehicles with seven different flow speeds. The emission intensity of each vehicle source is set as 85 dB. Traffic noise with different traffic flow speeds (30 km/h, 40 km/h, 50 km/h, 60 km/h, 70 km/h, 80 km/h, and 90 km/h) on each point was calculated. Light vehicle flow, middle traffic flow, heavy vehicle flow, and mixed traffic flow (with 30% heavy vehicles) were considered in this section.

With different type constituents and speeds of traffic flow, the insertion losses at three points are shown in Figure 8. From the results of the calculated L_D values, the analysis is as follows:

A. The insertion losses of barrier increase with the flow speeds grow at all chosen points, regardless of the composition of vehicle types. For example, when the speed is below 40 km/h, the insertion loss of light vehicle flow at R_1 is approximately 23.7 dB, and the insertion loss increases non-linearly as the speed increases. At the speed of 90 km/h, the L_D value reached approximately 26.3 dB.

B. Figure 8 clearly shows that the noise barrier has sound-shading effects in all situations. The trend of increase of insertion loss with speed is observed, and the trend is considerably more obvious for light vehicle flow, middle traffic flow, and mixed traffic flow compared to heavy vehicle flow. The added L_D value at R_1 of a 30–90 km/h speed range of light vehicle flow, middle traffic flow, and mixed traffic flow is 2.65 dB, 1.73 dB, and 2.33 dB, respectively. However, the added insertion loss value of the same speed range of heavy vehicle flow is only 0.36 dB, which is considerably smaller. The data at R_2 and R_3 present the same rule because the main acoustics of light vehicle and middle vehicle are shifted to higher frequencies, whereas the main acoustics of heavy vehicle are slightly changed in frequency.

C. From the values, the SPL behind a barrier of a common mixed traffic flow with high speed is approximately 2 dB lower than the level with a low speed with the same source emission intensity. For a heavy vehicle flow, the sound pressure levels behind a barrier are little different with high or low flow speed.

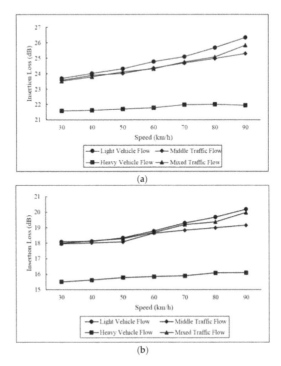

Figure 8. *Cont.*

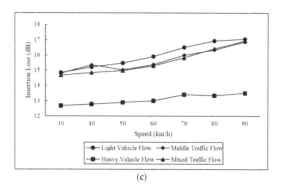

(c)

Figure 8. Insertion loss of barrier with different flow speeds. (**a**) Point R1; (**b**) Point R2; (**c**) Point R3.

5. Conclusions

A method for insertion loss calculation of a road noise barrier based on frequencies was presented in this paper. The proposed method realizes a more accurate calculation compared to the commonly used method based on a single 500-Hz sound. Several path length differences were preset to calculate the 1/3rd-octave-band diffraction and the total diffraction. Using different weights of spectrum according to the experimental noise data and traffic flow state, the insertion loss can be accurately calculated.

The method was verified by experiments with approximately 1.4 dB improvement in accuracy. The varied equivalent frequencies can be applied in the microscopic simulation of traffic noise, in which all different types of cars are considered to be the point source in the road network.

The noise protection of a road noise barrier with different flow states was analyzed based on a case study. The sound pressure level behind a barrier of a heavy vehicle flow is approximately 2.3 dB higher than that of a light vehicle flow with the same source emission intensity. The sound pressure level behind a barrier of a common mixed traffic flow with high speed is approximately 2 dB lower than the level with a low speed with the same source emission intensity. A road with high speed and high light vehicle proportion flow is recommended to establish a noise barrier.

Acknowledgments: This work was supported by the National Natural Science Foundation of China (No. 11574407) and the Fundamental Research Funds for the Central Universities (No. 17lgpy55).

Author Contributions: This paper is a result of the full collaboration of all the authors. Ming Cai and Haibo Wang conceived and designed the experiments; Peng Luo performed the experiments; Haibo Wang and Peng Luo analyzed the data; Haibo Wang wrote the paper.

Conflicts of Interest: The authors declare no conflict of interest.

References

1. Dehrashid, S.S.A.; Nassiri, P. Traffic noise assessment in the main roads of Sanandaj, Iran. *J. Low Freq. Noise Vib. Act. Control* **2015**, *34*, 39–48. [CrossRef]
2. Garacia, J.S.; Solano, J.J.P.; Serrano, M.C.; Camba, E.A.N.; Castel, S.F.; Asensi, A.S.; Suay, F.M. Spatial statistical analysis of urban noise data from a WASN gathered by an IoT system: Application to a small city. *Appl. Sci. Basel* **2016**, *6*, 380. [CrossRef]
3. Morel, J.; Marquisfavre, C.; Dubois, D.; Pierrette, M. Road traffic in urban areas: A perceptual and cognitive typology of pass-by noises. *Acta Acust. United Acust.* **2012**, *98*, 166–178. [CrossRef]
4. Di, G.Q.; Liu, X.Y.; Lin, Q.L.; Zheng, Y.; He, L.J. The relationship between urban combined traffic noise and annoyance: An investigation in Dalian, north of China. *Sci. Total Environ.* **2012**, *432*, 189–194. [CrossRef] [PubMed]

5. Roswall, N.; Raaschou-Nielsen, O.; Ketzel, M.; Overvad, K.; Halkjær, J.; Sørensen, M. Modeled traffic noise at the residence and colorectal cancer incidence: A cohort study. *Cancer Cause Control* **2017**, *28*, 745–753. [CrossRef] [PubMed]

6. Oiamo, T.H.; Luginaah, I.N.; Baxter, J. Cumulative effects of noise and odour annoyances on environmental and health related quality of life. *Soc. Sci. Med.* **2015**, *146*, 191–203. [CrossRef] [PubMed]

7. Monazzam, M.R.; Fard, S.M.B. Performance of passive and reactive profiled median barriers in traffic noise reduction. *Appl. Phys. Eng.* **2011**, *12*, 78–86. [CrossRef]

8. Reiter, P.; Wehr, R.; Ziegelwanger, H. Simulation and measurement of noise barrier sound-reflection properties. *Appl. Acoust.* **2017**, *123*, 133–142. [CrossRef]

9. Monazzam, M.R.; Lam, Y.W. Performance of profiled single noise barriers covered with quadratic residue diffusers. *Appl. Acoust.* **2005**, *66*, 709–730. [CrossRef]

10. Voropayev, S.I.; Ovenden, N.C.; Fernando, H.J.S.; Donovan, P.R. Finding optimal geometries for noise barrier tops using scaled experiments. *J. Acoust. Soc. Am.* **2017**, *141*, 722–736. [CrossRef] [PubMed]

11. Ishizuka, T.; Fujiwara, K. Performance of noise barriers with various edge shapes and acoustical conditions. *Appl. Acoust.* **2004**, *65*, 125–141. [CrossRef]

12. Arenas, J.P. Potential problems with environmental sound barriers when used in mitigating surface transportation noise. *Sci. Total Environ.* **2008**, *405*, 173–179. [CrossRef] [PubMed]

13. Zhao, W.C.; Chen, L.L.; Zheng, C.J.; Liu, C.; Chen, H.B. Design of absorbing material distribution for sound barrier using topology optimization. *Struct. Multidiscip. Optim.* **2017**, *56*, 315–329. [CrossRef]

14. Joynt, J. A Sustainable Approach to Environmental Noise Barrier Design. Ph.D. Thesis, University of Sheffield, Sheffield, UK, 2006.

15. Kotzen, B.; English, C. *Environmental Noise Barriers: A Guide to Their Acoustic and Visual Design*; E & FN Spon-Routledge: London, UK, 1999.

16. Sommerfeld, A. Mathematische theorie der diffraktion. *Math. Ann.* **1896**, *47*, 317–374. [CrossRef]

17. Wang, H.B.; Cai, M.; Zhong, S.Q.; Li, F. Sound field study of a building near a roadway via the boundary element method. *J. Low Freq. Noise Vib. Act. Control* **2017**, in press. [CrossRef]

18. He, Z.C.; Li, G.Y.; Liu, G.R.; Cheng, A.G.; Li, E. Numerical investigation of ES-FEM with various mass redistribution for acoustic problems. *Appl. Acoust.* **2015**, *89*, 222–233. [CrossRef]

19. Hiraishi, M.; Tsutahara, M.; Leung, R.C.K. Numerical simulation of sound generation in a mixing layer by the finite difference lattice Boltzmann method. *Comput. Math. Appl.* **2010**, *59*, 2403–2410. [CrossRef]

20. Ma, D.Y.; Shen, H. *Hankbook of Acoustics*; SciPress: Beijing, China, 2004.

21. Maekawa, Z. Noise reduction by screens. *Appl. Acoust.* **1968**, *1*, 157–173. [CrossRef]

22. Yamamoto, K.; Takagi, K. Expressions of Maekawa's chart for computation. *Appl. Acoust.* **1992**, *37*, 75–82. [CrossRef]

23. Menounou, P. A correction to Maekawa's curve for the insertion loss behind barriers. *J. Acoust. Soc. Am.* **2001**, *110*, 1828–1838. [CrossRef] [PubMed]

24. Kurze, U.J.; Anderson, G.S. Sound attenuation by barriers. *Appl. Acoust.* **1971**, *4*, 35–53. [CrossRef]

25. Cianfrini, C.; Corcione, M.; Fontana, L. Experimental verification of the acoustic performance of diffusive roadside noise barriers. *Appl. Acoust.* **2007**, *68*, 1357–1372. [CrossRef]

26. Li, K.M.; Wong, H.Y. A review of commonly used analytical and empirical formulae for predicting sound diffracted by a thin screen. *Appl. Acoust.* **2005**, *66*, 45–75. [CrossRef]

27. Anderson, G.S.; Menge, C.W.; Rossano, C.F.; Armstrong, R.E.; Ronning, S.A.; Fleming, G.G.; Lee, C.S.Y. FHWA traffic noise model, version 1.0: Introduction to its capacities and screen components. *Wall J.* **1996**, *22*, 14–17.

28. Shu, N.; Cohn, L.F.; Kim, T.K. Improving traffic-noise model insertion loss accuracy based on diffraction and reflection theories. *J. Transp. Eng.* **2007**, *133*, 281–287. [CrossRef]

29. Monazzam, M.R.; Nassiri, P. Contribution of quadratic residue diffusers to efficiency of tilted profile parallel highway noise barriers. *Iran. J. Environ. Health Sci. Eng.* **2009**, *6*, 271–284.

30. Grubeša, S.; Domitrović, H.; Jambrošić, K. Performance of traffic noise barriers with varying cross-section. *PROMET-Traffic Transp.* **2011**, *23*, 161–168. [CrossRef]

31. Huang, X.; Zou, H.; Qiu, X. A preliminary study on the performance of indoor active noise barriers based on 2D simulations. *Build. Environ.* **2015**, *94*, 891–899. [CrossRef]

32. Chen, S.M.; Wang, D.F.; Liang, J. Sound quality analysis and prediction of vehicle interior noise based on grey system theory. *Fluct. Noise Lett.* **2012**, *11*, 1250016. [CrossRef]
33. Ministry of Communications of the People's Republic of China. *Technical Specification for Construction of Asphalt Pavements*; JTG F40-2004; Ministry of Communications of the People's Republic of China: Beijing, China, 2005.
34. Cai, M.; Zhong, S.Q.; Wang, H.B.; Chen, Y.X.; Zeng, W.X. Study of the traffic noise source intensity emission model and the frequency characteristics for a wet asphalt road. *Appl. Acoust.* **2017**, *123*, 55–63. [CrossRef]
35. Zhang, Y.F. *Road Traffic Environmental Engineering*; China Communications Press: Beijing, China, 2001.
36. Ministry of Environmental Protection of the People's Republic of China. *Norm on Acoustical Design and Measurement of Noise Barriers*; HJ/T 90-2004; Ministry of Environmental Protection of the People's Republic of China: Beijing, China, 2004.

Review

The Boundary Element Method in Acoustics: A Survey

Stephen Kirkup

School of Engineering, University of Central Lancashire, Preston PR1 2HE, UK; smkirkup@uclan.ac.uk;
Tel.: +44-779-442-2554

Received: 23 January 2019; Accepted: 9 April 2019; Published: 19 April 2019

Abstract: The boundary element method (BEM) in the context of acoustics or Helmholtz problems is reviewed in this paper. The basis of the BEM is initially developed for Laplace's equation. The boundary integral equation formulations for the standard interior and exterior acoustic problems are stated and the boundary element methods are derived through collocation. It is shown how interior modal analysis can be carried out via the boundary element method. Further extensions in the BEM in acoustics are also reviewed, including half-space problems and modelling the acoustic field surrounding thin screens. Current research in linking the boundary element method to other methods in order to solve coupled vibro-acoustic and aero-acoustic problems and methods for solving inverse problems via the BEM are surveyed. Applications of the BEM in each area of acoustics are referenced. The computational complexity of the problem is considered and methods for improving its general efficiency are reviewed. The significant maintenance issues of the standard exterior acoustic solution are considered, in particular the weighting parameter in combined formulations such as Burton and Miller's equation. The commonality of the integral operators across formulations and hence the potential for development of a software library approach is emphasised.

Keywords: boundary element method; acoustics; Helmholtz equation

---

1. Introduction

Acoustics is the science of sound and has particular applications in sound reproduction, noise and sensing [1,2]. Acoustics may be interpreted as an area of applied mathematics [1,3]. In special cases, the mathematical equations governing acoustic problems can be solved analytically, for example if the equations are linear and the geometry is separable. However, for realistic acoustic problems, numerical methods provide a much more flexible means of solution [4,5]. Numerical methods are only useful in practice when they are implemented on computer. It is over the last sixty years or so that numerical methods have become increasingly developed and computers have become faster, with increasing data storage and more widespread. This brings us to the wide-ranging area of computational acoustics; the solution of acoustic problems on computer [6–11], which has significantly developed in this timescale.

In the context of this work, the acoustic field is assumed to be present within a fluid domain. If there is an obstacle within an existing acoustic field, then the disturbance it causes is termed *scattering*. If an object is vibrating and hence exciting an acoustic response, or contributing to an existing acoustic field, then this is termed *radiation*. If the fluid influences the vibration of the object or structure, and energy is exchanged in both directions between them, then this is termed coupled fluid-structure interaction [12] or, in the acoustic context, vibro-acoustics [13,14]. Alternatively, if the acoustic field is an outcome of a background flow, for example the generation of noise by turbulence, then this is termed aero-acoustics [11,15]. The determination of the properties of a vibrating or scattering object (for example, its shape, the surface impedance or the sound intensity) from acoustic measurements in the field is termed inverse acoustics [16].

Vibration analysis is not the focus of this work, but it clearly cannot be divorced from the study of acoustics. In vibration analysis, the domain of the structure oscillates, perhaps under variable loading or forcing. In the case of vibro-acoustics, the vibration can excite or be excited by the acoustic field and, in any analysis, both need to be modelled [17–19], and the equations coupled. The computational modelling of vibration is normally carried out by the finite element method (FEM) [20,21], and this method is significantly established in this application area. Vibrational modelling may be carried out in the time domain. However, with the nature of structural vibration in that it is normally dominated by modes and their corresponding resonant frequencies, and hence vibration is often close to periodic. Similarly, the excitation forces on a structure, such as the driver on a loudspeaker of the explosions in an engine can similarly be close to periodic in short time scales. It is, therefore, found that vibratory problems are routinely analysed in the frequency domain, both in the practical work and computational modelling.

An acoustic field is most straightforwardly interpreted as a sound pressure field, with the sound pressure varying over the extent of the domain, and with time. The transient acoustic field can be computationally modelled by the finite element method [22,23], the finite difference—time domain method (a particular finite difference method that was originally developed for electromagnetic simulation [24–26]) can also be applied in acoustics [27–30]) and the boundary element method [31–34], but acoustics and vibration are more often analysed and modelled in the frequency domain. In acoustic problems, the most likely fluid domains are air or water and, in many cases in these fluids, the linear wave equation is an acceptable model. By observing one frequency at a time, the wave equation can be simplified as a sequence of Helmholtz equations. Again, there are a variety of methods for the numerical solution of Helmholtz problems; the finite difference method, the finite element method [23,35–43] and, the subject of this paper, the boundary element method.

The boundary element method is one of several established numerical or computational methods for solving boundary-value problems that arise in mathematics, the physical sciences and engineering [44]. Typically, a boundary-value problem consists of a domain within or outside a boundary in which a variable of interest or physical property is governed by an identified partial differential equation (PDE). A computational method for solving a boundary-value problem is tasked with finding an approximation to the unknown variable within the domain. Mostly, this is carried out by domain methods, such as the finite element method [45], in which a mesh is applied to the domain. However, the boundary element method works in an entirely different way in that in the first stage further information is found on the boundary only; the solution at the domain points is found in the second stage by integrating over the known boundary data. The boundary element method requires a mesh of the boundary only, and hence is generally easier to apply than domain methods. The number of elements in the BEM is therefore expected to be much less than in the corresponding finite element method (for the same level of required accuracy or element size), and there is therefore often a potential for significant efficiency savings. The BEM is not as widely applicable as the domain methods; when problems are non-linear, for example, the application of the development of a suitable BEM requires significant further adaption [46–49]. Typically, the boundary element method has found application in sound reproduction modelling, such as loudspeakers [50,51], sonar transducers [52] and in modelling noise from vehicles [53–56] and, more recently, aircraft [57–59].

For the acoustic boundary element method to be accessible, and hence widely used, it has to be implemented in software. This was precisely the rationale behind the development of the software [60] and the monograph [61], the latter also serving as a manual. Several texts on the same theme were published at about the same time [62,63], following on from the earlier collection of works [64]. A chapter of a recent book contains a modern introduction to the acoustic boundary element method [65].

Implementing the acoustic BEM as software continues to be challenging in terms of scoping, the choice of sub-methods and efficiency. The method requires matrices to be formed with each component being the result of an integration, with some of the integrals being singular or hypersingular. The standard BEM requires the solution of a linear system of equations that is formed from the matrices, or, if the BEM is used for modal analysis, a non-linear eigenvalue problem. There have been significant reliability issues with the BEM for exterior problems. These challenges have been met, but much of the research is focused on future-proofing the method, with the increasing expectations in scaling-up and maintaining reasonable processing time. There is a continual desire to progress to higher resolutions of elements, particularly as this is necessary for modelling high frequency problems.

2. Acoustic Model

In this Section, the underlying acoustic model of the wave equation that governs the sound pressure in the domain is stated. It is shown how the model can be revised into a sequence of Helmholtz problems for periodic signals. The classes of domains that can be solved by boundary element methods are summarised and a generalised boundary condition is adopted. The other acoustic properties, such as sound intensity, radiation efficiency and sound power that that are mostly used in exterior 'noise' problems, are defined. Traditionally, acoustics properties are presented in the decibel scale, and it is shown how they are converted.

2.1. The Wave Equation and the Helmholtz Equation

The acoustic field is assumed to be present in the domain of a homogeneous isotropic fluid and it is modelled by the linear wave equation,

$$\nabla^2 \Psi(\mathbf{p}, t) = \frac{1}{c^2} \frac{\partial^2}{\partial t^2} \Psi(\mathbf{p}, t), \tag{1}$$

where $\Psi(\mathbf{p}, t)$ is the scalar time-dependent velocity potential related to the time-dependent particle velocity $V(\mathbf{p}, t)$ by $V(\mathbf{p}, t) = \nabla \Psi(\mathbf{p}, t)$ and c is the propagation velocity ($\mathbf{p}$ and t are the spatial and time variables). The time-dependent sound pressure $Q(\mathbf{p}, t)$ is given in terms of the velocity potential by $Q(\mathbf{p}, t) = -\rho \frac{\partial}{\partial t} \Psi(\mathbf{p}, t)$ where ρ is the density of the acoustic medium.

The time-dependent velocity potential $\Psi(\mathbf{p}, t)$ can be reduced to a sum of components each of the form

$$\Psi(\mathbf{p}, t) = Re\{\varphi(\mathbf{p}) e^{-i\omega t}\}, \tag{2}$$

where ω is the angular frequency ($\omega = 2\pi\nu$, where ν is the frequency in hertz) and $\varphi(\mathbf{p})$ is the (time-independent) velocity potential. The substitution of the above expression into the wave equation reduces it to the Helmholtz (reduced wave) equation:

$$\nabla^2 \varphi(\mathbf{p}) + k^2 \varphi(\mathbf{p}) = 0, \tag{3}$$

where $k^2 = \frac{\omega^2}{c^2}$ and k is the wavenumber. The complex-valued function φ relates the magnitude and phase of the potential field.

Similarly, the components of the particle velocity have the form $V(\mathbf{p}, t) = Re\{\nabla\varphi(\mathbf{p}) e^{-i\omega t}\}$. Often the boundary normal velocity $v(p)$ is given as a condition or is required and this is defined as follows,

$$v(\mathbf{p}) = \nabla\varphi(\mathbf{p}) \cdot n_p = \frac{\partial\varphi(\mathbf{p})}{\partial n_p}, \tag{4}$$

where n_p is the unit normal to the boundary at p.

2.2. Acoustic Properties

To carry out a complete solution, the wave equation is written as a series of Helmholtz problems, through expressing the boundary condition as a Fourier series with components of the form in Equation (2). For each wavenumber and its associated boundary and other conditions, the Helmholtz equation is then solved. The time-dependent velocity potential $\Psi(\mathbf{p}, t)$ can then be constituted from the separate solutions, but it is more usual that the results are considered in the frequency domain.

The sound pressure $p(\mathbf{p})$ at the point $\mathbf{p}$ in the acoustic domain is one of the most useful acoustic properties, and it is related to the velocity potential by the formula $p(\mathbf{p}) = i\omega\rho\varphi(\mathbf{p})$. In practice, the magnitude of the sound pressure is measured on the decibel scale in which it is evaluated as the sound pressure level as $20\log_{10}\left|\frac{p(\mathbf{p})}{\sqrt{2}p^*}\right|$, where p^* is the reference pressure of 2×10^{-5} Pa.

Particularly for 'noise' problems, the sound power, the time-averaged sound intensity and radiation efficiency are often considered to be useful properties. The normal sound intensity $I(\mathbf{p})$ at points p on a boundary is defined by the formula

$$I(\mathbf{p}) = \frac{1}{2}Re\{\bar{p}(\mathbf{p})v(\mathbf{p})\}, \tag{5}$$

where $\bar{p}$ represents the complex conjugate of p. The sound power W is an aggregation of the sound intensity to one value by direct integration,

$$W = \int_H I(\mathbf{q})dH_q. \tag{6}$$

where H is a boundary. The sound power is also often expressed in decibels as the sound power level as $10\log_{10}\left|\frac{W}{W^*}\right|$, where W^* is the reference sound power of 10^{-12} watts. The radiation ratio is defined as $\frac{W}{\frac{1}{2}\rho c \int_H v^*(q)v(q)dH_q}$.

2.3. The Scope of the Boundary Element Method in the Solution of Acoustic/Helmholtz Problems

In applying the boundary element method to acoustic or Helmholtz problems, the user is in effect adopting a model for the boundary/ies and domain(s), the nature of which determines the integral equation(s) that is/are employed. Traditionally, the BEM has been developed to solve the acoustic or Helmholtz problem interior or exterior to a closed boundary in the standard physical domains are two-dimensional, three-dimensional and axisymmetric three-dimensional space. The basis of the method is the integral equations that arise through applying Green's theorems (direct BEM) or by using layer potentials (indirect BEM). The exterior problem has received much more attention, because of its value in solving over an infinite domain from a surface mesh and because of the difficulty in resolving its reliability problem. Domain methods, such as the finite element method, may also be applied; more elements are required, but the matrices are sparse and structured and the FEM is more established than the BEM in general. If the FEM is applied to exterior problems then techniques such as infinite elements [66] can be used to complete the outer mesh or the perfectly matched layer may be used to absorb outgoing waves [67–69]. The solution of the interior and exterior acoustic problem by the BEM is considered in Sections 4.1 and 4.3.

An outline of the equations that arise in the finite element and related methods is given in Section 5.3.1. In domain methods like the FEM, whether it is applied to a structure or an enclosed fluid, a linear eigenvalue problem is a natural consequence. With the FEM, it is routine to extract the modes and resonant frequencies and use the modal basis to determine the response under excitation. From that perspective, it is natural to include the eigenvalue problem within the boundary element library. However, the application of the BEM leads to a non-linear eigenvalue problem, which requires special solution techniques and hence the construction of solutions through the modal basis is not a natural pathway in the BEM. Acoustic modal analysis via the BEM is considered in Section 4.2.

Although significantly restrictive, one of the simplest acoustic models is the Rayleigh integral [70]. The model is that of a vibrating plate set in an infinite rigid baffle and radiating into a half-space. The Rayleigh integral relates the velocity potential or sound pressure at any point to the velocity map on the plate. As the Rayleigh integral shares its one operator with the integral equation formulations in the boundary element method, the Rayleigh integral method (RIM) [71] can be adopted into the boundary element method fold. The Rayleigh integral is a special case of half-space problems and these are considered in Section 5.1

An acoustic model that uses the same operators—and hence fits in the BEM context—is that of a shell discontinuity. For example, this can be used to model the acoustic field around a thin screen or shield. The integral equation formulation for the shell are derived from those used in the traditional BEM, but taking the limit as the boundary becomes thinner [72,73], leaving a model that relates a discontinuity in the field across the shell. Shell elements in acoustics are considered in Section 5.2.

The development and application of the BEM from the models described are reviewed in this paper. Although this is a fairly exhaustive list of basic models as things stand, hybrid models can also be developed by using superposition or applying continuity and they will also be considered. The BEM may be applied in vibro-acoustics, aero-acoustics and inverse acoustics and these areas are discussed in Sections 5.3–5.5. Hybrid boundary element methods that include half-space formulation are covered in Section 5.1.2 and shells in Section 5.2.2.

There have been recent developments in the development of the BEM for periodic structures, with noise barriers being the usual application [74–81]. This special case will not be developed further in this paper.

2.4. Boundary Conditions

To maintain reasonable generality in the software, the author has generally worked with the boundary condition of the following general (Robin) form

$$\alpha(p)\varphi(p) + \beta(p)v(p) = f(p),\tag{7}$$

with the condition that $\alpha(p)$ and $\beta(p)$ cannot both be zero at any value of p. This model includes the Dirichlet boundary condition by setting $\beta(p) = 0$, the Neumann boundary condition by setting $\alpha(p) = 0$ and an impedance condition by setting $f(p) = 0$. In the boundary condition (7), p is any point on the boundary and α, β and f are complex-valued functions that may vary across the boundary. Although the boundary condition for the shell is an adaption of this, this generalised boundary condition model is apparently achievable and seems to cover most current expectations. An explanation of typical boundary conditions that occur in acoustics can be found in Marburg and Anderssohn [82].

For exterior problems, it is also necessary to introduce a condition at infinity, in order to ensure that all scattered and radiated waves are outgoing. This is termed the Sommerfeld radiation condition. In two-dimensional space the condition is

$$\lim_{r \to \infty} r^{\frac{1}{2}}\left(\frac{\partial \varphi}{\partial r} - ik\varphi\right) = 0\tag{8}$$

and

$$\lim_{r \to \infty} r\left(\frac{\partial \varphi}{\partial r} - ik\varphi\right) = 0.\tag{9}$$

in three dimensions. A thorough discussion on the Sommerfeld radiation condition can be found in Ihlenburg [43] (pp. 6–8).

3. The Boundary Element Method and the Laplace Equation Stem

Laplace's equation is the special case of the Helmholtz Equation (3) with $k = 0$,

$$\nabla^2\varphi(\mathbf{p}) = 0.\tag{10}$$

Although Laplace's equation models many phenomena, it is not directly useful in acoustic modelling. However, many of the issues that have to be tackled in solving the Helmholtz equation by the boundary element method are also found with Laplace's equation. The methods applied in developing the BEM for Laplace's equation also can be developed to be used in the Helmholtz problems. As such, applying the BEM to Laplace's equation is a useful entry point in initiating work on the Helmholtz equation. For example, much of the author's work in solving Helmholtz problems has been underpinned by corresponding work on Laplace's equation [83–87]. Thus, in this section, the BEM is communicated in its simplest, but still realistic and practical context as a stepping-stone on the journey to the full scope of the BEM in acoustics, the subject of this paper. The theoretical basis is set out in Kellogg [88] and development of integral equation methods in Jaswon and Symm [89]. In this Section, the boundary element method is derived for Laplace's equation, and this forms the foundation for its application to acoustic problems. In terms of comunication, it is helpful to sustain a notational conformance, that continues through this paper, and is helpful in relating mathematical expressions and their discrete equivalent as software components. Integration methods for finding the matrix components are considered. Interesting and useful properties of some of the operators and matrices are revealed.

3.1. Elementary Integral Equation Formulation for the Interior Problem

The boundary element method is not based on the direct solution of the PDE, but rather its reformulation as an integral equation. Historically, the integral equation reformulation of the PDE has followed two distinct routes, termed the direct method and the indirect method. The direct method is based on Green's second identity

$$\int_D \left(\varphi(q)\nabla^2\psi(q) - \psi(q)\nabla^2\varphi(q) \right) dV_q = \int_S \varphi(q)\frac{\partial\psi(q)}{\partial n_q} - \psi(q)\frac{\partial\varphi(q)}{\partial n_q} \, dS_q, \tag{11}$$

where φ and ψ are twice-differentiable scalar function in a domain D that is bounded by the closed surface S. If φ is a solution of Laplace's equation, $\nabla^2\varphi = 0$, then

$$\int_D \varphi(q)\nabla^2\psi(q) \, dV_q = \int_S \varphi(q)\frac{\partial\psi(q)}{\partial n_q} - \psi(q)\frac{\partial\varphi(q)}{\partial n_q} \, dS_q. \tag{12}$$

Green's third identity can be derived from the second identity by choosing $\psi(q) = G(p,q)$, where G is a Green's function. A Green's function is a fundamental solution of the partial differential equation, in this case Laplace's equation, that is the effect or influence a unit source at the point p has at the point q, and is defined by $\nabla^2 G(p,q) = -\delta(p-q)$. For the two-dimensional Laplace equation $G(p,q) = -\frac{1}{2\pi}\ln(r)$ and for three-dimensional problems $G(p,q) = \frac{1}{4\pi r}$ where $r = |p-q|$. The substitution of the Green's function into Equation (12) gives

$$\int_D \varphi(q)\nabla^2 G(p,q) \, dV_q = \int_S \varphi(q)\frac{\partial G(p,q)}{\partial n_q} - G(p,q)\frac{\partial\varphi(q)}{\partial n_q} \, dS_q$$

or

$$-\int_D \varphi(q)\delta(p-q) \, dV_q = \int_S \varphi(q)\frac{\partial G(p,q)}{\partial n_q} - G(p,q)\frac{\partial\varphi(q)}{\partial n_q} \, dS_q. \tag{13}$$

Hence, as a result of the properties of the Dirac delta function,

$$\int_S \varphi(q)\frac{\partial G(p,q)}{\partial n_q} - G(p,q)\frac{\partial\varphi(q)}{\partial n_q} \, dS_q = \begin{cases} -\varphi(p) \text{ if } p \in D \\ 0 \text{ if } p \in E \\ -c(p)\varphi(p) \text{ if } p \in S \end{cases}, \tag{14}$$

where $c(p)$ is the angle (in 2D, divided by 2π) or solid angle (in 3D, divided by 4π) subtended by the interior domain at the boundary point p. (Similarly, for exterior problems, $c(p)$ similarly relates to the exterior angle). If the boundary is smooth at p then $c(p) = \frac{1}{2}$, and, for simplicity, this is the value that will be used for the remainder of this paper.

In the indirect method, φ is presumed to be related by a layer potential σ on the boundary

$$\varphi(p) = \int_S G(p,q)\, \sigma(q)\, dS_q \quad (p \in D \cup S). \tag{15}$$

Differentiating Equation (15) with respect to a unit outward normal to a point on the boundary that that passes through p, gives the following equation:

$$\frac{\partial \varphi(p)}{\partial n_p} = \int_S \frac{\partial G(p,q)}{\partial n_p}\, \sigma(q)\, dS_q \quad (p \in D). \tag{16}$$

As p approaches the boundary, the integral operator has a 'jump' similar to the direct method (14) resulting in the following equation:

$$\frac{\partial \varphi(p)}{\partial n_p} = \int_S \frac{\partial G(p,q)}{\partial n_p}\, \sigma(q)\, dS_q + \frac{1}{2}\sigma(p) \quad (p \in S). \tag{17}$$

The BEM can be derived from the equations in this Section. The equation defined on the boundary (the last one in Equation (14) for the direct method, Equations (15) and (17) for the indirect method ($p \in S$)), the boundary integral equations, are used to find the unknown functions from the known data on the boundary. The corresponding integrals defined in the domain (the first one in Equation (14) for the direct method and Equation (15) for ($p \in D$) in the indirect method) return the solution at the chosen domain points.

3.2. Operator Notation and Further Integral Equations for the Laplace Problem

In this Subsection, operator notation is introduced and further integral equations are introduced in order to illuminate some useful properties. Operator notation provides a shorthand and is an aid to communication. The Laplace integral operators are defined as follows:

$$\{L\mu\}_\Gamma(p) = \int_\Gamma G(p,q)\, \mu(q) dS_q , \tag{18}$$

$$\{M\mu\}_\Gamma(p) = \int_\Gamma \frac{\partial G(p,q)}{\partial n_q}\, \mu(q) dS_q , \tag{19}$$

$$\{M^t\mu\}_\Gamma(p; v_p) = \frac{\partial}{\partial v_p} \int_\Gamma G(p,q)\, \mu(q) dS_q , \tag{20}$$

where Γ is a boundary (not necessarily closed), n_q is the unique unit normal vector to Γ at q, v_p is a unit directional vector passing through p and $\mu(q)$ is a function defined for $q \in \Gamma$. With this notation, the Equations (14), which form the basis of the elementary direct method for the interior problem, can be written as

$$\{M\varphi\}_S(p) - \{Lv\}_S(p) = \begin{cases} -\varphi(p) \text{ if } p \in D \\ 0 \text{ if } p \in E \\ -\frac{1}{2}\varphi(p) \text{ if } p \in S \end{cases} \tag{21}$$

Similarly, the equations for the indirect method ((15) and (17)) in operator notation for the interior problem are as follows:

$$\varphi(p) = \{L\sigma\}_S(p) \quad (p \in D \cup S), \tag{22}$$

$$v(p) = \{M^t\sigma\}_\Gamma(p; n_p) + \frac{1}{2}\sigma(p) \quad (p \in S). \tag{23}$$

Further integral equations for the direct formulation can be obtained by differentiating (21) (as in the derivation of Equations (16), (17)) and for the indirect formulation by introducing a *double layer potential*. Although they are generally unnecessary in solving Laplace problems, the equations resulting from differentiating the elementary integral equations or using double layer potentials are often used in the exterior acoustic problem, considered in the next Section. These equations are also useful in general and they will illuminate some important aspects. For example, differentiating Equation (21) with respect to the outward normal to the boundary,

$$\frac{\partial}{\partial n_p}\{M\varphi\}_S(p) - \frac{\partial}{\partial n_p}\{Lv\}_S(p) = -\frac{1}{2}\frac{\partial\varphi(p)}{\partial n_p}(p \in S), \tag{24}$$

which can be written in operator notation,

$$\{N\varphi\}_S(p; n_p) - \{M^t v\}_S(p; n_p) = -\frac{1}{2}v(p)(p \in S), \tag{25}$$

in which a new operator, N, has been introduced,

$$\{N\mu\}_\Gamma(p; v_p) = \frac{\partial}{\partial v_p}\int_\Gamma \frac{\partial G(p, q)}{\partial n_q}\mu(q)dS_q. \tag{26}$$

Further indirect integral equations may be obtained through presuming the field is the result of a double layer potential

$$\varphi(p) = \{M\zeta\}_S(p) \quad (p \in D), \tag{27}$$

$$v(p) = \{N\zeta\}_S(p; n_p) \quad (p \in S), \tag{28}$$

and, taking into account the jump discontinuity,

$$\varphi(p) = \{M\zeta\}_S(p) - \frac{1}{2}\zeta(p) \quad (p \in S). \tag{29}$$

The integral equations for the exterior Laplace problem may be derived in the same way. In general, the equations are the same as for the interior problem, except for a change of sign. The equations that make up the direct formulation are

$$\{M\varphi\}_S(p) - \{Lv\}_S(p) = \begin{cases} \varphi(p) \text{ if } p \in E \\ 0 \text{ if } p \in D \\ \frac{1}{2}\varphi(p) \text{ if } p \in S \end{cases} \tag{30}$$

$$\{N\varphi\}_S(p; n_p) - \{M^t v\}_S(p; n_p) = \frac{1}{2}v(p) \text{ if } p \in S. \tag{31}$$

The equivalent of equations for the indirect method ((15) and (17)) in operator notation for the exterior problem are similar, with just a couple of sign changes to indicate that the jump discontinuity in M and M^t in the limit from the exterior, rather than the interior:

$$\varphi(p) = \{L\sigma\}_S(p) \quad (p \in E \cup S), \tag{32}$$

$$v(p) = \{M^t\sigma\}_S(p; n_p) - \frac{1}{2}\sigma(p) \quad (p \in S). \tag{33}$$

$$\varphi(p) = \{M\zeta\}_S(p) \quad (p \in E), \tag{34}$$

$$v(p) = \{N\zeta\}_S(p; n_p) \quad (p \in S), \tag{35}$$

$$\varphi(p) = \{M\zeta\}_S(p) + \frac{1}{2}\zeta(p) \quad (p \in S). \tag{36}$$

3.3. Derivation of the Boundary Element Method

There are several integral equation methods that can be applied to transform integral equations into boundary element methods, but the most popular and straightforward method is that of collocation [90]. The development of the boundary element method from the selected integral equation requires that the boundary is represented by a set of *panels* or a mesh. An integral equation method is applied to solve the boundary integral equation. The solution in the domain can then be achieved by effecting the appropriate integration over the boundary.

3.3.1. Boundary Element Approximation

In the spirit of the author's previous work [61,91], in order to maintain generality, the discrete forms of the Laplace operators (18)–(20), (26) are sought, in order to effectively become a software component. Let the boundary Γ in Equation (18) be represented by the approximation $\tilde{\Gamma}$, a set of n panels:

$$\Gamma \approx \tilde{\Gamma} = \sum_{j=1}^{n} \Delta \tilde{\Gamma}_j , \tag{37}$$

The boundary function μ is replaced by its equivalent $\tilde{\mu}$ on the approximate boundary $\tilde{\Gamma}$:

$$\{L\mu\}_{\Gamma}(p) \approx \{L\tilde{\mu}\}_{\tilde{\Gamma}}(p) = \int_{\tilde{\Gamma}} G(p,q)\, \tilde{\mu}(q) dS_q = \sum_{j=1}^{n} \int_{\Delta\tilde{\Gamma}_j} G(p,q)\, \tilde{\mu}(q) dS_q. \tag{38}$$

In general, the function on the boundary is replaced by a sum of a set of basis functions. The simplest approximation is that of approximating the boundary functions by a constant on each panel:

$$\sum_{j=1}^{n} \int_{\Delta\tilde{\Gamma}_j} G(p,q)\, \tilde{\mu}(q) dS_q \approx \sum_{j=1}^{n} \int_{\Delta\tilde{\Gamma}_j} G(p,q)\, \tilde{\mu}_j dS_q = \sum_{j=1}^{n} \tilde{\mu}_j \{L\tilde{e}\}_{\Delta\tilde{\Gamma}_j}(p), \tag{39}$$

where $\tilde{e} = 1$.

For example, for the simple boundary integral Equation (22) with $p \in S$,

$$\varphi(p) = \{L\sigma\}_S(p) \approx \sum_{j=1}^{n} \tilde{\sigma}_j \{L\tilde{e}\}_{\Delta\tilde{S}_j}(p) . \tag{40}$$

The most common method of solving boundary integral equations is collocation, in which a linear system is formed through setting p to take the value of each collocation point in turn:

$$\varphi_{Si} = \varphi(p_{Si}) = \{L\sigma\}_S(p_{Si}) \approx \sum_{j=1}^{n} \tilde{\sigma}_j \{L\tilde{e}\}_{\Delta\tilde{S}_j}(p_{Si}) \text{ for } i = 1, 2, \ldots n . \tag{41}$$

3.3.2. Solution by Collocation

The linear system of approximations (41) may be written in matrix-vector form

$$\underline{\varphi}_S \approx L_{SS}\underline{\sigma}_S , \tag{42}$$

where $\underline{\varphi}_S = \begin{bmatrix} \varphi_{S1} \\ \varphi_{S2} \\ \vdots \\ \vdots \\ \varphi_{Sn} \end{bmatrix}$, $\underline{\sigma}_S = \begin{bmatrix} \sigma_{S1} \\ \sigma_{S2} \\ \vdots \\ \vdots \\ \sigma_{Sn} \end{bmatrix}$, and $[L_{SS}]_{ij} = \{L\tilde{e}\}_{\Delta\tilde{S}_j}(p_{Si})$. For example, for the Dirichlet problem in which $\underline{\varphi}_S$ is known, the solution of system (42) (now as an equation relating approximate values) returns an approximation $\widehat{\underline{\sigma}}_S$ to $\underline{\sigma}_S$. To find the solution at a set of m points in the domain the integral (22) is evaluated at the points $\boldsymbol{p}_{Di} \in D$ for $i = 1, 2, \ldots m$;

$$\varphi_{Di} = \varphi(\boldsymbol{p}_{Di}) = \{L\sigma\}_D(\boldsymbol{p}_{Di}) \approx \sum_{j=1}^{n} \tilde{\sigma}_j \{L\tilde{e}\}_{\Delta\tilde{S}_j}(\boldsymbol{p}_{Di}) \text{ for } i = 1, 2, \ldots m,$$

or, more concisely,

$$\underline{\varphi}_D \approx L_{DS}\underline{\sigma}_S, \tag{43}$$

where $\underline{\varphi}_D = \begin{bmatrix} \varphi_{D1} \\ \varphi_{D2} \\ \vdots \\ \vdots \\ \varphi_{Dn} \end{bmatrix}$ and $[L_{DS}]_{ij} = \{L\tilde{e}\}_{\Delta\tilde{S}_j}(\boldsymbol{p}_{Di})$. Hence approximations to the solution within the

domain $\widehat{\underline{\varphi}}_D$ may be found by the matrix-vector multiplication

$$\widehat{\underline{\varphi}}_D = L_{DS} \widehat{\underline{\sigma}}_S. \tag{44}$$

3.3.3. The Galerkin Method

Although most of the implementations of the boundary element method are derived through collocation, other techniques can be used, most typically the Galerkin method. The Galerkin method and collocation can both be derived from a more generalised approach that are termed weighted residual methods. For example, for the approximation (40), the residual is the difference between the approximation and the exact solution;

$$R(\underline{\sigma}_S; \boldsymbol{p}) = \sum_{j=1}^{n} \tilde{\sigma}_j \{L\tilde{e}\}_{\Delta\tilde{S}_j}(\boldsymbol{p}) - \varphi(\boldsymbol{p}) \quad (\boldsymbol{p} \in S).$$

In weighted residual methods, $R(\underline{\sigma}_S; \boldsymbol{p})$ is integrated with *test* basis functions $\chi_i(\boldsymbol{p})$ $(\boldsymbol{p} \cdot S)$ and the methods arise by setting the result to zero;

$$\int_S R(\widehat{\sigma}_S; \boldsymbol{p}_i)\chi_i(\boldsymbol{p}_i)dS - 0$$

at points $\boldsymbol{p}_i$ on the boundary, for $i = 1, 2, \ldots n_S$.

If $\chi_i(\boldsymbol{p}) = \delta(\boldsymbol{p} - \boldsymbol{p}_i)$, the Dirac delta function, the collocation method is derived;

$$\int_S R(\widehat{\sigma}_S; \boldsymbol{p})\delta(\boldsymbol{p} - \boldsymbol{p}_i) dS = R(\widehat{\sigma}_S; \boldsymbol{p}_i) = 0,$$

which leads to the methods outlined above. In the Galerkin method, the test functions are the same as the original basis functions. For example, for constant elements, the basis and the test functions are defined as

$$\chi_i(\boldsymbol{p}) = \begin{cases} 1 \text{ or } \tilde{e}, \text{ if } p \in \Delta\tilde{S}_i \\ 0, \text{ otherwise} \end{cases}.$$

Substituting this and the definition of the residual equation

$$\int_S R(\widehat{\sigma}_S; \boldsymbol{p}_i)\chi_i(\boldsymbol{p})dS = \int_{\Delta\tilde{S}_i} \sum_{j=1}^{n} \widehat{\sigma}_j \{L\tilde{e}\}_{\Delta\tilde{S}_j}(\boldsymbol{p}_i) - \varphi(\boldsymbol{p}_i) dS = 0.$$

The reason for the relative unpopularity of the Galerkin approach in boundary elements is that the matrix is now the result of a double integration, rather than the single integration in the collocation method. However, the matrix is understood to be symmetric.

3.4. Properties and Further Details

In the boundary element method, the boundary is represented by a mesh of *panels*. The boundary functions are approximated by a linear combination of basis functions on each panel. The boundary element is the combination the panel and the functional representation. In the previous Subsection, constant elements were mentioned as an example. In the finite element method, the degree of the approximation has to be at least half the order of the PDE. Fortunately, this is not the case for the application of the boundary element method, where constant elements are widely used. The panels that make up the boundaries are most simply represented by straight line panels in 2D, triangles in 3D and conic sections for axisymmetric 3D problems.

In this Subsection, important properties and further details of Laplace's equation, the related boundary integral equations and operators are discussed. An overview of methods for carrying out the integrations is included. The issue of non-uniquesness, that is an important feature of the boundary integral equation formulations for the exterior acoustic problem, is first addressed with Laplace's equation and useful outcomes from this are outlined.

3.4.1. Integration

In the previous Subsection, it was shown that up to four operators are involved in the boundary element method for Laplace problems, and these four operators extend to acoustic/Helmholtz problems. For the Laplace problem, it may be possible to develop analytic expressions for the integrals [92], but in general—and particularly for acoustic/Helmholtz problems—numerical integration is necessary. In the main, the integrals are regular, and these are most efficiently evaluated by Gauss-Legendre quadrature [93]. This is straightforward to apply to straight-line panels and in the generator and (typically in composite form) azimuthally. There are also published points and weights for Gaussian quadrature on a triangle [94].

However, the integrals corresponding to the diagonal components of the L_{SS} matrix are weakly singular and the diagonal components of the N_{SS} matrix are hypersingular, that is when the point lies on the panel. Moreover, if the point is close to the panel, for example at the centre of a neighbouring panel then the integrals are said to be nearly singular and may require a more accurate numerical integration rule, or special treatment [95]. Special numerical integration methods can be applied in order to evaluate the singular integrals and expressions for the hypersingular integrals may be found through limiting process. The weakly singular integrals, corresponding to the diagonal components of the L_{SS} matrix have a $O(\ln r)$ singularity in 2D and an $O\left(\frac{1}{r}\right)$ singularity in 3D, where r is the distance from the central collocation point. For the axisymmetric 3D case, the azimuthal integration resolves the $O\left(\frac{1}{r}\right)$ singularity to an $O(\ln r)$ singularity on the generator. For the simple elements discussed, analytic expressions for the diagonal components of the L_{SS} and N_{SS} matrices are available for the 2D and 3D (non-axisymmetic) cases for Laplace's equation. Further work on singular integration can be found in the following references [96–101].

Although for simplicity, the four operators are often lumped together as integral operators, N is not an integral operator: N is termed a pseudo-differential operator, and it therefore has distinct properties. The full expressions for the straight line and triangular panels are given in the author's book [61] (pages 49–50). For illustration the expressions are given for the straight-line panel of length h (2D) and for an equilateral triangular panel (3D) with each side of length h:

$$\{N\varphi\}_\Delta(p; n_p) = -\frac{2}{\pi h} \text{ for the straight-line panel of length } h \text{ (2D)} \tag{45}$$

$$\{N\varphi\}_\Delta\big(p;\boldsymbol{n_p}\big) = -\frac{3}{2\pi h} \text{ for the equilateral triangular panel with side of length } h \text{ (3D)} \qquad (46)$$

There is one simple but noteworthy remark to be made about the expressions (45) and (46). Normally, if the domain of integration is reduced in size, the integral is similarly reduced, at least in the limit as the domain size converges to zero. However, with these hypersingular integrals, the opposite is found to be the case! A significant consequence of this property is considered in Section 6.2.

3.4.2. Non-Uniqueness and Its Useful Outcomes

The non-uniqueness of some boundary integral equations in the exterior acoustic problem at a set of frequencies is well-known, and the issue and methods of resolution will be considered in the next Section. Because of the importance of the non-uniqueness of solution within the context of this paper, it is helpful to visit this early, with a simpler equation. Although the term non-uniqueness often prefixes the word problem, it is found that some very useful outcomes arise from this analysis.

The non-uniqueness is found with Laplace's equation itself; if φ is a solution of the interior Neumann problem then $\varphi + c$ is also solution, where c is any constant. The non-uniqueness in the interior Laplace problem with a Neumann boundary condition must be reflected in the boundary integral equations. It therefore follows from Equation (21) that the operator $M + \frac{1}{2}I$ is degenerate, and similarly for $M^t + \frac{1}{2}I$ from Equation (23) and N from Equations (25) and (28).

As we will see in the next Section on acoustic/Helmholtz problems, the operators from the interior problem equations are shared with the exterior problem, and this is shown for Laplace's equation in Section 3.2. For the exterior problem, the derivative direct formulation (31) is unsuitable for solving exterior Laplace problems as both the N and $M^t + \frac{1}{2}I$ operators are degenerate. Similarly, the indirect formulations, formed from double layer potentials, (35) and (36) are unsuitable. This correlation between the interior Neumann problem and the exterior derivative (direct) and double-layer potential (indirect) formulations carries through to the Helmholtz equation.

Most simply, if $v(p) = 0$ on the boundary then φ is any constant (e.g., $\varphi = 1$) throughout the domain is a solution. Substituting these values into the discrete form of the boundary integral Equation (21), the resulting matrix-vector equation is $\left[M_{SS} + \frac{1}{2}I\right]\underline{1} \approx \underline{0}$, where $\underline{0}$ is a vector of zeros and $\underline{1}$ is a vector of ones; every row of the M_{SS} matrix must approximately sum to $-\frac{1}{2}$. Similarly, from Equation (25), $N_{SS}\underline{1} \approx \underline{0}$; every row of the N_{SS} matrix must approximately sum to zero. A potentially useful outcome of this is that the hypersingular diagonal components of the N_{SS} matrix can be computed from the others, which are all 'regular' (at least for simple elements). This can be taken a step further and, the panel in question may be linked to a fictitious boundary made up of panels and the value of the hypersingular integral determined by summing the other integrals. The values of the singular and hypersingular integrals in the BEM solution of Laplace's equation may be stored and be used to subtract out the singular and hypersingular components of the same integrals for the Helmholtz equation. The method of inventing a fictitious surface to evaluate the hypersingular integrals is precisely the method used in the author's axisymmetric programs [60], as there were analytic expressions for the hypersingular integrals for the panels used in 2D and general 3D, as discussed in the previous Sub-subsection, but no other similar way forward for the axisymmetric panels.

The results in the previous paragraph also provide useful methods of validation. One row of the M_{SS} and the N_{SS} matrices need to be evaluated that that is when the point p is on the boundary. The result of summing the rows of M_{SS} can verify that a 'closed' boundary is actually closed, if their values are $-\frac{1}{2}$ and zero, and, if they are not then it indicated that the boundary may be open or there are errors in the mesh. Finally, substituting these values into Equation (30) gives the following,

$$\{Me\}_S(p) = \begin{cases} 1 \text{ if } p \in E \\ 0 \text{ if } p \in D \\ \frac{1}{2} \text{ if } p \in S \end{cases}, \qquad (47)$$

where e is the unit function; useful in validating that the solution points are within the domain.

These simple validation methods, arising from potential theory, are applicable to all BEMs and beyond; any computer simulation involving surface meshes may benefit from these simple techniques. The author includes these methods of validation routinely in his boundary element codes.

4. The Standard Boundary Element Method in Acoustics

In the author's book and software [60,61], three classes of acoustic problem were considered; the determination of the acoustic field within a closed boundary, outside of a closed boundary and the interior modal analysis problem. In this Section, the three methods are revisited and recent developments and applications are included. Recently, the software has been re-written in Python [102].

4.1. The Interior Acoustic Problem

In this Subsection, the BEM is developed for the solution of acoustic problems in a domain that is interior to a closed boundary [61,103]. The method has been applied to room acoustics [104–108], the interior of a vehicle [109–111], modelling sound in the human lung [112,113] and in biological cells [114].

4.1.1. Integral Equation Formulation

The direct integral equations for the interior acoustic problem, reformulating the Helmholtz Equation (10), but follows the same format as the formulation for Laplace's Equation (21);

$$\{M_k \varphi\}_S(p) - \{L_k v\}_S(p) = \begin{cases} -\varphi(p) \text{ if } p \in D \\ 0 \text{ if } p \in E \\ -\frac{1}{2}\varphi(p) \text{ if } p \in S \end{cases}. \tag{48}$$

Equation (48) introduces two Helmholtz integral operators that are analogous to the L and M operators for Laplace's equation, and are defined similarly;

$$\{L_k \mu\}_\Gamma(p) = \int_\Gamma G_k(p,q)\, \mu(q) dS_q, \tag{49}$$

$$\{M_k \mu\}_\Gamma(p) = \int_\Gamma \frac{\partial G_k(p,q)}{\partial n_q}\, \mu(q) dS_q, \tag{50}$$

where the Green's function is defined as follows:

$$G_k(p,q) = \frac{i}{4} H_0^{(1)}(kr), \text{ for two-dimensional problems, and} \tag{51}$$

$$G_k(p,q) = \frac{e^{jkr}}{4\pi r}, \text{ for three-dimensional problems.} \tag{52}$$

As with the interior Laplace Equations (22) and (23), the indirect integral equation is derived through presuming that the field is generated by a layer potential, defined on the boundary;

$$\varphi(p) = \{L_k \sigma\}_S(p) \quad (p \in D \cup S), \tag{53}$$

$$v(p) = \left\{\left(M_k^t + \frac{1}{2}I\right)\sigma\right\}_S(p; n_p) \quad (p \in S), \tag{54}$$

where, similarly with Equation (20), the operator M_k^t is defined as follows:

$$\{M_k^t \mu\}_\Gamma(p; v_p) = \frac{\partial}{\partial v_p} \int_\Gamma G_k(p,q)\, \mu(q) dS_q, \tag{55}$$

4.1.2. The Boundary Element Method for the Generalised Boundary Condition

Substituting the expressions (53) and (54) into the boundary condition (7) gives

$$\alpha(p)\{L_k\sigma\}_S(p) + \beta(p)\left\{\left(M_k^t + \frac{1}{2}I\right)\sigma\right\}_S(p; n_p) = f(p). \tag{56}$$

Following the derivation of the BEM through collocation, this resolves to a linear system of the form

$$[D_\alpha L_{SS,k} + D_\beta\left(M_{SS,k}^t + \frac{1}{2}I\right)]\widehat{\underline{\sigma}}_S = \underline{f}_S, \tag{57}$$

where D_α and D_β are diagonal matrices, with the values of α and β aligned along the diagonal. Equation (56) is solved in the primary stage of the boundary element method, yielding an approximation $\widehat{\underline{\sigma}}_S$ to $\underline{\sigma}_S$. In the secondary stage, the discrete equivalent of Equation (53) returns the solution at the interior points:

$$\widehat{\underline{\varphi}}_D = L_{DS,k}\,\widehat{\underline{\sigma}}_S. \tag{58}$$

For the direct formulation, the discrete equivalent of Equation (48) $(p \in S)$,

$$(M_{SS,k} + \frac{1}{2}I)\widehat{\underline{\varphi}}_S = L_{SS,k}\widehat{\underline{v}}_S, \tag{59}$$

which is solved with the discrete equivalent of the boundary condition (7),

$$D_\alpha\widehat{\underline{\varphi}}_S + D_\beta\widehat{\underline{v}}_S = \underline{f}_S. \tag{60}$$

Comparing the linear systems for the indirect method (56) with that of the direct method (59) and (60) illustrates an apparent significant advantage in the indirect approach, the number of unknowns in the system corresponding to the direct method is twice that of the indirect method. However, through exchanging columns, the system for the direct method can be reduced to an $n \times n$ system, and this method has been automated [115].

There have been several papers discussing the accuracy of the interior acoustic boundary element method near to the resonance frequencies. It is considered that real eigenvaulues are shifted into the complex plane following discretisation. This phenomenon is termed numerical damping [116–119].

4.1.3. Equations of the First and Second Kind

Integral equations with a fixed region of integration are termed Fredholm integral equations. Fredholm integral equations are categorised as first kind or second kind. Equation (53) is a typical first kind equation, in which we are solving over integral operator(s) alone, in this case the L_k operator, to find σ from φ. With second kind equations, we are solving not just over integral operators, but also the identity (or diagonal) operator. For example, Equation (54) is a second kind equation, solving over the $M_k^t + \frac{1}{2}I$ operator, in order to find σ from v.

In general, first kind equations have poor numerical properties and are avoided. Although pure first kind equations only occur in particular Dirichlet cases, they can be avoided, through using the derivatives of the integral equation formulations (direct) or double layer potentials (indirect). However, from experience, first kind equations with singular kernels, as we find with the L or L_k operator, do not have poor convergence properties. It follows, therefore, that the formulations and methods developed thus far in this Subsection are generally suitable in solving the interior acoustic problem.

4.1.4. Derivative and Double-Layer Potential Integral Equations for the Interior Helmholtz Problem

As with the equations listed in Section 4.1.1, the derivative equations are unnecessary in solving the interior Helmholtz problem. However, the derivative equations provide an alternative formulation and help with the general understanding. The form of the equations is analogous with the equations for the interior Laplace equation, developed in Section 3.2.

For the direct method, the derivative boundary integral equation is analogous to Equation (25):

$$\{N_k\varphi\}_S(p; n_p) - \{M_k^t v\}_S(p; n_p) = -\frac{1}{2}v(p) \quad (p \in S). \tag{61}$$

where

$$\{N_k\mu\}_\Gamma(p; v_p) = \frac{\partial}{\partial v_p} \int_\Gamma \frac{\partial G_k(p, q)}{\partial n_q} \mu(q) dS_q. \tag{62}$$

For the indirect method, the boundary integral equations follow the form of Equations (28) and (29):

$$\varphi(p) = \{(M_k - \frac{1}{2}I)\zeta\}_S(p) \quad (p \in S), \tag{63}$$

$$v(p) = \{N_k\zeta\}_S(p; n_p) \quad (p \in S). \tag{64}$$

4.2. Interior Acoustic Modal Analysis: The Helmholtz Eigenvalue Problem

An enclosed acoustic domain has resonance frequencies and associated mode shapes. Mathematically, these are the solutions k^* (the eigenvalues—that relate to the resonance frequencies) and φ^* (the eigenfunctions—that relate to the mode shapes) of the Helmholtz equation with the homogeneous form of the boundary condition (7) (i.e., with $f(p) = 0$). This problem is more typically solved by the finite element method and a finite element model of an acoustic or structural problem is developed in Section 5.3.1. In this Subsection the modal analysis of an enclosed fluid via the boundary element method is outlined.

4.2.1. The Generalised Non-Linear Eigenvalue Problem from the Boundary Element Method

In the indirect boundary element method, the eigenvalues k^* and the eigenfunctions σ^* are found through solving

$$\alpha(p)\{L_k\sigma\}_S(p) + \beta(p)\left\{\left(M_k^t + \frac{1}{2}I\right)\sigma\right\}_S(p; n_p) = 0, \tag{65}$$

that follows from Equation (56), with the true eigenfunctions φ^* can then be found with Equation (53). Through applying collocation, this is equivalent to solving the non-linear algebraic eigenvalue problem

$$[D_\alpha L_{SS,\hat{k}^*} + D_\beta(M_{SS,\hat{k}^*}^t + \frac{1}{2}I)]\widehat{\sigma}_S = \underline{0}, \tag{66}$$

which follows from Equation (57).

The eigen-solution of Equation (66) returns the approximations $\hat{k}^*$ to the eigenvalues and the approximation $\widehat{\sigma}_S$ to the layer potential eigenfunction. The approximations to the physical eigenfunctions at the chosen domain points can then be found with Equation (58). The method described can also be applied with the direct integral equation formulation but requiring the row-exchanging method mentioned in Section 4.1.2.

Although in this work, the focus is on the generalised boundary condition, most of the examples in the literature consider the Dirichlet and Neumann eigenvalues, and it is revealing to focus on those. The Dirichlet/Neumann interior eigenvalues and eigenfunctions are the solutions of the equations of this Subsection with $\varphi(p) = 0 / v(p) = 0$ ($p \in S$). Hence the Dirichlet eigenvalues are the eigenvalues of the L_k operator from Equations (48) and (53), of the $M_k - \frac{1}{2}I$ operator from Equation (63) and the $M_k^t - \frac{1}{2}I$ operator from Equation (61). Similarly, the Neumann eigenvalues are the eigenvalues of the $M_k + \frac{1}{2}I$ operator from Equation (48), $M_k^t + \frac{1}{2}I$ from Equation (54) and the N_k operator from Equations (61) and (64).

4.2.2. Solving the Non-linear Eigenvalue Problem

The standard algebraic eigenvalue problem has the form

$$A\underline{x} = \lambda \underline{x}, \tag{67}$$

where A is a square matrix. Because of the analogy between the Helmholtz Equation (3) and the standard algebraic eigenvalue problem (67), the Helmholtz eigenvalues are sometimes termed the eigenvalues of the Laplacian [120–124].

The generalised algebraic eigenvalue problem has the form

$$A\underline{x} = \lambda B\underline{x}, \tag{68}$$

where A and B are square matrices. Standard methods exist for solving these problems, termed the QR and QZ algorithm. As we have noted, the application of the boundary element method in Equation (66) results in a non-linear eigenvalue problem, and this has the form

$$A_k\underline{x} = \underline{0}. \tag{69}$$

Although this is a significantly complicated problem, at least—in the case of the boundary element matrices—the individual components of the components matrix are continuous. Earlier methods tended to find the eigenvalues by finding the zeros of $|A_k|$. Further developments and applications can be found in the following research papers [61,122–148].

4.3. The Exterior Acoustic Problem

The boundary element solution of the exterior acoustic problem is the most popular area of research in the context of this paper. A method that can solve over a theoretically infinite domain from data on a surface mesh has significant value in the context of acoustics. However, it was found more than half a century ago that the boundary integral equations, derived in the same way as the ones for the interior problem, resulted in unreliable boundary element methods. Much has been achieved on the road to developing a more successful outcome. However, to obtain a reliable exterior acoustic boundary element method remains a tantalising goal.

Two reasonably successful pathways for developing a boundary element solution of the exterior acoustic problem were developed about 50 years ago, and these remain today. In this paper, the first is termed the Schenck method [149] and the second is termed the combined boundary integral equation method (CBIEM). There are several references on a general review and evaluation of the methods [55,150–157] and of software implementations [158–160]. The methods have been used in a range of applications: loudspeakers [38,50,51,161–178], transducers [52,179–183], hearing aid [184], diffusers [185,186], sound within or around the human body [159,187–190], scattering by blood cells [191], engine/machine noise [53,56,192–208], aircraft noise [209–211], rail noise [212], non-destructive defect evaluation [142,213–215], noise barriers [74,75,216,217], environmental noise [218–224], underwater acoustics [225,226], detecting fish in the ocean [227,228] and perforated panels [229].

4.3.1. The Integral Equation Formulations of the Exterior Helmholtz Equation and Their Properties

The direct boundary element reformulation of the Helmholtz equation follows the format for Laplace's Equation (30) and is as follows:

$$\{M_k\varphi\}_S(\boldsymbol{p}) - \{L_k v\}_S(\boldsymbol{p}) = \begin{cases} 0 \text{ if } \boldsymbol{p} \in D \\ \varphi(\boldsymbol{p}) \text{ if } \boldsymbol{p} \in E \\ \frac{1}{2}\varphi(\boldsymbol{p}) \text{ if } \boldsymbol{p} \in S \end{cases}. \tag{70}$$

The indirect formulation is

$$\varphi(p) = \{L_k \sigma\}_S(p) \quad (p \in E \cup S), \tag{71}$$

$$v(p) = \left\{ M_k^t - \frac{1}{2}I \right\}_{\Gamma} \sigma(p; n_p) \quad (p \in S). \tag{72}$$

However, the L_k, $M_k - \frac{1}{2}I$ and $M_k^t - \frac{1}{2}I$ operators are degenerate at the eigenvalues of the interior Dirichlet problem (see Section 4.2.1). The issues in using the boundary integral equations in (70)–(72) as general purpose exterior acoustic/Helmholtz equation solvers has been known for over 50 years. The wavenumbers or frequencies in which these boundary integral equations are unsuitable are often termed the characteristic or irregular wavenumbers or frequencies in the literature. These characteristic wavenumbers are physical in the interior problem, but they are unphysical in the exterior problem in which they are not a property of the Helmholtz equation model but are manifest in the boundary integral equations. In Section 3.4.2 the effects of the non-uniqueness of the solution of the interior Laplace problem with a Neumann boundary condition were discussed. The boundary integral Equations (70)–(72) have a similar property at the characteristic wavenumbers, and this is also often termed the non-uniqueness problem.

Although it is 'unlikely' in practice that the value of wavenumber k is equal to a characteristic wavenumber k^*, it is shown in Amini and Kirkup [230] that the numerical error as a consequence of being 'close' to a characteristic wavenumber is $O(\frac{1}{|k-k^*|})$. Hence, one technique of resolving the problem is to use finer meshes in the neighbourhood of the characteristic wavenumber in order to offset this error [231]. However, this strategy is likely to be prohibitive, requiring the overhead of creating a range of meshes and increased computational cost. The values of the characteristic wavenumbers are also generally unknown, although they can be found (as discussed in Section 4.2), but at a substantial computational cost. It is also found that the characteristic wavenumbers cluster more and more as the frequency increases. In conclusion, therefore, the boundary integral equations are—in practice—only useful for frequencies reasonably below a conservatively estimated first characteristic wavenumber, severely restricting the methods to the low frequency range in practice.

4.3.2. The Derivative and Double Layer Potential Integral Equations

For the direct method, the derivative boundary integral equation is analogous to Equation (31):

$$\{N_k \varphi\}_S(p; n_p) = \left\{ (M_k^t + \frac{1}{2}I)v \right\}_S(p; n_p). \tag{73}$$

For the indirect method, the boundary integral equations follow the form of Equations (35) and (36):

$$\varphi(p) = \left\{ (M_k + \frac{1}{2}I)\zeta \right\}_S(p), \tag{74}$$

$$v(p) = \{N_k \zeta\}_S(p; n_p), \tag{75}$$

Again, some of the operators are shared with the interior formulation. The operators $M_k + \frac{1}{2}$, $M_k^t + \frac{1}{2}I$ and N_k are degenerate at the eigenvalues of the interior Neumann problem, and hence these equations are also unsuitable as the basis of methods of solution at those frequencies. These equations mirror the properties of the elementary equations discussed earlier and their straightforward solution does not result in an acceptable boundary element method.

The characteristic wavenumber for the derivative and double-layer potential Equations (73)–(75) (the interior Neumann eigenvalues) are generally different from those of the elementary Equations (70)–(72) (the interior Dirichlet eigenvalues), and at least therefore they provide alternative methods. However, a more useful path involves combining the elementary Equations (70)–(72) with these equations and these methods are considered in Section 4.3.4.

4.3.3. The Schenck Method

The Schenck method [149] is often termed the CHIEF method in the literature and it is a development of the standard direct method, based on Equation (70). Given the potential non-uniqueness, discussed in Section 4.3.1, the system of equations that form the discrete equivalent of the boundary integral equations are regarded as potentially underdetermined and they are augmented with equations related to a set of points in the interior D:

$$
\begin{bmatrix} M_{SS,k} - \frac{1}{2}I \\ M_{DS,k} \end{bmatrix} \hat{\varphi}_S = \begin{bmatrix} L_{SS,k} \\ L_{DS,k} \end{bmatrix} \hat{v}_S .
$$

By adding more equations, the expectation is to eliminate the non-uniqueness and determine a solution. The equations can be solved by the least-squares method for Dirichlet and Neumann problems, the column exchanging algorithm [115] could be used in the case of the general boundary condition (7). There are several reported implementations and applications of the Schenck method [216,232–234].

Obviously, there are immediate questions about the number and position of the interior CHIEF points. Juhl [235] develops an iterative method for selecting points and halting when the results are unchanged. Equation (47) can verify that CHIEF points are interior, so this could be usefully included in the method. There has been a significant number of implementations and testing of the Schenck method. In general, more and more points are required to offset the non-uniqueness, as k increases. This increases the computational overhead with respect to the wavenumber, additional to the potential need for more elements at higher wavenumbers. Several improvements in the method have been put forward and these are summarised in Marburg and Wu [236]. There have been several evaluations of the CHIEF and combined methods and their variants [157,237,238].

4.3.4. The Combined Integral Equation Method

In this method, a boundary integral equation is formed through a linear combination of the initial boundary integral equation and its corresponding derivative equation (for the direct method and double-layer potential for the indirect metod). This concept was initially derived for the indirect method and is attributed to Brakhage and Werner [239], Leis [240], Panich [241] and Kussmaul [242]. The corresponding direct integral equations are attributed to Burton and Miller [243].

The Burton and Miller method is based on a boundary integral equation that is a linear combination of the initial one (70) with the derivative (73),

$$
\left\{ (M_k - \frac{1}{2}I + \mu N_k)\varphi \right\}_S (p; n_p) = \left\{ (L_k + \mu(M_k^t + \frac{1}{2}I)v \right\}_S (p; n_p) \quad (p \in S), \tag{76}
$$

where μ is a complex number, a *coupling* or weighting parameter. Similarly, for the indirect method, the equation is based on writing φ as a linear combination of a single and double layer potential

$$
\varphi(p) = \{(L_k + \mu M_k)\eta\}(p) \; (p \in E). \tag{77}
$$

This returns the following boundary integral equations

$$
\varphi(p) = \{(L_k + \mu(M_k + \frac{1}{2}I))\eta\}_S (p) \quad (p \in S), \tag{78}
$$

$$
v(p) = \{(M_k^t - \frac{1}{2}I + \mu N_k)\eta\}_S (p; n_p) \quad (p \in S), \tag{79}
$$

The issue of the determination of the values for the coupling parameter will be revisited in Section 6.2.

4.3.5. Scattering

The boundary integral equations for the exterior acoustic problem are readily applicable to radiation problems. The equations can also be used for the scattering problem, but with extra terms involved that model the incident field. These same techniques can be applied in the interior problem. For example, for the indirect method, the exterior acoustic field (77) is presumed to be made up of the layer potentials, superposed with the (known) incident field:

$$\varphi(p) = \varphi_{\text{inc}}(p) + \{(L_k + \mu M_k)\eta\}(p) \ (p \in E)$$

This similarly adjusts Equations (78) and (79):

$$\varphi(p) = \varphi_{\text{inc}}(p) + \{(L_k + \mu(M_k - \frac{1}{2}I))\eta\}_S(p) \quad (p \in S),$$

$$v(p) = v_{\text{inc}}(p) + \{(M_k^t + \frac{1}{2}I + \mu N_k)\eta\}_S(p; n_p) \quad (p \in S).$$

5. Extending the Boundary Element Method in Acoustics

In this Section, extensions in the standard boundary element methods of the previous section are considered. These include the Rayleigh integral method for computing the acoustic field surrounding a vibrating plate set in an infinite reflecting baffle, as a case of the more general half-space methods. This Section also includes shell elements in which a revision in the boundary integral equation for the exterior problem returns a model for the acoustic field surrounding a thin screen. Through principles of continuity and superposition, hybrids of these models and the standard models also significantly extend the range of acoustic problems that come under the boundary element fold. Vibro-acoustic, aero-acoustic and inverse acoustic problems are also considered in this Section.

5.1. Half-Space Methods

In Section 3 it was stated that the boundary element solution of Laplace's equation was a useful entry to the BEM in acoustics. The Rayleigh integral method (RIM) is also a good a starting point, in that it required only one of the four Helmholtz operators, and, for Neumann problems, it is an integral, rather than an integral equation. In this Subsection, the Rayleigh integral method is defined and further developments are reviewed.

The Rayleigh integral method can be viewed as a solution to a half-space problem. If further boundaries are placed in the half-space, then the integral equation formulation, with a simple modification of the Green's function, forms the model with $\varphi = 0$ or $\frac{d\varphi}{dn} = 0$ on the plane. Further development of the Green's function have been researched in order that an impedance condition is modelled on the plane, a useful model in outdoor sound propagation.

On the other hand, if there is a cavity in the plane then the interior boundary element method can model the acoustic field within the cavity and this is coupled to the Rayleigh integral. The model is based on based on the interior formulation to model the cavity, applying a false flat boundary across the opening and coupling the interior formulation with that of the half-space. This method is a hybrid of the boundary element method and the Rayleigh integral method and is termed *BERIM*.

5.1.1. The Rayleigh Integral Method

In the operator notation of this paper, the Rayleigh integral is as follows:

$$\varphi(p) = -2\{L_k v\}_\Pi(p) \quad (p \in E^+ \cup \Pi), \tag{80}$$

where Π is the flat plate, radiating into the half-space E^+. The solution of the Neumann problem, finding φ from v, is simply the evaluation of an integral. For the general boundary condition of the form (7), the technique is again to solve the boundary integral Equation ((80) with $p \in \Pi$) to obtain v and then evaluating the integral to obtain φ at any point in E^+. In the Rayleigh integral method [71], the plate is divided into elements, as discussed and applying collocation or, the equivalent for the integral, product integration. Substituting (80) into the boundary condition (7) gives

$$-2\alpha(\boldsymbol{p})\{L_k v\}_{\Pi}(\boldsymbol{p}) + \beta(\boldsymbol{p})v(\boldsymbol{p}) = f(\boldsymbol{p}). \tag{81}$$

Using the discrete notation of this paper, the equation following the application of collocation is as follows:

$$\left(-2D_{\alpha}L_{\Pi\Pi,k} + D_{\beta}\right)\hat{\underline{v}}_{\Pi} = \underline{f}_{\Pi}. \tag{82}$$

There have been several reported implementations and developments of the Rayleigh integral method [244–254], including vibro-acoustics [197,255,256]. Applications of the method include sandwich panels [257–263], engine or machine noise [203], electrostatic speaker [264] and transducers [52,265,266].

5.1.2. Developments on Half-Space Problems

Integral equations for scattering or radiating boundaries above an infinite plane can be developed through altering the Green's function [247] in order that it also satisfies the condition on the plane. For the perfectly reflecting plane the Green's function must satisfy the Neumann condition on the plane; $\frac{\partial G^*}{\partial n} = 0$ and hence $G^*(\boldsymbol{p},\boldsymbol{q}) = G(\boldsymbol{p},\boldsymbol{q}) + G(\boldsymbol{p},\boldsymbol{q}^*)$, where $\boldsymbol{q}^*$ is the point that corresponds to $\boldsymbol{q}$, when reflected through the plane. Similarly, if there is a homogeneous Dirichlet condition on the plane then the revised Green's function is $G^*(\boldsymbol{p},\boldsymbol{q}) = G(\boldsymbol{p},\boldsymbol{q}) - G(\boldsymbol{p},\boldsymbol{q}^*)$. More generally, the modified Green's function is $G^*(\boldsymbol{p},\boldsymbol{q}) = G(\boldsymbol{p},\boldsymbol{q}) + R\, G(\boldsymbol{p},\boldsymbol{q}^*)$, with R representing the reflection of the plane $(-1 \le R \le 1)$. An impedance boundary condition is generally required in modelling environmental noise problems [220,267–270].

However, the methods in the previous paragraph are applicable when the acoustic scatterers or radiators are on or above the plane. Another set of methods apply if there is a cavity in the plane. Early examples of this type of problem have arisen in harbour modelling, in which the Helmholtz equation has been used to model the waves in a harbour (the cavity), which is open to the sea bounded by the coastline (the plane) [271,272]. The Boundary Element—Rayleigh Integral Method (BERIM) [169], is applicable to open cavity problems in acoustics. The interior boundary element method (Section 4.1) models the sealed interior and the Rayleigh integral method models the field exterior. The advantage in this model over the exterior model could be substantial; the mesh covers the interior of the cavity and the opening only, rather than the inside and outside. There is an important issue with the model, as it presumes that the cavity opens out on to an infinite reflecting baffle. However, this could be a small price to pay, and some problems—such as the loudspeaker problems considered by the author [169]—the mouth opens onto a front face of the cabinet. Motivated again by environmental noise problems, several methods have been developed for a cavity opening on to an impedance plane [273].

5.2. Shell Elements

The derivation of integral equations and methods for modelling thin shells (that is an open boundary modelling a discontinuity in the field) in the boundary element context takes us back to the works such as Krishnasamy [274], Gray [275], Terai [276] and Martin [277]. In these, and various other references, the shell is also termed a crack. For the Helmholtz equation, the derivation of the integral equations is set out in Warham [73] and Wu [72].

Shell problems may be solved using the standard boundary element methods already described in this paper. One method is to mesh the shell as a closed boundary, with a finite thickness. However, if the thickness of the shell is a fraction of the element size, then the boundary integral equations representing collocation points on either side of the shell are similar, and the equation approaches degeneracy. Alternatively, many elements may be required, to ensure that the elements on the shell are not disproportionate in comparison with those along the edges.

A more productive method of using existing methods is to artificially extend the shell to form a complete boundary, with the interior and exterior BEM applied to the inner and outer domains, and continuity applied over the artificial boundary. However, this approach, in many cases, would be prohibitive, as the number of elements would be substantially greater. However, analytic test problems for shell problems are difficult to develop and using the more-established interior and exterior BEM in this way can provide comparative solutions. As with the Rayleigh integral method of the previous Subsection, the same Helmholtz operators re-occur with the shell model. Hence, the inclusion of shells in boundary element software significantly extends the functionality of the library, at little extra cost.

5.2.1. Derivation of the BEM for Shells

In this model, φ is the solution of the exterior Helmholtz Equation (3) in the exterior domain, surrounding an open boundary Ω. The boundary Ω is presumed to be an infinitesimally thin discontinuity and so, at the points on the boundary, φ and its normal derivative v take two values, one at each side of the discontinuity. The two sides of the shell are labelled "+" and "-": it doesn't matter which way round this is, as long as it remains consistent. Hence, on the shell, four functions are modelled, $\varphi_+(p)$, $\varphi_-(p)$, $v_+(p)$ and $v_-(p)$ $(p \in \Omega)$, where the normal to the boundary, which orientates v_+ and v_-, is taken to point outward from the '+' side of the shell. However, rather than working with these functions, it is more straightforward to work with the difference and average functions;

$$\delta(p) = \varphi_+(p) - \varphi_-(p),$$

$$\Phi(p) = \frac{1}{2}(\varphi_+(p) + \varphi_-(p)),$$

$$v(p) = v_+(p) - v_-(p),$$

$$V(p) = \frac{1}{2}(v_+(p) + v_-(p)),$$

for $(p \in \Omega)$ and where, for simplicity, it is presumed that the boundary is smooth.

The boundary condition is defined in a similar way as in Equation (7), but as there are double the number of unknown function, two boundary conditions are required $(p \in \Omega)$:

$$\alpha(p)\delta(p) + \beta(p)v(p) = f(p), \tag{83}$$

$$A(p)\Phi(p) + B(p)V(p) = F(p), \tag{84}$$

The integral equations that govern the field around the shell discontinuities derived from the exterior direct formulation (70) [73], by taking the limit as the boundary becomes thinner, and they are as follows:

$$\Phi(p) = \{M_k\delta\}_\Gamma(p) - \{L_k v\}_\Gamma(p) \quad (p\epsilon\Gamma), \tag{85}$$

$$V(p) = \{N_k\delta\}_\Gamma(p; n_p) - \{M_k^t v\}_\Gamma(p; n_p) \quad (p\epsilon\Gamma), \tag{86}$$

$$\varphi(p) = \{M_k\delta\}_\Gamma(p) - \{L_k v\}_\Gamma(p) \quad (p\epsilon E). \tag{87}$$

There are few tests of methods based on these equations in the literature. The only known issue is at the edge, as it is likely that the solution is singular there. Therefore, mesh grading, using smaller and smaller elements close to the edge may improve efficiency.

5.2.2. Mixing Opem with Closed Boundaries

Again, using the superposition principle, discussed in Section 4.3.5, shell boundaries can be mixed with the traditional boundaries. For example, for the interior problem made up of a domain D

with boundary S and with shell discontinuities Γ within the domain, the superposition of the direct Equations (48) and (61) with Equations (85)–(87) returns the following equations:

$$\Phi(p) = \{L_k v\}_S(p) - \{M_k \varphi\}_S(p) + \{M_k \delta\}_\Gamma(p) - \{L_k v\}_\Gamma(p) \quad (p \epsilon \Gamma),$$

$$V(p) = \left\{M_k^t v\right\}_S(p) - \{N_k \varphi\}_S(p) + \{N_k \delta\}_\Gamma(p; n_p) - \left\{M_k^t v\right\}_\Gamma(p; n_p) \quad (p \epsilon \Gamma),$$

$$\varphi(p) = \{L_k v\}_S(p) - \{M_k \varphi\}_S(p) + \{M_k \delta\}_\Gamma(p) - \{L_k v\}_\Gamma(p) \quad (p \epsilon D),$$

$$\frac{1}{2}\varphi(p) = \{L_k v\}_S(p) - \{M_k \varphi\}_S(p) + \{M_k \delta\}_\Gamma(p) - \{L_k v\}_\Gamma(p) \quad (p \epsilon S).$$

Methods based on this analysis and equations were developed and demonstrated by the author for the interior Laplace equation [85], the exterior acoustic/Helmholtz problem [278], and for the interior acoustic/Helmholtz problem [135,279].

5.3. Vibro-Acoustics

Problems that involve structural vibration, as well as an acoustic response, are termed vibro-acoustics. In this Subsection, the modelling by a domain method, such as the finite element method is outlined. The FEM can be applied to the structure and/or the acoustic/fluid domain, but, in the context of this Subsection, the FEM is used as the structural model. When the structure and the fluid significantly influence each other's dynamic response then they are modelled as coupled fluid-structure interaction.

5.3.1. Discrete Structural or Acoustic (Finite Element) Model

The properties of the interior acoustic problem parallel the expected response an excited elastic structure, presuming no damping, or, more simply, simple harmonic motion. In this discussion, let us consider this further in order to bring context. With a mass M and a stiffness K, the equation of (unforced) simple harmonic motion is

$$M\ddot{q} + K q = 0,$$

where q is the displacement and $\ddot{q}$ is the acceleration. The same equation results from a system of masses or from the finite element method solution with M and K termed the mass and stiffness matrices and q and $\ddot{q}$ are vectors of displacement and acceleration of the individual masses or nodes in the FEM. The phasor solution $q = Qe^{j\omega t}$ returns the following equation,

$$-\omega^2 MQ + KQ = 0, \tag{88}$$

Hence, applying the finite element method to the structural or interior acoustic modal analysis problem returns a generalised algebraic eigenvalue problem (of the form of Equation (68)). The matrices are sparse and hence are amenable to more efficient methods of solution. Although the matrices are larger and the domain needs to be meshed, the BEM with its nonlinear eigenvalue problem struggles to compete with the FEM in acoustic modal analysis. Let ω_j^2 and Q_j for j = 1, 2, ... be the eigenvalues (natural frequencies) and corresponding eigenvectors (mode shapes) of Equation (88). It follows that $M^{-1}KQ_j = \omega_j^2 Q_j$.

Let us also now generalise the model in order to include and external excitation force g:

$$M\ddot{q} + K q = g. \tag{89}$$

Following the phasor substitution $g = Ge^{j\omega t}$ Equation (89) is modified as follows:

$$-\omega^2 MQ + KQ = G, \text{ or}$$

$$- \omega^2 \underline{Q} + M^{-1} K \underline{Q} = M^{-1} \underline{G} .$$

For the homogeneous equations ($\underline{G} = \underline{0}$), the above may be cast as a generalised or standard algebraic eigenvalue problems, as discussed in Section 4.2.2. Let us write the response and the excitation in terms of the modal basis

$$\underline{Q} = \sum_j \gamma_j Q_j \text{ and } M^{-1} \underline{G} = \sum_j a_j Q_j. \tag{90}$$

Considering each eigen-solution in turn relates the coefficients, so that

$$\gamma_j = \frac{a_j}{\omega_j^2 - \omega^2}.$$

In practice, a structure or an enclosed fluid experience damping, the simplest and usual model is to relate damping to the velocity:

$$M \ddot{\underline{q}} + C \dot{\underline{q}} + K \underline{q} = \underline{g}$$

where C is termed the damping matrix. Following the phasor substitutions, Equation (89) is modified as follows:

$$- \omega^2 M \underline{Q} + j \omega C \underline{Q} + K \underline{Q} = \underline{G}. \tag{91}$$

The response of the system to an applied boundary condition across a frequency sweep is to have a smoothed peak at the resonance frequency and a more gradual phase change. In general, the response of the system can be modelled as a sum of weighted modes (90) with the coefficients that are relatable as follows:

$$\gamma_j = d_j(\omega) a_i. \tag{92}$$

5.3.2. Coupled Fluid-Structure Interaction

The finite element model in the previous Sub-subsection is directly applicable to a structure when there is no significant coupling between the structure and a fluid. Similarly, the acoustic analysis methods of Sections 4 and 5.1 and Section 5.2 are directly applicable when there is no significant coupling between the fluid and a structure. However, for many dynamical systems, it is appropriate to couple the structural model, of the FEM form outlined in the previous Sub-subsection, with the acoustic model of the fluid with which it is in contact. The finite element method can be applied in both domains, with appropriate properties. In the context of this work, the boundary element method provides the computational acoustic model. The models are coupled together, through continuity of the particle velocity at the interface, and the resultant forcing on the structure is affected by the sound pressure. The discrete coupling is applied at the elements that coincide with the boundary.

The fluid-structure interaction model with the boundary element method modelling the acoustic domain has been developed and applied over several decades. Expansions on the general method can be found in the following works and the references therein [183,267,280–283]. Applications include the interaction of plates with fluids [23,245,284], sandwich panels/lightweight structures [257,261–263,285], sound insulation [283], screens [286], the passenger compartment of an automobile [111], hydrophones [284,287] and in seismo-acoustics [288].

In Section 4.2, the acoustic modal analysis of an enclosed fluid was discussed. Similarly, in the previous Section, structural modal analysis via the finite element method was outlined. Of course, when coupling occurs, the eigenvalue analysis need to be applied to the coupled system. For example, a structure typically exhibits different resonant frequencies in vacuo than it does when immersed in a fluid. In the literature, these are often termed the *dry* and *wet* modes (at least for structures placed outside of and in water). Modal analysis via the finite element method returns a standard and sparse eigenvalue problem (88), modal analysis by the boundary element method yields a non-linear eigenvalue problem (69) and hence the hybrid coupled FEM-BEM system is also non-linear. There are several reported implementations of coupled fluid-structure modal analysis using the boundary element method [19,255–257,265,284,287,289–301]. The coupled matrices are generally much larger than for the boundary element method alone and this can be resolved by determining the response in the dry structural modal basis and coupling that with the boundary element model [299,301].

5.4. Aero-Acoustics

While the boundary element method has been applied to coupled fluid-structure interaction problems for almost as long as the method has been around, the BEM in aero-acoustics is a much more recent development. The standard exterior BEM in a vibro-acoustic settting has been applied in aircraft noise [210,211]. The wave Equation (1) does not include convective flow and is only an adequate model in aero-acoustics in the special case of insignificant flow and the formulations and methods outlined in this work are no longer directly applicable. The Navier-Stokes equations model fluid flow and their solution by numerical methods is termed *computational fluid dynamics* (CFD). There are reports of applying the standard acoustic BEM and CFD to aircraft [47,48,302–304] and, similarly, to underwater vehicles [305,306].

To model the noise from aircraft, the domains are typically large and significant resolution is required to capture the higher frequencies, and hence domain methods come with a high computational cost. Computational aero-acoustics [11] has arisen for developing and applying numerical methods in this particular area. The attraction of the BEM in computational aero-acoustics is the same as it is in standard acoustics, a significant reduction in meshing and hence the potential for faster computation.

Work on the adaption of the BEM to a wider scope of problems has been developed, for example by the dual and multiple reciprocity method, which has also been applied to variants of the Helmholtz equation [46–48,103,131,148]. A generalisation of the boundary element method in acoustics that includes convection, is applicable to problems with uniform flow, but this can also form a useful approximation method with the mean flow rate substituting the value for the uniform flow rate [307]. Similar to the approach in half-pace problems, discussed in Section 5.1.2, aero-acoustic problems are adapted for the BEM by revising the Green's function [308–310]. Recently the BEM in acoustics has been adapted to model viscous and thermal losses [311–313].

Methods in aero-acoustics that use the BEM generally involve setting a fictitious surface, enclosing the significant effects such as noise generation and turbulence, within which typical methods of CFD are used to model the Navier-Stokes equations; the sound generation and sound propagation are modelled separately. The boundary element method models the outer domain, but requiring the fictitious surface to be meshed, modelling the uniform flow and the radiation condition from the boundary and into the far-field. There are several reported implementations and aero-acoustic and related applications of these methods [307,314–326].

5.5. Inverse Problems

An inverse problem in acoustics, involves measuring properties of the acoustic field and processing that information in order to determine something about its origin. Over recent decades, the main text that guides inverse acoustic (and electromagnetic) (scattering) problems is that of Colton and Kress [16], now in its third edition. Colton and Kress provide a mathematical analysis of inverse problems, and, in acoustics, their focus is on determining the shape of the boundary of a scatterer from the (disturbed) sound pressure at points in the far-field. An example of an application of this is identifying bodies on the sea floor [327].

Inverse problems are ill-posed. This means a range of solutions can give rise to the same measurements, in contrast to the forward problems, studied so far, which are usually well-posed, with a unique solution. In practice, the linear system that is returned by the boundary element method (or any standard method) is significantly ill-conditioned. If a conventional method is used to solve the linear system then the solution will not be acceptable. In the author's previous work on the inverse diffusion problem (classically, the backward heat conduction problem) [328], also included in Visser [237]) it was discussed that there is effectively insufficient information in the data to determine the solution of an inverse problem. The notion of an 'acceptable' solution, the bias of the observer, provides the final constraint that enables the ill-posed problem to be substituted by a nearby well-posed problem. This returns an acceptable solution, even though it is a less accurate solution of the discrete equations. In the literature, this technique is termed *regularisation*.

Acoustic holography—determining the sound field near the source from acoustic properties at a distance from the source—is one of the main application areas of the inverse boundary element method. In general, an array of pressure or velocity transducers provides the data and the inversion returns the acoustic properties on the surface. Acoustic holography can determine the sources of noise from a radiating structure, which can help guide a re-design. For example, starting with the discrete equivalent of Equation (71),

$$\underline{\varphi}_E = L_{ES,k}\,\widehat{\underline{\sigma}}_S \,, \tag{93}$$

the field data $\underline{\varphi}_E$ (effectively the sound pressure, see Section 2.1) is related to the layer potential σ_S. If we could find a discrete approximation to σ_S, then the approximation to the surface potential (sound pressure) and velocity could be found by the matrix-vector multiplication of the discrete equivalent of Equations (53) and (54) and the surface intensity could then be found by the Equation (5). However, as discussed, the solution of the linear system will not yield acceptable results. Even if the number of data points massively exceeds the number of elements, the system is stll effectively underdetermined.

Perhaps the most popular method of resolving the ill-posedness is to use *Tikonov regularisation*. This involves minimising the residual in system, along with a penalty function that constrains the variability of the solution in some sense. For example, Equation (93) is replaced by

$$min_{\,\widehat{\sigma}_S}\left\{\left\|\underline{\varphi}_E - L_{ES,k}\,\widehat{\underline{\sigma}}_S\right\| + \eta^2\left\|\widehat{\underline{\sigma}}_S\right\|^2\right\}, \tag{94}$$

where η is a parameter that can be 'tuned' to achieve and acceptable solution and the norm is the 2-norm.

A related regularisation method is termed truncated singular value decomposition (TSVD). For example, the singular value decomposition (SVD) of the $L_{ES,k}$ matrix factorises it as follows:

$$L_{ES,k} = U\Sigma V^H, \tag{95}$$

where U is a $n_E \times n_S$ matrix, V is $n_S \times n_S$ and Σ is a diagonal matrix containing n_S singular values s_i, non-negative values in non-decreasing order. In Equation (95), the H denotes the complex conjugate transposed. Let $U = \left[\underline{u}_1, \underline{u}_2, \ldots \underline{u}_n\right]$ and $V = \left[\underline{v}_1, \underline{v}_2, \ldots \underline{v}_n\right]$, where the $\underline{u}_i$ are the left singular vectors and the $\underline{v}_i$ are the right singular vectors. For ill-posed problems, the final singular vectors are oscillatory, and the corresponding singular values are very small. Hence, on solution of a system like (93), the oscillations dominate. In TSVD, the offending singular values are simply removed or filtered. The solution may then be determined as

$$\widehat{\underline{\sigma}}_S = \sum_{i=1}^{n^*} \frac{\underline{u}_i^H \cdot \underline{\varphi}_E}{s_i}\,\underline{v}_i \,,$$

where $n^* < \min{(n_S, n_E)}$.

There have been several reported implementation of acoustic holography using the methods discussed [237,329–334]. Similar methods have also been developed for finding the impedance of the surfaces in rooms from measured sound pressure data [335,336].

5.6. Meshless Methods

One of the main advantages of the boundary element method over domain methods, such as the finite element method is the reduced burden of meshing. With meshless methods, a mesh is not required at all. The simplest concept of a meshless method is the equivalent sources method (ESM) in which it is presumed that the acoustic field is equivalent to a field generated by a finite set of point sources:

$$\varphi(\boldsymbol{p}) \approx \sum_{j=1}^{m} \gamma_j G_k(\boldsymbol{p}, \boldsymbol{q}_j)$$

where the $\boldsymbol{q}_j$ are the positions of the sources that are outside of the acoustic domain and the γ_j are the unknown source strengths and G_k is the appropriate Green's function. For the Dirichlet boundary condition, by matching the Dirichlet data on the boundary $\varphi(\boldsymbol{p}_j)$ for $\boldsymbol{p}_j \in S$, gives a linear system of equations, provided there are at least as many source points as there are items of boundary data. Similarly, by differentiating the above equation with respect to the normal to the boundary

$$\frac{\partial \varphi(\boldsymbol{p})}{\partial n} \approx \sum_{j=1}^{m} \gamma_j \frac{\partial G_k}{\partial n}(\boldsymbol{p}, \boldsymbol{q}_j),$$

then approximate solution can be found by matching the values with Neumann data, and using a combination. Usually, more internal points than items of boundary data are chosen and a least-squares solution is sought. The meshless methods avoid the problem of singular integration. However, the ESM is not based on a boundary integral equation formulation and hence it is an alternative to the boundary element method and is therefore beyond the scope of this paper. Lee [337] provides a recent review of the ESM in acoustics.

An alternative method for developing meshless methods has more in common with the standard boundary element method. The method can be applied to the standard interior or exterior problem. For the exterior problem, the methods relates to the Schenck or CHIEF method, of Section 4.3.3, but only the integral equations for the internal points are applied, and the integrals are approximated by using only the midpoint value

$$\overline{M}_{DS,k}\underline{\phi}_S = \overline{L}_{DS,k}\hat{v}_S.$$

With the number of interior points exceeding the number of boundary points the solution can be found in the least-squares sense. For the exterior problem, the Burton and Miller form is expected to have superior numerical properties:

$$(\overline{M}_{DS,k} + \mu\overline{N}_{DS,k})\hat{\phi}_S = (\overline{L}_{DS,k} + \mu\overline{M}^t_{DS,k})\hat{v}_S.$$

These methods still avoid integration, particularly singular integration and can also be used on the modal problem. There are several reports of implementations of these methods [144,145,338–341].

As methods for solving acoustic problems, the meshless methods, are well behind the standard boundary element method in the sense of developing robust software. As with the Schenck or CHIEF method of Section 4.3.3, there is the added issue of determining the placement of the equivalent sources. The methods outlined in Section 3.4.2 could be used in determining whether source points are interior or exterior.

6. Areas of Discussion

Much of the development work on the boundary element method in acoustic has been on relatively simple shapes and relatively low frequencies. For practical problems, the existing methods must be applicable to more complicated domains and to high frequencies. In this Section, two significantly challenging areas in the acoustic BEM are surveyed. The first is that of efficiency, the relationship between the accuracy achieved and the computational effort. Error analysis of the acoustic/Helmholtz BEM has received significant attention [230,342,343]. Three categories of error arise in the BEM; the discretisation error due to the approximation of the boundary and boundary functions, the quadrature induced error resulting from the numerical integration method and the error in the solution of the linear system of equations. In general, efficiency is maximised by balancing the errors. Much of the focus in the development of the BEM in acoustics is on improving its efficiency so that a full frequency sweep, particularly including the resolution required to model high frequencies, can be achieved with reasonable computer time and memory requirements. In Section 6.1 the efficiency of the acoustic BEM is discussed and methods for improving efficiency are reviewed.

The most valuable acoustic boundary element method—the standard exterior problem, outlined in Section 5.3—has also been found the most difficult to maintain. The combined integral equations of Section 4.3.4 have held the most promise. The earlier work on the acoustic BEM suggested that the coupling parameter was somewhat arbitrary. In terms of scaling up the method, however, the coupling seems to have become a significant issue and, in Section 6.2, the choice of coupling parameter is revisited.

6.1. Efficiency

It is often stated that the boundary element method had a significant efficiency advantage over domain methods, such as the finite element method. However, efficiency concerns remain critical, not least for the acoustic boundary element method, in which it has been a focus of research for many decades. Efficiency, relates the accuracy of the output to the computational resources required to achieve it. The computational resources include the computer processing time and/or the memory requirements. One approach for improving the execution time of numerical methods is to use parallel processing, so that instructions are executed simultaneously, rather than sequentially, and such techniques have been applied in the boundary element method [304,344].

In this Subsection, the efficiency of the acoustic BEM is analysed and techniques for improvement are reviewed. Although a range of classes of boundary element methods have been outlined in this paper, the analysis is fairly generic, and can be applied to the boundary-value problems that arise in acoustics. The modal analysis or eigenvalue problem is not considered, but much of the analysis carries over. For the extended problems considered in Section 5, the analysis in this Subsection is only relevant insofar as the BEM is implemented with the wider context.

6.1.1. The Computational Cost of the Acoustic Boundary Element Method at One Frequency

In this Sub-subsection, the boundary element method in acoustics is developed in its most typical way for one frequency. The efficiency of the method is analysed and discussed. More disruptive methods for improving the efficiency are considered subsequently.

As discussed, the boundary element method is composed of two stages, the first stage involving determining the boundary functions and the domain solution in the second stage. In the first stage one or more $n \cdot n$ matrices are formed, where n is the number of boundary solution points (e.g., collocation points, or elements for simple constant elements). In the second stage, usually one $n_P \cdot n$ matrix is formed, where n_P is the number of domain points. Hence the storage requirement is $O(n^2 + n_P n)$ complex numbers.

The matrices are normally computed by numerical integration. In general, the number of quadrature points required on each panel would be varied with the size of the panel and the distance between the panel and the point [345]. For each quadrature point, the Green's function would be computed, together with the other geometric information, although the former is likely to incur the greater part of the computational cost. The discretisation could be carried out separately for each operator; however, for efficiency reasons, the computed values should be shared between the operators, as, for example, followed in the author's previous work [61,91]. The diagonal components of the square matrix (or matrices) in the initial stage may be the result of evaluating singular and hypersingular integrals, as discussed, and, although they require special treatment, it is usually more efficient if the quadrature points are similarly the same for all required discrete operators. Lumping together the matrices, for the $n \times n$ and $n_P \times n$ components, let the average number of quadrature points be N_Q

and let the average cost of the evaluation of the Green's function be C_G and the average cost of the geometrical properties be C_g, then the total cost (or computer time) of computing the matrices is

$$N_Q\left(n^2 + n_P n\right)\left(C_G + C_g\right).$$

Clearly C_g will be larger for the combined operators, used in the primary stage for exterior problems. Alternatively, for the Schenck method, the matrices are augmented by the discrete operators for the interior points.

Once, the matrix vector system is formed, the next step is to solve it. The most straightforward method of solution of a square system is the Gaussian elimination or LU factorisation method (re-writing a matrix as the product of a lower and upper triangular marices). The overall computer time for the acoustic boundary element method at one frequency can be summarised with the equation

$$N_Q\left(n^2 + n_P n\right)\left(C_G + C_g\right) + \left(\frac{4}{3}n^3 + O\left(n^2\right)\right)C_f$$

where C_f represents the cost of a floating-point operation, such as multiplication or division. The non-square system that arises in the Schenck method can also easily be resolved as a square system through pre-multiplying both sides by the transpose of the matrix over which the solution is sought and this is equivalent to the least-squares solution.

In its earlier development, the cost of computing the matrices generally far outweighed the cost of solving the system. However, in the first stage of the boundary element method, the computational cost of setting up the linear system is $O\left(n^2\right)$, whereas the cost of solving it (using the methods stated) is $O\left(n^3\right)$. For this reason, it has also been understood for a long time that LU factorisation or related methods for solving the linear system was unsustainable, as progress is made towards the finer meshes that are particularly required for high frequency problems. However, it is important also to state the particular advantages that LU factorisation has. LU factorisation is a robust method and, in the case when there are a range of inputs to a problem with a fixed boundary and boundary condition, once the $O\left(n^3\right)$ factorisation as been stored, it can be used repeatedly again with $O\left(n^2\right)$ cost.

The fast multiple method (FMM) is a popular method of speeding up the computation of the matrices in the boundary element method and it has been applied to acoustic/Helmholtz problems [220,227,254,284,287,290,320,346]. In this method the Green's function is approximated by local polar expansions that can be translated through the domain, rather than re-computed. The FMM is often used with iterative methods.

In this Subsection alternatives to standard LU factorisation are reviewed. Faster solution methods include the use of hierarchical matrices [190] or panel clustering [347] and iterative methods are also considered. If the solution of the linear system is potentially dominant, in terms of computational cost, then the attention also re-focuses on the integral equation method. For it may be advantageous to choose an integral equation method that results in a linear system that has faster convergence properties, rather than one that is the easiest to apply or the most efficient (such as the collocation method that is highlighted in this work).

6.1.2. The Frequency Factor, Multi-Frequency Methods and Wave Boundary Elements

Typically, in vibration and acoustics, the time-dependent signals are resolved into frequency components. For example, for air-acoustics, the audible range for human being is up to 20 kHz and typically his could be resolved into components with a 10Hz resolution; multiplying the computational cost, considered in the previous Sub-subsection, by around 2000. The prospect of solving acoustic problems over the full frequency range has led to another set of techniques focus on reducing the computational burden by solving a range of frequencies together and these are usually termed multi-frequency methods [297,348–355]. However, with the nature of an acoustic field, originating typically from structural vibration, is such that structural and acoustic resonances, and the driving profile, dominate the total acoustic response. From a computational point of view, effort should therefore be concentrated on these 'peak' frequencies. Hence, this adaption returns us to the standard mono-frequency problem, or applying the multifrequency method, centred on each peak.

Moreover, in general the frequency of observation is matched in the acoustic and vibratory properties. In practice, as the frequency increases, a higher elemental resolution is required and hence $n = O(k)$ for two-dimensional and axisymmetric problems and $n = O(k^2)$ in three dimensions. Although this seems to imply a revision of the mesh at every frequency, it is more likely that one mesh that is suitable for the highest frequency is used throughout the range, or separate meshes are applied in significant sub-ranges of the frequency range, in order that the burden of meshing is proportionate within the overall project.

A review of the actual number of elements required to capture the solution is provided by Marburg [356–358]. In general, for the simple elements that are often used, it is considered that 6-10 elements per wavelength is a reasonable guideline. Obviously, higher-order boundary element methods [100,359–363] would often return the same accuracy with fewer elements. There is significant interest in isogeometric elements, in which the boundary and the boundary functions are modelled with the same basis functions (typically splines), so that the BEM can be more easily integrated with computer-aided design [364–371]. For air acoustics, at high frequencies reaching 20 kHz, the wavelength is less than 2 cm. For example, even for a 10cm cube, the number of elements required to model the high frequencies is in the tens of thousands and for a 1m cube, over a million elements would be required. A variation on the boundary element method has arisen for developing sinusoidal basis functions or wave boundary elements, in order to more accurately model the acoustic functions and the term PUBEM or partition of unity BEM is often used to identify such methods [267,371–374], which has similarity with the application of the Treffetz method [292–294].

6.1.3. Iterative Methods and Preconditioning

The scalability of the standard BEM in acoustics has been in question for several decades. The computational estimate relates two areas of particular concern, the $O(n^3)$ nature of LU factorisation and the $O(n^2)$ cost of computing the matrices. Hence, much of the current research on the BEM in acoustics is focused on reducing its computational burden, so that the methods can be casually applied to more significant problems and high frequencies.

The conjugate gradient method was identified as a useful iterative solution method for the acoustic BEM [155,375,376]. The conjugate gradient method and related methods are generally termed Krylov subspace methods and these variants have been significantly tested on the linear systems arising from the boundary element method in acoustics [106,377–379]. In general, if the underlying operator is compact, corresponding to a clustering of the corresponding matrix eigenvalues, then iterative methods, such as the Krylov subspace methods, work well [380]. However, one of the operators, N_k, is not compact, the eigenvalues of its matrix equivalent are spread out, and hence the raw iterative methods are only applicable in particular cases, in which the hypersingular operator is not in play. Hence, in the spirit of generality that is sought in this work, the iterative methods are of limited value in solving the untreated equations. However, in the acoustic boundary element method, these iterative methods are usually applied following an intervention with the original linear system, termed preconditioning [381].

In the acoustic BEM, preconditioning has been advocated since its early days. The original concept can be derived from the integral equation formulations. By various substitutions with the boundary integral equations, operator identities can be derived. One of the most useful is the substitution of (74) and (75) into (70) ($p \in S$) (or substituting (63) or (64) into (48) (for $p \epsilon S$)) giving

$$L_k N_k = \left(M_k - \frac{1}{2}I\right)\left(M_k + \frac{1}{2}I\right) = \left(M_k + \frac{1}{2}I\right)\left(M_k - \frac{1}{2}I\right) = M_k^2 - \frac{1}{4}I*$$

and hence the left-hand side of the Burton and Miller Equation (76) may be re-written as follows

$$\left\{(M_k - \frac{1}{2}I + \mu L_k N_k)\varphi\right\}_S(\boldsymbol{p};\boldsymbol{n_p}) = \left\{(M_k - \frac{1}{2}I + \mu(M_k + \frac{1}{2}I)(M_k - \frac{1}{2}I)_k)\varphi\right\}_S(\boldsymbol{p};\boldsymbol{n_p})$$
$$= \left\{(L_k + \mu L_k(M_k^t + \frac{1}{2}I)v\right\}_S(\boldsymbol{p};\boldsymbol{n_p}),$$
(96)

The equation is preconditioned, as the N_k operator has been eliminated and all the operators are compact. However, the motivation for this technique can be with the elimination of the hypersingular integration, with incidental preconditioning. Methods based on (96) along with $L_k N_k = M_k^{t^2} - \frac{1}{4}$, that is obtained through the substitution of (53) and (54) into (61) or the substitution of (71) and (72) into (73) ($\boldsymbol{p} \in S$), are often termed the Calderon equations [382,383]. Methods that make use of this substitution have been developed [155,157,241]. Alternative approaches in developing well-conditioned integral formulations for iterative solution have also been the subject of research [384–388]. Langou [389] develops methods for the iterative solution of linear systems with multiple right-hand sides, echoing the stated advantage of the LU factorisation method, discussed in Section 6.1.1.

Matrix preconditioners are often based on constructing an approximate inverse and following a fixed point/defect correction/contraction method. Introducing a preconditioning matrix also introduces an $O(n^2)$ matrix multiplication at each step, but this can be reduced to $O(n)$ if the approximate inverse matrix is sparse or banded [378,380]. Methods based on an incomplete LU factorisation [390] have been tested in the acoustic BEM [391]. A similar approach is the construction of a *DtN* (or *DN*) map (Dirichlet to Neumann) or *NtD* (or *ND*) map, also termed an on surface radiation condition (OSRC), as the approximate inverse [190,392–394].

6.2. The Coupling Parameter

The direct and indirect integral formulations of the exterior problem that that were free of the characteristic frequencies or non-uniqueness, were introduced in Section 4.3.4. The equations were defined with a coupling parameter μ, which weighs the contribution of derivative equation with the original equation in the direct method, or between the double layer potential and the single layer potential equations in the indirect method. Mathematically, the coupling parameter has a non-zero imaginary part that ensures that the equations are free of characteristic wavenumbers, and, in general, μ is presumed to be an imaginary number. In accepting that μ is imaginary, the starting point is $-\infty < Im(\mu) < \infty$. If $|\mu|$ is very small or very large then one or the other of the original equations would dominate, and the issues with the dominant equation would be evident with the hybrid equation, and this narrows its range; $0 \ll |\mu| \ll \infty$.

The argument quickly shifted from introducing μ in order to avoid the potential catastrophic errors in the original equations of Section 4.3.1 to considering an optimal value. For several decades, based originally on the works of Kress [395–398] and the further works of Amini [399], there has been a strong recommendation, with supporting research, that $\mu = \frac{i}{k}$ is reasonably close to the 'optimal' choice, as least for the simple boundaries, like spheres. Obviously $\frac{i}{k}$ is unsuitable for low wavenumbers, as the parameter would be large, violating the condition $|\mu| \ll \infty$, and the second equation would dominate and a cap on its value is recommended, for example,

$$\mu = \begin{cases} 1 & \text{if } k \leq 1 \\ \frac{i}{k} & \text{if } k > 1 \end{cases}.$$

The term 'optimal' in these paragraphs, refers to the condition of the combined operators over which the equation is being solved. The stronger the condition of the operator, the less able it is to magnify the solution, or magnify its error. The original equations, on their own, have 'spikes' of ill-conditioning around their respective eigenvalues. The rationale is that in combining the equations, these spikes are significantly reduced, and the condition of the combined operator steers an even keel through the frequencies. It has been shown, for example for a sphere that $\frac{i}{k}$ provides a generally well-conditioned combined operator. A recent paper by Zheng [400] showed that each eigenvalue of the combined operator follows a loop-shaped locus as μ varies, joining the real axis when $|\mu| = 0, \infty$, and supporting $\mu = \frac{i}{k}$, as this reasonably approximated the point on the locus that was furthest from the real axis. Zheng terms the inclusion of the derivative equation as 'adding damping', relating the analogy with, for example, Equation (91), wherein the damping term similarly shifts the eigenvalue off the real axis; the combined integral equations are not free from irregular frequencies, rather they are simply moved from the real axis into the complex plane.

There is a simple case for the parameter with an inverse proportionality with the wavenumber. Considering the three-dimensional case, the Green's function (52) is the kernel of the L_k operator, and its magnitude does not change with k $\left(\left| e^{ikr} \right| = 1 \right)$. Hovever its derivative,

$$\frac{\partial G_k}{\partial r} = \frac{e^{ikr}}{4\pi r^2}(ikr - 1), \tag{97}$$

a factor in the kernel of the M_k and M_k^t operators, is $O(k)$. Hence, combining the operators of L_k with M_k or M_k^t, as in Section 4.3.4, a parameter that is inversely proportional to k equalises the contribution from the two operators. Similarly, differentiating again, the kernel of the N_k is $O(k^2)$, and the same parameter has a similar function, when M_k or M_k^t are combined with it.

A recent paper by Marburg [401] considered principally the sign of μ. Marburg noted that many papers had inadvertently used $\mu = -\frac{i}{k}$ as the coupling parameter. Given that its value was not thought to be critical, it might be thought that this would work as well as $\mu = \frac{i}{k}$. However, Marburg demonstrated that $\frac{i}{k}$ usually returns significantly more accurate results, and much faster convergence with the iterative methods for solving the systems of equations. The parameter $\mu = -\frac{i}{k}$ was not used out of choice in the papers reviewed therein, but rather through the confusion of the signs that underlie the basic definitions in the mathematical model. These results have been confirmed by Galkowski [116].

If L_k is coupled with its derivative, as it is in the Burton and Miller equation, then the following equation is obtained, for one side of the equation

$$\left\{ (L_k + \mu(M_k^t + \frac{1}{2}I)v \right\}_S (p; n_p) = \int_S (G_k + \mu \frac{\partial G_k}{\partial r} \frac{\partial r}{\partial n_p})v(q)dS + \frac{1}{2}\mu v(p)$$

$$= \int_S \frac{e^{ikr}}{4\pi r^2}\left(r + \mu ikr \frac{\partial r}{\partial n_p} - \mu \frac{\partial r}{\partial n_p} \right)v(q)\, dS + \frac{1}{2}\mu v(p).$$

With $\mu = \frac{i}{k}$ there is significant cancellation in $r + \mu ikr\frac{\partial r}{\partial n_p}$ when $\frac{\partial r}{\partial n_p} \approx 1$, possibly with a diagonalising effect on the operator, and this is also presumed for the other combinations of operators.

Another approach is to consider the condition of the matrices that arise in the BEM [402–404]. Earlier, it was discussed that the N_k operator was a pseudo-differential operator, rather than an integral operator. As a result of this, the $\|N_{SS,k}\|$ increases with the number of elements n_s, whereas for the other integral operators, the norm of the matrix is independent of the number of elements. This is also supported by Equations (45) and (46); the diagonal components of the $N_{SS,k}$ matrix are $O\left(\frac{1}{h}\right) = O(n_S)$ for two- dimensional and axisymmetric problems and $O\left(\sqrt{n_S}\right)$ for three-dimensional problems. It follows that $\|N_{SS,k}\| = O(n_S)$ or $O\left(\sqrt{n_S}\right)$. The scene is therefore set for two conflicting interests in choosing the coupling parameter $\mu = \frac{i}{k}$ is near-optimal in conditioning the combined operator, as a result of mathematical analysis whereas $\mu = \frac{i}{n_S}$ or $\mu = \frac{i}{\sqrt{n_S}}$ (put simply), balancing the matrix norms, as a result of numerical analysis. Without the latter correction, the $N_{SS,k}$ matrix will apparently provide an increasing dominant potential for (quadrature-induced) error as the number of elements increases.

However, in Section 6.1, the modern form of the combined method was outlined, involving pre-conditioning and iterative methods for solving the linear systems. For example, with the preconditioner in (96), the N_k operator is avoided and, in such cases, the analysis of the previous two paragraphs is not applicable.

7. Concluding Discussion

The main purpose of this paper is to encapsulate the modern scope of the boundary element method in acoustics; how it can be adopted for general 2D, 3D and axisymmetric dimensions, interior, exterior and half-space problems, thin shells and modal analysis. The standard BEM can be directly applied to an enclosed domain with cavities or to multi-boundary problems in an exterior domain. Boundary element methods are founded on a boundary integral equations formulations and these can be fused to form additional methods: thin shells can be used to model discontinuities in an interior or exterior domain defined by closed boundaries, half-space problems can be modelled by the Rayleigh integral or modifying the Green's function, domain methods can be linked to the BEM so that vibro-acoustic and aero-acoustic problems can be tackled and—through regularization of the BEM equations—inverse problems can be addressed.

The boundary element method is a numerical method that only becomes a potentially useful tool for solving real-world problems in acoustic engineering when it has been implemented in software. Clearly, it is important that the executing code fits within the memory of the computer and completes the computation in reasonable time, and efficiency issues have been considered in Section 6.1. There are issues of generality and, in that regard, this work has focused on the generalised Robin boundary condition and general topologies. Reliability and maintenance are also very important issues in software development, and hence much of the focus has been on the standard exterior acoustic problem, which has had issues and solution approaches documented for more than half a century. The robustness of the BEM, particularly considering the validity of a defined elemental boundary, has had little attention in the field, but methods for assisting with this are summarised in Section 3.4.2.

There is much commonality across the various topologies to which the boundary element method can be applied in acoustic/Helmholtz problems; the equations for different dimensional problems are literally the same, with a change in the definition of the Green's function and line integrals become surface integrals when we move from 2D to 3D. All the core equations, whether they are interior, exterior, shell problems or the straightforward half-space problems, have the same essential operators within a chosen dimensional space. Hence there is significant scope for component-wise development of software, adopting a 'library' approach and unifying the method.

Significant issues have always surrounded the N_k operator. This operator was initially included in the combined integral equation formulations for the exterior problem, outlined in Section 4.3.4. Firstly, for surface (collocation) points its integral definition is hypersingular, which is difficult to interpret and evaluate, and perhaps for that reason alone, the alternative Schenck/CHIEF method is the preference of many. Once this significant issue is surpassed, the N_k is not compact, it is a pseudo-differential operator and this causes further issues for solving over by iterative methods and hence preconditioning has been introduced to circumvent this. Preconditioning can eliminate the hypersingular operator N_k for the standard exterior problem (96). However, N_k is still required for shell elements and hence it has to be included in a general library.

Much of the current research theme in the area, put simply, is to shunt the BEM in acoustics from its traditional comfort-zone of problems with $\sim 10^3$ elements to modern problems, to include high frequencies, with $\sim 10^6$ elements and from straightforward problems to complicated applications [190,238,405]. The computational bottle-neck in the traditional BEM is in the use of LU factorisation or similar methods for solving the linear system. Hence the LU factorisation is the first necessary casualty of this move and iterative methods are favoured, although this has been presaged for several decades. The shift to $\sim 10^6$ elements could also have a significant demand on computer memory and hence the expectation of storing matrices must also be relaxed. With iterative methods, the effect of the approximation of the matrix components is more controllable and methods such as the fast multipole method and panel clustering are more able to broker the issues of storage and computation time to this end.

Probably the majority of research on the BEM in acoustics is on the seemingly intractable exterior problem. Our expectation that the solution in its infinite domain can be thread through the boundary is ultimately a questionable one, as the method is scaled up. The combined methods, and most prominently, the Burton and Miller method, have held centre-stage in this endeavour, seen by most as numerically superior to the Schenck/CHIEF method [238]. However, the two erstwhile competing methods may have to be married in our best efforts to achieve $\sim 10^6$ elements and high frequencies; using the equations from interior points to provide more stability in the Burton and Miller equation, as they did originally with the elementary equations to form the CHIEF method. The system could be augmented further with directional derivates of Equation (70) for points in the interior D.

The combined methods throw up two issues, one is the inclusion of the N_k operator, as discussed, and the other is the coupling parameter. The systems of equations arising in the BEM using combined methods require that the N_k operator is preconditioned in order to achieve convergence with iterative methods, as discussed in Section 6.1. The choice of coupling parameter is discussed in Section 6.2. However, the approach to choosing the coupling parameter is already to potentially optimise the system and hence it is itself a pre-conditioner, as pointed out in Betcke et al. [190]. In Harris and Amini [406], the coupling parameter is generalised from a singular value to a function over the surface. Hence, another marriage is proposed. As the coupling parameter and the preconditioner have a similar purpose, then the Burton and Miller Equation (76), for example, could take the following form

$$\left\{ (M_k - \frac{1}{2}I + U_k N_k)\varphi \right\}_S (p; n_p) = \left\{ (L_k + U_k(M_k^t + \frac{1}{2}I)v \right\}_S (p; n_p),$$

where U_k is the preconditioning operator, fusing the coupling parameter and the preconditioner into one entity. The objective then is to determine U_k, or an analogous preconditioning matrix, $U_{SS,k}$. For example, following the method in Equation (96), $U_k = \frac{i}{k}L_k$; however, in general, the goal is to set U_k or $U_{SS,k}$ to best-prepare the system for iterative solution.

Acknowledgments: The author would like to thank the anonymous reviewers for their enthusiasm for the paper. The author is also very grateful for the reviewers' many useful points and further references, that have helped to significantly improve on the original manuscript.

Conflicts of Interest: The authors declare no conflict of interest.

References

1. Crighton, D.G.; Dowling, A.P.; Ffowcs Williams, J.E.; Heckl, M.; Leppington, F.G. *Modern Methods in Analytical Acoustics: Lecture Notes*; Springer: Berlin/Heidelberg, Germany, 1992.
2. *Springer Handbook of Acoustics*, 2nd ed.; Rossing, T.D., Ed.; Springer Science+Business Media: New York, NY, USA, 2014.
3. Gulia, P.; Gupta, A. Mathematics and Acoustics. In *Mathematics Applied to Engineering*; Ram, M., Davin, J.P., Eds.; Academic Press: Cambridge, MA, USA; University of Aveiro: Aveiro, Portugal, 2017; pp. 55–82.
4. Pinsky, P.M.; Hughes, T.J.R. *Research in Computational Methods for Structural Acoustics*; Department of the Navy: Washington, DC, USA, 1996.
5. Petyt, M.; Jones, C.J.C. Numerical Methods in Acoustics. In *Advanced Applications in Acoustics, Noise & Vibration*; Fahy, F., Walker, J., Eds.; Spon Press: London, UK, 2004; pp. 53–99.
6. Bergman, D.R. *Computational Acoustics*; John Wiley & Sons: Chichester, UK, 2018.
7. Jensen, F.B.; Kuperman, W.A.; Porter, M.B.; Schmidt, H. *Computational Ocean Acoustics*, 2nd ed.; Springer Science & Business Media: New York, NY, USA, 2011.
8. Kaltenbacher, M. (Ed.) *Computational Acoustics*; Springer International Publishing: Cham, Switzerland, 2018.
9. Magoules, F. (Ed.) *Computational Methods for Acoustics Problems*; Saxe-Coburg Publications: Stirlingshire, UK, 2008.

10. Marburg, S.; Nolte, B. (Eds.) *Computational Acoustics of Noise propagation in Fluids*; Springer: Berlin/Heidelberg, Germany, 2008.
11. Tam, C.K.W. *Computatonal Aeroacoustics*; Cambridge University Press: Cambridge, UK, 2012.
12. Richter, T. *Fluid-structure Interaction: Models, Analysis and Finite Elements*; Springer International Publishing: Cham, Switzerland, 2017.
13. Hambric, S.A.; Sung, S.H.; Nefske, D.J. (Eds.) *Engineering Vibroacoustic Analysis: Methods and Applications*; John Wiley and Sons: Chichester, UK, 2016.
14. Hansen, C.H. *Foundations of Vibroacoustics*; CRC Press: Boca Raton, FL, USA, 2018.
15. Raman, G. (Ed.) *Jet Aeroacoustics*; Emerald Publishing Ltd.: Hockley, UK, 2008.
16. Colton, D.; Kress, R. *Inverse Acoustic and Electromagnetic Scattering Theory*; Springer: New York, NY, USA, 2013.
17. Wu, J.-S. *Analytical and Numerical Methods for Vibration Analyses*; John Wiley & Sons: Singapore, 2013.
18. Palazzolo, A. *Vibration Theory and Applications with Finite Elements and Active Vibration Control*; John Wiley & Sons: Chichester, UK, 2016.
19. Zhou, Y.; Lucey, A.D.; Liu, Y.; Huang, L. (Eds.) Fluid-Structure- Sound Interactions and Control. In *Proceedings of the 3rd Symposium on Fluid-Structure-Sound Interactions and Control*; Springer: Heidelberg, Germany, 2016.
20. Ross, C.T.F. *Finite Element Programs for Structural Vibrations*; Springer: London, UK, 2012.
21. Petyt, M. *Introduction to Finite Element Vibration Analysis*, 3rd ed.; Cambridge University Pess: New York, NY, USA, 2015.
22. Thompson, L.L. A review of finite-element methods for time-harmonic acoustics. *J. Acoust. Soc. Am.* **2006**, *119*, 1315–1330. [CrossRef]
23. Liu, G. *Formulation of Multifield Finite Element Models for Helmholtz Problems*; University of Hong Kong: Hong Kong, China, 2010.
24. Kirkup, S.M.; Mulla, I.; Ndou, G.; Yazdani, J. Electromagnetic Simulation by the FDTD method in Java. In Proceedings of the WSEAS MAMECTIS 2008, Corfu Island, Greece, 15–17 December 2019; pp. 370–375.
25. Kunz, K.S.; Luebbers, R.J. *The Finite Difference Time Domain Method for Electromagnetics*; CRC Press: Boca Raton, FL, USA, 2018.
26. Taflove, A.; Hagness, S.C. *Computational Electrodynamics: The Finite-Difference Time-Domain Method*, 3rd ed.; Artech House: Boston, MA, USA; London, UK, 2005.
27. Sheaffer, J.; Fazenda, B. FDTD/K-DWM Simulation of 3D Room Acoustics on General Purpose Graphics Hardware Using Compute Unified Device Architecture (CUDA). *Proc. Inst. Acoust.* **2010**, *32*. Available online: http://usir.salford.ac.uk/id/eprint/11568 (accessed on 15 April 2019).
28. Hargreaves, J. *Time Domain Boundary Element Method for Room Acoustics*; University of Salford: Salford, UK, 2007.
29. Bilbao, S. *Numerical Sound Synthesis: Finite Difference Schemes and Simulation in Musical Acoustics*; John Wiley & Sons: Chichester, UK, 2009.
30. Carvajal, C.S. *Time-Domain Numerical Methods in Room Acoustics Simulations*; Universitat Pompeu Fabra: Barcelona, Spain, 2009.
31. Parot, J.; Thirard, C.; Puillet, C. Elimination of a non-oscillatory instability in a retarded potential integral equation. *Eng. Anal. Bound. Elem.* **2007**, *31*, 133–151. [CrossRef]
32. Langer, S.; Schanz, M. Time Domain Boundary Element Method. In *Computational Acoustics of Noise Propagation in Fluids*; Marburg, S., Nolte, B., Eds.; Springer: Berlin/Heidelberg, Germany, 2008; pp. 495–518.
33. Chappell, D.J.; Harris, P.J.; Henwood, D.J.; Chakrabarti, R. A stable boundary element method for modeling transient acoustic radiation. *J. Acoust. Soc. Am.* **2006**, *120*, 74–80. [CrossRef]
34. Qiu, T. *Time Domain Boundary Integral Equation Methods in Acoustics, Heat Diffusion and Electromagnetism*; University of Delaware: Newark, DE, USA, 2016.
35. Tukac, M.; Vampola, T. Semi-analytic solution to planar Helmholtz equation. *Appl. Comput. Mech.* **2013**, *7*, 77–86.
36. Stewart, J.R.; Hughes, T.J.R. h-adaptive finite element computation of time-harmonic exterior acoustics problems in two dimensions. *Comput. Methods Appl. Mech. Eng.* **1997**, *146*, 65–89. [CrossRef]
37. Stewart, J.R.; Hughes, T.J.R. Adaptive Finite Element Methods for the Helmholtz Equation in Exterior Domains. In *Large-sale Structures in Acoustics and Electromagnetics: Proceedings of a Symposium*; National Academic Press: Washington, DC, USA, 1996; pp. 122–142.

38. Solgård, T.A. A Method of Designing Wide Dispersion Waveguides Using Finite Element Analysis. Master's Thesis, Norwegian University of Science and Technology, Trondheim, Norway, 2011.
39. Kechroud, R.; Soulaimani, A.; Antoine, X. A Performance Study of Plane Wave Finite Element Methods with a Padé-type Artificial Boundary Condition in Acoustic Scattering. *Adv. Eng. Softw.* **2009**, *40*, 738–750. [CrossRef]
40. Harari, I.; Hughes, T.J.R. Finite element methods for the helmholtz equation in an exterior domain: Model problems. *Comput. Methods Appl. Mech. Eng.* **1991**, *87*, 59–96. [CrossRef]
41. Harari, I.; Grosh, K.; Hughes, T.J.R.; Malhotra, M.; Pinsky, P.M.; Stewart, J.R.; Thompson, L.L. Recent developments in finite element methods for structural acoustics. *Arch. Comput. Methods Eng.* **1996**, *3*, 131–309. [CrossRef]
42. Borelli, D.; Schenone, C. A Finite Element Model to Predict Sound Attenuation in Lined and Parallel-baffle Rectangular Ducts. *HVAC R Res.* **2012**, *18*, 390–405.
43. Ihlenburg, F. *Finite Element Analysis of Acoustic Scattering*; Springer Science & Business Media: New York, NY, USA, 2006.
44. Wrobel, L.C.; Aliabadi, M.H. *The Boundary Element Method, Applications in Thermo-Fluids and Acoustics*; John Wiley & Sons: Chichester, UK, 2002.
45. Lindgren, L.E. From Weighted Residual Methods to Finite Element Methods. Available online: https://www.ltu.se/cms_fs/1.47275!/mwr_galerkin_fem.pdf (accessed on 10 April 2019).
46. Jumarhon, B.; Amini, S.; Chen, K. On the boundary element dual reciprocity method. *Eng. Anal. Bound. Elem.* **1997**, *20*, 205–211. [CrossRef]
47. Falletta, S.; Monegato, G. Exact Nonreflecting Boundary Conditions for Exterior Wave Equation Problems. *Publ. l'Institut Math.* **2014**, *96*, 103–123. [CrossRef]
48. Falletta, S.; Monegato, G.; Scuderi, L. A Space-time BIE Method for Nonhomogeneous Exterior Wave Equation Problems. The Dirichlet Case. *IMA J. Numer. Anal.* **2012**, *32*, 202–226. [CrossRef]
49. Bergman, D.R. Boundary Element Method in Refractive Media. In Proceedings of the 23rd International Congress on Sound and Vibration, Athens, Greece, 10–14 July 2016; pp. 1–5.
50. Jones, C.J.C. Finite Element Analysis of Loudspeaker Diaphragm Vibration and Prediction of the Resulting Sound Radiation. Ph.D. Thesis, University of Brighton (Polytechnic), Brighton, UK, 1986.
51. Feistel, S. *Modeling the Radiation of Modern Sound Reinforcement Systems in High Resolution*; Logos Verlag: Berlin, Germany, 2014.
52. Kocbach, J. Finite Element Modeling of Ultrasonic Piezoelectric Transducers. Ph.D. Thesis, University of Bergen, Bergen, Norway, 2000.
53. Kirkup, S.M.; Tyrell, R.J. Computer-aided analysis of engine noise. *Int. J. Veh. Des.* **1992**, *13*, 388–402.
54. Dowling, A.P. Automotive Noise. In *Proceedings of the Mathematics in the Automotive Industry*; Smith, J.R., Ed.; Clarenden Press: Oxford, UK, 1992; pp. 95–124.
55. Augusztinovicz, F. Calculation of Noise Control by Numerical methods—What We Can Do and What We Cannot Do Yet. In Proceedings of the INCE, Budapest, Hungary, 25 August 1997.
56. French, C.C.J. Advanced techniques for engine research and design. *Proc. Inst. Mech. Eng. Part D J. Automob. Eng.* **1989**, *203*, 169–183. [CrossRef]
57. Morris, P.J. Technical Evaluation Report. Available online: https://archive.org/details/DTIC_ADP014092 (accessed on 18 April 2019).
58. Dewitte, F.H. *V Aircraft Noise Shielding Assessment*; Delft University of Technology: Delft, The Netherlands, 2016.
59. Astley, R.J. Numerical methods for noise propagation in moving flows, with application to turbofan engines. *Acoust. Sci. Tech.* **2009**, *4*, 227–239. [CrossRef]
60. Kirkup, S.M. Boundary Element Method. Available online: www.boundary-element-method.com (accessed on 11 June 2018).
61. Kirkup, S.M. *The Boundary Element Method in Acoustics*; Integrated Sound Software: Hebden Bridge, UK, 1998.
62. Wu, T.W. *Boundary Element Acoustics: Fundamentals and Computer Codes*; WIT Press: Southampton, UK, 2000.
63. von Estorff, O. (Ed.) *Boundary Elements in Acoustics: Advances and Applications*; WIT Press: Southampton, UK, 2000.
64. Ciskowski, R.D.; Brebbia, C.A. (Eds.) *Boundary Element Methods in Acoustics*; Kluwer Academic Publishers Group: Dordrecht, The Netherlands, 1991.

65. Marburg, S. Boundary Element Method for Time-Harmonic Acoustic Problem. In *Computational Acoustics*; Kaltenbacher, M., Ed.; Springer International Publishing: Cham, Switzerland, 2018; pp. 69–158.

66. Astley, R.J. Infinite elements for wave problems: a review of current formulations and an assessment of accuracy. *Int. J. Numer. Methods Eng.* **2000**, *49*, 951–976. [CrossRef]

67. Qi, Q.; Geers, T. Evaluation of the Perfectly Matched Layer for Computational Acoustics. *J. Comput. Phys.* **1998**, *139*, 166–183. [CrossRef]

68. Liu, Q.; Tao, J. The perfectly matched layer for acoustic waves. *J. Acoust. Soc. Am.* **1997**, *102*, 2072–2082. [CrossRef]

69. Yang, J.; Zhang, X.; Liu, G.R.; Zhang, W. A compact perfectly matched layer algorithm for acoustic simulations in the time domain with smoothed particle hydrodynamic method. *J. Acoust. Soc. Am.* **2019**, *145*, 204–214. [CrossRef]

70. Rayleigh, J.W. Strutt, Lord. In *The Theory of Sound*; Dover: London, UK, 1945.

71. Kirkup, S.M. Computational solution of the acoustic field surrounding a baffled panel by the Rayleigh integral method. *Appl. Math. Model.* **1994**, *18*, 403–407. [CrossRef]

72. Wu, T.W.; Wan, G.C. Numerical modeling of acoustic radiation and scattering from thin bodies using a Cauchy principal integral equation. *J. Acoust. Soc. Am.* **1992**, *92*, 2900–2906. [CrossRef]

73. Warham, A.G.P. *The Helmholtz Integral Equation for a Thin Shell*; National Physical Laboratory: London, UK, 1988.

74. Koussa, F.; Defrance, J.; Jean, P.; Blanc-benon, P. Acoustic performance of gabions noise barriers: Numerical and experimental approaches. *Appl. Acoust.* **2013**, *74*, 189–197. [CrossRef]

75. Koussa, F.; Defrance, J.; Jean, P.; Blanc-Benon, P. Acoustical efficiency of a sonic crystal assisted noise barrier. *Acta Acust. United Acust.* **2013**, *99*, 399–409. [CrossRef]

76. Fard, S.M.B.; Peters, H.; Marburg, S.; Kessissoglou, N. Acoustic Performance of a Barrier Embedded With Helmholtz Resonators Using a Quasi-Periodic Boundary Element Technique Acoustic Performance of a Barrier Embedded With Helmholtz Resonators Using a Quasi-Periodic Boundary Element Technique. *Acta Acust. United Acust.* **2017**, *103*, 444–450. [CrossRef]

77. Karimi, M.; Croaker, P.; Kessissoglou, N. Boundary element solution for periodic acoustic problems. *J. Sound Vib.* **2016**, *360*, 129–139. [CrossRef]

78. Jean, P.; Defrance, J. Sound Propagation in Rows of Cylinders of Infinite Extent: Application to Sonic Crystals and Thickets Along Roads. *Acta Acust. United Acust.* **2015**, *101*, 474–483. [CrossRef]

79. Karimi, M.; Croaker, P.; Kessissoglou, N. Acoustic scattering for 3D multi-directional periodic structures using the boundary element method. *J. Acoust. Soc. Am.* **2017**, *141*, 313–323. [CrossRef]

80. Ziegelwanger, H. The Three dimensional Quasi-periodic Boundar Element Method: Implementation, Evaluation, and Use Cases. *Int. J. Comput. Methods Exp. Meas.* **2017**, *5*, 404–414.

81. Fard, S.M.B.; Peters, H.; Kessissoglou, N.; Marburg, S. Three-dimensional analysis of a noise barrier using a quasi-periodic boundary element method. *J. Acoust. Soc. Am.* **2015**, *137*, 3107–3114. [CrossRef]

82. Marburg, S.; Andersson, R. Fluid structure interaction and admittance boundary conditions: Setup of an analytical example. *J. Comput. Acoust.* **2011**, *19*, 63–74. [CrossRef]

83. Kirkup, S.M.; Yazdani, J. A Gentle Introduction to the Boundary Element Method in Matlab/Freemat. In Proceedings of the WSEAS MAMECTIS, Corfu, Greece, 26–28 October 2008.

84. Kirkup, S.M.; Henwood, D.J. An empirical error analysis of the boundary element method applied to Laplace's equation. *Appl. Math. Model.* **1994**, *18*, 32–38. [CrossRef]

85. Kirkup, S.M. The boundary and shell element method. *Appl. Math. Model.* **1994**, *18*, 418–422. [CrossRef]

86. Kirkup, S.M. DC Capacitor Simulation by the Boundary Element Method. *Commun. Numer. Methods Eng.* **2007**, *23*, 855–869. [CrossRef]

87. Kirkup, S.M. The Boundary Element Method in Excel for Teaching Vector Calculus and Simulation. *World Acad. Sci. Eng. Technol. Int. Sci. Int. J. Soc. Behav. Educ. Econ. Bus. Ind. Eng.* **2019**, *12*, 1605–1613.

88. Kellogg, O.D. *Foundations of Potential Theory*; Dover: New York, NY, USA, 1953.

89. Jaswon, M.A.; Symm, G.T. *Integral Equation Methods in Potential Theory and Elastostatics*; Academic Press: London, UK, 1977.

90. Baker, C.T.H. *The Numerical Treatment of Integral Equations*; Clarendon Press: Oxford, UK, 1977.

91. Kirkup, S.M. Fortran codes for computing the discrete Helmholtz integral operators. *Adv. Comput. Math.* **1998**, *9*, 391–404. [CrossRef]

92. Salvadori, A. Analytical integrations in 3D BEM for elliptic problems: Evaluation and implementation. *Int. J. Numer. Methods Eng.* **2010**, *84*, 505–542. [CrossRef]

93. Nodes and Weights of Gaussian Quadrature Calculator. Available online: https://keisan.casio.com/exec/system/1329114617 (accessed on 24 August 2018).

94. Laursen, M.E.; Gellert, M. Some criteria for numerically integrated matrices and quadrature formulas for triangles. *Int. J. Numer. Methods Eng.* **1978**, *12*, 67–76. [CrossRef]

95. Johnston, B.M.; Johnston, P.R.; Elliott, D. A new method for the numerical evaluation of nearly singular integrals on triangular elements in the 3D boundary element method. *J. Comput. Appl. Math.* **2013**, *245*, 148–161. [CrossRef]

96. Sikora, J.; Polakowski, K.; Pańczyk, B. Improper Integrals Calculations for Fourier Boundary Element Method. *Appl. Comput. Electromagn. Soc.* **2017**, *32*, 761–768.

97. Sikora, J.; Pańczyk, B.; Polakowski, K. Numerical calculation of singular integrals for different formulations of boundary element. *Prz. Elektrotechniczny* **2017**, 181–185. [CrossRef]

98. Gong, J.; Junying, A.; Ma, L.; Xu, H. Numerical quadrature for singular and near-singular integrals of boundary element method and its applications in large-scale acoustic problems. *Chinese J. Acoust.* **2017**, *36*, 289–301.

99. Chen, J.T.; Hong, H.-K. Review of Dual Integral Representations with Emphasis on Hypersingularity and Divergent Series. In Proceedings of the Fifith International Colloquium on Numerical Analysis, Plovdiv, Bulgaria, 18–23 August 1996; pp. 1–22.

100. do Rego Silva, J.J. *Acoustic and Elastic Wave Scattering Using Boundary Elements*, 1st ed.; Computational Mechanics Publications: Southampton, UK; Billerica, MA, USA, 1994.

101. Keuchel, S.; Hagelstein, N.C.; Zaleski, O.; von Estorff, O. Evaluation of hypersingular and nearly singular integrals in the Isogeometric Boundary Element Method for acoustics. *Comput. Methods Appl. Mech. Eng.* **2017**, *325*, 488–504. [CrossRef]

102. Jargstorff, F. Stephen Kirkup's Acoustic BEM codes written in Python. Available online: https://github.com/fjargsto/AcousticBEM (accessed on 7 September 2018).

103. Zhu, S.; Zhang, Y. A comparative study of the direct boundary element method and the dual reciprocity boundary element method in solving the Helmholtz equation. *ANZIAM J.* **2007**, *49*, 131–150. [CrossRef]

104. Cipriano, R.; Hersberger, R.; Hauser, G.; Noy, D.; Storyk, J. Low Frequency Behavior of Small Rooms. In Proceedings of the Audio Engineering Society, New York, NY, USA, 29 October–1 November 2015.

105. Chusov, A.A.; Statsenko, L.G.; Anisimov, P.N.; Mirgoronskaya, Y.V.; Cherkasova, N.A.; Bernavskaya, M.V. Computer Simulation of an Arbitrary Acoustical Field in Rooms. In Proceedings of the 2017 Asia Modelling Symposium, Kota Kinabalu, Malaysia, 4–6 December 2017; IEEE: Kota Kinabalu, Malaysia, 2017.

106. Ferreira, Á.C.; de Morais, M.V.G.; de Albuquerque, E.L.; Campos, L.S. Fast Boundary Element Simulations for for Complex Geometry Bidimensional Enclosures. In Proceedings of the CILAMCE; Avila, S.M., Ed.; ABMEC: Brasilia, Brazil, 2016.

107. Kuster, M. *Inverse Methods in Room Acoustics with Under-Determined Data and Applications to Virtual Acoustics*; Queen's University: Belfast, UK, 2007.

108. Konkel, F. *Sound Field in Small Fitted Enclosures*; Technical University of Berlin: Berlin, Germany, 2012.

109. Stringfellow, N.D.; Smith, R.N.L. The use of exact values at quadrature points in the boundary element method. *Trans. Model. Simul.* **1999**, *24*, 239–248.

110. Wozniak, Z.; Purcell, C. Variable separation in acoustic radiation problems using Chebyshev polynomials. *Trans. Model. Simul.* **1997**, *18*, 419–428.

111. Zhou, S.; Zhang, S. Structural-Acoustic Analysis of Automobile Passenger Compartment. *Appl. Mech. Mater.* **2012**, *236–237*, 175–179. [CrossRef]

112. Acikgoz, S.; Ozer, M.B.; Royston, T.J.; Mansy, H.A.; Sandler, R.H. Experimental and Computational Models for Simulating Sound Propagation Within the Lungs. *J. Vib. Acoust.* **2008**, *130*, 021010–021020. [CrossRef] [PubMed]

113. Ozer, M.B.; Acikgoz, S.; Royston, T.J.; Mansy, H.A.; Sandler, R.H. Boundary element model for simulating sound propagation and source localization within the lungs. *J. Acoust. Soc. Am.* **2007**, *122*, 657–671. [CrossRef]

114. Wijaya, F.B.; Mohapatra, A.R.; Sepehrirahnama, S.; Lim, K.M. Coupled acoustic-shell model for experimental study of cell stiffness under acoustophoresis. *Microfluid. Nanofluidics* **2016**, *20*, 1–15. [CrossRef]

115. Kirkup, S.M. Solving the Linear Systems of Equations in the Generalized Direct Boundary Element Method. In Proceedings of the 1st International Conference on Numerical Modelling in Engineering, Ghent, Belgium, 28–29 August 2018; Springer: Singapore, 2018.

116. Galkowski, J.; Muller, E.H.; Spence, E.A. Wavenumber-explicit analysis for the Helmholtz h -BEM: Error estimates and iteration counts for the Dirichlet problem. **2019**, not yet published. [CrossRef]

117. Baydoun, S.K.; Marburg, S. Quantification of Numerical Damping in the Acoustic Boundary Element Method for Two-Dimensional Duct Problems. *J. Theor. Comput. Acoust.* **2018**, *26*, 1850022. [CrossRef]

118. Fahnline, J.B. Numerical difficulties with boundary element solutions of interior acoustic problems. *J. Sound Vib.* **2009**, *319*, 1083–1096. [CrossRef]

119. Marburg, S. Numerical Damping in the Acoustic Boundary Element Method. *Acta Acust. United Acust.* **2016**, *102*, 415–418. [CrossRef]

120. Steinbach, O.; Unger, G. Convergence Analysis of a Galerkin Boundary Element Method for the Dirichlet Laplacian Eigenvalue Problem. *SIAM J. Numer. Anal.* **2012**, *50*, 710–728. [CrossRef]

121. Lu, Y.Y.; Yay, S.-T. Eigenvalues of the Laplacian through Boundary Integral Equations. *SIAM J. Matrix Anal. Appl.* **1991**, *12*, 597–609. [CrossRef]

122. Unger, G. *Analysis of Boundary Element Methods for Laplacian Eigenvalue Problems*; Brenn, G., Holzapfel, G.A., Schanz, M., Steinbach, O., Eds.; Monographi.: Graz, Austria, 2009; ISBN 9783851250817.

123. Zhao, L.; Barnett, A. Robust and efficient solution of the drum problem via Nystrom approximation of the Fredholm determinant. *SIAM J. Numer. Anal.* **2014**, *53*, 1–21. [CrossRef]

124. Barnett, A.H.; Hassell, A. Fast Computation of High Frequency Dirichlet Eigenmodes via the Spectral Flow of the Interior Neumann-to-Dirichlet Map. *Commun. Pure Appl. Math.* **2014**, *67*, 351–407. [CrossRef]

125. Kang, S.W.; Atluri, S.N. Application of Nondimensional Dynamic Influence Function Method for Eigenmode Analysis of Two-Dimensional Acoustic Cavities. *Adv. Mech. Eng.* **2015**. [CrossRef]

126. Kamiya, N.; Andoh, E.; Nogae, K. A new complex-valued formulation and eigenvalue analysis of the Helmholtz equation by boundary element method. *Adv. Eng. Softw.* **1996**, *26*, 219–227. [CrossRef]

127. Kamiya, N.; Andoh, E.; Nogae, K. Iterative local minimum search for eigenvalue determination of the Helmholtz equation by boundary element formulation. *Trans. Built Environ.* **1995**, *10*, 229–236.

128. Kamiya, N.; Andoh, E. Eigenvalue Analysis Schemes and Boundary Formulations: Recent Developments. *Trans. Built Environ.* **1995**, *10*, 489–496.

129. Ih, J.-G.; Kim, B.-K.; Choo, W.-S. Comparison of Eigenvalue Analysis Techniques in Acoustic Boundary Element Method. In Proceedings of the EuroNoise '95, Lyon, France, 21–23 March 1995; pp. 591–596.

130. Iemma, U.; Gennaretti, M. A boundary-field integral equation for analysis of cavity acoustic spectrum. *J. Fluids Struct.* **2006**, *22*, 261–272. [CrossRef]

131. Ghassemi, H.; Kohansal, A.R. Solving the Helmholtz Equation using Direct Boundary Element Method and Dual Reciprocity Boundary Element Method. *Int. J. Res. Curr. Dev.* **2016**, *2*, 81–85.

132. Gao, H.; Matsumoto, T.; Takahashi, T.; Isakari, H. Eigenvalue analysis for acoustic problem in 3D by boundary element method with the block Sakurai-Sugiura method. *Eng. Anal. Bound. Elem.* **2013**, *37*, 914–923. [CrossRef]

133. Effenberger, C.; Kressner, D. Chebyshev interpolation for nonlinear eigenvalue problems. *BIT Numer. Math.* **2012**, *52*, 933–951. [CrossRef]

134. Durán, M.; Nédélec, J.-C.; Ossandón, S. An Efficient Galerkin BEM to Compute High Acoustic Eigenfrequencies. *J. Vib. Acoust.* **2009**, *131*, 031001. [CrossRef]

135. Chen, J.T.; Lin, S.R.; Chen, K.H.; Chen, I.L.; Chyuan, S.W. Eigenanalysis for Membranes with Stringers using Conventional BEM in Conjunction with SVD Technique. *Comput. Methods Appl. Mech. Eng.* **2003**, *192*, 1299–1322. [CrossRef]

136. Chen, J.T.; Huang, C.X.; Chen, K.H. Determination of spurious eigenvalues and multiplicities of true eigenvalues using the real-part dual BEM. *Comput. Mech.* **1999**, *24*, 41–51. [CrossRef]

137. B&W Loudspeakers Ltd. *Development of the B&W 800D*; Worthing: West Sussex, UK. Available online: https://www.google.com.tw/url?sa=t&rct=j&q=&esrc=s&source=web& cd=1&ved=2ahUKEwjek72WqtvhAhXWfd4KHacjAFYQFjAAegQIBhAC&url=http%3A%2F% 2Fbwgroupsupport.com%2Fdownloads%2Freference%2Fbw%2F800_Development_Paper.pdf&usg= AOvVaw0MuUK5aLTcSXmKx2ph10K6 (accessed on 18 April 2019).

138. Ali, A.; Rajakumar, C.; Yunus, S.M. Advances in Acoustic Eigenvalue Analysis using Boundary Element Method. *Comput. Struct.* **1995**, *56*, 837–847. [CrossRef]

139. Ali, A.; Rajakumar, C. *The Boundary Element Method: Applications in Sound and Vibration*, 1st ed.; A.A. Balkenna: Leiden, The Netherlands, 2004.

140. Kirkup, S.M.; Amini, S. Solution of the Helmholtz eigenvalue problem via the boundary element method. *Int. J. Numer. Methods Eng.* **1993**, *36*, 321–330. [CrossRef]

141. Kirkup, S.M.; Jones, M.A. Computational methods for the acoustic modal analysis of an enclosed fluid with application to a loudspeaker cabinet. *Appl. Acoust.* **1996**, *48*, 275–299. [CrossRef]

142. Ossandon, S.; Klenner, J.; Reyes, C. Direct Nondestructive Algorithm for Shape Defects Evaluation. *J. Vib. Acoust.* **2011**, *133*, 031006. [CrossRef]

143. Ossandón, S.; Reyes, C.; Reyes, C.M. Neural network solution for an inverse problem associated with the Dirichlet eigenvalue. *Comput. Math. Appl.* **2016**, *72*, 1153–1163.

144. Leblanc, A. A Meshless Method for the Helmholtz Eigenvalue Problem Based on the Taylor Series of the 3-D Green's Function. *Acta Acust. United Acust.* **2013**, *99*, 770–776. [CrossRef]

145. Leblanc, A.; Lavie, A.; Vanhille, C. An Acoustic Resonance Study of Complex Three-Dimensional Cavities by a Particular Integral Method. *Acta Acust. United Acust.* **2005**, *91*, 873–879.

146. Xiao, J.; Meng, S.; Zhang, C.; Zheng, C. Contour integral based Rayleigh-Ritz method for large-scale nonlinear eigenvalue problems. *Comput. Methods Appl. Mech. Eng.* **2016**, *310*, 33–57. [CrossRef]

147. Xiao, J.; Zhang, C.; Huang, T.; Sakurai, T. Solving large-scale nonlinear eigenvalue problems by rational interpolation and resolvent sampling based Rayleigh—Ritz method. *Int. J. Numer. Methods Eng.* **2017**, *110*, 776–800. [CrossRef]

148. Yeih, W.; Chen, J.T.; Chen, K.H.; Wong, F.C. A study on the multiple reciprocity method and complex-valued formulation for the Helmholtz equation. *Adv. Eng. Softw.* **1998**, *29*, 1–6. [CrossRef]

149. Schenck, H.A. Improved Integral Formulation for Acoustic Radiation Problems. *J. Acoust. Soc. Am.* **1968**, *44*, 41–58. [CrossRef]

150. Zaman, S.I. A Comprehensive Review of Boundary Integral Formulations of Acoustic Scattering Problems. *Sci. Technol. Spec. Rev.* **2000**, 281–310. [CrossRef]

151. Wright, L.; Robinson, S.P.; Humphrey, V.F.; Harris, P.; Hayman, G. *The Application of Boundary Element Methods to Near- Field Acoustic Measurements on Cylindrical Surfaces at NPL*; National Physical Laboratory: London, UK, 2005.

152. Wright, L.; Robinson, S.P.; Humphrey, V.F. Prediction of acoustic radiation from axisymmetric surfaces with arbitrary boundary conditions using the boundary element method on a distributed computing system. *J. Acoust. Soc. Am.* **2009**, *125*, 1374–1383. [CrossRef] [PubMed]

153. Marburg, S. Conventional boundary element techniques: Recent developments and opportunities. In Proceedings of the Inter-Noise, Hong Kong, China, 27–30 August 2017; pp. 2673–2688.

154. Christensen, M.J. Using the Boundary Element Method for Prediction of Sound Radiated from an Arbitrarily Shaped Vibrating Body. Mater's Thesis, Western Michigan University, Kalamazoo, MI, USA, 2002.

155. Burton, A.J. *The Solution of Helmholtz Equation in Exterior Domains using Integral Equations*; NPL Report NAC30; National Physical Laboratory: London, UK, 1973.

156. Augusztinovicz, F. State of the Art of Practical Applications of Numerical Methods in Vibro-Acoustics. Available online: http://last.hit.bme.hu/download/fulop/Publikaciok/Iberoamericano_Statoftheart.pdf (accessed on 18 April 2019).

157. Amini, S.; Harris, P.J. A comparison between various boundary integral formulations of the exterior acoustic problem. *Comput. Methods Appl. Mech. Eng.* **1990**, *84*, 59–75. [CrossRef]

158. Margonari, M. The Solution of Exterior Acoustic Problems with Scilab. Available online: http://www.openeering.com/node/54 (accessed on 7 September 2018).

159. Fiala, P.; Rucz, P. NiHu: An open source C++ BEM library. *Adv. Eng. Softw.* **2014**, *75*, 101–112. [CrossRef]

160. Esward, T.J.; Lees, K.; Sayers, D.; Wright, L. *Testing Continuous Modelling Software: Three Case Studies*; National Physical Laboratory: London, UK, 2004.

161. Young, K.; Kearney, G.; Tew, A.I. Loudspeaker Positions with Sufficient Natural Channel Separation for Binaural Reproduction. In Proceedings of the 2018 AES International Conference on Spatial Reproduction - Aesthetics and Science, Tokyo, Japan, 7–9 August 2018; p. REB1-9.

162. Vanderkooy, J. The Acoustic Center: A New Concept for Loudspeakers at Low Frequencies. In Proceedings of the Audio Engineering Society, San Francisco, CA, USA, 5–8 October 2006; pp. 6859–6989.

163. Thompson, A. Line Array Splay Angle Optimisation. Available online: https://www.researchgate.net/publication/272490480 (accessed on 18 April 2019).

164. Sanalatii, M.; Herzog, P.; Melon, M.; Guillermin, R.; Le Roux, J.-C.; Poulain, N. Measurement of the Frequency and Angular Responses of Loudspeaker Systems Using Radiation Modes. In Proceedings of the Audio Engineering Society Convention 141, Los Angeles, CA, USA, 29 October–1 November 2016; Audio Engineering Society: New York, NY, USA, 2016; p. 9615.

165. Morgans, R.C. Optimisation Techniques for Horn Loaded Loudspeakers. Ph.D. Thesis, University of Adelaide, Adelaide, Australia, 2005.

166. Kolbrek, B.; Svensson, U.P. Using mode matching methods and edge diffraction in horn loudspeaker simulation. *Acta Acust. United Acust.* **2015**, *101*, 760–774. [CrossRef]

167. Kolbrek, B. Using Mode Matching Methods in Horn Loudspeaker Simulation. In Proceedings of the Forum Acousticum, Krakow, Poland, 8–12 September 2014.

168. Kolbrek, B. Extensions to the Mode Matching Method for Horn Loudspeaker Simulation. Ph.D. Thesis, Norwegian University of Science and Technology, Trodheim, Norway, 2016.

169. Kirkup, S.M.; Thompson, A.; Kolbrek, B.; Yazdani, J. Simulation of the acoustic field of a horn loudspeaker by the boundary element-Rayleigh integral method. *J. Comput. Acoust.* **2013**, *21*, 1250020. [CrossRef]

170. Henwood, D.J.; Vanderkooy, J. Polar Plots for Low Frequencies: The Acoustic Centre. In Proceedings of the Audio Engineering Society Convention 120, Paris, France, 20–23 May 2006.

171. Henwood, D.J. The Boundary Element Method and Horn Design. *J. Audio Eng. Soc.* **1993**, *41*, 486–496.

172. Hacihabiboğlu, H.; De Sena, E.; Cvetković, Z.; Johnston, J.; Smith, J.O. Perceptual Spatial Audio Recording, Simulation, and Rendering: An overview of spatial-audio techniques based on psychoacoustics. *IEEE Signal Process. Mag.* **2017**, *34*, 36–54. [CrossRef]

173. Grande, E.F. *Sound Radiation from a Loudspeaker Cabinet using the Boundary Element Method*; Technical University of Denmark: Copenhagen, Denmark, 2008.

174. Geaves, G.P.; Moore, J.P.; Henwood, D.J.; Fryer, P.A. Verification of an Approach for Transient Structural Simulation of Loudspeakers Incorporating Damping. In Proceedings of the Audio Engineering Society 100th Convention, Amsterdam, The Netherlands, 12–15 May 2001.

175. Fryer, P.A.; Henwood, D.; Moore, J.; Geaves, G. *Verification of an Approach for Transient Structural Simulation of Loudspeakers Incorporating Damping*; AES Convention Paper 5320; Audio Engineering Society: New York, NY, USA, 2001.

176. Feistel, S.; Thompson, A.; Ahnert, W. Methods and Limitations of Line Source Simulation. In Proceedings of the AES Convention 125, San Francisco, CA, USA, 2–5 October 2008; p. 7524.

177. Candy, J. Accurate Calculation of Radiation and Diffraction from Loudspeaker Enclosures at Low Frequency. *AES J. Audio Eng. Soc.* **2013**, *61*, 356–365.

178. Bastyr, K.J.; Capone, D.E. On the Acoustic Radiation from a Loudspeaker's Cabinet. *J. Audio Eng. Soc.* **2003**, *51*, 234–243.

179. Xu, Y.; Xu, L.; Li, X. The sound field analysis of piezoelectric micromachined ultrasound transducer array. In Proceedings of the 2009 International Conference on Mechatronics and Automation, Changechun, China, 9–12 August 2009.

180. Teng, D.; Chen, H.; Zhu, N. Computer Simulation of Sound Field Formed around Transducer Source Used in Underwater Acoustic Communication. In Proceedings of the ICACTE 3rd International Conference on Advanced Computer Theory and Engineering, Chengdu, China, 20–22 August 2010; pp. 144–148.

181. Kurowski, A.; Kotus, J.; Kostef, B.; Czyzewski, A. Numerical Modeling of Sound Intensity Distributions around Acoustic Transducer. In *Audio Engineering Society Convention 140*; Audio Engineering Society: Paris, France, 2016; p. 9525.

182. Kapuria, S.; Sengupta, S.; Dumir, P.C. Three-dimensional solution for simply-supported piezoelectric cylindrical shell for axisymmetric load. *Comput. Methods Appl. Mech. Eng.* **1997**, *140*, 139–155. [CrossRef]

183. Amini, S.; Harris, P.J.; Wilton, D.T. *Coupled Boundary and Finite Element Methods for the Solution of the Dynamic Fluid-Structure Interaction Problem*; Springer: Berlin, Germany, 1992; Volume 77, ISBN 978-3-540-55562-9.

184. Christensen, R. Acoustic Modeling Of Hearing Aid Components. Ph.D. Thesis, University of Southern Denmark, Odense, Denmark, 2010.

185. Lock, A. Development of a 2D Boundary Element Method to model Schroeder Acoustic Diffusers. Bachelor's Thesis, University of Tasmania, Tasmania, Australia, 2014.

186. Döşemeciler, A. A Study on Number Theoretic Construction and Prediction of Two Dimensional Acoustic Diffusers for Architectural Applications. Ph.D. Thesis, Izmir Institute of Technology, Izmir, Turkey, 2011.

187. Takane, S.; Matsuhashi, T.; Sone, T. Numerical estimation of individual HRTFs by using BEM. In Proceedings of the International Congress on Acoustics, Kyoto, Japan, 4–9 April 2004; pp. 1439–1440.

188. Jackson, P.J.B.; Desiraju, N. Use of 3D Head Shape for Personalized Binaural Audio. In Proceedings of the Audio Engineering Society Conference: 49th International Conference Audio for Games, London, UK, 6–8 February 2013; pp. 1–6.

189. Garcia, D.P.; Roozen, B.; Glorieux, C. Calculation of Human Echolocation Cues by Means of the Boundary Element Method. In Proceedings of the 19th International Conference on Auditory Display (ICAD2013), Lodz, Poland, 6–9 July 2013; pp. 253–259.

190. Betcke, T.; van 't Wout, E.; Gélat, P. Computationally Efficient Boundary Element Methods for High-Frequency Helmholtz Problems in Unbounded Domains. In *Modern Solvers for Helmholtz Problems*; Domenico, L., Tang, J., Vuik, K., Eds.; Springer International Publishing: Cham, Switzerland, 2017; ISBN 978-3-319-28832-1.

191. Wu, S.J.; Kuo, I.; Shung, K.K. Boundary element simulation of backscattering properties for red blood with high frequency ultrasonic transducers. *Ultrasonics* **2005**, *43*, 145–151. [CrossRef] [PubMed]

192. Zhang, B.; Chen, L.; Chen, J. Acoustic Analysis of a Structure Subjected to Stochastic Excitation Using Statistical Wave Superposition Approach. In Proceedings of the 2nd International Congress on Image and Signal Processing, Tianjin, China, 17–19 October 2009; IEEE: Tianjin, China, 2009; pp. 1–5.

193. Tillema, H.G. *Noise Reduction of Rotating Machinery by Viscoelastic Bearing Supports*; University of Twente: Enschede, The Netherlands, 2003.

194. Sorensen, J.D.; Frangopol, D.M. *Advances in Reliability and Optimization of Structural Systems*; Taylor & Francis: London, UK, 2006.

195. Saad, A.; El-Sebai, N. Combustion Noise Prediction Inside Diesel Engine. *SAE Trans.* **1999**, *108*, 2866–2872.

196. Roivainen, J. Unit-wave response-based modeling of electromechanical noise and vibration of electrical machines. Ph.D. Thesis, Helsinki University of Technology, Espoo, Finland, 2009.

197. Nijhuis, M.O. Analysis Tool for the Design of Active Structural Acoustic Control Systems. Ph.D. Thesis, University of Twente, Enschede, The Netherlands, 2003.

198. Mocsai, T.; Diwoky, F.; Hepberger, A.; Priebsch, H.-H.; Augusztinovicz, F. Application and analysis of an adaptive wave-based technique based on a boundary error indicator for the sound radiation simulation of a combustion engine model. *Comput. Assist. Methods Eng. Sci.* **2015**, *22*, 3–30.

199. Marburg, S. Developments in Structural – Acoustic Optimization for Passive Noise Control. *Arch. Comput. Methods Eng.* **2002**, *9*, 291–370. [CrossRef]

200. Lummer, M.; Akkermans, R.A.; Richter, C.; Pröber, C.; Delfs, J. Validation of a model for open rotor noise predictions and calculation of shielding effects using a fast BEM. In Proceedings of the 19th AIAA/CEAS Aeroacoustics Conference, Berlin, Germany, 27–29 May 2013; p. 2096.

201. Johnson, O.; Smith, A.V.; Morel, T. The Application of Advanced Analysis Methods to the Reduction of Noise from Air Compressors. In Proceedings of the International Compressor Engineering Conference, West Lafayette, IN, USA, 17–20 July 1990; p. 774.

202. Jaswon, M.A.; Zaman, S.I. A new BEM Formulation of Acoustic Scattering Problems. *Trans. Model. Simul.* **1993**, *1*, 11–21.

203. Fritze, D.; Marburg, S.; Hardtke, H.J. Estimation of radiated sound power: A case study on common approximation methods. *Acta Acust. United Acust.* **2009**, *95*, 833–842. [CrossRef]

204. Friot, E. *Limites et outils d'optimisation du contrôle acoustique actif, CNRS*; Laboratoire de Mécanique et d'Acoustique: Marseille, France, 2007.

205. Chauvicourt, F. *Vibro-Acoustics of Rotating Electric Machines Prediction, Validation and Solution*; Universite Libre de Bruxelles: Brussels, Belgium, 2018.

206. Bies, D.A.; Hansen, C.H. *Engineering Noise Control Theory and Practice*, 4th ed.; CRC Press: Abingdon, UK, 2009.

207. Ambrogio, M. *Virtual Acoustics for Product Design and Prototpying Process*; Polytecnico di Milano: Milan, Italy, 2012.

208. Nuraini, A.A.; Mohd Ihsan, A.K.A.; Nor, M.J.M.; Jamaluddin, N. Vibro-acoustic Analysis of Free Piston Engine Structure using Finite Element and Boundary Element Methods. *J. Mech. Sci. Technol.* **2012**, *26*, 2405–2411. [CrossRef]

209. Rossignol, K.; Lummer, M.; Delfs, J. Validation of DLR's sound shielding prediction tool using a novel sound source. In Proceedings of the 15th AIAA/CEAS Aeroacoustics Conference (30th AIAA Aeroacoustics Conference), Miami, FL, USA, 11–13 May 2009; AIAA 2009-3329.

210. Vlahopoulos, N.; Raveendra, S.T.; Vallance, C.; Messer, S. Numerical implementation and applications of a coupling algorithm for structural–acoustic models with unequal discretization and partially interfacing surfaces. *Finite Elem. Anal. Des.* **1999**, *32*, 257–277. [CrossRef]

211. Vlahopoulos, N.; Stark, R.D. Numerical Approach for Computing Noise-Induced Vibration from Launch Environments. *J. Spacecr. Rockets* **1998**, *35*, 355–360. [CrossRef]

212. Yeh, F.; Chang, X.; Sung, Y. Numerical and Experimental Study on Vibration and Noise of Embedded Rail System. *J. Appl. Math. Phys.* **2017**, *5*, 1629–1637. [CrossRef]

213. Prego-Borges, J.L. Lamb: a Simulation Tool for Air-Coupled Lamb Wave based Ultrasonic NDE Systems. Ph.D. Thesis, Polytechnic University of Catalonia, Barcelona, Splain, 2010.

214. Cao-Duc, T.; Nguyen-Dang, H. Inverse technique for detection of discontinuous geometry under acoustic field using boundary element method. In *Computation and Modeling in Structural adh Mechanical Engineering*; Ton Duc Thang University: Ho Chi Minh City, Vietnam, 2012.

215. Yan, N. Numerical Modelling and Condition Assessment of Timber Utility Poles using Stress Wave Techniques. Ph.D. Thesis, University of Tchnology, Sydney, Australia, 2015.

216. Grubeša, S.; Domitrović, H.; Jambrošić, K. Performance of Traffic Noise Barriers with Varying Cross-Section. *PROMET Traffic Transp.* **2011**, *23*, 161–168.

217. Hothersall, D.C.; Chandler-Wilde, S.N.; Hajmirzae, M.N. Efficiency of single noise barriers. *J. Sound Vib.* **1991**, *146*, 303–322. [CrossRef]

218. Georgiou, F. Modeling for auralization of urban environments: incorporation of directivity in sound propagation and analysis of a framework for auralizing a car pass-by. Ph.D. Thesis, Eindhoven University of Technology, Eindhoven, The Netherlaands, 2018.

219. Hargreaves, J.; Rendell, L.; Lam, Y.W. Simulation of acoustic environments for binaural reproduction using a combination of geometrical acoustics and Boundary Element Method. *J. Acoust. Soc. Am.* **2017**, *141*, 3783. [CrossRef]

220. Vuylsteke, X. *Development of a reference method based on the fast multipole boundary element method for sound propagation problems in urban environments: formalism, optimizations & applications*; Universite Paris-est: Paris, France, 2014.

221. Davis, D. A review of prediction methods for ground-borne noise due to tunnel construction activities. In Proceedings of the 14th Australasian Tunnelling Conference 2011: Development of Underground Space, Sydney, Australia, 8–10 March 2011; pp. 39–51.

222. Davis, D. A Review of Prediction Methods for Ground-Borne Noise due to Construction Activities. In Proceedings of the Proceedings of the 20th International Congress on Acoustics, Sydney, Australia, 23–27 August 2010; pp. 23–27.

223. Chandler-Wilde, S.N. The Boundary Element Method in Outdoor Noise Propagation. *Proc. Inst. Acoustcs* **1997**, *19*, 27–50.

224. Camargo, H.E.; Azman, A.S.; Peterson, J.S. Engineered Noise Controls for Miner Safety and Environmental Responsibility. In *Advances in Productive, Safe and Responsible Coal Mining*; Hirschi, J., Ed.; Elsevier B.V.: Amsterdam, The Netherlands, 2018; pp. 215–244.

225. Choi, J.-W.; Kim, K.-J. A New Sound Reception System using a Symmetrical Microphone Array and its Numerical Simulation. *J. Sh. Ocean Technol.* **2004**, *8*, 18–25.

226. Hunter, A.J. Underwater Acoustic Modelling for Synthetic Aperture Sonar. Ph.D. Thesis, University of Canterbury, Christchurch, New Zealand, 2006.

227. Wagenhoffer, N.; Moored, K.W.; Jaworski, J.W. Method and the Noise Generation of an Idealized School of Fish. In Proceedings of the International Conference on Flow Induced Noise and Vibration Issues and Aspects, Hong Kong, China, 27–28 April 2017; Ciappi, E., Ed.; Springer: Cham, Switzerland, 2017; pp. 157–178.

228. Mookerjee, A. Coherent Backscatter Enhancement from Finite Sized Aggregations of Scatterers. Ph.D. Thesis, University of Michigan, Ann Arbor, MI, USA, 2017.

229. Dimon, M.N.; Ula, M.; Hashim, A.W.I.M.; Hamid, S.Z.A.; Baharom, A.; Ahmad, A.H. *The Study of Normal Incidence Sound Absorption Coefficence (Sound Absorption) of Wood Circular Perforated Panel (CPP) Using Numerical Modelling Technique*; University of Malaysia: Kuala Lumpur, Malaysia, 2006.

230. Amini, S.; Kirkup, S.M. Solution of Helmholtz equation in the exterior domain by elementary boundary integral methods. *J. Comput. Phys.* **1995**, *118*, 208–221. [CrossRef]

231. Hall, W.S.; Robertson, W.H. Boundary element methods for acoustic wave scattering. In Proceedings of the Boundary Elements X, Southampton, UK, 6–9 September 1988; Computational Mechanics Publications: Southampton, UK, 1988; Volume 4, pp. 301–315.

232. Seybert, A.F.; Rengarajan, T.K. The use of CHIEF to obtain unique solutions for Acoustic Radiation using Boundary Integral Equations. *J. Acoust. Soc. Am.* **1987**, *81*, 1299–1306. [CrossRef]

233. Cutanda-Henriquez, V.; Juhl, P.M. OpenBEM - An open source Boundary Element Method software in Acoustics. In Proceedings of the Internoise, Lisbon, Portugal, 13–16 June 2010; pp. 1–10.

234. Mohsen, A.A.K.; Ochmann, M. Numerical experiments using CHIEF to treat the nonuniqueness in solving acoustic axisymmetric exterior problems via boundary integral equations. *J. Adv. Res.* **2010**, *1*, 227–232. [CrossRef]

235. Juhl, P.M. *The Boundary Element Method for Sound Field Calculations*; Technical University of Denmark: Copenhagen, Denmark, 1993.

236. Marburg, S.; Wu, T.W. Treating the Phenomenon of Irregular Frequencies. In *Computational Acoustics of Noise Propagation in Fluids - Finite and Boundary Element Methods*; Marburg, S., Nolte, B., Eds.; Springer: Berlin, Germany, 2008; pp. 411–434.

237. Visser, R. A Boundary Element Approach to Acoustic Radiation and Source Identification. Ph.D. Thesis, University of Twente, Enschede, The Netherlands, 2004.

238. Marburg, S.; Amini, S. Cat's Eye Radiation with Boundary Elements: Comparative Study on the Treatment of Irregular Frequencies. *J. Comput. Acoust.* **2005**, *13*, 21–45. [CrossRef]

239. Brakhage, H.; Werner, P. Ueber das Dirichletsche Au enraumproblem fuer die Helmholtzsche Schwingungsgleichung. *Arch. Math.* **1965**, *16*, 325–329. [CrossRef]

240. Leis, R. Zur Dirichletschen Randwertaufgabe des Auenraums der Schwingungsgleichung. *Math. Z.* **1965**, *90*, 205–211. [CrossRef]

241. Panich, O.I. On the Question of the Solvability of the Exterior Boundary Problem for the Wave Equation and Maxwell's Equation. *Uspeki Mat. Nauk* **1965**, *20*, 211–226.

242. Kussmaul, R. Ein numeriches Verfahren zur Losung des Neumannschen Au enraumproblems fuer die Helmholtsche Schwingungsgleichung. *Computing* **1979**, *4*, 246–273. [CrossRef]

243. Burton, A.J.; Miller, G.F. The Application of Integral Equations to the Numerical Solution of Some Exterior Boundary-Value Problems. *Proc. Roy. Soc. Lond.* **1971**, *323*, 201–210. [CrossRef]

244. Nowak, L.J.; Zielinski, T.G. Determination of the Free-Field Acoustic Radiation Characteristics of the Vibrating Plate Structures With Arbitrary Boundary Conditions. *J. Vib. Acoust.* **2015**, *137*, 051001. [CrossRef]

245. Kirkup, S.M. Fortran Codes for Computing the Acoustic Field Surrounding a Vibrating Plate by the Rayleigh Integral Method. In Proceedings of the Mathematical Methods, Computational Techniques, Non-Linerat Systems, Intelligent Systems, Corfu, Greece, 26–28 October 2008; pp. 364–369.

246. Khorshidi, K.; Akbari, F.; Ghadirian, H. Experimental and analytical modal studies of vibrating rectangular plates in contact with a bounded fluid. *Ocean Eng.* **2017**, *140*, 146–154. [CrossRef]

247. Hoernig, R.O.H. Green's Functions and Integral Equations for the Laplace and Helmholtz Operators in Impedance Half-Spaces. Ph.D. Thesis, Ecole Polytechnique, Paris, France, 2010.

248. Bose, T.; Mohanty, A.R. Sound Radiation Response of a Rectangular Plate Having a Side Crack of Arbitrary Length, Orientation, and Position. *J. Vib. Acoust.* **2015**, *137*, 021019. [CrossRef]

249. Arenas, J.P.; Ramis, J.; Alba, J. Estimation of the Sound Pressure Field of a Baffled Uniform Elliptically Shaped Transducer. *Appl. Acoust.* **2010**, *71*, 128–133. [CrossRef]

250. Arenas, J.P. Numerical Computation of the Sound Radiation From a Planar Baffled Vibrating Surface. *J. Comput. Acoust.* **2008**, *16*, 321–341. [CrossRef]

251. Arenas, J.P. Matrix Method for Estimating the Sound Power Radiated from a Vibrating Plate for Noise Control Engineering Applications. *Lat. Am. Appl. Res.* **2009**, *39*, 345–352.

252. Alia, A.; Soulie, Y. Simulation of Acoustical Response Using Rayleigh Method. In Proceedings of the Pressure Vessels and Piping/ICPVT-11 Conference, Vancouver, BC, Canada, 23–27 July 2006.

253. Liao, C.; Ma, C. Vibration characteristics of rectangular plate in compressible inviscid fluid. *J. Sound Vib.* **2016**, *362*, 228–251. [CrossRef]

254. Wu, H.; Jiang, W.; Liu, Y. Analyzing acoustic radiation modes of baffled plates with a fast multipole Boundary Element Method. *J. Vib. Acoust.* **2013**, *135*, 11007. [CrossRef]

255. Li, S. Modal models for vibro-acoustic response analysis of fluid-loaded plates. *J. Vib. Control* **2010**, *17*, 1540–1546.

256. Partha, B.; Atanu, S.; Arup, N.; Michael, R. A Novel FE_BE Approach for Free Field Vibro-acoustic Problem. In Proceedings of the Acoustics 2013 New Delhi, New Delhi, India, 10–15 November 2013; pp. 1367–1372.

257. Basten, T.G.H. Noise Reduction by Viscothermal Acousto-elastic Interaction in Double Wall Panels. Ph.D. Thesis, University of Twente, Enschede, The Netherlands, 2001.

258. Arunkumar, M.P.; Pitchaimani, J.; Gangadharan, K.V.; Lenin Babu, M.C. Effect of Core Topology on Vibro-acoustic Characteristics of Truss Core Sandwich Panels. *Procedia Eng.* **2016**, *144*, 1397–1402. [CrossRef]

259. Arunkumar, M.P.; Pitchaimani, J.; Gangadharan, K.V.; Lenin Babu, M.C. Influence of nature of core on vibro acoustic behavior of sandwich aerospace structures. *Aerosp. Sci. Technol.* **2016**, *56*, 155–167. [CrossRef]

260. Arunkumar, M.P.; Pitchaimani, J.; Gangadharan, K.V.; Lenin Babu, M.C. Sound transmission loss characteristics of sandwich aircraft panels: Influence of nature of core. *J. Sandw. Struct. Mater.* **2017**, *19*, 26–48. [CrossRef]

261. D'Alessandro, V.; Petrone, G.; Franco, F.; De Rosa, S. A Review of the Vibroacoustics of Sandwich Panels: Models and Experiments. *J. Sandw. Struct. Mater.* **2013**, *15*, 541–582. [CrossRef]

262. Denli, H.; Sun, J.Q. Structural-acoustic Optimization of Composite Sandwich Structures: A Review. *Shock Vib. Dig.* **2007**, *39*, 189–200. [CrossRef]

263. Galgalikar, R.; Thompson, L.L. Design Optimization of Honeycomb Core Sandwich Panels for Maximum Sound Transmission Loss. *J. Vib. Acoust.* **2016**, *138*, VIB-15-1339. [CrossRef]

264. Chiang, H.-Y.; Huang, Y.-H. Vibration and sound radiation of an electrostatic speaker based on circular diaphragm. *J. Acoust. Soc. Am.* **2015**, *137*, 1714–1721. [CrossRef] [PubMed]

265. Nowak, L.J.; Zielinski, T.G.; Meissner, M. Active vibroacoustic control of plate structures with arbitrary boundary conditions. *IPPT Reports Fundam. Technol. Res.* **2013**, *4*, 5–9.

266. Hasheminejad, S.M.; Keshavarzpour, H. Robust active sound radiation control of a piezo-laminated composite circular plate of arbitrary thickness based on the exact 3D elasticity model. *J. Low Freq. Noise Vib. Act. Control* **2016**, *35*, 101–127. [CrossRef]

267. Diwan, G.C. Partition of Unity Boundary Element and Finite Element Method: Overcoming Non-uniqueness and Coupling for Acoustic Scattering in Heterogeneous Media. Ph.D. Thesis, Durham University, Durham, UK, 2014.

268. Brick, H.; Ochmann, M. A Half-space BEM for the Simulation of Sound Propagation above an Impedance Plane. *J. Acoust. Soc. Am.* **2008**, *123*, 3418. [CrossRef]

269. Li, Y.L.; White, M.J.; Hwang, M.H. Green's Function for Wave Propagation above an Impedance Ground. *J. Acoust. Soc. Am.* **1994**, *96*, 2485–2490. [CrossRef]

270. Piscoya, R.; Ochmann, M. Acoustical Green's Function and Boundary Element Techniques for 3D Half-Space Problems. *J. Theor. Comput. Acoust.* **2017**, *25*, 1730001. [CrossRef]

271. Hwang, L.S.; Tuck, E.O. On the oscillations of harbours of arbitrary shape. *J. Fluid Mech.* **1970**, *42*, 447–464. [CrossRef]

272. Lee, J. Wave-induced oscillations in harbours of arbitrary geometry. *J. Fluid Mech.* **1971**, *45*, 375–394. [CrossRef]

273. Chandler-Wilde, S.N.; Peplow, A.T. A Boundary Integral Equation Formulation for the Helmholtz Equation in a Locally Perturbed Half-plane. *ZAMM J. Appl. Math. Mech.* **2005**, *85*, 79–88. [CrossRef]

274. Krishnasamy, G.; Schmer, L.W.; Rudolphi, T.J.; Rizzo, F.J. Hypersingular boundary integral equations: some applications in acoustic and elastic wave scattering. *J. Appl. Mech.* **1990**, *57*, 404–414. [CrossRef]

275. Gray, L.J. Boundary element method for regions with thin internal cavities. *Eng. Anal. Bound. Elem.* **1989**, *6*, 180–184. [CrossRef]

276. Terai, T. On calculation of sound fields around three dimensional objects by integral equation methods. *J. Sound Vib.* **1980**, *69*, 71–100. [CrossRef]

277. Martin, P.A.; Rizzo, F.J. On boundary integral equations for crack problems. *Proc. R. Soc. London, Ser. A-Mathematical Phys. Eng. Sci.* **1989**, *421*, 341–355. [CrossRef]

278. Kırkup, S.M. The Computational Modelling of Acoustic Shields by the Boundary and Shell Element Method. *Comput. Struct.* **1991**, *40*, 1177–1183. [CrossRef]

279. Kirkup, S.M. Solution of discontinuous interior Helmholtz problems by the boundary and shell element method. *Comput. Methods Appl. Mech. Eng.* **1997**, *140*, 393–404. [CrossRef]

280. Sedaghatjoo, Z.; Dehghan, M.; Hosseinzadeh, H. On uniqueness of numerical solution of boundary integral equations with 3-times monotone radial kernels. *J. Comput. Appl. Math.* **2017**, *311*, 664–681. [CrossRef]

281. Jeong, J.-H.; Ih, J.-G.; Lee, B.-C. A Guideline for Using the Multi-Domain BEM for Analyzing the Interior Acoustic Field. *J. Comput. Acoust.* **2003**, *11*, 403. [CrossRef]

282. Gennaretti, M.; Giordani, A.; Morino, L. A third-order boundary element method for exterior acoustics with applications to scattering by rigid and elastic shells. *J. Sound Vib.* **1999**, *222*, 699–722. [CrossRef]

283. Fahnline, J.B. Boundary-Element Analysis. In *Engineering Vibroacoustic Analysis: Methods and Applications*; Hambric, S.A., Sung, S.H., Nefske, D.J., Eds.; John Wiley & Sons: Chichester, UK, 2016; Chapter 7.

284. Wilkes, D.; Alexander, P.; Duncan, A. FMBEM analysis of sound scattering from a damping plate in the near field of a hydrophone. In Proceedings of the Proceedings of Acoustics, Fremantle, Australia, 21–23 November 2012; pp. 1–8.

285. Poblet-Puig, J. *Numerical Modelling of Sound Transmission in Lightweight Structures*; Universitat Politecnica de Catalunyu: Catalan, Spain, 2008.

286. Lee, J.-S.; Ih, J.-G. Reactive characteristics of partial screens for a sound source mounted in an infinite baffle. *J. Acoust. Soc. Am.* **1995**, *98*, 1008–1016. [CrossRef]

287. Wilkes, D.R. The Development of a Fast Multipole Boundary Element Method for Coupled Acoustic and Elastic Problems. Ph.D. Thesis, Curtin University, Perth, Australia, 2014.

288. Qian, Z.-H.; Yamanaka, H. An efficient approach for simulating seismoacoustic scattering due to an irregular fluid-solid interface in multilayered media. *Geophys. J. Int.* **2012**, *189*, 524–540. [CrossRef]

289. Zhou-Bowers, S.; Rizos, D.C. B-Spline Impulse Response Functions of Rigid Bodies for Fluid-Structure Interaction Analysis. *Hindawi Adv. Civ. Eng.* **2018**, *2018*. [CrossRef]

290. Zheng, C.-J.; Gao, H.F.; Du, L.; Chen, H.B.; Zhang, C. An accurate and efficient acoustic eigensolver based on a fast multipole BEM and a contour integral method. *J. Comput. Phys.* **2016**, *305*, 677–699. [CrossRef]

291. Qin, H.; Zheng, H.; Qin, W.; Zhang, Z. Lateral vibration control of a shafting-hull coupled system with electromagnetic bearings. *J. Low Freq. Noise, Vib. Act. Control* **2018**, 146134841881151. [CrossRef]

292. Pluymers, B.; Desmet, W.; Vanderpitte, D.; Sas, P. On the use of a wave based prediction technique for steady-state structural-acoustic radiation analysis. *CMES* **2005**, *7*, 173–183.

293. Pluymers, B. *Wave based Modelling Methods for Steady-State Vibro-acoustics*; Katholieke Universiteit Leuven: Leuven, Belgium, 2006.

294. Pluymers, B.; Desmet, W.; Vandepitte, D.; Sas, P. Application of an efficient wave-based prediction technique for the analysis of vibro-acoustic radiation problems. *J. Comput. Appl. Math.* **2004**, *168*, 353–364. [CrossRef]

295. Peters, H.; Kessissoglou, N.; Marburg, S. Modal decomposition of exterior acoustic-structure interaction. *J. Acoust. Soc. Am.* **2013**, *133*, 2668–2677. [CrossRef]

296. Peters, H.; Kessissoglou, N.; Marburg, S. Modal contributions to sound radiated from a fluid loaded cylinder. *J. Acoust. Soc. Am.* **2013**, *133*, 2668–2677. [CrossRef] [PubMed]

297. Liang, T.; Wang, J.; Xiao, J.; Wen, L. Coupled BE–FE based vibroacoustic modal analysis and frequency sweep using a generalized resolvent sampling method. *Comput. Methods Appl. Mech. Eng.* **2019**, *345*, 518–538. [CrossRef]

298. Lanzerath, H.; Waller, H. Computation of the time-history response of dynamic problems using the boundary element method and modal techniques. *Int. J. Numer. Methods Eng.* **1999**, *45*, 841–864. [CrossRef]

299. Kirkup, S.M.; Amini, S. Modal analysis of acoustically-loaded structures via integral equation methods. *Comput. Struct.* **1991**, *40*, 1279–1285. [CrossRef]

300. Jung, B.K.; Ryue, J.; Hong, C.; Jeong, W.B.; Shin, K.K. Estimation of dispersion curves of water-loaded structures by using approximated acoustic mass. *Ultrasonics* **2018**, *85*, 39–48. [CrossRef]

301. Brennan, D.P.; Chemuka, M.W. *Enhancements to AVAST*; National Defence, Research and Development Branch: Halifax, NS, Canada, 1996.

302. Prisacariu, V. CFD Analysis of UAV Flying Wing. *INCAS Bull.* **2016**, *8*, 65–72.

303. Colonius, T.; Lele, S.K. Computational aeroacoustics: Progress on nonlinear problems of sound generation. *Prog. Aerosp. Sci.* **2004**, *40*, 345–416. [CrossRef]

304. Alléon, G.; Champagneux, S.; Chevalier, G.; Giraud, L.; Sylvand, G. Parallel distributed numerical simulations in aeronautic applications. *Appl. Math. Model.* **2006**, *30*, 714–730. [CrossRef]

305. Shahbazi, M.; Mansourzadeh, S.; Pishevar, A.R. Hydrodynamic Analysis of Autonomous Underwater Vehicle (AUV) Flow Through Boundary Element Method and Computing Added-Mass Coefficients. *Int. J. Artif. Intell. Mechatron.* **2015**, *3*, 212–217.

306. Seol, H.; Pyo, S.; Suh, J.-C.; Lee, S. Numerical study of non-cavitating underwater propeller noise. *Noise Vib. Worldw.* **2004**, 11–26. [CrossRef]

307. Casenave, F.; Ern, A.; Sylvand, G. Coupled BEM – FEM for the Convected Helmholtz Equation with Non-uniform Flow in a Bounded Domain. *J. Comput. Phys.* **2014**, *257*, 627–644. [CrossRef]

308. Morfey, C.L.; Powles, C.J.; Wright, C.M. Green's Functions in Computational Aeroacoustics. *Int. J. Aeroacoustics* **2011**, *10*, 117–159. [CrossRef]

309. Mancini, S. *Boundary Integral Methods for Sound Propagation with Subsonic Potential Mean Flows*; University of Southampton: Southampton, UK, 2017.

310. Harwood, A.R.G. *Numerical Evaluation of Acoustic Green's Functions*; University of Manchester: Manchester, UK, 2014.

311. Andesen, P.R.; Cutanda-Henriquez, V.; Aage, N.; Marburg, S. A Two-Dimensional Acoustic Tangential Derivative Boundary Element Method Including Viscous and Thermal Losses Publication date: A Two-Dimensional Acoustic Tangential Derivative Boundary. *J. Theor. Comput. Acoust.* **2018**, *26*, 1850036. [CrossRef]

312. Cutanda-Henriquez, V.; Anderson, P.R.; Jensen, J.S.; Juhl, P.M.; Sanchez-Dehesa, J. A Numerical Model of an Acoustic Metamaterial Using the Boundary Element Method Including Viscous and Thermal Losses. *J. Comput. Acoust.* **2017**, *25*, 1750006. [CrossRef]

313. Cutanda-Henriquez, V.; Juhl, P.M. An axisymmetric boundary element formulation of sound wave propagation in fluids including viscous and thermal losses. *J. Acoust. Soc. Am.* **2013**, *134*, 3409–3418. [CrossRef] [PubMed]

314. Vieira, A.; Snellen, M.; Simons, D.G. Assessment of Engine Noise Shielding By the Wings of Current Turbofan Aircraft. In Proceedings of the 24th International Conference on Sound and Vibration, London, UK, 23–27 July 2017; pp. 1–8.

315. Salin, M.B.; Dosaev, A.S.; Konkov, A.I.; Salin, B.M. Numerical Simulation of Bragg Scattering of Sound by Surface Roughness for Different Values of the Rayleigh Parameter. *Acoust. Phys.* **2014**, *60*, 442–454. [CrossRef]

316. Mimani, A.; Croaker, P.; Karimi, M.; Doolan, C.J.; Kessissoglou, N. Hybrid CFD-BEM and Time-Reversal techniques applied to localise flow-induced noise sources generated by a flat-plate. In Proceedings of the 2nd Australasian Acoustical Societies Conference, ACOUSTICS 2016, Brisbane, Australia, 9–11 November 2016.

317. Martínez-lera, P.; Schram, C.; Bériot, H.; Hallez, R. An approach to aerodynamic sound prediction based on incompressible-flow pressure. *J. Sound Vib.* **2014**, *333*, 132–143. [CrossRef]

318. Kucukcoskun, K. *Prediction of Free and Scattered Acoustic Fields of Low-Speed Fans*; Ecole Central de Lyon: Lyon, France, 2012.

319. Heffernon, T.; Angland, D.; Zhang, X.; Smith, M. The Effect of Flow Circulation on the Scattering of Landing Gear Noise. In Proceedings of the 21st AIAA/CEAS Aeroacoustics Conference, Southampton, UK, 22–26 June 2015; University of Southampon: Southampton, UK, 2015; pp. 1–14.

320. Dürrwächter, L.; Kesslaer, M.; Kraemer, E. Numerical Assessment of CROR Noise Shielding with a Coupled Möhring Analogy and BEM Approach. In Proceedings of the A/AA/CEAS Aeroacoustics Conference, Atlanta, GA, USA, 25–29 June 2018; p. 2822.

321. Croaker, P.; Mimani, A.; Doolan, C.; Kessissoglou, N. A computational flow-induced noise and time-reversal technique for analysing aeroacoustic sources. *J. Acoust. Soc. Am.* **2018**, *143*, 2301–2312. [CrossRef] [PubMed]

322. Croaker, P.; Kessissoglou, N.; Marburg, S. Aeroacoustic Scattering Using a Particle Accelerated Computational Fluid Dynamics/Boundary Element Technique. *AIAA J.* **2016**, *54*, 1–18. [CrossRef]

323. Barhoumi, B. An improved axisymmetric convected boundary element method formulation with uniform flow. *Mech. Ind.* **2017**, *18*, 313. [CrossRef]

324. Sundkvist, E. *A High-Order Accurate, Collocated Boundary Element Method for Wave Propagation in Layered Media*; Uppsala University: Uppsala, Sweden, 2011.

325. Croaker, P.; Kessissoglou, N.; Marburg, S. Strongly singular and hypersingular integrals for aeroacoustic incident fields. *Int. J. Numer. Methods Fluids* **2015**, *7*, 274–318. [CrossRef]

326. Croaker, P.; Kessissoglou, N.; Kinns, R.; Marburg, S. Fast Low-Storage Method for Evaluating Lighthill's Volume Quadrupoles. *AIAA J.* **2013**, *51*, 867–884. [CrossRef]

327. Blondel, P.; Dobbins, P.F.; Jayasundere, N.; Cosci, M. High-frequency Bistatic Scattering Experiments using Proud and Buried Targets. In *Acoustic Sensing Techniques for the Shallow Water Environment: Inversion Methods and Experiments*; Caiti, A., Chapman, N.R., Hermand, J.-P., Jesus, S.M., Eds.; Springer: Dordrecht, The Netherlands, 2006; pp. 155–170, ISBN 1402043724.

328. Kirkup, S.M.; Wadsworth, M. Solution of Inverse Diffusion Problems by Operator-splitting Methods. *Appl. Math. Model.* **2002**, *26*, 1003–1018. [CrossRef]

329. Schuhmacher, A.; Hald, J.; Rasmussen, K.B.; Hansen, P.C. Sound source reconstruction using inverse boundary element calculations. *J. Acoust. Soc. Am.* **2003**, *113*, 114–127. [CrossRef] [PubMed]

330. Ko, B.; Lee, S.-Y. Enhancing the Reconsruction of Acoustic Source Field using Wavelet Transformation. *J. Mech. Sci. Technol.* **2005**, *19*, 680–686. [CrossRef]

331. Kim, G.T.; Lee, B.H. 3-D Source Reconstruction and Field Reprediction using the Helmholtz Integral Equation. *J. Sound Vib.* **2001**, *112*, 2645–2655. [CrossRef]

332. Bustamante, F.O.; Rodríguez, F.L.; Lopez, A.P. Experimental analysis of laptop fan noise radiation by acoustic source decomposition and inverse boundary element methods. In Proceedings of the NOISE-CON, Detroit, MI, USA, 28–31 July 2008.

333. Bai, M.R. Application of the BEM (Boundary Element Method)- based Acoustic Holography to Radiation Analysis of Sound Sources with Arbitrarily Shaped Geometries. *J. Acoust. Soc. Am.* **1992**, *92*, 533–549. [CrossRef]

334. van Wijngaarden, H.C.J. Prediction of Propeller-Induced Hull-Pressure Fluctuations Proefschrift, Delft. Ph.D. Thesis, Technische Universiteit, Delft, The Netherlands, 2011.

335. Nava, G.P. *Inverse sound rendering: In-situ estimation of surface acoustic impedance for acoustic simulation and design of real indoor environments*; University of Tokyo: Tokyo, Japan, 2006.

336. Piechowicz, J.; Czajka, I. Estimation of Acoustic Impedance for Surfaces Delimiting the Volume of an Enclosed Space. *Arch. Acoust.* **2012**, *37*, 97–102. [CrossRef]

337. Lee, S. Review: The Use of Equivalent Source Method in Computational Acoustics. *J. Comput. Acoust.* **2017**, *25*, 1630001. [CrossRef]

338. Fu, Z.J.; Chen, W.; Chen, J.T.; Qu, W.Z. Singular boundary method: Three regularization approaches and exterior wave applications. *Comput. Model. Eng. Sci.* **2014**, *99*, 417–443.

339. Fu, Z.J.; Chen, W.; Gu, Y. Burton-Miller-type singular boundary method for acoustic radiation and scattering. *J. Sound Vib.* **2014**, *333*, 3776–3793. [CrossRef]

340. Fu, Z.J.; Chen, W.; Qu, W. Numerical investigation on three treatments for eliminating the singularities of acoustic fundamental solutions in the singular boundary method. *WIT Trans. Model. Simul.* **2014**, *56*, 15–26.

341. Liu, L. Single layer regularized meshless method for three dimensional exterior acoustic problem. *Eng. Anal. Bound. Elem.* **2017**, *77*, 138–144. [CrossRef]

342. Treeby, B.E.; Pan, J. A practical examination of the errors arising in the direct collocation boundary element method for acoustic scattering. *Eng. Anal. Bound. Elem.* **2009**, *33*, 1302–1315. [CrossRef]

343. Juhl, P.M. A note on the convergence of the direct collocation boundary element method. *J. Sound Vib.* **1998**, *212*, 703–719. [CrossRef]

344. Menin, O.H.; Rolnik, V. Relation between accuracy and computational time for boundary element method applied to Laplace equation. *J. Comput. Interdiscip. Sci.* **2013**, *4*, 1–6.

345. Ramirez, I.H. Multilevel Multi-Integration Algorithm for Acoustics. Ph.D. Thesis, University of Twente, Enschede, The Netherlands, 2005.

346. Gumerov, N.A.; Duraiswami, R. A Broadband Fast Multipole Accelerated Boundary Element Method for the Three Dimensional Helmholtz Equation. *J. Acoust. Soc. Am.* **2009**, *125*, 191–205. [CrossRef]

347. Falletta, S.; Sauter, S.A. The panel-clustering method for the wave equation in two spatial dimensions. *J. Comput. Phys.* **2016**, *305*, 217–243. [CrossRef]

348. Zhang, Q.; Mao, Y.; Qi, D.; Gu, Y. An Improved Series Expansion Method to Accelerate the Multi-Frequency Acoustic Radiation Prediction. *J. Comput. Acoust.* **2015**, *23*, 1450015. [CrossRef]

349. Wang, Z.; Zhao, Z.G.; Liu, Z.X.; Huang, Q.B. A method for multi-frequency calculation of boundary integral equation in acoustics based on series expansion. *Appl. Acoust.* **2009**, *70*, 459–468. [CrossRef]

350. Wang, X.; Chen, H.; Zhang, J. An efficient boundary integral equation method for multi-frequency acoustics analysis. *Eng. Anal. Bound. Elem.* **2015**, *61*, 282–286. [CrossRef]

351. Schenck, H.A.; Benthien, G. The Application of a Coupled Finite-Element Boundary-Element technique to Large-Scale Structural Acoustic Problems. In Proceedings of the Eleventh International Conference on Boundary Element Methods, Advances in Boundary Elements, Vol. 2: Field and Fluid Flow; Brebbia, C.A., Connor, J.J., Eds.; Computational Mechanics Publications: Cambridge, MA, USA, 1989; pp. 309–318.

352. Lefteriu, S.; Souza Lenzi, M.; Bériot, H.; Tournour, M.; Desmet, W. Fast frequency sweep method for indirect boundary element models arising in acoustics. *Eng. Anal. Bound. Elem.* **2016**, *69*, 32–45. [CrossRef]

353. Kirkup, S.M.; Henwood, D.J. Methods for speeding up the boundary element solution of acoustic radiation problems. *J. Vib. Acoust. Trans. ASME* **1992**, *114*, 374–380. [CrossRef]

354. Amini, S. An Iterative Method for the Boundary Element Solution of the Exterior Acoustic Problem. *J. Comput. Appl. Math.* **1987**, *20*, 109–117. [CrossRef]

355. Coyette, J.-P.; Lecomte, C.; Migeot, J.-L.; Blanche, J.; Rochette, M.; Mirkovic, G. Calculation of Vibro-Acoustic Frequency Response Functions Using a Single Frequency Boundary Element Solution and a Padé Expansion. *Acta Acust. United Acust.* **1999**, *85*, 371–377.

356. Marburg, S. Discretization Requirements: How many Elements per Wavelength are Necessary? In *Computational Acoustics of Noise Propagation in Fluids: Finite and Boundary Element Methods*; Marburg, S., Nolte, B., Eds.; Springer: Berlin, Germany, 2010; Chapter 11.

357. Marburg, S. Six Boundary Elements per Wavelength: Is that Enough? *J. Comput. Acoust.* **2002**, *10*, 25–51. [CrossRef]

358. Marburg, S.; Schneider, S. Influence of Element Types on Numeric Error for Acoustic Boundary Elements. *J. Comput. Acoust.* **2003**, *11*, 363–386. [CrossRef]

359. Sun, Y.; Trevelyan, J.; Li, S. Evaluation of discontinuity in IGABEM modelling of 3D acoustic field. In Proceedings of the Eleventh UK Conference on Boundary Integral Methods (UKBIM 11), Nottingham, UK, 10–11 July 2017; Chappell, D.J., Ed.; Nottingham Trent University Publications: Nottingham, UK, 2017; pp. 177–185.

360. Amini, S.; Wilton, D.T. An investigation of boundary element methods for the exterior acoustic problem. *Comput. Methods Appl. Mech. Eng.* **1986**, *54*, 49–65. [CrossRef]

361. Zieniuk, E.; Szerszeń, K. Triangular Bézier Patches in Modelling Smooth Boundary Surface in exterior Helmholtz Problems Solved by the PIES. *Arch. Acoust.* **2009**, *34*, 51–61.

362. Zieniuk, E.; Szerszeń, K. A solution of 3D Helmholtz equation for boundary geometry modeled by Coons patches using the Parametric Integral Equation System. *Arch. Acoust.* **2006**, *31*, 99 111.

363. Zieniuk, E.; Boltuc, A. Bezier Curves in the Modeling of Boundary Gemetry for 2D Boundary Problems Defined by Helmholtz Equation. *J. Comput. Acoust.* **2006**, *14*, 353–367. [CrossRef]

364. Peake, M.J.; Trevelyan, J.; Coates, G. Extended isogeometric boundary element method (XIBEM) for three-dimensional medium-wave acoustic scattering problems. *Comp. Meth. Appl. Mech. Eng.* **2015**, *284*, 0762–780. [CrossRef]

365. Peake, M.J.; Trevelyan, J.; Coates, G. Extended isogeometric boundary element method (XIBEM) for two-dimensional Helmholtz problem. *Comp. Meth. Appl. Mech. Eng.* **2013**, *259*, 93–163. [CrossRef]

366. Liu, Z.; Majeed, M.; Cirak, F.; Simpson, R.N. Isogeometric FEM-BEM coupled structural-acoustic analysis of shells using subdivision surfaces. *Int. J. Numer. Methods Eng.* **2018**, *113*, 1507–1530. [CrossRef]

367. Coox, L.; Atak, O.; Vanderpitte, D.; Desmet, W. An isogeometric indirect boundary element method for solving acoustic problems in open-boundary domains. *Comput. Methods Appl. Mech. Eng.* **2017**, *316*, 186–208. [CrossRef]

368. Chen, L.; Liu, C.; Zhao, W.; Liu, L. An isogeometric approach of two dimensional acoustic design sensitivity analysis and topology optimization analysis for absorbing material distribution. *Comput. Methods Appl. Mech. Eng.* **2018**, *336*, 507–532. [CrossRef]

369. Khaki, H.; Trevelyan, J.; Hattori, G. Towards Isogeometric Boundary Element Method Based on Adaptive Hierarchical Refinement of NURBS for 3D Geometries. In Proceedings of the Boundary Integral Methods (UKBIM 11), Nottingham, UK, 10–11 July 2017; Chappell, D.J., Ed.; Nottingham Trent University Publications: Nottingham, UK, 2017; pp. 109–116.

370. Simpson, R.N.; Scott, M.A.; Taus, M.; Thomas, D. Acoustic isogeometric boundary element analysis. *Comput. Methods Appl. Mech. Eng.* **2014**, *269*, 265–290. [CrossRef]

371. Peake, M.J. *Enriched and Isogeometric Boundary Element Methods for Acoustic Wave Scattering*; Durham University: Durham, UK, 2014.

372. Melenk, J.M.; Babuška, I. The partition of unity finite element method: Basic theory and applications. *Comput. Methods Appl. Mech. Eng.* **1996**, *139*, 289–314. [CrossRef]

373. Perrey-Debain, E.; Trevelyan, J.; Bettess, P. Use of wave boundary elements for acoustic computations. *J. Comput. Acoust.* **2003**, *11*, 305–321. [CrossRef]

374. Perrey-Debain, E.; Trevelyan, J.; Bettess, P. Plane wave interpolation in direct collocation boundary element method for radiation and wave scattering: Numerical aspects and applications. *J. Sound Vib.* **2003**, *261*, 839–858. [CrossRef]

375. Amini, S.; Chen, K. Conjugate gradient method for second kind integral equations - applications to the exterior acoustic problem. *Eng. Anal. Bound. Elem.* **1993**, *6*, 72–77. [CrossRef]

376. Pocock, M.D. *Integral Equation Methods for Harmonic Wave Propagation*; University of London: London, UK, 1995.

377. Amini, S.; Maines, N.D.N. Preconditioned Krylov subspace methods for boundary element solution of the Helmholtz equation. *Int. J. Numer. Methods Eng.* **1998**, *41*, 875–898. [CrossRef]

378. Chen, K.; Harris, P.J. Efficient preconditioners for iterative solution of the boundary element equations for the three-dimensional Helmholtz equation. *Appl. Numer. Math.* **2001**, *36*, 475–489. [CrossRef]

379. Marburg, S.; Schneider, S. Performance of iterative solvers for acoustic problems. Part I. Solvers and effect of diagonal preconditioning. *Eng. Anal. Bound. Elem.* **2003**, *27*, 727–750. [CrossRef]

380. Carpentieri, B.; Duff, I.S.; Giraud, L. *Experiments with sparse preconditioning of dense problems from electromagnetic applications*; CERFACS: Toulouse, France, 2000.

381. Chen, K. *Matrix Preconditioning Techniques and Applications*, 1st ed.; Cambridge Univessity Press: Cambridge, UK, 2005.

382. Nedelec, J.C. *Acoustic and Electromagnetic Equations*; Springer: New York, NY, USA, 2001.

383. Steinbach, O. *Numerical approximation methods for elliptic boundary value problems*; Springer Science & Business Media: New York, NY, USA, 2008; ISBN 0190-535X.

384. Antoine, X.; Boubendir, Y. An Integral Preconditioner for Solving the Two-dimensional Scattering Transmission Problem using Integral Equations. *Int. J. Comput. Math.* **2008**, *85*, 1473–1490. [CrossRef]

385. Antoine, X.; Darbas, M. Integral equations and iterative schemes for acoustic scattering problems. In *Numerical Methods for Acoustics Problems*; Magoules, F., Ed.; Saxe-Coburg Publications, 2016. Available online: http://microwave.math.cnrs.fr/publications/files/chapterVersionFinale.pdf (accessed on 12 April 2019).

386. Antoine, X.; Darbas, M. Alternative Integral Equations for the Iterative Solution of Acoustic Scattering Problems. *Q. J. Mech. Appl. Math.* **1999**, *58*, 107–128. [CrossRef]

387. Antoine, X.; Darbas, M. Generalized Combined Field Integral Equations for te Iterative Solution of the Three-Dimensional Helmholtz Equation. *ESAIM Math. Model. Numer. Anal.* **2007**, *41*, 147–167. [CrossRef]

388. Darbas, M.; Darrigrand, E.; Lafranche, Y. Combining Analytic Preconditioner and Fast Multipole Method for the 3-D Helmholtz Equation. *J. Comput. Phys.* **2013**, *236*, 289–316. [CrossRef]

389. Langou, J. *Solving large linear systems with multiple right-hand sides*; L'Institut National des Sciences Appliquees de Toulouse: Toulouse, France, 2004.

390. Saad, Y. ILUT: A Dual Threshold Incomplete LU Factorization. *Numer. Linear Algebr. Appl.* **1994**, *1*, 387–402. [CrossRef]

391. Schneider, S.; Marburg, S. Performance of iterative solvers for acoustic problems. Part II. Acceleration by ILU-type preconditioner. *Eng. Anal. Bound. Elem.* **2003**, *27*, 751–757. [CrossRef]

392. Antoine, X. Advances in the On-Surface Radiation Condition Method: Theory, Numerics and Applications. In *Computational Methods for Acoustic Problems*; Magoules, F., Ed.; Saxe-Coburg Publications: Stirlingshire, UK, 2008.

393. Calvo, D.C. A wide-angle on-surface radiation condition applied to scattering by spheroids. *J. Acoust. Soc. Am.* **2004**, *116*, 1549. [CrossRef]

394. Chniti, C.; Eisa Ali Alhazmi, S.; Altoum, S.H.; Toujani, M. DtN and NtD surface radiation conditions for two-dimensional acoustic scattering: Formal derivation and numerical validation. *Appl. Numer. Math.* **2016**, *101*, 53–70. [CrossRef]

395. Kress, R. On the Condition Number of Boundary Integral Operators in Scattering Theory. In *Classsical Scattering*; Roach, G.F., Ed.; Shiva: Nantwich, UK, 1984.

396. Kress, R. Minimising the Condition Number of Boundary Integral Operators in Acoustic and Electromagnetic Scattering. *Q. J. Mech. Appl. Math.* **1985**, *38*, 324–342. [CrossRef]

397. Kress, R. On the Condition Number of Boundary Integal Equations in Acoustic Scattering using Combined Double- and Single- Layer Potentials. *Int. Ser. Numer. Math.* **1985**, *73*, 194–200.

398. Kress, R.; Spassov, W.T. On the condition number of boundary integral operators for the exterior Dirichlet problem for the Helmholtz equation. *Numer. Math.* **1983**, *95*, 77–95. [CrossRef]

399. Amini, S. Boundary integral solution of the exterior Helmholtz problem. *Comput. Mech.* **1993**, *13*, 2–11.

400. Zheng, C.J.; Chen, H.B.; Gao, H.F.; Du, L. Du Is the Burton – Miller formulation really free of fictitious eigenfrequencies? *Eng. Anal. Bound. Elem.* **2015**, *59*, 43–51. [CrossRef]

401. Marburg, S. The Burton and Miller Method: Unlocking Another Mystery of Its Coupling Parameter. *J. Comput. Acoust.* **2016**, *24*, 1550016. [CrossRef]

402. Kirkup, S.M. The influence of the weighting parameter on the improved boundary element solution of the exterior Helmholtz equation. *Wave Motion* **1992**, *15*, 93–101. [CrossRef]

403. Juhl, P.M. A numerical study of the coefficient matrix of the boundary element method near characteristic frequencies. *J. Sound Vib.* **1994**, *175*, 39–50. [CrossRef]

404. Dijkstra, W.; Mattheij, R.M.M. Condition Number of the BEM Matrix arising from the Stokes Equations in 2D. *Eng. Anal. Bound. Elem.* **2008**, *32*, 736–746. [CrossRef]

405. Hornikx, M.; Kaltenbacher, M.; Marburg, S. A platform for benchmark cases in computational acoustics. *Acta Acust. United Acust.* **2015**, *101*, 811–820. [CrossRef]

406. Harris, P.J.; Amini, S. On the Burton and Miller Boundary Integral Formulation of the Exterior Acoustic Problem. *ASME J. Vib. Acoust. Stress Reliab. Des.* **1992**, *114*, 540–546. [CrossRef]

Review

The Role of Powered Surgical Instruments in Ear Surgery: An Acoustical Blessing or a Curse?

Tsukasa Ito *, **Toshinori Kubota, Takatoshi Furukawa, Hirooki Matsui, Kazunori Futai, Melinda Hull and Seiji Kakehata**

Department of Otolaryngology, Head and Neck Surgery, Yamagata University Faculty of Medicine; Yamagata 990-9585, Japan; t-kubota@med.id.yamagata-u.ac.jp (T.K.); t-furukawa@med.id.yamagata-u.ac.jp (T.F.); matsuihirooki@gmail.com (H.M.); kfutai@gmail.com (K.F.); melindahull@nifty.com (M.H.); seijik06@gmail.com (S.K.)
* Correspondence: tuito@med.id.yamagata-u.ac.jp

Received: 23 January 2019; Accepted: 19 February 2019; Published: 21 February 2019

Abstract: Ear surgery in many ways lagged behind other surgical fields because of the delicate anatomical structures within the ear which leave surgeons with little room for error. Thus, while surgical instruments have long been available, their use in the ear would most often do more damage than good. This state of affairs remained the status quo well into the first half of the 20th century. However, the introduction of powered surgical instruments, specifically the electric drill used in conventional microscopic ear surgery (MES) and the ultrasonic aspirator, the Sonopet® Omni, in transcanal endoscopic ear surgery (TEES) marked major turning points. Yet, these breakthroughs have also raised concerns about whether the use of these powered surgical instruments within the confines of the ear generated so much noise and vibrations that patients could suffer sensorineural hearing loss as a result of the surgery itself. This paper reviews the intersection between the noise and vibrations generated during surgery; the history of surgical instruments, particularly powered surgical instruments, used in ear surgeries and the two main types of surgical procedures to determine whether these powered surgical instruments may pose a threat to postoperative hearing.

Keywords: noised-induced hearing loss; powered surgical instruments; ultrasonic aspirator; transcanal endoscopic ear surgery

1. Introduction

The internal anatomy of the ear is made up of extremely tiny, delicate, and interlocking anatomical structures that are surrounded by bone and muscle, with sound traveling through the external auditory canal as shown in Figure 1. In particular, the mastoid portion of the temporal bone lies behind the ear and serves as a solid, normally impenetrable, barrier protecting the internal ear. This bony barrier has made it particularly challenging to access the anatomical structures within the internal ear, which has primarily been accomplished by drilling straight through and removing the mastoid bone in a procedure called a mastoidectomy. This procedure has been the mainstay of ear surgery up until the turn of the 20th century. While most ear surgery procedures are performed today with the objective of either preserving or improving hearing, the potential exists for noise and vibration generating surgical instruments required in a mastoidectomy to damage hearing. Of special concern has been the adverse impact on hearing of the use of powered surgical instruments, particularly drills, and the more recently introduced ultrasonic devices which are used to remove bone and expose the internal anatomy of the ear. This paper discusses issues available in the literature which have reported on the effects on hearing of these powered surgical instruments.

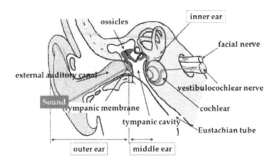

Figure 1. The internal anatomy of the ear.

The majority of current ear surgery procedures can be broken down into two broad approaches: conventional microscopic ear surgery (MES) and the more recently developed transcanal endoscopic ear surgery (TEES). While powered surgical instruments are a standard part of MES, such instruments are only used in a subset of TEES procedures referred to as powered TEES. As was noted above, TEES has not totally eliminated the need for a mastoidectomy, because some surgeons are more comfortable with MES, and some procedures are not indicated for TEES, such as the mastoid air cells, inner ear, and skull base that are beyond its reach.

This paper presents an overview on types and causes of hearing loss in the context of ear surgery; a brief history of surgical instruments used to access the internal anatomy of the ear focusing on the powered surgical instruments, specifically electric drills and ultrasonic devices; an overview of the two main surgical procedures of MES and TEES used in ear surgery to access the internal anatomy of the ear; and a review of the literature on the potential for hearing loss caused by noise and vibrations produced by powered surgical instruments used in MES and TEES procedures, together with the presentation of previously unpublished data on the use of the ultrasonic aspirator, a powered surgical instrument, used in TEES.

2. Types and Causes of Hearing Loss

2.1. Types of Hearing Loss

Hearing loss falls into three broad categories: conductive hearing loss, sensorineural hearing loss and mixed hearing loss. Conductive hearing loss generally occurs when a physical impediment or barrier prevents the transmission of sound waves through the pathway from the outer ear through to the middle ear. Such impediments or barriers can range from a simple buildup of ear wax, accumulation of fluid within the ear due to an infection or an abnormal growth such as bony tissue or a tumor. The other type of hearing loss, sensorineural hearing loss, can be attributed to problems within the inner ear, primarily the cochlea and associated hair cells or the vestibulocochlear nerve (cranial nerve VIII). Sensorineural hearing loss can be caused by either intrinsic factors such as genetics resulting in congenital abnormalities, or extrinsic factors such as inner ear infections; ototoxic drugs such as aminoglycosides and cisplatin; or exposure to high noise levels both over an extended period of time such as in an industrial workplace, prolonged use of headphones or concert/entertainment venues or a single discrete event such as a blast of noise from equipment, gun shot, or bomb blast. The third type of hearing loss, mixed hearing loss, as the name implies, is a combination of the other two types of hearing loss [1].

2.2. Hearing Loss and Powered Surgical Instruments

Hearing loss as related to powered surgical instruments has primarily been studied from two perspectives: noise levels (air-conducted) and vibrations (bone-conducted). The vast majority of studies have focused on noised-induced hearing loss, which is a clear subcategory of sensorineural hearing loss, while vibrations, or more precisely, skull vibrations, have garnered much less attention until recently and deserve further study and consideration. Such hearing loss is measured based on the degree to which the hearing threshold sensitivity has risen and is classified as either a permanent threshold shift or a temporary threshold shift [2]. Most sensorineural hearing loss caused by powered surgical instruments fortunately falls into the temporary threshold shift category.

Separate from these two types of hearing loss caused by powered surgical instruments is the physical contact of an instrument with the ossicular chain. Such drill-to-bone contact results in vibrations which are transmitted via the ossicular chain to the cochlea. The subsequent damage to the cochlea generally results in permanent hearing loss. However, this hearing loss can be attributed to surgeon error rather than conventional use of the surgical instrument itself.

2.2.1. Noised-Induced Hearing Loss

Noised-induced hearing loss can, as alluded to above, be caused by either chronic, accumulative, and gradual exposure, or an acute, one-time event. The chronic, accumulative, and gradual exposure is generally defined in terms of daily exposure over years. Specifically, the National Institute for Occupational Safety and Health (NIOSH) has set down the recommended exposure limit (REL) as 85 decibels, A-weighted (dBA) for an 8-h time-weighted average (TWA) [3] while the acute, one-time event is generally set at 140 dB or higher [4]. However, any noise-induced hearing loss caused by powered surgical instruments falls into an undefined category between these two defined categories. While the noise levels tend to fall within chronic category, surgery time is only measured in minutes up to a couple of hours on a single day, rather than accumulated hours over multiple days. Moreover, while the noise levels generated by powered surgical instruments fall below noise levels defined as dangerous for acute one-time events, the fact that these instruments are used directly within the anatomy of the ear needs to be factored in.

The connection between possible noise-induced hearing loss and powered surgical instruments used during ear surgery has long been a source of concern and study in the case of surgical drills [5–8], as well as a recent target of research in the case of ultrasonic devices [9].

2.2.2. Vibration-Induced Hearing Loss

The second potential way that powered surgical instruments can cause sensorineural hearing loss is by skull vibrations generated by these instruments. This cause has not been totally ignored, but has not received the same amount of attention over the years as noise levels. Moreover, occupational standards are not widely codified, particularly in regard to hearing, as opposed to damage to the circulatory system through the use of hand-held heavy equipment which can result in what is known as hand-arm vibration syndrome (HAVS), a type of Raynaud's syndrome [10]. Vibrations have been posited to cause damage to hearing through inner ear damage and Seki et al. [11] and Miyasaka [12] have both posited that morphological changes, specifically permeability, occur in stria vascularis capillaries, when the auditory ossicles or mastoid are subjected to vibrations. Again, as was the case with noised-induced hearing loss, vibration-induced hearing loss has been studied with surgical drills [10,13,14], but less so with ultrasonic devices [15].

3. History of Otological Surgical Instruments

Progress in ear surgery has, like other surgical fields, been driven primarily by progress in the development of appropriate instruments. These instruments were often originally designed with another purpose in mind, often dentistry, but were eventually adapted by enterprising ear surgeons who co-opted them for their own purposes. Mudry has reported on the history of instruments in ear surgery and divided this history into three periods: trepans (Figure 2a); chisels and gouges (Figure 2b): and electrical drills (Figure 2c [16]. The introduction of powered TEES has added a fourth period of ultrasonic devices (Figure 2d) [17].

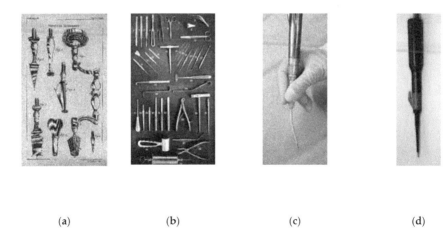

(a)   (b)   (c)   (d)

Figure 2. (a) Surgical instruments for trepanning. Engraving with etching by T. Jefferys. Credit: Wellcome Collection. CC BY (b) Tools for mastoidectomy [18]. (c) Visao® High-Speed Otologic Drill, Medtronic, Shinagawa, Tokyo, Japan). (d) Sonopet® Omni, (model UST-2001 Ultrasonic Surgical Aspirator, Stryker Corporation, Kalamazoo, Michigan, USA).

3.1. Pre-Powered Surgical Instruments

Though scattered references have been found throughout history that may be describing a mastoidectomy-like procedure such as by Galen of Pergamon (129 A.D–210 A.D.) [19]. Surgical instruments had only developed to the point where a mastoidectomy could seriously be contemplated in the 18th century. These developments were attributed to advances made in metallurgy and machine tooling that fueled the Industrial Revolution and underscore the critical importance of technological advances outside medicine, leading to breakthroughs in medical procedures.

3.2. Powered Surgical Drills

The powered surgical drill, as opposed to a hand drill type instrument such as a trepan, was first used in the 1880s. However, its use did not catch on, and like many new approaches, was a bit ahead of its time. Instead, the electric drill was reintroduced into otological practice in the late 1920s by Julius Lempert, who is commonly recognized as the father of the use of the electric drill in ear surgery [16]. The electrical drill has been the workhorse in ear surgery up until the present day.

3.3. Ultrasonic Devices

A new type of device technology began to make an appearance in the medical literature around the 1970s: ultrasonic devices. Broadly speaking, three types of ultrasonic devices used in surgery have appeared on the market: cavitation ultrasonic aspirators or CUSA devices; piezoelectric devices; and ultrasonic aspirators equipped with longitudinal and torsional oscillation.

A CUSA instrument selectively targets the fluid in soft tissue such as tumors, but leaves hard tissue such as bone untouched [20]. CUSAs are quite frequently used in neurosurgery [21] and renal surgery [22] and have enabled surgeons in these fields to remove tumor tissue located in areas with little room for error with less possibility of damaging anatomical structures in comparison to electric drills.

Piezoelectric devices are another type of ultrasonic device that can cut both bone and soft tissue depending on the frequency setting [23]. These devices are another example of borrowing a tool primarily used in dental surgery and expanding its reach into a wide variety of fields, including craniofacial surgery, to perform osteotomies [24] and ear surgery to perform mastoidectomies [25]. However, some surgeons have found that piezoelectric devices compare unfavorably to electric drills in terms of bone cutting power and speed [9].

A third type of ultrasonic device offers the reverse of the CUSA by targeting hard tissue such as bone and leaving soft tissue relatively untouched. Both the CUSA and this new type of ultrasonic aspirator that removes only bone are often referred to in the literature as an ultrasonic bone curette (UBC) because they can both use the same handpiece to which a specialized tip is attached. These third types of ultrasonic aspirators remove bone using a longitudinal oscillation (L mode) or longitudinal and torsional oscillation (LT mode) to emulsify bone and these L mode/LT mode UBCs have been used in fields such as neurosurgery [26], paranasal sinus surgery [27], maxillofacial surgery [28], and spinal surgery [29].

Our original paper on powered TEES referred to the ultrasonic aspirator which we used as a Sonopet® UBC [17]. However, this ultrasonic aspirator that we used and still use today, emerged on the market in the early 2000s and has passed through a number of companies [30]. This ultrasonic aspirator is today more properly called the Sonopet® Omni and uses an H101 tip (model UST-2001 Ultrasonic Surgical Aspirator, Stryker Corporation, Kalamazoo, Michigan, USA) (Figure 3).

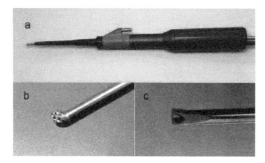

Figure 3. (a) Sonopet® Omni, (model UST-2001 Ultrasonic Surgical Aspirator, Stryker Corporation, Kalamazoo, Michigan, USA). (b) Bone removal view (c) Aspirator view.

It is the Sonopet® Omni together with its H101 tip that has allowed us to perform powered TEES. The ability of this Sonopet® Omni to target bone while generally avoiding soft tissue made it the perfect tool for working within the narrow confines of the external auditory canal and for removing bone from the ear canal. In addition, the Sonopet® Omni not only removes bone, but also has irrigation and aspiration functions that were critical in opening the door to powered TEES, as will be described below. However, little information has been reported on the safety of the Sonopet® Omni in terms of noise levels and vibrations [15], which will also be discussed below.

4. History of Otological Surgical Procedures

The primary otological surgical procedure has been the mastoidectomy. A mastoidectomy involves the removal of the bone behind the ear in order to access the internal anatomy of the ear. However, the tools used in performing a mastoidectomy and related objectives and considerations in terms of safety have changed dramatically over the years. Moreover, since around the turn of the 20th century, a new surgical approach, TEES, has emerged that circumvents the need for a mastoidectomy. It should be noted that TEES has not totally eliminated the need for a mastoidectomy; some surgeons have yet to adopt the procedure and some procedures are not indicated for TEES. Moreover, TEES has raised its own safety issues that need to be addressed.

4.1. Pre-MES Procedures

Once tools were available for removing bone, humans opened holes in skulls for various purposes. The objective in prehistoric and early historic times could have been to release evil spirits, while other procedures may have been to relieve a buildup of pressure within the skull. This pressure buildup could be due to the presence of excess fluids caused by internal hemorrhaging or infection. Prior to the development of antibiotics and the appropriate tools, the main concern of otologists was the treatment of middle ear infections that resulted in pus accumulating within the ear, and having nowhere to go, were potentially life threatening. Such procedures often involved simple incisions of the tympanic membrane or abscesses located inside or outside of the ear. The French physician Ambroise Paré was the first to be credited in the 16th century with making a surgical incision to drain such an infection [19].

The first confirmed mastoidectomy is credited to the French surgeon Jean Louis Petit in 1736 [16] and thereafter the mastoidectomy became a standard, but not universally practiced part of ear surgery from around 1860 [19]. A mastoidectomy was sometimes performed to allow draining of the pus and cleaning out of the infected area. However, removing the mastoid bone using the hand-drill-like trepan or by chipping away at the bone using chisels and gouges did not offer the level of control needed. Some patients were reported to have developed post-operative meningitis and subsequently died, an outcome which would have been due to removing too much bone and tissue. The inadequacy of the pre-powered age surgical instruments led to the mastoidectomy falling out of favor.

It should be noted that since these mastoidectomies were often performed as a life saving measure with little to no regard to the potential damage to internal ear anatomy and hearing caused by the surgical instruments. It is not a stretch, however, to assume that those patients who survived probably did have hearing loss. Eventually, Gustave Bondy developed improvements to the mastoidectomy with the aim of preserving the middle ear anatomical structures and better results were achieved. Bondy's improvements and the introduction of the electric drill resulted in the mastoidectomy becoming a mainstay of ear surgery. This situation improved even further with the introduction of antibiotics in the 1940s, which dramatically cut down on the severity of middle ear infections [19].

4.2. MES Procedures

While the introduction of an electric drill was a major breakthrough in ear surgery, an equally important breakthrough was the introduction of the microscope. This breakthrough is generally attributed to Dr. William House in the mid-1950s [31]. The combination of the widespread use of antibiotics and the microscope led to the development of new surgical possibilities and techniques under the general category of microscopic ear surgery (MES). A majority of these procedures still involve the mastoidectomy as a first step to opening up the internal anatomy of the ear to the enhanced visualization afforded by the microscope. At the same time, higher standards emerged in terms of preserving hearing and preventing hearing loss. Concerns thus began to be raised, and are still raised today, about the potential for noised-induced hearing loss due to the use of these electric drills.

Figure 4a shows a close-up view of a mastoidectomy in progress while Figure 4b shows the typical set up of the operating room and positioning of surgeons when performing an MES procedure.

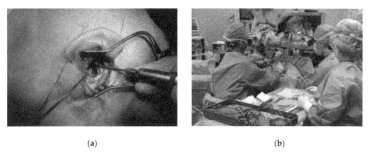

(a)   (b)

Figure 4. Microscopic ear surgery (MES) (**a**) A mastoidectomy in progress. (**b**) Surgeons performing MES.

4.3. TEES Procedures

While MES is still, by far, the common surgical approach employed in ear surgery, transcanal endoscopic ear surgery (TEES) has emerged as a viable alternative in the last twenty years. The microscope is not used in TEES, but instead an endoscope is employed to access and visualize the internal anatomy of the ear through the external auditory canal instead of the more invasive mastoidectomy approach. Even though the endoscope has long been a standard part of surgery in other surgical fields, ear surgeons have had to face a unique set of anatomical and technological challenges that delayed the use of the endoscope within the ear. Endoscopes equipped with cameras were first used together with the microscope when performing mastoidectomies in the 1990s as a way to get a better view of the internal anatomy of the ear [32]. The first surgeries performed completely via the external auditory canal with the endoscope alone were reported on by Dr. Muaaz Tarabichi in 1997 [33] and once again in 1999 [34]. TEES really took off after 2008 with the further development of the 3-charged-coupled device (CCD) camera connected to high definition (HD) monitors, which resulted in high-resolution images during surgery of the tiny structures of the ear [35,36]. Figure 5a illustrates how the endoscope and forceps can be simultaneously inserted into the external auditory canal to perform TEES, and Figure 5b shows the typical set up of the operating room and positioning of surgeons when performing a TEES procedure.

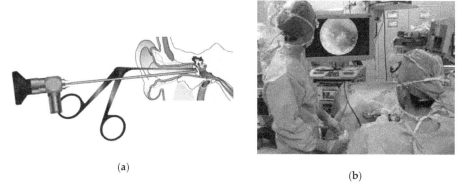

(a)

(b)

Figure 5. Transcanal endoscopic ear surgery (TEES) (**a**) Accessing the internal anatomy of the ear with an endoscope via the external auditory canal. (**b**) A surgeon performing TEES.

TEES offers a number of advantages over MES, including no need to perform an invasive mastoidectomy which requires bone loss, better visualization of the surgical field, the ability to see into deep recesses within the ear, no disfiguring retroauricular scarring, and a quicker recovery time. Moreover, most TEES procedures are performed entirely without powered surgical instruments, which eliminates the potential for sensorineural hearing loss resulting from noise levels or vibrations. However, the "conventional" or what we have unofficially dubbed "non-powered" TEES can be performed only so far into the middle ear, but a subset of TEES procedures use the Sonopet® Omni to remove bone within the middle ear and enable access to the antrum which we call powered TEES. Figure 6 illustrates the internal anatomy of the middle ear (Figure 6a); the scope of the indications for non-powered TEES which can only reach into the inferior portion of the attic (Figure 6b); and the scope of the indications for powered TEES that can reach into the antrum (Figure 6c). This figure underscores that the Sonopet® Omni is reaching deep into the middle ear and closer to the ossicular chain and cochlea, which raises the specter of sensorineural hearing loss resulting from noise levels or vibrations, and is addressed below.

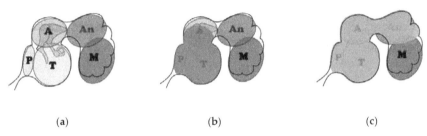

(a)   (b)   (c)

Figure 6. Revised system for staging and classifying of middle ear cholesteatomas to be treated by non-TEES based on the PTAM System for Staging and Classification of Middle Ear Cholesteatomas as proposed by the Japan Otological Society [37] (**a**) The middle ear (P: protympanum; T: tympanic cavity; A: attic; An: antrum; M: mastoid cells). (**b**) Non-powered TEES (orange overlay). (**c**) Powered TEES (yellow overlay).

5. Generation of Noise and Vibrations by Otological Surgical Instruments and Hearing Loss

5.1. Pre-MES Procedures

One can reliably posit that the surgical instruments and procedures used in the pre-MES period were so crude that sensorineural hearing loss was common, but no data is available.

5.2. MES Procedures

5.2.1. Drill Generated Noise Levels

Many researchers have conducted basic research targeting the problem of measuring the noise levels generated by surgical drills in a non-clinical setting. Among the earliest studies of drill generated noise levels and most frequently cited is that of Kylén and Arlinger [5], published in 1976. They measured vibrations generated by drills using isolated temporal bones and cadavers, whereafter the data was then converted into equivalent air-borne noise levels. They found that the isolated temporal bones produced lower noise levels than cadavers, and thus concluded that cadavers better simulated real surgical conditions. The maximum equivalent air-borne noise levels were found to be around 100 dB, which falls below the 130 to 140 dB threshold for causing permanent hearing loss from an acute one-time exposure The same group reported additional results in 1977 [6], looking at different variables focusing on the size of the burr; the type of the burr; (sizes: 2 mm, 4-mm and 6-mm burrs; types: diamond versus cutting burrs); drill rotation speed; and location of drilling. They found that the smaller the burr, the lower the noise levels, as well as that lower noise levels were obtained with diamond burrs, as opposed to cutting burrs with the highest noise level of 108 dB discovered with a 6-mm cutting burr. They further found that drill rotation speed and location of drilling had little effect on noise levels.

Several researchers subsequently collected intraoperative in vivo data with the objective of getting a better picture of drill generated noise during real-life conditions. Holmquist et al. [38] reported higher in vivo noise levels in 1979 from the contralateral ear (the ear which was not being operated on) of six patients while undergoing a mastoidectomy on the ipsilateral ear (the ear which was being operated on). They reported noise levels in excess of the 108 dB recorded by the Kylén group at 116 dBA for 8-mm burrs; 109 dBA for 4-mm burrs; and surprisingly, a dangerously high 125 dBA for 2-mm burrs. Hickey and Fitzgerald O'Connor [39] conducted an in vivo study in 1991 in which they attempted to directly measure the drill-generated noise levels by monitoring intraoperative *electrocochleographic* (ECoG) noise levels and calculating these levels by using a masking technique. They were able to determine that drill-generated noise levels were present in excess of 90 dBHL at the level of the cochlea, but were unable to determine peak values.

Other researchers collected data on postoperative sensorineural hearing loss in the contralateral ear in order to attempt to eliminate surgical trauma as a possible cause of hearing loss from surgery, while at the same time acknowledging the added distance between the contralateral and ipsilateral ears. Man and Winerman [40] conducted a study on 62 patients and reported in 1985 that they found a minimal difference between the peak noise levels which did not exceed 84 dB in the ipsilateral ear and 82 dB in the contralateral ear. Moreover, they found no hearing loss in the contralateral ear, but did find loss in 16 out of the 62 patients in the ipsilateral ear. They suggested that these results indicated that drill-generated noise levels did not cause sensorineural hearing loss. daCruz et al. [41] reported on drill-induced hearing loss in the contralateral ear in 1997 based on determining outer hair cell (OHC) function using distortion-product otoacoustic emissions (DPOAE). They found a change in the amplitude of the intraoperative DPOAE in 2 out of 12 patients undergoing temporal bone surgery, indicating a transient OHC dysfunction that subsequently returned to normal. This transient but reversed dysfunction was attributed to drill-generated noise levels. Goyal et al. [42] reported on the effect of mastoid bone drilling on the contralateral ear in 2013 based on otoacoustic emissions (OAE). Their study looked at the results for 30 patients and they stated that 15 of these patients exhibited a reduction in postoperative OAE levels out of which only 10 out of 15 completely recovered. However, data was only collected for up to 72 h, which makes a definite conclusion that the hearing loss was permanent a bit premature. In contrast, in two similar studies, Baradaranfar et al. in 2015 [43] and Latheef et al. [44] in 2018 reported transient hearing losses that all had disappeared by 72 h, while Badarudeen et al. [45] in 2018 reported transient hearing losses that were still present on the 7th day after surgery.

The above findings thus paint a mixed picture on transient hearing loss in the contralateral ear caused by drilling.

5.2.2. Drill Generated Vibrations

Recent studies which have examined skull vibrations have indicated that this factor should not be discounted. Suits et al. reported on a guinea pig model in 1993 which was used to measure both noise and vibration levels based on the auditory brainstem response (ABR). They did find that a temporary threshold shift occurred, but that the shift had disappeared by three weeks [13]. In 2001, Zou et al. also created a guinea pig model and compared the results for younger animals versus older animals when they were exposed to both noise and noise + vibrations. They reported that the older animals were more vulnerable to a threshold shift and that the sound-induced threshold shift was significantly less than the vibration + sound-induced threshold shift at three days after exposure [14]. The same group of researchers reported in 2007 that temporal bone vibration in a guinea pig model showed that vibrations at high frequencies caused more severe hearing loss than at lower frequencies, but that the threshold shift was generally temporary [10]. However, Hilmi et al. contended that high speed drills do not produce sufficient high levels of high frequency skull vibrations to result in damage to hearing [8].

5.2.3. Hearing Loss after MES Procedures

A commonly accepted range for the incidence of sensorineural hearing loss in the ipsilateral ear after ear surgery is from 1.2% to 4.5% of patients. The higher figure of 4.5% is from a study of 1680 ear surgeries reported by Palva et al. [46] in 1973, and the lower figure of 1.2% is from a study of 2,303 ear surgeries performed from 1965 to 1980, reported by Tos et al. [47] in 1984. What is notable about these papers is that they are among the first large scale studies published in the literature and that they are from more than close to 35 years ago. Both authors attributed some of these hearing losses to the surgeon being too aggressive in the area of the ossicular chain, and Tos et al. [47] in particular, noted that the incidence of sensorineural hearing loss was lower in patients treated from 1974 to 1980. They attributed this drop to better technique and better drills. In 1989, Doménech et al. [48] reported what they characterize as an important sensorineural hearing loss after tympanoplasty in a larger percentage of patients at 16.7% than previously reported in literature, but the hearing loss was restricted to over 8000 Hz, which is typically not measured. Urquhart et al. [7] reported in 1992 that they found no evidence of an even temporary threshold shift after ear surgery in a patient group of 40; however, they only tested up to 4000 Hz. A recent study by Kent et al. [49] in 2017 looked at factors of the experience of the surgeon and the use of a powered drill in hearing outcomes for patients undergoing a tympanoplasty which required drilling in the ear canal, a less invasive procedure than a mastoidectomy, and they found that neither factor exhibited a correlation with high-frequency hearing loss. In contrast, Al Anazy et al. [50] found in 2016 that the experience of the surgeon was a significant factor, but the use of a drill was not significant in the incidence of postoperative sensorineural hearing loss between 500 Hz to 4000 Hz after tympanoplasty which was 7% for residents, but only 1% for more experienced surgeons. However, they could not identify any obvious errors on the part of the residents.

Thus, the literature as it relates to the role of powered surgical instruments, specifically electric drills in postoperative sensorineural hearing loss, is inconclusive at best, and depends on the study design and definition of postoperative sensorineural hearing loss.

5.3. TEES Procedures

While a considerable amount of study has been done on noise-induced and vibration-induced hearing loss caused by surgical drills when performing MES, ultrasonic devices have not been studied in similar depth. CUSA devices are not used in ear surgery, and a search of the literature did not reveal any related studies from outside of ear surgery. Kramer et al. [9] look at the potential for noise trauma caused by piezoelectric devices in craniofacial osteotomies and concluded that piezoelectric devices offer no advantage over regular drills in acoustic properties. They ultimately recommended using a drill because of the slower speed of the piezoelectric device. Research on TEES and the Sonopet® Omni has, to our knowledge, only been conducted by our own group. The reason for this difference is that, as stated above, TEES procedures do not usually require the use of electric drills and only the small, but important subset of powered TEES, employ an ultrasonic aspirator together with a standard drill. The Sonopet® Omni is inserted via the external auditory canal and used to remove part of the canal wall to expose the antrum while the surgical drill is used to polish a bony shelf which is designed to protect the facial nerve.

We took it upon ourselves to collect data on noise levels and vibration levels produced by the Sonopet® Omni and compare it to data collected for standard surgical drills. The data on noise levels is presented herein for the first time while the data on skull vibrations was previously published in 2013 [15].

5.3.1. Ultrasonic Aspirator Generated Noise Levels

Our study was designed to confirm that the noise levels generated by an ultrasonic aspirator during powered TEES fall within safe levels and should not induce sensorineural hearing loss. The study was conducted from September 2014 to February 2015 and data was collected during surgery from patients undergoing a powered TEES procedure to remove a cholesteatoma with a total of 14 patients (5 males/9 females) ranging in age from 15 to 76 and a median age of 50. All patients underwent a transcanal atticoantrotomy which was performed using a Sonopet® Omni ultrasonic aspirator (model UST-2001 Ultrasonic Surgical Aspirator, Stryker Corporation, Kalamazoo, Michigan, USA) in the LT (longitudinal-torsion) mode at 25 kHz (Figure 3a) and a high-speed drill with a curved burr at 80,000 rpm/1333 Hz (Visao® High-Speed Otologic Drill, Medtronic, Shinagawa, Tokyo, Japan) (Figure 7).

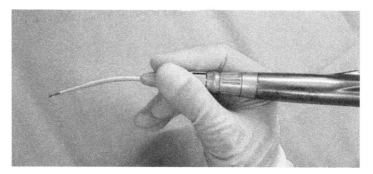

Figure 7. The high-speed drill with a curved burr at 80,000 rpm/1333 Hz (Visao® High-Speed Otologic Drill, Medtronic, Shinagawa, Tokyo, Japan) used for polishing bone during the transcanal atticoantrotomy.

The noise levels were measured at 10 cm (Figure 8a and 70 cm (Figure 8b) from the devices from 0.5 kHz to 16 kHz. These distances were selected because 10 cm is the closest that it was physically possible to measure noise levels generated within the external auditory canal and 70 cm represents the distance to the surgeon's ear. The noise level at 70 cm was measured in our original study because when new powered instruments are introduced, it is important to take into consideration the potential for hearing loss for anyone in the operating room due to long term noise exposure. Furthermore, we only measured up to 16 kHz despite the Sonopet® LT mode generating frequencies of up to 25 kHz, because standard noise measurement equipment can only measure up to 16 kHz, which is also close to the maximum auditory threshold of a normal adult. The results for both the Sonopet® Omni and the Visao® drill at 10 cm were at 85 dB or below with the noise levels lower for the Sonopet® Omni up to 2000 Hz and higher for the Sonopet® Omni from 4000 Hz to 16,000 Hz. The results for both the Sonopet® Omni and the Visao® drill at 70 cm were below 70 dB with the noise levels for the Sonopet® Omni lower or equal to the Visao® drill up to 8000 Hz and higher for the Sonopet® Omni from 8000 Hz to 16,000 Hz.

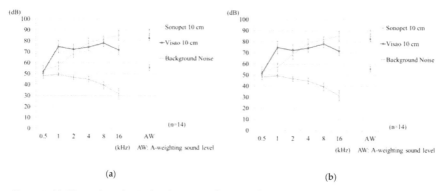

(a)   (b)

Figure 8. (**a**) Air-conducted noise levels measured at 10 cm from the tip of the tool; (**b**) Air-conducted noise levels measured at 70 cm from the tip of the tool.

5.3.2. Ultrasonic Aspirator Generated Vibration Levels

Our group conducted a study designed to determine the vibration levels generated by an ultrasonic aspirator and compare the ultrasonic aspirator vibration levels to those of two surgical drills: an Osteon® drill at 20,000 rpm/333 Hz (Zimmer Biomet, Warsaw, Indiana, USA) and a Visao® High-speed Otologic Drill at 40,000 rpm/667 Hz and 80,000 rpm/1333 Hz (Medtronic, Shinagawa,

Tokyo, Japan) [15]. All measurements were taken during an MES mastoidectomy and the skull bone vibrations were measured with a polyvinylidene difluoride (PVDF) film taped to the forehead as shown in Figure 9. PVDF is a piezoelectric material and the charge builds up in the PVDF film in response to any applied mechanical stress.

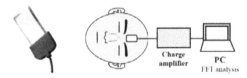

PolyVinylidene DiFluoride (PVDF) film

Figure 9. The setup for measuring skull vibrations using PolyVinylidene DiFluoride film.

Figure 10 shows the mean values of the measured skull vibrations and the background noise level at four frequency bands. In the frequency bands of 500–2000 Hz and 2000–8000 Hz, the mean values of the Sonopet® Omni with an LT-vibration tip did not exceed the values for the Visao® revolution speed of 40k rpm or 80k rpm as well as the Osteon® drill. The peak values of skull vibrations by the Sonopet® Omni with an LT-vibration tip were lower than the vibrations of Visao® at 40 k rpm in the band of 500–2000 Hz; those of Visao® at 80k rpm in the bands of 500–2000 Hz and 2000–8000 Hz; and those of Osteon® in the bands of 500–2000 Hz and 2000–8000 Hz. No significant differences in the skull vibrations were observed among the three instruments below 500 Hz or above 8000 Hz.

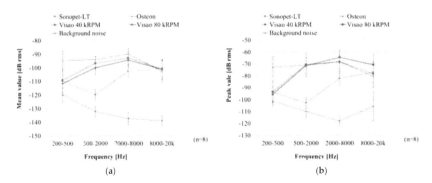

Figure 10. (**a**) Mean values for vibration levels. (**b**) Peak values for vibration levels.

Figure 11 shows the power spectrum produced by the Sonopet® at 25 kHz in LT-mode versus background noise for the purpose of reference.

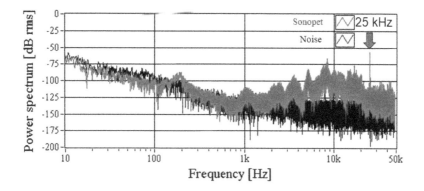

Sonopet® LT 25,000 Hz

Figure 11. Comparison of the power spectrum produced by the Sonopet® at 25 kHz in LT mode versus background noise.

5.3.3. Hearing loss and TEES procedures

Our group has performed powered TEES since approximately 2011, and we have yet to record any postoperative sensorineural hearing loss in any patient up to a frequency of 8000 Hz. Thus, powered TEES can be performed, when indicated, without the worry of postoperative sensorineural hearing loss up to the highest measurable frequency.

6. Conclusions

The early days of ear surgery often focused on saving the lives of patients with little regard to postoperative hearing loss. However, the end of the 19th century and into the 20th century saw the introduction and steady improvement of surgical approaches and tools, particularly powered surgical instruments in the form of electric drills for MES and, more recently, ultrasonic aspirators for powered TEES. Concerns were raised about the impact that these powered surgical instruments could have on sensorineural hearing loss, and research has been done looking at both noise-induced hearing loss and vibration-induced hearing loss caused by powered surgical instruments when performing MES. The results of this research are long and often contradictory, as related to conventional MES and postoperative sensorineural hearing loss, particularly as related to high frequency hearing loss. Thus, effort and research should continue to be expended toward improving both MES techniques and electric drill noise level specifications, because MES, as stated above, will continue to be a standard and essential part of ear surgery, particularly in the areas of the mastoid air cells, inner ear, and skull base.

The introduction of TEES in the late 20th century eliminated the issue of sensorineural hearing loss for a circumscribed set of middle ear procedures because non-powered TEES does not require the use of powered surgical instruments. However, the emergence of powered TEES in the 21th century, once again, required that the research be conducted anew related to the use of powered surgical instruments, specifically the ultrasonic aspirator and curved burr, directly within the confines of the EAC and postoperative sensorineural hearing loss. We presented our research herein on whether powered TEES could be shown to cause either noise-induced hearing loss or vibration-induced hearing loss, and we found that powered TEES is as safe as, if not safer, in regard to the potential for postoperative sensorineural hearing loss than MES based on the data which we collected on noise levels, vibration levels, and the occurrence of postoperative sensorineural hearing loss for powered TEES. Thus, while MES will continue to be an essential part of ear surgery, surgeons can now rest assured that, when indicated, powered TEES can be performed safely and does not present any more of a risk of postoperative sensorineural hearing loss than MES.

Author Contributions: Conceptualization, T.I.; Investigation, T.I., T.K., T.F., H.M. and K.F.; Project administration, S.K.; Writing—original draft, M.H.

Funding: This research received no external funding.

Conflicts of Interest: The authors declare no conflict of interest.

References

1. Isaacson, B. Hearing loss. *Med. Clin. N. Am.* **2010**, *94*, 973–988. [CrossRef] [PubMed]
2. Ryan, A.F.; Kujawa, S.G.; Hammill, T.; Le Prell, C.; Kil, J. Temporary and Permanent Noise-induced Threshold Shifts: A Review of Basic and Clinical Observations. *Otol. Neurotol.* **2016**, *37*, e271–e275. [CrossRef] [PubMed]
3. National Institute for Occupational Safety Health (NIOSH). *Occupational Noise Exposure. Revised Criteria*; Department of Health and Human Services: Washington, DC, USA, 1998.
4. Occupational Safety and Health Administration (OSHA). *Occupational Safety and Health Standards, Occupational Noise Exposure*; Standard Number 1910.95; Department of Labor: New York, NY, USA.
5. Kylen, P.; Arlinger, S. Drill-generated noise levels in ear surgery. *Acta Otolaryngol.* **1976**, *82*, 402–409. [CrossRef] [PubMed]
6. Kylen, P.; Stjernvall, J.E.; Arlinger, S. Variables affecting the drill-generated noise levels in ear surgery. *Acta Otolaryngol.* **1977**, *84*, 252–259. [CrossRef] [PubMed]
7. Urquhart, A.C.; McIntosh, W.A.; Bodenstein, N.P. Drill-generated sensorineural hearing loss following mastoid surgery. *Laryngoscope* **1992**, *102*, 689–692. [CrossRef] [PubMed]
8. Hilmi, O.J.; McKee, R.H.; Abel, E.W.; Spielmann, P.M.; Hussain, S.S. Do high-speed drills generate high-frequency noise in mastoid surgery? *Otol. Neurotol.* **2012**, *33*, 2–5. [CrossRef] [PubMed]
9. Kramer, F.J.; Bornitz, M.; Zahnert, T.; Schliephake, H. Can piezoelectric ultrasound osteotomies result in serious noise trauma? *Int. J. Oral Maxillofac. Surg.* **2015**, *44*, 1355–1361. [CrossRef] [PubMed]
10. Sutinen, P.; Zou, J.; Hunter, L.L.; Toppila, E.; Pyykko, I. Vibration-induced hearing loss: Mechanical and physiological aspects. *Otol. Neurotol.* **2007**, *28*, 171–177. [CrossRef]
11. Seki, M.; Miyasaka, H.; Edamatsu, H.; Watanabe, K. Changes in permeability of strial vessels following vibration given to auditory ossicle by drill. *Ann. Otol. Rhinol. Laryngol.* **2001**, *110*, 122–126. [CrossRef]
12. Miyasaka, H. Morphological changes in the stria vascularis and hair cells after mastoid-vibration using a cutting bur. *Nihon Jibiinkoka Gakkai Kaiho* **1999**, *102*, 1249–1257. [CrossRef]
13. Suits, G.W.; Brummett, R.E.; Nunley, J. Effect of otologic drill noise on ABR thresholds in a guinea pig model. *Otolaryngol. Head Neck Surg.* **1993**, *109*, 660–667. [CrossRef] [PubMed]
14. Zou, J.; Bretlau, P.; Pyykko, I.; Starck, J.; Toppila, E. Sensorineural hearing loss after vibration: An animal model for evaluating prevention and treatment of inner ear hearing loss. *Acta Otolaryngol.* **2001**, *121*, 143–148. [PubMed]
15. Ito, T.; Mochizuki, H.; Watanabe, T.; Kubota, T.; Furukawa, T.; Koike, T.; Kakehata, S. Safety of ultrasonic bone curette in ear surgery by measuring skull bone vibrations. *Otol. Neurotol.* **2014**, *35*, e135–e139. [CrossRef] [PubMed]
16. Mudry, A. History of instruments used for mastoidectomy. *J. Laryngol. Otol.* **2009**, *123*, 583–589. [CrossRef] [PubMed]
17. Kakehata, S.; Watanabe, T.; Ito, T.; Kubota, T.; Furukawa, T. Extension of indications for transcanal endoscopic ear surgery using an ultrasonic bone curette for cholesteatomas. *Otol. Neurotol.* **2014**, *35*, 101–107. [CrossRef] [PubMed]
18. West, C.E.; Scott, S.R. *The Operations of Aural Surgery. Together with Those for the Relief of the Intracranial Complications of Suppurative Otitis Media*; P. Blakiston: Philadelphia, PA, USA, 1909.
19. Bento, R.F.; Fonseca, A.C. A brief history of mastoidectomy. *Int. Arch. Otorhinolaryngol.* **2013**, *17*, 168–178. [PubMed]
20. Epstein, F. The Cavitron ultrasonic aspirator in tumor surgery. *Clin. Neurosurg.* **1983**, *31*, 497–505. [CrossRef] [PubMed]
21. Brock, M.; Ingwersen, I.; Roggendorf, W. Ultrasonic aspiration in neurosurgery. *Neurosurg. Rev.* **1984**, *7*, 173–177. [CrossRef] [PubMed]
22. Chopp, R.T.; Shah, B.B.; Addonizio, J.C. Use of ultrasonic surgical aspirator in renal surgery. *Urology* **1983**, *22*, 157–159. [CrossRef]

23. Zhang, Y.; Wang, C.; Zhou, S.; Jiang, W.; Liu, Z.; Xu, L. A comparison review on orthopedic surgery using piezosurgery and conventional tools. *Procedia Cirp* **2017**, *65*, 99–104. [CrossRef]

24. Gleizal, A.; Bera, J.C.; Lavandier, B.; Beziat, J.L. Piezoelectric osteotomy: A new technique for bone surgery-advantages in craniofacial surgery. *Child's Nerv. Syst.* **2007**, *23*, 509–513. [CrossRef] [PubMed]

25. Dellepiane, M.; Mora, R.; Salzano, F.A.; Salami, A. Clinical evaluation of piezoelectric ear surgery. *Ear Nose Throat J.* **2008**, *87*, 212–213, 216. [PubMed]

26. Hadeishi, H.; Suzuki, A.; Yasui, N.; Satou, Y. Anterior clinoidectomy and opening of the internal auditory canal using an ultrasonic bone curette. *Neurosurgery* **2003**, *52*, 867–870; discussion 870-1. [CrossRef] [PubMed]

27. Pagella, F.; Giourgos, G.; Matti, E.; Colombo, A.; Carena, P. Removal of a fronto-ethmoidal osteoma using the sonopet omni ultrasonic bone curette: First impressions. *Laryngoscope* **2008**, *118*, 307–309. [CrossRef] [PubMed]

28. Ueki, K.; Nakagawa, K.; Marukawa, K.; Yamamoto, E. Le Fort I osteotomy using an ultrasonic bone curette to fracture the pterygoid plates. *J. Cranio-Maxillo-Facial Surg.* **2004**, *32*, 381–386. [CrossRef] [PubMed]

29. Nakagawa, H.; Kim, S.D.; Mizuno, J.; Ohara, Y.; Ito, K. Technical advantages of an ultrasonic bone curette in spinal surgery. *J. Neurosurg. Spine* **2005**, *2*, 431–435. [CrossRef] [PubMed]

30. Kakehata, S.; Ito, T.; Yamauchi, D. *Innovations in Endoscopic Ear Surgery*; Springer: Cham, Switzerland, in press.

31. House, W.F. *The Struggles of a Medical Innovator: Cochlear Implants and Other Ear Surgeries: A Memoir*; Better Hearing Institute: Washington, DC, USA, 2011.

32. Thomassin, J.M.; Korchia, D.; Doris, J.M. Endoscopic-guided otosurgery in the prevention of residual cholesteatomas. *Laryngoscope* **1993**, *103*, 939–943. [CrossRef] [PubMed]

33. Tarabichi, M. Endoscopic management of acquired cholesteatoma. *Am. J. Otol.* **1997**, *18*, 544–549. [PubMed]

34. Tarabichi, M. Endoscopic middle ear surgery. *Ann. Otol. Rhinol. Laryngol.* **1999**, *108*, 39–46. [CrossRef] [PubMed]

35. Marchioni, D.; Mattioli, F.; Alicandri-Ciufelli, M.; Presutti, L. Transcanal endoscopic approach to the sinus tympani: A clinical report. *Otol. Neurotol.* **2009**, *30*, 758–765. [CrossRef] [PubMed]

36. Marchioni, D.; Mattioli, F.; Alicandri-Ciufelli, M.; Presutti, L. Endoscopic approach to tensor fold in patients with attic cholesteatoma. *Acta Otolaryngol.* **2009**, *129*, 946–954. [CrossRef] [PubMed]

37. Tono, T.; Sakagami, M.; Kojima, H.; Yamamoto, Y.; Matsuda, K.; Komori, M.; Hato, N.; Morita, Y.; Hashimoto, S. Staging and classification criteria for middle ear cholesteatoma proposed by the Japan Otological Society. *Auris Nasus Larynx* **2017**, *44*, 135–140. [CrossRef] [PubMed]

38. Holmquist, J.; Oleander, R.; Hallen, O. Peroperative drill-generated noise levels in ear surgery. *Acta Otolaryngol.* **1979**, *87*, 458–460. [CrossRef] [PubMed]

39. Hickey, S.A.; O'Connor, A.F. Measurement of drill-generated noise levels during ear surgery. *J. Laryngol. Otol.* **1991**, *105*, 732–735. [CrossRef] [PubMed]

40. Man, A.; Winerman, I. Does drill noise during mastoid surgery affect the contralateral ear? *Am. J. Otol.* **1985**, *6*, 334–335. [PubMed]

41. da Cruz, M.J.; Fagan, P.; Atlas, M.; McNeill, C. Drill-induced hearing loss in the nonoperated ear. *Otolaryngol. Head Neck Surg.* **1997**, *117*, 555–558. [CrossRef]

42. Goyal, A.; Singh, P.P.; Vashishth, A. Effect of mastoid drilling on hearing of the contralateral ear. *J. Laryngol. Otol.* **2013**, *127*, 952–956. [CrossRef] [PubMed]

43. Baradaranfar, M.H.; Shahbazian, H.; Behniafard, N.; Atighechi, S.; Dadgarnia, M.H.; Mirvakili, A.; Mollasadeghi, A.; Baradaranfar, A. The effect of drill-generated noise in the contralateral healthy ear following mastoid surgery: The emphasis on hearing threshold recovery time. *Noise Health* **2015**, *17*, 209–215. [PubMed]

44. Latheef, M.N.; Karthikeyan, P.; Coumare, V.N. Effect of Mastoid Drilling on Hearing of the Contralateral Normal Ear in Mastoidectomy. *Indian J. Otolaryngol. Head Neck Surg.* **2018**, *70*, 205–210. [CrossRef]

45. Badarudeen, S.; Somayaji, G. Influence of mastoid drilling on otoacoustic emissions of the nonoperated ear. *Indian J. Otol.* **2018**, *24*, 95.

46. Palva, T.; Karja, J.; Palva, A. High-tone sensorineural losses following chronic ear surgery. *Arch. Otolaryngol.* **1973**, *98*, 176–178. [CrossRef] [PubMed]

47. Tos, M.; Lau, T.; Plate, S. Sensorineural hearing loss following chronic ear surgery. *Ann. Otol. Rhinol. Laryngol.* **1984**, *93*, 403–409. [CrossRef] [PubMed]

48. Domenech, J.; Carulla, M.; Traserra, J. Sensorineural high-frequency hearing loss after drill-generated acoustic trauma in tympanoplasty. *Arch. Oto-Rhino-Laryngol.* **1989**, *246*, 280–282. [CrossRef]
49. Kent, D.T.; Chi, D.H.; Kitsko, D.J. Surgical trainees and powered-drill use do not affect type I tympanoplasty hearing outcomes. *Ear Nose Throat J.* **2017**, *96*, 366–371. [PubMed]
50. Al Anazy, F.H.; Alobaid, F.A.; Alshiha, W.S. Sensorineural hearing loss following tympanoplasty surgery: A prospective cohort study. *Egypt. J. Otolaryngol.* **2016**, *32*, 93–97. [CrossRef]

MDPI

St. Alban-Anlage 66

4052 Basel

Switzerland

Tel. +41 61 683 77 34

Fax +41 61 302 89 18

www.mdpi.com

Applied Sciences Editorial Office

E-mail: applsci@mdpi.com

www.mdpi.com/journal/applsci

CPSIA information can be obtained
at www.ICGtesting.com
Printed in the USA
LVHW072114190720
661007LV00012B/132